A Themed Issue in Memory of Academician Duzheng Ye (1916–2013)

A Themed Issue in Memory of Academician Duzheng Ye (1916–2013)

Editors

Xiaolei Zou
Guoxiong Wu
Zhemin Tan

MDPI

Basel • Beijing • Wuhan • Barcelona • Belgrade • Novi Sad • Cluj • Manchester

Editors

Xiaolei Zou
School of
Atmosphere Sciences
Nanjing University of
Information Science &
Technology
Nanjing, China

Guoxiong Wu
Institute of
Atmospheric Physics
Chinese Academy of Sciences
Beijing, China

Zhemin Tan
Department of Meteorology
Nanjing University
Nanjing, China

Editorial Office
MDPI
St. Alban-Anlage 66
4052 Basel, Switzerland

This is a reprint of articles from the Topic published online in the open access journals *Atmosphere* (ISSN 2073-4433), *Climate* (ISSN 2225-1154), and *Remote Sensing* (ISSN 2072-4292) (available at: https://www.mdpi.com/topics/Memory_of_Academician_Duzheng_Yeh).

For citation purposes, cite each article independently as indicated on the article page online and as indicated below:

Lastname, A.A.; Lastname, B.B. Article Title. *Journal Name* **Year**, *Volume Number*, Page Range.

ISBN 978-3-0365-8932-9 (Hbk)
ISBN 978-3-0365-8933-6 (PDF)
doi.org/10.3390/books978-3-0365-8933-6

Cover image courtesy of Xiaolei Zou

Contents

About the Editors

Xiaolei Zou

Dr. Xiaolei Zou is currently a Professor at the School of Atmosphere Sciences of Nanjing University of Information Science & Technology, China. She received her Ph.D. degree from the Institute of Atmospheric Physics, Academia Sinica, Beijing, China, in 1988. She developed the National Meteorological Center (now the National Center for Environmental Prediction) medium-range global forecast model 4D-Var system with full physics during 1989–1993, and the fifth-generation Penn State/NCAR Mesoscale Model (MM5) 4D-Var system during 1993–1997. She has worked on GPS radio occultation data assimilation since 1993. From 1997 to 2014, she was a Professor in the Department of Earth, Ocean, and Atmospheric Science, Florida State University, USA. During 2014 to 2018, she worked mainly on satellite data assimilation for quantitative precipitation forecasts and hurricane tracking and intensity forecasts at the Earth System Science Interdisciplinary Center (ESSIC), University of Maryland, USA. She has published over 220 papers in peer-reviewed journals. Dr. Zou was the recipient of the 2008 American Meteorological Society Fellow Award for her outstanding contributions to the applications of satellite data in numerical weather prediction models and education in data assimilation.

Guoxiong Wu

Dr. Guoxiong Wu is a Professor at the Institute of Atmospheric Physics (IAP) of Chinese Academy of Sciences (CAS), China. He received his Ph.D. in meteorology from Imperial College of Science and Technology, London University, UK, in 1983. He served as Director of Key Laboratory of Numerical Experiment for Atmospheric Sciences and Geophysical Fluid Dynamics (LASG), IAP, from 1993 to 2000. He has a solid background in both mathematics and physics and rich practical experiences. His scientific contributions have been multidimensional, ranging from advances in theory, numerical experimentation, and physical understanding of the large-scale dynamical processes. He served PIs on many major climate-related projects in China. He was elected the Academician of CAS and Life Tenure Professor in 1997, Honorary Fellow of the Royal Meteorological Society (UK) in 2012, Honorary Fellow of Chinese-American Oceanic and Atmospheric Association (COAA) and Tan Kah Kee Science Award in 2014, International Union of Geodesy and Geophysics (IUGG) Fellow and American Geophysical Union (AGU) Fellow in 2015, and International Science Council Fellow (ISC) Fellow in 2022.

Zhemin Tan

Dr. Zhemin Tan is a Professor at the School of Atmospheric Sciences of Nanjing University, China. He received his Ph.D. degree from Nanjing University, Nanjing, China in 2000. Since 1999, he has been a Professor in the Department of Meteorology, Nanjing University. Dr. Tan was elected as Academician of Chinese Academy of Sciences in 2021. He was previously the Vice President of the Chinese Meteorological Society and member of International Committee for Dynamic Meteorology (ICDM). Dr. Tan is currently the chair of the China National Committee for the World Weather Research Programme (WWRP), Vice President of the China National Committee of the International Association of Meteorological and Atmospheric Science (IAMAS), and a member of the National Expert Committee on Climate Change. Dr. Tan conducts research in atmospheric dynamics, tropical cyclone dynamics and atmospheric predictability. By bridging theoretical analysis with applied research, Dr. Tan sought new methods for model parameterization and data assimilation that are objective yet dynamics based.

Preface

Professor Duzheng Ye was an internationally acclaimed scientist and a primary founder of modern meteorology in China. His profound contributions encompassed a wide spectrum of Earth sciences, notably including pioneering work in Tibetan Plateau meteorology, the theory of atmospheric longwave energy dispersion, and climate change adaptation theories. Recognized for his groundbreaking insights, Prof. Ye was honored with prestigious accolades, including the 2005 State Preeminent Science and Technology Award as well as the 48th International Meteorological Organization (IMO) Prize in 2003. The passing of this luminary in 2013 left an immense void that continues to resonate.

This Topic, dedicated to the memory of Academician Duzheng Ye (1916–2013), serves a dual purpose: to allow those who hold Prof. Ye in high regard not only to pay homage to his academic legacy and charismatic personality but also to transmit their admiration for him to the forthcoming generations.

In celebration of Professor Duzheng Ye's remarkable contributions to Earth sciences, this Topic comprises 31 original research manuscripts and reviews from 30 invited authors spanning atmospheric dynamics and physics, synoptic weather patterns, climate change, and remote sensing observations for weather and climate. The 30 distinguished invited authors are a harmonious blend of luminaries from the Institute of Atmospheric Physics (IAP), Chinese Academy of Sciences, including Academicians Guoxiong Wu and Ronghui Huang, as well as Drs. Hongbin Chen, Bin Wang, Yiming Liu, and Zhongwei Yan. Additionally, this contingent features five esteemed professors and a scientist from the United States: Professors William K. M. Lau, Zhengyu Liu, Zhanqin Li, Ming Cai, and Zhaohua Wu, alongside scientist Xiaoxu Tian.

Complementing this assemblage are four notable scientists from the China Meteorological Administration (CMA), including Academician Yihui Ding and Drs. Hongli Ren, Juan Li, and Xiaoyong Zhuge. Further enriching this tapestry are three distinguished professors from Nanjing University, China: Academicians Congbin Fu and Zhemin Tan, and Professor Kun Zhao. Fudan University's contribution features two prominent professors, Academician Mu Mu and Professors Zhiwei Wu. Further diversity is introduced by Professor Jianfang Fei from the National University of Defense Technology, China; Professor Jianhua Lu from Sun Yat-sen University; and Professor Shiqiu Peng from the South China Sea Institute of Oceanology, Chinese Academy of Sciences, China.

Concluding this remarkable ensemble are five professors and an assistant professor from Nanjing University of Information Science & Technology: Professors Xiaolei Zou, Yubao Liu, Wenjun Zhang, Li Guan, Zhengkun Qin, and Assistant Professor Meirong Wang. It is a testament to their collective dedication and expertise that they accomplished the formidable task of contributing to this endeavor within a remarkably short span, from 27 June 2022 to 30 June 2023. To each of these esteemed invited authors and their collaborators, we extend our heartfelt gratitude.

This book was supported by the National Key R&D Program of China (91937302).

Xiaolei Zou, Guoxiong Wu, and Zhemin Tan
Editors

remote sensing

Editorial

Editorial for the Topic "A Themed Issue in Memory of Academician Duzheng Ye (1916–2013)"

Xiaolei Zou [1,*], Ming Cai [2], Guoxiong Wu [3] and Zhemin Tan [4]

1 Joint Center of Data Assimilation for Research and Application, Nanjing University of Information and Science & Technology, Nanjing 211544, China
2 Department of Earth, Ocean & Atmospheric Sciences, Florida State University, Tallahassee, FL 32304, USA; mcai@fsu.edu
3 State Key Laboratory of Numerical Modelling for Atmospheric Sciences and Geophysical Fluid Dynamics (LASG), Institute of Atmospheric Physics (IAP), Chinese Academy of Sciences (CAS), Beijing 100029, China; gxwu@lasg.iap.ac.cn
4 School of Atmospheric Sciences, Nanjing University, Nanjing 210023, China; zmtan@nju.edu.cn
* Correspondence: xzou@nuist.edu.cn

Citation: Zou, X.; Cai, M.; Wu, G.; Tan, Z. Editorial for the Topic "A Themed Issue in Memory of Academician Duzheng Ye (1916–2013)". *Remote Sens.* **2023**, *15*, 4586. https://doi.org/10.3390/rs15184586

Received: 13 September 2023
Accepted: 15 September 2023
Published: 18 September 2023

This Topic covers a wide range of topics, including atmospheric dynamics and physics, synoptic weather, climate variability, climate change, and remote sensing observations for weather and climate studies. There are a total of 29 research papers and two review papers published in "A Themed Issue in Memory of Academician Duzheng Ye (1916–2013)", comprising 14 in *Atmosphere* [1–14], 2 in *Climate* [15,16], and 15 in *Remote Sensing* [17–31].

The first paper (Lu [1]) presents a chronological overview of Professor Ye's remarkable 79-year research journey, spanning from preparatory years and graduate studies, through pioneering breakthroughs in atmosphere dynamics, to a pivotal transition that encompasses global climate sciences and delves into the intricate interplay of climate change's societal impacts, culminating in a globally influential era, offering a comprehensive portrayal of his lifelong scholarly expedition. These prominent accomplishments encompass pioneering seminal research on the Rossby wave dispersion theory and scale-dependent theory of geostrophic adjustment; the establishment of a new meteorology subdiscipline, Qinghai–Tibetan Plateau meteorology; groundbreaking research on the atmospheric general circulation across East Asia and the globe; and visionary research on global changes and their implications for socioeconomic and ecological welfare.

The following five papers in this themed Topic collectively further explore the impacts of the Qinghai–Tibetan Plateau on shaping weather patterns, climate variability, and climate changes across China, East Asia, and beyond. These contributions exemplify the continuous growth and enrichment of our understanding of Qinghai–Tibetan Plateau meteorology, a subdiscipline founded by Professor Ye. The study of Liu et al. [2] introduces a novel circulation index to characterize the orographic potential vorticity (PV) forcing across the Tibetan Plateau. The PV-based index demonstrates a connection to the interannual variability of the East Asian monsoon. The underlying mechanism for this linkage is the downstream influence of the orographic PV forcing across the Tibetan Plateau through the vertical differential advection of PV, reinforcing upward motions and producing extensive precipitation along the Jianghuai Meiyu front. Liu et al. [3] reveal the frequent occurrence of large-scale Kármán vortex streets on the leeward side of the Tibetan Plateau under the seasonally varying zonal mean flow. These vortex streets significantly impact the timing of the seasonal migration and the strength of the rain band along the Meiyu Front to the east of the Tibetan Plateau. Their findings add to the previously discovered roles of the Tibetan Plateau in shaping the weather and climate in eastern Asia. Huang et al. [4] provide a comprehensive review on the characteristics of the two prominent teleconnection patterns, the East Asia/Pacific (EAP) pattern for meridionally propagating wave trains and the "Silk Road" pattern for zonally propagating wave trains, and their impacts on the

East Asian summer monsoon system and summer precipitation variability in China. This paper also reviews the dynamic processes responsible for interdecadal oscillations of boreal quasi-stationary planetary waves and their impacts on the variability of the East Asian winter monsoon system. The study by Ding et al. [5] found that the Tibetan Plateau is the source region of Rossby waves for synoptical disturbances downstream in boreal summer through both vortex stretching and vorticity advection. The interdecadal variability of the Rossby source over the Tibetan Plateau exerts a significant influence on the variability of precipitation in the Huang-huai River Basin, South Korea, and Japan. Sun and Mu [6] demonstrate the effectiveness of the conditional nonlinear optimal parameter perturbation method in discerning the nonlinear characteristics of terrestrial ecosystems and its utility in identifying physical parameters in ecosystem models that exhibit high sensitivity to climate forcing, consequently addressing sources of significant uncertainties in ecosystem models. Their study areas of ecosystems include the Inner Mongolia region, the north–south transect of eastern China, and the Qinghai–Tibet Plateau region.

The common theme of the next four papers revolves around the presentation of observational and modeling evidence highlighting climate changes across diverse regions and varying temporal scales. The contribution from Liu [7] provides a comprehensive review on the recent progress towards a better understanding of thermohaline instability in observations and model simulations, which is a primary mechanism for abrupt climate change as evidenced in paleoclimate proxy data. Lau et al. [8] investigated changes in the characteristics of tropical extreme precipitation–cloud regimes in response to anthropogenic radiative forcing, pinpointing increases in the surface moisture (temperature) as the key driver leading to enhanced convective instability over tropical oceans (land). The paper by Secor et a. [9] is an observational study that quantifies the change in amplitude of synoptic-scale surface temperature variability across the U.S., finding a surge in the surface temperature variability in the Rockies and surrounding regions but a reduction over low land regions. Zhu et al. [10] report a comprehensive observational study on the seasonal variation of the diurnal cycle of the surface sensible heat flux at different stations over the Tibetan Plateau region, finding that the standard heat transfer coefficient used in models tends to overestimate surface sensible heat flux.

The last four papers published in *Atmosphere* are devoted to new techniques for diagnosing the climate variability of the stratospheric polar vortex using a parametric model of elliptic orbits [11], the development of a four-dimensional variational data assimilation system using a neural network-based machine learning technique [12], improving tropical cyclone track forecasts through assimilating Advanced Microwave Sounding Unit data derived from satellites and large-scale flows derived from a global model [13], and improving surface weather forecasts through applying the Grid-to-Point Deep Learning Error Correction method [14]. The two papers published in *Climate* investigate the impacts of El Niño–Southern Oscillation (ENSO) on the subseasonal variations in the North Atlantic Oscillation (NAO) in winter seasons [15] and how the diversity of ENSO types further complicates the ENSO-NAO relation [16].

The 15 papers published in *Remote Sensing* span a wide range of applications of remote sensing observations. Utilizing high-density PM2.5 stations across eastern China and machine learning models, Li et al. [17] demonstrate the indispensable utility of satellite-derived aerosol optical thickness in accurately estimating surface PM2.5 levels, especially in situations when monitoring sites are sparsely distributed. Via combining satellite-derived vegetation images, meteorological observations, and reanalysis datasets, Mao et al. [18] reveal that the normalized difference vegetation index (NDVI) of the vegetation on the Tibetan Plateau is positively correlated with both the Indian summer monsoon (ISM) and precipitation. The correlation between NDVI and ISM is higher than that between NDVI and precipitation since the ISM can also influence the surface thermal conditions favoring vegetation growth on the Tibetan Plateau. Based on the TRMM and ERA5 datasets, Gao et al. [19] found that the spatial pattern of tropical latent heating (TLH) associated with equatorially symmetric tropical surface temperature (TST) is nearly identical to the spatial pattern of TLH associated with equatorially antisymmetric TST. The results suggest an urgent need for taking a whole-tropics perspective in understanding the variability of tropical air–sea–land system.

The next three papers focus on applications of ground-based and satellite radar observations. The main finding of Teng et al. [20] is that turbulent mixing is an additional cause of clear-air echoes besides atmobios migration in Chinese Doppler S-band Weather Radar (CINRAD/SA) observations in Beijing. Specifically, about 58% of dual-wavelength ratios between the CINRAD/SA S-band and the X-band observations from three X-band dual-polarization (X-POL) radars falls within the range of 18 dB and 24 dB, which is quite consistent with the Villars–Weisskopf turbulence theory. The diurnal variation in clear-air echoes is also associated with turbulent mixing since excessive turbulence mixing weakens clear-air echo signals. Yu et al. [21] explore the utility of a deep learning-based radar echo reconstruction model for accurately monitoring severe convective weather over oceans using Advanced Himawari-8 Imager (AHI) brightness temperature observations. Using the precipitation products derived from the dual-frequency precipitation radar (DPR) onboard the Global Precipitation Measurement core observatory (GPM) satellite, Wang et al. [22] conducted a composite analysis of the precipitation structures and microphysical properties in 6432 cases of the northeastern China cold vortex (NCCV) from 2014 to 2019. Their main conclusion is that the precipitation associated with NCCV has a different mechanism for the growth of hydrometeors above and below ~4 km altitude. The lower-level particles grow through collision, whereas the upper hydrometeors grow through the Bergeron process.

The papers [23,24] are on the satellite data assimilation of microwave temperature sounding observations with two different focuses. Zhu et al. [23] improve the methodology for assimilating long-existing instruments in a research data assimilation system, whereas Li et al. [24] conduct an exploratory study on a newly launched satellite in an operational system. Zhu et al. [23] propose a weighted average hypsometry as an observation space localization method to improve AMSU-A radiance assimilation in an ensemble four-dimensional variational (En4DVar) data assimilation system. As a result, the forecast skill of the En4DVar-initialized model becomes comparable to the 4DVar-initialized forecast skill in the southern extratropics and tropics. Li et al. [24] demonstrate that a direct assimilation of Chinese early-morning polar-orbiting satellite FY-3E Microwave Temperature Sounder-3 (MWTS-3) radiance observations has a positive contribution to the China Meteorological Administration Global Forecast System's (CMA-GFS's) global analysis and forecasts for the Southern Hemisphere.

The next three papers deal with satellite data quality. Chen and Guan [25] assess the quality of observations from the Hyperspectral Infrared Atmospheric Sounder-II (HIRAS-II), which is an infrared hyperspectral instrument onboard the world's first early-morning polar-orbiting satellite FY-3E. They concluded that the performance of the HIRAS-II instrument is consistently stable. The Geostationary Interferometric Infrared Sounder (GIIRS) onboard FY-4A is the first infrared hyperspectral atmospheric vertical sounder onboard a geostationary satellite. Yao and Guan [26] derive atmospheric temperature and humidity profiles under all sky conditions from GIIRS-observed brightness temperatures using three deep machine learning algorithms. As expected, the accuracy of the temperature retrieval, validated against radiosonde observations, is the highest when the field of view is completely clear. Shen et al. [27] apply an iterative principal component analysis method for filling the data gaps of Advanced Microwave Scanning Radiometer (AMSR)-2 C-band data in areas where the effect of radio frequency interference (RFI) is strong and thus removed.

Hu and Zou [28] employ brightness temperature (TB) observations at an infrared channel (10.3 μm) of the Advanced Baseline Imager (ABI) onboard the U.S. 16th Geostationary Operational Environmental Satellite (GOES-16) to directly determine tropical cyclone (TC) center positions as well as the radii of inner (R_{IR}) and outer (R_{OR}) rainbands. The root-mean-square differences for 108 cases between ABI-determined centers and the best track centers are 45.35 and 29.06 km for tropical storms and hurricanes, respectively. Their estimates of R_{IR} and R_{OR} can be used to determine storms' annulus, which is defined as the area of asymmetric rainbands dominated by the azimuthal wavenumbers 1–3 from R_{IR} and R_{OR}.

Aiming at applying satellite data for climate studies, Dong and Zou [29] examine 20 years of observations from 2003 to 2023 from the special sensor microwave imager sounder (SSMIS) onboard Defense Meteorological Satellite Program (DMSP) F16, F17, and F18. F16 was launched on 18 October 2003 and carried the first conical-scanning radiometer SSMIS that combines the special sensor microwave/imager (SSM/I), the special sensor microwave/temperature sounder (SSM/T), and the special sensor microwave/water vapor sounder (SSM/T2) together. Having access to nearly two decades of F16 SSMIS data offers abundant opportunities for studying the atmosphere at both the synoptic and decadal scales. Unfortunately, data noise of complicated structures occurred in brightness temperature (TB) observations at the lower atmospheric sounding (LAS) channels since 25 March 2013. Due to noise interference, TB observations reflecting rain, clouds, tropical cyclone warm core, temperature and water vapor distributions are not readily distinguishable, especial in channels above the middle troposphere (channels 4–7 and 24), whose dynamic ranges of TB are smaller than low tropospheric channels 1–3. Dong and Zou [29] not only found that the data noise is around 1–2 K, occurs at certain cross-track wavelengths, and has a latitudinal variation, but also developed an effective noise mitigation scheme. It was shown that TB observations from conical-scanning radiometer SSM/T can directly capture hurricane 3D structures. We may also investigate the decadal change in many features of tropical cyclones derivable from these TB observations once the noise in F16 SSMIS LAS channels from April 25 to the present are eliminated.

The second-to-last paper is on oceanic remote sensing data. Liu et al. [30] conducted a study to find the decorrelation length scale of background errors is a key factor for the two-dimensional variational method (2D-Var) to generate multi-satellite merged maps of altimeters with an effective resolution capable of capturing meso-scale eddies in the ocean. They conclude that having a higher proportion of small-scale signals and a smaller proportion of large-scale dynamic signals result in a smaller decorrelation length scale of background errors and thus a higher effective resolution of the merged altimeter data.

Finally, the paper from Zhao et al. [31] conducts an observation-based investigation of the dynamics and microphysical characteristics of the extreme heavy rainfall on 20 July 2021 in Zhengzhou, China. The record-breaking hourly rainfall of 201.9 mm caused severe urban flooding and human casualties. The synoptic flow was characterized by an erect updraft at the low levels and an enhanced easterly inflow, which brought abundant moisture from the oceans and converged at Zhengzhou. A slow-moving convective storm persisted for an hour-long time over Zhengzhou and produced the extreme rainfall. The disdrometer data reveal unusually high concentrations of all sizes of raindrops with both maritime and continental convection properties, and the polarimetric radar data confirm the important contributions of both ice-based and warm rain processes. Therefore, the extremely heavy rainfall resulted from an interaction among convective-scale, mesoscale dynamics and microphysical processes under favorable synoptic conditions.

In conclusion, many novel developments and advances in meteorology and oceanography are represented in the publications in this Topic. We hope that this collection of papers will stimulate further research in atmospheric and oceanographic sciences.

Author Contributions: All authors contributed equally to all aspects of this editorial. All authors have read and agreed to the published version of the manuscript.

Acknowledgments: The topic editors would like to thank all authors who contributed to this Topic for sharing their scientific findings. They would also like to thank the reviewers for their valuable comments and suggestions to improve the quality of papers published within this Topic, the academic editors, and the editorial teams of *Remote Sensing*, *Atmosphere*, and *Climate* for all the support in the process.

Conflicts of Interest: The authors declare no conflict of interest.

References

1. Lu, J. From General Circulation to Global Change: The Evolution, Achievements, and Influences of Duzheng Ye's Scientific Research. *Atmosphere* **2023**, *14*, 1202. [CrossRef]
2. Liu, Y.; Luan, L.; Wu, G.; Ma, T. Impacts of the Surface Potential Vorticity Circulation over the Tibetan Plateau on the East Asian Monsoon in July. *Atmosphere* **2023**, *14*, 1038. [CrossRef]
3. Liu, Q.; Wu, Z.; Tan, Z.-M.; Yang, F.; Fu, C. The Atmospheric Vortex Streets and Their Impact on Precipitation in the Wake of the Tibetan Plateau. *Atmosphere* **2023**, *14*, 1096. [CrossRef]
4. Huang, R.; Huangfu, J.; Liu, Y.; Lu, R. The Guiding Role of Rossby Wave Energy Dispersion Theory for Studying East Asian Monsoon System Dynamics. *Atmosphere* **2023**, *14*, 962. [CrossRef]
5. Ding, Y.; Sun, X.; Li, Q.; Song, Y. Interdecadal Variation in Rossby Wave Source over the Tibetan Plateau and Its Impact on the East Asia Circulation Pattern during Boreal Summer. *Atmosphere* **2023**, *14*, 541. [CrossRef]
6. Sun, G.; Mu, M. Applications of CNOP-P Method to Predictability Studies of Terrestrial Ecosystems. *Atmosphere* **2023**, *14*, 617. [CrossRef]
7. Liu, Z. Instability of Atlantic Meridional Overturning Circulation: Observations, Modelling and Relevance to Present and Future. *Atmosphere* **2023**, *14*, 1011. [CrossRef]
8. Lau, W.K.M.; Kim, K.-M.; Harrop, B.; Leung, L.R. Changing Characteristics of Tropical Extreme Precipitation–Cloud Regimes in Warmer Climates. *Atmosphere* **2023**, *14*, 995. [CrossRef]
9. Secor, M.; Sun, J.; Yang, F.; Zou, X.; Wu, Z. Global Warming and Topography Impact the Amplitude of the Synoptic-Scale Surface Temperature Variability across the US. *Atmosphere* **2023**, *14*, 979. [CrossRef]
10. Zhu, Z.; Wang, M.; Wang, J.; Ma, X.; Luo, J.; Yao, X. Diurnal Variation Characteristics of the Surface Sensible Heat Flux over the Tibetan Plateau. *Atmosphere* **2023**, *14*, 128. [CrossRef]
11. Yu, Y.; Sun, J.; Secor, M.; Cai, M.; Luo, X. A Parametric Model of Elliptic Orbits for Annual Evolutions of Northern Hemisphere Stratospheric Polar Vortex and Their Interannual Variability. *Atmosphere* **2023**, *14*, 870. [CrossRef]
12. Tian, X.; Conibear, L.; Steward, J. A Neural-Network Based MPAS—Shallow Water Model and Its 4D-Var Data Assimilation System. *Atmosphere* **2023**, *14*, 157. [CrossRef]
13. Lai, Z.; Peng, S. The Effect of Assimilating AMSU-A Radiance Data from Satellites and Large-Scale Flows from GFS on Improving Tropical Cyclone Track Forecast. *Atmosphere* **2022**, *13*, 1988. [CrossRef]
14. Qin, Y.; Liu, Y.; Jiang, X.; Yang, L.; Xu, H.; Shi, Y.; Huo, Z. Grid-to-Point Deep-Learning Error Correction for the Surface Weather Forecasts of a Fine-Scale Numerical Weather Prediction System. *Atmosphere* **2023**, *14*, 145. [CrossRef]
15. Zhang, W.; Jiang, F. Subseasonal Variation in the Winter ENSO-NAO Relationship and the Modulation of Tropical North Atlantic SST Variability. *Climate* **2023**, *11*, 47. [CrossRef]
16. Zhang, P.; Wu, Z. Insight into Asymmetry in the Impact of Different Types of ENSO on the NAO. *Climate* **2023**, *11*, 136. [CrossRef]
17. Tian, Z.; Wei, J.; Li, Z. How important is satellite-retrieved aerosol optical depth in deriving surface $PM_{2.5}$ using machine learning? *Remote Sens.* **2023**, *15*, 3780. [CrossRef]
18. Mao, X.; Ren, H.-L.; Liu, G.; Su, B.; Sang, Y. Influence of the Indian summer monsoon on inter-annual variability of the Tibetan-plateau NDVI in its main growing season. *Remote Sens.* **2023**, *15*, 3612. [CrossRef]
19. Gao, Y.; Liu, X.; Lu, J. Tropical surface temperature and atmospheric latent heating: A whole-tropics perspective based on TRMM and ERA5 datasets. *Remote Sens.* **2023**, *15*, 2746. [CrossRef]
20. Teng, Y.; Li, T.; Ma, S.; Chen, H. Turbulence: A significant role in clear-air echoes of CINRAD/SA at night. *Remote Sens.* **2023**, *15*, 1781. [CrossRef]
21. Yu, X.; Lou, X.; Yan, Y.; Yan, Z.; Cheng, W.; Wang, Z.; Zhao, D.; Xia, J. Radar echo reconstruction in oceanic area via deep learning of satellite data. *Remote Sens.* **2023**, *15*, 3065. [CrossRef]
22. Wang, J.; Zhuge, X.; Chen, F.; Chen, X.; Wang, Y. A preliminary analysis of typical structures and microphysical characteristics of precipitation in northeastern China cold vortexes. *Remote Sens.* **2023**, *15*, 3399. [CrossRef]
23. Zhu, S.; Wang, B.; Zhang, L.; Liu, J.; Liu, Y.; Gong, J.; Xu, S.; Wang, Y.; Huang, W.; Liu, L.; et al. Assimilating AMSU-A radiance observations with an ensemble four-dimensional variational (En4DVar) hybrid data assimilation system. *Remote Sens.* **2023**, *15*, 3476. [CrossRef]
24. Li, J.; Qian, X.; Qin, Z.; Liu, G. Direct assimilation of Chinese FY-3E Microwave Temperature Sounder-3 Radiances in the CMA-GFS: An Initial Study. *Remote Sens.* **2022**, *14*, 5943. [CrossRef]
25. Chen, H.; Guan, L. Assessing FY-3E HIRAS-II radiance accuracy using AHI and MERSI-LL. *Remote Sens.* **2022**, *14*, 4309. [CrossRef]
26. Yao, S.; Guan, L. Comparison of three convolution neural network schemes to retrieve temperature and humidity profiles from the FY4A GIIRS observations. *Remote Sens.* **2022**, *14*, 5112. [CrossRef]
27. Shen, W.; Lin, Z.; Qin, Z.; Bai, X. Improved estimation of O-B bias and standard deviation by an RFI restoration method for AMSR-2 C-Band observations over North America. *Remote Sens.* **2022**, *14*, 5558. [CrossRef]
28. Hu, Y.; Zou, X. Determining tropical cyclone center and rainband size in geostationary satellite imagery. *Remote Sens.* **2022**, *14*, 3499. [CrossRef]
29. Dong, H.; Zou, X. Mitigation of systematic noise in F16 SSMIS LAS channels observations for tropical cyclone applications. *Remote Sens.* **2022**, *14*, 3171. [CrossRef]

30. Liu, L.; Zhang, X.; Fei, J.; Li, Z.; Shi, W.; Wang, H.; Jiang, X.; Zhang, Z.; Lv, X. Key for improving resolution capability of sea surface height merged maps from multi-satellite altimeters' data in the South China Sea. *Remote Sens.* **2023**, *15*, 4275. [CrossRef]
31. Zhao, K.; Xu, X.; Xue, M.; Tan, Z.-M.; Huang, H.; Zhou, A.; Fan, X.; Liu, Q.; Wang, C.; Fang, J.; et al. The dynamics and microphysical characteristics of convection producing the Record-breaking Hourly Precipitation on 20 July 2021 in Zhengzhou, China. *Remote Sens.* **2023**, *15*, 4511. [CrossRef]

atmosphere

MDPI

Review

From General Circulation to Global Change: The Evolution, Achievements, and Influences of Duzheng Ye's Scientific Research

Jianhua Lu

School of Atmospheric Sciences, Southern Marine Science and Engineering Guangdong Laboratory (Zhuhai), Sun Yat-sen University (SYSU), Zhuhai 519082, China; lvjianhua@mail.sysu.edu.cn

Abstract: Duzheng Ye (Tu-cheng Yeh) was an active member of Rossby's Chicago School, one of the founders of modern meteorology in China since the 1950s, and a pioneer of global change science in China and over the world. His achievements have been central to the development of atmospheric and climate dynamics and global change studies in China, and many of them remain to be fundamental in the context of global climate change. In this review, his lifelong research career is divided into five periods: (1) the preparatory period (1935–1944); (2) the Chicago period (1945–1950); (3) the period of breaking ground (1950–1966); (4) the period of transition (1972–1983); and (5) the period of global change (1984–2013). The evolution of Yeh's main achievements is described in the context of the historical background of both China and the world. These well-known achievements included the theory of energy dispersion in the atmosphere, the general circulation of the atmosphere (GCA) over East Asia and the globe, Qinghai–Tibetan Plateau meteorology, the scale-dependence theory of geostrophic adaptation (adjustment), and his pioneering ideas on global change. Special emphases are put on some of Yeh's investigations that were well ahead of his time, such as his investigations on trade inversion, the GCA as an internally consistent whole, abrupt seasonal changes in the GCA, the physical mechanism of atmospheric blocking, and orderly human activities.

Keywords: Duzheng Ye; energy dispersion; trade inversion; the general circulation of the atmosphere; Qinghai–Tibetan Plateau meteorology; geostrophic adjustment; atmospheric blocking; global change; orderly human activities

Citation: Lu, J. From General Circulation to Global Change: The Evolution, Achievements, and Influences of Duzheng Ye's Scientific Research. *Atmosphere* **2023**, *14*, 1202. https://doi.org/10.3390/atmos14081202

Academic Editors: Xiaolei Zou, Guoxiong Wu and Zhemin Tan

Received: 30 June 2023
Revised: 23 July 2023
Accepted: 24 July 2023
Published: 26 July 2023

1. Introduction

Professor Duzheng Ye (21 February 1916–16 October 2013; also spelled as Tu-cheng Yeh, and Yeh shall be used hereafter) was one of the active members in Rossby's Chicago School before his return to China and one of the founders of modern meteorology in China since the 1950s. He was also one of the pioneers who initiated and steered international global change research in the early 1980s [1–4].

The late Prof. Shiyan Tao, a lifelong colleague of Yeh's since 1950, together with Prof. Zhongxiang Hong, briefly summarized Yeh's achievements from three perspectives: (1) achievements in research and training; (2) the establishment of the Institute of Atmospheric Physics (IAP) in the Chinese Academy of Sciences (CAS); (3) coordinating international cooperation [5]. In terms of achievements in research and training, they mentioned energy dispersion, the adaptation (adjustment) of atmospheric motions, atmospheric general circulation, Tibetan Plateau meteorology, and climate dynamics and global change. This review addresses Yeh's achievements in research only.

Although Yeh's many accomplishments are well-known by scientific communities and his cause is developing well both in China and worldwide, there is still a need to provide a relatively thorough retrospective on the evolution of his scientific ideas and investigations via a meta-analysis of his published works. The necessity comes from the fact that many of Yeh's investigations were highly prospective, with some of them well ahead of his time, and hence the aim of this review is not so much to honor the past achievements of a giant in atmospheric sciences, but to inspire new scientific findings from Yeh's previous works.

As an esteemed scientist, Yeh never put himself in a lofty position. On the contrary, he always considered that these accomplishments were not just his, but also belonged to his colleagues and the meteorological and climatic communities at large in China. He was most proud not of the numerous prizes and honors he obtained, but that he could dedicate his life and talent with his scientific endeavors to the advancement of science in China and the civilization of the world. Therefore, it is appropriate to place this review in the broader context of the historical background of both China and the world.

The organization of the review follows chronologically the five periods in Yeh's lifelong research career, which are named by the author of the review as follows: (I) the preparatory period (1935–1944) in Section 2; (II) the Chicago period (1945–1950) in Section 3; (III) the period of breaking ground (1950–1966) in Section 4; (IV) the period of transition (1972–1983) in Section 5; (V) the period of global change (1984–2013) in Section 6. Each of the main fields to which Yeh contributed will be described in one of the five periods for the sake of narrative clarity. Section 7 is a brief summary.

2. Period I: The Preparatory Period (1935–1944)

During this period, Yeh initially enrolled as a Physics student at Tsinghua University in 1935; then, he shifted his interest to meteorology in 1938 due to its crucial role in the war against Japan's invasion of China. In 1940, Yeh successfully graduated from the Southwestern Associated University in Kunming, a united institution comprising Peking University, Tsinghua University, and Nankai University. After a very short period of teaching at a high school, Yeh became a graduate student of meteorology at Zhejiang University, which moved westward to Guizhou Province in southwestern China during the war. During this time, Yeh studied atmospheric electricity under the guidance of Changwang Tu (1902–1962), a renowned meteorologist and central figure of modern meteorological research and operational systems, and Ganchang Wang (1907–1998), a former student of Lise Meitner and a pioneer of China's nuclear science. Meanwhile, Yeh published in 1942 his first research paper on isentropic analysis (Figure 1).

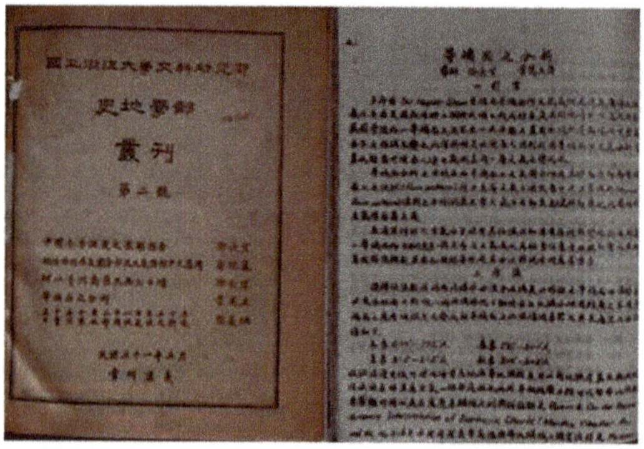

Figure 1. A picture of Yeh's first paper, "Analysis on Isentropic Surfaces" in *Memoir of the Division of History and Geography of Zhejiang University*, No. 2, 1942.

After completing his studies at Zhejiang University in the summer of 1943, Yeh joined the Meteorological Institute, which was one of the first eight institutes in Academia Sinica established in 1928, situated in Beibei, Chongqing during the war. Yeh served as a research assistant there until his arriving at the University of Chicago in early 1945 as a doctoral student under the supervision of C.-G. Rossby.

During this preparatory period, Yeh underwent comprehensive training in physics and meteorology under the guidance of China's finest mentors. His strong background in scientific education would play a pivotal role in shaping his subsequent career and achievements.

3. Period II: The Chicago Period (1945–1950)

The Chicago School established by C.-G. Rossby, since his move to the University of Chicago in 1940, was undoubtedly at the center of the explosive development of modern meteorology after World War II. During this period, J. G. Charney, who had just finished his Ph.D. thesis at UCLA and was to develop his quasi-geostrophic theory and baroclinic instability theory and lead the Princeton project of numerical weather prediction, had stayed in Chicago from 1946 to 47 and lectured hydrodynamics there [6]. E. Palmén, during his extended stay in Chicago, engaged in stimulating debates with V. Starr on the respective roles of the meridional Hadley circulation and transient eddies in momentum and energy transport [7]. Prominent European meteorologists such as Tor Bergeron, Alf Nyberg, and Halvor Solberg visited Chicago and sometimes taught classes [6]. Rossby, while being occupied with his organizing duties [8], advanced his theory of the westerly jet based on the mixing of vorticity [9]. He was also working on theoretical explanations for the general circulation of the atmosphere (referred to as the GCA hereafter). David Fultz was conducting his noteworthy dishpan experiment; during this period, Herbert Riehl, returning from his work at the Institute of Tropical Meteorology located in Puerto Rico, taught tropical meteorology in Chicago [7]. It was during these "great days of the Chicago School" [8] that Yeh came to the very center of world meteorology and, as an active member of the Chicago School, soon contributed to the explosive development of the field. Indeed, this may also be considered as the formative period of Yeh's entire research career.

3.1. *On Energy Dispersion in the Atmosphere*

Rossby initially introduced the concept of group velocity into meteorology [10] in 1945, the same year that Yeh arrived at Chicago, and Yeh chose the topic of energy dispersion in the atmosphere as his doctoral thesis. Before his graduation in 1948, part of his results had been included in the seminal 1947paper, "On the General Circulation of the Atmosphere in Middle Latitudes", which was led by Rossby and Palmén, with Yeh being one of the participants [11] (note: the reference numbers in bold indicate that Yeh was the author or one of the authors of the publication). Figure 23 of [11], which was also included in Yeh's "energy dispersion" paper published one and a half years later (Figure 9 in [12]), illustrated the establishing process of a one-dimensional stationary wave as a result of the injection of cyclonic vorticity (wave source) at a prescribed longitude, based on the theoretical calculation in [12]. More than thirty years later, Brian Hoskins and David Karoly developed a theory for stationary Rossby wave propagation in a spherical atmosphere [13], which can be considered a two-dimensional extension of Yeh's one-dimensional theory. Interested readers may refer to [14,15] for a better understanding of the role of energy dispersion theory in atmospheric and climatic dynamics.

Indeed, the immediate and long-term applications of Yeh's energy dispersion theory went beyond this in stationary wave dynamics. It is interesting to consider Rossby's evaluation of Yeh's thesis, which was submitted to the *Journal of Meteorology* on 27 April 1948, in his letter to George Platzman (Figure 2) on 29 October 1948:

> *"Who has gone through the analytical part of Yeh's mathematics? Personally, I preferred to have his paper devoted as much as possible to basic questions, relatively less attention to detailed mathematical computations. The breakdown, dispersion, of a solitary ridge, is of importance. The coastal effects should not be included. Please let Yeh understand that for his own sake, overloading of the paper must be avoided. It will merely result in nobody reading the paper. It must be recognized that Yeh's thesis deals with the very heart of the Princeton project and must be written so as to promote this development. Best wishes, C. G. R."*

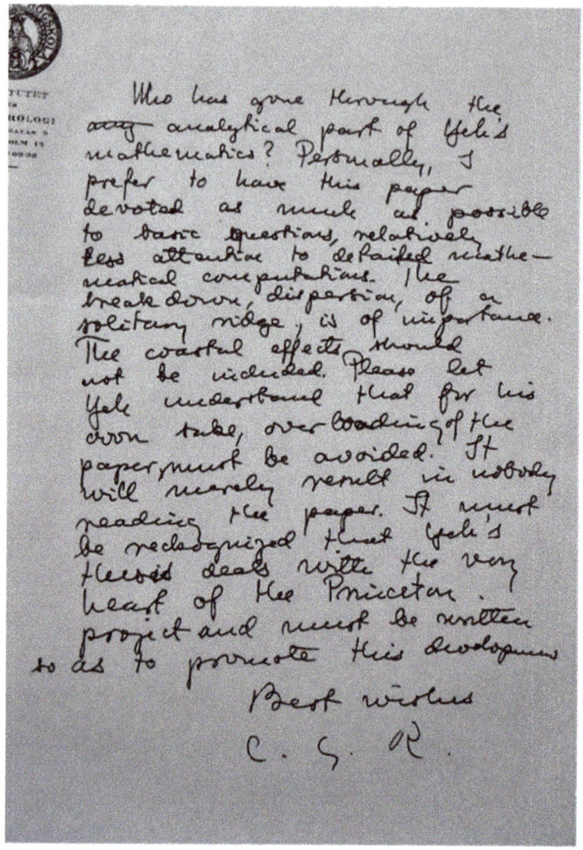

Who has gone through the analytical part of Yeh's mathematics? Personally, I prefer to have this paper devoted as much as possible to basic questions, relatively less attention to detailed mathematical computations. The breakdown, dispersion, of a solitary ridge, is of importance. The coastal effects should not be included. Please let Yeh understand that for his own sake, overloading of the paper must be avoided. It will merely result in nobody reading the paper. It must be recognized that Yeh's thesis deals with the very heart of the Princeton project and must be written so as to promote this development.

Best wishes

C. G. R.

Figure 2. The letter of C.-G. Rossby to G. W. Platzman regarding Yeh's "energy dispersion" thesis. The date of the letter was 29 October 1948. (Photo courtesy of Anders Persson.)

'The Princeton project' in Rossby's letter refers to the numerical weather prediction (NWP) project led by Charney and von Neumann [16]. The reason why "Yeh's thesis deals with the very heart of the Princeton project" was because the results in Yeh's thesis were essential to Charney's choice of a minimum forecast area that could ensure the success of the NWP project. N. A. Phillips, who was one of the team members of the project, later further stated that if too small an area had been chosen, the failure of the NWP project would have led to a 'tremendous setback' for Charney's quasi-geostrophic theory [17] (p. 4). Indeed, Rossby not only asked Charney to write a brief note to accompany Yeh's paper [18] (p. 126), but also wrote by himself another note in the first issue of the first volume of *Tellus*, which was launched by Rossby in Sweden, to introduce Yeh's paper to a broader audience [19].

The importance of "the breakdown, dispersion, of a solitary ridge" that Rossby emphasized in his letter refers to Yeh's theoretical analyses and calculations on the different dispersive features of solitary waves at low, middle, and high latitudes (see Section 5 and Figures 10–12 in [12]), which were used to understand fast dispersion at low latitudes and the preferred occurrence and maintenance of blocking highs over high latitudes, as was stated by Yeh himself in [12].

The abstract and main contents of [12] indicated that Yeh understand the potential applications of his work well. The first application, "the formation of a new trough over North America following an intense cyclogenesis in the Gulf of Alaska" [12], was soon visualized by E. Hovmöller in his troughs-and-ridges diagram [20] and was used to explain recent extreme cold events [21]. In the second application, Yeh explained the formation of blocking action at high latitudes by the dispersion of an initial solitary wave, in contrast to Rossby's earlier hypothesis based on the convergence of group velocity [10]. Later, during the late 1950s and early 1960s, Yeh and his Chinese colleagues would conduct a systematic survey and series of theoretical and numerical studies on blocking highs (see Section 4.4).

3.2. On Hadley Circulation, Trade Inversion, and the Motion of Tropical Storms

Yeh continued his pursuit of the general circulation over middle latitudes during the period 1949–1950 after the energy dispersion paper was published. For example, he discussed the maintenance of zonal circulation from the perspective of vertical and meridional vorticity transport [22]. During this period (1949–1950), while Rossby spent most of his time at Stockholm, Yeh joined Riehl's group on tropical meteorology, in which Yeh was second to Riehl [7]. Together with Riehl and the other team members, Yeh made significant contributions to the field of tropical meteorology at its very beginning. One of these contributions was the proof of the existence of the Hadley circulation. To settle the debate between Palmén and Starr on the role of the Hadley circulation in the global circulation system [7], Riehl and Yeh proved for the first time, by computations based on data obtained from the climatic atlas, the existence of meridional Hadley circulation [23].

Following their work on the Hadley cell, they further explored the spatial structure and the processes governing the balance of radiation, surface sensible and latent heat, and momentum in the north-east trade wind of the Pacific Ocean. The resultant 1951 publication by Riehl, Yeh, J. Malkus (later J. Simpson), and N. La Seur [24] is widely considered a seminal paper on the physics of tropical trade inversion and is one of the most celebrated contributions to meteorology of the twentieth century [7]. Interested readers may find a detailed observational and historical background for [24] in [7]. By carefully reading the 29-page paper containing 24 figures, one cannot help feeling the comprehensive and physically penetrating nature of the analyses that Riehl, Yeh, and their colleagues conducted on the observations obtained from the five weather ships. They concluded that the trade inversion is not a discontinuity separating the upper-layer dry air and lower-layer moist air, and that downward mass transfer occurs through the inversion. Their conclusions inspired the later development of theoretical and numerical models of trade cumuli, which are essential to the parameterization of shallow convections in climate models [25,26]. Indeed, in the mind of the author of this review, the broad view presented by Riehl, Yeh, and their colleagues in [24] may well help provide a better perspective on the relations between clouds, circulation, the hydrological cycle, and global climate change.

Before Yeh's return to China in 1950, he also conducted an investigation on the mechanism leading to the oscillating trace of tropical storms [27] and a survey on the rainfall over the islands of Hawaii [28], which originated from the collaborative research between Rossby's School and the Pineapple Research Institute in Hawaii [8].

4. Period III: The Period of Breaking Ground (1950–1966)

The education and research in modern meteorology in China was relatively limited in terms of scope and depth before 1949 [29]. The situation underwent a significant transformation after the establishment of the People's Republic of China. The former Institute of Meteorology in Academia Sinica was reorganized as the Institute of Geophysics and Meteorology (IGM) under the Chinese Academy of Sciences (CAS) in 1950, which was the main body of meteorological and geophysical studies in China until 1966, when it was separated into several institutes. Yeh and Chen-chao Koo [30], both protégés of Rossby who returned to China in 1950 (Koo from Stockholm), were the two directors responsible for leading the meteorological studies in the institute, under the leadership of their former teacher, J. J. Jaw, who was the director-general of the IGM.

Indeed, in 1973, two renowned atmospheric dynamicists from the U.S., W. Blumen and W. M. Washington, noted that, based on translated meteorological publications from

China, there was 'accumulated evidence' suggesting that "the field of meteorology (in China) had become a well established and continually growing scientific activity" between 1949 and 1966 [31]. They acknowledged the need for "a more exhaustive overview of the contributions made by Chinese meteorologists to the theory of the general circulation" [31]. It is worth mentioning that the groundbreaking works undertaken by Yeh and his colleagues at the IGM and in the universities played a pivotal role in this remarkable development. It was their efforts that realized Rossby's expectations [32] of an "extremely vigorous development of Chinese meteorology" and "many significant realistic contributions" in a relatively short span of approximately 16 years.

4.1. *On the General Circulation over East Asia*

Even prior to and, presumably, for his return to China, Yeh paid attention to the GCA over China and East Asia and published in *Tellus* a paper titled "The circulation of the high troposphere over China in the winter of 1945–46" [33]. Yeh unveiled, for the first time, the profound influence of the Qinghai–Tibetan Plateau on the upper tropospheric circulation, utilizing the best available data at that time. Within this seminal paper, Yeh observed the abrupt appearance of a new westerly jet at the southern flank of the Qinghai–Tibetan Plateau during mid-October, which merged downstream with the northern westerly jet into one singular and robust jet stream extending beyond the edge of the continent [33].

Given the pivotal importance of comprehending the general circulation over East Asia, which remained predominantly unexplored in 1950 both within China and worldwide, Yeh and his colleagues (Figure 3) in the IGM embarked on a systematical investigation of its three-dimensional structures, seasonal cycle, associated synoptic phenomena, and underlying principles. This dedicated pursuit began with the establishment of the Section of Synoptic and Dynamic Meteorology in the institute in 1950. A series of high-quality papers were published, most of them in Chinese and some in English, during this groundbreaking period.

Figure 3. Photos of Ye–Gu–Tao–Yang (Yeh–Koo–Dao–Yang), a legendary team leading the meteorological research in the IGM, under the leadership of J.J. Jaw, the director-general of the institute, during the period 1950–1966. Yeh and Koo were the two directors of the meteorological section in the IGM. (Photos courtesy of the family members of Ye, Gu, Tao, and Yang.)

A synthesis of their findings was initially presented by Chen-chao Koo to the international meteorological community on the evening of 4 June 1957, during the Numerical Weather Prediction Conference in Stockholm, through a seminar organized to showcase the meteorological achievements in China [34]. The audience of the seminar comprised C.-G. Rossby, J. G. Charney, B. Bolin, E. N. Lorenz, N. A. Phillips, and A. M. Obukhov, among others. Parts I and II of a manuscript titled "On the general circulation over Eastern Asia" [35,36] were submitted to *Tellus* on the following day of the seminar, with Part III [37] being submitted four months later. In a groundbreaking move, Yeh and his colleagues presented the international meteorological community with a systematic and physically coherent depiction of the GCA and weather systems over East Asia. This included elucidating their three-dimensional structures, seasonal disparities, oft-abrupt seasonal transitions, the development of associated weather systems, and the discernable influence of the Qinghai–Tibetan Plateau on circulation patterns and weather systems. These features were vividly illustrated through the 27 meticulously crafted figures presented in Part I and II of the three *Tellus* papers [35,36].

Yeh and his colleagues did not limit their attention to merely descriptive studies, but sharply realized fundamental theoretical problems while analyzing the phenomenological aspects of the general circulation over East Asia. These theoretical problems included, but were not limited to, the thermal and dynamical effects of huge orographic features on the general circulation, the role of the heat sources and sinks associated with the land–sea distributions, the role of orography in determining the positioning of mean troughs and ridges (now referred to as stationary waves), and the mechanism underlying abrupt seasonal changes in the general circulation. For instance, Figures 3 and 4 in Part III [37] of the three *Tellus* papers, obtained from a calculation applying the Green's function method to a steady, linearized Sawyer–Bushby model [38] forced by the topography of the Qinghai–Tibetan Plateau and the distribution of heat sources and sinks over the Northern Hemisphere, distinctively illustrated the respective roles of the topography and diabatic heating in the formation of a deep wintertime trough along the East Asian coast.

4.2. On the Meteorology of the Qinghai–Tibetan Plateau

While previous research had predominantly focused on the mechanical influence of topography on atmospheric circulation [39], Yeh and his colleague C.-C. Koo were quick to recognize and emphasize, as early as 1955, the potential thermal impact of large-scale topographic features, such as the Qinghai–Tibetan Plateau [40]. Through meticulous calculations using available aerological and surface observations from 1955–1956, Yeh and his two associates unequivocally stated in 1957 that the Qinghai–Tibetan Plateau acts as a heat source during summer and likely a heat sink (except in the southeastern region of the plateau) during winter [41]. This pioneering study stands as the first reliable calculation of the thermal effect of the Qinghai–Tibetan Plateau. Subsequently, Yeh and his colleagues devoted themselves to studying various aspects of the Qinghai–Tibetan Plateau's role in the GCA, climate, and weather, not only over China and East Asia, but also over the entire globe. Their research ultimately led to the establishment of the field known as "the Meteorology of the Tibetan Plateau" [42] or "the Meteorology of the Qinghai–Tibetan Plateau" [43], as suggested by the two monographs published in 1960 and 1979.

The investigation of Qinghai–Tibetan Plateau Meteorology permeated Yeh's lifelong career, and he made many other contributions to this field. For the sake of narrative clarity, I will briefly outline here his significant later contributions in this area, so as to focus in the subsequent sections on the new fields he pioneered and influenced. Yeh himself summarized the studies conducted by himself and his colleagues in this field in English-language journals such as the *Bulletin of the American Meteorological Society* (BAMS) in 1981 [44] and *Meteorology and Atmospheric Physics* in 1998 [45]. These review papers provide in-depth insights into their research, and interested readers are encouraged to refer to them for further details.

During the mid-1970s, Yeh and his colleagues conducted experiments by constructing a rotating annulus with a heated "plateau" to investigate the thermal effects of the Tibetan Plateau on the general circulation. Their research focused on various aspects such as the vertical structures of horizontal circulation, the zonal and meridional overturning circulations, the convective activities, and the movements of the Tibetan High over the Plateau, as is well summarized in [46]. In 1979, Yeh and his colleague Youxi Gao led the publication of [43], which provided a comprehensive summary of the achievements in this field by the Chinese meteorological community up to that time.

During the 1980s, Yeh, along with his colleagues and students, expanded his research to explore the influences of the Qinghai–Tibetan Plateau on the regions beyond East Asia [45]. Their work became particularly significant for current studies on polar climate change and the global effects of the Tibetan Plateau. In her doctoral thesis supervised by Yeh and G. X. Wu, X. L. Zou discovered that wintertime planetary waves with low wavenumbers (wave 1 and wave 2), induced by the mechanical forcing of the Tibetan Plateau (but not the Rocky Mountains), could propagate northward into high latitudes and vertically into the stratosphere [47,48]. This finding implied that the Tibetan Plateau plays a crucial role in shaping the spatial structure of the climate over the high latitudes of the Northern Hemisphere, affecting both its mean state and variations.

4.3. On the Scale-Dependence Theory of Geostrophic Adjustment

Geostrophic adjustment, known as "geostrophic adaptation" in China, has long been an important theoretical issue in atmospheric and oceanic dynamics [49]. In contrast to traditional wisdom, both Rossby in the 1930s [49] and Obukhov in 1949 [50] demonstrated that during the process of geostrophic adjustment, the mass (pressure) field rapidly aligns itself with the velocity field until the geostrophic balance is achieved. In 1957, Yeh was the first to highlight that the direction of mutual adjustment between the velocity and mass fields depends on the spatial scale of the initial geostrophic imbalance, as explicitly stated in [51]:

> *"From the foregoing discussions we may give the following statement about the production of quasigeostrophic motion: When due to some reason or other the quasigeostrophic equilibrium breaks down, then for small scale motion (not so small that the earth's rotation may be neglected) it is the pressure field to fit the new velocity field to attain new quasigeostrophic motion; for very large scale it is the velocity field which changes more to give new quasigeostrophic motion; and for intermediate scale both fields will change."*

Later, Q. C. Zeng (T.-S. Tseng) mathematically proved that the initial scale of a non-geostrophic disturbance determining the direction of geostrophic adjustment (adaptation) between the mass and velocity fields is the Rossby radius of deformation [52,53]. Yeh's research served as inspiration for Chinese scientists, including some who worked directly under Yeh's influences, leading to a uniquely systematic investigation of this topic. The findings of their research were summarized in a 1965 monograph titled "The Problems of Adaptation in Atmospheric Motion", authored by Yeh and his student M.-T. Li [54]. In this monograph, they also explored the adjustment (adaptation) in meso-scale atmospheric motion and the hydrostatic adjustment.

4.4. On the Dynamics of Atmospheric Blocking Highs

The slower dispersion of solitary waves at high latitudes, as described in the fifth section of Yeh's energy dispersion thesis in 1949 [12], remains a fundamental theoretical paradigm for the dynamics of blocking highs, a concept that continues to be supported by recent studies [55,56]. In the late 1950s and early 1960s, Yeh and his colleagues systematically investigated the climatology, synoptic features, numerical simulations, and dynamics of atmospheric blocking over the Northern Hemisphere, based on a thorough survey of 85 wintertime blocking events (54 events over the region from the North Atlantic to the Ural Mountains and 31 events over the North Pacific) that occurred between 1955 and 1960. Two monographs were published in 1962 and 1963, respectively: one focusing on synoptic and dynamical investigations [57], and the other on numerical simulations [58]. From their survey and simulations, Yeh and his colleagues developed synoptic and conceptual models for the onset, maintenance, and decay of blocking highs. This systematic investigation into blocking highs was truly exceptional, as it took more than twenty years for another comprehensive volume on the topic to be published internationally [59].

Yeh further discussed a possible physical mechanism for the onset and decay of blocking highs. He emphasized the significant role played by the baroclinic instability of long waves (approximately 5000 km or above in scale) in the onset of the meridional-type circulation. Furthermore, he highlighted the importance of local conditions favoring the appearance of highly non-geostrophic conditions in the cut-off process of a closed high cell. Additionally, positive vorticity advection toward the blocking high leads to its decay.

Furthermore, in an article published in 1963, Yeh explicitly emphasized the essential role of nonlinear wave–wave interaction in the formation of blocking highs, as explicitly stated in the abstract of [60]:

> *"...we shall keep in the vorticity and thermodynamic equations the nonlinear terms which are dropped generally. This enables us to study the mutual interaction of the disturbances. It will be shown theoretically that this mutual interaction is very important in the formation of Ω-shaped blocking highs."*

4.5. On the Fundamental Problems of Global Atmospheric Circulation

Yeh's theoretical interest in the GCA was not limited to that over East Asia, but he also considered a more fundamental problem, i.e., the very nature of the global atmospheric circulation as a whole. While the debates and investigations within Rossby's Chicago School, in which Yeh actively participated [7,23], influenced his early interest and perspective on this matter, Yeh and his colleagues in China soon independently developed their own views through systematic research. Due to space limitations, I will only highlight two examples of his work: his monograph on some fundamental problems of the GCA [61] and his investigation on abrupt seasonal changes in the GCA [62].

4.5.1. On an Internally Consistent Picture of the General Circulation

From July to September 1957, only about one year after N. A. Phillips published his seminal numerical experiment on the GCA [63], Yeh and his colleague Pao-chen Chu conducted a series of seminars in Beijing. The content of these seminars revolved around their initial draft on some fundamental problems of the GCA. Just one year later in 1958, approximately nine years before E. N. Lorenz published his monograph on the nature and theory of the GCA [64], Yeh and Chu's monograph (a 159-page volume) was published. Although the text was in Chinese, it included at the end of the book 16 pages of chapter abstracts in English [61] (p. 144–159). B. Hoskins commented, "The chapter headings give an idea of the broad sweep and ground-breaking nature of his (Yeh's) ideas at this time" [1], and the detailed chapter and section headings can be found in Figure 1 of [65]. A comparison between Yeh and Chu's classic monograph and Lorenz's indicates the different styles and visions of the three authors regarding the GCA [66], while both books demonstrated a penetrating depth of physical thinking.

In the final chapter of [61], Yeh and Chu made an effort to present an internally consistent picture of the GCA. They emphasized that the GCA is a coherent system in which various components and physical processes, influenced by external factors such as radiation and the Earth's rotation, are interconnected. Particularly, they emphasized the central role of large-scale eddies in connecting these fundamental elements of, and key physical processes in, the GCA (Figure 4, excerpts from Chapter XI). They analyzed how and why the large-scale-eddy-induced angular momentum transport in the upper layer and heat transport in the lower layer of the atmosphere should be considered an integrated entity, and their analyses were qualitatively consistent with the Eliassen–Palm flux obtained from the theory of wave–mean flow interaction by Andrews and McIntyre [67,68], which was developed during the mid-1970s.

In this chapter ***an attempt is made to give an internally consistent picture of the general circulation.*** *The way to reach this goal is to try to relate the main physical mechanism with the physical processes operating in the general circulation. The following is the main discussion of this chapter.*

The basic state of general circulation may be described as follows: In the mean meridional plane there are three cells, two direct and one indirect. (This is what we call mean meridional circulation). In the horizontal plane there are planetary wind belts (i.e. westerlies and easterlies). The wind distribution is not uniform. There is the so-called jet stream. Superimposed on these wind belts are lows and highs, troughs and ridges (large-scale disturbances). We call these the basic elements of the general circulation. ***They are not independent. They are mutually related and form an internally consistent integrity. In the course of formation of this integrity, ... large-scale disturbances play a basic role.***

From time to time the large-scale disturbances transport and redistribute the physical properties (as heat, angular momentum etc.) of the atmosphere. ***Through these transports and redistributions the large-scale disturbances tie up the elements of the general circulation.*** ...

……

In the formation of jet stream large-scale disturbances are also important.

……

Not only the main elements of general circulation are mutually constrained. The main physical processes of the general circulation are also mutually related. By main physical processes we mean those processes that are operating in the balance of important physical properties of the atmosphere, such as heat, water vapour, kinetic energy and angular momentum. ……From the above discussion we see that the main physical processes operating in the atmosphere are connected with one another. ***In connecting these physical processes the unstable disturbances play an important role.***

……

Figure 4. Excerpts from the chapter abstract, originally in English, from Chapter XI in [61] (pp. 153–156).

4.5.2. On Abrupt Seasonal changes of the General Circulation

While abrupt seasonal changes of the GCA had been observed over the southern Asian monsoon regions and the Middle East, Yeh and his colleagues extended their studies to East Asia and other regions of the Northern Hemisphere. They made a significant discovery, which they explicitly stated in the *Rossby Memorial Volume*: that the abrupt seasonal change could well be a global phenomenon [62]. Moreover, Yeh and his colleagues conjectured that a certain type of *'instability'* in the atmosphere could be the underlying mechanism:

> *"Concerning the cause of the abrupt changes, we shall only propose the following reasons as a conjecture: From winter to summer the inclination of the sun over the Northern Hemisphere gradually increases. With this increase the temperature contrast between the equator and pole gradually decreases. When it has decreased to a certain value, a certain type of 'instability' in the atmosphere appears and the abrupt change of the upper-air circulation takes place. From summer to winter the reversed sequence of events would occur."* [62]

Yeh further suggested a possible model experiment "to answer conclusively the above conjecture": "In a rotating half sphere or two coaxial cylinders [. . .] we heat differently the inner and outer part, then gradually decrease or increase the heating difference and observe whether we get the abrupt transition of the circulation as observed in the atmosphere.".

It is remarkable that they took such a broad perspective on abrupt seasonal changes of the GCA during a time when data were limited and model experiments were still in their very early stages. Notably, their conjecture did not mention topography or land–sea contrasts. Meanwhile, Yeh emphasized the central role of large-scale eddies in the maintenance of the GCA (Figure 4); therefore, it is very likely that Yeh and his colleagues believed that eddies would play a part in the speculated instability of the GCA. Interestingly, only half a century later, similarly idealized experiments using general circulation models (GCMs) were conducted, replacing the rotating half sphere or two coaxial cylinders used in Yeh and colleagues' thought the experiment [69,70]. These GCM experiments revealed that the abrupt seasonal change is caused by the rapid transition between two different circulation regimes: the equinox regime controlled by eddy momentum transport, and the monsoon regime driven directly by thermal forcing. If we interpret the "instability" mentioned in Yeh and his colleagues' conjecture as the instability of circulation regimes, then Yeh's idea aligns qualitatively with the recent findings in [69,70]. For a more comprehensive perspective on the evolution of the theory regarding abrupt seasonal changes in the GCA, refer to [65].

5. Period IV: The Period of Transition (1972–1983)

In 1966, the Institute of Atmospheric Physics (IAP) was established from the meteorological section of the IGM, and Yeh's research was stopped for several years due to the Cultural Revolution. It was not until 1972 that Yeh resumed his research activities, and his first project was the study of the GCA through the construction of a laboratory for rotating fluid physics. The significant achievements of these experiments were discussed in the section on Qinghai–Tibetan Plateau Meteorology (Section 4.2). During the period from 1972 to 1983, China underwent a transition from the turmoil of the Cultural Revolution to the era of Reform and Opening-up. Yeh became the director of the IAP in 1978 and later the vice-president of the CAS in 1981. With these roles, he took on more responsibilities in leading scientific research and promoting international cooperation in China [5]. Simultaneously, there was also a transition in the focus of Yeh's investigations from atmospheric dynamics and general circulation to climate dynamics.

During this period, Yeh maintained his theoretical interests in atmospheric dynamics, such as in his investigation of the multiple time-scale theory of atmospheric motions, conducted collaboratively with his former student M.-T. Li [71]. As a collaborator with younger theoreticians at the IAP, Yeh was supportive of the development of Q. C. Zeng's rotational adaptation (adjustment) theory, which focused on the adjustment of atmospheric motions slower than the geostrophic adjustment [72]. Additionally, Yeh contributed to J. P. Chao's spiral-like planetary wave theory in a barotropic atmosphere [73]. Furthermore, he continued to lead the research into Qinghai–Tibetan Plateau Meteorology in China [43].

During the late 1970s and onwards, Yeh's primary focus shifted from atmospheric dynamics towards the field of climate dynamics. This transition coincided with the emergence of global climate models in the 1970s; the publication of the *Charney Report* in 1979 [74]; and the establishment of the World Climate Research Programme (WCRP) in 1980, of whose joint scientific committee (JSC) Yeh became a member from 1982 to 1988 [3].

In 1981, during his visit to the Geophysical Fluid Dynamics Laboratory (GFDL) in Princeton, Yeh collaborated with S. Manabe and R. T. Wetherald. Together, they conducted idealized GCM simulations to investigate the short-term climate effects of snow-cover removal and irrigation [75,76]. Their findings revealed that soil moisture anomalies resulting from these land surface modifications can induce climate anomalies that persist for several months, creating a cross-seasonal "memory" in the general circulation and climate. It is worth noting that Yeh had already gained familiarity with surface and atmospheric energy balance analysis through his earlier work on the northeast trade wind and his subsequent studies on Qinghai–Tibetan Plateau Meteorology and the dynamics of the global circulation as an internally consistent whole. Therefore, it is not surprising to find a unique clarity in the physical insights derived from Yeh and his collaborators' analyses of the surface energy balance and the response of the global circulation to surface anomalies.

6. Period V: The Period of Global Change (1984–2013)

During the early 1980s, the international scientific community recognized the importance of the interaction between global biogeochemical processes and physical processes in shaping global environmental and climate change. In response to this realization, the International Geosphere-Biosphere Programme (IGBP) was launched in 1988, symbolizing the establishment of global change science, also known as Earth system science [77].

The concept of "global change" encompasses a wide spectrum of transformations in the Earth's environment, spanning changes in the solid earth, oceans, atmosphere, biosphere, cryosphere, and more [77]. Yeh played a key role from 1984 as one of the pioneering figures who led the launch of the IGBP [78], and he is regarded as the founder of global change studies in China. His ideas, such as those on the sensitive zones of global change [3], were reflected in the early projects of the IGBP. For a more comprehensive account of Yeh's involvement in the establishment of international global change studies, interested readers may refer to [3,79]. However, in this discussion, the focus is only placed on two aspects of Yeh's contributions: the inherent connection between adaptation to global change and the principles of sustainable development, and the concept of "orderly human activities".

At the turn of the 21st century, Yeh recognized the imperative of establishing a strong connection between adaptation to global change and sustainable development. In this context, adaptation refers to the necessary adjustments in natural and human systems to effectively respond to anticipated changes and their resulting impacts, with the aim of minimizing harm and capitalizing on beneficial opportunities [4]. Yeh believed that the lack of awareness among policymakers and the general public regarding global change could impede the success of sustainable development in China. Therefore, he organized a conference in Beijing to discuss how to link sustainable development and the adaptation to global change, inviting about 30 leading scientists of various fields from China, and Yeh's thoughts on the issue were further distilled in [80] and in an interview featured in the *WMO Bulletin* [4].

Yeh conveyed his views in [4,80] regarding the close connection between adaptation to global change and sustainable development. He claimed that the adaptation to global change must align with the principles of sustainable development. Otherwise, the adaptation for temporary or local interests may only result in greater destructive change. On the other hand, sustainable development will not achieve its goals without taking future global change into account. Yeh also analyzed the systematic nature of both sustainable development and adaptation to global change. Yeh emphasized that adaptation to global change must transcend the boundaries between regions, organizations, and business sectors; meanwhile, the trend towards the integration of regional and global economies concisely exemplifies the systematic nature of sustainable development [79].

Yeh and his colleagues further put forward the idea of "orderly human activities", which were defined as human activities that could ensure the maintenance of the life-supporting environment as a whole without notable degeneration, or even with some improvement, while meeting the demands of socio-economic development [81]. It is evident that sustainable development serves not only as the objective of orderly human activities but also as the criterion by which the orderliness of large-scale human activities is measured [3]. Yeh and his colleagues analyzed the attributes of orderly human activities, highlighting their alignment with sustainable development, hierarchical structure, systematic nature, and scale effect [81]. They further suggested an approach to investigating orderly human activities in which scientists, policymakers, and stakeholders at different levels should be closely integrated.

From a retrospective viewpoint twenty years since the publication of [80,81], one can easily identify the visionary nature of Yeh's thoughts on global change, sustainable development, and orderly human activities.

7. Conclusions

In this review, Prof. Duzheng Ye's (Tu-cheng Yeh's) research career was divided into five periods, namely: (I) the preparatory period (1935–1944); (II) the Chicago period (1945–1950); (III) the period of breaking ground (1950–1966); (IV) the period of transition (1972–1983); and (V) the period of global change (1984–2013). Figure 5 summarizes the timeline of Yeh's research career with the main foci and selected key publications for each period included. The evolution of his main achievements is provided based on a meta-analysis of his published works. Some of the evaluations of his work by his contemporaries are provided, and the influences of his achievements are reflected in the literature mentioned here, which represents the advances in the relevant research fields.

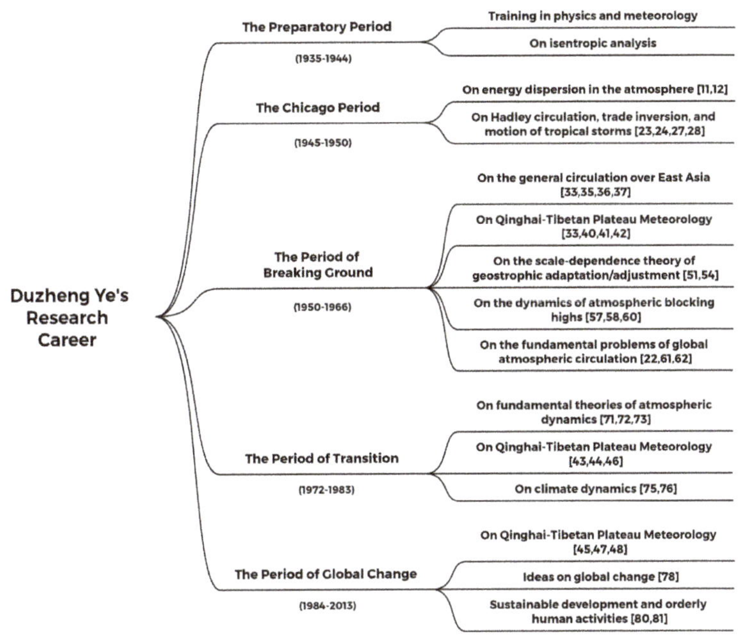

Figure 5. The timeline of Yeh's research career and the main foci and selected key publications for each period.

Needless to say, his achievements after he returned to China in 1950 were not only his, but also a part of 'many significant realistic contributions' [32] by the Chinese meteorological community at large [30,82]. In the long course, Yeh played a central role in the explosive development of atmospheric dynamics in China during the 1950s and 1960s, as well as in climate change and global change science in China after the late 1970s. Yeh's dedicated efforts and leadership in the field of global change led to the establishment of a nationwide and decade-long global change research program in China, and he also provided many science-based suggestions on climate change policy to the leaders of China.

It is emphasized that many of Yeh's ideas were very prospective, and hence this review aimed to not only honor past achievements, but also inspire more new scientific findings based on Yeh's visionary ideas. Indeed, Yeh's investigations on energy dispersion in the atmosphere [12] and on the physical mechanism of atmospheric blocking [57,60] are highly relevant to current research endeavors on the possible changes in Rossby wave propagation [83–85] under global warming and the mechanisms of blocking and extreme events [21,86–88]. His study on trade inversion in collaboration with Riehl et al. [24] remains to be used as a mechanistic basis for very recent theoretical and modeling investigations on the dynamics of trade cumuli [25,26]. Yeh's proposal regarding the cause of abrupt seasonal changes of the GCA [62] have been echoed by recent theoretical studies [69,70], and his perspective on the atmospheric general circulation as an internally consistent whole [61] remains the goal of theoretical efforts seeking a unified theory of the GCA [65]. Recent progress on the global effects of the Qinghai–Tibetan Plateau can be found in [89,90].

While many of Yeh's investigations were theory-oriented, Yeh's theoretical investigations were firmly rooted in observations. The availability of observations was indeed very limited in his time, such as during his investigations on the upper-tropospheric circulation over China during the winter of 1945–46 [33], trade inversion [24], and the thermal effects of the Tibetan Plateau [41]. However, the combination of his physical intuition and insights based on these limited data often led to groundbreaking new findings.

Finally, it should be noted that, due to the space limitations, many of Yeh's scientific contributions were not discussed here, and the emphases in the current review inevitably reflected—and were limited by—the author's understanding of Yeh's works. Suffice it to say, however, many of his scientific ideas are still very relevant to current pivotal issues, such as the need for a universal law of the general circulations of the atmosphere, and human responses to climate change. Therefore, there is no doubt that Prof. Duzheng Ye's works will continue to inspire generations of young scientists.

Funding: This research was funded by the National Natural Science Foundation of China, grant number 42042011.

Acknowledgments: The author appreciates the assistance in typing the references and reading the manuscript provided by his graduate students Tao Wang, Zhixiang Li, and Yue Gao, and also the assistance of Yue Gao in designing Figure 5. The suggestions from Weijiang Yeh and Weijian Yeh are also appreciated.

Conflicts of Interest: The author declares no conflict of interest.

References

1. Hoskins, B. Obituary: Professor Duzheng Ye, 1916–2013. *Weather* **2014**, *69*, 82–83. [CrossRef]
2. WMO. Obituary for Prof. YE Duzheng. 2013. Available online: https://public.wmo.int/en/resources/meteoworld/obituary (accessed on 29 June 2023).
3. Fu, C. From climate to global change: Following the footprint of Prof. Duzheng YE's Research. *Adv. Atmos. Sci.* **2017**, *34*, 1159–1168. [CrossRef]
4. Taba, H. The Bulletin interviews: Ye Duzheng. *WMO Bull.* **2003**, *52*, 7–16.
5. Tao, S.Y.; Hong, Z.X. Preface. In *From Atmospheric Circulation to Global Change: Celebration of the 80th Birthday of Professor YE Duzheng*; The Institute of Atmospheric Physics, Chinese Academy of Sciences, Eds.; China Meteorological Press: Beijing, China, 1996; pp. 1–3.
6. Platzman, G.W. The Atmosphere—A Challenge: Charney's Recollections. In *The Atmosphere—A Challenge: The Science of Jule Gregory Charney*; Lindzen, R.S., Lorenz, E.N., Platzman, G.W., Eds.; American Meteorological Society: Boston, MA, USA, 1990; pp. 11–82.
7. Lewis, J.M.; Fearon, M.G.; Klieforth, H.E. Herbert Riehl: Intrepid and Enigmatic Scholar. *Bull. Am. Meteorol. Soc.* **2012**, *93*, 963–985. [CrossRef]
8. Byers, H.R. Carl-Gustaf Rossby the Organizer. In *The Atmosphere and the Sea in Motion*; Bolin, B., Ed.; Rockefeller Institute Press: New York, NY, USA, 1959; pp. 56–59.
9. Rossby, C.-G. On the Distribution of Angular Velocity in Gaseous Envelopes Under the Influence of Large-Scale Horizontal Mixing Processes. *Bull. Am. Meteorol. Soc.* **1947**, *28*, 53–68. [CrossRef]
10. Rossby, C.G. On the propagation of frequencies and energy in certain types of oceanic and atmospheric waves. *J. Meteorol.* **1945**, *2*, 187–204. [CrossRef]
11. Staff Members of the Department of Meteorology of the University of Chicago. On the General Circulation of the Atmosphere in Middle Latitudes: A Preliminary Summary Report on Certain Investigations Conducted at the University of Chicago during the Academic Year 1946–1947. *Bull. Am. Meteorol. Soc.* **1947**, *28*, 255–280. [CrossRef]
12. Yeh, T.C. On energy dispersion in the atmosphere. *J. Meteorol.* **1949**, *6*, 1–16. [CrossRef]
13. Hoskins, B.J.; Karoly, D.J. The steady linear response of a spherical atmosphere to thermal and orographic forcing. *J. Atmos. Sci.* **1981**, *38*, 1179–1196. [CrossRef]
14. Lau, N.C. The pioneering works of Professor Duzheng YE on atmospheric dispersion, Tibetan Plateau meteorology, and air–sea interaction. *Adv. Atmos. Sci.* **2017**, *34*, 1137–1149. [CrossRef]
15. Huang, R.; Huangfu, J.; Liu, Y.; Lu, R. The guiding role of Rossby wave energy dispersion theory for studying East Asian monsoon system dynamics. *Atmosphere* **2023**, *14*, 962. [CrossRef]
16. Charney, J.G.; Fjørtoft, R.; von Neumann, J. Numerical integration of the barotropic vorticity equation. *Tellus* **1950**, *2*, 237–254. [CrossRef]
17. Phillips, N.A. *Dispersion Processes in Large-Scale Weather Prediction*; WMO-No.700; WMO: Geneva, Switzerland, 1990; 126p. Available online: https://library.wmo.int/index.php?lvl=notice_display&id=1678 (accessed on 29 June 2023).
18. Harper, K.C. *Weather by the Numbers*; MIT Press: Cambridge, MA, USA, 2008; 308p.
19. Rossby, C.G. On the dispersion of planetary waves in a barotropic atmosphere. *Tellus* **1949**, *1*, 54–58. [CrossRef]
20. Hovmöller, E. The Trough-and-Ridge Diagram. *Tellus* **1949**, *1*, 62–66. [CrossRef]

21. Yao, Y.; Zhuo, W.; Gong, Z.; Luo, B.; Luo, D.; Zheng, F.; Zhong, L.; Huang, F.; Ma, S.; Zhu, C.; et al. Extreme Cold Events in North America and Eurasia in November−December 2022: A Potential Vorticity Gradient Perspective. *Adv. Atmos. Sci.* **2023**, *40*, 953–962. [CrossRef]
22. Yeh, T.C. On the maintenance of zonal circulation in the atmosphere. *J. Meteorol.* **1951**, *8*, 146–150. [CrossRef]
23. Riehl, H.; Yeh, T.C. The intensity of the net meridional circulation. *Q. J. R. Meteorol. Soc.* **1950**, *76*, 182–188. [CrossRef]
24. Riehl, H.; Yeh, T.C.; Malkus, J.S.; la Seur, N.E. The north-east trade of the Pacific Ocean. *Q. J. R. Meteorol. Soc.* **1951**, *77*, 598–626. [CrossRef]
25. Stevens, B. On the growth of layers of nonprecipitating cumulus convection. *J. Atmos. Sci.* **2007**, *64*, 2916–2931. [CrossRef]
26. Albright, A.L.; Stevens, B.; Bony, S.; Vogel, R. A new conceptual picture of the trade wind transition layer. *J. Atmos. Sci.* **2023**, *80*, 1547–1563. [CrossRef]
27. Yeh, T.C. The motion of tropical storms under the influence of a super-imposed southerly current. *J. Meteorol.* **1950**, *7*, 108–113. [CrossRef]
28. Yeh, T.C.; Wallén, C.C.; Carson, J.E. A study of rainfall over Oahu. *Meteorol. Monographs.* **1951**, *1*, 34–46. [CrossRef]
29. Jaw, J.J. Development of research on meteorology in China in the past ten years. *Acta Meteorol. Sin.* **1959**, *30*, 206–211. (In Chinese) [CrossRef]
30. Lu, J. Chen-Chao Koo and the early numerical weather prediction experiments in China. *Adv. Atmos. Sci.* **2021**, *38*, 707–716. [CrossRef]
31. Blumen, W.; Washington, W.M. Atmospheric dynamics and numerical weather prediction in the People's Republic of China 1949–1966. *Bull. Am. Meteorol. Soc.* **1973**, *54*, 502–518. [CrossRef]
32. Rossby, C.-G. Note on cooperative research projects. *Tellus* **1951**, *3*, 212–216. [CrossRef]
33. Yeh, T.C. The circulation of the high troposphere over china in the winter of 1945–46. *Tellus* **1950**, *2*, 173–183. [CrossRef]
34. Koo, C.-C. Report on the numerical weather forecasting symposium held in Stockholm. *Chin. Sci. Bull.* **1957**, *8*, 480. (In Chinese) [CrossRef]
35. Staff Members of Academia Sinica. On the general circulation over Eastern Asia (I). *Tellus* **1957**, *9*, 432–446. [CrossRef]
36. Staff Members of Academia Sinica. On the general circulation over Eastern Asia (II). *Tellus* **1958**, *10*, 58–75. [CrossRef]
37. Staff Members of Academia Sinica. On the general circulation over Eastern Asia (III). *Tellus* **1958**, *10*, 299–312. [CrossRef]
38. Sawyer, J.S.; Bushby, F.H. A baroclinic model atmosphere suitable for numerical integration. *J. Meteorol.* **1953**, *10*, 54–59. [CrossRef]
39. Bolin, B. On the influence of the earth's orography on the general circulation of the Westerlies. *Tellus* **1950**, *2*, 184–195. [CrossRef]
40. Yeh, T.C.; Koo, C.C. On the influence of Tibetan Plateau on the circulation over Eastern Asia and weather in China. *Sci. Sin.* **1955**, *29*, 33. (In Chinese)
41. Yeh, T.C.; Lo, S.W.; Chu, P.C. On the heat balance and circulation structure in troposphere over Tibetan Plateau and its vicinity. *Acta Meteorol. Sin.* **1957**, *28*, 108–121. (In Chinese) [CrossRef]
42. Yang, C.-C.; Tao, S.-Y.; Koo, C.C.; Yeh, T.-C. *The Meteorology of Tibetan Plateau*; Science Press: Beijing, China, 1959. (In Chinese)
43. Yeh, T.-C.; Kao, Y.-S. *The Meteorology of Qinghai-Tibetan Plateau*; Science Press: Beijing, China, 1979; 278p. (In Chinese)
44. Ye, D. Some characteristics of the summer circulation over the Qinghai-Xizang (Tibet) Plateau and its neighborhood. *Bull. Am. Meteorol. Soc.* **1981**, *62*, 14–19. [CrossRef]
45. Ye, D.Z.; Wu, G.X. The role of the heat source of the Tibetan Plateau in the general circulation. *Meteorl. Atmos. Phys.* **1998**, *67*, 181–198. [CrossRef]
46. Yeh, T.C.; Chang, C.C. A preliminary experimental simulation on the heating effect of the Tibetan Plateau on the general circulation over Eastern Asia in summer. *Sci. Sin.* **1974**, *17*, 397–420. [CrossRef]
47. Zou, X.L.; Ye, D.Z.; Wu, G.X. Analysis of the dynamic effects on winter circulation of the two main mountains in the Northern Hemisphere (I): Relationship among general circulation, teleconnection and stationary waves. *J. Meteorol. Res.* **1992**, *6*, 395–407.
48. Zou, X.L.; Wu, G.X.; Ye, D.Z. Analysis of the dynamic effects on winter circulation of the two main mountains in the Northern Hemisphere (II): Vertical propagation of planetary waves. *J. Meteorol. Res.* **1992**, *6*, 408–420.
49. Rossby, C.G. On the mutual adjustment of pressure and velocity distributions in certain simple current systems, II. *J. Mar. Res.* **1938**, *1*, 239–263. Available online: https://elischolar.library.yale.edu/journal_of_marine_research/532 (accessed on 29 June 2023). [CrossRef]
50. Obukhov, A.M. On the question of the geostrophic wind. *Izv. Akad. Nauk SSSR Ser. Geogr. Geofiz.* **1949**, *13*, 281–306. (In Russian)
51. Yeh, T.C. On the formation of quasi-geostrophic motion in the atmosphere. *J. Meteorol. Soc. Japan.* **1957**, *35A*, 130–134. [CrossRef]
52. Zeng, Q.C. The influence of disturbance on the process of adaption and the problems of application of the observed wind fields. *Acta Meteorol. Sin.* **1963**, *33*, 37–50. (In Chinese) [CrossRef]
53. Zeng, Q.C. The adjustment and evolutional process in the atmosphere. *Acta Meteorol. Sin.* **1963**, *33*, 281–289. (In Chinese) [CrossRef]
54. Yeh, T.-C.; Li, M.-T. *The Problems of Adaptation in Atmospheric Motion*; Science Press: Beijing, China, 1965; 126p. (In Chinese)
55. Lupo, A.R. Atmospheric blocking events: A review. *Ann. N. Y. Acad. Sci.* **2021**, *1504*, 5–24. [CrossRef] [PubMed]
56. Luo, D. Planetary-scale baroclinic envelope Rossby solitons in a two-layer model and their interaction with synoptic-scale eddies. *Dyn. Atmos. Ocean.* **2000**, *32*, 27–74. [CrossRef]
57. Yeh, T.-C.; Tao, S.-Y.; Chu, P.-C.; Chen, L.S. *Studies on the Blocking Situations in Northern Hemisphere Winter*; Science Press: Beijing, China, 1962; 135p. (In Chinese)

58. Yeh, T.C. (Ed.) *Numerical Studies on the Developmental Processes of Blocking Situations. Papers in Dynamical Meteorology (II)*; Memoir of the Institute of Geophysics and Meteorology, Academia Sinica Science Press: Beijing, China, 1963; 152p.
59. Saltzman, B.; Benzi, R.; Wiin-Nielsen, A.C. (Eds.) *Anomalous Atmospheric Flows and Blocking*; Advances in Geophysics; Elsevier: Amsterdam, The Netherlands, 1986; Volume 29, 459p.
60. Yeh, T.C.; Chen, X.S. On the nonlinear effect in the formation of blocking high. *Sci. Sin.* **1963**, *12*, 391–402.
61. Yeh, T.-C.; Chu, P.-C. *Some Fundamental Problems of the General Circulation of the Atmosphere*; Science Press: Beijing, China, 1958; 159p. (In Chinese)
62. Yeh, T.C.; Dao, S.; Li, M.T. The Abrupt Change of Circulation Over the Northern Hemisphere during June and October. In *The Atmosphere and the Sea in Motion*; Bolin, B., Ed.; Rockefeller Institute Press: New York, NY, USA, 1959; pp. 249–267.
63. Phillips, N.A. The general circulation of the atmosphere: A numerical experiment. *Q. J. R. Meteorol. Soc.* **1956**, *82*, 123–164. [CrossRef]
64. Lorenz, E.N. *The Nature and Theory of the General Circulation of the Atmosphere*; World Meteorological Organization: Geneva, Switzerland, 1967; 161p, Available online: https://library.wmo.int/index.php?lvl=notice_display&id=5571 (accessed on 29 June 2023).
65. Lu, J.; Schneider, T. Evolving perspectives on abrupt seasonal changes of the general circulation. *Adv. Atmos. Sci.* **2017**, *34*, 1185–1194. [CrossRef]
66. Lu, J.H. On "some fundamental problems of the general circulation of the atmosphere". *Chin. J. Atmos. Sci.* **2016**, *40*, 78–85. (In Chinese) [CrossRef]
67. Andrews, D.G.; McIntyre, M.E. Planetary waves in horizontal and vertical shear: The generalized Eliassen–Palm relation and the mean zonal acceleration. *J. Atmos. Sci.* **1976**, *33*, 2031–2048. [CrossRef]
68. Andrews, D.G.; McIntyre, M.E. An exact theory of nonlinear waves on a Lagrangian-mean flow. *J. Fluid Mech.* **1978**, *89*, 609–646. [CrossRef]
69. Schneider, T.; Bordoni, S. Eddy-mediated regime transitions in the seasonal cycle of a Hadley circulation and implications for monsoon dynamics. *J. Atmos. Sci.* **2008**, *65*, 915–934. [CrossRef]
70. Bordoni, S.; Schneider, T. Monsoons as eddy-mediated regime transitions of the tropical overturning circulation. *Nat. Geosci.* **2008**, *1*, 515–519. [CrossRef]
71. Yeh, T.C.; Li, M.T. On the characteristics of the scales of the atmospheric motions. *J. Meteorol. Soc. Jpn.* **1982**, *60*, 16–23. [CrossRef]
72. Zeng, Q.C.; Ye, D.Z. The adaptative processes in the rotating atmospheric motion. *J. Mech.* **1980**, *16*, 1–11. (In Chinese)
73. Chao, J.P.; Ye, D.Z. The spiral-like planetary waves in the barotropic atmosphere. *Sci. Atmos. Sin.* **1977**, *1*, 81–88. (In Chinese) [CrossRef]
74. Charney, J.G. Members of Ad Hoc Study Group on Carbon Dioxide and Climate. In *Carbon Dioxide and Climate: A Scientific Assessment*; National Academy of Sciences: Washington, DC, USA, 1979; 33p. [CrossRef]
75. Yeh, T.-C.; Wetherald, R.T.; Manabe, S. A model study of the short-term climatic and hydrologic effects of sudden snow-cover removal. *Mon. Weather Rev.* **1983**, *111*, 1013–1024. [CrossRef]
76. Yeh, T.-C.; Wetherald, R.T.; Manabe, S. The effect of soil moisture on the short-term climate and hydrology change—A numerical experiment. *Mon. Weather Rev.* **1984**, *112*, 474–490. [CrossRef]
77. IGBP. *The International Geosphere-Biosphere Programme: A Study of Global Change*; Final Report of the Ad-Hoc Planning Group; Current Science Association: Berne, Switzerland, 1986; 21p.
78. Yeh, T.-C.; Fu, C.B. Climatic Change—A Global and Multidisciplinary Theme. In *Global Change*; Malone, T.F., Roederer, J.G., Eds.; Cambridge University Press: Cambridge, UK, 1985; pp. 127–145.
79. Lu, J. On the role of global change science in sustainable development: Reflecting on Ye Duzheng's contributions. *Cult. Sci.* **2021**, *4*, 55–62. [CrossRef]
80. Ye, D.Z.; Lu, J.H. On sustainable development and the adaptation to the impact of global change. *Bull. Chin. Acad. Sci.* **2000**, *14*, 164–168.
81. Ye, D.Z.; Fu, C.B.; Ji, J.J.; Dong, W.J.; Lu, J.H.; Wen, G.; Yan, X.D. Orderly human activities and subsistence environment. *Adv. Earth Sci.* **2001**, *16*, 453–460. (In Chinese) [CrossRef]
82. Li, T.; Wang, L.; Peng, M.; Wang, B.; Zhang, C.; Lau, W.; Kuo, H.-C. A Paper on the Tropical Intraseasonal Oscillation Published in 1963 in a Chinese Journal. *Bull. Am. Meteorol. Soc.* **2018**, *99*, 1765–1779. [CrossRef]
83. Wirth, V.; Riemer, M.; Chang, E.K.M.; Martius, O. Rossby wave packets on the midlatitude waveguide—A review. *Mon. Weather Rev.* **2018**, *146*, 1965–2001. [CrossRef]
84. Blackport, R.; Screen, J.A. Insignificant effect of Arctic amplification on the amplitude of midlatitude atmospheric waves. *Sci. Adv.* **2020**, *6*, eaay2880. [CrossRef] [PubMed]
85. Wu, Y.S.; Lu, J.H. A quantitative method of detecting transient Rossby wave phase speed: No evidence of slowing down with global warming. *Adv. Atmos. Sci.* **2023**, *40*, 251–261. [CrossRef]
86. Hoskins, B.; Woollings, T. Persistent extratropical regimes and climate extremes. *Curr. Clim. Chang. Rep.* **2015**, *1*, 115–124. [CrossRef]
87. Luo, D.H.; Luo, B.H.; Zhang, W.Q. A perspective on the evolution of atmospheric blocking theories: From eddy-mean flow Interaction to nonlinear multiscale interaction. *Adv. Atmos. Sci.* **2023**, *40*, 553–569. [CrossRef]

88. Kautz, L.-A.; Martius, O.; Pfahl, S.; Pinto, J.G.; Ramos, A.M.; Sousa, P.M.; Woollings, T. Atmospheric blocking and weather extremes over the Euro-Atlantic sector—A review. *Weather Clim. Dynam.* **2022**, *3*, 305–336. [CrossRef]
89. Wu, G.X.; Duan, A.M.; Liu, Y.M.; Mao, J.Y.; Ren, R.C.; Bao, Q.; He, B.; Liu, B.Q.; Hu, W.T. Tibetan Plateau climate dynamics: Recent research progress and outlook. *Natl. Sci. Rev.* **2015**, *2*, 100–116. [CrossRef]
90. Liu, Y.M.; Lu, M.M.; Yang, H.J.; Duan, A.M.; He, B.; Yang, S.; Wu, G.X. Land-atmosphere-ocean coupling associated with the Tibetan Plateau and its climate impacts. *Natl. Sci. Rev.* **2020**, *7*, 534–552. [CrossRef] [PubMed]

atmosphere

Article

Impacts of the Surface Potential Vorticity Circulation over the Tibetan Plateau on the East Asian Monsoon in July

Yimin Liu [1,2], Lulu Luan [1], Guoxiong Wu [1,2,*] and Tingting Ma [1,*]

1 State Key Laboratory of Numerical Modelling for Atmospheric Sciences and Geophysical Fluid Dynamics (LASG), Institute of Atmospheric Physics (IAP), Chinese Academy of Sciences (CAS), Beijing 100029, China; lym@lasg.iap.ac.cn (Y.L.); luanlulu@lasg.iap.ac.cn (L.L.)
2 College of Earth and Planetary Sciences, University of Chinese Academy of Sciences, Beijing 100049, China
* Correspondence: gxwu@lasg.iap.ac.cn (G.W.); matingting@lasg.iap.ac.cn (T.M.)

Abstract: Based on the definition of potential vorticity substance (W) and its equation, an index "iPV" representing the leading mode of the surface potential vorticity circulation (PVC) over the Tibetan Plateau is defined to characterize the orographic potential vorticity (PV) forcing on the atmospheric general circulation. The relationships between the iPV index and the East Asian monsoon in July, as well as the Silk Road pattern in Eurasia, are investigated on an interannual time scale. Results show that the iPV in July is closely related to the interannual variability of the East Asian monsoon. Corresponding to the positive phase of iPV with negative (positive) PVC over the north (south) of the plateau, strong positive PV anomalies and westerly flows develop in the troposphere over the plateau. Consequently, in the downstream region, the zonal PV advection increases with height just above the Jianghuai Meiyu front, which is conducive to the generation of upward movement. Over the East Asian area, the upper troposphere is controlled by the eastward shifted South Asian High. In the lower troposphere, the southwesterly flow anomaly on the northwestern side of the strengthened western Pacific subtropical high transports abundant water vapor to the north, forming a convergence in the Jianghuai area, leading to the formation of large-scale precipitation along the Meiyu front. Results from partial correlation analysis also demonstrate that the link between the variability of the East Asian monsoon in July and the plateau PV forcing is affected very little by the Silk Road pattern, whereas the plateau PV forcing plays a key "bridging" role in the influence of the Silk Road pattern on the East Asian monsoon.

Keywords: potential vorticity substance; potential vorticity circulation; East Asian monsoon; Tibetan Plateau

Citation: Liu, Y.; Luan, L.; Wu, G.; Ma, T. Impacts of the Surface Potential Vorticity Circulation over the Tibetan Plateau on the East Asian Monsoon in July. *Atmosphere* **2023**, *14*, 1038. https://doi.org/10.3390/atmos14061038

Academic Editor: Massimiliano Burlando

Received: 18 May 2023
Revised: 8 June 2023
Accepted: 15 June 2023
Published: 16 June 2023

1. Introduction

The East Asian summer monsoon (EASM) is a complex monsoon system composed of multiple important components such as tropospheric upper circulations, lower circulations, and precipitation. The seasonal evolution of the EASM is accompanied by the advance of the rain belt from south to north and has significant differences in interannual variation during different periods [1]. According to the definition of East Asian monsoon intensity [2], there exist significant differences in the intensity and pattern of EASM in different months of summer. Other studies even show that some factors that influence the interannual variability of the EASM also exhibit intra-seasonal differences [3,4].

Previous studies have shown that the Tibetan Plateau has significant impacts on the EASM through dynamical and thermodynamical effects. The dynamical forcing of the plateau has a very important impact on the establishment, development, evolution, and precipitation of the monsoon. The thermodynamic forcing mainly comes from the heating of the plateau, which is the main driver of EASM precipitation [5]. Numerous studies have been contributed to investigate the separate influence of either the dynamical forcing of the Tibetan Plateau [6,7] or its thermal forcing [8–10] on the EASM. For example, Okajima and Xie [7] find that the uplifted terrain plays an important role in the formation of the northwestern Pacific monsoon, while Duan and Wu [10] emphasize the role of the thermal forcing of the Tibetan Plateau in influencing the summer climate patterns over subtropical Asia. The study of the influence on the EASM of the combined mechanical and thermal forcing of the Tibetan Plateau is rare.

As we all know, PV is defined as $P = \frac{1}{\rho}\vec{\zeta}_a \cdot \nabla\theta$, where ρ is air density, $\vec{\zeta}_a$ is three-dimensional absolute vorticity, and θ is potential temperature. It is a physical quantity that combines the intrinsic dynamic and thermodynamic characteristics of the atmosphere. The huge uplift Tibetan Plateau has been proven the most important PV source in the world. The surface PV over the Tibetan Plateau may better represent the plateau forcing and have a closer relationship with the EASM. Sheng et al. [11,12] and He et al. [13] have demonstrated that the surface PV over the Tibetan Plateau can have a significant impact on the interannual variation of summer precipitation in East Asia via changing the atmospheric circulations. In this study, we study the PV circulation (PVC) at the surface of the Tibetan Plateau, rather than the PV within the surface layer, and reveal the relationship between the surface PVC of the plateau and the EASM.

Interestingly, the circulation structure and precipitation anomalies over the EASM region in July induced by the above-mentioned PVC forcing over the Tibetan Plateau look similar to those associated with the "Silk Road" pattern (SRP) in summer over the troposphere of Eurasia, which has been considered the main mode of the meridional wind anomaly [14]. The quasi-stationary Rossby remote-correlated wave train over the Asian troposphere in July in the Northern Hemisphere can affect the climate of a vast area along the axis of the Asian subtropical jet stream (about 40° N) [15–17] and has a significant impact on the circulation and precipitation of the tripolar type of the EASM [4,18,19]. The Tibetan Plateau is located along the Rossby wave train induced by the SRP and upstream of the EASM region. It is still unclear whether the Tibetan Plateau can modulate the effect of the SRP on EASM or not. What role the Tibetan Plateau plays in the relationship between the SRP and EASM needs clarification.

This paper aims to explore the influence on the EASM of the surface PVC forcing over the Tibetan Plateau in July, and the relationship between the circulation anomalies induced by the PVC forcing over the Tibetan Plateau and by the SRP. The context of the study is arranged as follows: Section 2 introduces the data used for the study and the concept of surface potential vorticity circulation (SPV). Section 3 demonstrates the July-mean global distributions of SPV and its zonal deviation, and an index presenting the SPV forcing of the Tibetan Plateau, i.e., iPV, is thereby introduced. By using this iPV index, the impact of the SPV forcing of the Tibetan Plateau on the EASM is studied by using regression analysis in Section 4. In Section 5, the influence on the EASM of the SPV forcing of the Tibetan Plateau is compared with the impact on the EASM of the SRP forcing by using partial correlation analysis. Conclusions are provided in Section 6.

2. Data and Methods

2.1. Data

The data of atmospheric variables used to calculate the surface PVC over the Tibetan Plateau were obtained from the daily reanalysis data at the bottom (level = 72, σ = 0.993) of the MERRA-2 (Modern-Era Retrospective Analysis for Research and Applications Version 2) published by NASA [20]. The variables include the horizontal wind, temperature, and pressure with a horizontal resolution of 0.625 × 0.5° (longitude × latitude). The atmospheric data used in the diagnostic analysis of the atmospheric circulation are from monthly data under the MERRA2 barometric coordinate system, including Ertel potential vorticity (EPV), horizontal wind, geopotential height, specific humidity, and surface pressure with a horizontal resolution of 0.625 × 0.5° (longitude × latitude). Precipitation data are monthly averaged data provided by the Global Precipitation Climatology Project, version 2.3 [21] with a horizontal resolution of 2.5 × 2.5° (longitude × latitude). The time span is from 1980 to 2019 for all datasets.

To focus on the relationship between interannual variations, all variable data were preempted with detrending and 2–9 year Lanczos bandpass filtering to preserve interannual signals.

2.2. Partial Correlation Analysis

Partial correlation refers to the method of studying the correlation between only two variables while another variable remains unchanged in a multivariate study [22]. In actual research work, it is common to find that one of the three variables remains unchanged and the correlation between the other two variables is explored. When the influence of the sequence Z(t) is excluded, the partial correlation coefficient between the sequence X(t) and Y(t) is calculated as follows:

$$r_{xy,z} = \frac{r_{xy} - r_{xz}r_{yz}}{\sqrt{(1-r_{xz}^2)(1-r_{yz}^2)}}$$

where r_{xy}, r_{xz}, and r_{yz} are, respectively, the correlation coefficients between sequence X(t) and Y(t), X(t) and Z(t), and Y(t) and Z(t); and $r_{xy,z}$ is the partial correlation coefficient, which represents the correlation coefficient between sequence X(t) and sequence Y(t) after excluding the influence of time series Z(t). Student's t-test is used to determine the confidence level of the partial correlation coefficient, and the test statistic quantity t with a sample size n is:

$$\mathrm{t} = \frac{r\sqrt{n-3}}{\sqrt{1-r^2}}$$

2.3. Earth's Surface PVC Forcing-SPV

PV substance (W) is defined as Refs. [23,24]:

$$\mathrm{W} = \rho P = \nabla \cdot (\vec{\zeta}_a \theta) \tag{1}$$

The local variation equation for PV substance (also known as potential vorticity density) can be rewritten in flux form:

$$\frac{\partial \mathrm{W}}{\partial \mathrm{t}} = -\nabla \cdot \left[\mathrm{W}\vec{V} - (Q\vec{\zeta}_a + \theta\vec{F})\right] = -\nabla \cdot \vec{J} \tag{2}$$

From Equations (1) and (2),

$$\nabla \cdot \frac{\partial}{\partial \mathrm{t}}(\vec{\zeta}_a \theta) = -\nabla \cdot \vec{J} \tag{3}$$

Equation (2) shows that the local variation of PV substance is related to its advection, diabatic heating, and friction. In Equation (3), $\vec{J}$ represents the "effective potential vorticity flux" that affects the local change of W.

By defining the PV circulation (PVC) as [25]:

$$\vec{J_C} = -\vec{\zeta_a}\theta = \int_{\mathrm{t_0}}^{\mathrm{t_0}+\Delta \mathrm{t}} \vec{J}\,\mathrm{dt} + \vec{C} \tag{4}$$

Then we have:

$$\mathrm{W} = -\nabla \cdot (\vec{J_C}) \tag{5}$$

In the pressure coordinate system with unit vectors $(\vec{i}, \vec{j}, \vec{k})$ pointing eastward, northward, and downward, respectively,

$$\vec{J_C} = (J_C^x, J_C^y, J_C^p) = -\frac{\partial v}{\partial p}\theta\,\vec{i} + \frac{\partial u}{\partial p}\theta\,\vec{j} + (f + \frac{\partial v}{\partial x} - \frac{\partial u}{\partial y})\theta\,\vec{k} \tag{6}$$

In Equation (4), the physical interpretation of PVC is the time accumulated effective potential vorticity flux. Equation (5) indicates that the convergence (divergence) of PVC corresponds to a positive (negative) PV substance (W). When we consider the total PV substance of a global atmosphere covered by an upper boundary "top" of an isentropic surface θ_T (e.g., 390 K), we need to perform a global integration for W. By employing the Gaussian theorem to convert the global volume integration of W into a surface integration of the crossing boundary PVC over the boundary surfaces (S) enclosing the global volume, we can obtain:

$$\underbrace{\iiint}_{\text{Globe}} \mathrm{W}dv = \underbrace{\iiint}_{Globe} -\nabla\cdot(\vec{J_C})dxdydp = \oiint_S -(\vec{J_C}\cdot\vec{n})ds = \underbrace{\iint}_{top} J_C^p dxdy + \underbrace{\iint}_{bot} -J_C^p dxdy \quad (7)$$

where S is the complete surface that surrounds the volume "Globe", and $\vec{n}$ is the outward normal unit vector of the surface S. The lower boundary "bot" of the global column is the surface of the Earth, and the upper boundary θ_T is a complete isentropic surface. According to Stokes' theorem, the integration on a surface of the normal component of relative vorticity equals the closed-loop circulation along the boundary surrounding the surface. Since there is no boundary for the upper surface θ_T, the first term on the right-hand side of Equation (7) vanishes, and the total amount of W in the entire bounded atmosphere becomes:

$$\underbrace{\iiint}_{\text{Globe}} \mathrm{W}dv = \underbrace{\iint}_{bot} -J_C^p dxdy = \oiint [-(f+\zeta_\mathrm{s})\theta_\mathrm{s}]ds \quad (8)$$

where the subscript "s" indicates the Earth's surface, and $[(f+\zeta_\mathrm{s})\theta_\mathrm{s}]$ represents PVC at the surface. Taking time difference on Equation (8) leads to:

$$\underbrace{\iiint}_{\text{Globe}} \frac{\partial}{\partial t}\mathrm{W}dv = \oiint \frac{\partial}{\partial t}[-(f+\zeta_\mathrm{s})\theta_\mathrm{s}]ds \quad (9)$$

Equation (9) indicates that increase in surface $[-(f+\zeta_\mathrm{s})\theta_\mathrm{s}]$ will lead to the increase in the PV substance (W) of the atmosphere. Therefore, a new variable:

$$\mathrm{SPV} = [-(f+\zeta_\mathrm{s})\theta_\mathrm{s}] \quad (10)$$

can be used as a factor to represent the PV forcing at the Earth's surface on the PV substance of the atmosphere.

3. Distributions of SPV and Index of the PV Forcing of the Tibetan Plateau

3.1. Global Distributions of SPV and Its Zonal Deviation

Figure 1 shows the spatial distributions of the climatic mean and the zonal deviation of SPV $[-(f+\zeta_\mathrm{s})\theta_\mathrm{s}]$ in July, respectively. The climatic mean distribution (Figure 1a) is mainly determined by the distribution of the Coriolis parameter f presenting a zonal orientation feature and gradually decreasing from the south pole to the north pole. The zonal deviation distribution (Figure 1b) is characterized by a weak anomaly over the ocean and a strong anomaly over large terrain and plateaus, such as Greenland and the Rocky Mountains in North America, the Andes Mountains in South America, the Alps in Europe, and the Mongolia Plateau, Iranian Plateau, and Tibetan Plateau in Asia. Such a distribution of SPV is related to the fact that elevated mountains penetrate more isentropic surfaces in the lower troposphere and produce additional PV sources for the atmosphere [12]. The most remarkable negative area is on the Tibetan Plateau, indicating that the summertime negative SPV forcing is over land, particularly over the Tibetan Plateau.

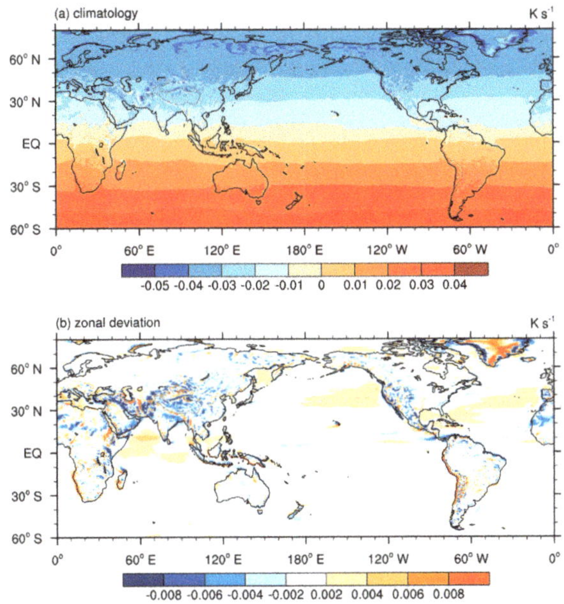

Figure 1. (**a**) Climatological July mean distribution of surface potential vorticity circulation (SPV, $[-(f+\zeta_s)\theta_s]$) and (**b**) its zonal deviation. Unit: K s^{-1}.

3.2. Index of the PV Forcing over the Tibetan Plateau

Figure 2 shows the first two empirical orthogonal function (EOF) modes of SPV on the Tibetan Plateau area higher than 3 km in July. The explanatory variance of the first EOF mode (EOF1) is 22.6%, showing a triple anomaly pattern of SPV with three anomalous centers over the northern, central, and southern Tibetan Plateau. The explanatory variance of EOF2 is 13.2%, which approximately presents a north-south reversed distribution across the plateau midline, implying a cyclonic circulation anomaly on the north and an anticyclonic circulation anomaly on the south, or vice versa. This mode mainly reflects the seesaw effect of SPV between the northern and southern Tibetan Plateau. To evaluate the relationships between these modes and the EASM, the East Asian monsoon intensity index is defined with reference to Wang et al. [2]: taking MV-EOF analysis on the three-dimensional circulation and precipitation in the East Asia region (0–50° N, 100–140° E), then define the principal component (PC) of the main mode as the intensity index of the East Asian monsoon (iEAM) in July. As both PC1 and PC2 of the SPV exhibit interannual variations, interannual relationships will be investigated in the following study.

The correlation coefficient between PC1 of the SPV and iEAM in July is only 0.05, beyond the 90% confidence level. As a result, PC1 and iEAM are approximately independent, whereas the correlation coefficient between PC2 of the SPV and iEAM is 0.54, exceeding the 99% confidence level based on Student's *t*-test. The high correlation between PC2 and iEAM indicates that the interannual variability of EASM is closely linked to the dipole mode of SPV over the Tibetan Plateau. Therefore, PC2 of the SPV over the Tibetan Plateau can be used to investigate the impacts of the PV forcing of the Tibetan Plateau on the EASM. According to Equation (9), the change in $[-(f+\zeta_s)\theta_s]$ will lead to the change of the PV substance W in the atmosphere. PC2 of SPV over the Tibetan Plateau is thus defined as the intrinsic PV forcing index of the plateau for the following study, which is abbreviated as "iPV":

$$iPV = PC2\,of\;[-(f+\zeta_s)\theta_s]over\;the\;Tibetan\;Plateau\;(>3\,\mathrm{km}) \quad (11)$$

A positive iPV then corresponds to a positive (negative) PVC over the northern (southern) Tibetan Plateau, with surface westerly flows passing through the central plateau and affecting downstream circulations.

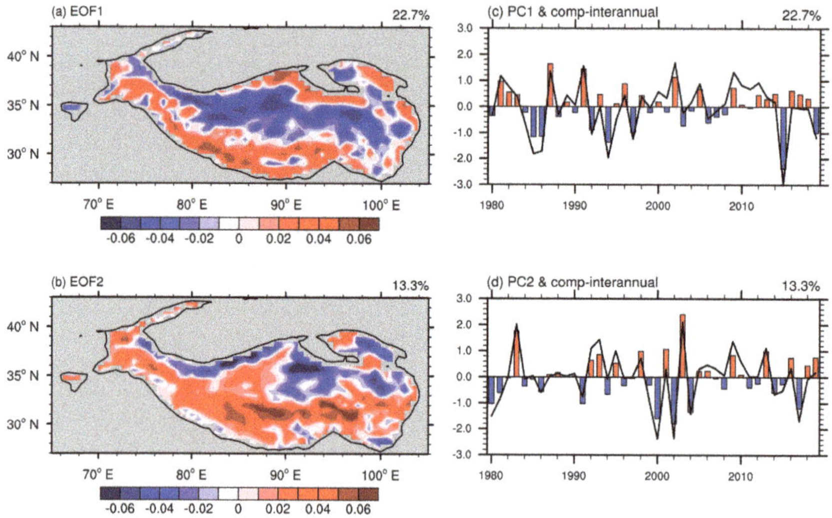

Figure 2. Spatial patterns of the first (**a**) and second (**b**) EOF modes of SPV on the Tibetan Plateau area higher than 3 km in July. (**c**,**d**) are the corresponding principal components, respectively, to (**a**,**b**) (black curve). The histogram in (**c**,**d**) represents the interannual variation after detrending during 1980–2019 and 2–9 years of bandpass filtering.

4. PV Forcing over the Tibetan Plateau on the EASM

The left column of Figure 3 shows the anomalies of the 200 hPa circulation and geopotential height (Figure 3a) and the 850 hPa circulation and precipitation (Figure 3b) regressed onto the iEAM index, representing the characteristic anomalous precipitation and circulation in the upper and lower troposphere in July for the strong EASM years. Its main characteristics are the presenting of a ray of circulation anomalies with anticyclone and cyclone centers located, respectively, over Mid Asia, northern Tibetan Plateau, the Jianghuai Basin, and Japan in the upper troposphere. The anticyclonic circulation anomaly centered over the Jianghuai Basin indicates the eastward shifting of the South Asian High at 200 hPa. In climatology, PV generally increases with latitude. The anomalous northerly flow associated with the anticyclonic circulation anomaly over East Asia thus favors the transport of positive PV anomaly to the Jianghuai region.

In the lower troposphere, the Jianghuai area is influenced by the enhanced Western Pacific subtropical high. The associated southwesterly anomaly transports much more water vapor and contributes to the positive precipitation anomaly over the Jianghuai region. In addition, the associated southwesterly anomaly transports negative PV anomaly to the Jianghuai area. Then the circulation background of PV advection increases with height forms, which is conducive to the development of air ascent [26–28], whereas the precipitation over the northern part of the South China Sea and near the Philippine Sea shows significant negative anomalies. The right column of Figure 3 is for the same variables but regressed onto the iPV index. The spatial anomalous patterns are very similar to their counterparts in the left column, particularly in their prominent circulation anomalies in the upper troposphere. This implies that there is an inextricable relationship between the in situ PV forcing over the Tibetan Plateau and the intensity of the EASM.

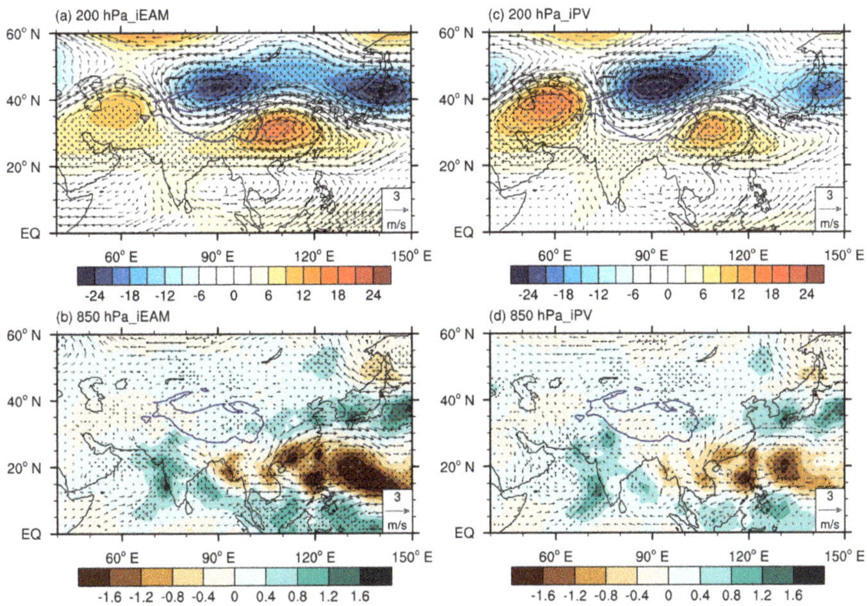

Figure 3. The 200 hPa wind (vector, unit: m s^{-1}) and geopotential height (shading, unit: gpm) anomaly (**a**,**c**) and the 850 hPa wind (vector, unit: m s^{-1}) and precipitation (shading, unit: mm day^{-1}) anomaly (**b**,**d**) regressed onto the iEAM index (**a**,**b**) and onto the iPV index (**c**,**d**).

To explore this relationship, we first investigate how the PV in the atmosphere over the plateau and the zonal PV flux downstream vary associated with the SPV over the plateau by regressing these fields onto the iPV index. Because the sign of EOF2 of SPV over the plateau is inverted on the northern and southern parts of the plateau (Figure 2b), the analysis is performed in two zonal-vertical cross sections located along 34–36° N in the north and 30–32° N in the south, respectively. As shown in Figure 4a,b, corresponding to a positive iPV, positive PV develops over the whole plateau in the north (Figure 4a) and its western part in the south (Figure 4b), presenting an upward and northward intensification of PV. Meanwhile, westerly flow predominates over the whole plateau, which will result in positive zonal PV advection in the upper troposphere and form a structure of PV advection increasing with height over the EASM region (Figure 4c,d). According to Hoskins et al. [26,27] and Wu et al. [28], air ascent will develop where PV advection increases with increasing height. Consequently, upward motion and precipitation develop over the EASM region.

Figure 5 presents the regression onto the iPV index of the vertically integrated (from 1000 to 300 hPa) water vapor flux and its divergence. The water vapor in the troposphere in the whole East Asia region is mainly transported through the anomalous southwesterly and westerly flow along the abnormal subtropical anticyclone over the northwestern Pacific. Significant water vapor divergence anomaly occurs within the anomalous subtropical anticyclone, whereas the anomalous southwesterly and westerly winds on the northwestern side of the anomalous subtropical anticyclone bring much more water vapor to the vicinity of the Jianghuai Basin, forming a large-scale water vapor convergence anomaly. As mentioned above, this anomalous pattern indicates the stronger EASM years with the strengthened western Pacific subtropical high which provides abundant water vapor conditions for the positive precipitation anomaly downstream of the Tibetan Plateau.

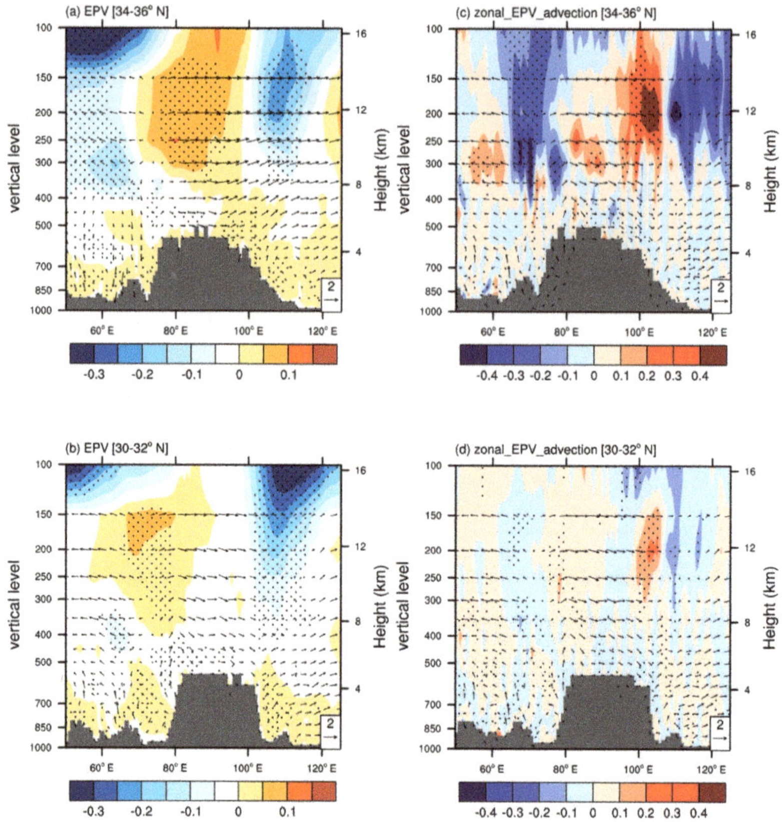

Figure 4. Zonal cross sections averaged over 34–36° N (**a**,**c**) and 30–32° N (**b**,**d**) of zonal circulation (vector, unit: m s^{-1}), and PV ((**a**,**b**); shading, unit: PVU, 1PVU = 10^{-6} K m^2 s^{-1} kg^{-1}) and zonal PV advection ((**c**,**d**); shading, unit: 10^{-5} PVU s^{-1}) regressed onto the iPV index.

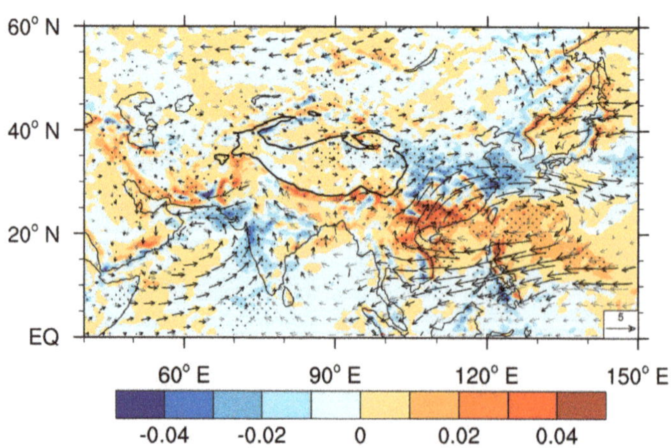

Figure 5. The vertically integrated (from 1000 to 300 hPa) water vapor flux (vector, kg m^{-1} s^{-1}) and its divergence (shading, unit: 10^{-4} kg m^{-2} s^{-1}) regressed onto the iPV index.

The above analysis demonstrates that when the surface PVC over the Tibetan Plateau is in its positive phase of EOF2 with positive SPV on its south and negative SPV on its north (Figure 2b), prominent positive PV anomaly develops in the troposphere above the plateau platform accompanied with abnormally intensified westerly flows, forming a structure of zonal PV advection increasing with height in the troposphere of the downstream area, which is conducive to the generation of upward movement over the Jianghuai region. The upper troposphere over East Asia is controlled by the eastward-shifted South Asian High. The southwesterly wind anomaly on the northwest side of the Western Pacific subtropical high in the lower atmosphere transports abundant water vapor to the north, forming a convergence above the Jianghuai region. This not only enhances the typical anomaly of the three-dimensional circulation of the East Asian monsoon in July, but also further strengthens the formation of the Meiyu front, the main component of the EASM in July.

5. Relationship with the Overland Silk Road Pattern

Figures 3c and 4a demonstrate that the in situ positive PV forcing over the Tibetan Plateau corresponds well with the positive PV anomaly in the atmosphere over the plateau. To identify the link of this positive PV anomaly with the upper layer general circulation, the 200 hPa meridional wind distribution regressed onto the iPV index is shown in Figure 6. Significant southerly and northerly anomalies are concentrated in the mid-latitudes of Eurasia from the western Eurasian continent to East Asia. The structure of the meridional wind anomaly is similar to the SRP which also appears as alternate southerly and northerly anomalies along the mid-latitude Asian westerly jet from western Europe to East Asia [14,16]. The correlation coefficient between the iPV and the index of SRP (SRPI) is as high as 0.59, exceeding the significance level of 0.01.

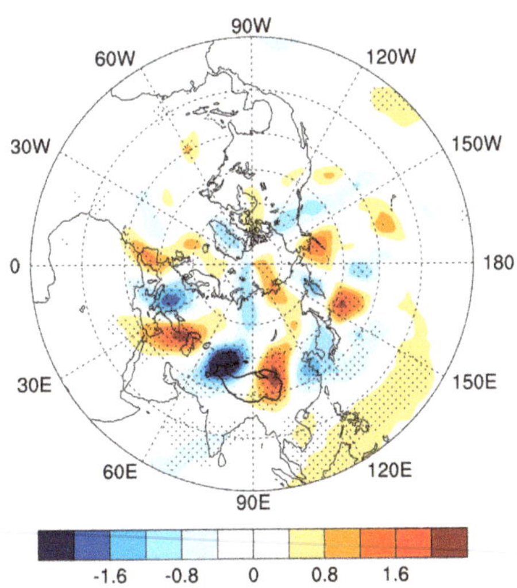

Figure 6. The distribution of the meridional wind anomaly at 200 hPa (shading, unit: m s^{-1}) regressed onto the iPV index. Areas exceeding the 0.05 significance level are highlighted by black dots.

Previous studies have shown that, through the associated anomalous northerly wind over East Asia, the summertime SRP has varying degrees of influence on the circulation and the precipitation in the East Asian areas [4,15,18,19,29]. In order to better understand the relationship between plateau PV forcing and the onland SRP in the influence of the interannual intensity of the EASM, we conduct two sets of partial correlation analysis: one by removing the linear influence of SRPI from that of iPV and the other by removing the linear influence of iPV from that of SRPI, and the results are presented in Figures 7 and 8, respectively.

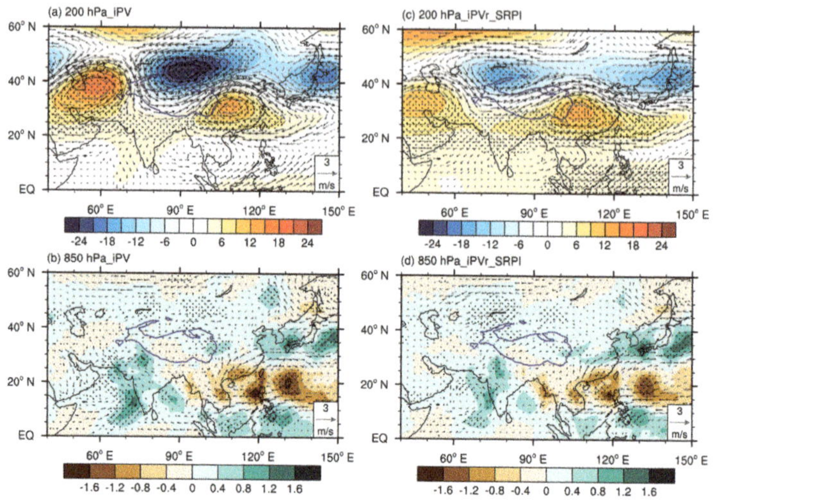

Figure 7. The distributions of circulation (vector, unit: m s^{-1}) and geopotential height (shading, unit: gpm) at 200 hPa (**a**,**c**) and the 850 hPa circulation (vector, unit: m s^{-1}) and precipitation (shading, unit: m day^{-1}) (**b**,**d**) regressed onto the iPV index (**a**,**b**) and regressed onto the iPVr index (**c**,**d**). iPVr represents the remaining time series after removing the linear correlation with SRPI from the iPV index. Dotted regions indicate the geopotential height and precipitation exceeding the 95% confidence level.

It can be seen from Figure 7 that when the linear influence of the SRP is removed, the distributions of circulation and precipitation anomalies in the lower tropospheric layer change little (Figure 7d) compared with the original distributions (Figure 7b). The main changes in the upper troposphere circulation are the weakening and westward shifting of the anomalous cyclone to the north of the Tibetan Plateau and its upstream anticyclonic anomaly (Figure 7c), while the position and intensity of the anomalous anticyclone over East Asia (Figure 7a) is almost unchanged (Figure 7c). The associated anomalous northerly wind still prevails over East Asia. As mentioned in Section 4 and previous studies [4,15], the northerly wind anomaly is a key factor that induces the precipitation anomaly over the Jianghuai region. Consequently, the results shown in Figure 7 indicate that no matter whether there is the influence of the SRP or not, PV forcing over the Tibetan Plateau can directly influence the intensity of EASM. The correlation coefficient between iPVr and iEAM is 0.49, which is only 0.05 lower than that between iPV and iEAM, but still reaches a significance level of 0.01. On the contrary, when the influence of plateau PV forcing is removed from the Silk Road tele-correlation, the regressed upper tropospheric circulation undergoes remarkable changes (Figure 8c versus Figure 8a). The strong cyclone anomaly to the north of the plateau (Figure 8a) becomes significantly weak and shifts southward (Figure 8c). The anomalous westerly wind prevailing over the entire plateau platform (Figure 8a) becomes much weaker and moves southward to the south of the plateau (Figure 8c). The downstream anticyclone anomaly originally located over the Jianghuai region moves northeastward and is located over Northeast China. The associated anomalous northerly wind moves northeastward correspondingly. The Jianghuai region is controlled by the easterly anomaly in the upper troposphere (Figure 8c). In the lower troposphere (Figure 8b,d), the subtropical anticyclone circulation over the northwestern Pacific is weakened, and no apparent precipitation anomalies occur in the Jianghuai area. The correlation coefficient between SRPIr and the overall East Asian monsoon intensity index iEAM drops to 0.01, which is almost independent. These results indicate that the SRP cannot influence the EASM directly in July. PV forcing over the Tibetan Plateau may play a role in "bridging" the connection between the EASM and the SRP.

The above results suggest that PV forcing over the Tibetan Plateau can directly influence the East Asian monsoon's variability. Such connection between the EASM and the plateau PV forcing in July is affected little by the SRP, whereas the plateau PV forcing plays a key role in "bridging" the SRP and the EASM precipitation. If the plateau PV forcing disappears, the upstream SRP may not have a significant effect on the East Asian summer monsoon.

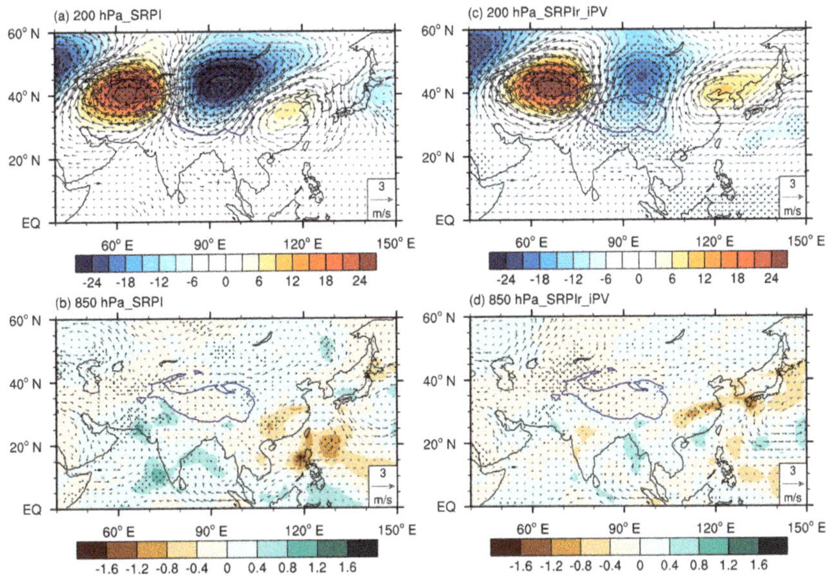

Figure 8. The distributions of circulation (vector, unit: m s^{-1}) and geopotential height (shading, unit: gpm) at 200 hPa (**a**,**c**) and the 850 hPa circulation (vector, unit: m s^{-1}) and precipitation (shading, unit: m day^{-1}) (**b**,**d**) regressed onto the SRPI index (**a**,**b**) and regressed onto the SRPIr index (**c**,**d**). SRPIr represents the remaining time series after removing the linear correlation with iPV from the SRPI index. Dotted regions indicate the geopotential height and precipitation exceeding the 95% confidence level.

6. Conclusions

By integrating the PV substance and its local change equation over the global atmospheric volume bounded by an enclosed isentropic surface as the upper boundary, it shows that the global gross PV substance equals the integral of the PV circulation (PVC) at the earth's surface of the whole globe. That means the gross source of PV substance of the global atmosphere is located at the Earth's surface. EOF analysis of the surface PV circulation (SPV) over the Tibetan Plateau higher than 3 km in July indicates that PC2 can be used as an index to characterize effects on the EASM of the intrinsic plateau PV forcing. By making partial correlation and regression analysis, this paper further studies the influence mechanism of the plateau PV forcing on the interannual variability of the EASM and its relationship with that of the onland SRP over Eurasia. The main conclusions can be summarized as follows:

(1) When the second mode of SPV on the surface of the Tibetan Plateau platform is in the positive phase (Figure 2b), with positive SPV on its south and negative SPV on its north, a strong positive PV anomaly and strengthened westerly flow will develop in the troposphere over the plateau, forming a structure of zonal PV advection increasing with height in the troposphere over the downstream Jianghuai region, which is conducive to the generation of air ascent. The upper troposphere over East Asia is controlled by the strong positive anomaly of geopotential height due to

the eastward shifting of the South Asian High. The associated northerly anomaly favors the transport of positive PV anomaly to the Jianghuai region; whereas, in the lower troposphere, the anomalous southwesterly flow on the northwestern side of the enhanced western Pacific subtropical high transports not only abundant water vapor, but also negative PV anomaly to the Jianghuai region, forming a circulation background of PV advection increasing with height. This not only enhances the three-dimensional circulation anomaly of the East Asian monsoon in July, but also facilitates stronger precipitation along the Meiyu front;

(2) The link between East Asian monsoon variability and plateau PV forcing in July is influenced very little by the SRP. The latter mainly impacts the wave position and intensity in the upper troposphere to the west of the plateau, but has limited effect on the spatial distributions of circulation and precipitation downstream of the plateau. However, when the linear signal of plateau PV forcing is removed from the SRP sequence, the cyclone anomaly to the north of the plateau is significantly weakened and shifts southward. The westerly wind originally over the plateau becomes much weaker and shifts to the south of the plateau as well. At the same time, the anticyclone anomaly originally located over the Jianghuai region shifts to Northeast China. The Jianghuai region is controlled by the easterly anomaly in the upper troposphere, which weakens the circulation background of PV advection increasing with height. Consequently, the positive precipitation anomaly over the Jianghuai region becomes weak and the interannual variability of the SRP and the East Asian monsoon are no longer correlated. These results indicate that the SRP cannot influence the EASM directly in July. The plateau PV forcing plays a key role in "bridging" the influence of the SRP to the East Asian summer monsoon: the PV forcing over the Tibetan Plateau can modulate the influence of the SRP on the EASM by changing the position of the anticyclone anomaly in the upper troposphere downstream of the Tibetan Plateau which is critical for the development of air ascent and precipitation of the EASM. When the influence of plateau PV forcing is removed, this anticyclone anomaly is located over Northeast China, which has little impact on the EASM. However, when the influence of plateau PV forcing is considered, the anticyclone anomaly shifts to central China, contributing to a stronger EASM year. In other words, the influence of the SRP in the Eurasian region on the East Asian monsoon in July is inseparable from the involvement of the PV forcing over the Tibetan Plateau. In summary, it is the surface PV forcing of the Tibetan Plateau that directly and significantly affects the interannual variability of the EASM over the Jianghuai region.

Author Contributions: Y.L.: Conceptualization; validation; Funding acquisition; supervision; writing—review and editing. L.L.: Data curation; formal analysis; investigation; writing—original draft. G.W.: Conceptualization; Funding acquisition; writing—review and editing. T.M.: writing—review and editing. All authors have read and agreed to the published version of the manuscript.

Funding: The study was supported by the National Natural Science Foundation of China Projects (91937302) and the Strategic Priority Research Program of Chinese Academy of Sciences (XDB40030205).

Data Availability Statement: All datasets used in this study are publicly available. We would like to thank the Global Modeling and Assimilation Office and the Goddard Earth Sciences Data and Information Services Center for the dissemination of MERRA-2 reanalysis data (https://climatedataguide.ucar.edu/climate-data/nasas-merra2-reanalysis; accessed on 18 May 2023). The GPCP precipitation data are available from https://psl.noaa.gov/data/gridded/data.gpcp.html (accessed on 18 May 2023).

Conflicts of Interest: The authors declare no conflict of interest.

References

1. Liu, F.; Ouyang, Y.; Wang, B.; Yang, J.; Ling, J.; Hsu, P.C. Seasonal evolution of the intraseasonal variability of China summer precipitation. *Clim. Dyn.* **2020**, *54*, 4641–4655. [CrossRef]
2. Wang, B.; Wu, Z.W.; Li, J.P.; Liu, J.; Chang, C.P.; Ding, Y.H.; Wu, G.X. How to measure the strength of the East Asian summer monsoon. *J. Clim.* **2008**, *21*, 4449–4463. [CrossRef]

3. Hong, X.W.; Lu, R.Y.; Li, S.L. Differences in the Silk Road Pattern and Its Relationship to the North Atlantic Oscillation between Early and Late Summers. *J. Clim.* **2018**, *31*, 9283–9292. [CrossRef]
4. Li, X.Y.; Lu, R.Y. Subseasonal change in the seesaw Pattern of Precipitation between the Yangtze River Basin and the Tropical Western North Pacific during Summer. *Adv. Atmos. Sci.* **2018**, *35*, 1231–1242. [CrossRef]
5. Wu, G.X.; Liu, Y.M.; Wang, T.M.; Wan, R.J.; Liu, X.; Li, W.P.; Wang, Z.Z.; Zhang, Q.; Duan, A.M.; Liang, X.Y. The influence of Mechanical and thermal forcing by the Tibetan Plateau on Asian climate. *J. Hydrometeorol.* **2007**, *8*, 770–789. [CrossRef]
6. Yeh, T.C. The circulation of the high troposphere over China in the winter of 1945–1946. *Tellus* **1950**, *2*, 173–183. [CrossRef]
7. Okajima, H.; Xie, S.P. Orographic effects on the northwestern Pacific monsoon: Role of air-sea interaction. *Geophys. Res. Lett.* **2007**, *34*, L21708. [CrossRef]
8. Yanai, M.; Li, C.; Song, Z. Seasonal heating of the Tibetan Plateau and its effects of the evolution of the Asian summer monsoon. *J. Meteor. Soc. Jpn.* **1992**, *70*, 319–351. [CrossRef]
9. Kitoh, A. Effects of large-scale mountains on surface climate—A coupled ocean atmosphere general circulation model study. *J. Meteorol. Soc. Jpn.* **2002**, *80*, 1165–1181. [CrossRef]
10. Duan, A.M.; Wu, G.X. Role of the Tibetan Plateau thermal forcing in the summer climate patterns over Subtropical Asia. *Clim. Dyn.* **2005**, *24*, 793–807. [CrossRef]
11. Sheng, C.; He, B.; Wu, G.X.; Liu, Y.M.; Zhang, S.Y. Interannual influences of the surface potential vorticity forcing over the Tibetan Plateau on East Asian summer rainfall. *Adv. Atmos. Sci.* **2022**, *39*, 1050–1061. [CrossRef]
12. Sheng, C.; Wu, G.X.; Tang, Y.Q.; He, B.; Xie, Y.K.; Ma, T.T.; Ma, T.; Li, J.X.; Bao, Q.; Liu, Y.M. Characteristics of the potential vorticity and its budget in the surface layer over the Tibetan Plateau. *Int. J. Climatol.* **2021**, *41*, 439–455. [CrossRef]
13. He, B.; Sheng, C.; Wu, G.X.; Liu, Y.M.; Tang, Y.Q. Quantification of seasonal and interannual variations of the Tibetan Plateau surface thermodynamic forcing based on the potential vorticity. *Geophys. Res. Lett.* **2022**, *49*, e2021GL097222. [CrossRef]
14. Lu, R.Y.; Oh, J.H.; Kim, B.J. A teleconnection pattern in upper-level meridional wind over the North African and Eurasian continent in summer. *Tellus* **2002**, *54*, 44–55. [CrossRef]
15. Ding, Y.H.; Wang, B. Circumglobal Teleconnection in the Northern Hemisphere Summer. *J. Clim.* **2005**, *18*, 3483–3505. [CrossRef]
16. Enomoto, T. Interannual Variability of the Bonin High Associated with the Propagation of Rossby Waves along the Asian Jet. *J. Meteorol. Soc. Jpn.* **2003**, *82*, 1019–1034. [CrossRef]
17. Hong, X.W.; Xue, S.H.; Lu, R.Y.; Liu, Y.Y. Comparison between the interannual and decadal components of the Silk Road pattern. *Atmos. Ocean. Sci. Lett.* **2018**, *11*, 270–274. [CrossRef]
18. Huang, R.H.; Chen, J.L.; Wang, L.; Lin, Z.D. Characteristics, processes, and causes of the spatio-temporal variabilities of the East Asian monsoon system. *Adv. Atmos. Sci.* **2012**, *29*, 910–942. [CrossRef]
19. Lin, S.M.; Hsu, H.H. Asymmetry of the Tripole Rainfall Pattern during the East Asian Summer. *J. Clim.* **2007**, *20*, 4443–4458.
20. Gelaro, R.; McCarty, W.; Suarez, M.J.; Todling, R.; Molod, A.; Takacs, L.; Randles, C.A.; Darmenov, A.; Bosilovich, M.G.; Reichle, R.; et al. The Modern-Era Retrospective Analysis for Research and Applications, Version 2 (MERRA-2). *J. Clim.* **2017**, *30*, 5419–5454. [CrossRef]
21. Adler, R.F.; Huffman, G.J.; Chang, A.; Ferraro, R.; Xie, S.P.; Janowiak, J.; Rudolf, B.; Schneider, U.; Curtis, S.; Bolvin, D.; et al. The Version-2 Global Precipitation Climatology Project (GPCP) Monthly Precipitation Analysis (1979–Present). *J. Hydrometeorol.* **2003**, *4*, 1147–1167. [CrossRef]
22. Huang, J.Y. *Statistical Analysis Method of Meteorological Data*, 4th ed.; China Meteorological Press: Beijing, China, 2016; p. 25, ISBN 978-7-5029-6346-0. (In Chinese)
23. Haynes, P.H.; McIntyre, M.E. On the evolution of vorticity and potential vorticity in the presence of diabatic heating and frictional or other forces. *J. Atmos. Sci.* **1987**, *44*, 828–841. [CrossRef]
24. Haynes, P.H.; McIntyre, M.E. On the conservation and impermeability theorems for potential vorticity. *J. Atmos. Sci.* **1990**, *47*, 2021–2031. [CrossRef]
25. Sheng, C.; Wu, G.X.; He, B.; Liu, Y.M.; Ma, T.T. Linkage between cross-equatorial potential vorticity flux and surface air temperature over the mid-high latitudes of Eurasia during boreal spring. *Clim. Dyn.* **2022**, *59*, 3247–3263. [CrossRef]
26. Hoskins, B.J.; McIntyre, M.E.; Robertson, A.W. On the use and significance of isentropic potential vorticity maps. *Q. J. R. Meteorol. Soc.* **1985**, *111*, 877–946. [CrossRef]
27. Hoskins, B.J.; Pedder, M.; Jones, D.W. The omega equation and potential vorticity. *Q. J. R. Meteorol. Soc.* **2003**, *129*, 3277–3303. [CrossRef]
28. Wu, G.X.; Ma, T.T.; Liu, Y.M.; Jiang, Z.H. PV-Q Perspective of Cyclogenesis and Vertical Velocity Development Downstream of the Tibetan Plateau. *J. Geophys. Res. Atmos.* **2020**, *125*, e2019JD030912. [CrossRef]
29. Lu, R.Y.; Li, X.Y. Extratropical Factors Affecting the Variability in Summer Precipitation over the Yangtze River Basin, China. *J. Clim.* **2019**, *30*, 8357–8374.

atmosphere

Article

The Atmospheric Vortex Streets and Their Impact on Precipitation in the Wake of the Tibetan Plateau

Qi Liu [1,2,3], Zhaohua Wu [4,*], Zhe-Min Tan [1], Fucheng Yang [4] and Congbin Fu [1,2,3,*]

1 School of Atmospheric Sciences, Nanjing University, Nanjing 210023, China; qiliu@nju.edu.cn (Q.L.); zmtan@nju.edu.cn (Z.-M.T.)
2 Jiangsu Collaborative Innovation Center for Climate Change, Nanjing University, Nanjing 210023, China
3 Institute for Climate and Global Change Research, Nanjing University, Nanjing 210023, China
4 Department of Earth, Ocean, and Atmospheric Science, and Center for Ocean-Atmospheric Prediction Studies, Florida State University, Tallahassee, FL 32306, USA
* Correspondence: zwu@fsu.edu (Z.W.); fcb@nju.edu.cn (C.F.)

Abstract: Atmospheric vortex streets (AVSs) are often observed in the wake of the leeward side of mountainous islands and are considered atmospheric analogs of the classic Kármán vortex street when a fluid flows past a cylindrical obstacle. The prevailing westerlies were observed year-round around the Tibetan Plateau. However, it remains to be understood whether the wake on the leeward side of the Tibetan Plateau exhibits a stable AVS and how the AVS impacts precipitation over the downstream region. In this study, the environmental meteorological factors, spatiotemporal characteristics, and various properties of the AVS on the leeward side of the Tibetan Plateau were examined for the period of 1979–2018 using global reanalysis datasets. The results show that the spatial structure of these AVSs closely resembles that of the classic Kármán vortex street observed in the laboratory. The meteorological factors satisfy the conditions in which a stable AVS can exist year-round. Moreover, various properties of these AVSs, including the aspect ratio and Strouhal number, are similar to those in previous studies of smaller obstacle caused AVS. Thus, these AVSs on the leeward side of the Tibetan Plateau can be interpreted as the atmospheric analog of classic Kármán vortex streets. The results further show that the spatiotemporal structure of precipitation over the wake of the Tibetan Plateau was largely shared by the cyclonic activities in the AVS. Approximately 80–90% of the total precipitation and heavy rain days in the main rainband over the wake of the Tibetan Plateau are closely tied to the seasonal evolution of the AVS.

Keywords: atmospheric vortex street; Tibetan Plateau; precipitation; East Asia; Kármán vortex street

Citation: Liu, Q.; Wu, Z.; Tan, Z.-M.; Yang, F.; Fu, C. The Atmospheric Vortex Streets and Their Impact on Precipitation in the Wake of the Tibetan Plateau. *Atmosphere* **2023**, *14*, 1096. https://doi.org/10.3390/atmos14071096

Academic Editor: Yoshihiro Tomikawa

Received: 29 May 2023
Revised: 26 June 2023
Accepted: 28 June 2023
Published: 30 June 2023

1. Introduction

When steady wind flows around an isolated obstacle, such as a mountain or a mountainous island, atmospheric vortex streets (AVSs) can be generated on the leeward side of the obstacle under favorable meteorological conditions. The AVS pattern exhibits a double row of counter-rotating vortex pairs shedding alternately and resembles the classic von Kármán vortex street [1–3], as schematically shown in Figure 1. Vortex streets have been frequently observed in the atmosphere [4,5] and ocean [6–11]. These types of vortex streets have significant weather and climate implications. Oceanic vortex trains could enhance biological production and turbulent mixing, impacting fishing activities [12–15]. The atmospheric vortex streets may modulate the cloud and wind patterns over the downstream region [16].

The studies of AVS can be traced back to as early as the 1930s. Lettau [17] suggested that AVSs could be shed by large islands. However, it was not until the early 1960s that researchers observed AVS in cloud images taken by the first generation of earth-orbiting satellites (e.g., [18–20]). These studies also revealed the properties of AVS, such as the rate of vortex shedding eddy lifetime, eddy viscosity, and obstacle drag, from satellite imagery and suggested that the AVS on the lee side of obstacles can be interpreted as the atmospheric analog of classic von Kármán vortex streets [18,20–26]. Further studies analyzed the mechanisms and meteorological conditions under which a stable vortex on the lee side of an obstacle can develop, e.g., Etling [27] conducted laboratory experiments and found that a stable vortex street on the leeward side of an obstacle can exist when the Reynolds number (Re) is larger than a particular value and the Froude number is smaller than 0.4.

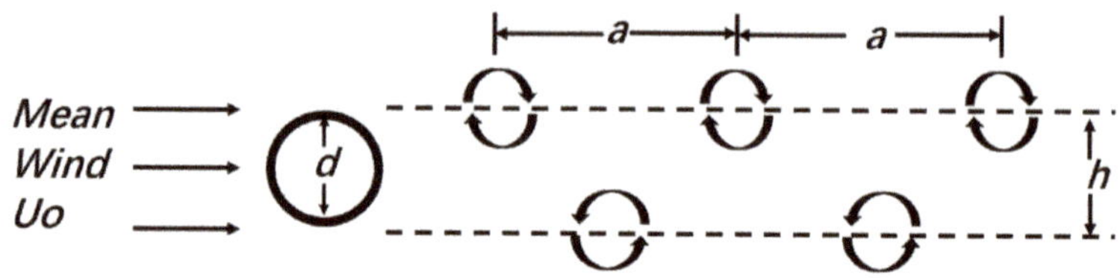

Figure 1. Schematic plot of a Kármán vortex street generated by wind passing a cylindrical obstacle of diameter d. a is the vortex spacing and h is the width of the Kármán vortex street.

In addition to observational and laboratory studies, numerical studies have been carried out to reveal more details of the formation of a vortex pair on the leeward side of the island and its separation from the island. These studies are documented in the review paper of Young and Zawislak [7]. To date, most numerical simulations of Kármán vortex shedding in the real atmosphere have focused on islands in the Northeast Atlantic [5,28–30] or in the Northwest Pacific [31]. Based on a numerical simulation model with a horizontal resolution of 2 km, Nunalee and Basu [5] revealed that the whole region in the satellite image with a cellular stratocumulus cloud pattern was disturbed by Kármán vortex shedding. Li et al. [31] simulated the observed AVS using an MM5 model and estimated a vortex-shedding rate of 1 h.

It is noted here that previous observations of AVS, including those mentioned above, focused on AVS at a spatial scale of approximately a few hundred kilometers that can be captured in a non-merged satellite image. For AVS of these spatial scales, the Reynolds number, the ratio of inertial forces to viscous forces within a fluid, is usually between 50 and 500 [25]. From a theoretical perspective, given that the Reynolds number is proportional to the spatial size of the obstacle, a stable vortex street can exist when the Reynolds number is as high as 10^5 [25,32], as demonstrated in laboratory experiments. However, there have been few observational studies on that scale. It is indeed this hidden possibility that is one of the reasons for us to explore whether the Tibetan Plateau, standing in strong seasonally varying westerlies, can cause AVS of similar spatial scales on its leeward side.

The Tibetan Plateau, located over South-Central Asia, is the world's highest plateau above sea level, with an average elevation of approximately 4500 m (approximately one-third of the tropospheric height). Prevailing year-round lower tropospheric westerlies flow over or flow around the Tibetan Plateau and are divided into two branches after passing by the Tibetan Plateau [33,34], with the portion of flow around being dominant in the low- to mid-tropospheric region [35–37]. Yeh noticed, as early as the 1950s, that a pair of vortices appeared frequently on the east sides of the Tibetan Plateau [33]. The southern vortex is cyclonic and associated with low surface pressure and is termed the southwest China vortex [38]. The downstream propagation of southwest China vortices can result in substantial precipitation in downstream regions [39,40]. The northern vortex is anticyclonic and called the little northwest high [41]. However, whether the downstream mesoscale vortices are indeed AVS and how they are directly tied to seasonally varying westerlies have not received much attention from researchers.

This study aims to answer the following questions by analyzing high-resolution reanalysis data: (1) Can the mesoscale systems on the leeward side of the Tibetan Plateau be interpreted as the atmospheric analog of classic von Kármán vortex streets? Do the surrounding meteorological factors and properties of the AVS satisfy conditions in which a stable vortex on the lee side of an obstacle can exist? (2) If so, does the AVS on the lee side of the Tibetan Plateau impact the precipitation and heavy rain days over the wake of the Tibetan Plateau?

The paper is organized as follows: Section 2 describes the reanalysis datasets and precipitation data used. Section 3.1 gives an overview of the seasonal variability of the meteorological situation of the Tibetan Plateau and demonstrates whether the meteorological factors satisfy the conditions under which a stable vortex AVS can exist. Section 3.2 demonstrates the similarity between the AVS on the lee side of the Tibetan Plateau and the vortex street recorded in the laboratory experiment. Section 3.3 calculates some AVS properties, such as the aspect ratio and the Strouhal number identified directly from the reanalysis datasets. Section 3.4 reveals the impact of the AVS on precipitation over the wake of the Tibetan Plateau. Finally, Section 4 summarizes our results and discusses the relationship between AVS and the meteorological systems controlling precipitation over the wake of the Tibetan Plateau.

2. Materials and Methods

2.1. Data

Daily and 6-h horizontal winds and air temperature were derived from the reanalysis of the European Center for Medium-Range Weather Forecasts (ECMWF) interim reanalysis (ERA-Interim) products [42], spanning the 40-yr period from 1979 to 2018. The dataset we obtained is a spatially gridded one that has a fixed horizontal resolution of $1^\circ \times 1^\circ$ and 37 vertical levels. Daily and 6-h horizontal winds and air temperature ($2.5^\circ \times 2.5^\circ$ and 17 vertical levels) were also obtained from the National Centers for Environmental Prediction and National Center for Atmospheric Research (NCEP/NCAR) [43] for the period of January 1979–December 2018. It is noted that the key results presented later in this paper are not sensitive to the selection of reanalysis datasets. Following Curio et al. [44], the relative vorticity was calculated from the wind field and was employed to represent the vortices. Interpolation methods were used to obtain a spatial resolution of $1^\circ \times 1^\circ$ and a temporal resolution of 1 h, if necessary.

The daily precipitation in East Asia for the period of 1 January 1979 to 31 December 2018 was obtained from the CPC precipitation dataset [45]. We selected this period to match the duration of the ERA-Interim and NCEP/NCAR datasets. The APHRO_MA_025deg_V1003R1 product based on the APHRODITE rain gauge data precipitation dataset for the 1979–2015 period was also used in this study [46]. Both precipitation datasets have a fixed horizontal resolution of $0.5^\circ \times 0.5^\circ$. In the following section, we will demonstrate that the results relevant to precipitation are not sensitive to the selection of the precipitation datasets.

2.2. Spatial Fourier Transform to Derive the AVS Pattern

The relative vorticity and horizontal wind field were divided into the AVS-related component and other wave-related components using the Fourier transform, which has been widely applied in wave analysis [47,48]. In this study, the spatial lowpass filtered relative vorticity captured the AVS related to the Tibetan Plateau. The spatial domain was 60° E–120° W, 10° S–80° N. The components with a meridional wavelength shorter than 7–9° and a zonal wavelength shorter than 14–18° were filtered out. We chose these thresholds due to the distances between neighboring vortices in the AVS being longer than these wavelength thresholds.

2.3. The AVS-Related Precipitation and Heavy Rain Days

Mesoscale cyclonic activities could trigger substantial precipitation. Previous studies have attempted to determine the precipitation associated with mesoscale cyclonic activities. Some studies use a fixed-radius scheme to identify the precipitation related to mesoscale cyclonic activities [44,49]. However, other studies use the outermost closed contour (OCC) to detect the affected precipitation [50]. The domain of mesoscale cyclonic activities is naturally defined by the region covered by the outermost closed contour (OCC) in the potential height or relative vorticity fields. In addition, the precipitation within the coverage of the OCC is defined as related to mesoscale cyclonic activities.

To determine the precipitation associated with the AVS, in this study, following Hanley and Caballero [50], we used the region covered by the OCC instead of a fixed radius to detect the precipitation affected by cyclonic activities in the AVS. The OCC of the AVS was defined as the contour where the spatial Fourier-filtered relative vorticity was zero. The AVS-related precipitation was determined over 110–120° E, 125–130° E, and 130–145° E, respectively. These three regions represent East China, the Korean Peninsula, and Japan, respectively.

3. Results

3.1. Topography of the Tibetan Plateau and Surrounding Meteorological Conditions

The Tibetan Plateau spans the region of approximately 26~40° N, 73~105° E. The grid of (40° E, 32° N) is located upstream of the Tibetan Plateau with an elevation of only 500 m and serves as a reference coordinate grid for the Tibetan Plateau. Figure 2 displays the mean seasonal cycle of the zonal wind at that grid. Strong westerlies prevail over the upstream region and over almost the whole troposphere from 1000 hPa to 100 hPa for a whole year. The minima of the westerlies occurred over the boreal summer.

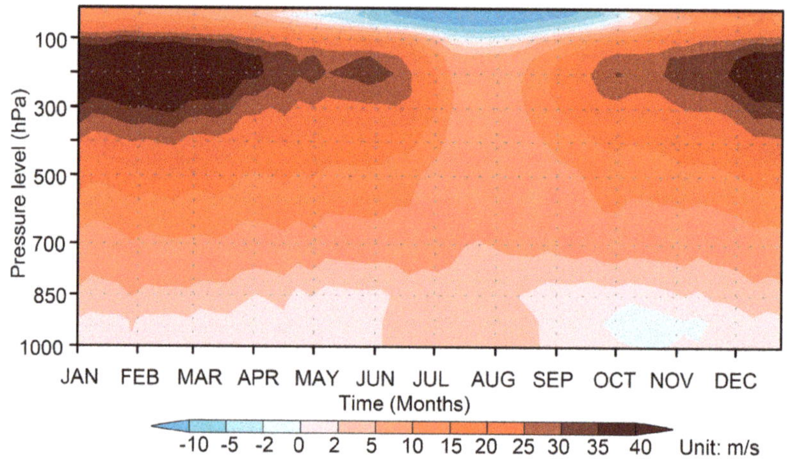

Figure 2. Vertical-temporal distributions of climatological mean values of the zonal wind at the grid (40° E, 32° N) during 1979–2018 (unit: m/s).

To determine whether meteorological conditions around the Tibetan Plateau favor the generation of vortex street shedding, two dimensionless indices were employed here to measure the basic flow parameters: the Froude number (*Fr*) and Reynolds number (*Re*). Etling [27] showed that a stable vortex street on the lee side of an obstacle can exist when the Froude number (*Fr*) is smaller than 0.4 and the Reynolds number (*Re*) falls in a particular range. The particular range of Reynolds numbers (*Re*) for the generation of vortex street shedding was suggested to be 50–10^5 for different horizontal sizes of the obstacle [25].

The Froude number is the dimensionless ratio of flow inertia to gravitational forces. The parameter is relevant in the description of stratified atmospheric flows. Here, we calculate the streamlines of the air parcels that flow over and around the obstacle separately [51]. The flow below the level of the partitioning streamline, referred to as dividing streamline height, is regarded as a quasi-2D streamline in horizontal planes.

When the vertical profile of wind speed and stratification is given, the dividing streamline height h_c can be calculated based on an implicit expression derived by Snyder et al. [51]:

$$0.5U^2(h_c) = \int_{h_c}^{h_m} N^2(z)(h_m - z)dz \quad (1)$$

where $U(z)$ is the vertical profile of upstream wind speed, h_m is the average height of the Tibetan Plateau that is felt by the large-scale flow (average height of Tibetan Plateau plus its boundary layer depth), which is approximately 5000 m, and $N(z)$ is the Brunt-Väisälä frequency. The dividing streamline height h_c is then used to calculate the Froude number in the slowly varying (approximated as constant) upstream velocity and stratification [27]:

$$Fr = 1 - \frac{h_c}{h_m} \quad (2)$$

Note that this equation of Fr is different from its glossary definition; this equation of Fr is derived based on laboratory experiments [51] and was recommended to calculate Fr in the real atmosphere [27]. Figure 3 plots the climatological mean of the dividing streamline height h_c for various seasons. The dividing streamline height h_c falls in the range of 3500 m to 4000 m (approximately 600–700 hPa), indicating that the Froude number varies from 0.2 to 0.3 and falls in the range of Froude numbers that could support vortex shedding year-round. The value of dividing streamline height coincides well with the results revealed by numerical experiments, which illustrates that the westerly flowing around the Tibetan Plateau dominates the flowing over in the middle-low troposphere [52]. At the dividing streamline height level, the diameter of the Tibetan Plateau is approximately 1000 km year-round. The upstream velocity for the Tibetan Plateau at that level is observed to be approximately 8 m/s in summer and 15 m/s in other seasons.

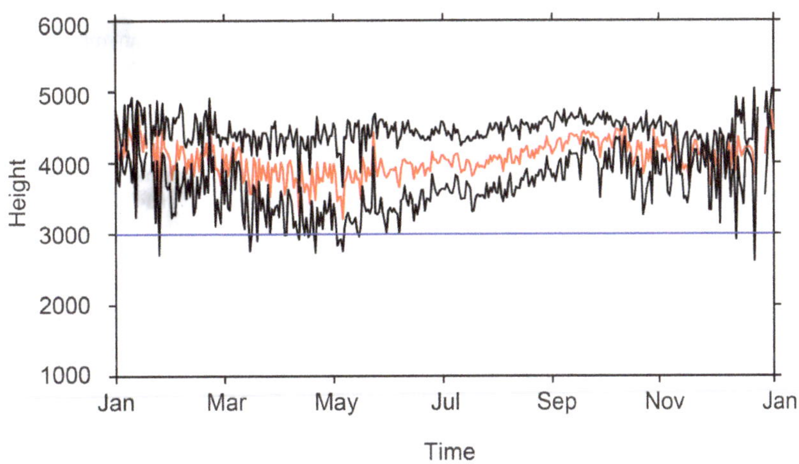

Figure 3. Red line: The seasonal variation in the climatological mean of the height of the dividing streamline h_c of the Tibetan Plateau (unit: m). Black lines mark 1 standard deviation. The blue line is the threshold above which the stable AVS could exist.

Using the above values, we can estimate the Reynolds number, the dimensionless ratio of inertial force to viscous force, for various seasons. The Reynolds number is defined as:

$$Re = \frac{U_o d}{v} \quad (3)$$

where U_0 is the upstream velocity, d is the obstacle (cylinder) diameter, and ν is the kinematic viscosity of the fluid. A value of $\nu = 1000\ m^2/s$ was used in this study, as suggested by Thomson et al. [25]. The estimated Re is approximately 1.5×10^4 in winter and 0.8×10^4 in summer. The estimated Re is below the upper threshold (10^5) at which the vortex street with a predominant frequency can exist, suggesting that the Reynolds number falls in the range of Reynolds numbers that could support vortex shedding year-round.

3.2. Characteristics of the AVS on the Leeward Side of the Tibetan Plateau

The analysis in Section 3.1 revealed that the meteorological conditions around the Tibetan Plateau are favorable for the generation of vortex street shedding year-round. What are the characteristics of the AVS on the leeward side of the Tibetan Plateau? In this study, the focus is placed on the downstream AVS, far beyond the sharp downhill region.

To answer this question, in Figure 4a–d, we display four cases of the spatial structure of the Fourier-filtered relative vorticity and horizontal wind over the leeward side of the Tibetan Plateau. These four cases are displayed for the spring, summer, autumn, and winter seasons. Clearly, the Fourier-filtered relative vorticity fields in various seasons all bear close resemblance to that of the classic von Kármán vortex-street patterns observed in laboratory flow experiments and illustrated in Figure 1. The double row of counter-rotating vortex pairs appears on the leeward side of the Tibetan Plateau, confirming a well-defined AVS pattern related to the Tibetan Plateau. The cyclonic vortices and anticyclonic vortices in the wake of the Tibetan Plateau are generated over the Sichuan Basin and Gansu Province, respectively, and then both propagate eastward toward the Pacific Ocean, with the trail persisting a considerable distance downstream of the Tibetan Plateau. Note that the AVSs discussed here are mostly generated by the portion of the mean flow split horizontally by the Tibetan Plateau at the upstream and merged downstream, not by the portion that overflows the Tibetan Plateau.

The spatial structures of the AVS showed seasonal variations. A common feature that emerged between spring and winter is that the centerlines connecting the centers of a cyclonic-anticyclonic vortex pair all have an approximately west-east orientation. The propagating direction of rotating vortex pairs in these seasons follows that of westerlies, consistent with the results in Horváth et al. [16]. However, the centerlines connecting the centers of a cyclonic-anticyclonic vortex pair in summer and early autumn have a southwest-northeast orientation. Notably, a strong anticyclonic circulation was located over the western Pacific (Figure 4b,c), which was termed the western Pacific subtropical anticyclone (WPSA). The WPSA penetrates northwards in summer, and the strong southwesterlies along its western edge may favor the southwest-northeast propagation direction of the AVS in summer and early autumn. Thus, the differences in the centerline orientation among various seasons may be related to the difference in the position of the WPSA.

A question that naturally follows is whether the AVS pattern was simply a coincidence or a consequence of using Fourier filtering. Our further analysis of the historical data without filtering eliminates the possibility of these artifacts. Figure 4e,h displays the spatial structure of the raw relative vorticity field of the four cases listed in Figure 4a–d. The AVS structure can still be distinguished in the raw field of these cases. The spatial patterns of unfiltered relative vorticity share similar spatial structures of their corresponding filtered field, with the filtered field explaining approximately 66% to 76% of the raw field among various seasons (Figure 4e–h vs. Figure 4a–d).

Figure 5a–c displays the spatiotemporal evolution of daily relative vorticity at 700 hPa (without filtering) over the downstream region of the Tibetan Plateau (110–120° E) from February to April of 1981, 1984, and 1992. The AVS structure remains robust in the raw data; the double row of counter-rotating vortex pairs alternately appeared in the region between 25–40° N, with the centerline located at 32° N, which is the central line of the Tibetan Plateau (Figure 5a–c). In this double-row pattern, each vortex is opposite the center of the spacing between the two vortices in the other row, and the lateral spacing is roughly equal to the cross-stream diameter of the Tibetan Plateau. Another noticeable feature is that the vortex pairs are generated at a predominant period of approximately 3–4 days and then propagate downstream to the Pacific at a similar, steady speed (Figure 5a–f). The above features suggest that the double row of counter-rotating vortex pairs over the leeward side of the Tibetan Plateau in the unfiltered field exists in various seasons, and this structure also bears a close resemblance to that of the classic von Kármán vortex-street patterns observed in laboratory flow experiments and illustrated in Figure 1.

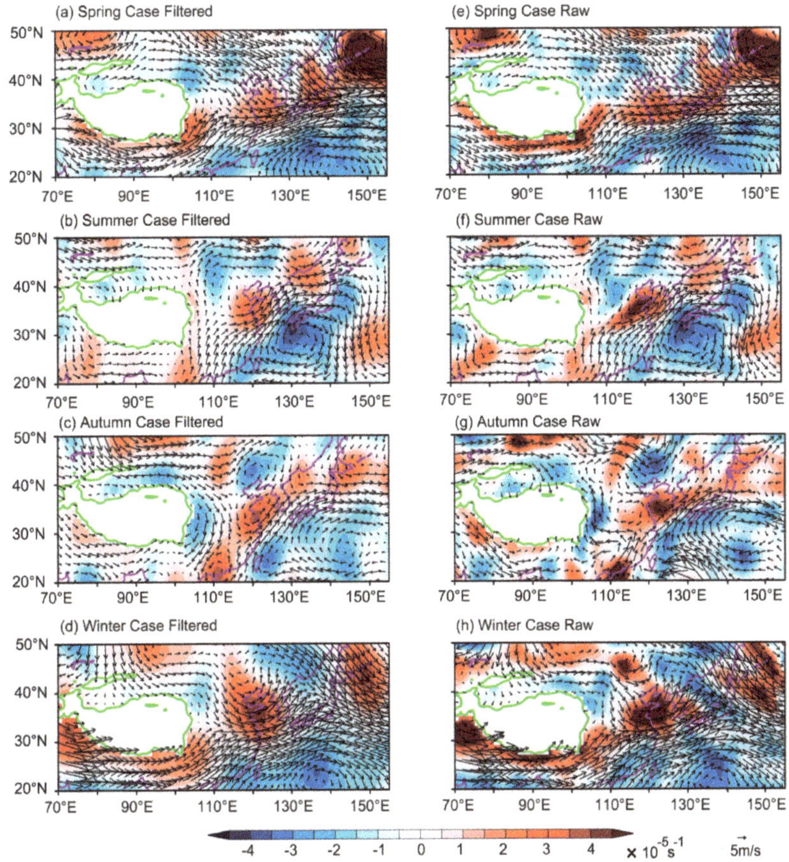

Figure 4. (**a**–**d**) Fourier filtered horizontal structure of 4 cases in daily relative vorticity at 700 hPa for the period of boreal (**a**) spring (16 March 1992), (**b**) summer (1 August 1991), (**c**) autumn (15 September 2002), and (**d**) winter (15 February 2005) (unit: s^{-1}). The green line marks the topography of the Tibetan Plateau. The vector exhibited the Fourier filtered 700 hPa horizontal wind field. (**e**–**h**) Same as (**a**–**d**) but for the unfiltered field replacing the Fourier filtered field.

3.3. The Properties of the AVS on the Leeward Side of the Tibetan Plateau

To further confirm that the AVSs developing on the leeward side of the Tibetan Plateau can be interpreted as the atmospheric analog of the classic von Kármán vortex street, various properties of AVSs were calculated to compare with those in previous studies of smaller obstacle caused AVS. These properties are summarized in Tables 1 and 2.

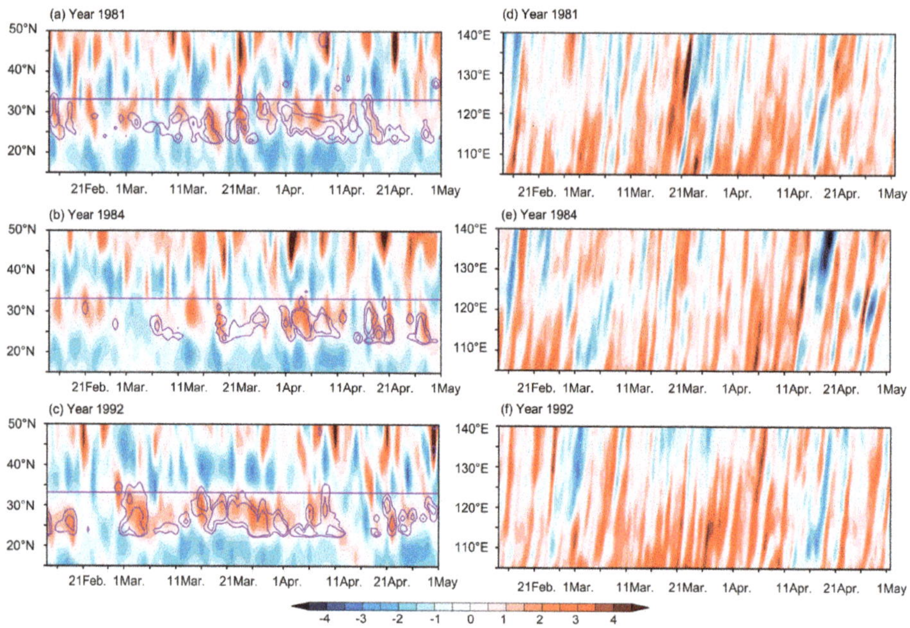

Figure 5. (**a–c**) Hovmöller diagram of daily relative vorticity at 700 hPa averaged over 110–120° E (unit: 10^{-5} s^{-1}) for the period of February to April (winter to spring) in (**a**) 1981, (**b**) 1984, and (**c**) 1992. The straight line in panel a is used to represent 32° N, which is roughly the central line of the Tibetan Plateau. Precipitation averaged over 110–120° E was also exhibited by the purple lines (2 levels of 5 mm/day and 15 mm/day). (**d–f**) Hovmöller diagram of daily relative vorticity at 700 hPa averaged over 28–32° N from March to April in (**a**) 1981, (**b**) 1984, and (**c**) 1992 (unit: 10^{-5} s^{-1}).

Table 1. Characteristic values for the vortex streets of the leeward side of the Tibetan Plateau for 12 cases in spring and winter. The characteristic values include vortex spacing *a*, vortex width *h*, and aspect ratio *h/a*, AVS vortex shedding period T_e, the propagation velocity of the AVS patterns U_e, the undisturbed wind velocity U_o, the Reynolds number *Re*, and the Strouhal number *S*.

Date	*a* (km)	*h* (km)	*h*/*a*	U_e (m/s)	U_o (m/s)	T_e (hour)	*Re*	*S*
1981/3/18	2791	1196	0.462	5.66	7.28	137	7.28×10^3	0.279
1981/4/8	3074	525	0.201	10.40	19.40	82	1.24×10^4	0.273
1981/4/19	2478	1458	0.588	8.39	9.99	82	1.56×10^4	0.216
1981/4/24	3440	741	0.217	11.12	20.74	86	1.59×10^4	0.204
1984/3/14	3087	1083	0.381	7.20	10.27	119	8.58×10^3	0.272
1984/4/7	3195	1088	0.343	17.78	28.38	50	2.27×10^4	0.245
1984/4/21	2622	982	0.366	9.11	13.27	80	1.70×10^4	0.205
1984/4/25	3036	989	0.326	10.27	17.29	82	1.22×10^4	0.278
1992/3/23	3538	1023	0.327	13.83	23.28	71	1.97×10^4	0.198
1992/2/15	3807	1298	0.241	12.95	24.14	86	2.41×10^4	0.133
1992/4/1	3576	1066	0.313	9.20	17.15	108	1.71×10^4	0.150
2004/2/15	2987	1211	0.268	8.43	15.72	108	1.57×10^4	0.164

The aspect ratio *h/a* (see Figure 1) is a basic property of an AVS. Laboratory experiments show that a stable vortex formed on the lee side of an obstacle is characterized by $0.28 < h/a < 0.52$ [19]. The aspect ratio *h/a* of AVS recorded in previous studies falls in the range of 0.15–0.8 [25]. Rows 2–4 in Table 1 display the vortex spacing a, vortex width h, and aspect ratio h/a, respectively. In this study, the dimensions of the AVS can be measured directly based on the position of extreme points. Considering that the longer shedding period may cause an increasingly disordered AVS structure, only the two westernmost vortex pairs in each case were used to calculate the aspect ratio h/a. The aspect ratio h/a in the 12 AVS cases listed in Table 1 mainly falls into the range of 0.2 to 0.59, which is comparable to that observed by previous studies. From Table 1, one can see that the vortex spacing a varies from 2478 km to 3807 km. Moreover, the average AVS width h is 1055 km, which is roughly equal to the cross-stream diameter of the Tibetan Plateau.

Another property that can be identified directly is the AVS vortex shedding period T_e (Figure 5a–c). T_e was defined as the difference between the timing of the maximal value in consecutive cyclonic vortices in the relative vorticity field at 700 hPa in the latitude-time diagram (row 7 of Table 1). Three properties of AVS can be calculated based on T_e: (1) The propagation velocity of the AVS patterns, U_e, (2) the ratio between the vortex propagation velocity U_e and the undisturbed wind velocity U_o (referred to as U_e/U_o below), and (3) the Strouhal number S.

Following Chopra and Hubert [20], The AVS propagation velocity U_e is defined as:

$$U_e = \frac{a}{T_e} = af \tag{4}$$

where f is the vortex shedding rate (or frequency), U_e is the vortex propagation velocity and is defined as a divided by T_e, and T_e is the vortex shedding period. From Table 1, one can see that the vortex shedding period T_e falls in the range from 50 to 137 h, and these values are also approximately one to two orders of magnitude greater than those observed by previous studies. From Table 1, U_e was estimated to fall into the range between 5.7 m/s and 17.8 m/s. These values are comparable to those observed in previous studies.

Table 2. Characteristic values for the vortex streets of the leeward side of the Tibetan Plateau for 12 cases in summer. The characteristic values include vortex spacing a, vortex width h, and aspect ratio h/a, AVS vortex shedding period T_e, the propagation velocity of the AVS patterns U_e, the undisturbed wind velocity U_o, the Reynolds number Re, and the Strouhal number S.

Date	a (km)	h (km)	h/a	U_e (m/s)	U_o (m/s)	T_e (hour)	Re	S
1981/8/12	1392	825	0.594	5.95	7.08	65	8.66×10^3	0.493
1981/8/16	2270	842	0.376	5.84	8.33	108	7.45×10^3	0.345
1981/8/20	1978	911	0.465	6.96	8.87	79	1.17×10^4	0.300
1981/8/24	1795	1070	0.600	5.60	6.65	89	1.04×10^4	0.299
1984/7/2	2520	1472	0.591	6.09	7.25	115	1.02×10^4	0.236
1984/7/6	1842	841	0.475	3.60	4.56	142	4.63×10^3	0.422
1984/7/10	1790	826	0.469	3.50	4.46	142	4.82×10^3	0.406
1984/7/14	2024	563	0.291	5.98	11.14	94	7.25×10^3	0.407
1992/7/2	2204	1008	0.457	4.90	6.30	125	9.13×10^3	0.243
1992/7/6	1483	824	0.552	4.21	5.10	98	4.92×10^3	0.576
1992/7/10	1383	877	0.635	2.49	2.92	154	3.49×10^3	0.517
1992/7/14	1687	1293	0.767	3.04	3.45	154	3.45×10^3	0.523

The ratio U_e/U_o is a common choice to estimate U_o when directly measuring U_o is difficult. Following Chopra and Hubert [20], as well as Li et al. [4], the ratio U_e/U_o and the aspect ratio h/a satisfy the following equation:

$$(2B - A)(U_e/U_o)^2 + (2A - 3B)\left(\frac{U_e}{U_o}\right) + \left(B - A + \frac{B}{4A}\right) = 0 \tag{5}$$

where $A = coth\ (\pi h/a)$ and $B = \pi h/a$. The analytical solution shows that the ratio U_e/U_o equals 0.75 if the aspect ratio h/a is close to 0.39. In previous studies, the ratio U_e/U_o fell in a range of 0.7 to 0.85 [25]. In our study, the ratio U_e/U_o in the 12 AVS cases mainly falls into the range of 0.54 to 0.88, and the estimated U_o varies from 7.3 m/s to 28.4 m/s (Table 1).

The Strouhal number, S, is an essential dimensionless quantity in the description of oscillating flows. It can be considered a normalized shedding frequency, defined as:

$$S = \frac{d}{T_e U_o} \tag{6}$$

where U_o is the upstream velocity, d is the crosswind island diameter at dividing streamline height h_c, and T_e is the shedding period between two consecutive like-rotating vortices. Laboratory experiments show that the Strouhal number S varies from 0.12 to 0.21 when the Reynolds number Re is smaller than 10^4 [53]. In this study, the Strouhal number S fluctuated within the broad range of 0.13–0.28, which is consistent with the conclusions in previous studies.

In a nonrotating, unstratified fluid, the nature of the wake only depends on the Reynolds number, which is the dimensionless ratio of inertial force to viscous force. In Section 3.1, the Reynolds number, Re, was estimated to be O (10^4). In this study, the Reynolds number Re fluctuated within the broad range of 0.73×10^4–2.41×10^4, which is consistent with our previous estimation.

When the wake on the leeward side of the Tibetan Plateau was characterized by an AVS with a southwest-northeast orientation, many AVS properties changed considerably (Table 2). The distinctions in the AVS properties for these two periods can be summarized as follows: (1) The aspect ratio h/a in the subtropical AVS increased to 0.38–0.77. The increases in the aspect ratio h/a were mainly caused by the shortening of vortex spacing a. (2) The AVS propagation velocity U_e declined to 2.5–7.0 m/s, which mainly resulted from the decrease in upstream velocity in boreal summer. The Reynolds number, Re, was thereby decreased to 0.35×10^4–1.17×10^4, and the average AVS vortex shedding period T_e increased to 114 h.

Thus, the above results indicate that the AVS on the leeward side of the Tibetan Plateau can be interpreted as the atmospheric analog of classic von Kármán vortex streets in various seasons.

3.4. Impacts of the AVS on Precipitation over the Wake of the Tibetan Plateau

Obviously, cyclonic activities in the AVS caused substantial precipitation (Figure 5a–c). A question that naturally follows is how much precipitation over the wake of the Tibetan Plateau can be closely tied to the AVS. To answer this question, we compared the spatiotemporal evolution of precipitation and cyclonic activities and calculated the relationship between heavy rain days and AVS-related cyclonic activities over East China, Japan, and the Korean Peninsula.

Figure 6 exhibits the seasonal variation in the climatological mean of the rainband and positive vorticity over 110–120° E, 125–130° E, and 130–145° E for the whole year. The rainband with precipitation exceeding 4 mm/day over these three regions began in the temporal span of March to September and the spatial span of the south to 32° N, and then the rainband penetrated northwards in the subtropics, propagating from 25° N to 40° N. Such a spatiotemporal structure was largely shared by that of positive relative vorticity. The correlation coefficients for the spatiotemporal domain were 0.178 (5691 samples, $p < 0.01$), 0.133 ($p < 0.01$), and 0.124 ($p < 0.01$).

The similarity between the seasonal evolution of the climatological mean was further supported by a single case. Figure 7 presents the seasonal meridional evolution of the rainband and positive vorticity over 110–120° E for the period from February to September 1983. The spatiotemporal structure of the rainband bears a close resemblance to that of positive vorticity. The correlation coefficients for the spatiotemporal domain were 0.164 (5691 samples, $p < 0.01$).

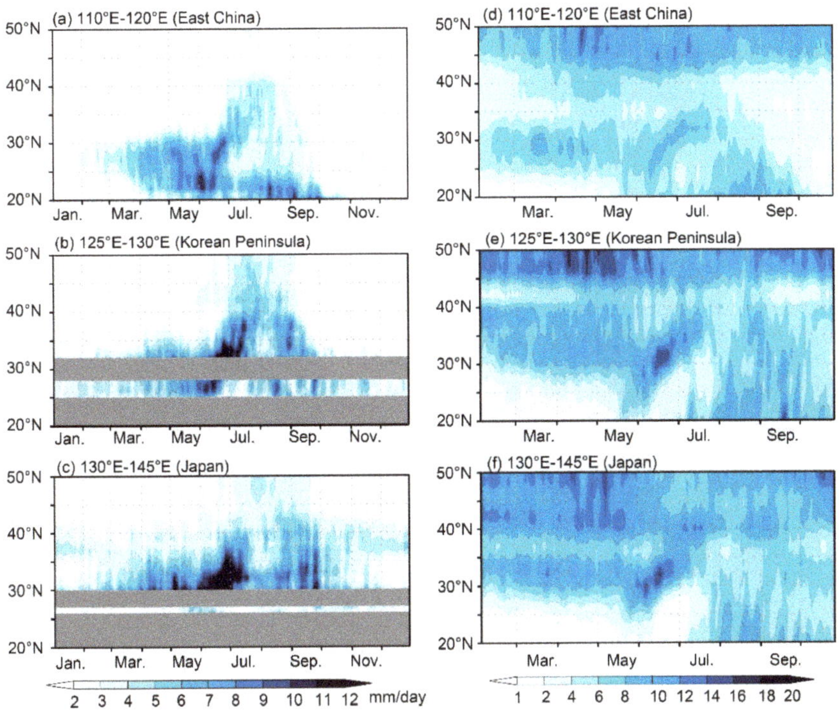

Figure 6. Hovmöller diagram of (**a–c**) daily precipitation derived from CPC datasets (unit: mm) and (**d–f**) daily relative vorticity at 850 hPa (unit: 10^{-6} s^{-1}) averaged over (**a**,**d**) 110–120° E, (**b**,**e**) 125–130° E, and (**c**,**f**) 130–145° E. The results were smoothed by a pentad temporal domain. Gray represents missing records.

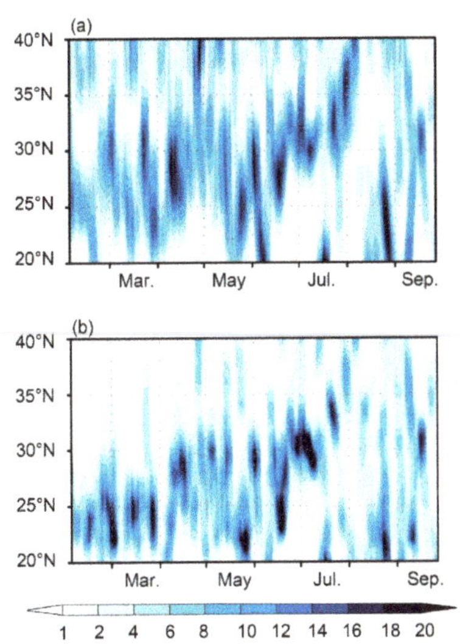

Figure 7. Hovmöller diagram of (**a**) daily relative vorticity at 850 hPa (unit: 10^{-6} s^{-1}) and (**b**) daily precipitation derived from CPC datasets (unit: mm) for the period from February to September 1983 averaged over 110–120° E. The results were smoothed by a pentad temporal domain.

How much precipitation and how many heavy rain days can be closely tied to the AVS in the leeside wake of the Tibetan Plateau? Figures 8 and 9 present the proportion of precipitation (Figure 8) and heavy rain days (Figure 9, defined as daily precipitation exceeding 8 mm/day) that can be closely tied to AVS (characterized by positive vorticity in the Fourier filtered field). A common feature that emerged in these three regions is that 80–90% of heavy rain days were accompanied by positive vorticity in AVS, which means that the seasonal variations in AVS are closely tied to the heavy rain days in the main rainband. Moreover, 80–90% of the total precipitation in the main rainband is closely tied to the seasonal variations in AVS. Our study reveals that the impact of the Tibetan Plateau on precipitation can be in a larger region (a scale of a few thousand kilometers) downstream of the Tibetan Plateau and that this impact is facilitated by the AVS. The AVS provides a favorable cyclonic environment for precipitation. When other low-value weather systems march to the cyclic vortex region of the AVS, precipitation can significantly increase.

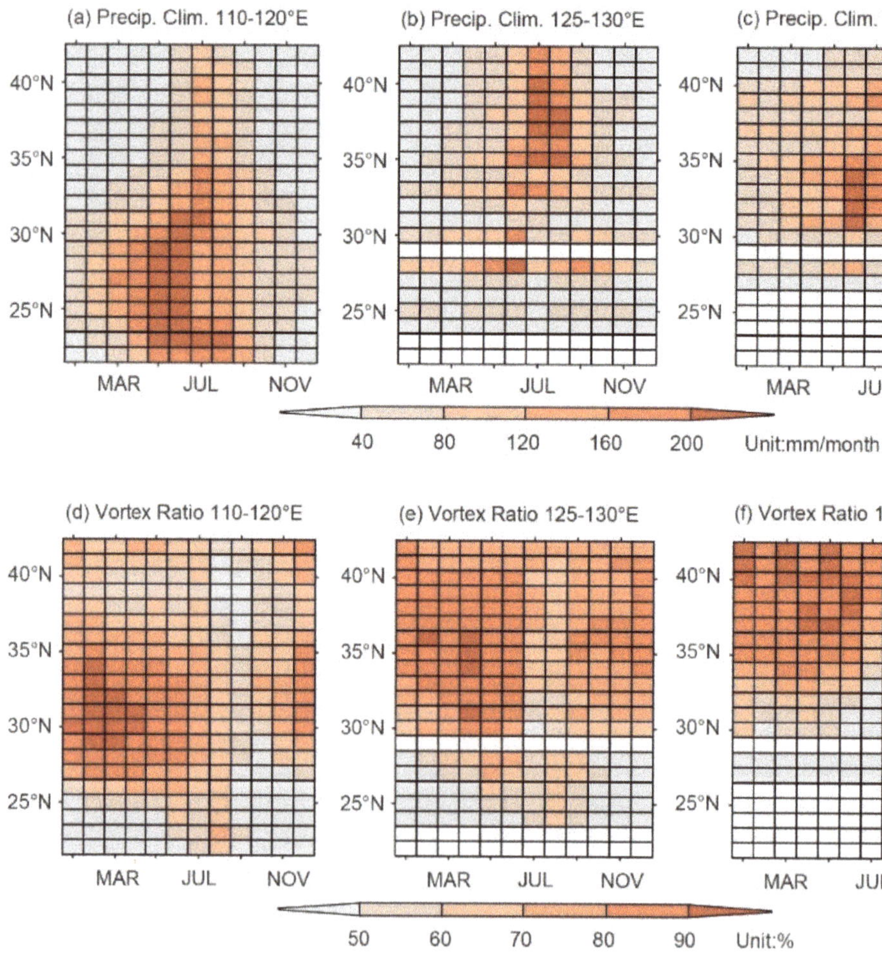

Figure 8. Hovmöller diagram of the (**a–c**) monthly precipitation and (**d–f**) the ratio of the monthly precipitation closely tied to AVS to the total monthly precipitation averaged over (**a,d**) 110–120° E, (**b,e**) 125–130° E, and (**c,f**) 130–145° E. White represents missing records of land precipitation.

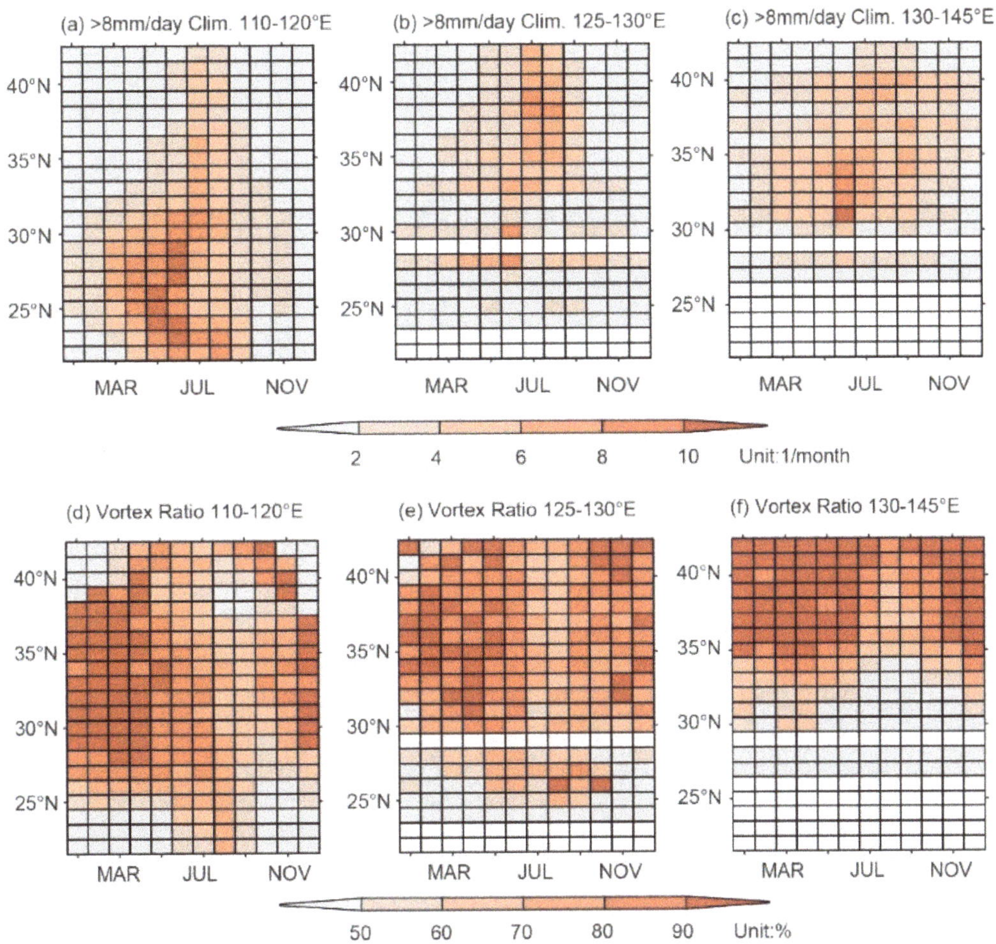

Figure 9. Hovmöller diagram of (**a–c**) the yearly frequency of heavy rain days (defined as daily precipitation exceeding 8 mm/day) and the ratio of (**d–f**) the number of heavy rain days closely tied to AVS to the total number of heavy rain days averaged over (**a**,**d**) 110–120° E, (**b**,**e**) 125–130° E, and (**c**,**f**) 130–145° E. White represents missing records of land precipitation.

4. Conclusions

The atmospheric vortex street (AVS) is a common phenomenon recorded on the leeward side of mountainous islands. The prevailing westerlies were observed year-round around the Tibetan Plateau, which is the world's highest plateau above sea level. However, it remains unknown whether the wake on the leeward side of the Tibetan Plateau exhibits a stable AVS and how the AVS impacts precipitation over the downstream region. In this study, we present evidence that the wake on the leeward side of the Tibetan Plateau can be interpreted as the atmospheric analog of classic von Kármán vortex streets. Further analysis measured the percentage of precipitation over the wake of the Tibetan Plateau closely tied to seasonal variations in AVS. The major findings are summarized below.

(1) The meteorological factors around the Tibetan Plateau satisfy conditions in which a stable vortex street on the lee side of an obstacle can exist for the whole year. The Froude number varies from 0.2 to 0.3 and falls in the range of Froude numbers that could support vortex shedding for the whole year, whereas the Reynolds number

was estimated to be 0.7×10^4–2.4×10^4 in winter and 0.4×10^4–1.2×10^4 in summer. Both of these dimensionless indices fall in the range of meteorological conditions summarized by previous studies [27].

(2) The spatiotemporal structures indicate that the wake on the leeward side of the Tibetan Plateau showed seasonal variations. The wake was characterized by a stable vortex street with a southwest-northeast orientation in summer and early autumn but with a west-east orientation in other seasons. The differences in the centerline orientation among various seasons may be related to the difference in the position of the Western Pacific Subtropical Anticyclone. The wake in the Tibetan Plateau bears a close resemblance to that of the classic von Kármán vortex-street patterns observed in laboratory flow experiments. Moreover, various properties, including aspect ratio, Strouhal number, etc., calculated for these AVSs are in the same range as previous studies. Thus, the wake on the leeward side of the Tibetan Plateau can be interpreted as the atmospheric analog of classic von Kármán vortex streets in various seasons.

(3) We further show that the spatiotemporal structure of precipitation was largely shared by that of cyclonic activities in the AVS, both in the climatological mean and case study. Approximately 80–90% of the precipitation and heavy rain days in the main rainband over the wake of the Tibetan Plateau are closely tied to the seasonal evolution of the AVS.

Previous studies have suggested that the subtropical convergence zone (the Meiyu-Baiu-Changma frontal zone) is a key circulation system dominating summer precipitation over the wake of the Tibetan Plateau. In this study, the AVS in the subtropics was found on the leeward side of the Tibetan Plateau, spatiotemporally coinciding well with the subtropical convergence zone. Thus, the AVS can be regarded as the dominating mechanism of the subtropical convergence zone. Exploring the underlying relationship between the large-scale circulation and the AVS would help to improve the simulating skill of precipitation over East Asia in numerical simulations in the future.

Author Contributions: Conceptualization, Q.L., Z.W. and C.F.; methodology, Q.L.; validation, Z.W. and F.Y.; formal analysis, Q.L.; investigation, Q.L. and Z.W.; data curation, Q.L.; writing—original draft preparation, Q.L.; writing—review and editing, Z.W., C.F., Z.-M.T. and F.Y.; visualization, Q.L. and Z.W.; supervision, Z.W., C.F. and Z.-M.T.; project administration, Z.W. and C.F.; funding acquisition, C.F. All authors have read and agreed to the published version of the manuscript.

Funding: This research was funded by the Chinese Jiangsu Collaborative Innovation Center for Climate Change, the Frontiers Science Center for Critical Earth Material Cycling of Nanjing University, and the High-Performance Computing Centers of Nanjing University.

Institutional Review Board Statement: Not applicable.

Informed Consent Statement: Not applicable.

Data Availability Statement: The ERA-Interim dataset can be downloaded from the website https://apps.ecmwf.int/datasets/data/interim-full-daily/levtype=sfc/ (accessed on 1 January 2022). The NCEP/NCAR dataset can be downloaded from https://psl.noaa.gov/data/gridded/data.ncep.reanalysis.html (accessed on 1 January 2022). The CPC precipitation data can be downloaded from the website https://psl.noaa.gov/data/gridded/data.cpc.globalprecip.html (accessed on 1 January 2022). The APHRODITE precipitation data can be downloaded from the website http://aphrodite.st.hirosaki-u.ac.jp/products.html (accessed on 1 January 2022).

Conflicts of Interest: The authors declare no conflict of interest.

References

1. von Kármán, T. Über den Mechanismus des Widerstandes, den ein bewegter Körper in einer Flüssigkeit erfärt. 1. Teil, Nachr. Ges. Wiss. Göttingen. *Math.-Phys. Kl.* **1911**, *1911*, 509–517.
2. von Kármán, T. Über den Mechanismus des Widerstandes, den ein bewegter Körper in einer Flüssigkeit erfärt. 2. Teil, Nachr. Ges. Wiss. Göttingen. *Math.-Phys. Kl.* **1912**, *1912*, 547–556.
3. Kundu, P.K. *Fluid Dynamics*; Academic Press: San Diego, CA, USA, 1990; pp. 321–338.

4. Nunalee, C.G.; Basu, S. On the periodicity of atmospheric von Kármán vortex streets. *Environ. Fluid Mech.* **2014**, *14*, 1335–1355. [CrossRef]
5. Barton, E.D.; Basterretxea, G.; Flament, P.; Mitchelson-Jacob, E.G.; Jones, B.; Arîstegui, J.; Herrera, F. Lee region of gran canaria. *J. Geophys. Res.-Oceans* **2000**, *105*, 17173–17193. [CrossRef]
6. Li, X.; Clemente-Colón, P.; Pichel, W.G. Atmospheric vortex streets on a RADARSAT SAR image. *Geophys. Res. Lett.* **2000**, *27*, 1655–1658. [CrossRef]
7. Young, G.S.; Zawislak, J. An observational study of vortex spacing in island wake vortex streets. *Mon. Weather Rev.* **2006**, *134*, 2285–2294. [CrossRef]
8. Dong, C.; McWilliams, J.C.; Shchepetkin, A.F. Island wakes in deep water. *J. Phys. Oceanogr.* **2007**, *37*, 962–981. [CrossRef]
9. Zheng, Q.; Lin, H.; Meng, J.; Hu, X.; Song, Y.T.; Zhang, Y.; Li, C. Sub-mesoscale ocean vortex trains in the Luzon Strait. *J. Geophys. Res.-Oceans* **2008**, *113*, C04032. [CrossRef]
10. Teinturier, S.; Stegner, A.; Didelle, H.; Viboud, S. Small-scale instabilities of an island wake flow in a rotating shallow-water layer. *Dyn. Atmos. Oceans* **2010**, *49*, 1–24. [CrossRef]
11. Topouzelis, K.; Kitsiou, D. Detection and classification of mesoscale atmospheric phenomena above sea in SAR imagery. *Remote Sens. Environ.* **2015**, *160*, 263–272. [CrossRef]
12. Caldeira, R.M.A.; Groom, S.; Miller, P.; Pilgrim, D.; Nezlin, N.P. Sea-surface signatures of the island mass effect phenomena around Madeira Island, Northeast Atlantic. *Remote Sens. Environ.* **2002**, *80*, 336–360. [CrossRef]
13. Caldeira, R.M.A.; Marchesiello, P.; Nezlin, N.P.; DiGiacomo, P.M.; McWilliams, J.C. Island wakes in theSouthern California Bight. *J. Geophys. Res.-Oceans* **2005**, *110*, C11012. [CrossRef]
14. Hasegawa, D.; Yamazaki, H.; Lueck, R.G.; Seuront, L. How islands stir and fertilize the upper ocean. *Geophys. Res. Lett.* **2004**, *31*, L16303. [CrossRef]
15. Hasegawa, D.; Lewis, M.R.; Gangopadhyay, A. How islands cause phytoplankton to bloom in their wakes. *Geophys. Res. Lett.* **2009**, *36*, L20605. [CrossRef]
16. Horváth, Á.; Bresky, W.; Daniels, J.; Vogelzang, J.; Stoffelen, A.; Carr, J.L.; Wu, D.L.; Seethala, C.; Günther, T.; Buehler, S.A. Evolution of an Atmospheric Kármán Vortex Street from High-Resolution Satellite Winds: Guadalupe Island Case Study. *J. Geophys. Res.-Atmos.* **2020**, *125*, e2019JD032121. [CrossRef]
17. Lettau, H. Atmosphärische Turbulenz. *Akad. Verl. Leipz.* **1939**, 283.
18. Hubert, L.F.; Krueger, A.F. Satellite Pictures of Mesoscale Eddies. *Mon. Weather Rev.* **1962**, *90*, 457–463. [CrossRef]
19. Chopra, K.P.; Hubert, L.F. Kármán vortex streets in earth's atmosphere. *Nature* **1964**, *203*, 1341–1343. [CrossRef]
20. Chopra, K.P.; Hubert, L.F. Mesoscale Eddies in Wake of Islands. *J. Atmos. Sci.* **1965**, *22*, 652–657. [CrossRef]
21. Lyons, W.A.; Fujita, T. Mesoscale motions in oceanic stratus as revealed by satellite data. *Mon. Weather Rev.* **1968**, *96*, 304–314. [CrossRef]
22. Tsuchiya, K. The clouds with the shape of Kármán vortex street in the wake of Cheju Island, Korea. *J. Meteorol. Soc. JPN Ser. II* **1969**, *47*, 457–465. [CrossRef]
23. Zimmerman, L.I. Atmospheric wake phenomena near the Canary Islands. *J. Appl. Meteorol. Clim.* **1969**, *8*, 896–907. [CrossRef]
24. Chopra, K.P. Atmospheric and Oceanic Flow Problems Introduced by Islands. In *Advances in Geophysics*; Academic Press: New York, NY, USA, 1973.
25. Thomson, R.E.; Gower, J.; Bowker, N.W. Vortex Streets in the Wake of the Aleutian Islands. *Mon. Weather Rev.* **1977**, *105*, 873–884. [CrossRef]
26. Jensen, N.O.; Agee, E.M. Vortex cloud street during AMTEX 75. *Tellus* **1978**, *30*, 517–523. [CrossRef]
27. Etling, D. On atmospheric vortex streets in the wake of large islands. *Meteorol. Atmos. Phys.* **1989**, *41*, 157–164. [CrossRef]
28. Li, X.; Zheng, W.; Zou, C.Z.; Pichel, W.G. A SAR Observation and Numerical Study on Ocean Surface Imprints of Atmospheric Vortex Streets. *Sensors* **2008**, *8*, 3321–3334. [CrossRef] [PubMed]
29. Couvelard, X.; Caldeira, R.M.A.; Araújo, I.B.; Tomé, R. Wind mediated vorticity-generation and eddy-confinement, leeward of the Madeira Island: 2008 numerical case study. *Dynam. Atmos. Oceans* **2012**, *58*, 128–149. [CrossRef]
30. Caldeira, R.M.; Tomé, R. Wake response to an ocean-feedback mechanism: Madeira Island case study. *Bound.-Lay. Meteorol.* **2013**, *148*, 419–436. [CrossRef]
31. Ito, J.; Niino, H. Atmospheric Kármán Vortex Shedding from Jeju Island, East China Sea: A Numerical Study. *Mon. Weather Rev.* **2015**, *144*, 139–148. [CrossRef]
32. Papailiou, D.D.; Lykoudis, P.S. Turbulent vortex streets and the entrainment mechanism of the turbulent wake. *J. Fluid Mech.* **1974**, *62*, 11–31. [CrossRef]
33. Yeh, T. The circulation of the high troposphere over China in the winter of 1945–46. *Tellus* **1950**, *2*, 173–183. [CrossRef]
34. Bolin, B. On the Influence of the Earth's Orography on the General Character of the Westerlies. *Tellus* **2016**, *2*, 184–195. [CrossRef]
35. Queney, P. The problem of air flow over mountains: A summary of theoretical studies. *Bull. Am. Meteorol. Soc.* **1948**, *29*, 16–26. [CrossRef]
36. Charney, J.G.; Eliassen, A. A numerical method for predicting the perturbations of the middle latitude westerlies. *Tellus* **1949**, *1*, 38–54. [CrossRef]
37. Wu, G.X. The nonlinear response of the atmosphere to large-scale mechanical and thermal forcing. *J. Atmos. Sci.* **1984**, *41*, 2456–2476. [CrossRef]

38. Wang, Q.; Tan, Z.-M. Multi-scale topographic control of southwest vortex formation in Tibetan Plateau region in an idealized simulation. *J. Geophys. Res.-Atmos.* **2014**, *119*, 11543–11561. [CrossRef]
39. Zhang, G.; Mao, J.; Liu, Y.; Wu, G. PV perspective of impacts on downstream extreme rainfall event of a Tibetan Plateau vortex collaborating with a southwest China vortex. *Adv. Atmos. Sci.* **2021**, *38*, 1835–1851. [CrossRef]
40. Wu, G.; Tang, Y.; He, B.; Liu, Y.; Mao, J.; Ma, T.; Ma, T. Potential vorticity perspective of the genesis of a Tibetan Plateau vortex in June 2016. *Clim. Dyn.* **2022**, *58*, 3351–3367. [CrossRef]
41. Zhang, J.; Li, J.; Guo, B.; Ma, Z.; Li, X.; Ye, X.; Yu, H.; Liu, Y.; Yang, C.; Zhang, S.; et al. Magnetostratigraphic age and monsoonal evolution recorded by the thickest Quaternary loess deposit of the Lanzhou region, western Chinese Loess Plateau. *Quat. Sci. Rev.* **2016**, *139*, 17–29. [CrossRef]
42. Dee, D.P.; Uppalaa, S.M.; Simmonsa, A.J.; Berrisforda, P.; Polia, P.; Kobayashib, S.; Andraec, U.; Balmasedaa, M.A.; Balsamoa, G.; Bauera, P.; et al. The ERA-Interim reanalysis: Configuration and performance of the data assimilation system. *Q. J. R. Meteorol. Soc.* **2011**, *137*, 553–597. [CrossRef]
43. Kalnay, E.; Kanamitsu, M.; Kistler, R.; Collins, W.; Deaven, D.; Gandin, L.; Iredell, M.; Saha, S.; White, G.; Woollen, J.; et al. The NCEP/NCAR 40-Year Reanalysis Project. *Bull. Am. Meteorol. Soc.* **1996**, *77*, 437–471. [CrossRef]
44. Curio, J.; Schiemann, R.; Hodges, K.I.; Turner, A.G. Climatology of Tibetan Plateau vortices in reanalysis data and a high-resolution global climate model. *J. Clim.* **2019**, *32*, 1933–1950. [CrossRef]
45. Chen, M.; Shi, W.; Xie, P.; Silva, V.B.S.; Kousky, V.E.; Wayne Higgins, R.; Janowiak, J.E. Assessing objective techniques for gauge-based analyses of global daily precipitation. *J. Geophys. Res.-Atmos.* **2008**, *113*, D04110. [CrossRef]
46. Yatagai, A.; Arakawa, O.; Kamiguchi, K.; Kawamoto, H.; Nodzu, M.I.; Hamada, A. A 44-Year Daily Gridded Precipitation Dataset for Asia Based on a Dense Network of Rain Gauges. *Sola* **2009**, *5*, 137–140. [CrossRef]
47. Wheeler, M.; Kiladis, G.N. Convectively coupled equatorial waves:analysis of clouds and temperature in the wavenumber-frequency domain. *J. Atmos. Sci.* **1999**, *56*, 374–399. [CrossRef]
48. Kiladis, G.N.; Wheeler, M.C.; Haertel, P.T.; Straub, K.H.; Roundy, P.E. Convectively coupled equatorial waves. *Rev. Geophys.* **2009**, *47*, 2008RG000266. [CrossRef]
49. Hawcroft, M.K.; Shaffrey, L.C.; Hodges, K.I.; Dacre, H.F. How much Northern Hemisphere precipitation is associated with extratropical cyclones? *Geophys. Res. Lett.* **2012**, *39*, L24809. [CrossRef]
50. Hanley, J.; Caballero, R. Objective identification and tracking of multicentre cyclones in the ERA-Interim reanalysis dataset. *Q. J. R. Meteorol. Soc.* **2012**, *138*, 612–625. [CrossRef]
51. Snyder, W.H.; Thomson, R.S.; Eskridge, R.E.; Lawson, R.E.; Castro, I.P.; Lee, S.T.; Hunt, J.C.R.; Ogawa, J. The structure of strongly stratified flows over hills: Dividing streamline concept. *J. Fluid Mech.* **1985**, *152*, 249–288. [CrossRef]
52. Trenberth, K.E.; Chen, S.C. Planetary waves kinematically forced by Himalayan orography. *J. Atmos. Sci.* **2010**, *45*, 2934–2948. [CrossRef]
53. Roshko, A. On the Development of Turbulent Wakes from Vortex Streets (No. NACA-TR-1191). Ph.D. Thesis, California Institute of Technology, Pasadena, CA, USA, 1954.

atmosphere

Review

The Guiding Role of Rossby Wave Energy Dispersion Theory for Studying East Asian Monsoon System Dynamics

Ronghui Huang [1,2], Jingliang Huangfu [1,*], Yong Liu [1] and Riyu Lu [2]

1 Center for Monsoon System Research, Institute of Atmospheric Physics, Chinese Academy of Sciences, Beijing 100029, China; hrh@mail.iap.ac.cn (R.H.); liuyong@mail.iap.ac.cn (Y.L.)
2 State Key Laboratory of Numerical Modeling for Atmospheric Sciences and Geophysical Fluid Dynamics (LASG), Institute of Atmospheric Physics, Chinese Academy of Sciences, Beijing 100029, China; lr@mail.iap.ac.cn
* Correspondence: hfjl@mail.iap.ac.cn

Abstract: This paper is written to commemorate the 10th anniversary of academician Ye Duzheng (Yeh T.C.) pass away and his great contributions to the development of atmospheric dynamics. Under the inspiration and guidance of the theory of Rossby wave energy dispersion, remarkable progresses have been made in research on planetary wave dynamics and teleconnections of atmospheric circulation anomalies. This paper aims to make a brief review of the studies on the propagating characteristics of quasi-stationary planetary waves in a three-dimensional spherical atmosphere and the dynamic processes of the interannual and interdecadal variabilities of the East Asian summer and winter monsoon systems. Especially, this paper systematically reviews the progresses of the studies on the impacts of the interannual and interdecadal variabilities of the East Asia/Pacific (EAP) pattern teleconnection wave train propagating along the meridional direction over East Asia and the "Silk Road" pattern teleconnection wave train propagating along the zonal direction within the subtropical jet from West Asia to East Asia on the East Asian summer monsoon system and the summer precipitation variability in China, under the guidance of the theory of Rossby wave energy dispersion. Moreover, this paper reviews the dynamic processes of the impact of the interannual and interdecadal oscillations of the propagating waveguides of boreal quasi-stationary planetary waves on the variability of the East Asian winter monsoon system.

Keywords: Rossby waves; energy dispersion; East Asian monsoon system; dynamic processes; planetary wave train

Citation: Huang, R.; Huangfu, J.; Liu, Y.; Lu, R. The Guiding Role of Rossby Wave Energy Dispersion Theory for Studying East Asian Monsoon System Dynamics. *Atmosphere* **2023**, *14*, 962. https://doi.org/10.3390/atmos14060962

Academic Editor: Anthony R. Lupo

Received: 14 April 2023
Revised: 24 May 2023
Accepted: 28 May 2023
Published: 31 May 2023

1. Introduction

It has been 10 years since academician Duzheng Ye left us. His pass away was a great loss not only to the atmospheric science community in China but also to the international atmospheric science community. This article is written to commemorate the 10th anniversary of his pass away and to remember his great contributions to the development of atmospheric dynamics.

It is well known that the theory of Rossby wave energy dispersion [1] proposed by Duzheng Ye more than 70 years ago is one of the classical theories of atmospheric dynamics. The theory of Rossby wave energy dispersion is not only widely used in weather forecasting but also inspire the development of planetary wave dynamics, especially the study on the propagating characteristics of quasi-stationary planetary waves in two- and three-dimensional spherical atmosphere [2–11] and the teleconnection mechanism of global atmospheric circulation anomalies [12–20]. Under the guidance of the theory of Rossby wave energy dispersion proposed by academician Ye, the research on the dynamic processes of the variability of the East Asian monsoon system has been made in recent years. In particular, our research group studied the dynamic processes of the impacts of the East Asian/Pacific (EAP) and Silk Road (SR) pattern teleconnection wave trains on the variability of the East Asian summer monsoon system [16,21]. In addition, significant progress has achieved in the studies on the dynamic processes of the variability of the East Asian winter monsoon system due to the oscillations of the propagating waveguides of quasi-stationary planetary waves in the three-dimensional spherical atmosphere [22–25]. Thus, this paper presents a brief review of the recent studies on the dynamic processes of the interannual and interdecadal variabilities of the East Asian summer monsoon system and the dynamical impact of the interannual and interdecadal oscillations of the propagating waveguides of quasi-stationary planetary waves on the East Asian winter monsoon system. Besides, under the guidance of the theory of Rossby wave energy dispersion, some studies in China and abroad related to the dynamic processes of the East Asian monsoon system are simply reviewed in this paper. Many figures in this paper are the result of our recent analysis using 60-year reanalysis data and summer precipitation data in eastern China for 1961–2020.

2. The Guiding Role of the Theory of Rossby Wave Energy Dispersion in the Study on Quasi-Stationary Planetary Wave Propagation in a Three-Dimensional Spherical Atmosphere

The theory of Rossby wave energy dispersion proposed by academician Ye has not only benefited the study on the two-dimensional propagation of quasi-stationary planetary waves in the spherical atmosphere but also played a guiding role in the study on the three-dimensional propagation of quasi-stationary planetary waves in the spherical atmosphere during boreal winter.

2.1. Study on the Propagating Waveguides of Quasi-Stationary Planetary Waves in the Spherical Atmosphere during Boreal Winter

After academician Duzheng Ye proposed the theory of Rossby wave energy dispersion, many meteorologists focused on the energy dispersion of planetary waves in the vertical direction of the atmosphere. For example, several meteorologists [26,27] applied the concepts of the wave refraction index and energy, respectively, to study the vertical propagating characteristics of quasi-stationary planetary waves in the basic flow with a vertical wind shear. Later, Dickinson [5] applied the concept of waveguides to study the vertical propagating characteristics of quasi-stationary planetary waves in the ideal basic flow in boreal winter and proposed that quasi-stationary planetary waves can propagate from the troposphere to the stratosphere over high latitudes, which is called the polar waveguide. After the Dickinson's study, Matsuno [6,7] studied the vertical propagation of quasi-stationary planetary waves from the troposphere to the stratosphere in the actual basic flow in boreal winter and proposed the dynamic mechanism of the stratospheric sudden warming. He noted that the propagation of tropospheric quasi-stationary planetary waves into the stratosphere and their interaction with the stratospheric basic flow leads to the stratospheric sudden warming in boreal winter. Thus, the theory of Rossby wave energy dispersion guided the study on the propagating characteristics of quasi-stationary planetary waves in the atmosphere from the troposphere to the stratosphere, which established the theoretical basis for the study on the mechanism of tropospheric-stratospheric interactions.

Guided by the theory of Rossby wave energy dispersion and based on the studies made by Matsuno [6,7], Huang and Gambo [10,11] systematically studied the characteristics of the three-dimensional propagations of quasi-stationary planetary waves in a spherical atmosphere using the wave refraction index and the E-P flux, respectively.

Huang and Gambo [11] studied the characteristics of three-dimensional propagation of quasi-stationary planetary waves in the actual basic flow using the wave refraction index. Figure 1 shows the vertical distribution of boreal winter Q_0 and a schematic diagram of the propagating waveguides of quasi-stationary planetary waves in the three-dimensional atmosphere, where $Q_0 = Q_k + k^2/\cos^2\varphi$ (Q_k is the square of the refraction index for wavenumber k, calculated from the basic flow and related parameters; k is the wavenumber, and φ is latitude). In Figure 1, we can see that the propagations of quasi-stationary planetary waves in the three-dimensional atmosphere in boreal winter exist two waveguides. The first waveguide is called the polar waveguide [5], i.e., quasi-stationary planetary waves can propagate from the troposphere to the stratosphere at high latitudes through this waveguide. Similarly, Figure 1 shows that quasi-stationary planetary waves can also propagate from the lower troposphere at mid-latitudes to the upper troposphere at low latitudes through the second waveguide. This propagating waveguide over low latitudes is called the "low-latitude waveguide" or "alternate waveguide". As shown in Figure 1, if there is a quasi-stationary planetary wave generated by a forcing source at low or mid-latitudes, it cannot propagate directly from the troposphere to the stratosphere at mid-latitudes but can propagate quasi-horizontally from the troposphere at low or mid-latitudes to high latitudes and then to the stratosphere at high latitudes through the polar waveguide, and it can also propagate to the upper troposphere at low latitudes through the low-latitude waveguide. This schematic picture indicates that the propagations of quasi-stationary planetary waves in the three-dimensional atmosphere in boreal winter are not limited to the propagation through polar waveguide but also exist the propagation through the low-latitude waveguide.

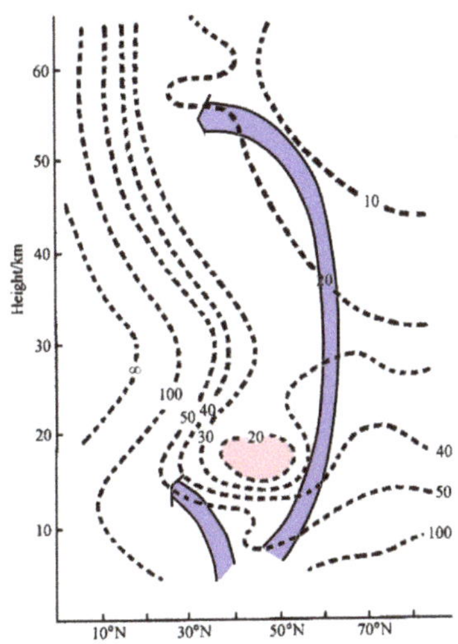

Figure 1. The vertical distribution of Q_0 (refraction index, dashed lines) and a schematic diagram of the waveguides of the three-dimensional propagations of quasi-stationary planetary waves in boreal winter (from Huang and Gambo [11]).

2.2. *Waveguides of Quasi-Stationary Planetary Wave Propagation Characterized by the E-P Flux*

The characteristics of three-dimensional propagations of quasi-stationary planetary waves in the atmosphere over the Northern Hemisphere can also be studied in terms of the wave energy flux. Eliassen and Palm [27] studied the vertical propagation of waves using the concept of energy flux, which was generalized by Andrews and McIntyre [28] in the 1970s, and they proposed the E-P flux under the β-plane approximation. Later, Edmon et al. [29] extended it to a sphere under the assumption that $\Delta f/f$ is small within the Rossby deformation radius.

Since the E-P flux of quasi-stationary planetary waves is parallel to the group velocity of the wave, meaning the E-P flux can represent the propagation of wave energy. Therefore, the E-P flux can be used to graphically characterize the propagations of quasi-stationary planetary waves in the three-dimensional atmosphere. Figure 2 shows the distributions of the E-P fluxes of quasi-stationary planetary waves averaged for 60 winters which were recently calculated by using the NCEP/NCAR reanalysis data from 1961 to 2020. As clearly shown in Figure 2, there are two propagating waveguides of quasi-stationary planetary waves in the three-dimensional atmosphere in boreal winter. The first waveguide is the propagation of planetary waves from the troposphere to the stratosphere at high-latitudes via the polar waveguide, while the other is the propagation of planetary waves from the mid-latitude region to the upper troposphere at low latitudes via the low-latitude waveguide. Comparing Figures 1 and 2, the propagating characteristics of quasi-stationary planetary waves in the spherical atmospherie during boreal winter (as shown in Figure 2) are consistent with the results analyzed earlier using the wave refraction index.

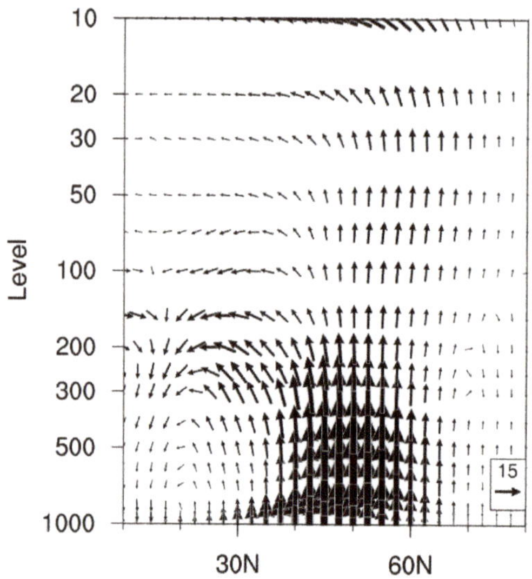

Figure 2. E-P flux cross sections (vector scale: m^3 s^{-2}; *Y*-axis denotes vertical levels, units: hPa) of quasi-stationary planetary waves averaged for 60 winters from the NCEP/NCAR [30] reanalysis data for 1961–2020.

2.3. Study on the Propagation of Quasi-Stationary Planetary Waves in the Three-Dimensional Atmosphere during Boreal Summer

There are few studies on the propagating characteristics of quasi-stationary planetary waves in the three-dimensional atmosphere during boreal summer. Huang and Gambo [31] studied the three-dimensional propagating characteristics of quasi-stationary planetary waves in the actual flow during boreal summer using the wave refraction index and E-P flux and noted that quasi-stationary planetary waves can propagate quasi-horizontally from the subtropical region to the middle and high latitudes during boreal summer, but not to the stratosphere.

3. The Guiding Role of Theory of Rossby Wave Energy Dispersion in the Study on the Dynamical Processes of the Variabilities of the East Asian Summer Monsoon System

Under the guidance of the theory of Rossby wave energy dispersion, the research on the dynamical processes of the variabilities of the East Asian summer monsoon system has been in recent years, especially on the interannual and interdecadal variabilities of the EAP pattern and "Silk Road" teleconnection wave trains and their effects on the variability of summer monsoon precipitation in eastern China.

3.1. The Guiding Role of the Theory of Rossby Wave Energy Dispersion in the Study on the Dynamical Processes of the Interannual Variability of the East Asian Summer Monsoon System

3.1.1. Dynamical Influence of the EAP Pattern Teleconnection Wave Train on the Interannual Variability of the East Asian Summer Monsoon System

In summer, the monsoon is prevalent in East Asia. Due to the obvious interannual variability of the monsoon, severe droughts and floods often occur in the eastern part of China.

Therefore, the causes of the interannual variability of the East Asian summer monsoon system and summer monsoon precipitation in China are important research topics.

The dynamical processes of the interannual variability of the East Asian summer monsoon system have been an interesting research topic for many scholars in East Asia since the 1980s. Nitta [13] firstly proposed that there is an anti-correlated oscillation, or the Pacific-Japan (PJ) oscillation, between the summer atmospheric circulation anomalies over the tropical western Pacific around the Philippines and those around Japan. Kosaka and Nakamura [20,32] also investigated the dynamics of the PJ oscillation. Under the guidance of the theory of Rossby wave energy dispersion, our research group [14–17,33] systematically studied the influence of strong heat sources formed by convective activities and the SST in the tropical western Pacific Ocean around the Philippines on the East Asian summer monsoon system and summer precipitation in eastern China from observational facts, theory and numerical simulations, respectively. These studies show that there is a distribution of teleconnection wave train of atmospheric circulation anomalies over the tropical western Pacific Ocean through East Asia and the Okhotsk Sea to Alaska and the west coast of North America in summer, which is referred as the East Asia/Pacific (EAP) pattern teleconnection wave train. These studies revealed that, as shown in Figure 3a, when the western Pacific warm pool is in a warm state and the convective activities over the tropical western Pacific around the Philippines are strong, there is a cyclonic circulation anomaly over the tropical western Pacific around the Philippines. Meanwhile, there is an anticyclonic circulation anomaly over Japan and the Yangtze and Huaihe River basins in China, i.e., the western Pacific subtropical high shifts northward. There is a cyclonic circulation anomaly that extends from north China towards the Okhotsk Sea. Under this condition, the plum rains over the Yangtze and Huaihe River basins are weak. In contrast, when the western Pacific warm pool is in a cold state and the convective activities over the tropical western Pacific around the Philippines are weak, as shown in Figure 3b, there is an anticyclonic circulation anomaly around the Philippines. There is a cyclonic circulation anomaly over Japan and the Yangtze and Huaihe River basins in China; that is, the West Pacific subtropical high shifts southward. In this configuration, strong plum rains often cause floods in the Yangtze River and Huaihe River basins. Due to the EAP pattern teleconnection wave train, the summer monsoon precipitation anomalies in eastern China often show a distribution of "+, −, +" or "−, +, −" tripole pattern wave train from south to north.

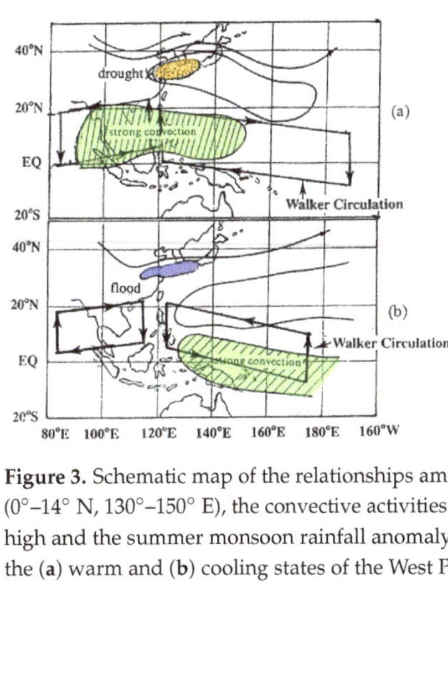

Figure 3. Schematic map of the relationships among the thermal state of the West Pacific warm pool (0°–14° N, 130°–150° E), the convective activities around the Philippines, the West Pacific subtropical high and the summer monsoon rainfall anomaly in the Yangtze River and the Huai River valleys in the (**a**) warm and (**b**) cooling states of the West Pacific warm pool (From Huang and Sun [17]).

It can also be studied from the EOF analysis of summer monsoon precipitation that the distributions of summer precipitation anomalies in eastern China are similar to the EAP pattern teleconnection. Our research group recently conducted the EOF analysis of the summer precipitation in eastern China using data for 60 years from 1961 to 2020. Figure 4a–d show the spatial distribution and corresponding temporal coefficients of summer precipitation in eastern China from 1961 to 2020, respectively. As shown in Figure 4a, the first mode of summer precipitation in eastern China shows a meridional tripole structure, which is similar to the EAP pattern. Additionally, as shown in Figure 4c, the first main mode of the variability of summer monsoon precipitation in eastern China shows a 2–3 year period. This is closely related to the meridional tripole distribution of the first mode of zonal water vapor transports in East Asia in summer [34].

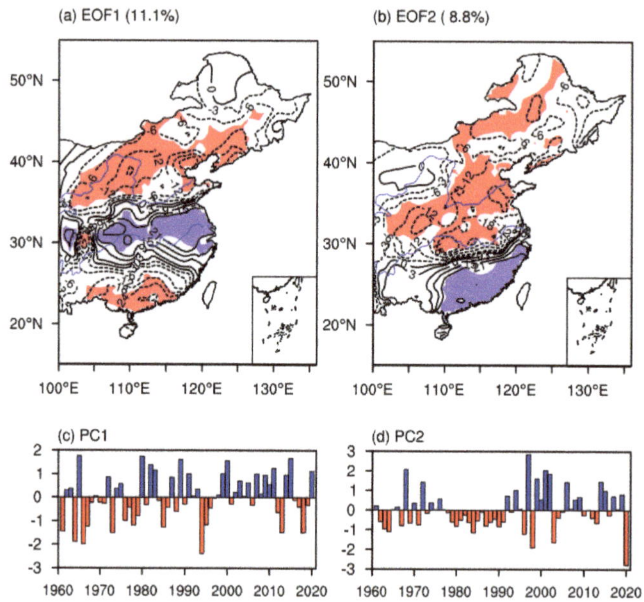

Figure 4. (**a**,**b**) Spatial distributions and (**c**,**d**) corresponding time coefficient series of the first and second components of EOF analysis of summertime (JJA) rainfall in eastern China for 1961–2020. The sold and dashed lines in Figure 4a,b indicate positive and negative values, respectively, and EOF1 and EOF2 explain 11.1% and 8.8%, respectively.

Huang et al. [35] used the EAP pattern teleconnection of atmospheric circulation anamolies to further explain the causes of the interannual variability of the summer precipitation in eastern China. This study showed that the spatial distribution of the variabilities of summer precipitation and East Asian summer monsoon water vapor transport fluxes in eastern China appears a meridional tripole distribution with a quasi-biennial period. The interannual variability with the quasi-biennial period maybe caused by the planetary wave train (i.e., EAP pattern teleconnection wave train) generated by the interannual variation of thermal forcing over the tropical western Pacific warm pool. Figure 5 shows the distribution of 500 hPa geopotential height anomalies regressed from the PC1. As shown in Figure 5, we can see that the summer monsoon circulation anomalies over East Asia and the tropical western Pacific exhibit a meridional tripole structure from south to north, and this distribution resembles the EAP pattern teleconnection wave train.

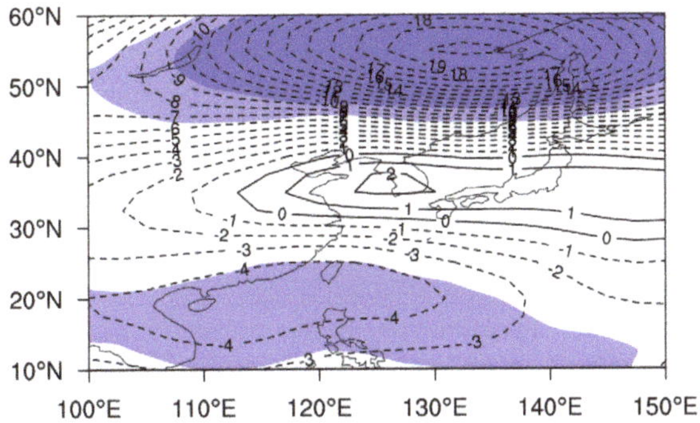

Figure 5. Distribution of summertime 500 hPa height anomalies over East Asia regressed by the time coefficients of EOF1 of summertime 500 hPa height for 1961–2020. Solid and dashed lines indicate positive and negative height anomalies, respectively, and areas with confidence level over 95% are shaded. Data of geopotential height fields is from the NCEP/NCAR reanalysis data [30].

To reflect the interannual variability of the EAP pattern teleconnection wave train and its influence on the interannual variability of summer monsoon precipitation in the Yangtze and Huaihe River basins, Huang G. [36] defined an EAP index from the distribution of the EAP pattern teleconnection wave train. This index is calculated with the standardized seasonal-mean (June-July-August) 500 hPa height anomaly at three different grid points (20° N, 125° E; 40° N, 125° E; 60° N, 125° E), which can well measure the strength of the East Asian summer monsoon and describe the interannual variability of summer rainfall and surface air temperature in East Asia. This index has a significantly negative correlation not only with summer monsoon precipitation in the Yangtze and Huaihe River basins in China but also with summer precipitation in Korea. Moreover, this index has a positive correlation with summer precipitation in North and South China. This shows that this index can describe the interannual variability of precipitation caused by East Asian summer monsoon. Figure 6 shows the wavelet analysis of the EAP index for 60 summers (June to August) recently calculated by using the definition of Huang G. [36] and the NCEP/NCAR reanalysis data from 1961 to 2020. From Figure 6, it can be seen that this index shows the interannual variability with the 2–3 year period from the 1970s to the late 1990s and from 2015 onward, which indicates that the interannual variability of East Asian summer monsoon precipitation with the 2–3 year period is closely related to the interannual variability of the EAP pattern teleconnection wave train.

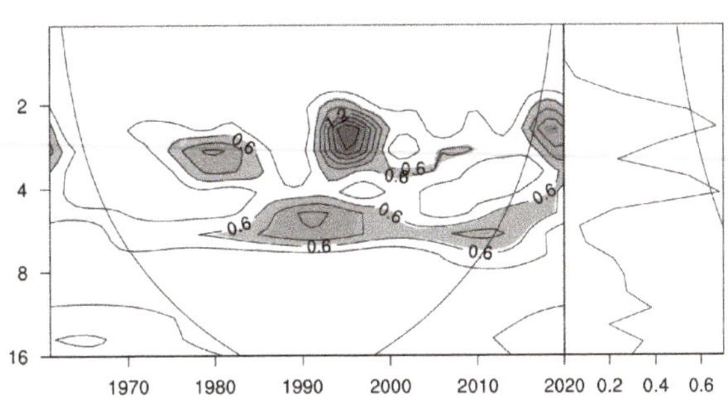

Figure 6. Wavelet analysis (*Y*-axis denotes periods, units: year; the curve on the right panel denotes the global wavelet spectrum; shading depicts power spectrum significant beyond 95% level based on the Chi-square test) of the EAP index calculated by using the definition in Huang G. [36] and 500 hPa height fields of the NCEP/NCAR reanalysis data [30] for 1961–2020.

3.1.2. Dynamical Influence of the "Silk Road" Pattern Teleconnection Wave Train on the Interannual Variability of the East Asian Summer Monsoon System

The variability of the East Asian summer monsoon system is influenced not only by the heat of the tropical western Pacific but also by the subtropical jet over Asia. The results studied by previous studies [18,19,21] demonstrated the existence of a wave train in the meridional wind variability in the upper troposphere from West Asia to East Asia. Later, Enomoto et al. [37] referred to this teleconnection wave train as the "Silk Road" pattern teleconnection. Therefore, the anomalies of East Asian summer monsoon precipitation with the meridional tripole or meridional dipole distribution may be influenced not only by the EAP pattern teleconnection but also by the Silk Road pattern teleconnection propagating along the upper tropospheric subtropical jet stream from West Asia to East Asia. Tao and Wei [38] suggested that the "Silk Road" pattern teleconnection has an important influence on the northward or southward movement of the western Pacific subtropical high and the East Asian summer monsoon rainfall belt. Hsu and Lin [39] and Kosaka et al. [40] also suggested that the Silk Road pattern teleconnection has an important influence on the western Pacific subtropical high and the northward or southward movement of and the summer monsoon rainfall belt. Hsu and Lin [39] and Kosaka et al. [40] further noted that the distribution of the summer precipitation anomalies with the tripole pattern is related not only to the EAP pattern teleconnection but also to the Silk Road pattern teleconnection propagating along the subtropical jet over Asia. Huang et al. [41] studied the dynamical processes of the variability of East Asian summer monsoon precipitation in terms of water vapor transports and showed that the variability of the East Asian summer monsoon precipitation is driven by a combination of the EAP pattern teleconnection wave train and the Silk Road pattern teleconnection wave train.

3.2. The Guiding Role of the Theory of Rossby Wave Energy Dispersion in the Study on the Dynamical Processes of the Interdecadal Variability of the East Asian Summer Monsoon System

The East Asian summer monsoon system not only has significant interannual variability but also obvious interdecadal variability. Due to the influence of interdecadal variability of the East Asian summer monsoon system, there is significant interdecadal variability in summer precipitation in eastern China. The result studied by Huang et al. [41] showed that the summer monsoon precipitation in eastern China experienced three significant interdecadal variations from the 1970s to the beginning of the 21st century. Recently, our research group analyzed the interdecadal variability of summer precipitation in eastern China using the summer precipitation data from 1961 to 2020 (Figure 7). The result shows that during 1961–1976, the summer precipitation anomalies in eastern China exhibited a "+, −, +" meridional tripole distribution from south to north, i.e., there was more precipitation in southern and northern China and less precipitation in the Yangtze River basin. However, during 1978–1992, the summer precipitation anomalies in eastern China appeared to have the opposite meridional distribution to those during 1961–1976. In the period from 1977 to 1992, the summer precipitation anomalies in eastern China appeared to have the opposite distribution of the meridional tripole pattern from 1961 to 1977, i.e., the meridional tripole pattern of "−, +, −", when the summer precipitation in southern and northern China decreased, while the summer precipitation in the Yangtze River basin increased. However, from 1999 to 2010, the summer monsoon precipitation anomaly in eastern China changed from a meridional tripole distribution to a meridional dipole distribution, i.e., appeared a feature with floods in southern China and droughts in northern China. Recently, our analysis also shows that there has been an additional variation in the interdecadal variability of summer precipitation pattern in eastern China since 2011, the summer precipition anomalies appear a meridional "+, −, +" tripole, i.e., more summer precipitation in northern and southern China and less summer precipitation in the Yangtze and Huaihe River basins.

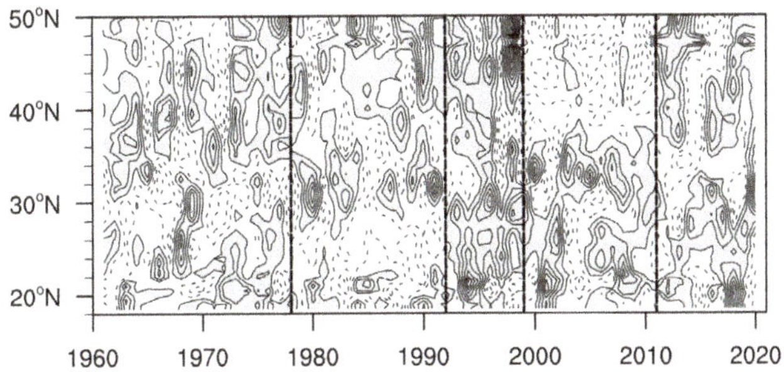

Figure 7. Distribution of summertime (JJA) rainfall anomalies (percentage) averaged for 110°–120° E in eastern China with latitude and time. Solid and dashed lines indicate positive and negative anomalies, respectively, and positive anomalies are shaded. Data of precipitation is from the dataset of precipitations at 822 observational stations in China.

Huang et al. [41] investigated the dynamic process of the interdecadal variability of summer precipitation in eastern China and noted that the interdecadal variability of summer precipitation in eastern China is closely related to the interdecadal variability of the teleconnection wave trains of the summer atmospheric circulation anomalies over East Asia, especially in regards to the interdecadal variability of the EAP pattern teleconnection wave train, which plays an important role in the variabilities of summer precipitation in eastern China. Recently, it is analyzed that the interdecadal variability of summer whole-layer water vapor transport fluxes between 1000 and 300 hPa using the NCEP/NCAR reanalysis data of water vapor transport fluxes over the Eurasian region from 1961 to 2020. Figure 8a–d show the 1000–300 hPa whole-layer water vapor transport fluxes averaged for the summers of 1977–1992, 1993–1998, 1999–2010, and 2011–2020, respectively. In Figure 8a, the anomalies of water vapor transport fluxes over Southeast Asia and East Asia appeared as a "cyclone-anticyclone-cyclone" tripole anomaly pattern during 1977–1992, i.e., there was an EAP-like pattern teleconnection wave train. And there was a strong southward water vapor transport anomaly in the eastern part of China. This finding indicates that the East Asian summer monsoon weakened. In addition, the anomalies of water vapor transport fluxes from the Caspian Sea through central Asia to North China appeared the distribution of "cyclone-anticyclone-cyclone", those are similar to the distribution of the Silk Road-like pattern teleconnection wave train. In contrast, the anomalies of summer water vapor transport fluxes over East Asia during 1993–1998 (see Figure 8b) showed a somewhat different distribution from Figure 8a, i.e., anomalies of water vapour transport fluxes over Southeast Asia and East Asia appeared the dipole pattern distribution of "anticyclone-cyclone". This finding indicates that the East Asian summer monsoon became stronger during this period. Moreover, the "anticyclone-cyclone-anticyclone-cyclone" pattern anomalies of water vapour transport fluxes from the Caspian Sea to North China were similar to the Silk Road pattern teleconnection wave train. However, comparing Figure 8c with Figure 8a,b, it is obvious that the distributions of the anomalies of water vapor transport fluxes along the zonal direction over Eurasia or along the meridional direction over East Asia significantly changed during the period from 1999 to 2010. During this period, the anomaly distribution of water vapor fluxes over East Asia appeared the meridional dipole pattern, and the anomaly distribution of water vapor fluxes over Eurasia showed an "anticyclone-cyclone-anticyclone-cyclone" pattern, which is similar to the Silk Road pattern teleconnection wave train. In addition, Figure 8c shows that there was the southward water vapor transport flux anomalies extending from the northeast region to the southwest region of China, which indicates that the East Asian summer monsoon became weak again during this period, which weakens the water vapor transport to the northeast and north China. Therefore, the persistent droughts occurred in these regions in different degrees during summer. However, there was a strong northward water vapor transport flux anomaly from the southeast coastal area of China to the east of the Huaihe River basin, which caused significant increase of the summer precipitation and severe floods in these regions during this period. Recently, the result analyzed by our research group shows that during 2011–2020, as shown in Figure 8d, the southward water vapor transport flux anomalies over the northeast to southwest region of China were smaller than those during 1999–2011, while the northward water vapor transport fluxes over the southeast coast of China to the Yellow River and Huaihe River regions in China were strengthened. This caused the increase of the summer precipitation in northern China and the water vapor transport flux anomalies from Southeast Asia to the northeast region of China along the eastern part of China. The water vapor transport flux anomalies from Southeast Asia along the eastern part of China to the northeast part of China appeared the distribution of "anticyclone-cyclone-anticyclone", which is similar to the EAP-like pattern teleconnection wave train. And the water vapor transport flux anomalies from the Caspian Sea to East Asia appeared the "anticyclone-cyclone-anticyclone-cyclone" distribution, which is the Silk Road-like pattern teleconnection distribution.

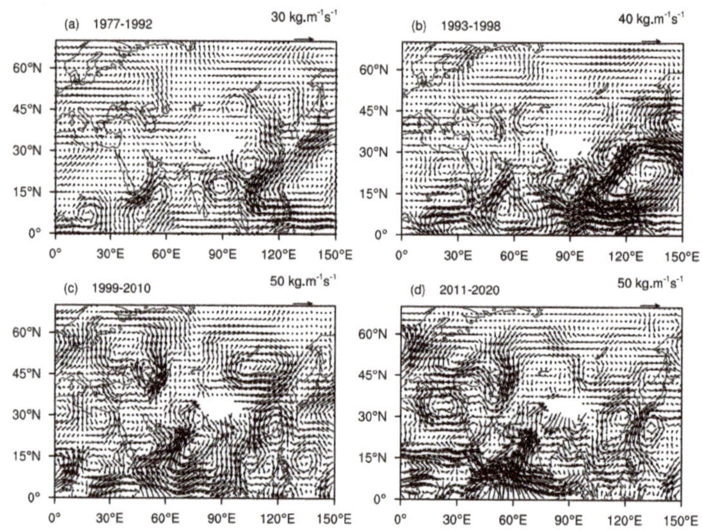

Figure 8. Anomaly distributions of summertime (JJA) water vapor transport fluxes integrated from 1000 hPa to 300 hPa over the Eurasian continent, averaged for (**a**) 1977–1992, (**b**) 1993–1998, (**c**) 1999–2010, and (**d**) 2011–2020. Data of water vapor and wind fields are from the NCEP/NCAR reanalysis data for 1961–2020 [30].

From the above analyses of the spatial distribution of the interdecadal variability of summertime water vapor transport fluxes over Eurasia, we can see that the interdecadal variability in summer monsoon precipitation over eastern China since the 1970s is closely related to not only the interdecadal variability of the distribution of the EAP pattern teleconnection wave train propagating along the meridional direction in East Asia but also related to the interdecadal variability of the Silk Road teleconnection wave train propagating along the subtropical jet in the upper troposphere over the Eurasian subtropics. Both wave trains are formed by the propagations of quasi-stationary planetary waves in the spherical atmosphere.

In summary, the interannual and interdecadal variabilities of summer monsoon precipitation and whole-layer water vapor transport in eastern China may be result of joint action of the interannual and interdecadal variabilities of the EAP pattern teleconnection wave train propagating along the meridional direction over East Asia and the Silk Road pattern teleconnection wave train propagating along the subtropical regions from West

Asia to East Asia. Therefore, the theory of Rossby wave energy dispersion proposed by academician Ye provides an important scientific basis for short- and medium-term weather forecasting but also a new idea for short-term summer climate prediction.

4. The Guiding Role of the Theory of Rossby Wave Energy Dispersion in the Study on the Dynamic Processes of the Variability of the East Asian Winter Monsoon System

The variability of the winter climate in China is controlled by the East Asian winter monsoon system. The interannual and interdecadal variabilities of the East Asian winter monsoon system is the main cause of winter rain, snow, and ice disasters in China [42]. However, there are few studies on the dynamic processes of interannual and interdecadal variabilities in East Asian winter monsoon system. Under the guidance of the theory of Rossby wave energy dispersion, some studies on the dynamic processes of the East Asian winter monsoon system have been carried out in recent years. Especially, some meteorogists in China have paid a special attention to the impacts of the interannual and interdecadal variations of quasi-stationary planetary waves in the three-dimensional spherical atmosphere on the East Asian winter monsoon system and the low-temperature rain, snow, and ice disasters in China.

4.1. The Guiding Role of the Theory of Rossby Wave Dispersion in the Study on the Dynamical Processes of the Interannual Variability of the East Asian Winter Monsoon System

The winter monsoon prevails in East Asia, and due to the interannual variability of the winter monsoon system, winter temperature and precipitation in China show the significant interannual variability. Occasionally, severe low-temperature and snow disasters frequently occur in winter, such as the freezing event occurred in January 2008. Due to the anomalously strong East Asian winter monsoon, the severe low-temperature and snow disasters occurred in southwest, central and southern China, which caused economic losses of more than 150 billion yuan [42].

The dynamical processes of interannual variability of the East Asian winter monsoon system are closely related to the interannual oscillations of these two propagating waveguides of quasi-stationary planetary waves during boreal winter. Previous studies [22,23,43] proposed that the variability of these two waveguides exists an opposite oscillation on the interannual time scale. The above studies suggest that the interannual variability of the East Asian winter monsoon system is significantly related to the interannual oscillation of the propagating waveguides of quasi-stationary planetary waves. When the polar waveguide strengthens, then the low-latitude waveguide weakens, and anomalously strong quasi-stationary planetary waves in the troposphere will propagate towards the stratosphere through the polar waveguide, while the propagation of quasi-stationary planetary waves towards the upper troposphere near the low latitudes through the low-latitude waveguide will be weakened. In contrast, when the polar waveguide weakens, then, the low-latitude waveguide strengthens, and anomalously strong quasi-stationary planetary waves in the troposphere will propagate from middle latitudes towards the upper troposphere at low latitudes through the low-latitude waveguide. In this case, the propagation of anomalously strong quasi-stationary planetary waves in the troposphere from middle to high latitudes towards the top of the troposphere over low-latitude region via the low-latitude waveguide is stronger, while the propagation of quasi-stationary planetary waves to the stratosphere through the polar waveguide is significantly weaker.

Huang et al. [24] studied the influence of the interannual oscillations of the propagating waveguides of quasi-stationary planetary waves in the three-dimensional atmosphere on the interannual variability of the East Asian winter monsoon system and analyzed the relationship between the East Asian climate anomalies and boreal quasi-stationary planetary wave activity during the winters of 2005 and 2006. In the winter of 2005, the temperature from Western Europe through the Urals to Siberia and East Asia was lower than normal, and the temperature in the east side of the Urals and the northwest side of the Mongolian Plateau was more than 2 °C lower than normal, which resulted in the cold winter and frequent cold waves outbreaked in China. This in turn led to anomalously strong snow in Northwest and Northeast China as well as strong precipitation in the Yangtze River basin. However, in the winter of 2006, the temperature in the Siberia and East Asia was higher than normal, and the temperature in Europe from the Urals to Baikal Lake was more than 2 °C higher than normal, resulting in a warm winter in China.

The obvious difference between the winter climate in Eurasia in 2005 and that in 2006 was closely related to the oscillation of the propagating waveguides of quasi-stationary planetary waves in the Northern Hemisphere during these two winters. Figure 9a,b show the E-P fluxes and the scatterplot distributions of the quasi-stationary planetary waves during the boreal winters of 2005 and 2006, respectively. As shown in Figure 9a, the polar waveguide of the boreal quasi-stationary planetary wave propagation in the winter of 2005 was strong, while the low-latitude waveguide was weak, i.e., the propagation of quasi-stationary planetary wave along the polar waveguide to the stratosphere was enhanced. This caused that the convergence of planetary wave E-P fluxes was enhanced in the upper troposphere at high latitudes and weakened the convergence of planetary wave E-P fluxes in the upper troposphere in the subtropics, i.e., enhanced divergence of the E-P fluxes in this region, which caused the weakening of the boreal polar frontal jet and the strengthening of the subtropical jet. This can facilitate the development of high pressure systems over the Siberia and the strengthening of the East Asian winter monsoon. As shown in Figure 9b, the polar waveguide of quasi-stationary planetary wave propagation in the winter of 2006 was weak, while the low-latitude waveguide was strong, i.e., the propagation of quasi-stationary planetary waves along the low-latitude waveguide was enhanced in the upper troposphere at low latitudes. This caused that the convergence of planetary wave E-P fluxes was enhanced in the upper troposphere in the subtropical region, while the convergence of planetary wave E-P fluxes in the upper troposphere at high latitudes was weakened, i.e., the divergence of the E-P fluxes was enhanced. This caused the strengthening of the boreal polar jet and the weakening of the subtropical jet, which was detrimental to the development of the Siberian high and brought about the weakening of the East Asian winter monsoon. As shown in Figure 9c, the difference between the propagations of quasi-stationary planetary waves in these two boreal winters was also evident.

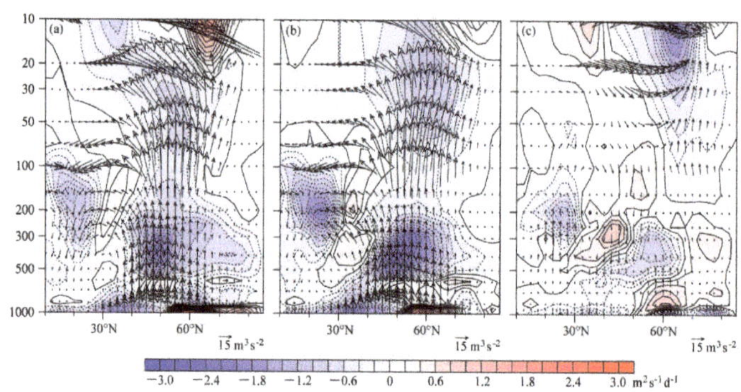

Figure 9. Composite distributions of the E-P fluxes (multiplied by ρ^{-1} for displaying purpose) (vectors, units: $m^3\ s^{-2}$) of quasi-stationary planetary waves for wavenumbers 1–3 and their divergence (shaded, units: $m^3\ s^{-1}\ d^{-1}$; *Y*-axis denotes vertical levels, units: hPa) over the Northern Hemisphere in the winters of (**a**) 2005 (December 2005 to February 2006) and (**b**) 2006 (December 2006 to February 2007), and (c) the difference between them. Solid and dashed lines indicate positive (divergence) and negative (convergence) of planetary wave E-P fluxes, respectively. And the divergence/convergence regions of the E-P flux are shaded with red/blue colors (from Huang et al. [24]).

4.2. The Guiding Role of the Theory of Rossby Wave Energy Dispersion in the Study on the Dynamic Processes of the Interdecadal Variability of the East Asian Winter Monsoon System

Wang et al. [25] have proposed that the East Asian winter monsoon system has not only significant interannual variability but also significant interdecadal variability. From the 1970s to the beginning of the 21st century, winter temperatures in China experienced two significant interdecadal variations. During the period from 1976 to 1987, the East Asian winter monsoon was strong, the winter temperature in China was generally low, and the frequency of cold wave outbreaks in China was high. However, during the period from 1988 to 1998, the East Asian winter monsoon was weak, the winter temperature in China was generally high, the frequency of cold wave outbreaks was significantly low, then, the warm winter frequently occurred. The result studied by Huang et al. [42] showed that during 1999–2010, winter temperatures in China changed significantly again, the colder temperature occurred in north China and warmer temperature appeared in south China. Moreover, temperature changed to a meridional oscillation pattern, and the interannual variability of winter temperatures changed from a 3–4 year cycle to a 2–8 year cycle. Our recent analysis shows that the variability of the East Asian winter monsoon during 2011–2020 was roughly similar to that during 1999–2010, and no a significant interdecadal variability occurred. During this period, several cold waves occurred in China, such as 21–25 January 2016, and 28–31 December 2020, where many regions in China experienced severe cooling and occurred low-temperature and snow disasters. These disasters caused severe economic losses.

The dynamic processes of interdecadal variability of winter climate in China and the East Asian winter monsoon system have been studied [25,42]. The results showed that the interdecadal variability of the East Asian winter monsoon system occurred in the mid-late 1980s and late 1990s, which was closely related to the interdecadal oscillation of the propagating waveguides of quasi-stationary planetary waves in boreal winter. Recently, it is analyzed that the interdecadal oscillations of propagating waveguides of quasi-stationary planetary waves in the Northern Hemisphere using the NCEP/NCAR reanalysis data from 1961 to 2020. Figure 10a–c show the composite distributions of the E-P fluxes of quasi-stationary planetary waves for wavenumber 1–3 and their divergences over the Northern Hemisphere averaged for the winters of 1976–1987, 1988–1998, and 1999–2020, respectively. From Figure 10a, it can be seen that during the period of 1976–1987, the polar waveguide of quasi-stationary planetary waves for the boreal winters was strong, i.e., the propagation of planetary waves was strong along the polar waveguide up to the stratosphere over 60° N and weak along the low-latitude waveguide to the upper troposphere over low latitudes. These caused strong convergence of the E-P flux of planetary waves in the troposphere and stratosphere over high latitudes of the Northern Hemisphere and weak divergence of the E-P fluxes in the middle and upper troposphere over the subtropical region near 30° N. Moreover, as shown in Figure 10b, during the period of 1988–1998, the propagation of the boreal winter quasi-stationary planetary waves changed. Compared with Figure 10a, the polar waveguide of the boreal winter quasi-stationary planetary waves was weak during this period, while the low-latitude waveguide became strong. In other words, the propagation of planetary waves towards the stratosphere along the polar waveguide over high latitudes became weaker in the winters of 1988–1998, comparing with that in the winters of 1976–1987. And the propagation of planetary waves to the upper troposphere over low latitudes along the low waveguide became stronger, comparing with that in the winters of 1976–1987, which caused the E-P fluxes of the quasi-stationary planetary waves over high latitudes during winters of 1988–1998 were weaker than those during 1976–1987. This means that there was a positive difference of the divergence. But the divergence of the E-P fluxes of quasi-stationary planetary waves in the upper troposphere over the subtropical region became stronger in the winters of 1988–1998 than that in the winters of 1976–1987. In addition, as shown in Figure 10c, the polar waveguide of propagation of the boreal quasi-stationary planetary waves was again stronger and the low-latitude waveguide was weaker in the winters from 1999 to 2020. This means that the propagation of planetary waves became stronger along the polar waveguide up to the stratosphere at high latitudes and weaker towards the upper troposphere of the subtropical region near 30° N via the low-latitude waveguide. The convergence of E-P fluxes of planetary waves in the upper troposphere and stratosphere at high latitudes was stronger than that in the winters of 1988–1998, but the divergence of planetary wave E-P fluxes in the upper troposphere over the subtropical region near 30° N was weaker than that in the winters of 1988–1998.

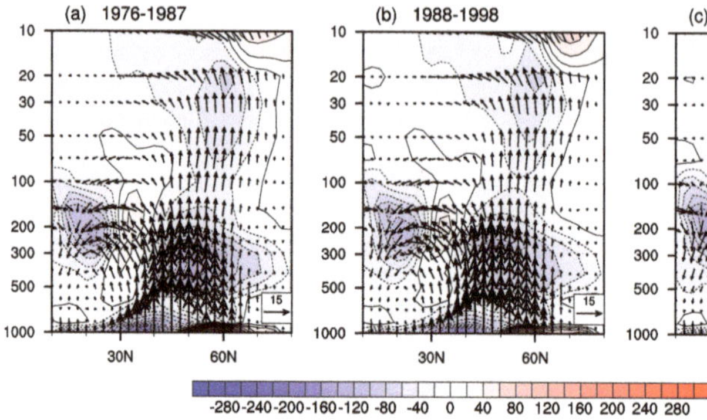

Figure 10. Composite distributions of E-P fluxes of quasi-stationary planetary waves for wave numbers 1–3 and their divergences (shaded, units: $m^3\ s^{-1}\ d^{-1}$; *Y*-axis denotes vertical levels, units: hPa) over the Northern Hemisphere averaged for the winters of (**a**) 1976–1987, (**b**) 1988–1998 and (**c**) 1999–2020. Solid and dashed lines indicate positive (divergence) and negative (convergence) divergence of E-P fluxes. Data of wind fields and temperature are from the NCEP/NCAR reanalysis data [30].

4.3. *Dynamic Effect of the Propagating Waveguide Oscillations of Quasi-Stationary Planetary Waves on the Variability of the East Asian Winter Monsoons*

The above results show that the boreal winter quasi-stationary planetary waveguides in the three-dimensional atmosphere have not only significant interannual oscillations but also significant interdecadal oscillations. Moreover, two significant interdecadal oscillations of the propagating waveguides of the boreal winter quasi-stationary planetary waves occurred since the 1970s. These oscillations of the propagating waveguides of quasi-stationary planetary waves caused the variability of the divergence or convergence of the E-P fluxes of quasi-stationary planetary waves. According to the wave-flow interaction equation for the spherical atmospheric planetary waves derived by Edmon et al. [29], the variation of the divergence of the E-P fluxes of quasi-stationary planetary waves will cause the variation of the zonal mean flow during boreal winter and the variation of the Arctic Oscillation (AO) index. According to previous studies [44,45], if the AO index is negative in a winter, the winter monsoon in East Asia is strong in the winter; conversely, if the AO index is positive in a winter, then the winter monsoon in East Asia is weak in the winter. Therefore, the interannual and interdecadal oscillations of the propagating waveguides of quasi-stationary planetary waves in the boreal winter will affect the interannual and interdecadal variabilities of the East Asian winter monsoon system.

It may see from the above studies that under the guidance of academician Ye's theory of Rossby wave dispersion, some studies on the dynamic processes of the variabilities of the East Asian winter monsoon system have been carried out. In particular, the study on the dynamic processes of the influence on the interannual and interdecadal oscillations of the propagating waveguides of quasi-stationary planetary waves on the East Asian winter monsoon variations has achieved an important progress.

5. Conclusions and Discussion

The theory of Rossby wave energy dispersion proposed by academician Ye in the 1940s not only improves the study on the characteristics of two-dimensional and three-dimensional propagations of quasi-stationary planetary waves in the spherical atmosphere but also provides a scientific basis for the study on the mechanisms related to the anomalies of global atmospheric circulation. And this theory also lays the theoretical basis for the study on the dynamic processes of tropospheric-stratospheric interactions and their mechanisms. This paper reviews the guiding role of the theory of Rossby wave energy dispersion in the studies on the characteristics of three-dimensional propagation of quasi-stationary planetary waves in the spherical atmosphere and their impacts on the dynamic processes of the interannual and interdecadal variabilities of the East Asian summer and winter monsoon systems. In particular, this paper reviews the impacts of the interannual and interdecadal variabilities of the EAP pattern teleconnection wave train propagating along the meridional direction over East Asia and the Silk Road pattern teleconnection wave train propagating along the zonal direction in the subtropical jet from West Asia to East Asia on the variabilities of the East Asian summer monsoon system. This paper also reviews the studies on the dynamical processes of the impacts of the interannual and interdecadal oscillations of the propagating waveguides of quasi-stationary planetary waves on the variabilities of the East Asian winter monsoon system.

The energy dispersion of waves is an important theoretical problem in the fluid dynamics, academician Ye first applied it to the study on the mechanisms of atmospheric circulation variability. His research inspired the study on the dynamics of the two- and three-dimensional spherical propagations of quasi-stationary planetary waves, and guides the study on the dynamical processes of the variabilities of the East Asian winter and summer monsoon systems.

Under the guidance of academician Yeh's theory of Rossby wave energy dispersion, our research group has investigated the dynamical processes of the interannual and interdecadal variabilities of the East Asian winter and summer monsoon systems. The results show that the variabilities of the East Asian winter and summer monsoon systems are the variablities of circulation seeing from their phenomena, but these variabilities are also the variations of quasi-stationary planetary waves according to their mechanisms. Currently, this theory is developing and expanding, and it has wide applications not only in the study on atmospheric circulation dynamics at middle and high latitudes but also in the study on the dynamics of typhoon genesis and evolution in tropical regions [46–51]. Moreover, the nonlinear effects of the energy dispersion of Rossby wave in atmosphere and the nonlinear interactions of different quasi-stationary planetary wave trains also need to be further investigated.

Author Contributions: Conceptualization, R.H.; validation, R.H., J.H., Y.L. and R.L.; writing—original draft preparation, R.H. and J.H.; writing—review and editing, R.H., J.H. and Y.L.; visualization, Y.L., J.H. and R.H.; supervision, R.H. and R.L. All authors have read and agreed to the published version of the manuscript.

Funding: This work was supported by the National Natural Science Foundation of China (41721004).

Data Availability Statement: The NCEP/NCAR reanalysis data can be found in the website: [https://psl.noaa.gov/data/gridded/data.ncep.reanalysis.html (accessed on 5 July 2021)]. The daily observational precipitation datasets in mainland China are provided by the National Meteorological Information Center of Chinese Meteorological Administration, which can be achived in the website: [https://data.cma.cn/ (accessed on 20 March 2021)].

Acknowledgments: We are thankful to the NCEP/NCAR and other scientific agencies for providing the datasets for our analysis. The authors are grateful for the comments and suggestions provided by the editor and anonymous reviewers.

Conflicts of Interest: The authors declare no conflict of interest.

References

1. Yeh, T.C. On energy dispersion in the atmosphere. *J. Meteor.* **1949**, *6*, 1–16. [CrossRef]
2. Longuest-Higgins, M.S. On group velocity and energy flux in planetary wave motions. *Deep-Sea Res.* **1964**, *11*, 35–42.
3. Longuest-Higgins, M.S. Planetary waves on a sphere. *Proc. R. Soc.* **1964**, *279*, 446–473.
4. Longuest-Higgins, M.S. Planetary waves on a rotating sphere, Part I. *Proc. R. Soc.* **1965**, *284*, 40–68.
5. Dickinson, R.E. Planetary Rossby waves propagating vertically through weak westerly wind wave guide. *J. Atmos. Sci.* **1968**, *25*, 981–1002. [CrossRef]
6. Matsuno, T. Vertical propagation of stationary planetary waves in the winter Northern Hemisphere. *J. Atmos. Sci.* **1970**, *27*, 871–883. [CrossRef]
7. Matsuno, T. A dynamical model of the stratospheric sudden warming. *J. Atmos. Sci.* **1971**, *28*, 1479–1494. [CrossRef]
8. Hoskins, B.J.; Simmons, A.H.; Andrews, D.G. Energy dispersion in a barotropic atmosphere. *Quart. J. R. Meteor. Soc.* **1977**, *103*, 553–567. [CrossRef]
9. Hoskins, B.J.; Karoly, D.J. The steady linear response of a spherical atmsphere to thermal and orographic forcing. *J. Atmos. Sci.* **1981**, *38*, 1179–1196. [CrossRef]
10. Huang, R.H.; Gambo, K. The response of a hemispheric multi-level model atmosphere to forcing by topography and stationary heat sources. Parts I, II. *J. Meteor. Soc. Jpn.* **1982**, *60*, 78–108. [CrossRef]
11. Huang, R.H.; Gambo, K. On other wave guide in stationary planetary wave propagations in the winter Northern Hemisphere. *Sci. China* **1983**, *26*, 940–950.
12. Wallace, J.M.; Gutzler, D.S. Teleconnections in the geopotential height field during the Northern Hemisphere winter. *Mon. Weather Rev.* **1981**, *109*, 748–812. [CrossRef]
13. Nitta, T.S. Convective activities in the tropical western Pacific and their impact on the Northern Hemisphere summer circulation. *J. Meteor. Soc. Jpn.* **1987**, *64*, 373–390. [CrossRef]
14. Huang, R.H.; Li, W.J. Influence of the heat source anomaly over the western tropical Pacific on the subtropical high over East Asia. In Proceedings of the International Conference on the General Circulation of East Asia, Chengdu, China, 10–15 April 1987.
15. Huang, R.H.; Li, W.J. Influence of heat source anomaly over the western tropical Pacific on the subtropical high over East Asia and its physical mechanism. *Chin. J. Atmos. Sci.* **1988**, *14*, 107–116. (In Chinese)
16. Huang, R.H. The East/Pacific pattern teleconnection of summer circulation and climate anomaly in East Asia. *Acta Meteorol. Sin.* **1992**, *6*, 25–37.
17. Huang, R.H.; Sun, F.Y. Impact of the tropical western Pacific on the East Asian summer monsoon. *J. Meteor. Soc. Jpn.* **1992**, *70*, 243–256. [CrossRef]
18. Lu, R.Y.; Oh, J.H.; Kim, B.J. A teleconnection pattern in upper-level meridional wind over the North African and Eurasian continent in summer. *Tellus* **2002**, *54*, 44–55. [CrossRef]
19. Lin, Z.D.; Lu, R.Y. Interannual meridional displacement of the Asian upper-tropospheric jet stream in summer. *Adv. Atmos. Sci.* **2005**, *22*, 199–211.
20. Kosaka, Y.; Nakamura, H. Structure and dynamics of the summertime Pacific-Japan teleconnection pattern. *Quart. J. R. Meteor. Soc.* **2006**, *132*, 2000–2030. [CrossRef]
21. Lu, R.Y.; Kim, B.J. The climatological Rossby wave source over the STCZs in the summer Northern Hemisphere. *J. Meteor. Soc. Jpn.* **2004**, *82*, 657–669. [CrossRef]
22. Chen, W.; Takahashi, M.; Graf, H.F. Interannual variations of stationary planetary waves activity in the northern winter troposphere and stratosphere and their relations to NAM and SST. *J. Geophys. Res.* **2003**, *108*, 4797. [CrossRef]
23. Chen, W.; Yang, S.; Huang, R.H. Relationship between stationary planetary wave activity and the East Asian winter monsoon. *J. Geophys. Res.* **2005**, *110*, D14110. [CrossRef]
24. Huang, R.H.; Wei, K.; Chen, J.L.; Chen, W. The East Asian winter monsoon anomalies in the winters of 2005 and 2006 and their relations to the quasi-stationary planetary wave activity in the Northern Hemisphere. *Chin. J. Atmos. Sci.* **2007**, *31*, 1033–1048. (In Chinese)
25. Wang, L.; Huang, R.H.; Gu, L.; Chen, W.; Kang, L.H. Interdecadal variations of the East Asian winter monsoon and their association with quasi-stationary planetary wave activity. *J. Clim.* **2009**, *22*, 4850–4872. [CrossRef]
26. Charney, J.G.; Drazin, G. Propagation of planetary-scale disturbances from the lower into the upper atmosphere. *J. Geophys. Res.* **1961**, *66*, 83–109. [CrossRef]
27. Eliassen, A.; Palm, E. On the transfer of energy in stationary mountain waves. *Geophys. Publ.* **1961**, *22*, 1–23.
28. Andrews, D.G.; McIntyre, M.E. Planetary waves in horizontal and vertical shear: The generalized Eliassen-Palm relation and the mean zonal acceleration. *J. Atmos. Sci.* **1976**, *33*, 2031–2048. [CrossRef]
29. Edmon, M.J.; Hoskins, B.J.; McIntyre, M.E. Eliassen-Palm sections for the troposphere. *J. Atmos. Sci.* **1980**, *37*, 2600–2617. [CrossRef]
30. Kalnay, E.; Kanamitsu, M.; Kistler, R.; Collins, W.; Deaven, D.; Gandin, L.; Iredell, M.; Saha, S.; White, G.; Woollen, J.; et al. The NCEP/NCAR 40-year reanalysis project. *Bull. Am. Meteor. Soc.* **1996**, *77*, 437–471. [CrossRef]
31. Huang, R.H.; Gambo, K. The response of a hemispheric multi-level model atmosphere to forcing by topograph and stationary heat sources in summer. *J. Meteor. Soc. Jpn.* **1983**, *61*, 495–509. [CrossRef]

32. Kosaka, Y.; Nakamura, H. Mechanisms of meridional teleconnection observed between a summer monsoon system and a subtropical anticyclone. Part I: The Pacific-Japan. *J. Clim.* **2010**, *23*, 5085–5108. [CrossRef]
33. Huang, R.H.; Lu, L. Numerical simulation of the relationship between the anomaly of the subtropical high over East Asia and the convective activities in the western tropical Pacific. *Adv. Atmos. Sci.* **1989**, *6*, 202–214.
34. Huang, R.H.; Chen, J.L.; Huang, G.; Zhang, Q.L. The quasi-biennial oscillation of summer monsoon rainfall in China and its cause. *Chin. J. Atmos. Sci.* **2006**, *30*, 545–560. (In Chinese)
35. Huang, R.H.; Chen, J.L.; Huang, G. Characteristics and variations of the East Asian monsoon system and its impacts on climate di sasters in China. *Adv. Atmos. Sci.* **2007**, *24*, 993–1023. [CrossRef]
36. Huang, G. An index measuring the interannual variation of the East Asian summer monsoon-The EAP index. *Adv. Atmos. Sci.* **2004**, *21*, 41–52. [CrossRef]
37. Enomoto, T.; Hoskins, B.J.; Matsuda, Y. The formation mechanism of the Bonin high in August. *Quart. J. R. Meteor. Soc.* **2003**, *129*, 157–178. [CrossRef]
38. Tao, S.Y.; Wei, J. The westward, northward advance of the subtropical high over the west Pacific in summer. *J. Appl. Meteor.* **2006**, *17*, 513–525. (In Chinese)
39. Hsu, H.H.; Lin, S.M. Asymmetry of the tripole rainfall pattern during the East Asian summer. *J. Clim.* **2007**, *20*, 4443–4458. [CrossRef]
40. Kosaka, Y.; Xie, S.P.; Nakamura, H. Dynamics of internannual variability in summer precipitation over East Asia. *J. Clim.* **2011**, *24*, 5435–5453. [CrossRef]
41. Huang, R.H.; Liu, Y.; Feng, T. Interdecadal change of summer precipitation over eastern China around the late-1990s and associated circulation anomalies, internal dynamical causes. *Chin. Sci. Bull.* **2013**, *58*, 1339–1349. [CrossRef]
42. Huang, R.H.; Liu, Y.; Huangfu, J.L.; Feng, T. Characteristics and interanal dynamical causes of the interdecadal variability of East Asian winter monsoon near the late 1990s. *Chin. J. Atmos. Sci.* **2014**, *38*, 627–644. (In Chinese)
43. Chen, W.; Huang, R.H. The three-dimensional propagation of quasi-stationary waves in the northern hemisphere winter and its interannual variations. *Chin. J. Atmos. Sci.* **2005**, *29*, 139–146. (In Chinese)
44. Gong, D.Y.; Wang, S.W.; Zhu, J.H. East Asian winter monsoon and Arctic oscillation. *Geophys. Res. Lett.* **2001**, *28*, 2073–2076. [CrossRef]
45. Wu, B.Y.; Wang, J. Winter Arctic Oscillation, Shiberian high and East Asian winter monsoon. *Geophys. Res. Lett.* **2002**, *29*, 1897. [CrossRef]
46. Li, T.; Fu, B. Tropical Cyclogenesis Associated with Rossby Wave Energy Dispersion of a Preexisting Typhoon. Part I: Satellite Data Analyses. *J. Atmos. Sci.* **2006**, *63*, 1377–1389. [CrossRef]
47. Ge, X.; Li, T.; Peng, M.S. Cyclogenesis Simulation of Typhoon Prapiroon (2000) Associated with Rossby Wave Energy Dispersion. *Mon. Weather Rev.* **2010**, *138*, 42–54. [CrossRef]
48. Luo, D.; Zhang, W. A Nonlinear Multiscale Theory of Atmospheric Blocking: Dynamical and Thermodynamic Effects of Meridional Potential Vorticity Gradient. *J. Atmos. Sci.* **2020**, *77*, 2471–2500. [CrossRef]
49. Yang, M.; Li, C.; Li, X.; Tan, Y.; Wu, Y.; Chen, X.; Xia, F. Association of the zonal migration of North Pacific storm track with the East Asian monsoon in boreal wintertime. *J. Geophys. Res.* **2021**, *126*, e2020JD033790. [CrossRef]
50. Li, Y.; Chao, J.; Kang, Y. Variations in Amplitudes and Wave Energy along the Energy Dispersion Paths for Rossby Waves in the Quasigeostrophic Barotropic Model. *Adv. Atmos. Sci.* **2022**, *39*, 876–888. [CrossRef]
51. Narinesingh, V.; Booth, J.F.; Ming, Y. Blocking and general circulation in GFDL comprehensive climate models. *J. Clim.* **2022**, *35*, 3687–3703. [CrossRef]

atmosphere

MDPI

Article

Interdecadal Variation in Rossby Wave Source over the Tibetan Plateau and Its Impact on the East Asia Circulation Pattern during Boreal Summer

Yihui Ding [1], Xiaoting Sun [1,*], Qingquan Li [1,2] and Yafang Song [1]

1 China Meteorological Administration Key Laboratory for Climate Prediction Studies, National Climate Center, Beijing 100081, China
2 Collaborative Innovation Center on Forecast and Evaluation of Meteorological Disasters, Nanjing University of Information Science and Technology, Nanjing 210044, China
* Correspondence: sunxt@cma.gov.cn

Abstract: The wave activity flux representing the energy propagation direction of planetary Rossby wave generally originates from a large wave source area. This study investigates the interdecadal variability and formation mechanism of Rossby wave source over the Tibetan Plateau (TP-RWS) and its impact on the atmospheric circulation and precipitation pattern in East Asia based on the ERA-20C reanalysis dataset in summer (June–July–August) during 1900 to 2010. Results show that the region with the maximum variabilities of Rossby wave source (RWS) in the past 110 years appears over the Tibetan Plateau (TP) during boreal summer, and the TP-RWS shows prominent characteristics of interdecadal oscillation. Secondly, the TP-RWS is mainly composed of the vortex stretching term (RWS-S1) and the absolute vorticity advection term (RWS-S2). The interdecadal TP-RWS is a synergistic result of the snow cover over northwestern TP associated with the RWS-S1, and the deep convection over southeastern TP associated with the RWS-S2. Furthermore, the interdecadal TP-RWS can lead to an alternatively positive and negative pattern of geopotential height anomalies from the northwestern TP to the North Pacific, which has a great climate effect on the precipitation in Huang-huai River Basin, South Korea and Japan Island. Under the guidance of the anomalous cyclonic circulation in East Asia, the prevailing southerly and easterly winds occur over the West Pacific and the Huang-huai River Basin, which lead to the water vapor convergence and upward movement at middle and lower troposphere.

Keywords: Rossby wave source; Tibetan Plateau; interdecadal variation; precipitation anomaly in East Asia

Citation: Ding, Y.; Sun, X.; Li, Q.; Song, Y. Interdecadal Variation in Rossby Wave Source over the Tibetan Plateau and Its Impact on the East Asia Circulation Pattern during Boreal Summer. *Atmosphere* **2023**, *14*, 541. https://doi.org/10.3390/atmos14030541

Academic Editors: Xiaolei Zou, Guoxiong Wu and Zhemin Tan

Received: 8 February 2023
Revised: 7 March 2023
Accepted: 8 March 2023
Published: 11 March 2023

1. Introduction

The Rossby wave (RW) propagation caused by external forcing is one of the mechanisms leading to the configuration of wave trains called teleconnection. In the northern hemisphere (NH), a favorable area with RW-forming dynamic conditions is located in East Asia [1,2]. The Rossby wave pattern at the upper troposphere generally manifests as Rossby wave train of atmospheric response to one or more local wave sources [3–5]. In the NH midlatitudes, the source of RW may be related to topography, marine–terrestrial contrast, or transient baroclinic systems [6]. Ye and Zhu [7] believe that both topographic and diabatic heating play an important role in the climatological atmospheric teleconnections along the westerly jet stream, especially the large troughs over East Asia and the east coast of North America. Previous studies of Rossby wave source (RWS) have shown that most of the strong RWS are located in the subtropical zone on the seasonal timescales, although most of the upper divergent flow regions are located near the equator. Because the divergent flow is greater at the edge of the greatest divergence zone and the absolute vorticity gradient is larger at higher latitudes [2,8,9], these all provide conditions for RWS formation in the subtropics. Therefore, RWS may be associated with both local forcing and absolute vorticity advection caused by atmospheric diabatic heating. Plumb [10] shows that the main forcing of the quasi-stationary wave originates from the topographic influence over the Tibetan Plateau (TP), as well as interaction of the diabatic heating with transient flows over the Pacific, Northwest Atlantic, and Siberia by diagnosing the wave activity flux in the NH winter.

Ye [11] first applied the RW energy dispersion theory to the study of the atmospheric circulation change mechanism. Previous studies have shown that the energy propagation process of Rossby waves along the westerly jet in the NH summer is an important dynamic mechanism for the development of high-latitude trough in East Asia, and its downstream effect is also an important driver of flood disasters in China [12–15]. On the interdecadal timescale, there is a Rossby wave train in the midlatitudes. The eastward propagation of wave energy has a significant impact on the precipitation pattern in the middle and lower reaches of the Yangtze River basin [16,17]. The Tibetan Plateau, as the steepest and most complex terrain on the Earth and the region with the strongest land–atmosphere interaction in the NH midlatitudes, has a great influence on regional and global climate [18–20]. Moreover, its local thermal forcing can directly affect the downstream atmospheric circulation and rainfall pattern [21–23]. Previous studies have shown that the surface turbulent heat flux in the southeastern TP plays an important role in regulating Meiyu and rainstorm in the Yangtze River Basin [24–27]. There is also a close relationship between the variation of snow cover in different regions of the TP and the atmospheric circulation [28–31]. In addition, the anomalous diabatic heating over the TP can trigger Rossby waves propagating westward and eastward along the extratropical westerly jet to change the large-scale climate on the interannual scale [32–34].

The above studies on the impact of RWS mainly focus on the seasonal and interannual timescales and discuss the RW role on persistent circulation anomaly and corresponding precipitation pattern in East Asia. However, where is the key area of RW energy dispersion in the NH summer on interdecadal scale? What is the physical mechanism of the Rossby wave train excited from the TP on the precipitation anomaly in East Asia? All these are worthy of further discussion. Thus, we analyze the interdecadal variability and spatial distribution of the TP-RWS in summer during 1900 to 2010. In addition, the causes of the TP-RWS and its effects on the anomalous circulation in East Asia are discussed so as to understand the physical mechanism of precipitation anomalies associated with TP-RWS. Section 2 of the manuscript introduces the data and methods used. Section 3 analyzes the interdecadal variability of the TP-RWS and its impact on the East Asia Circulation Pattern. Conclusions and discussion are shown in Sections 4 and 5, respectively.

2. Data and Methods

The atmospheric data used in this paper are from the ERA-20th Century (ERA-20C) monthly reanalysis dataset provided by the European Center for Medium Range Weather Forecasting (ECMWF) for 1900 to 2010 [35]. In order to compare the calculation results, we also use monthly atmospheric variables of the 20th Century Reanalysis (NCEP-20C) from the National Center for Environmental Prediction/National Center for Atmospheric Research (NCEP/NCAR) for 1836 to 2015 [36]. The horizontal resolution of the two datasets is $1^\circ \times 1^\circ$, and the rectilinear grid numbers are 181×360. The data include wind field (V), temperature (T), geopotential height (H), vertical velocity (ω), and specific humidity (q) of 27 layers from 1000 hPa to 100 hPa. In addition, the surface sensible heat flux (SHTFL), total cloud cover (TCC), snow albedo (ASN), and surface pressure are used. The centennial precipitation data used in our study are based on the global terrestrial meteorological grid dataset established by the University of East Anglia (CRU_ts4) from 1901 to 2010, with a horizontal resolution of $0.5^\circ \times 0.5^\circ$. The monthly land precipitation data provided by the Global Precipitation Climatology Centre (GPCC) for 1900–2010 are also used, with a horizontal resolution of $0.25^\circ \times 0.25^\circ$. The 2500 m terrain height is selected as the criteria for calculating the regional average over the TP. In this study, interdecadal variation denotes the time series of TP-RWS and other physical quantities remove the linear trend during 1900 to 2010 in summer, and then perform 10-year low-pass filtering.

According to the derivation of Sardeshmukh and Hoskins [4,8] based on the nonlinear vorticity equation, the Rossby wave source (RWS) on a horizontal level can be calculated as:

$$RWS = -\nabla\cdot(f+\zeta)V_\chi = -(f+\zeta)D - V_\chi\cdot\nabla(f+\zeta) \quad (1)$$

where the RWS represents Rossby wave source. The f and ζ are planetary vorticity and relative vorticity, respectively, and absolute vorticity is the sum of the two. D represents the horizontal divergence $\frac{\partial u}{\partial x} + \frac{\partial v}{\partial y}$ and V_χ is the divergent component of the horizontal wind. The divergent wind component is calculated by inverting the Laplacian operator in spherical harmonic space after computing the divergence. The Formula (1) shows that time-averaged vorticity can be considered as a combination of the vortex stretching term (RWS-S1) produced by local strong divergence and the absolute vorticity advection term (RWS-S2) caused by large-scale divergent flow [2]. When an anomalous vorticity diverges outward, the local vorticity decreases, and this vorticity divergence center is called the vortex source region. The converse is the vortex sink area. Therefore, analyzing the RWS distribution and its variability can help to understand the origin and physical mechanism of planetary wave generation and atmospheric changes [37].

The method of calculating Q1 in this study is based on the inverted algorithm of Yanai et al. [38]:

$$Q_1 = C_P\left[\frac{\partial T}{\partial t} + V\cdot\nabla T + \left(\frac{P}{P_0}\right)^k \omega\frac{\partial\theta}{\partial P}\right] \quad (2)$$

where T is temperature, V is horizontal wind vector, ω is vertical velocity, and θ is the potential temperature. $k = R/C_p$, where R and C_P are dry atmospheric constant and isobaric specific heat capacity, respectively. All these variables are in p co-ordinates; thus, Q_1 at each isobaric layer can be calculated.

In addition, the three-dimensional T-N wave flux derived by Takaya and Nakamura [39,40] based on the Plumb wave flux [10] is used in our study to describe the Rossby wave energy propagation. These elements are able to better describe the Rossby wave disturbance along the westerly jet in zonal inhomogeneous flow [41]. The formula is expressed as follows:

$$W = \frac{p\cos\varphi}{2|U|}\begin{Bmatrix} \frac{U}{a^2\cos^2\varphi}\left[\left(\frac{\partial\psi'}{\partial\lambda}\right)^2 - \psi'\frac{\partial^2\psi'}{\partial\lambda^2}\right] + \frac{V}{a^2\cos\varphi}\left[\frac{\partial\psi'}{\partial\lambda}\frac{\partial\psi'}{\partial\varphi} - \psi'\frac{\partial^2\psi'}{\partial\lambda\partial\varphi}\right] \\ \frac{U}{a^2\cos\varphi}\left[\frac{\partial\psi'}{\partial\lambda}\frac{\partial\psi'}{\partial\varphi} - \psi'\frac{\partial^2\psi'}{\partial\lambda\partial\varphi}\right] + \frac{V}{a^2}\left[\left(\frac{\partial\psi'}{\partial\varphi}\right)^2 - \psi'\frac{\partial^2\psi'}{\partial\varphi^2}\right] \\ \frac{f_0^2}{N^2}\left\{\frac{U}{a\cos\varphi}\left[\frac{\partial\psi'}{\partial\lambda}\frac{\partial\psi'}{\partial z} - \psi'\frac{\partial^2\psi'}{\partial\lambda\partial z}\right] + \frac{V}{a}\left[\frac{\partial\psi'}{\partial\varphi}\frac{\partial\psi'}{\partial z} - \psi'\frac{\partial^2\psi'}{\partial\varphi\partial z}\right]\right\} \end{Bmatrix} \quad (3)$$

where the φ, λ, Φ, and a represent the latitude, longitude, geopotential, and radius of the Earth, respectively. z is the vertical co-ordinates of the $\log p$. $\psi' = \frac{\Phi'}{f}$ is the disturbance of quasi-geostrophic stream function relative to the climatology. The basic flow U represents the climate average.

3. Results

3.1. Interdecadal Variation in Rossby Wave Source over the Tibetan Plateau

In the June–July–August (JJA) period, a Rossby wave source clearly appears along the upper tropospheric westerly jet over East Asia and the extratropical Pacific, which is much stronger than that in the tropics (Figure 1a). In the North Hemisphere (NH), the Asian Monsoon–Tibet Plateau region exhibits a strong negative RWS and the North Africa and Mediterranean Sea are positive RWS regions. In addition, the eastern subtropical Pacific and the Atlantic Ocean showed weak RWSs. The RWSs in the Southern Hemisphere are mainly located over the South Indian Convergence Zone, South Pacific, and South Atlantic. These are consistent with previous studies [2,9,42]. For the JJA mean, the intertropical convergence zone moves towards the Northern Hemisphere, and its associated heavy rainfall (Figure 1b) extends northeast from East Asia and the Northwest Pacific. Meanwhile, there is a strong divergent flow and velocity potential center over the Asian monsoon region, which corresponds to its strong convection. The divergence flow radiates outward from East Asia and extends into Eurasia and the Pacific. The concurrence of large-scale divergence at 200 hPa and deep convection indicates that the velocity potential center at the upper troposphere in the NH is largely caused by diabatic heating over the Asian monsoon region [43], while North Africa occurs as a net radiation sink in summer, resulting in continuous cooling as a powerful cold source [44]. This distribution of heat and cold sources forms a divergent/convergent field center over the Asian monsoon region and the Mediterranean Sea. According to Formula (1), negative RWS sources over the Asian summer monsoon region in Figure 1a are formed. Previous studies have shown that these large-scale heating forcings and their RWSs play important roles in forming and maintaining atmospheric circulation over East Asia and the North Pacific [45,46].

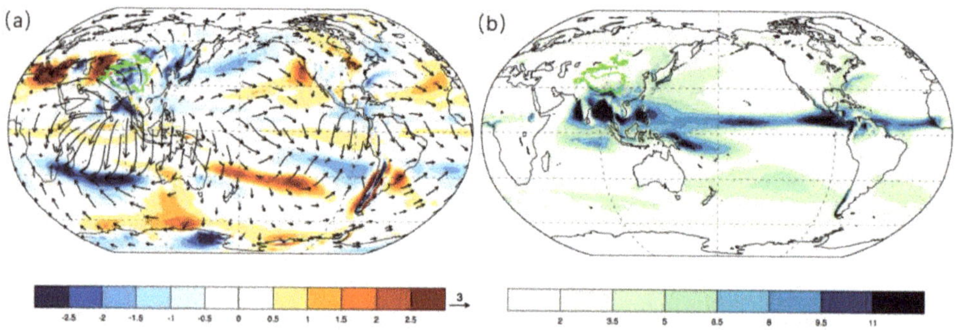

Figure 1. Distribution of global (**a**) Rossby wave source (shading; unit: 10^{-10} s^{-2}) and divergent wind component (vectors; unit: $m{\cdot}s^{-1}$), and (**b**) precipitation (unit: $mm{\cdot}day^{-1}$) in June–August of 1900–2010. The green contours in (**a**,**b**) represent the 2.5km topography height of the Tibetan Plateau.

How the RWSs have changed during boreal summer over the past 110 years and what physical processes they are associated with are the main concerns in this study. The NH summer RWS variability has exhibited prominent differences, as exemplified by the RWS variance patterns on interannual and interdecadal timescales in Figure 2. Notably, the maximum variability of RWSs on both timescales occurs over the northwestern TP, and the fluctuations are also large over the northeast of China and near the Mediterranean Sea. The interdecadal spatial distributions of NH RWS variabilities are very consistent with those on the interannual timescale. Furthermore, the TP-RWS variance on interdecadal timescale accounts for 22.3% of that in raw from 1900 to 2010. All these indicate that the northwestern TP, including most parts of the Eurasian continent, is the anomalous RWS fluctuation region. As the largest plateau and strongest heat source in boreal summer, the changes in TP thermal conditions may lead to the interdecadal RWS anomalies over this region.

Figure 3 shows the JJA interannual and interdecadal RWS time-series over the TP (topographic height above 2500 m) of 200 hPa from 1900 to 2010. It can be seen that the TP-RWS in the upper troposphere is characterized by multidecadal variations. From the perspective of intensity changes, the TP-RWS showed continuous strong periods during the early 20th century, the 1920s to 1940s, the 1950s to 1970s, and after 2000. Accordingly, the temporal series turned into weak periods during the 1910s to 1920s, 1940s to 1950s, and from the 1970s to the end of the 20th century. Moreover, the TP-RWS variation shows multidecadal differences in the recent 110 years. In the first half of the 20th century, it oscillated in longer periods and, after the late 1950s, decadal variations were more evident. These may be affected by the joint influence of various internal forcing oscillations, as well as the increasing instability of the climate system since global warming [47,48]. The NCEP-20C datasets from 1900 to 2015 are also used to calculate TP-RWS time evolution in order to reduce the uncertainty of calculation results from different datasets. The temporal correlation of TP-RWS between ERA-20C and NCEP-20C datasets reaches 0.72 on the interannual scale, which is 0.69 on the interdecadal scale, and they both pass the 99% confidence test. Therefore, the selected ERA-20C data in this paper can accurately reflect the variations in TP-RWS. Furthermore, in order to explore the spatial distribution characteristics with the TP-RWS interdecadal evolution in the past 110 years, we calculate one-point correlation between the TP-RWS anomalies and large-scale divergent flow in the NH summer (Figure 4). When the interdecadal intensity of TP-RWS increases, the corresponding maximum RWS area appears over the northwestern TP, which is accompanied by a strong divergence center at the upper troposphere. Meanwhile, the RWS in the south of the TP, especially in the northwestern Indo–China Peninsula, shows a weakly negative correlation with the RWS in the whole plateau. The above results show that, as a key area of the interdecadal RWS fluctuations in boreal summer, the causes of the TP-RWS and its influences are worth studying.

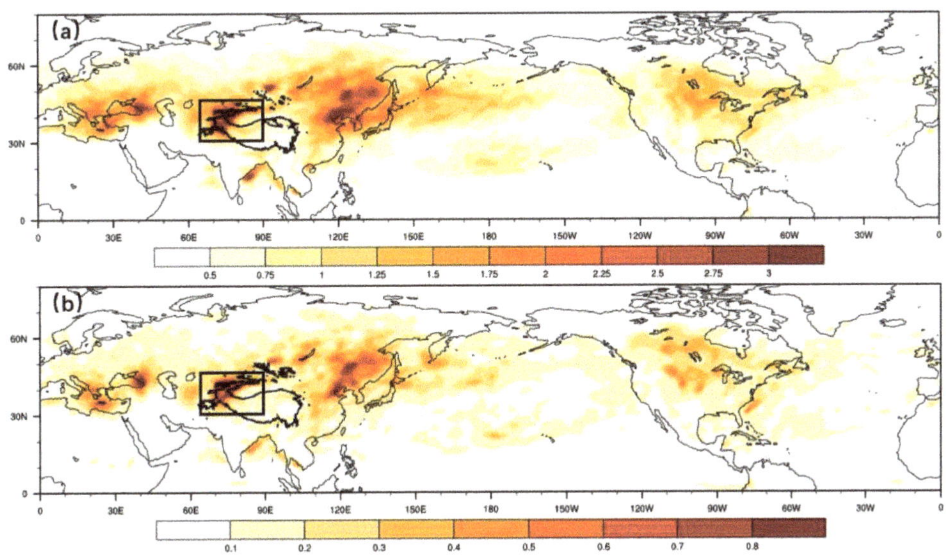

Figure 2. Spatial distribution of the RWS variance (unit: 10^{-10} s^{-4}) during boreal summer of 1900–2010 (**a**) on an interannual timescale and (**b**) on an interdecadal timescale (black box indicates the key area of the RWS variance 65–90° E, 30–45° N).

3.2. *The Mechanisms Contributing to Interdecadal Variations in the TP-RWS*

Since the large-scale divergent flow is directly related to anomalous diabatic heating of the atmosphere, we calculate the correlation map of the atmospheric heat source (Q1) with the TP-RWS in order to explore the possible causes of the TP-RWS interdecadal variations (Figure 5). It is worth noting that two Q1 key areas occur over the plateau, which are located in the northwestern TP (Pamirs plateau) and southeastern TP. When the interdecadal TP-RWS intensifies, the atmospheric heating characteristics over the two regions are significantly different, that is, the anomalous Q1 in northwestern TP is mainly found between the near-ground plateau and 300 hPa in the troposphere, while the Q1 in southeastern TP is more pronounced from 500 hPa to 200 hPa. The regional climate of Pamir Plateau and Tarim Basin are mainly characterized by drought, with less rainfall due to the faint water vapor imported from the tropical Indian Ocean. Meanwhile, the giant mountains in southeastern TP are mainly affected by the monsoon and moist water vapor from the tropical ocean in summer, resulting in abundant water vapor being lifted here, which is conducive to the occurrence of deep convection [49,50]. Therefore, dynamical mechanisms contributing to the interdecadal variations in TP-RWS may be different in northwestern and southeastern TP.

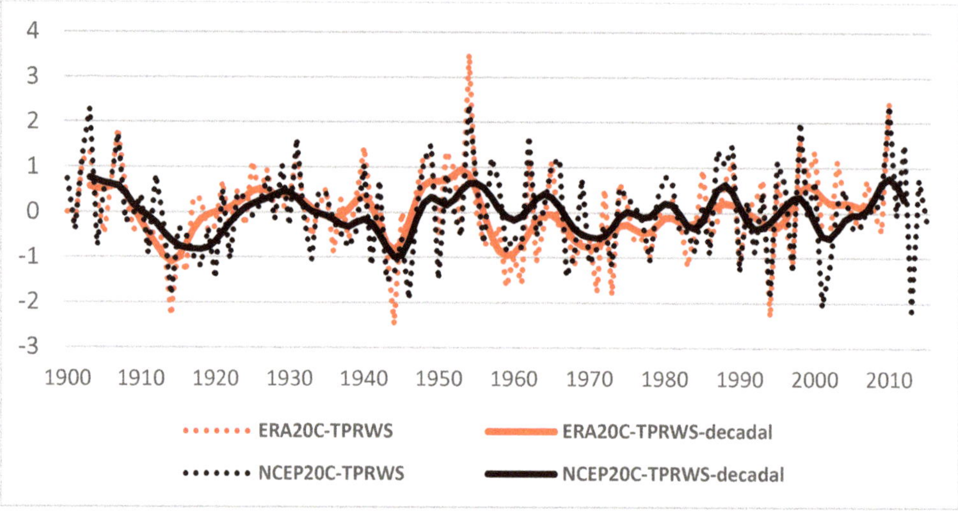

Figure 3. The standardized time series of Rossby wave source over the TP (*−1; dotted lines) and its 10−year low−pass filtering value (solid lines) in summer of 1900−2010. The red and black lines are results from the ERA−20C and NCEP−20C reanalysis datasets, respectively. The RWS*−1 represents the intensity of negative Rossby wave source over the TP.

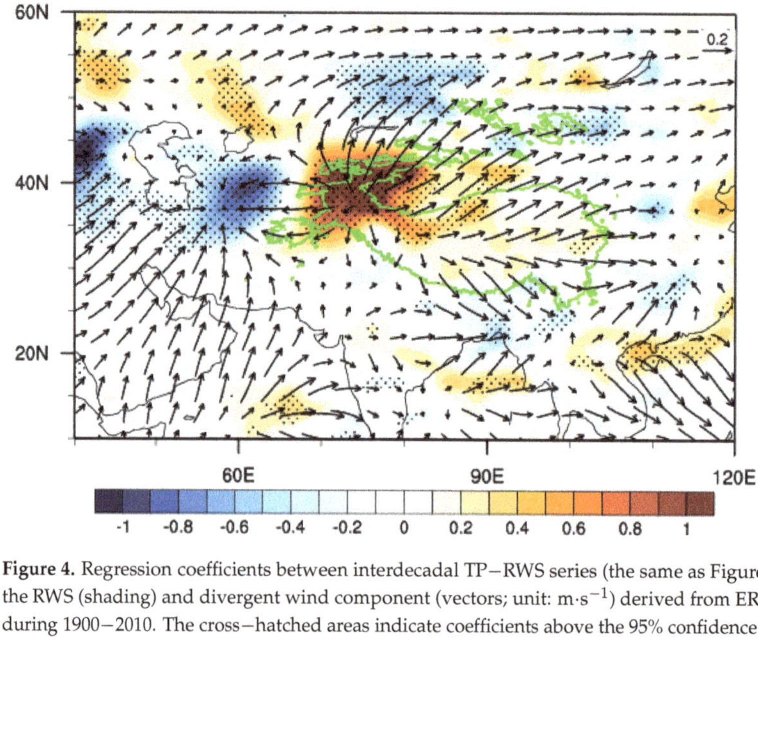

Figure 4. Regression coefficients between interdecadal TP−RWS series (the same as Figure 3) and the RWS (shading) and divergent wind component (vectors; unit: m·s^{-1}) derived from ERA−20C during 1900−2010. The cross−hatched areas indicate coefficients above the 95% confidence level.

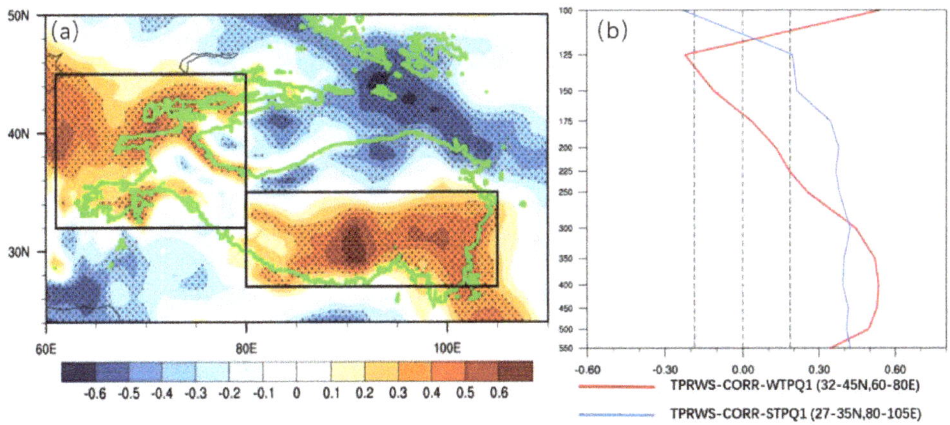

Figure 5. (**a**) Spatial correlations between interdecadal TP−RWS series (the same as Figure 3) and the atmospheric heat source (Q1) derived from ERA−20C during 1900−2010. (**b**) Vertical profile of correlation coefficients between the TP−RWS and the Q1 over northwestern TP (red solid line) and the Q1 over southeastern TP (blue solid line). The cross−hatched areas indicate coefficients above the 95% confidence level. The black rectangles in (**a**) denote the northwestern and southeastern TP.

The components of RWS, i.e., the vortex stretching term (RWS-S1) and the absolute vorticity advection term caused by divergent flow (RWS-S2), can be used to determine the causes of interdecadal TP-RWS during 1900 to 2010. From Formula (1), RWS-S1 is mainly determined by absolute vorticity and strong divergence, and the RWS-S2 is directly related to absolute vorticity gradient and the divergent wind component. In general, strong atmospheric heating can lead to circulation changes, which can be understood as the local influence of the RWS-S1 term on Rossby wave source. Moreover, other regions also can affect the RWS through large-scale divergent flow, which can be used as a horizontal distribution rebalancing of absolute vorticity by RWS-S2 term. The time series of TP-RWS sub-items have obvious interdecadal changes in summer from 1900 to 2010 (Figure 6). Both RWS-S1 and RWS-S2 are closely related to TP-RWS on the interdecadal scale, with correlation coefficients of 0.81 and 0.67, respectively. By comparison, RWS-S1 seems to play a more dominant role. However, the correlation coefficient between RWS-S1 and RWS-S2 is only 0.14, which indicates that their interdecadal evolutions are independent of each other. Furthermore, we find that the interdecadal Q1 fluctuations over northwestern and southeastern TP also represent significant differences with correlation coefficient of −0.05 only. Meanwhile, similar interdecadal temporal variations exist between TPRWS-S1 and northwestern TP-Q1 and between TPRWS-S2 and southeastern TP-Q1, with the correlation coefficients reaching 0.60 and 0.51, respectively. These results show that the TPRWS-S1 and TPRWS-S2, which cause interdecadal variation of TP-RWS, have independent changes along with the related atmospheric diabatic heating in different regions over the TP. Therefore, the interdecadal TP-RWS can be regarded as a result of the joint contribution of Q1 over northwestern and southeastern TP by different dynamical mechanisms.

Next, we calculate correlation maps between the two sub-terms of RWS and multiple meteorological elements on the interdecadal scale. As shown in Figure 7, the contribution of TPRWS-S1 to TP-RWS mainly occurs over the Pamir Plateau, which is characterized by both strong divergence and anomalous absolute vorticity. In addition, significant negative snow albedo and increased surface sensible heat flux are observed along the northwestern TP. Since the albedo of snow cover is significantly greater than that of other underlying surfaces (e.g., soil, forest, grass, and water), the albedo decreases when the snow melts, resulting in more solar radiation absorbed by the ground and the net radiation increases. The processes enhance surface sensible heat flux and lead to an increased Q1 over northwestern TP in the middle and lower troposphere. Therefore, the main contribution component of TP-RWS in northwestern TP is TPRWS-S1, which is caused by interdecadal variations in snow cover. In addition, we find that the snow albedo associated with the TPRWS-S1 shows a spatial east–west dipole-like pattern over the TP. This feature has also been observed in the study of the plateau snow cover [51,52]. In contrast, the main effect of TPRWS-S2 on TP-RWS is manifested in southeastern TP (Figure 8) due to the strong absolute vorticity advection caused by deep convection. A strong divergence is accompanied by the meridional northward flow in southeastern TP. This process excites disturbance in vorticity field, which is conducive to the RWS-S2 formation at the upper troposphere. Thus, the interdecadal TP-RWS variations during boreal summer are a synergistic result of the snow cover related to the vortex stretching term in northwestern TP and the deep convection related to the absolute vorticity advection term in southeastern TP.

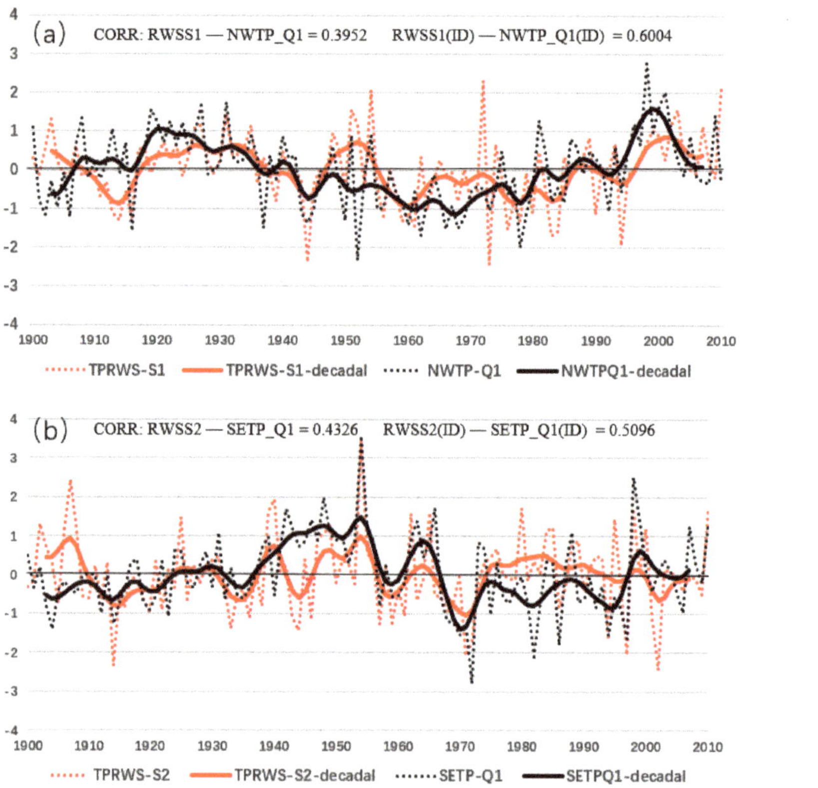

Figure 6. The standardized interannual and interdecadal time series of TP−RWS components (*−1; red lines) and the Q1 over the northwestern and southeastern TP (black lines) in summer of 1900−2010. (**a**) The vortex stretching term RWS−S1 and the northwestern TP−Q1 (60–80° E, 32–45° N), (**b**) the absolute vorticity advection term RWS−S2 and the southeastern TP-Q1 (80–105° E, 27–35° N).

In summary, the interdecadal variation in TP-RWS is closely related to atmospheric diabatic heating over northwestern and southeastern TP. The interdecadal variations in snow cover over the TP result in divergence and absolute vorticity anomalies by affecting atmospheric heating from TP surface to the middle troposphere, and then contribute to the TP-RWS by vortex stretching term. On the other hand, due to latent heat release of deep convection in southeastern TP, the TP-RWS can be formed and maintained through absolute vorticity advection caused by large-scale divergent flow in the middle and upper troposphere. Therefore, although the TP-RWS manifests as a whole, the causes of its interdecadal variations are not the same in northwestern and southeastern TP.

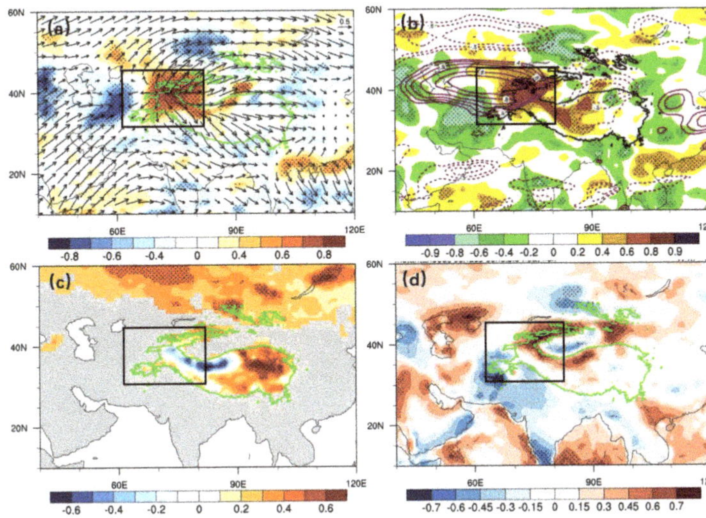

Figure 7. The correlation coefficients between interdecadal TPRWS−S1 series (the same as Figure 6a) and (**a**) the RWS (shading) and divergent wind component (vectors), (**b**) the divergence (shading) and absolute vorticity (contours) at 200 hPa, (**c**) snow albedo, (**d**) surface sensible heat flux derived from ERA−20C during 1900−2010. The cross-hatched areas indicate coefficients above the 95% confidence level. The black rectangles represent the northwestern TP.

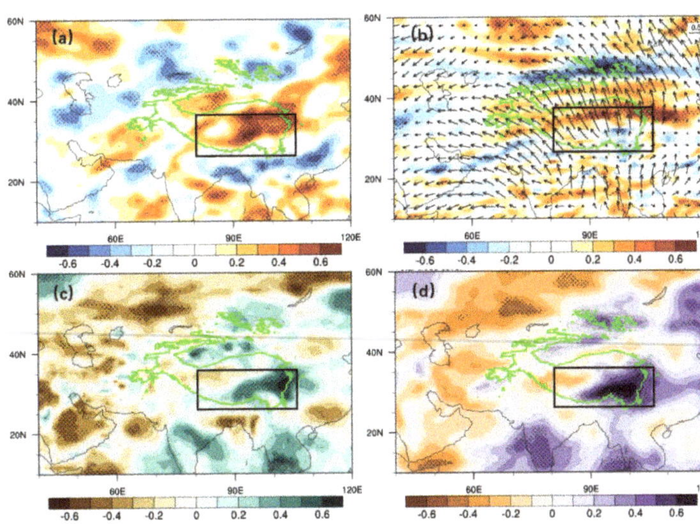

Figure 8. The correlation coefficients between interdecadal TPRWS−S2 series (the same as Figure 6b) and (**a**) the RWS, (**b**) divergent wind component (vectors) and absolute vorticity gradient (shading) at 200 hPa, (**c**) precipitation, (**d**) total cloud cover derived from ERA−20C during 1900−2010. The cross-hatched areas indicate coefficients above the 95% confidence level. The black rectangles represent the southeastern TP.

3.3. *Impacts of the Interdecadal TP-RWS on East Asia Circulation Pattern*

The interdecadal TP-RWS in JJA have a significant impact on the downstream atmospheric circulation and precipitation. From the correlation maps between TP-RWS and the land precipitation (Figure 9), an enhanced RWS corresponds to more JJA rainfall over the TP, especially in the southeastern and northwestern TP. These two regions are consistent with the atmospheric diabatic heating areas in the above section. For East Asia, the TP-RWS interdecadal variation corresponds to a clear tripole rainfall pattern. When the intensity of TP-RWS is relatively strong, the regions from the Huang-huai River Basin to the southern Korean Peninsula and the southern part of the Japan Island are accompanied by moisture anomalies, while the South China and North China are usually dry. By comparing different reanalysis results of TP-RWS and precipitation datasets, we find that there are significant relationships between the interdecadal TP-RWS and the precipitation in the Huang-huai River Basin. All the results show that the JJA rainfall in the north of the Yangtze River increases, while the precipitation in South China is suppressed.

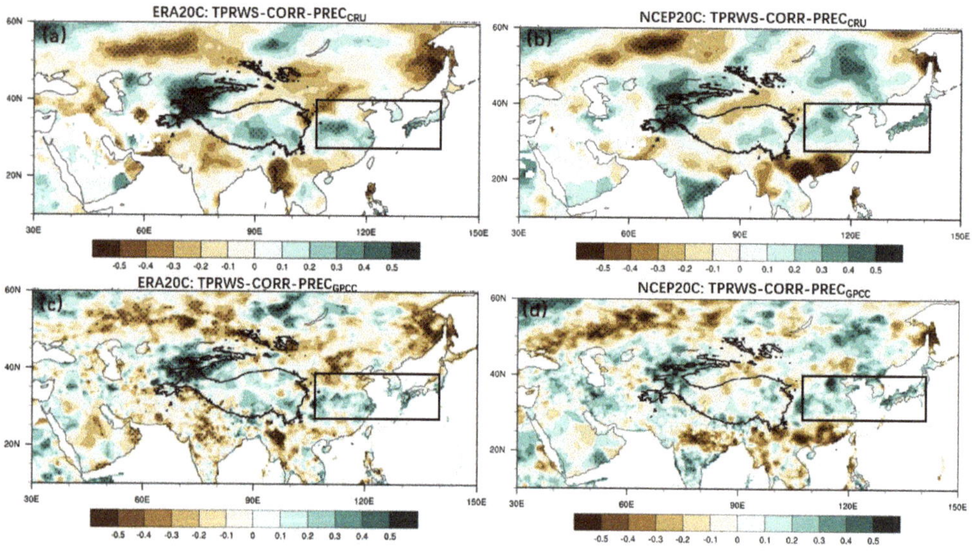

Figure 9. The correlation coefficients between interdecadal TP−RWS (same as Figure 3) and global terrestrial precipitation. The interdecadal TP−RWS in (**a**,**c**) and (**b**,**d**) are from ERA−20CR and NCEP−20C, respectively. The precipitation data in (**a**,**b**) and (**c**,**d**) are from CRU_ts4 and GPCC, respectively. The cross−hatched areas indicate coefficients above the 95% confidence level. Black boxes indicate the key areas of interdecadal precipitation.

In order to reveal the causes of the close relationship between TP-RWS and precipitation in East Asia, the Asia–Pacific atmospheric circulation anomaly associated with the TP-RWS is further discussed and analyzed in this study. As shown in Figure 10, significant negative geopotential height anomalies are observed at the middle and low troposphere over northwestern TP, Huang-huai River Basin, and the southern part of Japan, while positive geopotential heights appear over the southeastern TP and the northwestern Pacific Ocean. The stream function field at 850 hPa in the southern TP and East Asia is dominated by anomalous cyclonic circulation, and the centers are located in eastern India and Bangladesh, the Indochina Peninsula, and the Huang-huai River Basin. The prevailing southerly and easterly winds in the northwestern Pacific Ocean and the Huang-huai River Basin are attributed to the anomalous cyclonic circulation at the lower troposphere in East Asia, which lead to the water vapor convergence and upward movement at the middle and lower troposphere. The above circulation pattern provides favorable water vapor transmission and dynamic conditions for the occurrence of precipitation in these regions, which increase the precipitation in the Huang-huai River Basin, South Korea, and Japan.

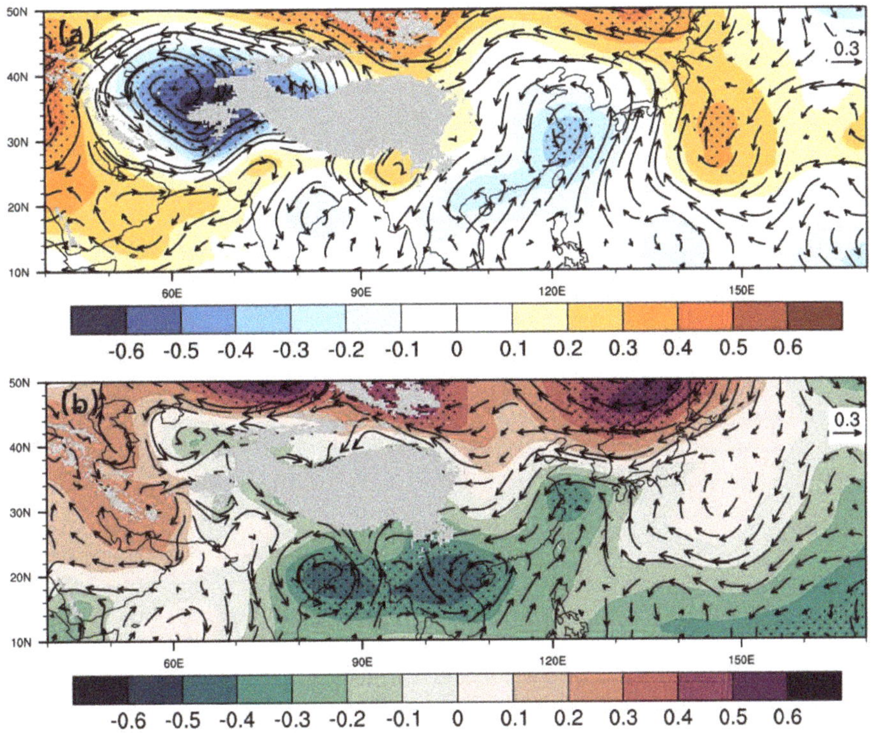

Figure 10. The correlation coefficients between interdecadal TP-RWS (same as Figure 3) and (**a**) geopotential height (shading) and water vapor transport flux (vectors) at 600 hPa, (**b**) stream function (shading) and horizontal wind field (vectors) at 850 hPa from 1900 to 2010. The cross–hatched areas indicate coefficients above the 95% confidence level.

We further discuss the spatial distribution of the relationship between interdecadal TP-RWS and related wave-activity flux. A prominent characteristic is that the TP-RWS triggers a zonal wave train along the northwestern TP to the North Pacific (Figure 11a). The geopotential height in northwestern TP, eastern TP, East Asia, and the northwestern Pacific Ocean correspond to negative, positive, negative, and positive wave train anomalies, respectively. As shown in Figure 11b, strengthening TP-RWS is accompanied by an anomalous wave activity flux originating over the northwestern TP and propagating eastward along the westerly jet stream to the East Asia and northwestern Pacific. This anomalous wave flux also has a tendency to spread southward but to a lesser extent. Moreover, we investigate the vertical structure to further explore the teleconnection pattern associated with interdecadal TP-RWS. A significant zonal wave train can be seen from the TP to the northwestern Pacific, and the strongest wave flux is located over the western TP, accompanied by a Rossby wave spreading downstream to East Asia and the North Pacific. In addition, the TP-RWS teleconnection pattern exhibits an equivalent barotropic structure at the vertical profile. It is worth noting that the anomalous wave activity excites from the TP has obvious characteristics of propagating from the upper troposphere to the lower troposphere over East Asia. Accompanied by the low pressure over the Huang-huai River Basin to Japan Island, the TPRWS-excited teleconnection is conducive to the strengthening ascent movement at the middle and lower troposphere.

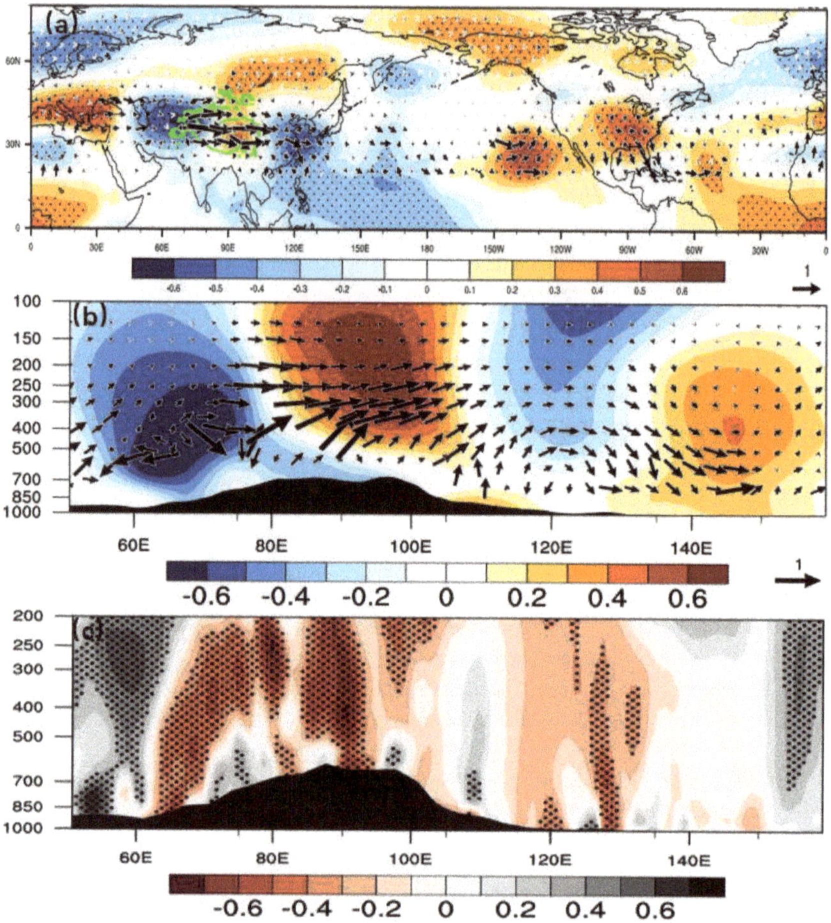

Figure 11. (**a**) The correlation coefficients between interdecadal TP−RWS (same as Figure 3) and the geopotential height (shading) and T−N wave flux (vectors) at 200 hPa. Vertical profiles of correlation coefficients between interdecadal TP-RWS and (**b**) the geopotential height and T−N wave flux, (**c**) the vertical velocity averaged along 28–45° N from 1900 to 2010. The cross-hatched areas indicate coefficients above the 95% confidence level.

Previous studies have shown that the Atlantic multidecadal oscillation (AMO) can modulate the intensity of the East Asian summer monsoon by exciting teleconnection wave train [53–56]. The interannual variation in deep convection in the southeastern TP is also closely associated with the AMO [57,58]. In order to distinguish the effects of TP-RWS and AMO on East Asia, we calculate the partial correlation between interdecadal TP-RWS and the atmospheric circulation without AMO series in the Northern Hemisphere (Figure 12). Consistent with Figure 11, the zonal wave train excited by interdecadal TP-RWS propagating from northwestern TP to East Asia is still clear and the corresponding circulation intensity over East Asia and Northwest Pacific is significantly strengthened. Therefore, it is reasonable to believe that the TP-RWS during 1900 to 2010 can stimulate an equivalent barotropic zonal wave train that travels eastward along the westerly jet to the northwest Pacific and exhibits "low-high-low-high" geopotential height anomalies at the vertical profile. Furthermore, the pattern is conducive to the upward movement and precipitation over the Huang-huai River Basin.

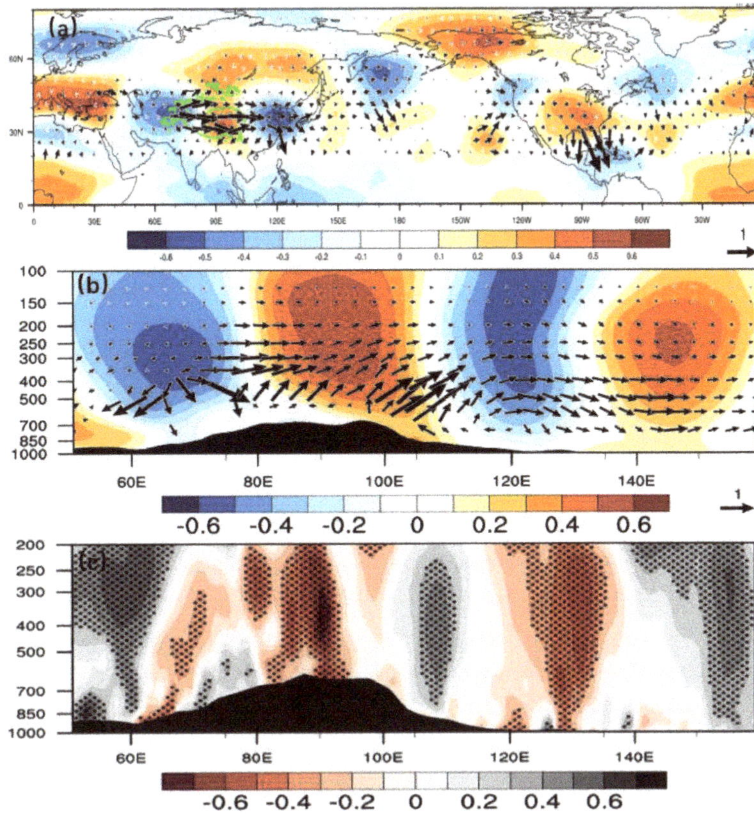

Figure 12. (**a**) Partial correlation coefficients between interdecadal TP−RWS (same as Figure 3) and the geopotential height (shading) and T−N wave flux (vectors) at 200 hPa (remove interdecadal AMO). Vertical profiles of partial correlation coefficients between interdecadal TP−RWS and (**b**) the geopotential height and T−N wave flux, (**c**) the vertical velocity averaged along 28−45° N from 1900 to 2010. The cross−hatched areas indicate coefficients above the 95% confidence level.

4. Conclusions

Using the ERA-20C monthly reanalysis datasets in summer (June–July–August) from 1900 to 2010 and a variety of global precipitation data, our manuscript studies the multidecadal variations in Rossby wave source over the Tibetan Plateau, the corresponding energy transmission, and its impact on atmospheric circulation pattern over East Asia. The conclusions are summarized as follows:

(1) There is a strong negative RWS over the TP during boreal summer. In the recent 110 years, the RWS variability has shown great regional differences in the NH midlatitudes. The regions with large interannual and interdecadal variability occur over the northwestern TP, the northeast of China to the Korean Peninsula, and areas near the Mediterranean Sea along the westerly jet, among which the variability over the TP is the largest.

(2) The TP-RWS during 1900 to 2010 is characterized by multidecadal variations. From the perspective of intensity changes, it showed continuous strong periods during the early 20th century, the 1920s to 1940s, the 1950s to 1970s, and after 2000. Correspondingly, the temporal series turned into weak periods during the 1910s to 1920s, 1940s to 1950s, and from 1970s to the end of the 20th century. When the interdecadal TP-RWS intensifies, it is accompanied by strong divergence flow at the upper troposphere.

(3) The two sub-items TPRWS-S1 and TPRWS-S2, which constitute the TPRWS have independent interdecadal variations. Meanwhile, similar interdecadal temporal variations exist between TPRWS-S1 and northwestern TP-Q1 and between TPRWS-S2 and southeastern TP-Q1. The interdecadal variations in snow cover over the TP cause the divergence and absolute vorticity anomalies by affecting atmospheric heating from TP surface to the middle troposphere, and then contribute to the TP-RWS. On the other hand, due to latent heat release of deep convection in southeastern TP, the TP-RWS can be formed through absolute vorticity advection caused by large-scale divergent flow in the middle and upper troposphere. Therefore, although the TP-RWS manifests as a whole, the causes of its interdecadal variations are not the same in northwestern and southeastern TP.

(4) The interdecadal TP-RWS has a great climate effect on the atmospheric circulation and precipitation pattern in Huang-huai River Basin, South Korea, and Japan Island. The enhanced TP-RWS can excite an anomalous wave active flux originating from the northwestern TP and propagating eastward along the westerly jet to East Asia and the Northwest Pacific. The TP-RWS teleconnection exhibits an alternatively positive and negative pattern of geopotential height anomalies from the northwestern TP to the North Pacific, which has an equivalent barotropic structure at vertical profile. Under the guidance of the anomalous cyclonic circulation in East Asia, the prevailing southerly and easterly winds over the West Pacific and Huang-huai River Basin are conducive to the water vapor convergence and precipitation anomaly.

5. Discussion

By calculating the components of RWS equation and the three-dimensional T-N wave flux, we analyze the causes of TP-RWS and discuss the influence of TP-RWS on the precipitation and atmospheric circulation pattern in East Asia on an interdecadal timescale. The above conclusions are of great significance for further understanding the variabilities and maintenance mechanisms of Rossby wave sources in the NH summer. Considering that the linear detrending may not fully remove the influence of external forcing, the variations in TP-RWS are also modulated by a variety of external forcing factors. While the tropical cumulus convection and heavy rainfall also play a role in the formation of the steady vortex source and the RWS energy propagation during boreal summer. How the interdecadal variations in sea surface temperature in the three oceans will affect the TP-RWS and how the corresponding physical mechanism is are questions worth studying. We will use numerical models to further explore this issue in future work. Due to the limited space, this paper does not elaborate too much.

Author Contributions: Formal analysis, X.S. and Y.D.; data curation, X.S. and Q.L.; writing—original draft preparation, X.S.; writing—review and editing, X.S., Y.D., Q.L. and Y.S.; supervision, Y.D. and Y.S. All authors have read and agreed to the published version of the manuscript.

Funding: This work was funded by the Strategic Priority Research Program of the Chinese Academy of Sciences (XDA20100304), the Second Tibetan Plateau Scientific Expedition and Research Program of China (2019QZKK0208), the National Key Research and Development Program of China (2022YFE0136000), the National Natural Science Foundation of China (U2242207, 41965005, 41790471), the Natural Science Foundation of Anhui Province of China (2208085UQ08), and the Innovative Development Special Project of China Meteorological Administration (CXFZ2022J039).

Institutional Review Board Statement: Not applicable.

Informed Consent Statement: Not applicable.

Data Availability Statement: The ERA-20C data can be downloaded from the website https://www.ecmwf.int/en/forecasts/datasets/reanalysis-datasets/era-20c (accessed on 1 January 2022). The NCEP-20C data can be downloaded from the website https://www.psl.noaa.gov/data/gridded/data.20thC_ReanV3.html (accessed on 1 January 2022).

Conflicts of Interest: The authors declare no conflict of interest.

References

1. Rossby, C.G. Relation between variations in the intensity of the zonal circulation of the atmosphere and the displacements of the semi-permanent centers of action. *J. Mar. Res.* **1939**, *2*, 38–55. [CrossRef]
2. Shimizu, M.H.; Cavalcanti, I. Variability patterns of Rossby wave source. *Clim. Dyn.* **2011**, *37*, 441–454. [CrossRef]
3. Hoskins, B.J.; Karoly, D.J. The steady linear response of a spherical atmosphere to thermal and orographical forcing. *J. Atmos. Sci.* **1981**, *38*, 1179–1196. [CrossRef]
4. Sardeshmukh, P.D.; Hoskins, B.J. The generation of global rotational flow by steady idealized tropical divergence. *J. Atmos. Sci.* **1988**, *45*, 1228–1251. [CrossRef]
5. Geng, Q.; Chen, S. Vorticity Source by Stationary Divergent Wind and Energy Propagation of Stationary Waves. *Chin. J. Atmos. Sci.* **1996**, *20*, 298–305.
6. James, I.N. *Introduction to Circulating Atmospheres*; Cambridge University Press: Cambridge, UK, 1994.
7. Ye, D.Z.; Zhu, B.Z. *Some Basic Problems of Atmospheric Circulation*; Science Press: Beijing, China, 1958.
8. Sardeshmukh, P.D.; Hoskins, B.J. Vorticity balances in the tropics during the 1982-83 El Niño-Southern Oscillation event. *Quart. J. Roy. Meteor. Soc.* **1985**, *111*, 261–278. [CrossRef]
9. Lu, R.Y.; Baek-Jo, K.I.M. The Climatological Rossby Wave Source over the STCZs in the Summer Northern Hemisphere. *J. Meteorol. Soc. Japan. Ser. II* **2004**, *82*, 2, 657–669. [CrossRef]
10. Plumb, R.A. On The three-dimensional propagation of stationary waves. *J. Atmos. Sci* **1985**, *42*, 217–229. [CrossRef]
11. Yeh, T.C. On energy dispersion in the atmosphere. *J. Atmos. Sci.* **2010**, *6*, 1–16. [CrossRef]
12. Tao, S.Y.; Wei, J.; Liang, F.; Zhang, X.L. Analysis of High Impact Weather Induced by the Downstream Effect of Rossby Waves. *Meteor Mon.* **2010**, *36*, 81–93.
13. Xu, J.P.; Wang, W.; Cai, X.J.; Xu, Z.L.; Xu, J.X. A comparison of the Rossby wave activities and circulation features of the drought in winter-spring of 2011 and in summer of 2013 over mid-lower reaches of the Yangtze river basin. *J. Trop. Meteorol.* **2017**, *33*, 992–999.
14. Wang, X.W.; Li, Y.; Zhou, B. An Applied Research About Spread Effects of Rossby Wave in Medium Range Forecast on Heavy Rain During Meiyu Flood Period. *Meteor. Mon.* **2012**, *38*, 1070–1077.
15. Sun, X.; Ding, Q.; Wang, S.Y.S.; Topál, D.; Li, Q.; Castro, C.; Teng, H.; Luo, R.; Ding, Y. Enhanced jet stream waviness induced by suppressed tropical Pacific convection during boreal summer. *Nat. Commun.* **2022**, *13*, 1288. [CrossRef] [PubMed]
16. Li, M.; Guan, Z.; Mei, S. Interannual and Interdecadal Variations of Summer Rainfall Duration over the Middle and Lower Reaches of the Yangtze River in Association with Anomalous Circulation and Rossby Wave Activities. *Chin. J. Atmos. Sci.* **2016**, *40*, 1199–1214.
17. Yang, N.; Jin, R.; Xiao, T.; Sun, X. Propagation of Rossby Wave Along Asian Subtropical Westerly Jet in Summer and Its Impact on Anomaly of Precipitation in China. *Meteor. Mon.* **2020**, *46*, 1–14.
18. Duan, A.M.; Li, F.; Wang, M.R.; Wu, G. Persistent weakening trend in the spring sensible heat source over the Tibetan Plateau and its impact on the Asian summer monsoon. *J. Clim.* **2011**, *24*, 5671–5682. [CrossRef]
19. Wu, G.X.; Liu, Y.M.; He, B.; Bao, Q.; Duan, A.; Jin, F.F. Thermal controls on the Asian summer monsoon. *Sci. Rep.* **2012**, *2*, 404. [CrossRef] [PubMed]
20. Xu, X.D.; Lu, C.G.; Ding, Y.H.; Shi, X.; Guo, Y.; Zhu, W. What is the relationship between China summer precipitation and the change of apparent heat source over the Tibetan Plateau? *Atmos. Sci. Lett.* **2013**, 14. [CrossRef]
21. Lu, M.M.; Yang, S.; Li, Z.N.; He, B.; He, S.; Wang, Z. Possible effect of the Tibetan Plateau on the 'upstream' climate over West Asia, North Africa, South Europe and the North Atlantic. *Clim. Dyn.* **2018**, *51*, 1485–1498. [CrossRef]
22. Wu, G.X.; He, B.; Liu, Y.M.; Bao, Q.; Ren, R. Location and variation of the summertime upper-troposphere temperature maximum over South Asia. *Clim. Dyn.* **2015**, *45*, 2757–2774. [CrossRef]
23. Sun, R.Z.; Duan, A.M.; Chen, L.L.; Li, Y.; Xie, Z.; Zhao, Y. Interannual variability of the North Pacific mixed layer associated with the spring Tibetan Plateau thermal forcing. *J. Clim.* **2019**, *32*, 3109–3130. [CrossRef]
24. Dong, L.; Xu, X.; Zhao, T.; Ren, H. Linkage between moisture transport over the Yangtze River Basin and a critical area of the Tibetan Plateau during the Meiyu. *Clim. Dyn.* **2019**, *53*, 2643–2662. [CrossRef]
25. Duan, A.; Wang, M.; Lei, Y.; Cui, Y. Trends in summer rainfall over China associated with the Tibetan Plateau sensible heat source during 1980–2008. *J. Clim.* **2013**, *26*, 261–275. [CrossRef]
26. Duan, A.; Wu, G. Role of the Tibetan Plateau thermal forcing in the summer climate patterns over subtropical Asia. *Clim. Dyn.* **2005**, *24*, 793–807. [CrossRef]
27. Liu, Y.; Lu, M.; Yang, H.; Duan, A.; He, B.; Yang, S.; Wu, G. Land–atmosphere–ocean coupling associated with the Tibetan Plateau and its climate impacts. *Nat. Sci. Rev.* **2020**, *7*, 534–552. [CrossRef] [PubMed]
28. Song, L.; Wu, R.; An, L. Diferent sources of 10- to 30-day intraseasonal variations of autumn snow over western and eastern Tibetan Plateau. *Geophy. Res. Lett.* **2019**, *46*, 9118–9125. [CrossRef]
29. Zhao, P.; Zhou, Z.; Liu, J. Variability of Tibetan spring snow and its associations with the hemispheric extratropical circulation and East Asian summer monsoon rainfall: An observational investigation. *J. Clim.* **2007**, *20*, 3942–3955. [CrossRef]
30. Wu, R.; Zhao, P.; Liu, G. Change in the contribution of spring snow cover and remote oceans to summer air temperature anomaly over Northeast China around 1990. *J. Geophy. Res. Atmos.* **2014**, *119*, 663–676. [CrossRef]

31. Si, D.; Ding, Y. Decadal change in the correlation pattern between the Tibetan Plateau winter snow and the East Asian summer precipitation during 1979–2011. *J. Clim.* **2013**, *26*, 7622–7634. [CrossRef]
32. Zhou, X.J.; Zhao, P.; Chen, J.; Chen, L.; Li, W. Impacts of thermodynamic processes over the Tibetan Plateau on the Northern Hemispheric climate. *Sci. China Ser. D Earth Sci.* **2009**, *52*, 1679–1693. [CrossRef]
33. Wu, B.; Lin, J.; Zhou, T. Interdecadal circumglobal teleconnection pattern during boreal summer. *Atmos. Sci. Lett.* **2016**, *17*, 446–452. [CrossRef]
34. Li, Q.Q.; Zhao, M.C.; Yang, S.; Shen, X.; Dong, L.; Liu, Z. A zonally-oriented teleconnection pattern induced by heating of the western Tibetan Plateau in boreal summer. *Clim. Dyn.* **2021**, *57*, 2823–2842. [CrossRef]
35. Poli, P.; Hersbach, H.; Dee, D.P.; Berrisford, P.; Simmons, A.J.; Vitart, F.; Laloyaux, P.; Tan, D.G.; Peubey, C.; Thépaut, J.N.; et al. ERA-20C: An Atmospheric Reanalysis of the Twentieth Century. *J. Clim.* **2016**, *29*, 4083–4097. [CrossRef]
36. Compo, G.P.; Whitaker, J.S.; Sardeshmukh, P.D.; Matsui, N.; Allan, R.J.; Yin, X.; Gleason, B.E.; Vose, R.S.; Rutledge, G.; Bessemoulin, P.; et al. The Twentieth Century Reanalysis Project. *Q. J. R. Meteorol. Soc.* **2011**, *137*, 1–28. [CrossRef]
37. Xu, G.Q.; Zhu, Q.G. Source or Sink Features of Atmospheric Low Frequency Oscillation over the Tibetan Plateau. *Trans Atmos Sci.* **2002**, *3*, 358–365.
38. Yanai, M.; Li, C.F.; Song, Z.S. Seasonal heating of the Tibetan Plateau and its effects on the evolution of the Asian summer monsoon. *J. Meteor. Soc. Jpn.* **1992**, *70*, 319–350. [CrossRef]
39. Takaya, K.; Nakamura, H. A formulation of a wave-activity flux for stationary Rossby waves on a zonally varying basic flow. *Geophys. Res. Lett.* **1997**, *24*, 2985–2988. [CrossRef]
40. Takaya, K.; Nakamura, H. A formulation of a phase-independent wave-activity flux for stationary and migratory quasigeostrophic eddies on a zonally varying basic flow. *J. Atmos. Sci.* **2001**, *58*, 608–627. [CrossRef]
41. Shi, C.H.; Jin, X.; Liu, R.Q. The differences in characteristics and applicability among three types of Rossby wave activity flux in atmospheric dynamics. *Trans. Atmos. Sci.* **2017**, *40*, 850–855.
42. Moon, J.Y.; Ha, K.J. Association between tropical convection and boreal wintertime extratropical circulation in 1982/83 and 1988/89. *Adv. Atmos. Sci.* **2003**, *20*, 593–603. [CrossRef]
43. Sun, X.T.; Ding, Y.H.; Li, Q.Q. Interdecadal Variation of the Atmospheric Heat Source over the Tibetan Plateau and Surrounding Asian Monsoon Region: Impact on the Northern Hemisphere Summer Circulation. *J. Meteor. Res.* **2021**, *35*, 1–20. [CrossRef]
44. Webster., P.J. *The Large-Scale Structure of the Tropical Atmosphere, Large-Scale Dynamical Processes in the Atmosphere*; Hoskins, B.J., Pearce, E.P., Eds.; Academic Press: London, UK, 1983; pp. 235–273.
45. Kato, K. Seasonal transition of the lower-level circulation systems around the Baiu front in China in 1979 and its relation to the northern summer monsoon. *J. Meteor. Soc. Jpn.* **1989**, *67*, 249–265. [CrossRef]
46. Ose, T. Seasonal change of Asian summer monsoon circulation and its heat source. *J. Meteor. Soc. Jpn.* **1998**, *76*, 1045–1063. [CrossRef]
47. Yao, T.; Bolch, T.; Chen, D.; Gao, J.; Immerzeel, W.; Piao, S.; Su, F.; Thompson, L.; Wada, Y.; Wang, L.; et al. The imbalance of the Asian water tower. *Nat. Rev. Earth Environ.* **2022**, *3*, 618–632. [CrossRef]
48. Zhang, Z.; Sun, X.; Yang, X.Q. Understanding the interdecadal variability of East Asian summer monsoon precipitation: Joint influence of three oceanic signals. *J. Clim.* **2018**, *31*, 5485–5506. [CrossRef]
49. Cane, M. A moist model monsoon. *Nature* **2010**, *463*, 163–164. [CrossRef] [PubMed]
50. Wu, G.; Duanm, A.; Lium, Y.; Mao, J.; Ren, R.; Bao, Q.; He, B.; Liu, B.; Hu, W. Tibetan Plateau climate dynamics: Recent research progress and outlook. *Natl. Sci. Rev.* **2014**, *2*, 100–116. [CrossRef]
51. Zhao, H.; Moore, G.W.K. On the relationship between Tibetan snow cover, the Tibetan plateau monsoon and the Indian summer monsoon, Geophys. *Res. Lett.* **2004**, *31*, L14204. [CrossRef]
52. Zhang, C.; Guo, Y.; Wen, Z. Interdecadal change in the effect of Tibetan Plateau snow cover on spring precipitation over Eastern China around the early 1990s. *Clim Dyn.* **2022**, *58*, 2807–2824. [CrossRef]
53. Sutton, R.T.; Hodson, D.L.R. Atlantic Ocean forcing of North American and European summer climate. *Science* **2005**, *309*, 115–118. [CrossRef] [PubMed]
54. Knight, J.R.; Folland, C.K.; Scaife, A.A. Climate impacts of the Atlantic multidecadal oscillation. *Geophys. Res. Lett.* **2006**, *33*, L17706. [CrossRef]
55. Wang, Y.; Li, S.; Luo, D. Seasonal response of Asian monsoonal climate to the Atlantic multidecadal oscillation. *J. Geophys. Res.* **2009**, *114*, D02112. [CrossRef]
56. Miao, J.; Jiang, D. Multidecadal variations in the East Asian winter monsoon and their relationship with the Atlantic Multidecadal Oscillation since 1850. *J. Clim.* **2021**, *34*, 7525–7539. [CrossRef]
57. Zhang, P.; Duan, A.; Hu, J. Combined effect of the tropical Indian Ocean and tropical North Atlantic sea surface temperature anomaly on the Tibetan Plateau precipitation anomaly in late summer. *J. Clim.* **2022**, *35*, 3899–3918. [CrossRef]
58. Wang, L.; Yu, J.-Y.; Paek, H. Enhanced biennial variability in the Pacific due to Atlantic capacitor effect. *Nat. Commun.* **2017**, *8*, 1–7. [CrossRef] [PubMed]

atmosphere

MDPI

Communication

Applications of CNOP-P Method to Predictability Studies of Terrestrial Ecosystems

Guodong Sun [1,2] and Mu Mu [3,4,5,*]

1 State Key Laboratory of Numerical Modelling for Atmospheric Sciences and Geophysical Fluid Dynamics (LASG), Institute of Atmospheric Physics, Chinese Academy of Sciences, Beijing 100029, China
2 University of Chinese Academy of Sciences, Beijing 100049, China
3 Department of Atmospheric and Oceanic Sciences and Institute of Atmospheric Sciences, Fudan University, Shanghai 200438, China
4 CMA-FDU Joint Laboratory of Marine Meteorology, Shanghai 200438, China
5 Shanghai Frontiers Science Center of Atmosphere-Ocean Interaction, Shanghai 200438, China
* Correspondence: mumu@fudan.edu.cn

Abstract: In this paper, recent research on terrestrial ecosystem predictability using the conditional nonlinear optimal parameter perturbation (CNOP-P) method is summarized. The main findings include the impacts of uncertainties in climate change on uncertainties in simulated terrestrial ecosystems, the identification of key physical parameters that lead to large uncertainties in terrestrial ecosystem modeling and prediction, and the evaluation of the simulation ability and prediction skill of terrestrial ecosystems by reducing key physical parameter errors. The study areas included the Inner Mongolia region, north–south transect of eastern China, and Qinghai–Tibet Plateau region. The periods of the studies were from 1961 to 1970 for the impacts of uncertainties in climate change on uncertainties in simulated terrestrial ecosystems, and from 1951 to 2000 for the identification of the most sensitive combinations of physical parameters. Climatic Research Unit (CRU) data were employed. The numerical results indicate the important role of nonlinear changes in climate variability due to the occurrences of extreme events characterized by CNOP-P in the abrupt grassland ecosystem equilibrium state and formation of carbon sinks in China. Second, the most sensitive combinations of physical parameters to the uncertainties in simulations and predictions of terrestrial ecosystems identified by the CNOP-P method were more sensitive than those obtained by traditional methods (e.g., one-at-a-time (OAT) and stochastic methods). Furthermore, the improvement extent of the simulation ability and prediction skill of terrestrial ecosystems by reducing the errors of the sensitive physical parameter combinations identified by the CNOP-P method was higher than that by the traditional methods.

Keywords: CNOP-P; model uncertainties; predictability; terrestrial ecosystem

Citation: Sun, G.; Mu, M. Applications of CNOP-P Method to Predictability Studies of Terrestrial Ecosystems. *Atmosphere* **2023**, *14*, 617. https://doi.org/10.3390/atmos14040617

Academic Editors: Alfredo Rocha, Xiaolei Zou, Guoxiong Wu and Zhemin Tan

Received: 9 February 2023
Revised: 16 March 2023
Accepted: 18 March 2023
Published: 24 March 2023

1. Introduction

As a part of the Earth system, terrestrial ecosystems interact and couple with the atmosphere through water cycling and energy exchange, so terrestrial ecosystems have important impacts on weather and climate systems [1–3]. However, there are large uncertainties in current terrestrial ecosystem simulations and predictions, and these uncertainties affect our quantitative estimates of terrestrial ecosystem carbon flux and carbon storage and are an obstacle to the simulation and prediction of weather and climate events. Therefore, it is essential to conduct studies on uncertainties in terrestrial ecosystem simulations and predictions [4–8].

Model errors are one of the factors that contribute to uncertainties in the simulation and prediction of terrestrial ecosystems [9,10]. Model errors include climate forcing errors, uncertainties in the physical processes of models, and errors in the physical parameters of models. Climate change is an important factor that can induce variations in terrestrial ecosystems, especially under the background of global warming [11–13]. Climate change is reflected not only in variations in climatology but also in climate variability. However, in previous studies, linearly increased temperature and precipitation changes were employed to assess the impacts of uncertainties in climate change on uncertainties in simulated terrestrial ecosystems [14].

Recently, many studies have found that climate variability plays a key role in the variation in terrestrial ecosystems [15]. For example, Botta and Foley [16] demonstrated that climate variability resulted in changes in ecosystem structure, soil carbon, and vegetation carbon. Mitchell and Csillag [17] also emphasized that climate variability could influence the stability of grasslands and result in high uncertainty in estimating the net primary production (NPP) of grasslands. Zaghloul et al. [18] investigated the impact of climate change on river flow and showed that early spring warming caused water flow to increase in cold climate regions of Canada due to snowpack melting and gradual glacier melting. Li et al. [19] explored the climatic impact of vegetation spring phenology in China and provided important support for modeling vegetation phenology and growth in northern China. Dastour et al. [20] showed that the seasonal cycles of vegetation and climate were generally coherent but there was a time delay. Their wavelet methods also considered the observational uncertainties. Although the effects of climate variability change on terrestrial ecosystems have been investigated, the extreme effects of uncertainties in climate variability change on uncertainties in simulated terrestrial ecosystems are often neglected [21–23].

Moreover, the uncertainties of physical parameters in numerical models are a major factor contributing to the uncertainties in terrestrial ecosystem simulations and predictions. Reducing the errors of physical parameters in numerical models is an effective way to improve the simulation ability and forecasting skills of terrestrial ecosystems. The simulation ability and forecasting skill of terrestrial ecosystems can be improved by adjusting the model parameters. For example, by assimilating the parameters in the model, Rayner et al. [24] found that the model could match the seasonal cycle and annual variation in CO_2 well with the observation with the Biosphere Energy Transfer Hydrology (BETH) model. Mo et al. [25] optimized the physical parameters of the boreal ecosystem productivity simulator (BEPS) model using the ensemble Kaman filter and found that the simulation abilities of total primary productivity, total ecosystem respiration, and net ecosystem productivity were improved. From these results, it was found that the simulation capability of terrestrial ecosystems could be improved by adjusting the parameters in numerical models.

Numerical models contain a large number of parameters in dynamic vegetation models, which simulate carbon storage and cycling in terrestrial ecosystems. There are three categories for the above parameters in numerical models. The first is related to the discrete format of the model, which is independent of observations; the second is for parameters that can be determined from direct observations; and the third is for parameters that can be determined from indirect observations. For example, the random number seed parameter in the Lund–Potsdam–Jena (LPJ) numerical model [26] belongs to the first type; the co-limitation shape parameter obtained directly from observations belongs to the second type [27]; and the temperature sensitivity parameter to the Q10 obtained from indirect observations belongs to the third type [28]. The latter two types of physical parameters determined by direct and (or) indirect observation (PDOs) are the focus of attention in the above studies.

The numerical model contains a large number of PDOs, and reducing the errors of all PDOs at the same time would be very costly. Identifying which PDO errors should be reduced first is critical, and this question involves identifying the sensitivity and importance of the physical parameters. There has been ample research on how to identify the sensitivities of physical parameters in numerical models. For example, Pitman [29] analyzed the sensitivities of 18 physical parameters in the Biosphere Atmosphere Transfer Scheme (BATS) model using the one-at-a-time (OAT) method. When the sensitivity of one of the parameters was analyzed, the remaining 17 physical parameters remained unchanged.

However, the OAT approach ignores the interaction of physical processes characterized by physical parameters [30,31].

The above sensitivity analysis method was also used to analyze the sensitivity of the parameters. However, this method is based on the assumption of linearity and can be used to explore only small parameter errors and short integration times and is not valid for large parameter errors and long integration times. To consider the interaction of physical processes, some scholars have conducted sensitivity analysis of parameters with finite parameter error samples using the multiobjective generalized sensitivity analysis (MOGSA) method, Monte Carlo method, and extended Fourier amplitude sensitivity test (EFAST) method [32]. Zaehle et al. [28] applied the Monte Carlo hierarchical sample method to identify the sensitivity of model parameters. Bastidas et al. [33] used the MOGSA method to analyze the sensitivity of parameters according to different significance levels. These aforementioned methods were characterized by their low computational cost due to the use of limited samples in the parameter space to identify the sensitivities of physical parameters. However, there were certain limitations; for example, either the interaction among all physical parameters was not considered, or the sensitivity of physical parameters was identified within the parameter space using finite samples.

The responses of terrestrial ecosystems to uncertainties in climate change and physical parameters are a component of predictability studies. Although many studies have been conducted on the uncertainties of terrestrial ecosystem simulations and predictions in terms of uncertainties in climate change and physical parameters, the maximum extent of their uncertainty has rarely been determined. The conditional nonlinear optimal perturbation (CNOP) approach [34,35] is a powerful tool to study predictability. The CNOP approach is related to initial errors (CNOP-I) and model errors (CNOP-P) and has been widely applied to predictability studies in atmospheric and oceanic sciences [36–42].

In this study, the applications of the CNOP-P method to predictability studies of terrestrial ecosystems are introduced. The content includes the maximum extent of uncertainties in climate change on the simulation uncertainties in terrestrial ecosystems using the CNOP-P method. Second, key physical parameters and combinations of physical parameters that lead to uncertainties in terrestrial ecosystem simulations and predictions are identified using the CNOP-P method. Furthermore, the degree of improvement in terrestrial ecosystem simulations and projections is assessed by reducing the errors of sensitive physical parameter combinations identified by the CNOP-P method. These works are reviewed mainly to demonstrate the usefulness and adaptability of nonlinear optimization methods (e.g., the CNOP-P method) in terrestrial ecosystem predictability studies. Furthermore, it provides an outlook for more scholars to use this method to conduct uncertainty studies on numerical simulations and predictions of terrestrial ecosystems using the method.

This paper is organized as follows: studies on the influence of grassland ecosystem equilibrium on moisture index perturbation are introduced in Section 2.1. The impact of uncertainties in climate change on the uncertainties in simulated terrestrial ecosystems is presented in Section 2.2. In Section 2.3, the impact of uncertainties in physical parameters on the terrestrial ecosystem is introduced; in Section 3, the summary and conclusion are provided.

2. Results of Reviews

2.1. The Impact of Moisture Index Perturbation on the Stability of Grassland Ecosystem Equilibrium

To investigate the stability of grassland ecosystem equilibrium to climate perturbation, Sun and Mu [43] used the CNOP-P method and a five-variable grassland ecosystem model. For a grassland equilibrium state (GES) and a desert equilibrium state (DES) within the five-variable grassland ecosystem model, moisture index perturbations were generated using the CNOP-P method, and these perturbations represented the climate perturbation. They first found that the variations in the moisture index resulting from CNOP-P showed nonlinear characteristics. For instance, for the GES, the humidity index of CNOP-P gradually decreased when the amplitude of the moisture indices was small, while when the amplitude of the moisture indices was large, the humidity index of CNOP-P showed a "decreasing–increasing–decreasing" pattern and changed sharply at the end of the period. The variation in the GES also exhibited nonlinear characteristics due to the above humidity index variations.

With the small amplitude of moisture indices, grassland ecosystems returned to the grassland equilibrium state under the influence of the CNOP-P-type humidity index. There were different times required for recovery for different amplitudes of moisture indices. However, grassland ecosystems gradually evolved toward the desert equilibrium state with abrupt changes in the larger amplitude of moisture indices. Numerical results indicated that grassland ecosystems eventually evolved toward a desert state with nonlinear instability when subjected to sufficiently large climate changes. For the DES, Sun and Mu [43] also demonstrated a nonlinear character similar to that of the GES.

To further explore the nonlinear characteristics of the stability of the GES and DES to different types of climatic disturbances, Sun and Mu [43] analyzed the nonlinear evolution of grassland ecosystems under the influence of nonlinear and linear climatic disturbances (Table 1). To interpret the differences between the two, they created two linear climate perturbations that could be distinguished in light of their linear slopes, which were zero or nonzero. For the GES, they found that nonlinear climate change had a severe impact on grassland ecosystems. Grassland ecosystems degraded to a desert equilibrium state and tended to be nonlinearly unstable under the influence of the CNOP-P-type moisture indices. For the DES, they found that nonlinear moisture indices had a severe impact on desert ecosystems. The desert ecosystem influenced by the CNOP-P-type moisture index degenerated into the grassland equilibrium state and became nonlinearly unstable. All of the above work suggests that nonlinear changes in climate variability play an important role in abrupt changes in the equilibrium state of grassland ecosystems.

2.2. The Impact of Uncertainties in Climate Change on the Uncertainties in Simulated Terrestrial Ecosystems

Soil carbon, as a large carbon sink, plays an important role in the carbon cycle in terrestrial ecosystems [14,44]. Changes in soil carbon can cause large changes in atmospheric CO_2, which may further accelerate global warming. It is therefore necessary to determine the uncertainty in modeled soil carbon. Sun and Mu [45] used the CNOP-P method to analyze the maximum degree of uncertainty in the contribution of soil carbon to climate change uncertainty (both climatological change and climate variability) in China (Table 1).

Table 1. Summary of the studies of terrestrial ecosystem predictability using the CNOP-P method.

Sources of Uncertainty	Descriptions/Limitations	Reference
Moisture index	Stability analysis of grassland ecosystem equilibrium was shown due to moisture index perturbation using CNOP-P method. A theoretical model was employed.	Sun and Mu [43]
Climate condition	Uncertainties in simulated soil carbon due to temperature and precipitation perturbations were estimated using the CNOP-P method.	Sun and Mu [45]
Physical parameters	A new parameter sensitivity analysis method based on CNOP-P was proposed. The new method was applied to identify the most sensitive physical parameters set to uncertainties in simulated NPP in China. The improvement extent by reducing the errors of sensitive physical parameters set determined by the new method was evaluated.	Sun and Mu [46]
Physical parameters	The new parameter sensitivity analysis method based on CNOP-P was applied to identify the most sensitive physical parameters set to uncertainties in simulated soil carbon in China.	Sun and Mu [47]
Physical parameters	The new parameter sensitivity analysis method based on CNOP-P was applied to identify the most sensitive physical parameters set to uncertainties in simulated ET over the TP. The improvement extent by reducing the errors of sensitive physical parameters set determined by the new method was evaluated.	Sun et al. [48]

Under the background of global warming, they provided a nonlinear climate change, i.e., CNOP-P-type climate change, and a linear climate change. The key difference between the CNOP-P-type climate change and the linear climate change was whether there was a change in temperature or precipitation variability compared to a reference temperature or precipitation variability. Sun and Mu [45] showed that there were different regional responses to uncertainties in simulated soil carbon caused by CNOP-P-type and linear temperature changes.

By exploring three components of soil carbon in the LPJ model, namely, rapidly decomposing soil carbon, slowly decomposing soil carbon, and subsurface apoplastic material, they found that the decrease in subsurface apoplastic matter was probably the main reason for the decrease in soil carbon in arid and semiarid zones as a result of the two temperature climate changes. The different effects of the two temperature climate changes in southern China may be caused mainly by the rapid decomposition of soil carbon. The uncertainties in simulated soil carbon caused by the two precipitation climate changes were similar. In the arid and semiarid zones, both precipitation and climate changes led to increased uncertainty in the simulated soil carbon. This research implied that the variation in temperature variability played a crucial role in the variations in soil carbon and its components in the study region.

2.3. *The Impact of Uncertainties in Physical Parameters on the Terrestrial Ecosystem*

2.3.1. The Sensitivity Analysis Method Based on CNOP-P

The numerical model contains a large number of physical parameters. Finding the key physical processes and physical parameters in the numerical model is an important way to improve simulation capabilities and prediction skills. To find the most sensitive physical parameters, Sun and Mu [46] proposed a sensitivity analysis (SA) method based on CNOP-P (Figure 1, Table 1). For the SA method based on CNOP-P, there were two steps. First, some insensitive physical parameters were eliminated using the CNOP-P method. Next, among the remaining physical parameters, the combination of relatively sensitive and important physical parameters was judged using the idea of combination and the CNOP-P method. In the second step, the sensitivity of a single parameter was identified using the CNOP-P approach, which in theory was the optimal way to ensure the ranking of every parameter in terms of its sensitivity. Obviously, this method fully considered the nonlinear synergistic effects between physical parameters. Moreover, this method identified relatively sensitive and important combinations of physical parameters in the whole physical parameter space.

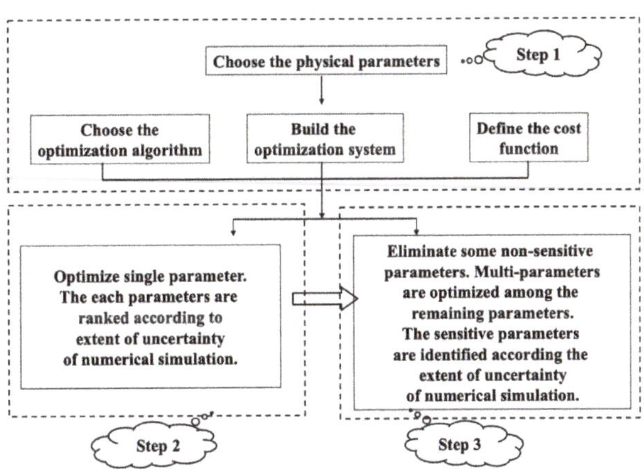

Figure 1. Flowchart depicting the steps involved in the SA method based on CNOP-P (From research findings by Sun and Mu [46]).

2.3.2. Identification of Sensitive Physical Parameters

Model errors are a critical source affecting the uncertainty in simulated terrestrial ecosystems. It is important to determine which parameter errors should be reduced to improve the simulation ability of terrestrial ecosystems. Sun and Mu [47] used the SA method based on CNOP-P to identify the most sensitive physical parameters to soil carbon. To compare the sensitivity of the parameter combination, the one-at-a-time (OAT) approach was also applied to judge the sensitivity of each parameter.

Sun and Mu [47] noted that the most sensitive parameters to soil carbon varied between plant functional types (Figure 2, Table 1, and the physical meanings of the parameters can be found in Table S1). For example, for C3 perennial grasses under semiarid conditions, the uncertainty in hydrological processes was also critical for modeling soil carbon.C3 perennial grasses are cool season grasses and are great at fixing CO_2 at cooler temperatures. However, at higher temperatures, e.g., above 90 degrees F, they are not as efficient. The most sensitive parameter combinations using the SA method based on CNOP-P differed from the highest rank of sensitivity for each parameter using the OAT method. This difference suggested that the nonlinear effects of parameter combinations were key to determining sensitive parameter combinations (Figure 3, and the physical meanings of the parameters can be found in Table S1).

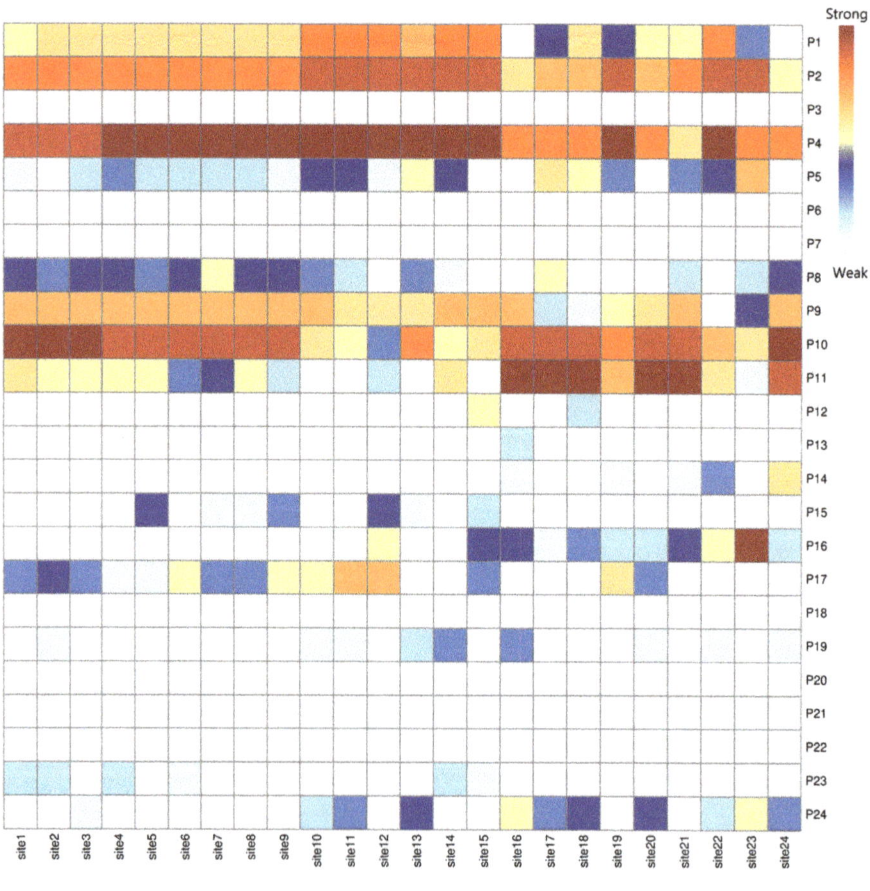

Figure 2. The sensitivity of each parameter for the simulated soil carbon using the CNOP-P method (From research findings by Sun and Mu [47]. Parameter corresponding to the number can be found in studies by Sun and Mu [47] and Table S1 in Supplementary Materials).

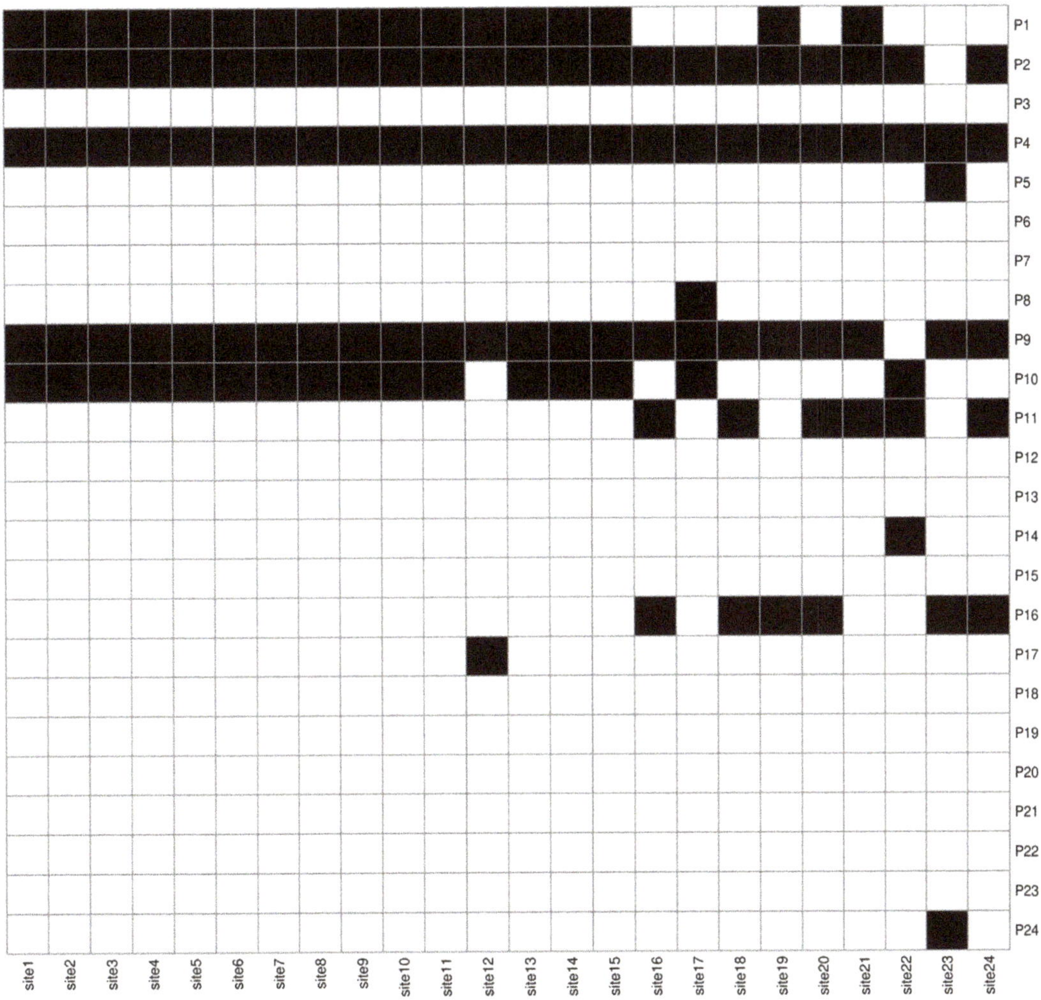

Figure 3. The sensitive parameter combination for the simulated soil carbon using the SA method based on CNOP-P (From research findings by Sun and Mu [47]. Parameter corresponding to the number can be found in studies by Sun and Mu [47] and Table S1 in Supplementary Materials).

Numerical simulations and predictions of carbon fluxes (net primary production, NPP) on the Qinghai–Tibet Plateau (TP) are still subject to large uncertainties. To reduce the uncertainty in numerical simulations and improve the predictive power of simulated NPP, Sun et al. [48] identified the key physical processes associated with uncertainty at nine stations on the TP using the SA method based on CNOP-P. In the mid-precipitation region of the Tibetan Plateau, the parameters related to photosynthesis were the main factors contributing to the large uncertainty in the NPP simulations; in regions with low and high precipitation on the Tibetan Plateau, the combined effects of the parameters related to hydrological processes and photosynthesis played an important role (Figures 4 and 5, and the physical meanings of the parameters can be found in Table S2). All the above results showed that the SA based on the CNOP-P method could reasonably identify relatively sensitive and important combinations of parameters and was more physically meaningful.

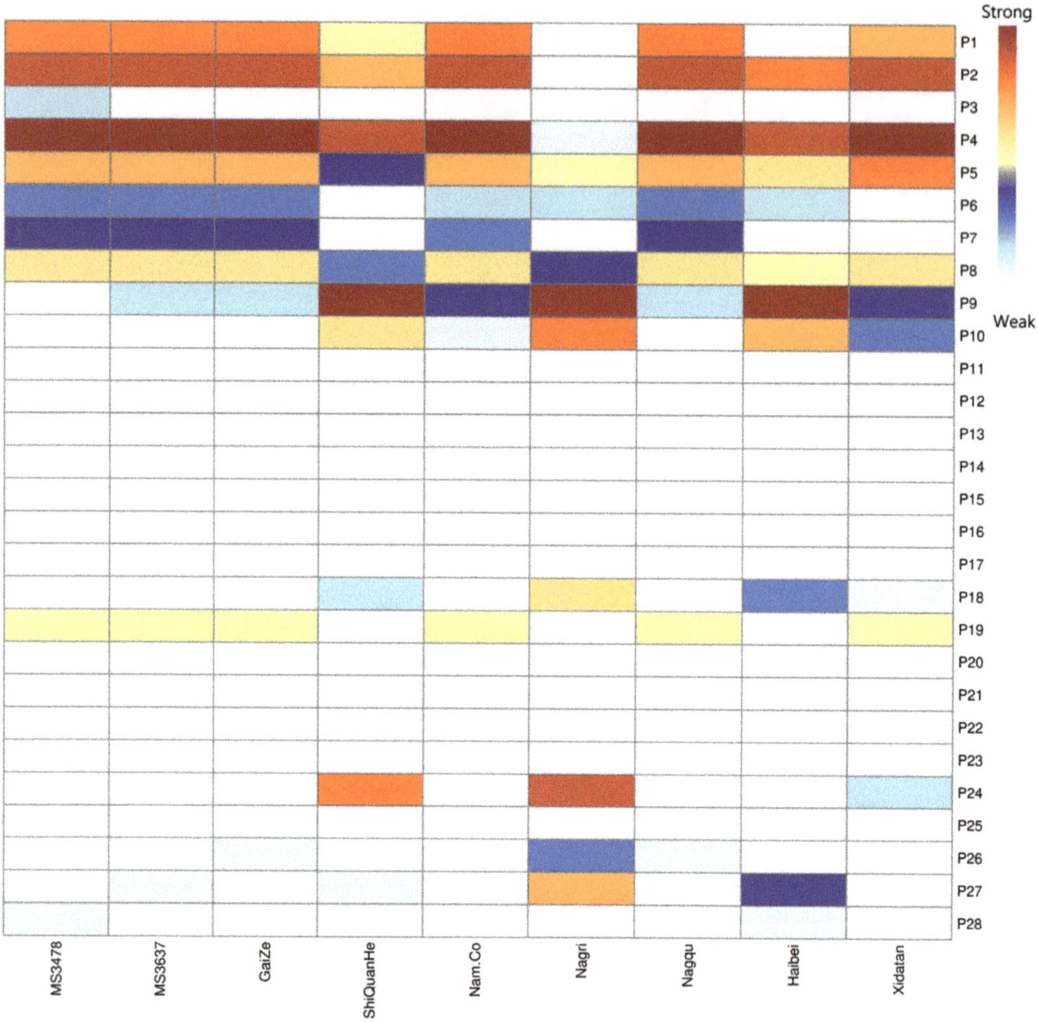

Figure 4. The sensitivity of each parameter using CNOP-P method over the TP (From research findings by Sun et al. [48]. Parameter corresponding to the number can be found in studies by Sun et al. [48] and Table S2 in Supplementary Materials).

2.3.3. Evaluation of Simulation Ability and Prediction Skill by Reducing the Errors of Sensitive Physical Parameters

An important objective of finding the sensitive parameter subset is to improve the simulation ability and prediction skill of terrestrial ecosystems. Sun et al. [48] designed an ideal numerical experiment to reduce the uncertainty in the simulation of NPP over the TP (Table 1). To explore the benefits of modeling NPP while reducing the parameter errors associated with the most sensitive parameter subset, an experiment was implemented as follows:

$$\tau = \frac{\|M_T(\boldsymbol{U}_0, \boldsymbol{P}+\boldsymbol{p}) - M_T(\boldsymbol{U}_0, \boldsymbol{P})\| - \|M_T(\boldsymbol{U}_0, \boldsymbol{P}+(1-\alpha)\boldsymbol{p}) - M_T(\boldsymbol{U}_0, \boldsymbol{P})\|}{\|M_T(\boldsymbol{U}_0, \boldsymbol{P}+\boldsymbol{p}) - M_T(\boldsymbol{U}_0, \boldsymbol{P})\|} \times 100\% \quad (1)$$

where τ represents the benefit of modeling NPP based on reducing the parameter errors of the sensitive parameter subset. A larger τ value indicates a better improvement in the NPP simulation. $\boldsymbol{P}$ is the reference state of the sensitive parameter subsets. $\boldsymbol{p}$ is the CNOP-P, which is related to the errors of five sensitive parameter subsets. α (=0.2, 0.4, 0.6, and 0.8) represents the extent of the error reduction for the correct parameters due to data assimilation or observation.

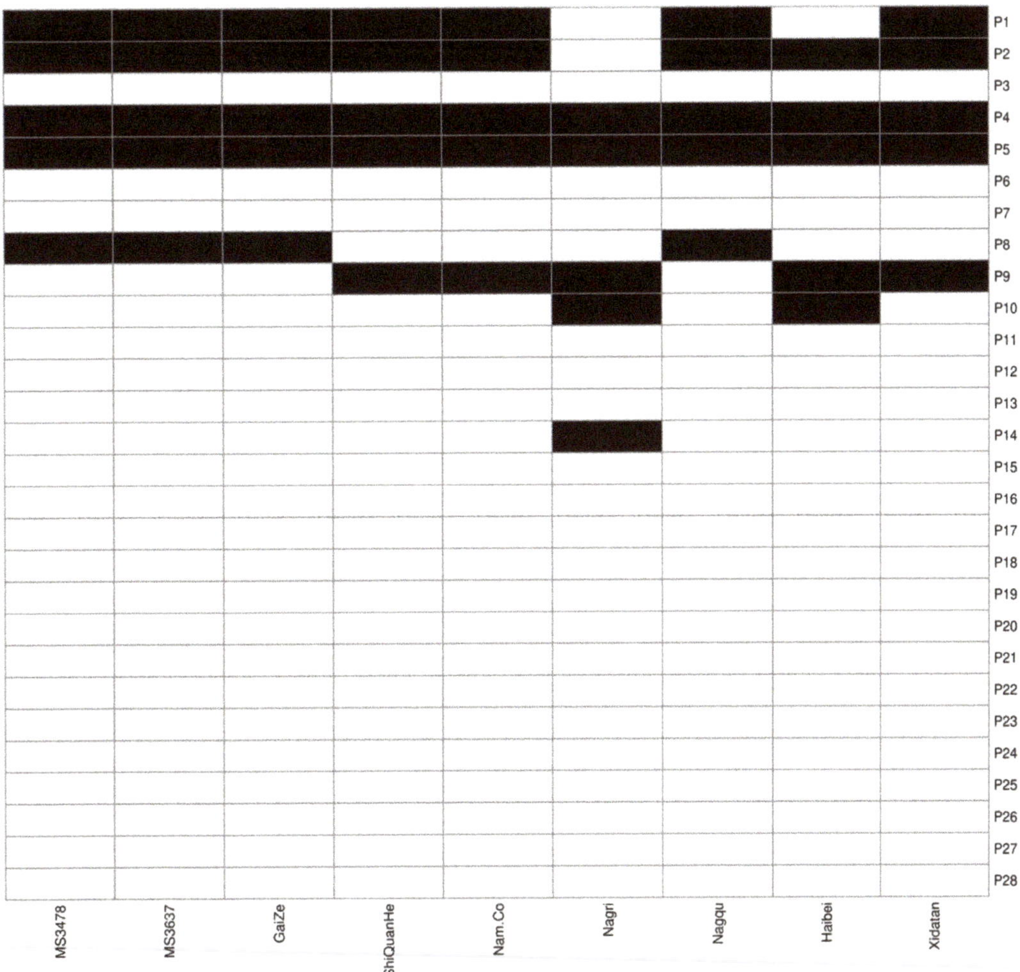

Figure 5. The sensitive parameter combination using the SA method based on CNOP-P over the TP (From research findings by Sun et al. [48]. Parameter corresponding to the number can be found in studies by Sun et al. [48] and Table S2 in Supplementary Materials).

Sun et al. [48] demonstrated that eliminating the errors associated with the most sensitive and important parameter subset with the SA method based on CNOP-P led to the maximum benefit in terms of reducing the uncertainty of simulated NPP when compared to that obtained using the traditional method. For all cases over the TP in the studies of Sun et al. [48], the numerical results showed that the simulation abilities of NPP were improved by reducing the uncertainties in sensitive physical parameters identified by the CNOP-P method compared to the OAT method. In addition, for some cases, the extent of improvement in the simulated NPP by reducing the uncertainties in sensitive physical parameters identified by the CNOP-P method was distinctly better than that by the OAT method [48]. For example, for the Ngari site, the extent of the improvement in the simulated NPP was 34.3% using the CNOP-P method and 28.6% using the OAT method. This study suggested that we should prioritize reducing the uncertainty of relatively sensitive parameter combinations among all physical parameters to improve the prediction or simulation ability of NPP over the TP. Sun et al. [48] also emphasized the importance of nonlinear interactions among sensitive parameter sets for uncertainties in the simulation ability and prediction skill of terrestrial ecosystems.

3. Discussion

Although the CNOP-P method has been studied in terms of uncertainties in terrestrial ecosystem modeling and prediction, more research should be conducted. It is not enough for studies to consider only the effects of a 2 °C temperature increase on terrestrial ecosystem variations. Climate change with multimodel prediction results should be considered. Additionally, ideal numerical experiments are implemented when studying sensitive combinations of physical parameters. In the future, studies of sensitive physical parameter combinations can be conducted with observational data. Finally, the study of the CNOP-P method in terrestrial ecosystem predictability is not limited to the above two aspects.

On the one hand, ensemble forecasting is one of the methods that can be used to improve the simulation and prediction of terrestrial ecosystems, and research on the CNOP-P method is worth exploring land carbon cycle ensemble predictions (LEPS). On the other hand, the impacts of extreme events (e.g., droughts, high temperatures, and fires) on terrestrial ecosystems have received increasing attention from scholars. Studies of terrestrial ecosystem responses to climate change imply that this approach can be used to carry out research on the effects of extreme events on terrestrial ecosystems.As the underlying surface of the Earth system, terrestrial ecosystems affect local and global climate change through land–atmosphere interactions. The impact of terrestrial ecosystems on regional and global climate change will be discussed in the future using the CNOP-P method, especially for studies of extreme events. The results reviewed in this article may not be sufficient to conclude significant findings that are part of uncertainties in simulated and predicted terrestrial ecosystems over multiple years. In this study, uncertainties in simulated and predicted terrestrial ecosystems were shown using the nonlinear optimization method (CNOP-P method). These results encourage us to further research the uncertainty and predictability of terrestrial ecosystems.

4. Conclusions

In this paper, the applications of CNOP methods in terrestrial ecosystem predictability studies are reviewed. The paper contained two main parts. First, using the CNOP method, climate changes were given where both climate state changes and climate variability changes were considered. The numerical results showed that the nonlinear changes in climate variability were considered to show more significant changes in terrestrial ecosystems. This result shows the important role of nonlinear variations in climate variability in terrestrial ecosystem changes.

Additionally, to overcome the limitations of traditional methods in studying the identification of key physical parameters for terrestrial ecosystem simulation and prediction uncertainty, a CNOP-P-based SA method for identifying combinations of sensitive physical parameters was proposed. This method can consider both the nonlinear interactions among physical parameters and the sensitivity of the parameters in the whole physical parameter error space. The sensitive physical parameter combinations identified by the CNOP-P-based SA method for identifying sensitive physical parameter combinations were more sensitive than those identified by the traditional methods. Furthermore, reducing the errors of sensitive physical parameters identified by the CNOP-P-based SA method resulted in a higher degree of improvement in terrestrial ecosystem simulation and prediction. All of these applications imply that the CNOP method is an important theoretical tool that can be used to study the uncertainties in terrestrial ecosystem simulations and predictions.

Supplementary Materials: The following supporting information can be downloaded at: https://www.mdpi.com/article/10.3390/atmos14040617/s1, Table S1: The chosen physical parameters in studies of Sun and Mu [46,47]; Table S2: The chosen physical parameters in studies of Sun et al. [48].

Author Contributions: Conceptualization, M.M.; methodology, M.M.; software, G.S.; validation, M.M.; formal analysis, G.S.; investigation, M.M. and G.S.; resources, G.S.; data curation, G.S.; writing—original draft preparation, G.S.; writing—review and editing, M.M.; visualization, G.S.; supervision, M.M.; project administration, M.M.; funding acquisition, M.M. and G.S. All authors have read and agreed to the published version of the manuscript.

Funding: Funding was provided by the National Natural Science Foundation of China (Nos. 42288101), by the Guangdong Major Project of Basic and Applied Basic Research (No. 2020B0301030004), and by the National Natural Science Foundation of China (Nos.41975132, 42175077).

Institutional Review Board Statement: Not applicable.

Informed Consent Statement: Not applicable.

Data Availability Statement: Not applicable.

Acknowledgments: The authors would like to acknowledge the helpful and constructive comments that improved the manuscript substantially from Editor and three anonymous reviewers.

Conflicts of Interest: The authors declare no conflict of interest.

References

1. Yuan, Q.; Wu, S.; Zhao, D.; Dai, E.; Chen, L.; Zhang, L. Modeling net primary productivity of the terrestrial ecosystem in China from 1961 to 2005. *J. Geogr. Sci.* **2014**, *24*, 3–17. [CrossRef]
2. Chen, Z.; Chen, J.M.; Zheng, X.; Jiang, F.; Qin, J.; Zhang, S.; Yuan, W.; Ju, W.; Mo, G. Optimizing photosynthetic and respiratory parameters based on the seasonal variation pattern in regional net ecosystem productivity obtained from atmospheric inversion. *Sci. Bull.* **2015**, *60*, 1943–1950. [CrossRef]
3. Friedlingstein, P.; Jones, M.W.; O'Sullivan, M.; Andrew, R.M.; Hauck, J.; Peters, G.P.; Peters, W.; Pongratz, J.; Sitch, S.; Le Quéré, C.; et al. Global Carbon Budget 2019. *Earth Syst. Sci. Data* **2019**, *11*, 1783–1838. [CrossRef]
4. Tian, H.; Melillo, J.; Lu, C.; Kicklighter, D.; Liu, M.; Ren, W.; Xu, X.; Chen, G.; Zhang, C.; Pan, S.; et al. China's terrestrial carbon balance: Contributions from multiple global change factors. *Glob. Biogeochem. Cycles* **2011**, *25*, GB1007. [CrossRef]
5. Luo, Y.; Keenan, T.F.; Smith, M. Predictability of the terrestrial carbon cycle. *Glob. Change Biol.* **2015**, *21*, 1737–1751. [CrossRef]
6. Roberts, A.M.I.; Tansey, C.; Smithers, R.J.; Phillimore, A.B. Predicting a change in the order of spring phenology in temperate forests. *Glob. Change Biol.* **2015**, *21*, 2603–2611. [CrossRef]
7. Tagesson, T.; Schurgers, G.; Horion, S.; Ciais, P.; Tian, F.; Brandt, M.; Ahlström, A.; Wigneron, J.-P.; Ardö, J.; Olin, S.; et al. Recent divergence in the contributions of tropical and boreal forests to the terrestrial carbon sink. *Nat. Ecol. Evol.* **2020**, *4*, 202–209. [CrossRef] [PubMed]
8. Liddicoat, S.K.; Wiltshire, A.J.; Jones, C.D.; Arora, V.K.; Brovkin, V.; Cadule, P.; Hajima, T.; Lawrence, D.M.; Pongratz, J.; Schwinger, J.; et al. Compatible fossil fuel CO_2 emissions in the CMIP6 earth system models' historical and shared socioeconomic pathway experiments of the twenty-first century. *J. Clim.* **2021**, *34*, 2853–2875. [CrossRef]
9. Lawrence, D.M.; Fisher, R.A.; Koven, C.D.; Oleson, K.W.; Swenson, S.C.; Bonan, G.; Collier, N.; Ghimire, B.; van Kampenhout, L.; Kennedy, D.; et al. The community land model version 5: Description of new features, benchmarking, and impact of forcing uncertainty. *J. Adv. Model. Earth Syst.* **2019**, *11*, 4245–4287. [CrossRef]
10. Lombardozzi, D.L.; Lu, Y.; Lawrence, P.J.; Lawrence, D.M.; Swenson, S.; Oleson, K.W.; Wieder, W.R.; Ainsworth, E.A. Simulating agriculture in the community land model version 5. *J. Geophys. Res. Biogeosci.* **2020**, *125*, e2019JG005529. [CrossRef]
11. Xia, J.; Chen, J.; Piao, S.; Ciais, P.; Luo, Y.; Wan, S. Terrestrial carbon cycle affected by non-uniform climate warming. *Nat. Geosci.* **2014**, *7*, 173–180. [CrossRef]
12. Shen, X.; Jiang, M.; Lu, X.; Liu, X.; Liu, B.; Zhang, J.; Wang, X.; Tong, S.; Lei, G.; Wang, S.; et al. Aboveground biomass and its spatial distribution pattern of herbaceous marsh vegetation in China. *Sci. China Earth Sci.* **2021**, *64*, 1115–1125. [CrossRef]
13. Shen, X.; Liu, B.; Henderson, M.; Wang, L.; Jiang, M.; Lu, X. Vegetation Greening, Extended Growing Seasons, and Temperature Feedbacks in Warming Temperate Grasslands of China. *J. Clim.* **2022**, *35*, 5103–5117. [CrossRef]
14. Post, J.; Hattermann, F.F.; Krysanova, V.; Suckow, F. Parameter and input data uncertainty estimation for the assessment of long-term soil organic carbon dynamics. *Environ. Model. Softw.* **2008**, *23*, 125–138. [CrossRef]
15. Cao, M.K.; Prince, S.D.; Li, K.R.; Tao, B.; Small, J.; Shao, X.M. Response of terrestrial carbon uptake to climate interannual variability in China. *Glob. Chang. Biol.* **2003**, *9*, 536–546. [CrossRef]

16. Botta, A.; Foley, J.A. Effects of climate variability and disturbances on the Amazonian terrestrial ecosystem dynamics. *Glob. Biogeochem. Cycles* **2002**, *16*, 1070. [CrossRef]
17. Mitchell, S.W.; Csillag, F. Assessing the stability and uncertainty of predicted vegetation growth under climatic variability: Northern mixed grass prairie. *Ecol. Model.* **2001**, *139*, 101–121. [CrossRef]
18. Zaghloul, M.S.; Ghaderpour, E.; Dastour, H.; Farjad, B.; Gupta, A.; Eum, H.; Achari, G.; Hassan, Q.K. Long Term Trend Analysis of River Flow and Climate in Northern Canada. *Hydrology* **2022**, *9*, 197. [CrossRef]
19. Li, Z.; Wu, Y.; Wang, R.; Liu, B.; Qian, Z.; Li, C. Assessment of Climatic Impact on Vegetation Spring Phenology in Northern China. *Atmosphere* **2023**, *14*, 117. [CrossRef]
20. Dastour, H.; Ghaderpour, E.; Zaghloul, M.S.; Farjad, B.; Gupta, A.; Eum, H.; Achari, G.; Hassan, Q.K. Wavelet-based spatiotemporal analyses of climate and vegetation for the Athabasca river basin in Canada. *Int. J. Appl. Earth Obs. Geoinf.* **2022**, *114*, 103044. [CrossRef]
21. Wang, S.; Tian, H.; Liu, J.; Pan, S. Pattern and change of soil organic carbon storage in China: 1960s–1980s. *Tellus B* **2003**, *55*, 416–427. [CrossRef]
22. Eglin, T.; Ciais, P.; Piao, S.L.; Barre, P.; Bellassen, V.; Cadule, P.; Chenu, C.; Gasser, T.; Koven, C.; Reichstein, M.; et al. Historical and future perspectives of global soil carbon response to climate and land-use changes. *Tellus B* **2010**, *62*, 700–718. [CrossRef]
23. Pan, Z.; Andrade, D.; Segal, M.; Wimberley, J.; McKinney, N.; Takle, E.S. Uncertainty in future soil carbon trends at a central U.S. site under an ensemble of GCM scenario climates. *Ecol. Model.* **2010**, *221*, 876–881. [CrossRef]
24. Rayner, P.J.; Scholze, M.; Knorr, W.; Kaminski, T.; Giering, R.; Widmann, H. Two decades of terrestrial carbon fluxes from a carbon cycle data assimilation system (CCDAS). *Glob. Biogeochem. Cycles* **2005**, *19*, GB2026. [CrossRef]
25. Mo, X.; Chen, J.M.; Ju, W.; Black, T.A. Optimization of ecosystem model parameters through assimilating eddy covariance flux data with an ensemble Kalman filter. *Ecol. Model.* **2008**, *217*, 157–173. [CrossRef]
26. Sitch, S.; Smith, B.; Prentice, I.C.; Arneth, A.; Bondeau, A.; Cramer, W.; Kaplan, J.O.; Levis, S.; Lucht, W.; Sykes, M.T.; et al. Evaluation of ecosystem dynamics, plant geography and terrestrial carbon cycling in the LPJ dynamic vegetation model. *Glob. Chang. Biol.* **2003**, *9*, 161–185. [CrossRef]
27. Collatz, G.J.; Berry, J.A.; Farquhar, G.D.; Pierce, J. The relationship between the Rubisco reaction mechanism and models of photosynthesis. *Plant Cell Environ.* **1990**, *13*, 219–225. [CrossRef]
28. Zaehle, S.; Sitch, S.; Smith, B.; Hatterman, F. Effects of parameter uncertainties on the modeling of terrestrial biosphere dynamics. *Glob. Biogeochem. Cycles* **2005**, *19*, GB3020. [CrossRef]
29. Pitman, A.J. Assesing the sensitivity of a land-surface scheme to the parameter values using a single column model. *J. Clim.* **1994**, *7*, 1856–1869. [CrossRef]
30. Jackson, C.; Xia, Y.; Sen, M.K.; Stoffa, P.L. Optimal parameter and uncertainty estimation of a land surface model: A case study using data from Cabauw, Netherlands. *J. Geophys. Res.* **2003**, *108*, 4583. [CrossRef]
31. Bastidas, L.A.; Hogue, T.S.; Sorooshian, S.; Gupta, H.V.; Shuttleworth, W.J. Parameter sensitivity analysis for different complexity land surface models using multicriteria methods. *J. Geophys. Res.* **2006**, *111*, D20101. [CrossRef]
32. Henderson-Sellers, A. Assessing the Sensitivity of a Land-Surface Scheme to Parameters Used In Tropical-Deforestation Experiments. *Q.J.R. Meteorol. Soc.* **1992**, *118*, 1101–1116. [CrossRef]
33. Bastidas, L.A.; Gupta, H.V.; Sorooshian, S.; Shuttleworth, W.J.; Yang, Z.L. Sensitivity analysis of a land surface scheme using multicriteria methods. *J. Geophys. Res.* **1999**, *104*, 19481–19490. [CrossRef]
34. Mu, M.; Duan, W.S.; Wang, B. Conditional nonlinear optimal perturbation and its applications. *Nonlinear. Proc. Geoph.* **2003**, *10*, 493–501. [CrossRef]
35. Mu, M.; Duan, W.S.; Wang, Q.; Zhang, R. An extension of conditional nonlinear optimal perturbation approach and its applications. *Nonlinear Proc. Geoph* **2010**, *17*, 211–220. [CrossRef]
36. Mu, M.; Xu, H.; Duan, W.S. A kind of initial errors related to spring predictability barrier for El Niño events in Zebiak-Cane model. *Geophys. Res. Lett.* **2007**, *34*, L03709. [CrossRef]
37. Mu, M.; Jiang, Z.N. A method to find perturbations that trigger blocking onset: Conditional nonlinear optimal perturbations. *J. Atmos. Sci.* **2008**, *65*, 3935–3946. [CrossRef]
38. Mu, M.; Zhou, F.F.; Wang, H.L. A method for identifying the sensitive areas in targeted observations for tropical cyclone prediction: Conditional nonlinear optimal perturbation. *Mon. Weather Rev.* **2009**, *137*, 1623–1639. [CrossRef]
39. Mu, M.; Duan, W.S.; Tang, Y.M. The predictability of atmospheric and oceanic motions: Retrospect and prospects. *Sci. China Earth Sci.* **2017**, *60*, 2001–2012. [CrossRef]
40. Wang, Q.; Mu, M.; Sun, G.D. A useful approach to sensitivity and predictability studies in geophysical fluid dynamics: Conditional non-linear optimal perturbation. *Natl. Sci. Rev.* **2019**, *7*, 214–223. [CrossRef]
41. Mu, M. Methods, current status, and prospect of targeted observation. *Sci. China Earth Sci.* **2013**, *56*, 1997–2005. [CrossRef]
42. Mu, M.; Wang, Q. Applications of nonlinear optimization approach to atmospheric and oceanic sciences. *Sci. Sin. Math.* **2017**, *47*, 1207–1222. (In Chinese) [CrossRef]
43. Sun, G.D.; Mu, M. Response of a grassland ecosystem to climate change in a theoretical model. *Adv. Atmos. Sci.* **2011**, *28*, 1266–1278. [CrossRef]
44. Lal, R. Soil carbon sequestration impacts on global climate change and food security. *Science* **2004**, *304*, 1623–1627. [CrossRef] [PubMed]

45. Sun, G.D.; Mu, M. Responses of soil carbon variation to climate variability in China using the LPJ model. *Theor. Appl. Climatol.* **2012**, *110*, 143–153. [CrossRef]
46. Sun, G.D.; Mu, M. A new approach to identify the sensitivity and importance of physical parameters combination within numerical models, using the Lund–Potsdam–Jena (LPJ) model as an example. *Theor. Appl. Climatol.* **2017**, *128*, 587–601. [CrossRef]
47. Sun, G.D.; Mu, M. A flexible method to determine the sensitive physical parameter combination for soil carbon under five plant types. *Ecosphere* **2017**, *8*, e01920. [CrossRef]
48. Sun, G.; Mu, M.; You, Q. Identification of key physical processes and improvements for simulating and predicting net primary production over the Tibetan Plateau. *J. Geophys. Res. Atmos.* **2020**, *125*, e2020JD033128. [CrossRef]

atmosphere

Review

Instability of Atlantic Meridional Overturning Circulation: Observations, Modelling and Relevance to Present and Future

Zhengyu Liu [1,2]

[1] Department of Geography, The Ohio State University, Columbus, OH 43210, USA; liu.7022@osu.edu
[2] School of Geography Science, Nanjing Normal University, Nanjing 210024, China

Abstract: The Atlantic Meridional Overturning Circulation (AMOC) has changed dramatically during the glacial–interglacial cycle. One leading hypothesis for these abrupt changes is thermohaline instability. Here, I review recent progress towards understanding thermohaline instability in both observations and modelling. Proxy records available seem to favor thermohaline instability as the cause of the abrupt climate changes during the glacial–deglacial period because the deep North Atlantic water mass and AMOC seemed to have changed before the North Atlantic climate. However, most fully Coupled General Circulation Models (CGCMs) so far seem to exhibit monostable AMOC, because (1) these models have failed to simulate abrupt AMOC changes unless they are forced by an abrupt change of external forcing and, (2) these models have shown opposite freshwater convergence from the current observations. This potential model bias in the AMOC stability leaves the model projection of the future AMOC change uncertain.

Keywords: thermohaline instability; model bias; abrupt changes

Citation: Liu, Z. Instability of Atlantic Meridional Overturning Circulation: Observations, Modelling and Relevance to Present and Future. *Atmosphere* **2023**, *14*, 1011. https://doi.org/10.3390/atmos14061011

Academic Editors: Xiaolei Zou, Guoxiong Wu and Zhemin Tan

Received: 18 May 2023
Revised: 7 June 2023
Accepted: 7 June 2023
Published: 12 June 2023

1. Introduction

Paleoclimate proxies suggest that the Atlantic Meridional Overturning Circulation (AMOC) and global climate have experienced abrupt changes of millennial time scales [1,2]. Figure 1 shows multiple proxies relevant to the AMOC, its forcing and its climate impact, for the last 80,000 years [3]. The meltwater flux to the North Atlantic is characterized by strong millennial changes superimposed on the last glacial–deglacial cycle, as seen in the oxygen isotope ratio $\delta^{18}O$ ($\sim^{18}O/^{16}O$) over the North Atlantic surface (Figure 1c). A more depleted (lower) surface water $\delta^{18}O$ indicates a greater meltwater flux, due to the contribution of more meltwater from ice sheets that have all been very depleted $\delta^{18}O$ relative to the average ocean water (−30‰ vs. 0‰). Similarly, large variability is also evident in two proxies of deep water masses ($\delta^{13}C$, ε_{Nd} in Figure 1d,e) and three proxies of circulation ($^{231}Pa/^{230}Th, \delta^{18}O$, grain size, Figure 1f–h). The carbon isotope composition $^{13}C/^{12}C$, expressed as $\delta^{13}C$, and the neodymium isotopic composition $^{143}Nd/^{144}Nd$, expressed as ε_{Nd}, from the shells of benthic foraminifera are two proxies for deep water masses, because their value in the deep water is determined by the competition between the high $\delta^{13}C$/low ε_{Nd} North Atlantic Deep Water (NADW) source water and the low $\delta^{13}C$/high ε_{Nd} of the Antarctic Bottom Water (AABW) source water [4–6]. The different behavior of uranium decay-series nuclides of protactinium and thorium, expressed as $^{231}Pa/^{230}Th$, from sediments in the North Atlantic is considered a proxy of the rate of deep Atlantic circulation, because ^{231}Pa has a longer residence time than ^{230}Th (111 years vs. 26 years) such that the sediment $^{231}Pa/^{230}Th$ decreases with increased export of Atlantic deep water [7]. The variability of benthic calcite $\delta^{18}O_c$ is related to sea water temperature and salinity, and is therefore related to density anomaly and, in turn, ocean current strength through the thermal wind relation [8]. The grain size in the deep western boundary current region (DWBC) is a proxy of DWBC strength because a coarser grain size is likely to be caused by a stronger current. Finally, large variability is also found accompanying deep ocean changes in the proxy for surface air temperature in Greenland and Antarctica ice cores as indicated in the stable water isotope composition in precipitation $\delta^{18}O$, which represents annual temperature via the "temperature effect" associated with Rayleigh distillation [9] (Figure 1a,b).

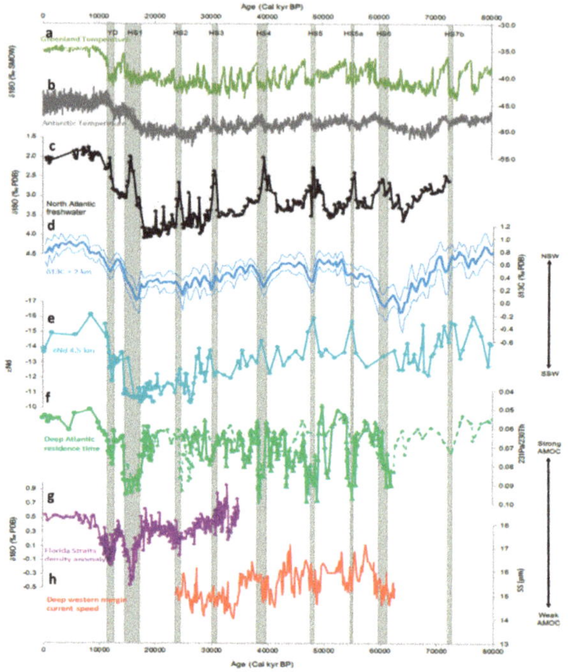

Figure 1. North Atlantic water-mass and circulation proxy records over the Heinrich events, along with temperature proxies over Greenland and Antarctica. (**a**) Greenland ice core $\delta^{18}O$. (**b**) Antarctica $\delta^{18}O$. (**c**) The oxygen isotope ratio in the planktonic foraminifer Neogloboquadrina pachyderma from the western North Atlantic; low values reflect the presence of glacial meltwater. (**d**) Mean and 2-sigma standard-error $\delta^{13}C$ values (thick and thin lines, respectively) for the deep North Atlantic (>2 km). (**e**) The deep (4.55 km) Nd isotope ratio (εNd) at the Bermuda Rise. (**f**) The Pa^{231}/Th^{230} ratio, which can reflect changes in deepwater residence time, from the deep North Atlantic at the Bermuda Rise. (**g**) The ice-volume-corrected oxygen isotope ratio of benthic foraminifera on the Florida Margin, which can reflect changes in the density structure in the Florida Straits and the strength of the upper branch of the Atlantic meridional overturning circulation (AMOC). (**h**) The mean sortable silt grain size along the western boundary of the North Atlantic at the depth of today's North Atlantic Deep Water, which can indicate the current speed on the deep western margin; a large mean size indicates a vigorous flow. See [3] Figure 4 for more details on the figure. (Courtesy of J. Lynch-Stieglitz).

After the discovery of abrupt changes in events of Dansgaard–Oeschger (DO) oscillation in Greenland ice cores and their links to North Atlantic water masses in the 1980s, AMOC instability associated with basin-wide positive salinity feedback [10,11] has been proposed as one leading mechanism for the millennial climate variability [12–14]. Greenland stadial and NADW reduction have been further found to be associated with massive ice rafting of the Heinrich events [15,16]. The direct link of abrupt climate changes with the AMOC intensity was further found in the proxies sensitive to the AMOC intensity, notably sediment $^{231}Pa/^{230}Th$ [7,17,18], $\delta^{18}O_c$ gradient for Florida Current [8] and sortable grain size for DWBC current speed [19]. Furthermore, the greatest AMOC reduction and the coldest stadial intervals were found to be concurrent with the largest iceberg discharges during Heinrich events [3,18,20,21] and the AMOC reductions associated with the Heinrich stadial also coincide with intervals of rising atmospheric CO_2 [22], suggesting a potential influence of deep ocean circulation on atmospheric CO_2 [23]. Within the chronology uncertainty, these AMOC changes were found to be concurrent with climate variability in most records. In the real world, abrupt climate change could have been caused by ice-sheet instability as well as thermohaline instability. If the thermohaline instability is the cause of abrupt changes, AMOC changes likely occurred before climate changes. On the other hand, if the ice-sheet instability is the cause, the climate likely has changed with ice sheet before the AMOC. This observed concurrence of AMOC variability and climate variability, therefore, has left the origin of the abrupt climate change still wide open: is it of the ocean origin from AMOC instability, or the ice-sheet origin from ice-sheet instability?

Observations of abrupt climate changes have stimulated intensified modelling studies on AMOC instability first in ocean-alone models with mixed boundary conditions [24–27] and then a Coupled General Circulation Model (CGCM) [28] and, most comprehensively, in earth system models of intermediate complexity (EMICs) [29,30]. The AMOCs in most EMICs have been found in a bistable regime [31], associated with the positive salinity feedback [10] in the Atlantic basin [32,33]. Therefore, abrupt climate change events similar to DO and Henrich events have been simulated in these models even under a gradual change in external forcing [27,29,30,34]. In the state-of-the-art CGCMs, however, what is the current status of AMOC instability and how is the model AMOC instability compared with observations?

In this paper, I will review recent progress in observations, modelling and understanding of AMOC instability. Given extensive reviews on this topic, especially related to the theoretical aspect [35–38], I will focus on some recent progress in paleo observations and potential bias in CGCMs.

The review is arranged as follows. In Section 3, I will review the current status of AMOC instability in CGCMs. In Section 4, I discuss some personal perspective on the relevance of AMOC instability to the present and future climate changes. A summary and perspective are given in Section 5.

2. Materials and Methods

All materials are derived from published papers where the detailed methods are described.

3. Results

3.1. Paleo Evidence of AMOC Instability

Evidence of an ocean origin of the abrupt change in AMOC, or thermohaline instability, requires a clear lead of the AMOC change over the ice sheet and the resulting meltwater flux and ice rafting. Similarly, an ice-sheet origin of abrupt change via ice-sheet instability requires a clear lead of the ice-sheet and meltwater changes over the AMOC. Recent paleo observations seemed to favor an ocean origin, although not yet unambiguously. These analyses were carried out on multiple proxies in the same sediment core, such that the relative phasing was accurate among different variables, saving the potentially different preservation times of different proxies. Ref. [21] analyzed ice rafting debris and AMOC strength in the subpolar North Atlantic and showed a lead of ~200–300 years of the AMOC weakening prior to ice rafting events for most Heinrich events, demonstrating clearly that AMOC collapses were not triggered by icebergs. This result was consistent with reconstructions of the subsurface temperature in the North Atlantic, which showed warming leading to ice rafting events by hundreds of years for the most recent Henrich events [39,40].

However, it remains possible that the AMOC weakening was triggered by the initial meltwater flux from ice-sheet instability [41] if the iceberg export occurred later than the initial melting. There are no records that can clarify this point unambiguously. However, ref. [18] so far likely provides the clearest evidence suggesting AMOC change preceding North Atlantic climate change. Ref. [18] performed a lagged correlation between multiple proxies and the Greenland temperature proxy ($\delta^{18}O_p$) in a sediment core over the Bermuda

Rise. Relative to the Greenland temperature, this analysis showed a clear lead of 200 years for the benthic $\delta^{13}C$ (purple), accompanied by an almost in phase AMOC strength proxy $-^{231}Pa/^{230}Th$ (green, the legend should be of a minus sign) and North Atlantic SST (red) (Figure 2). At face value, this analysis suggested that the NADW was enhanced (represented by a positive $\delta^{13}C$ anomaly) before the enhancement of the AMOC intensity (represented by more negative $^{231}Pa/^{230}Th$ or more positive $-^{231}Pa/^{230}Th$), North Atlantic SST and Greenland temperature. Allowing for a century scale response time of sediment $^{231}Pa/^{230}Th$ after the AMOC change, the AMOC strength likely has changed preceding Pa/Th with its phase close to the NADW [18]. Thus, the AMOC likely has changed before the temperatures over the North Atlantic region, precluding the AMOC change as a response to climate change, which would otherwise have led to an AMOC response lagging surface temperature. This analysis so far provides the strongest evidence favoring the oceanic origin of AMOC collapse. This analysis, however, has left the question open as to why the temperature changed after the AMOC, as opposed to synchronously (within data resolution time scale of decades) with the AMOC and the associated heat transport.

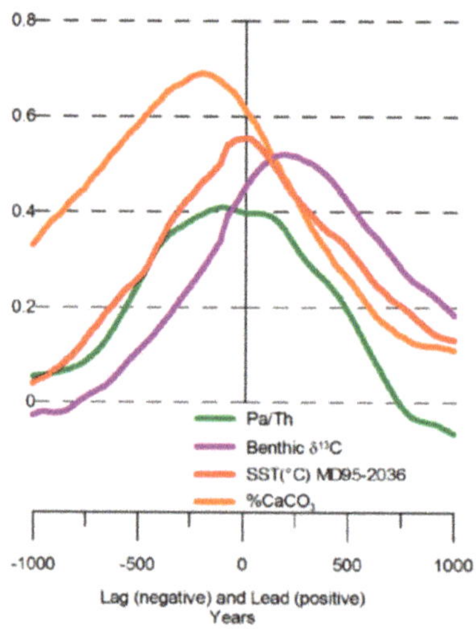

Figure 2. Cross-correlation of NGRIP ice core $\delta^{18}O$ with CDH[19] CaCO3 flux (orange), Pa/Th of bulk sediment from CDH[19] (green), $\delta^{13}C_{BF}$ from CDH19 (purple), Sea Surface Temperature (SST, °C) from Bermuda Rise sediment core MD95-2036 (red). A lead (positive) refers to the lead of the proxy relative to Northern Greenland Ice Project (NGRIP) ice core. See [25] for details. Adapted from [18]. (Note: the legend for Pa/Th should be of minus sign −Pa/Th, J. McManus, personal communication.)

Regardless of the AMOC instability, even the concurrent variability of the AMOC and climate implies a robust positive feedback between the AMOC and ice-sheet change. Proxy evidence [39,40] and modelling studies [42–44] suggested a subsurface warming accompanying the surface cooling in the subpolar North Atlantic in response to a meltwater flux, weakened deep convection and AMOC. This subsurface warming would promote ice-sheet calving and further melting and, in turn, a further weakening of the AMOC [44]. This positive feedback suggests a coupled AMOC–ice-sheet instability that depends on the coupling of both components, even if either component alone is stable. This is analogous to the case of El Nino, which is caused by a positive coupled ocean–atmosphere feedback, or instability, while the atmosphere or ocean alone is stable when the other component is fixed.

Finally, it should be noted that some abrupt transition can occur within decades. For example, the termination of the Younger Dryas event (YD, 12,900–11,600 years ago) has been observed to occur within 40 years in the stable water isotope signals in Chinese cave records [45,46]. It remains challenging to interpret such a rapid transition in terms of AMOC instability. Other mechanisms, notably those related to sea ice changes, may be responsible for such rapid transitions.

3.2. AMOC Instability in CGCMs

In most CGCMs so far, the AMOC seemed stable. This statement, however, remains highly uncertain, because, unlike EMICs, CGCMs are computationally too expensive to perform comprehensive sensitivity experiments over thousands of years under different perturbation forcing. Instead, so far, AMOC stability in most CGCMs has been assessed in two approaches. The first approach is the direct perturbation experiment, but mostly of a short duration of hundreds of years. In response to a meltwater pulse, the AMOCs in most CGCMs have been found to recover after the termination of the freshwater forcing for the Present Day (PD) [47] and LGM [48]. In response to long term CO_2 rise, CGCMs so far also show gradual responses [49]. Similarly, in response to a rapid rise of atmospheric CO_2, the AMOC also tended to decline gradually (instead of abruptly), including the simulations in IPCC reports [50], followed by slow recovery at millennial time scales [49,50]. These short experiments have led to the impression of a monostable AMOC, although it is possible that the perturbations might not have been strong or long enough to push the AMOC out of the stability attractor [32]. This impression was corroborated by a few long simulations of past climate changes in a small subset of models. Therefore, in most current CGCMs, abrupt climate changes could be generated only by abrupt changes in the forcing ([51–55]).

There are a few CGCMs (without flux adjustment) that exhibited AMOC instability in idealized paleo climate modelling. In a coarse resolution model, the AMOC seemed to be locked in a broad bistability regime such that the AMOC changed abruptly to another state in response to a gradual change in ice-sheet topography, atmospheric CO_2 or orbital forcing [56–58]. Furthermore, AMOC instability and the resulting millennial variability tended to occur for intermediate levels of glacial ice sheet and CO_2, consistent with the observations that the strongest millennial variability occurred in Stage 3 in the last glacial–interglacial cycle (Figure 1). Nevertheless, temperature variability over Greenland was underestimated in this model relative to reconstructions [59]. In several CGCMs, the AMOC seemed to transition to an unstable state of strong millennial variability in response to perturbation climate forcing [60–64]. In one version of the Community Earth System Model version 1 of the modified ocean diffusivity coefficient, the AMOC in the LGM setting exhibited DO-like variability of about 700 years of duration after being triggered by a strong meltwater pulse [60,61]. Why the change in diffusivity caused the change in AMOC stability, however, remains unclear. Oceanic diapycnal mixing has been suggested to enhance AMOC instability in some theoretical models and EMICs [65,66], but to suppress AMOC instability in a CGCM [67] and not to qualitatively affect the AMOC stability in an EMIC [68]. Overall, it remains unclear why these few CGCMs that do not exhibit monostable AMOC are different from most CGCMs that seem to exhibit monostable AMOC, given their comparable resolution and complexity of model physics.

The second approach was an indirect approach in which AMOC stability was inferred from a diagnostic indicator $M_{ov,S}$ that represents the southward freshwater export by the zonal mean overturning circulation across ~30° S in the Atlantic [11]. The AMOC is inferred bistable for freshwater exporting $M_{ov,S} < 0$, because a weakening AMOC leads to a freshwater pileup in the Atlantic, and then a further weakening of the AMOC. This indicator has been found to be remarkably successful in EMICs [32,33,69–71], and even in a few CGCMs [72,73], in the sense that its sign was consistent with the AMOC stability tested in direct simulations. It should also be noted that the fidelity of the indicator seemed to be improved substantially after a refinement to include the AMOC freshwater transport across the northern boundary of the North Atlantic $M_{ov,N}$ as the net freshwater import $\Delta M_{ov} = M_{ov,S} - M_{ov,N}$ [68,74]. For example, almost all CMIP5 models responded to increased CO_2 with a gradual slowdown, likely implying monostable AMOC, even though 40% of the models were inferred as bistable in terms of $M_{ov,S}(< 0)$ [75–78]. However, in terms of ΔM_{ov}, most models were indeed inferred as stable [79] (Figure 3), suggesting the non-negligible role of $M_{ov,N}$. An additional improvement can be further made to take into account the slow temporal evolution of the AMOC [76,77]. The most valuable aspect of this indicator was its applicability to present observations. All the observational data sets showed a net freshwater export $M_{ov,S} < 0$ or $\Delta M_{ov} < 0$, implying a bistable regime of the present AMOC [24,32,38,72,78–81] (Figure 3). If this indicator is indeed correct, the real-world AMOC would be unstable at the present and most CGCMs are too stable and thus might have underestimated the likelihood of abrupt climate changes [80,81].

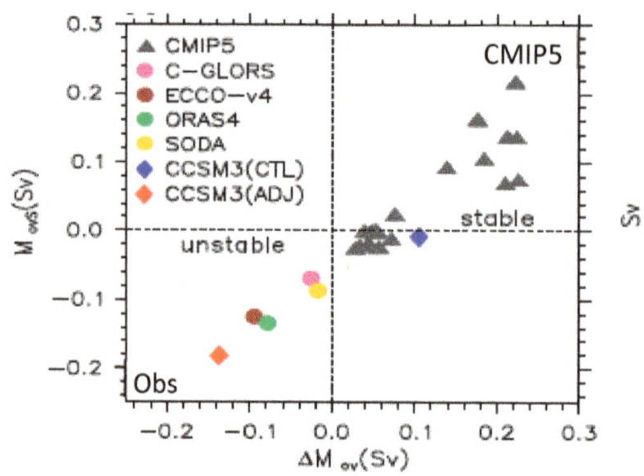

Figure 3. AMOC stability in the modern climate as assessed by stability indicators. The indicator values in the modern climate are shown in solid circles with color for 4 reanalysis data sets, in gray triangles for the CMIP5 model simulations, and in blue and red diamonds for the CCSM3 before (CTL) and after (ADJ) flux adjustment. A positive or negative indicator implies that the AMOC resides in a stable or bistable regime. Note that for CMIP models, about 40% shows $M_{ov,S} < 0$, but all show $\Delta M_{ov} > 0$. Adapted from [72].

Given the potential importance of this stability indicator, it is interesting to ask why most CGCMs have shown a net AMOC freshwater import, opposite to current observations, while EMICs have shown an AMOC freshwater export as in current observations. The biased freshwater transport in CGCMs has been found to be caused mainly by the salinity bias in the South Atlantic, especially the fresh bias of the surface subtropical water [38,82,83]. From the oceanic perspective, this surface fresh bias could be caused by the deficient ocean model, such as the coarse model resolution and the deficient Agulhas Retroflection [37,38,84]. Yet, even with similar coarse resolution ocean models, EMICs were able to simulate the freshwater export as in observations. This seemed to suggest that the salinity bias in the CGCMs were caused by certain model biases common in CGCMs, but not in EMICs. One such bias was suggested to be the tropical bias of the double Intertropical Convergence Zone (ITCZ) [81], which has been a persistent bias in almost all CGCMs [85,86], especially over the tropical Atlantic sector [87]. This southward bias of ITCZ would lead to excessive rainfall and surface freshening in the surface subtropical South Atlantic [81,83]. This hypothesis has been tested systematically in one CGCM using flux-adjustment sensitivity experiments [81]. While the original CCSM3 model was indeed monostable under direct meltwater perturbation, consistent with $\Delta M_{ov} > 0$, the model after flux adjustment is changed to being bistable in terms of both direct perturbation experiment and $\Delta M_{ov} < 0$ (Figure 3). This hypothesis appeared to also be consistent in another two CGCMs, in which the model without flux adjustment is monostable but the model with flux adjustment is bistable [88,89]. However, a systematic study is still lacking across models on the role of flux adjustment on AMOC stability.

The tropical bias hypothesis seemed to offer an explanation as to why EMICs tended to be more unstable than CGCMs. Due to the simplified atmospheric physics in EMICs, their model climatology is generated with some information on the present climatology of wind and/or SST. Thus, all EMICs could be considered as models after flux adjustment. Another hypothesis is that the salinity bias was contributed by the model bias of the Antarctica Intermediate Water (AAIW) [90]. However, for most CGCMs, the AAIW was more saline than observations, which would have favored freshwater export, instead of import, in these CGCMs [91].

The key question still remains open as to how correct the indicator is, in CGCMs and ultimately the real world. Physically, the freshwater budget of the North Atlantic and the final salinity feedback with the AMOC can be affected by the AMOC freshwater transport as well as other processes, such as the azonal gyre transport M_{az}, atmospheric feedback on evaporation minus precipitation and wind stress (and in turn gyre transport) and oceanic mixing processes (see [37,84] for comprehensive reviews). Indeed, there are at least two CGCMs showing AMOC instability, but without flux adjustment, they are inconsistent with this indicator ([72], but see an alternative explanation by [92,93]).

3.3. *Relevance to the Present and Future*

AMOC instability is highly relevant to the future climate state and abrupt changes [94], which remain highly uncertain [95]. AMOC instability can be state dependent such that the AMOC stability differs in different times, glacial periods, Holocene, the present and future. What is then needed is the understanding of how the AMOC stability changes with climate state. In spite of the uncertainty, the stability indicator can serve as a useful starting point. The indicator implies an unstable AMOC in observations at the present time. This unstable AMOC, however, seems to contradict the lack of abrupt events in the Holocene (except for the modest 8.2 ka event, [96]), if the Holocene is taken as an analogue of the present. Alternatively, this lack of large abrupt changes in the Holocene could be caused by the lack of a strong trigger in the Holocene.

If we take the evidence of a likely unstable AMOC in the glacial cycle and assume the indicator is correct for the present period, most current CGCMs would be too stable, implying an underestimation of the possibility of abrupt climate changes in the future [79]. This over-stable AMOC is consistent for most recent CGCMs without flux adjustment, in which the AMOC responded gradually to the future rise of CO_2 [49]. There are, nevertheless, three exceptions for CGCMs without flux adjustment that show abrupt collapse in hundreds of years, as presented in ref. [57,93,97]. The evolution of the AMOC in the future will be affected further by the melting of ice sheets in Greenland and Antarctica in the long run. It therefore remains highly uncertain how the AMOC will change in the future. It should be kept in mind that abrupt changes in models with flux adjustment should be treated with great caution [37], because of the potential distortion of the AMOC stability by flux adjustment, as analyzed in simple models [98]. Equally, however, it should be realized that there is no reason to trust the projections more from those current models without flux adjustment, as long as these models still suffer from severe salinity bias and, in turn, the AMOC freshwater transport, even if the stability indicator may not be perfectly correct.

4. Discussion

Recent progress in paleoceanographic proxies seem to favor the oceanic origin of AMOC instability as the cause of the abrupt climate changes during the glacial–interglacial period. Most CGCMs, however, seem to be over-stable, judging from the limited sensitivity experiments available, as well as the stability indicator ΔM_{ov}, although it remains uncertain how correct this indicator is across CGCMs and in the real world.

Further paleo proxy records, especially those with high-temporal resolution, are needed to distinguish the AMOC instability from the ice-sheet instability as the origin for abrupt climate changes. These records may further include those outside the North Atlantic, say, in the North Pacific [99], because of their potential links to the abrupt changes in the AMOC.

Even more challenging is the assessment of the AMOC instability in the real world for the present and future. For the present, it has remained difficult to detect the AMOC response to the global warming of rising CO_2. This is because the direct instrument measurement of AMOC transport has only been available for two decades. This short record can be significantly distorted by multidecadal variability and therefore is too short to detect the trend response to CO_2 rise. Observational evidence of deep warming in the Atlantic and Southern Ocean [100–104] is not good evidence of the AMOC response either. Besides also being too short, the deep warming could be caused simply by the advection of the mean circulation, notably, the deep western boundary current [105], instead of a change in the AMOC circulation. Nevertheless, a recent study of two AMOC fingerprints in the North Atlantic surface temperature [106,107] and South Atlantic surface salinity [108,109] seemed to provide the clearest evidence so far of the AMOC slowdown response to global warming. This slowdown response, if true, could be simply the forced response of the AMOC, even without instability.

Finally, it is certainly worrisome, to say the least, that the state-of-the-art CGCMs still show the opposite AMOC freshwater transport, which is potentially related to the salinity feedback and in turn, AMOC instability. A diagnostic indicator, even if imperfect, provides the only way that the AMOC stability can be assessed for the present day real world, which then can be compared with models. Given all the odds of potential feedback beyond a simple conceptual model, it is already surprising that the indicator ΔM_{ov} even works in many EMICs and some CGCMs. In EMICs, this indicator has been shown to represent the physical process of basin-wide salinity feedback associated with perturbation flow on mean salinity, while the gyre-induced freshwater transport is not sensitive to AMOC changes [32,33]. These feedback processes may be altered in CGCMs, especially in high-resolution models, leading to inconsistency between the indicator and AMOC stability [37,73,110]. Is it then possible to derive an improved stability indicator? For example, should the AMOC freshwater transport be calculated at a latitude other than 30° S, such as the intergyre boundary where the gyre transport change seems to be weak [73,90]?

5. Conclusions

Ultimately, AMOC instability, including any potential instability indicator, should be studied in the most realistic models without flux adjustment: high-resolution models with little bias in model climatology. This poses several challenges. First, the high-computational cost for the eddy resolving high-resolution models makes it difficult to perform extensive and long simulations that are needed to test any stability indicator. Second, if the indicator is related to a certain model bias, such as the tropical bias, these biases need to be significantly reduced in these CGCMs for a credible test of the indicator. The reduction of this bias, however, will be challenging because some biases are stubborn, notably the tropical bias which has been one of the most stubborn biases in CGCMs. Finally, AMOC instability may involve different feedback on different time scales, which may also be related to various transient behaviors of the AMOC responses, the latter being more relevant in the near future of climate change [93,97,110–112]. The different transient behaviors may be related to the basin-wide salinity feedback [35], as well as other feedback, such as the local convective feedback in the subpolar North Atlantic [113], feedback with atmosphere and sea ice [114], and the salinity feedback between the tropical and North Atlantic [89].

Funding: LSKJ202203303: NOAA20AR4310403, DOE ImPaCT.

Institutional Review Board Statement: Not applicable.

Informed Consent Statement: Not applicable.

Data Availability Statement: Not applicable.

Acknowledgments: I would like to thank W. Liu, J. Lynch-Steiglitz, P. Clark, J. Mecking and J. McManus for our discussions. I would also like to thank three anonymous reviewers for their comments.

Conflicts of Interest: The authors declare no conflict of interest.

References

1. Broecker, W.S. Paleocean circulation during the Last Deglaciation: A bipolar seesaw? *Paleoceanography* **1998**, *13*, 119–121. [CrossRef]
2. Pedro, J.B.; Jochum, M.; Buizert, C.; He, F.; Barker, S.; Rasmussen, S.O. Beyond the bipolar seesaw: Toward a process understanding of interhemispheric coupling. *Quat. Sci. Rev.* **2018**, *192*, 27–46. [CrossRef]
3. Lynch-Stieglitz, J. The Atlantic Meridional Overturning Circulation and Abrupt Climate Change. *Annu. Rev. Mar. Sci.* **2017**, *9*, 83–104. [CrossRef] [PubMed]
4. Duplessy, J.C.; Shackleton, N.J.; Fairbanks, R.G.; Labeyrie, L.; Oppo, D.; Kallel, N. Deepwater source variations during the last climatic cycle and their impact on the global deepwater circulation. *Paleoceanography* **1988**, *3*, 343–360. [CrossRef]
5. Goldstein, S.L.; Hemming, S.R. Long-lived Isotopic Tracers in Oceanography, Paleoceanography, and Ice-sheet Dynamics. *Treatise Geochem.* **2003**, *6*, 453–483.
6. Lynch-Stieglitz, J.; Adkins, J.F.; Curry, W.B.; Dokken, T.M.; Hall, I.R.; Herguera, J.C.; Hirschi, J.J.-M.; Ivanova, E.; Kissel, C.; Marchal, O.; et al. Atlantic Meridional Overturning Circulation During the Last Glacial Maximum. *Science* **2007**, *316*, 66–69. [CrossRef]
7. McManus, J.F.; Francois, R.; Gherardi, J.M.; Keigwin, L.D.; Brown-Leger, S. Collapse and rapid resumption of Atlantic meridional circulation linked to deglacial climate changes. *Nature* **2004**, *428*, 834–837. [CrossRef]
8. Lynch-Stieglitz, J.; Curry, W.B.; Slowey, N. Weaker Gulf Stream in the Florida Straits during the Last Glacial Maximum. *Nature* **1999**, *402*, 644–648. [CrossRef]
9. Dansgaard, W. Stable isotopes in precipitation. *Tellus* **1964**, *16*, 436–468. [CrossRef]
10. Stommel, H. Thermohaline Convection with Two Stable Regimes of Flow. *Tellus* **1961**, *13*, 224–230. [CrossRef]
11. Rahmstorf, S. On the freshwater forcing and transport of the Atlantic thermohaline circulation. *Clim. Dyn.* **1996**, *12*, 799–811. [CrossRef]
12. Dansgaard, W.; Johnsen, S.J.; Clausen, H.B.; Dahl-Jensen, D.; Gundestrup, N.; Hammer, C.U.; Oeschger, H. North Atlantic Climatic Oscillations Revealed by Deep Greenland Ice Cores. In *Climate Processes and Climate Sensitivity*; Wiley Online Library: New York, NY, USA, 1984; pp. 288–298. [CrossRef]
13. Oeschger, H.; Beer, J.; Siegenthaler, U.; Stauffer, B.; Dansgaard, W.; Langway, C.C. Late Glacial Climate History from Ice Cores. In *Climate Processes and Climate Sensitivity*; Wiley Online Library: New York, NY, USA, 1984; pp. 299–306. [CrossRef]
14. Broecker, W.S.; Peteet, D.M.; Rind, D. Does the ocean–atmosphere system have more than one stable mode of operation? *Nature* **1985**, *315*, 21–26. [CrossRef]
15. Bond, G.; Heinrich, H.; Broecker, W.; Labeyrie, L.; McManus, J.; Andrews, J.; Huon, S.; Jantschik, R.; Clasen, S.; Simet, C.; et al. Evidence for massive discharges of icebergs into the North Atlantic ocean during the last glacial period. *Nature* **1992**, *360*, 245–249. [CrossRef]
16. Bond, G.; Showers, W.; Cheseby, M.; Lotti, R.; Almasi, P.; deMenocal, P.; Priore, P.; Cullen, H.; Hajdas, I.; Bonani, G. A Pervasive Millennial-Scale Cycle in North Atlantic Holocene and Glacial Climates. *Science* **1997**, *278*, 1257–1266. [CrossRef]
17. Gherardi, J.-M.; Labeyrie, L.; Nave, S.; Francois, R.; McManus, J.F.; Cortijo, E. Glacial-interglacial circulation changes inferred from 231Pa/230Th sedimentary record in the North Atlantic region. *Paleoceanography* **2009**, *24*, PA2204. [CrossRef]
18. Henry, L.G.; McManus, J.F.; Curry, W.B.; Roberts, N.L.; Piotrowski, A.M.; Keigwin, L.D. North Atlantic ocean circulation and abrupt climate change during the last glaciation. *Science* **2016**, *353*, 470–474. [CrossRef]
19. Hoogakker, B.A.A.; McCave, I.N.; Vautravers, M.J. Antarctic link to deep flow speed variation during Marine Isotope Stage 3 in the western North Atlantic. *Earth Planet. Sci. Lett.* **2007**, *257*, 463–473. [CrossRef]
20. Clark, P.U.; Pisias, N.G.; Stocker, T.F.; Weaver, A.J. The role of the thermohaline circulation in abrupt climate change. *Nature* **2002**, *415*, 863–869. [CrossRef]
21. Barker, S.; Chen, J.; Gong, X.; Jonkers, L.; Knorr, G.; Thornalley, D. Icebergs not the trigger for North Atlantic cold events. *Nature* **2015**, *520*, 333–336. [CrossRef]
22. Ahn, J.; Brook, E.J. Atmospheric CO_2 and Climate on Millennial Time Scales During the Last Glacial Period. *Science* **2008**, *322*, 83–85. [CrossRef]
23. Schmittner, A.; Galbraith, E.D. Glacial greenhouse-gas fluctuations controlled by ocean circulation changes. *Nature* **2008**, *456*, 373–376. [CrossRef] [PubMed]
24. Bryden, H.L.; King, B.A.; McCarthy, G.D. South Atlantic overturning circulation at 24°S. *J. Mar. Res.* **2011**, *69*, 38–55. [CrossRef]
25. Marotzke, J.; Welander, P.; Willebrand, J. Instability and multiple steady states in a meridional-plane model of the thermohaline circulation. *Tellus A Dyn. Meteorol. Oceanogr.* **1988**, *40*, 162–172. [CrossRef]
26. Stocker, T.F.; Wright, D.G.; Broecker, W.S. The influence of high-latitude surface forcing on the global thermohaline circulation. *Paleoceanography* **1992**, *7*, 529–541. [CrossRef]
27. Weaver, A.J.; Hughes, T.M.C. Rapid interglacial climate fluctuations driven by North Atlantic ocean circulation. *Nature* **1994**, *367*, 447–450. [CrossRef]
28. Manabe, S.; Stouffer, R.J. Two Stable Equilibria of a Coupled Ocean-Atmosphere Model. *J. Clim.* **1988**, *1*, 841–866. [CrossRef]
29. Ganopolski, A.; Rahmstorf, S. Rapid changes of glacial climate simulated in a coupled climate model. *Nature* **2001**, *409*, 153–158. [CrossRef]

30. Weaver, A.J.; Saenko, O.A.; Clark, P.U.; Mitrovica, J.X. Meltwater pulse 1A from Antarctica as a trigger of the Bølling-Allerød warm interval. *Science* **2003**, *299*, 1709–1713. [CrossRef]
31. Rahmstorf, S.; Crucifix, M.; Ganopolski, A.; Goosse, H.; Kamenkovich, I.; Knutti, R.; Lohmann, G.; Marsh, R.; Mysak, L.A.; Wang, Z.; et al. Thermohaline circulation hysteresis: A model intercomparison. *Geophys. Res. Lett.* **2005**, *32*, L23605. [CrossRef]
32. Huisman, S.E.; den Toom, M.; Dijkstra, H.A.; Drijfhout, S. An Indicator of the Multiple Equilibria Regime of the Atlantic Meridional Overturning Circulation. *J. Phys. Oceanogr.* **2010**, *40*, 551–567. [CrossRef]
33. Sijp, W.P. Characterising meridional overturning bistability using a minimal set of state variables. *Clim. Dyn.* **2012**, *39*, 2127–2142. [CrossRef]
34. Yin, Q.Z.; Wu, Z.P.; Berger, A.; Goosse, H.; Hodell, D. Insolation triggered abrupt weakening of Atlantic circulation at the end of interglacials. *Science* **2021**, *373*, 1035–1040. [CrossRef] [PubMed]
35. Marotzke, J. Abrupt climate change and thermohaline circulation: Mechanisms and predictability. *Proc. Natl. Acad. Sci. USA* **2000**, *97*, 1347–1350. [CrossRef] [PubMed]
36. Liu, W.; Liu, Z. Assessing the stability of the Atlantic meridional overturning circulation of the past, present, and future. *J. Meteorol. Res.* **2014**, *28*, 803–819. [CrossRef]
37. Gent, P.R. A commentary on the Atlantic meridional overturning circulation stability in climate models. *Ocean Model.* **2018**, *122*, 57–66. [CrossRef]
38. Weijer, W.; de Ruijter, W.P.M.; Dijkstra, H.A.; van Leeuwen, P.J. Impact of Interbasin Exchange on the Atlantic Overturning Circulation. *J. Phys. Oceanogr.* **1999**, *29*, 2266–2284. [CrossRef]
39. Marcott, S.A.; Clark, P.U.; Padman, L.; Klinkhammer, G.P.; Springer, S.R.; Liu, Z.; Otto-Bliesner, B.L.; Carlson, A.E.; Ungerer, A.; Padman, J.; et al. Ice-shelf collapse from subsurface warming as a trigger for Heinrich events. *Proc. Natl. Acad. Sci. USA* **2011**, *108*, 13415–13419. [CrossRef]
40. Max, L.; Nürnberg, D.; Chiessi, C.M.; Lenz, M.M.; Mulitza, S. Subsurface ocean warming preceded Heinrich Events. *Nat. Commun.* **2022**, *13*, 4217. [CrossRef]
41. MacAyeal, D.R. Binge/purge oscillations of the Laurentide Ice Sheet as a cause of the North Atlantic's Heinrich events. *Paleoceanography* **1993**, *8*, 775–784. [CrossRef]
42. Mignot, J.; Ganopolski, A.; Levermann, A. Atlantic Subsurface Temperatures: Response to a Shutdown of the Overturning Circulation and Consequences for Its Recovery. *J. Clim.* **2007**, *20*, 4884–4898. [CrossRef]
43. He, C.; Liu, Z.; Zhu, J.; Zhang, J.; Gu, S.; Otto-Bliesner, B.L.; Brady, E.; Zhu, C.; Jin, Y.; Sun, J. North Atlantic subsurface temperature response controlled by effective freshwater input in "Heinrich" events. *Earth Planet. Sci. Lett.* **2020**, *539*, 116247. [CrossRef]
44. Shaffer, G.; Olsen, S.M.; Bjerrum, C.J. Ocean subsurface warming as a mechanism for coupling Dansgaard-Oeschger climate cycles and ice-rafting events. *Geophys. Res. Lett.* **2004**, *31*, L24202. [CrossRef]
45. Ma, Z.-B.; Cheng, H.; Tan, M.; Edwards, R.L.; Li, H.-C.; You, C.-F.; Duan, W.-H.; Wang, X.; Kelly, M.J. Timing and structure of the Younger Dryas event in northern China. *Quat. Sci. Rev.* **2012**, *41*, 83–93. [CrossRef]
46. Cheng, H.; Zhang, H.; Spötl, C.; Baker, J.; Sinha, A.; Li, H.; Bartolomé, M.; Moreno, A.; Kathayat, G.; Zhao, J.; et al. Timing and structure of the Younger Dryas event and its underlying climate dynamics. *Proc. Natl. Acad. Sci. USA* **2020**, *117*, 23408–23417. [CrossRef]
47. Stouffer, R.J.; Yin, J.-j.; Gregory, J.M.; Dixon, K.W.; Spelman, M.J.; Hurlin, W.J.; Weaver, A.J.; Eby, M.; Flato, G.; Hasumi, H.; et al. Investigating the Causes of the Response of the Thermohaline Circulation to Past and Future Climate Changes. *J. Clim.* **2006**, *19*, 1365–1387. [CrossRef]
48. Kageyama, M.; Merkel, U.; Otto-Bliesner, B.; Prange, M.; Abe-Ouchi, A.; Lohmann, G.; Ohgaito, R.; Roche, D.M.; Singarayer, J.; Swingedouw, D.; et al. Climatic impacts of fresh water hosing under Last Glacial Maximum conditions: A multi-model study. *Clim. Past* **2013**, *9*, 935–953. [CrossRef]
49. Bonan, D.B.; Thompson, A.F.; Newsom, E.R.; Sun, S.; Rugenstein, M. Transient and Equilibrium Responses of the Atlantic Overturning Circulation to Warming in Coupled Climate Models: The Role of Temperature and Salinity. *J. Clim.* **2022**, *35*, 5173–5193. [CrossRef]
50. Collins, M.; Knutti, R.; Arblaster, J.M.; Dufresne, J.L.; Fichefet, T.; Friedlingstein, P.; Gao, X.; Gutowski, W.J.; Johns, T.C.; Krinner, G.; et al. Long-term Climate Change: Projections, Commitments and Irreversibility. In *Climate Change 2013: The Physical Science Basis*; Cambridge University Press: Cambridge, UK, 2013.
51. Liu, Z.; Otto-Bliesner, B.L.; He, F.; Brady, E.C.; Tomas, R.; Clark, P.U.; Carlson, A.E.; Lynch-Stieglitz, J.; Curry, W.B.; Brook, E.J.; et al. Transient Simulation of Last Deglaciation with a New Mechanism for Bølling-Allerød Warming. *Science* **2009**, *325*, 310–314. [CrossRef]
52. Zhu, J.; Liu, Z.; Zhang, X.; Eisenman, I.; Liu, W. Linear weakening of the AMOC in response to receding glacial ice sheets in CCSM3. *Geophys. Res. Lett.* **2014**, *41*, 6252–6258. [CrossRef]
53. Ivanovic, R.F.; Gregoire, L.J.; Kageyama, M.; Roche, D.M.; Valdes, P.J.; Burke, A.; Drummond, R.; Peltier, W.R.; Tarasov, L. Transient climate simulations of the deglaciation 21–9 thousand years before present. PMIP 4 Core experiment design and boundary conditions. *Geosci. Model Dev.* **2016**, *9*, 2563–2587. [CrossRef]
54. Ivanovic, R.F.; Gregoire, L.J.; Wickert, A.D.; Valdes, P.J.; Burke, A. Collapse of the North American ice saddle 14,500 years ago caused widespread cooling and reduced ocean overturning circulation. *Geophys. Res. Lett.* **2017**, *44*, 383–392. [CrossRef]

55. He, C.; Liu, Z.; Otto-Bliesner, B.L.; Brady, E.C.; Zhu, C.; Tomas, R.; Clark, P.U.; Zhu, J.; Jahn, A.; Gu, S.; et al. Hydroclimate footprint of pan-Asian monsoon water isotope during the last deglaciation. *Sci. Adv.* **2021**, *7*, eabe2611. [CrossRef] [PubMed]
56. Zhang, X.; Lohmann, G.; Knorr, G.; Purcell, C. Abrupt glacial climate shifts controlled by ice sheet changes. *Nature* **2014**, *512*, 290–294. [CrossRef]
57. Zhang, X.; Knorr, G.; Lohmann, G.; Barker, S. Abrupt North Atlantic circulation changes in response to gradual CO_2 forcing in a glacial climate state. *Nat. Geosci.* **2017**, *10*, 518–523. [CrossRef]
58. Zhang, X.; Barker, S.; Knorr, G.; Lohmann, G.; Drysdale, R.; Sun, Y.; Hodell, D.; Chen, F. Direct astronomical influence on abrupt climate variability. *Nat. Geosci.* **2021**, *14*, 819–826. [CrossRef]
59. Zhang, X.; Prange, M. Stability of the Atlantic overturning circulation under intermediate (MIS3) and full glacial (LGM) conditions and its relationship with Dansgaard-Oeschger climate variability. *Quat. Sci. Rev.* **2020**, *242*, 106443. [CrossRef]
60. Peltier, W.R.; Vettoretti, G. Dansgaard-Oeschger oscillations predicted in a comprehensive model of glacial climate: A "kicked" salt oscillator in the Atlantic. *Geophys. Res. Lett.* **2014**, *41*, 7306–7313. [CrossRef]
61. Vettoretti, G.; Peltier, W.R. Fast Physics and Slow Physics in the Nonlinear Dansgaard–Oeschger Relaxation Oscillation. *J. Clim.* **2018**, *31*, 3423–3449. [CrossRef]
62. Kuniyoshi, Y.; Abe-Ouchi, A.; Sherriff-Tadano, S.; Chan, W.-L.; Saito, F. Effect of Climatic Precession on Dansgaard-Oeschger-Like Oscillations. *Geophys. Res. Lett.* **2022**, *49*, e2021GL095695. [CrossRef]
63. Izumi, K.; Armstrong, E.; Valdes, P.J. Global footprints of dansgaard-oeschger oscillations in a GCM. *Quat. Sci. Rev.* **2023**, *305*, 108016.
64. Klockmann, M.; Mikolajewicz, U.; Marotzke, J. Two AMOC States in Response to Decreasing Greenhouse Gas Concentrations in the Coupled Climate Model MPI-ESM. *J. Clim.* **2018**, *31*, 7969–7984. [CrossRef]
65. Prange, M.; Lohmann, G.; Paul, A. Influence of Vertical Mixing on the Thermohaline Hysteresis: Analyses of an OGCM. *J. Phys. Oceanogr.* **2003**, *33*, 1707–1721. [CrossRef]
66. Nof, D.; Van Gorder, S.; de Boer, A. Does the Atlantic meridional overturning cell really have more than one stable steady state? *Deep Sea Res. Part I Oceanogr. Res. Pap.* **2007**, *54*, 2005–2021. [CrossRef]
67. Manabe, S.; Stouffer, R.J. Are two modes of thermohaline circulation stable. *Tellus A* **1999**, *51*, 400–411. [CrossRef]
68. Dijkstra, H.A. Characterization of the multiple equilibria regime in a global ocean model. *Tellus A Dyn. Meteorol. Oceanogr.* **2007**, *59*, 695–705. [CrossRef]
69. de Vries, P.; Weber, S.L. The Atlantic freshwater budget as a diagnostic for the existence of a stable shut down of the meridional overturning circulation. *Geophys. Res. Lett.* **2005**, *32*, L09606. [CrossRef]
70. Oka, A.; Abe-Ouchi, A.; Sherriff-Tadano, S.; Yokoyama, Y.; Kawamura, K.; Hasumi, H. Glacial mode shift of the Atlantic meridional overturning circulation by warming over the Southern Ocean. *Commun. Earth Environ.* **2021**, *2*, 169. [CrossRef]
71. Drijfhout, S.S.; Weber, S.L.; van der Swaluw, E. The stability of the MOC as diagnosed from model projections for pre-industrial, present and future climates. *Clim. Dyn.* **2011**, *37*, 1575–1586. [CrossRef]
72. Hawkins, E.; Smith, R.S.; Allison, L.C.; Gregory, J.M.; Woollings, T.J.; Pohlmann, H.; de Cuevas, B. Bistability of the Atlantic overturning circulation in a global climate model and links to ocean freshwater transport. *Geophys. Res. Lett.* **2011**, *38*, L10605. [CrossRef]
73. Mecking, J.V.; Drijfhout, S.S.; Jackson, L.C.; Graham, T. Stable AMOC off state in an eddy-permitting coupled climate model. *Clim. Dyn.* **2016**, *47*, 2455–2470. [CrossRef]
74. Boers, N. Observation-based early-warning signals for a collapse of the Atlantic Meridional Overturning Circulation. *Nat. Clim. Chang.* **2021**, *11*, 680–688. [CrossRef]
75. Weaver, A.J.; Sedláček, J.; Eby, M.; Alexander, K.; Crespin, E.; Fichefet, T.; Philippon-Berthier, G.; Joos, F.; Kawamiya, M.; Matsumoto, K.; et al. Stability of the Atlantic meridional overturning circulation: A model intercomparison. *Geophys. Res. Lett.* **2012**, *39*, L20709. [CrossRef]
76. Liu, W.; Liu, Z.; Hu, A. The stability of an evolving Atlantic meridional overturning circulation. *Geophys. Res. Lett.* **2013**, *40*, 1562–1568. [CrossRef]
77. Liu, W.; Liu, Z. A Note on the Stability Indicator of the Atlantic Meridional Overturning Circulation. *J. Clim.* **2014**, *27*, 969–975. [CrossRef]
78. Garzoli, S.L.; Baringer, M.O.; Dong, S.; Perez, R.C.; Yao, Q. South Atlantic meridional fluxes. *Deep Sea Res. Part I Oceanogr. Res. Pap.* **2013**, *71*, 21–32. [CrossRef]
79. Liu, W.; Xie, S.-P.; Liu, Z.; Zhu, J. Overlooked possibility of a collapsed Atlantic Meridional Overturning Circulation in warming climate. *Sci. Adv.* **2017**, *3*, e1601666. [CrossRef]
80. Weber, S.L.; Drijfhout, S.S.; Abe-Ouchi, A.; Crucifix, M.; Eby, M.; Ganopolski, A.; Murakami, S.; Otto-Bliesner, B.; Peltier, W.R. The modern and glacial overturning circulation in the Atlantic ocean in PMIP coupled model simulations. *Clim. Past* **2007**, *3*, 51–64. [CrossRef]
81. Liu, W.; Liu, Z.; Brady, E.C. Why is the AMOC Monostable in Coupled General Circulation Models? *J. Clim.* **2014**, *27*, 2427–2443. [CrossRef]
82. Liu, W. *Why Is the AMOC Monostable in Coupled General Circulation Models?* American Geophysical Union: Washington, DC, USA, 2016; p. PO54A-3227.
83. Mecking, J.V.; Drijfhout, S.S.; Jackson, L.C.; Andrews, M.B. The effect of model bias on Atlantic freshwater transport and implications for AMOC bi-stability. *Tellus A Dyn. Meteorol. Oceanogr.* **2017**, *69*, 1299910. [CrossRef]

84. Weijer, W.; Cheng, W.; Drijfhout, S.S.; Fedorov, A.V.; Hu, A.; Jackson, L.C.; Liu, W.; McDonagh, E.L.; Mecking, J.V.; Zhang, J. Stability of the Atlantic Meridional Overturning Circulation: A Review and Synthesis. *J. Geophys. Res. Ocean.* **2019**, *124*, 5336–5375. [CrossRef]
85. Mechoso, C.R.; Robertson, A.W.; Barth, N.; Davey, M.K.; Delecluse, P.; Gent, P.R.; Ineson, S.; Kirtman, B.; Latif, M.; Treut, H.L.; et al. The Seasonal Cycle over the Tropical Pacific in Coupled Ocean–Atmosphere General Circulation Models. *Mon. Weather Rev.* **1995**, *123*, 2825–2838. [CrossRef]
86. Lin, J.-L. The Double-ITCZ Problem in IPCC AR4 Coupled GCMs: Ocean–Atmosphere Feedback Analysis. *J. Clim.* **2007**, *20*, 4497–4525. [CrossRef]
87. Davey, M.; Huddleston, M.; Sperber, K.; Braconnot, P.; Bryan, F.; Chen, D.; Colman, R.; Cooper, C.; Cubasch, U.; Delecluse, P.; et al. STOIC: A study of coupled model climatology and variability in tropical ocean regions. *Clim. Dyn.* **2002**, *18*, 403–420. [CrossRef]
88. Stouffer, R.J.; Manabe, S. Equilibrium response of thermohaline circulation to large changes in atmospheric CO_2 concentration. *Clim. Dyn.* **2003**, *20*, 759–773. [CrossRef]
89. Yin, J.; Stouffer, R.J. Comparison of the Stability of the Atlantic Thermohaline Circulation in Two Coupled Atmosphere–Ocean General Circulation Models. *J. Clim.* **2007**, *20*, 4293–4315. [CrossRef]
90. Mignac, D.; Ferreira, D.; Haines, K. Decoupled Freshwater Transport and Meridional Overturning in the South Atlantic. *Geophys. Res. Lett.* **2019**, *46*, 2178–2186. [CrossRef]
91. Zhu, C.; Liu, Z.; Gu, S. Model bias for South Atlantic Antarctic intermediate water in CMIP5. *Clim. Dyn.* **2018**, *50*, 3613–3624. [CrossRef]
92. Jackson, L.C.; Smith, R.S.; Wood, R.A. Ocean and atmosphere feedbacks affecting AMOC hysteresis in a GCM. *Clim. Dyn.* **2017**, *49*, 173–191. [CrossRef]
93. Rind, D.; Schmidt, G.A.; Jonas, J.; Miller, R.; Nazarenko, L.; Kelley, M.; Romanski, J. Multicentury Instability of the Atlantic Meridional Circulation in Rapid Warming Simulations With GISS ModelE2. *J. Geophys. Res.* **2018**, *123*, 6331–6355. [CrossRef]
94. Alley, R.B.; Marotzke, J.; Nordhaus, W.; Overpeck, J.T.; Peteet, D.M.; Pielke, R.A.; Pierrehumbert, R.T.; Rhines, P.B.; Stocker, T.F.; Talley, L.D.; et al. Abrupt climate change: Inevitable surprises. *Science* **2003**, *299*, 2005–2010. [CrossRef]
95. Fox-Kemper, G.; Hewitt, H.T.; Xiao, C.; Aðalgeirsdóttir, G.; Drijfhout, S.S.; Edwards, T.L.; Golledge, N.R.; Hemer, M.; Kopp, R.E.; Krinner, G.; et al. 2021: Ocean, Cryosphere and Sea Level Change. In *Climate Change 2021: The Physical Science Basis. Contribution of Working Group I to the Sixth Assessment Report of the Intergovernmental Panel on Climate Change*; Masson-Delmotte, V.P., Zhai, A., Pirani, S.L., Connors, C., Péan, S., Berger, N., Caud, Y., Chen, L., Goldfarb, M.I., Gomis, M., et al., Eds.; Cambridge University Press: Cambridge, UK; New York, NY, USA, 2021; pp. 1211–1362. [CrossRef]
96. Alley, R.B.; Mayewski, P.A.; Sowers, T.; Stuiver, M.; Taylor, K.C.; Clark, P.U. Holocene climatic instability: A prominent, widespread event 8200 yr ago. *Geology* **1997**, *25*, 483–486. [CrossRef]
97. Romanou, A.; Rind, D.; Jonas, J.; Miller, R.; Kelley, M.; Russell, G.; Orbe, C.; Nazarenko, L.; Latto, R.; Schmidt, G.A. Stochastic Bifurcation of the North Atlantic Circulation Under A Mid-Range Future Climate Scenario With The NASA-GISS ModelE. *J. Clim.* **2023**, *1*, 1–49. [CrossRef]
98. Dijkstra, H.A.; Neelin, J.D. Imperfections of the Thermohaline Circulation: Multiple Equilibria andFlux Correction. *J. Clim.* **1999**, *12*, 1382–1392. [CrossRef]
99. Walczak, M.H.; Mix, A.C.; Cowan, E.A.; Fallon, S.; Fifield, L.K.; Alder, J.R.; Du, J.; Haley, B.; Hobern, T.; Padman, J.; et al. Phasing of millennial-scale climate variability in the Pacific and Atlantic Oceans. *Science* **2020**, *370*, 716–720. [CrossRef] [PubMed]
100. Campos, E.J.D.; van Caspel, M.C.; Zenk, W.; Morozov, E.G.; Frey, D.I.; Piola, A.R.; Meinen, C.S.; Sato, O.T.; Perez, R.C.; Dong, S. Warming Trend in Antarctic Bottom Water in the Vema Channel in the South Atlantic. *Geophys. Res. Lett.* **2021**, *48*, e2021GL094709. [CrossRef]
101. Johnson, G.C.; Cadot, C.; Lyman, J.M.; McTaggart, K.E.; Steffen, E.L. Antarctic Bottom Water Warming in the Brazil Basin: 1990s Through 2020, From WOCE to Deep Argo. *Geophys. Res. Lett.* **2020**, *47*, e2020GL089191. [CrossRef]
102. Johnson, G.C.; Lyman, J.M. Warming trends increasingly dominate global ocean. *Nat. Clim. Chang.* **2020**, *10*, 757–761. [CrossRef]
103. Frey, D.I.; Krechik, V.A.; Bashirova, L.D.; Ostroumova, S.A.; Smirnova, D.A.; Kuleshova, L.A.; Ponomarenko, E.P.; Morozov, E.G.; Ligi, M.; Dudkov, I.Y.; et al. Multiple Abyssal Jets Flowing Into the Vema Deep, Romanche Fracture Zone. *J. Geophys. Res. Ocean.* **2023**, *128*, e2022JC019434. [CrossRef]
104. Purkey, S.G.; Johnson, G.C. Warming of Global Abyssal and Deep Southern Ocean Waters between the 1990s and 2000s: Contributions to Global Heat and Sea Level Rise Budgets. *J. Clim.* **2010**, *23*, 6336–6351. [CrossRef]
105. Zhang, M.; Wu, Z.; Qiao, F. Deep Atlantic Ocean Warming Facilitated by the Deep Western Boundary Current and Equatorial Kelvin Waves. *J. Clim.* **2018**, *31*, 8541–8555. [CrossRef]
106. Rahmstorf, S.; Box, J.E.; Feulner, G.; Mann, M.E.; Robinson, A.; Rutherford, S.; Schaffernicht, E.J. Exceptional twentieth-century slowdown in Atlantic Ocean overturning circulation. *Nat. Clim. Chang.* **2015**, *5*, 475–480. [CrossRef]
107. Caesar, L.; Rahmstorf, S.; Robinson, A.; Feulner, G.; Saba, V. Observed fingerprint of a weakening Atlantic Ocean overturning circulation. *Nature* **2018**, *556*, 191–196. [CrossRef] [PubMed]
108. Zhu, C.; Liu, Z. Weakening Atlantic overturning circulation causes South Atlantic salinity pile-up. *Nat. Clim. Chang.* **2020**, *10*, 998–1003. [CrossRef]
109. Zhu, C.; Liu, Z.; Zhang, S.; Wu, L. Likely accelerated weakening of Atlantic overturning circulation emerges in optimal salinity fingerprint. *Nat. Commun.* **2023**, *14*, 1245. [CrossRef] [PubMed]

110. Weijer, W.; Maltrud, M.E.; Hecht, M.W.; Dijkstra, H.A.; Kliphuis, M.A. Response of the Atlantic Ocean circulation to Greenland Ice Sheet melting in a strongly-eddying ocean model. *Geophys. Res. Lett.* **2012**, *39*, L09606. [CrossRef]
111. Winton, M.; Anderson, W.G.; Delworth, T.L.; Griffies, S.M.; Hurlin, W.J.; Rosati, A.J. Has coarse ocean resolution biased simulations of transient climate sensitivity? *Geophys. Res. Lett.* **2014**, *41*, 8522–8529. [CrossRef]
112. Jackson, L.C.; Wood, R.A. Hysteresis and Resilience of the AMOC in an Eddy-Permitting GCM. *Geophys. Res. Lett.* **2018**, *45*, 8547–8556. [CrossRef]
113. Lenderink, G.; Haarsma, R.J. Variability and Multiple Equilibria of the Thermohaline Circulation Associated with Deep-Water Formation. *J. Phys. Oceanogr.* **1994**, *24*, 1480–1493. [CrossRef]
114. Jayne, S.R.; Marotzke, J. A Destabilizing Thermohaline Circulation-Atmosphere-Sea Ice Feedback. *J. Clim.* **1999**, *12*, 642–651. [CrossRef]

atmosphere

MDPI

Article

Changing Characteristics of Tropical Extreme Precipitation–Cloud Regimes in Warmer Climates

William K. M. Lau [1,*], Kyu-Myong Kim [2], Bryce Harrop [3] and L. Ruby Leung [3]

1 Earth System Science Interdisciplinary Center, University of Maryland, College Park, MD 20740, USA
2 Climate and Radiation Laboratory, NASA/Goddard Space Flight Center, Greenbelt, MD 20771, USA; kyu-myong.kim@nasa.gov
3 Pacific Northwest National Laboratory, Atmospheric Sciences and Global Change Division, Richland, WA 99352, USA; bryce.harrop@pnnl.gov (B.H.); ruby.leung@pnnl.gov (L.R.L.)
* Correspondence: wkmlau@umd.edu

Abstract: In this study, we investigated the changing characteristics of climatic scale (monthly) tropical extreme precipitation in warming climates using the Energy Exascale Earth System Model (E3SM). The results are from Atmospheric Model Intercomparison Project (AMIP)-type simulations driven by (a) a control experiment with the present-day sea surface temperature (SST) and CO_2 concentration, (b) P4K, the same as in (a) but with a uniform increase of 4K in the SST globally, and (c) the same as in (a), but with an imposed SST and CO_2 concentration from the outputs of the coupled E3SM forced by a 4xCO_2 concentration. We found that as the surface warmed under P4K and 4xCO_2, both convective and stratiform rain increased. Importantly, there was an increasing fractional contribution of stratiform rain as a function of the precipitation intensity, with the most extreme but rare events occurring preferentially over land more than the ocean, and more so under 4xCO_2 than P4K. Extreme precipitation was facilitated by increased precipitation efficiency, reflecting accelerated rates of recycling of precipitation cloud water (both liquid and ice phases) in regions with colder anvil cloud tops. Changes in the vertical profiles of clouds, condensation heating, and vertical motions indicate increasing precipitation–cloud–circulation organization from the control and P4K to 4xCO_2. The results suggest that large-scale ocean warming, that is, P4K, was the primary cause contributing to an organization structure resembling the well-known mesoscale convective system (MCS), with increased extreme precipitation on shorter (hourly to daily) time scales. Additional 4xCO_2 atmospheric radiative heating and dynamically consistent anomalous SST further amplified the MCS organization under P4K. Analyses of the surface moist static energy distribution show that increases in the surface moisture (temperature) under P4K and 4xCO_2 was the key driver leading to enhanced convective instability over tropical ocean (land). However, a fast and large increase in the land surface temperature and lack of available local moisture resulted in a strong reduction in the land surface relative humidity, reflecting severe drying and enhanced convective inhibition (CIN). It is argued that very extreme and rare "record-breaking" precipitation events found over land under P4K, and more so under 4xCO_2, are likely due to the delayed onset of deep convection, that is, the longer the suppression of deep convection by CIN, the more severe the extreme precipitation when it eventually occurs, due to the release of a large amount of stored surplus convective available potential energy in the lower troposphere during prolonged CIN.

Keywords: climate-scale extreme tropical precipitation; stratiform and convective precipitation; precipitation efficiency; meso-scale convective complex; surface warming vs. moistening; convective inhibition over land

Citation: Lau, W.K.M.; Kim, K.-M.; Harrop, B.; Leung, L.R. Changing Characteristics of Tropical Extreme Precipitation–Cloud Regimes in Warmer Climates. *Atmosphere* **2023**, *14*, 995. https://doi.org/10.3390/atmos14060995

Academic Editors: Xiaolei Zou, Guoxiong Wu and Zhemin Tan

Received: 28 April 2023
Revised: 28 May 2023
Accepted: 31 May 2023
Published: 8 June 2023

1. Introduction

Recent reports of devastation resulting from record-breaking heavy precipitation around the world have provided strong indications that humanity is already experiencing the disastrous effects of increased extreme precipitation, i.e., flash floods, soil erosion, landslide, degradation of the eco-system, destruction of properties and loss of lives, attributable to anthropogenic greenhouse warming. Without a timely reduction in the emissions of greenhouse gases, the current trend in extreme precipitation will continue, and adverse impacts on the socio-economic system are likely to become worst [1]. Extreme precipitation in the tropics not only adversely affects the livelihood of more than 40% of the world population but is also a primary driver of global climate variability and change [2–5]. Hence, a better understanding of the physical processes underlying tropical extreme precipitation and its global impacts is paramount for the development and implementation of effective adaptation and mitigation strategies for global climate variability and change.

In the tropics and subtropics, climatologically strong surface heating and low-level moisture convergence lead to increased convective instability, enhancing heavy precipitation preferentially in regions with a warm surface temperature, i.e., the Inter-Tropical Convergence Zone (ITCZ), monsoon regions, and the maritime continent [6–9]. Changes in precipitation under global warming generally follow a geographic distribution pattern of "wet-gets-wetter" and "warmer-gets-wetter" [10–15]. A necessary condition for precipitation is the formation of clouds. Both precipitation and clouds, and their associated temporal and spatial distributions, are strong functions of atmospheric heating/cooling and moistening/drying processes, modulated by the surface temperature, heat and moisture fluxes, cloud microphysics, convection, and large-scale circulation [16–22]. Previous research on precipitation and clouds under climate variability and change have emphasized: (a) regional extreme precipitation events, cloud microphysics, and latent heating and forcing by a mesoscale convective system (MCS) [23–29], and (b) radiation heating feedback by various cloud types in determining global climate sensitivities [30–37]. While much knowledge has been gained and both approaches need to be continued in order to narrow down uncertainties, an emerging paradigm is that a deeper understanding of the myriad factors leading to extreme precipitation under climate change is predicated on a more comprehensive approach based on the broader context of interactions and enhanced by feedback processes involving cloud radiation, convection, and large-scale circulation [38–44].

Previous observational and climate modeling studies have shown that under global warming, the rate of increase in the top 0.1% of tropical daily precipitation has been estimated to be near 10% K^{-1}, significantly higher than those in the extratropics, which is limited by a thermodynamic rate of 6–7% K^{-1}, governed by the Clausius–Clapeyron relationship for atmospheric saturated moisture and temperature [2–5]. Models and observations have also shown that as Earth's surface and the atmosphere warm up under anthropogenic CO_2 radiative forcing, convection becomes more vigorous, and clouds grow faster, wider, and taller, producing more extreme precipitation [45]. An increasing number of recent studies [5,46–48] have shown that extreme precipitation events attributable to GHG warming tend to occur preferentially in tropical/subtropical regions with a strong and sustained organization of deep convection embedded in extended areas of high anvil clouds associated with long-lived strong mesoscale convective systems (MCS). Even though such long-lived MCS occur in less than 5% of the tropical precipitation events in preferred climatological wet regions, they account for more than 40% of the extreme precipitation amount [49]. This could mean that extreme precipitation, which occurs on hourly/daily time scales, could have organization signals on monthly and longer time scales over specific land or oceanic regions, and even over the entire tropics.

In spite of the increasing reports on devastating and destructive impacts on populated land regions, the scientific question of whether extreme cloud–precipitation organization is (a) fundamentally different and (b) more or less intense and/or frequent over land vs. ocean on climatic time scales remains uncertain. In this paper, we focus on addressing these questions and the scientific rationales underlying them based on general circulation model (GCM) simulations. However, because of the GCM's coarse resolution (>50–100km), MCS are not explicitly resolved and not well simulated in traditional GCM cloud–precipitation parameterizations. More recently, MCS-like features have been simulated and tracked in a moderate resolution (50 km) GCM, with improved physical cloud–precipitation parameterization that includes organization features occurring across the scales [50]. In this study, we conducted AMIP (Atmospheric Model Intercomparison Project)-type [51] simulations using the Department of Energy's Exascale Energy Earth System Model (E3SM), which includes an improved unified parameterization of clouds and precipitation types, to examine its capability in simulating MCS-like features and contributing to extreme tropical precipitation on climatic time scales. See a further discussion on the E3SM model's physics in Section 2.

Specifically, we disentangled the effects of surface warming vs. atmospheric heating and moistening by increased CO_2 radiation forcing, leading to an occurrence of extreme precipitation–cloud regimes, with respect to changes in the stratiform vs. convective precipitation, precipitation efficiency, and thermodynamic vs. dynamical forcing over land and ocean. The organization of the paper is as follows. In Section 2, we describe the methodology, including the key physical parameterizations of the clouds and precipitation processes and the experimental design of the E3SM model experiments. In Section 3, we present the key results of the experiments. The conclusions and scope of continuing work are discussed in Section 4.

2. Model Description and Methodology

The U.S. Department of Energy (DOE)'s Energy Exascale Earth System Model Version 1 (E3SMv1) [52] was developed with the aim of addressing the grand challenge of actionable prediction of the Earth system's variability and changes to meet scientific and societal needs. The E3SMv1 is a fully coupled ocean–atmosphere–land–biosphere model, developed on the foundation of the Community Earth System Model version 1 (CESM1), but it includes adaptations and improvements to optimize the computational performance and science/application requirements of the DOE.

For clouds and precipitation, the E3SM atmospheric model (EAM) uses an improved version of Cloud Layers Unified by Binormals (CLUBB), which includes a third-order turbulence closure parameterization that unifies the treatment of boundary-layer clouds, shallow and deep convection, and cloud microphysics [53,54]. In the E3SM, improving the model of shallow cumulus clouds and stratocumulus clouds and precipitation was achieved by optimizing the scale dependence of the CLUBB parameterization for a diurnal cycle of precipitation over land [55]. Deep convective clouds and precipitation are based on the improved version of the Zhang and McFarlane (1995) [56] scheme, which included a recent update on the bulk parameterization of updraft processes (entrainment, detrainment, condensation, and precipitation) and downdraft processes (entrainment and evaporation of falling rain) from both liquid- and ice-phase precipitation [57]. Aerosol and cloud microphysics interactions in stratiform clouds are included in an updated version of the Modal Aerosol Module (MAM4) [58], which predicts the concentrations of major aerosol species (sulfate, black carbon, primary and secondary organic matter, mineral dust, and sea spray). The Morrison and Gettelman Version 2 [59] aerosol–cloud microphysics parameterization, coupled with CLUBB and MAM4, was used for the generation of shallow and stratiform clouds. The implementation of a convective gustiness adjustment to CLUBB significantly improved the simulation of stratiform and shallow clouds over the tropical ocean, where the climatological surface mean winds are weak [60]. Radiation–cloud–convection–circulation interaction (RC3I) processes in the EAM have also been significantly improved by better microphysics-based treatment of wet scavenging and re-suspension of evaporating precipitation, which affect the abundance and size of cloud condensation nuclei for liquid- and ice-phase precipitation, respectively [61]. In addition, this study used a variant of EAMv1 that adopted a consistent set of parameter adjustments, including sub-grid scale wind variance, resulting in better simulations of cloud properties [55].

Based on the Community Land Model (4.5) of CESM2, the land model (ELM) of the E3SM includes improvements in the representation of the water cycle processes of soil hydrology, river routing, coastal erosion, and biogeochemistry fluxes [52]. A new river routing Model for Scale Adaptive River Transport (MOSART) was implemented, with particular emphases on human activities, including the management of water availability from river flow and the mitigation of floodplain inundation [62–64]. Two-way coupling between the MOSART and ELM was implemented to estimate the amount of water available from precipitation, river run-off, and storage in reservoirs for irrigation.

A key motivation for our analytical approach was to assess the degree to which the E3SM parameterization of fast and subgrid-scale cloud microphysical processes reflect the important contribution of mesoscale convective systems (MCS) to extreme cloud–precipitation organization on climatic (monthly and longer) time scales. Key features of MCS producing heavy precipitation over the ocean and land have been well documented [24,65]. During peak MCS development, a deep core with intense convective precipitation is coupled with extensive anvil clouds in the downwind regions, where stratiform precipitation dominates (Figure 1a). In the stratiform region, condensation heating associated with increased precipitation by active ice-phase microphysics (deposition, riming, and aggregation) causes a large-scale ascent in the upper troposphere above the freezing level (0 °C isotherm) near 500 hPa. At the same time, evaporative cooling by falling rain results in a large-scale mean descent in the lower troposphere. For tropical extreme precipitation events, the associated MCS life cycle may consist of multiple clusters of MCS complexes at various stages of development, starting with predominant convective precipitation and evolving to an increased contribution from stratiform precipitation. The results of cloud-resolving model simulations have shown that for non-MCS (100% convective) precipitation, the heating profile has a maximum near 500 hPa, while for "pure" stratiform precipitation, the heating profile shows a dipole structure with maximum heating (cooling) in the upper (lower) troposphere (Figure 1b). As a result, the degree of MCS development is reflected in the elevation of the level of maximum condensation heating, relative to that of convective (non-MCS) precipitation. However, it is important to note that the presence of an active convective core coupled with a substantial fraction of the stratiform (anvil cloud) region is essential for the development and maintenance of an MCS. A stratiform anvil cloud–precipitation regime decoupled from its convective core lacking in a sustained supply of ice-phase condensates from the convective core region represents a decaying MCS that readily dissipates and ceases to rain [24]. Hence, while the proportion of stratiform rain in a developing and active MCS is expected to increase with increasing extreme precipitation, it is unlikely to be close to 100% over the life cycle of the development of multiple organized MCS systems in extreme precipitation [65].

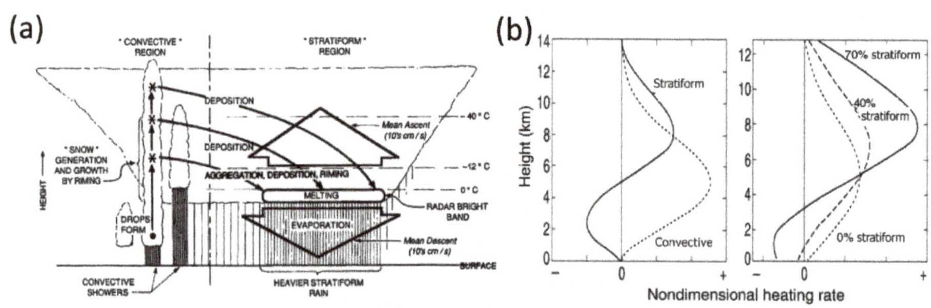

Figure 1. (**a**) Schematic showing organization of convective and stratiform cloud precipitation associated with a mesoscale convective system (adopted from Houze et al., 1989 [24]). Formation of water droplet and precipitation ice are denoted as • and *, respectively. (**b**) Idealized vertical profiles of latent heating as a function of stratiform precipitation fraction from simulations of cloud-resolving models (adopted from Sui et al., 2020 [66]).

For model integration, the control experiment was represented by an equilibrium solution of an AMIP-type 10-year integration of the E3SM with a 100 km latitude–longitude resolution, and 72 layers with variable thickness in the vertical direction, with a top at 60 km, under present-day sea surface temperature (SST), sea-ice conditions, and an atmospheric concentration of CO_2. To disentangle the effects of surface warming vs. CO_2 radiative forcing, equilibrium solutions based on two separate AMIP simulations identical to the control were conducted. First, the SST was increased by including an idealized plus-4K (P4K) anomaly uniformly across the globe. Second, an SST anomaly (SSTA) and CO_2 radiative forcing were imposed based on the climatology of the last 30-year simulation of an abrupt 4 times CO_2 (4xCO_2) experiment with the coupled ocean–atmosphere version of the E3SM, as part of the Coupled Model Intercomparison Project phase 6 (CMIP6) [67,68]. Changes in the tropical cloud precipitation characteristics for extreme precipitation were compared among the control, P4K, and 4xCO_2 simulations. Emphases were placed on a better understanding of the forcing and competing influences and feedback arising from surface warming vs. atmospheric heating and moistening processes. The realism of the model parameterization of the MCS and extreme precipitation, in terms of the changes in stratiform vs. convective precipitation, precipitation intensity, and large-scale circulation, was evaluated.

3. Results

3.1. *Stratiform vs. Convective Precipitation*

Since climate models do not resolve clouds explicitly because of their coarse resolution, model precipitation is classified as "convective" if they are produced by the subgrid-scale parameterization of deep convection, and as "stratiform" precipitation if they are produced by condensation processes of the large-scale (LS) circulation represented by the cloud microphysics parameterization. In this paper, for convenience, we use the term LS rain fraction (LSRF) in the model as synonymous with stratiform rain fraction. The variations in the LSRF (Figure 2a) as a function of the monthly precipitation from January to December (J2D) show that there was an increasing contribution of LS rain as a function of the precipitation intensity (P) over the entire tropics in the control, P4K, and 4xCO_2 simulations, respectively. The LS rain fraction increased faster (steeper rise) in the order of control, P4K, and 4xCO_2. For very extreme precipitation (P > 30 mm/day), the LS rain fraction rose to above 50%, reaching a maximum of ~70%, for P > 40 mm/day under P4K and 4xCO_2. For convenience, the unit for P is omitted hereafter. In comparing the same plots but separated into ocean and land, it is clear that for P < 30, most of the increase in the LSRF came from the ocean (Figure 2b). This is not surprising because of the much larger area of ocean compared to land in providing precipitable water to the atmosphere. Over the ocean, increases in the LSRF as a function of P were more robust under 4xCO_2 compared to P4K, with the former showing a steady increase up to P < 30, and the latter showing a peak in the LSRF at P~20. As explained in later sections, the stronger signal under 4xCO_2 was likely due to stronger dynamical feedback under a physically consistent SSTA and additional CO_2-induced radiative forcing and response compared to the idealized uniform SSTA-only forcing under P4K. Most interestingly, in comparing the LSRF variation over land (Figure 2c) to those over land and ocean (Figure 2a) and over ocean only (Figure 2b), it is clear that almost all of the very extreme precipitation (P > 30), while occurring rarely, were found only over tropical land regions. Worth noting is that the LSRF seldom reached above 0.7 even over land, indicating that the generation and transport of ice-phase condensate by deep convection to the upper troposphere are essential in order for ice-phase microphysics generating stratiform rain to take place under the extended anvil clouds (see Figure 1a). Anvil clouds will dissipate quickly without the sustained generation of ice-phase condensation from the convective core [65,66].

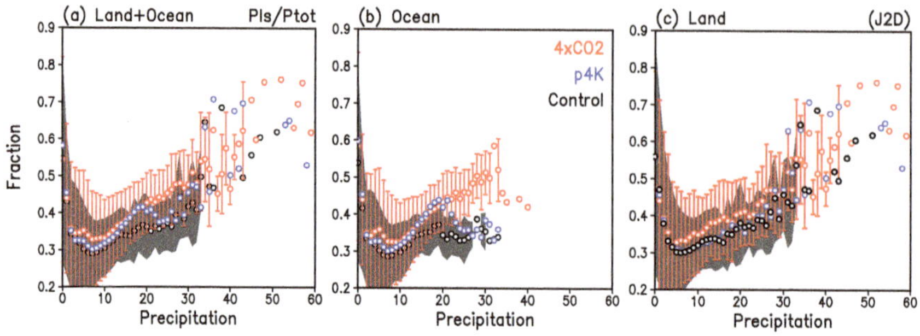

Figure 2. Stratiform rain fraction as a function of precipitation rate (mm day^{-1}) over (**a**) entire tropics, (**b**) ocean-only, and (**c**) land-only, for control (black), 4xCO_2 (red), and P4K (blue). Gray shading and red vertical bars represent a 1-s standard deviation for control, and 4xCO_2, respectively. Standard deviations for P4K are similar to 4xCO_2, and are not shown for clarity. J2D stands for monthly data taken from January to December.

A breakdown of the cumulative frequency of occurrence (FOC) of extreme monthly precipitation in terms of the total number of model grid points exceeding a given precipitation threshold (Table 1) shows that there was a rapid drop-off in the FOC with increasingly extreme precipitation. In the control climate, the FOCs of very extreme precipitation (P > 25–35) were indeed rarely (fewer than 1 in 1000) occurring preferentially over land, and rare or absent over the ocean. The FOCs of P > 30 increased by 3–5-fold under P4K and 4xCO_2 compared to the control and were stronger for the latter than the former. Analysis of the precipitation intensity threshold as a function of the top-percentile (PCT) rain rates showed similar signals, indicating more extreme heavy rain over land than the ocean (see Table S1). The preference for very extreme precipitation over land compared to the ocean appears to be an intrinsic property of the tropical ocean–land–atmosphere system, which was already present in the control, amplified under P4K, and even more so under 4xCO_2.

Table 1. Frequency of occurrence (FOC) measured in total number of model grid points over the entire tropics (30° S–30° N) as a function extreme monthly precipitation (P) exceeding a given threshold intensity, for control, P4K, and 4xCO_2 over land, ocean, and land + ocean, respectively. Quantities in bracket in first column show total number of grid points over the entire tropics (30° S–30° N). Unit of P is mm day^{-1}.

		P > 15	P > 20	P > 25	P > 30	P > 35
Ocean (192984)	Control	555	149	27	6	0
	P4K	2357	322	43	6	0
	4xCO_2	5555	1012	161	36	3
Land (66216)	Control	840	283	104	39	13
	P4K	1172	276	111	27	13
	4xCO_2	1739	581	244	117	51
Ocean+ Land (259200)	Control	1395	432	131	45	13
	P4K	3529	598	154	33	13
	4xCO_2	7294	1593	405	153	54

The spatial distributions of the frequency of occurrence (FOC) of extreme precipitation based on the rain rate for the top 1 percentile (PCT01) and top 5 percentile (PCT05) rainfall were computed. To facilitate comparison, the thresholds for the control for ocean + land were used to compute the FOC geographical distributions for P4K and 4xCO_2. The PCT01 (P > 13) rains (Figure 3a) occurred over limited areas within the climatological rainy regions of the Asian monsoon, the maritime continent/Pacific warm pool (SST > 302K), and the equatorial East Pacific ITCZ, with isolated signals over land regions in equatorial South

America and Africa. Under P4K (Figure 3b), the warm pool areas expanded substantially, covering much of the tropics. The PCT01 rain areas also expanded, but were still anchored to the climatological wet regions within the much warmer SST (SST > 304K). The increased PCT01 precipitation over the equatorial land region was more prominent compared to the control. Worth noting is that under P4K, except for the expansion of wetter areas, there were no fundamental changes in the spatial structure of tropical rainfall distribution compared to the control, suggesting a strong wet-getting-wetter (WeGW) scenario [10,11]. Under 4xCO_2 (Figure 3c), the areal extent of the Pacific warm pool was further expanded compared to P4K, covering the entire tropical ocean (25° S–25° N). Over the aforementioned WeGW regions, PCT01 rain FOCs were further enhanced and expanded compared to P4K. Additionally, prominent centers of action for PCT01 precipitation were found over the equatorial Indian Ocean, and over the equatorial Atlantic Ocean under 4xCO_2. that is, the expanded PCT01 rain areas exhibited not only WeGW but also a warmer-getting-wetter (WaGW) pattern [12]. Overall, the tropical SST was warmer by 1.85 K under 4xCO_2 compared to P4K.

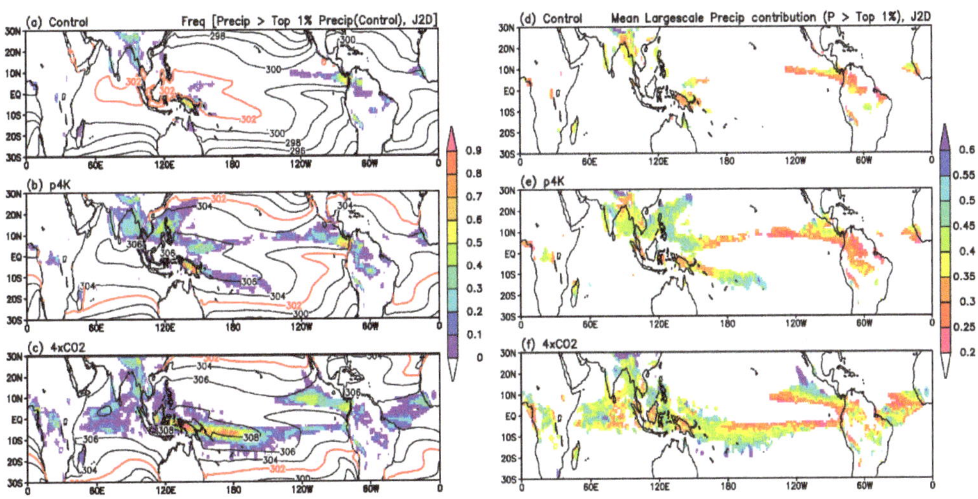

Figure 3. Spatial distribution of SST and frequency of occurrence (FOC) for top 1% precipitation (PCT01) in fractional units, based on monthly rainfall data from January through December (J2D), for (**a**) control, (**b**) P4K, and (**c**) 4xCO_2, with warm pool SST (302K) outlined in red. Corresponding distributions for stratiform (large-scale) rain in fractional units are shown in (**d**–**f**), respectively.

Under the control climate, close matches between the areas of PCT01 rain (Figure 3a) and regions of an enhanced LSRF were discernable over the Asian monsoon land, the maritime continent, and the eastern Pacific ITCZ (Figure 3d). The sparse spatial extent of the PCT01 LSRF signals the rarity of such events in the control. Under P4K, the regions of enhanced FOC (Figure 3b) were well co-located with those with a large LSRF (>45–50%) over the Asian monsoon region, the maritime continent, the SPCZ, and the northern edge of the ITCZ over the eastern Pacific (Figure 3e). Under 4xCO_2, the co-location of high FOCs of PCT01 rain (Figure 3c) with an increased LSRF (Figure 3f) could be seen over the aforementioned regions, as well as the land regions of equatorial Africa and the Amazon, consistent with the WeGW and WaGW patterns. Similar patterns of the FOC for PCT05 (P > 10) and an enhanced LS rain fraction were computed, indicating increasing contributions from LS (stratiform) rain types in more expansive regions of a high FOC compared to PCT01 (see Figure S1).

3.2. Precipitation Efficiency and MCS Organization

Recent model simulations and observations have shown that increased precipitation intensity is highly correlated with enhanced precipitation efficiency (PE), that is, an enhanced rate of microphysical auto-conversion of cloud water (liquid and ice phase), as the surface temperature rises [4,66,69–71], and it is a key driver of the large-scale circulation sustaining tropical heavy precipitation under global warming [72,73]. Here, we define the PE as the ratio of precipitation to the column integration of the total cloud water (TCW), including liquid and ice, as simulated by the microphysics parameterization of clouds and precipitation used in the E3SM (see discussion in Section 2).

$$PE = P/TCW \text{ (in units of } s^{-1}). \quad (1)$$

Physically, the inverse of PE ($\tau = PE^{-1}$) represents a characteristic residence time scale for the total condensed cloud water in an atmospheric column undergoing precipitation for a given precipitation rate. A high value of PE (low value of t) reflects vigorous water recycling within the atmosphere, converting cloud liquid and ice water into precipitation, while maintaining an abundant stock of the TCW in the atmosphere through enhanced surface moisture flux and low-level moisture convergence [66,74].

Figure 4a shows a nearly linear increase in the PE as a function of P over the entire tropics, with a faster rate (steeper gradient) of increase in the PE for extreme precipitation from the control to P4K to 4xCO_2. The typical range of values of PE (0.02–0.2) is from τ = 50 to 5 minutes, that is, there is a 10-fold reduction in the residence time scale of the TCW in the atmosphere, from light to the most extreme precipitation in the tropics. These values of τ can be considered a crude estimate of increasingly fast cloud–water–precipitation recycling time scales in MCS-like organization systems, contributing to the extreme precipitation in the E3SM model. Compared to the ocean-only plot (Figure 4b), it can be seen that most of the PE increase for P < 30 represents contributions mainly from oceanic precipitation, with a faster increase in the order of control, P4K, and 4xCO_2. However, very extreme precipitation (P > 30) with high PE (PE > 0.1) was not found over the ocean. In contrast, the rate of increase in the PE as a function of P (Figure 4c) was faster over land than over the ocean for all precipitation rates. For extreme precipitation (P > 30) over land, the rate of increase in the PE was clearly accelerated compared to lower rain rates (Figure 4c). Comparing Figure 4a–c, it can be seen that almost all of the very extreme tropical precipitation (P > 30) and high PE (>0.1) events came from the land.

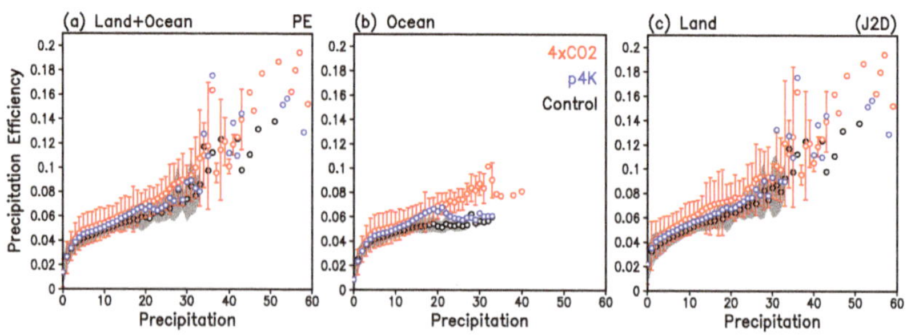

Figure 4. Precipitation efficiency ($minute^{-1}$) as a function of the precipitation rate (mm day^{-1}) for January through December, for (**a**) land + ocean, (**b**) ocean-only, and (**c**) land only, based on monthly data from January through December (J2D). Gray shading and red vertical bars represent 1 s standard deviation for control and 4xCO_2, respectively. Standard deviations for P4K are similar to 4xCO_2, but are not shown for clarity.

The geographic distributions of the PE of PCT01 rainfall for the control, P4K, and 4xCO_2 (Figure 5a–c) show strong similarities to the pattern of outgoing longwave radiation (OLR), indicating an abundance of cold anvil clouds with low OLR (<190 Wm^{-2}) in regions with enhanced PE (Figure 5d–f). Under P4K and $4xCO_2$, more so in the latter than the former, higher PE with lower OLR (more elevated clouds with colder tops) were found over the Asian monsoon, maritime continent, and equatorial Africa and South America regions. In contrast, over the open oceans of the Pacific ITCZ, the tropical western Pacific, and the South Pacific Convergence Zone (SPCZ), extreme precipitation was derived mostly from increased PE in regions with OLR >215 Wm^{-2}, consistent with an increased abundance of warm rain as a key signal of climate warming [69,75]. For moderately extreme precipitation (PCT05), the areal extent of high PE and low OLR increased substantially in conjunction with the expansion of the tropical SST warm pool (see Figure 3). Overall, the PE and OLR distributions for PCT01 and PCT05 exhibited the WeGW pattern under the control and P4K, and the WeGW + WaGW under $4xCO_2$, similar to those for the LSRF (see discussion about Figure 3).

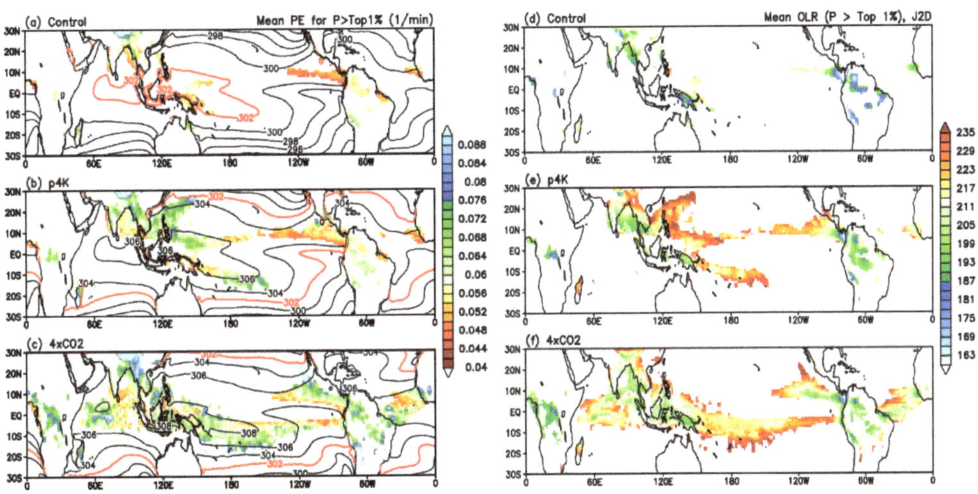

Figure 5. Spatial distribution of SST and precipitation efficiency (PE) for top 1% precipitation (PCT01) in units of $minute^{-1}$, based on monthly rainfall data from January to December (J2D) for (**a**) control, (**b**) P4K, and (**c**) $4xCO_2$, with respective warm pool SST outlined in red. Corresponding distributions for outgoing longwave radiation (OLR) in units of Watt m^{-2} are shown in (**d**), (**e**), and (**f**) respectively.

Next, we explored the capability of the E3SM in simulating MCS-like extreme precipitation organization, with regard to an increased contribution from stratiform (anvil) rain, enhanced PE in the production by freezing, and the removal of cloud ice by melting and precipitation fallout. Specifically, we computed composite change patterns of P4K and $4XCO_2$ relative to the control and that of $4xCO_2$ relative to P4K in the vertical profiles of key MCS quantities, i.e., cloud ice concentration, condensation heating, and large-scale vertical velocity, as a function of the precipitation intensity of the entire tropics, separately for land and ocean. Over the ocean, the level of maximum cloud ice can be seen to shift upward relative to the control as precipitation increases under P4K (Figure 6a), starting at P~10 and continuing up to P > 25–30. The negative (positive) values of cloud ice signals accelerated the removal (accretion) of cloud ice below (above) 300 hPa by enhanced precipitation (condensation) relative to the control. Given the co-location of the regions of enhanced precipitation (PCT01) and the increased LSRF (Figure 3), as well as the increased PE and low OLR values (Figure 5), the cloud ice features are consistent with the enhanced model of MCS-like organization compared to the control and analogous to those shown in Figure 1a. Under $4xCO_2$ (Figure 6b), the cloud ice anomaly pattern is similar to that under P4K, indicating the primary importance of ocean warming in initiating the MCS organization. However, the MCS structure appears to be more robust under $4xCO_2$ compared to under P4K. The stronger organized MCS development under 4xCO_2 can also be seen in the difference plot of 4xCO_2-minus-P4K (Figure 6c), indicating a stronger removal of cloud ice by precipitation near 400–250 hPa, and increased melting due to the warming of the middle and lower troposphere, coupled with enhanced cloud ice formation near 250–150 hPa. These likely reflect the effect of increased CO_2 radiative heating in the lower troposphere, enhancing convective instability in the upper troposphere [76]. Over land (Figure 6d,e), the changes in the cloud ice in the upper troposphere reflecting the increasing MCS organization under P4K and 4xCO_2 are similar to those in the ocean, as is evident by the strong removal of cloud ice near 500–350 hPa and the accumulation of cloud ice above (250–150 hPa) associated with anvil cloud development. Under P4K and 4xCO_2 (Figure 6d,e), the MCS organization over land shows less cloud ice loading (solid contours) but a more vertically confined region of negative anomalies, indicating stronger cloud ice removal compared to over the ocean. However, very extreme precipitation $P \geq 30$–35 occurred only over land in 4xCO_2, but not over the ocean (solid contours in Figure 6b,e). The additional radiative heating effect due 4xCO_2 further enhanced the MCS precipitation organization over land compared to P4K (Figure 6f).

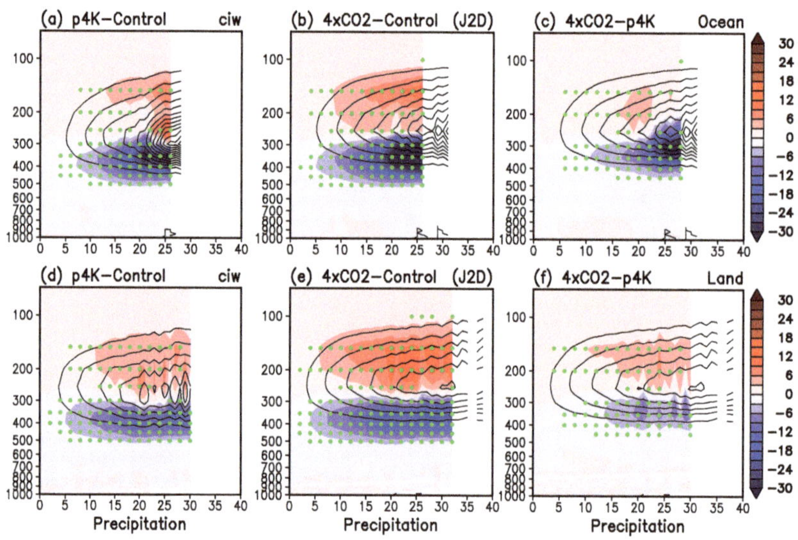

Figure 6. Vertical profiles of anomalous cloud ice contents (10^{-6} kg/kg, ice mass per kilogram of air mass) as a function of precipitation intensity (mm day^{-1}) over ocean for (**a**) P4K-minus-control, (**b**) 4xCO_2-minus-control, and (**c**) 4xCO_2-minus-P4K, from January to December (J2D). Panels (**d**), (**e**), and (**f**) are the same as (**a**), (**b**), and (**c**), respectively, but over land. Contours show the mean profiles of condensation heating for the minuend (first term of the subtraction) indicated in the respective subpanel labels. Regions with statistical significance exceeding 95% confidence are highlighted by green dots.

Over the ocean, the condensation heating profiles as a function of P for P4K and 4xCO_2 (Figure 7a,b) reveal an essential feature of MCS organization, that is, the elevation of the level of condensation heating is characterized by positive (negative) anomalies above (below) 300 hPa as the precipitation intensifies (cf. Figure 1b). This is consistent with the increase in the LSRF (see Figure 2) and PE (see Figure 4), as discussed previously. Strong cooling found near the freezing level at 500 hPa and regions slightly above signals enhanced melting and evaporation of falling rain. The MCS organization appears to be stronger under 4xCO_2 relative to P4K, with more condensation heating above (below) 250hPa (Figure 7c). Over land, the condensation heating profiles (Figure 7d,e) exhibit similar features to their ocean counterparts, but with more robust MCS-like features, that is, elevated condensation heating, strong cooling at the mid-troposphere freezing level and regions below (Figure 7d,e), and a stronger response in 4xCO_2 compared to P4K (Figure 7f) due to the additional radiative heating of the atmospheric CO_2.

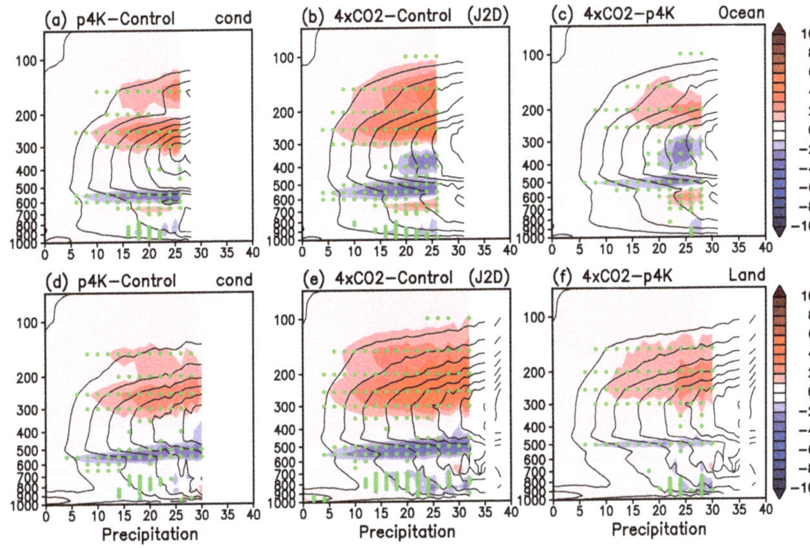

Figure 7. Vertical profiles of anomalous condensational heating (K day^{-1}) as a function of precipitation intensity (mm day^{-1}) over ocean for (**a**) P4K-minus-control, (**b**) 4xCO_2-minus-control, and (**c**) 4xCO_2-minus-P4K, from January to December (J2D). Panels (**d**), (**e**), and (**f**) are the same as (**a**), (**b**), and (**c**), respectively, but over land. Contours show the profiles of condensation of the minuend (first term of the subtraction) indicated in the respective subpanel labels. Regions with statistical significance exceeding 95% confidence are highlighted by green dots.

For the large-scale vertical velocity over the ocean under P4K and 4xCO_2 (Figure 8a,b), increased upward (downward) motions in the upper (middle-and-lower) troposphere are evident and consistent with the condensation heating (cooling) changes (see Figure 7). The decrease in the upward vertical motion in the mid-troposphere is indicative of the MCS organization, pertaining to an increased melting of cloud ice at the distinctive freezing level near 500 hPa and increased downdraft associated with evaporative cooling in the regions of falling rain (cf. Figure 1a). Again, the effects are stronger under 4xCO_2 compared to P4K (Figure 8c). Over land (Figure 8d,e), the changes in the large-scale vertical motions are similar to those in the ocean, except they appear more muted under both P4K and 4xCO_2, with the latter only slightly stronger than the former (Figure 8f). The stronger MCS-like signals over the ocean, especially the strong, distinctive cooling at the freezing level compared to over land, reflect the direct effects of stronger forcing over the ocean from the SSTA, as well as positive feedback from changes in the large-scale circulation. Importantly, the anomalous large-scale vertical motions over land shown here are likely attributable to not only the MCS organization but also changes in the large-scale Walker Circulation, driven by an east–west SST gradient and the land–sea thermal contrast, further modulating changes in the MCS convective updraft over land [77,78].

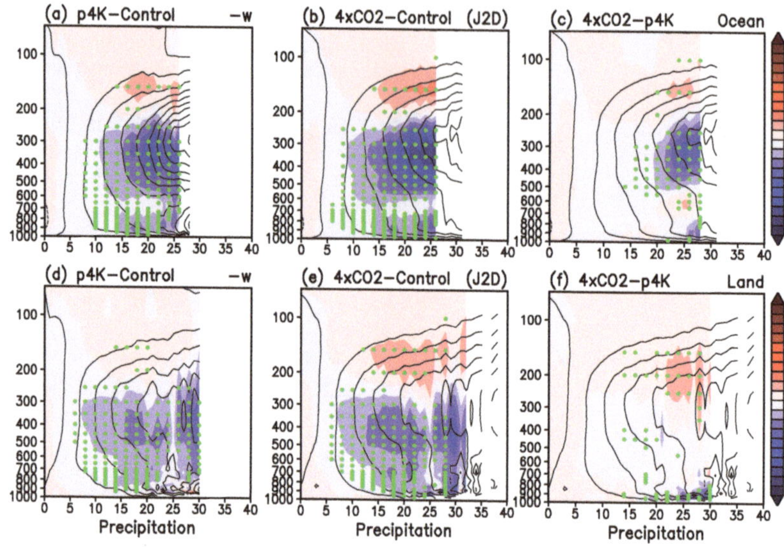

Figure 8. Vertical profiles of anomalous upward vertical velocity (Pa/s) as a function of precipitation intensity (mm day^{-1}) over ocean for (**a**) P4K-minus-control, (**b**) 4xCO_2-minus-control, and (**c**) 4xCO_2-minus-P4K, from January to December (J2D). Panels (**d**), (**e**), and (**f**) are the same as (**a**), (**b**), and (**c**), respectively, but over land. Contours show the profiles of condensation of the minuend (first term of the subtraction) shown in the respective subpanel labels. Regions with statistical significance exceeding 95% confidence are highlighted by green dots.

3.3. *Convective Inhibition (CIN) and Extreme Precipitation*

In this subsection, we explore further the fundamental reason why very extreme but rare (record-breaking) precipitation tends to occur over land rather than the ocean. Under GHG warming, the convective available potential energy (CAPE) is expected to increase due to the relative fast rate (~7% K^{-1}) of increase in the atmospheric saturated moisture with higher temperature. However, convective inhibition (CIN), that is, near-surface negative buoyancy, is known to be enhanced under global warming over land, resulting in increased drying (sub-saturation) of the near-surface air due to a lack of moisture supply relative to the fast land warming [79,80]. CIN drying is reflected in reduced low-level relative humidity, a higher lifting condensation level (LCL), and an elevated level of free convection (LFC), inhibiting deep convection [81].

To illustrate the effect of CIN under P4K and 4xCO_2 and its relationship with extreme precipitation, an analysis of the surface moist energy budget follows. The convective instability of the atmosphere is controlled by the vertical gradient of the moist static energy (MSE), with

$$MSE = C_pT + Lq + gz, \quad (2)$$

where C_p is the thermal capacity at a constant pressure, T is the surface air temperature, L is the latent heat of condensation, q is the specific humidity, g is the gravitation constant, and z is the geopotential height. A negative MSE vertical gradient (high—below, low—above) implies convective instability and vice versa for stability. For CIN, we focused on the first two terms (1) near the surface, that is, the lowest model level, where gz is negligibly small.

Under P4K (Figure 9a), the near-surface C_pT anomalies (relative to the control) increased nearly uniformly (~4–5 kJ/kg) over the entire tropical ocean, following closely that of the imposed idealized 4K uniform SST warming. The C_pT increase over land was stronger (~5–7kJ/Kg) compared to over the ocean because land has a lower thermal capacity than water. As a result, the land temperature rises faster and higher than that of the ocean with the same amount of heat input. In addition, the lack of land moisture sources results in less evaporative cooling. Hence, under global warming, the land temperature has to rise much higher compared to the ocean temperature to enhance outgoing longwave radiative cooling, which is needed to balance the land heating from the CO_2 greenhouse effect and increased downward solar radiation from reduced clouds due to drying. The surface Lq (Figure 9b) followed a similar change pattern to C_pT and was clearly the dominant forcing (~8–15kJ/kg), stronger than that of C_pT by 2–3 times. This is because of the well-known exponential increase in the atmospheric saturated moisture content as a function of temperature, governed by the Clausius–Clapeyron relationship. Over the ocean, due to the readily available moisture from below, the near-surface relative humidity (RH) remained close to the saturation values. As a result, the anomalous relative humidity under P4K vs. the control was positive but small (<2–4%) over the ocean (Figure 9c). However, over land, because of the larger increase in the C_pT, the additional moisture required to reach saturation far exceeded that which could be derived from local moisture sources. As a result, there was a distinctive reduction in the RH (~3–6 %), indicating drying over land under P4K relative to the control (Figure 9c) and signaling increased CIN [81]. However, as the land temperature rises and CIN increases under P4K, the triggering of convection induced by mesoscale convergence, episodic outflow from land–sea breeze, and forced lifting from surface inhomogeneity and/or orographic may lead to an explosive growth of convection, releasing a large amount of stored CAPE during CIN [81]. The delayed onset of deep convection due to increased CIN could facilitate the occurrence of very extreme but rare precipitation in a warming climate, specifically over land. The stronger and longer-lasting the CIN, the more CAPE builds up in the lower troposphere, and the more extreme the precipitation when it eventually occurs, releasing a large amount of built-up CAPE during CIN [82–85].

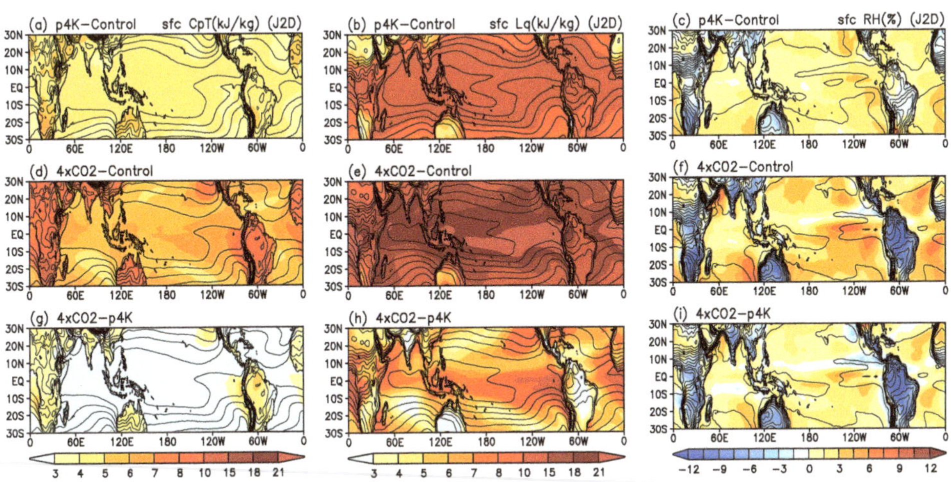

Figure 9. Spatial distribution of surface anomalous (**a**) C_pT (kJ/kg), (**b**) Lq (kJ/kg), and (**c**) relative humidity (%) for January to December (J2D) under P4K. Panels (**d**), (**e**), and (**f**) are the same as in (**a**), (**b**), and (**c**), respectively, but under $4xCO_2$. Panels (**g–i**) are the same as (**a–c**) but for $4xCO_2$-minus-P4K.

Under $4xCO_2$ (Figure 9d), the near-surface anomalous C_pT warming over land was much larger (>8–15 kJ/kg) than over the ocean (~5–7 kJ/kg). Again, the increase in the surface Lq (Figure 9e) over the tropical oceans followed the corresponding increase in the SST, consistent with $4xCO_2$ forcing and indicating enhanced warming and moistening of the surface air over the tropical ocean, following the Clausius–Clapeyron relationship. The differential magnitude of the anomalous Lq and C_pT resulted in a large contrast in the relative humidity (RH) between the ocean and land, that is, increased RH over the tropical ocean, and decreased RH over land (Figure 9f), further enhancing the land–sea contrast, as noted in P4K (Figure 9c). Judging from the 4xCO_2-minus-P4K pattern in C_pT (Figure 9g), it can be seen from the near uniform and small values (<3 kJ/kg) over the ocean that the P4K SST warming was a reasonable analog of the SST surface thermal forcing under 4xCO_2. However, the surface moisture forcing Lq reveals more regional features with higher values (>6–10 kJ/kg) over oceanic regions near the equator and subtropical monsoon regions adjacent to land in 4xCO_2 compared to P4K (Figure 9h). Clearly, enhanced atmospheric warming by 4xCO_2 further exacerbated the surface RH reduction over land compared to P4K (Figure 9i). This is likely due to enhanced land–atmosphere feedback arising from 4xCO_2, radiative forcing of the atmosphere and facilitated by cloud–convection–circulation interactions [19] under dynamically consistent SSTA forcing, increasing CIN, and the occurrence of very extreme but rare precipitation already operative under the control, but enhanced by P4K, and further amplified under 4xCO_2.

4. Concluding Remarks

Based on the AMIP-type model simulations using the Exascale Energy Earth System Model (E3SM), we investigated the changing characteristics of climate-scale (monthly) tropical extreme precipitation in a warming climate. Three ten-year-long AMIP-type model simulations were carried out: (1) a control, with the present-day SST, and CO_2 atmospheric concentration, (2) P4K, the same as the control but with a forced idealized uniform 4K increase in the SST globally, and (3) 4xCO_2, the same as the control but with SSTA derived from coupled model simulations under a four-times-higher atmospheric CO_2 concentration, including the corresponding 4xCO_2 radiative heating of the atmosphere. The key results of this study include the following:

- In a warming tropical climate, while both convective and stratiform rain increase, there is an increasing contribution from the stratiform rain fraction to extreme precipitation, with the most extreme but rare precipitation occurring preferentially over land compared to the ocean. However, the stratiform rain fraction approaches an upper limit of approximately 0.7, indicating that a deep convection core is essential to provide ice-phase condensate for stratiform rain even for the most extreme precipitation.
- The distributions of extreme precipitation (top 1% and 5%) generally follow the paradigms of wet-getting-wetter (WeGW) under the control and P4K, but both show WeGE and warmer-getting-wetter (WaGW) within an expanded tropical SST warm pool, and regional SST warming under 4xCO_2.
- Extreme precipitation is facilitated by increased precipitation efficiency (PE), reflecting an accelerated rate of recycling of precipitation and total cloud water (both liquid and ice phases) in regions of strongly reduced outgoing longwave radiation (<190Wm^{-2}), associated with colder (higher) anvil cloud tops.
- The increase in PE associated with the extreme precipitation under P4K and 4xCO_2 is reflected in a more MCS-like organization structure over land and ocean compared to the control, including (a) increased ice-phase upper-level clouds, (b) an elevated level of condensation heating in the upper troposphere and strong cooling from the enhanced melting of ice near the freezing level and altitudes below from the evaporation of falling rain, and (c) an increased ascent (descent) of large-scale vertical motion in the upper (lower) troposphere.
- Analysis of the surface moist static energy distribution revealed that moisture forcing (Lq) from an increased higher SST is the primary driver of extreme precipitation over the ocean, in accordance with the Clausius–Clapeyron relationship. However, surface temperature forcing (C_pT) is more important over land, as reflected in the much higher land surface temperature due to the smaller heat capacity of land and a lack of moisture sources from land.
- The high surface temperature over land leads to enhanced convective inhibition (CIN), that is, the drying of the land surface, reflected in reduced relative humidity of the near-surface air over land under P4K and 4xCO_2, more so in the latter than the former. We argue that the very extreme but rare precipitation over land is likely due

to increased CIN, delaying the triggering of deep convection, while building up the convective available energy in the lower atmosphere associated with the warming climate. When deep convection is triggered eventually through moisture advection from episodic small-scale atmospheric eddy processes associated with land–sea breeze, thunderstorms, and orography, an explosive growth of MCS-like organization occurs preferentially over land, releasing extra amounts of convective available potential energy (CAPE) stored during CIN, and resulting in very extreme "record-breaking" precipitation over land, as global climate warming continues unabated.

The similarities in MCS extreme precipitation development over ocean and land and between $4xCO_2$ and P4K underscore the importance of SST warming as the primary forcing in the development of MCS-like organization, leading to extreme precipitation. However, non-uniform SSTA based on ensemble coupled models together with dynamically consistent CO_2 radiative forcing of the atmosphere is needed to produce stronger and presumably more realistic regional characteristics of extreme precipitation in the warming climate of a future world through dynamical adjustments and feedback processes in the coupled atmosphere–ocean–land system. For a better understanding of the effects of CIN in staging very extreme "record-breaking" regional precipitation events over land, intrinsic land–atmosphere feedback processes and impacts by concomitant changes in the tropical large-scale circulation, land–sea contrast, and under P4K and $4xCO_2$, comparisons with CMIP6 model outputs, and multiple sources of precipitation and cloud observations are being investigated in our ongoing research.

Finally, we note that high-resolution MCS resolving meso-scale (10–20 km) and cloud-scale (<5–10 km) models are required to conduct studies of extreme precipitation events on hourly/daily time scales over limited spatial/time domains. Cloud-scale GCM and coupled GCMs are certainly desirable for better simulations of MCS over the global domain. However, such GCM simulations are highly labor-intensive and expensive for climate-scale long-term integrations. That is why most long-term GCM climate experiments, such as in CMIP6, are still expected to run at moderate-to-low resolution (>50–100 km) in the foreseeable future. Here, we show important results indicating that improved cumulus parameterization in a state-of-the-art GCM with moderate resolution can show MCS-like organization features for extreme tropical precipitation, on monthly time scales. Such an approach allows for the physics of extreme precipitation, such as MCS-like organization, to be explored and evaluated by precipitation and cloud observations on a global climatic scale, bridging the gap between meso-scale and low-resolution climate models.

Supplementary Materials: The following supporting information can be downloaded at: https://www.mdpi.com/article/10.3390/atmos14060995/s1, Figure S1. Spatial distribution of SST, Frequency of Occurrence (FOC), and the fraction of stratiform (large-scale) rain for top 5% monthly precipiptation; Figure S2. Spatial distribution of SST, precipitation efficiency (PE), outgoing longwave radiation (OLR) for top 5% monthly precipitation; Table S1. Extreme tropical monthly precipitation intensity threshold (mm/day) as a function of top-percentile rain rate for the entire tropics (30° S–30° N).

Author Contributions: Conceptualization, W.K.M.L. and L.R.L.; methodology, W.K.M.L., B.H. and K.-M.K.; software, B.H. and K.-M.K.; validation, W.K.M.L., B.H. and K.-M.K.; formal analysis, K.-M.K.; investigation, W.K.M.L. and K.-M.K.; resources, L.R.L., B.H. and K.-M.K.; data curation, B.H. and K.-M.K.; writing—original draft preparation, W.K.M.L.; writing—review and editing, L.R.L., B.H. and K.-M.K.; visualization, K.-M.K. and W.K.M.L.; supervision, W.K.M.L.; funding acquisition, W.K.M.L. and L.R.L. All authors have read and agreed to the published version of the manuscript.

Funding: This research was funded by the U.S. Department of Energy (DOE), the Office of Science, Biological, and Environmental Research as part of the Regional and Global Model Analysis Program Area under Grant Award #300426-00001, awarded to the University of Maryland from the Pacific Northwest National Laboratory (PNNL). The PNNL is operated for the DOE by the Battelle Memorial Institute under contract DE-AC05-76RL01830. Partial support was also provided by the NASA Modeling and Analysis Program, Federal Award Identification # 80NSSC21K1800, awarded to the University of Maryland and to the NASA Goddard Space Flight Center.

Institutional Review Board Statement: Not applicable.

Informed Consent Statement: Not applicable.

Data Availability Statement: The data used in this research were based on model outputs of the DOE/PNNL, available at https://portal.nersc.gov/archive/home/b/beharrop/www/e3sm_v1_pd_and_warming_experiments/e3sm_v1_control_simulations.tar (assessed on 1 June 2023).

Conflicts of Interest: The authors declare no conflict of interest.

References

1. IPCC. *Summary for Policymakers*; The Physical Science Basis. Contribution of Working Group I to the Sixth Assessment Report of the Intergovernmental Panel on Climate Change; Cambridge University Press: Cambridge, UK, 2021; pp. 3–32.
2. Trenberth, K.E.; Dai, A.; Rasmussen, R.M.; Parsons, D.B. The Changing Character of Precipitation. *Bull. Am. Meteor. Soc.* **2003**, *84*, 1205–1218. [CrossRef]
3. O'Gorman, P.A. Sensitivity of tropical precipitation extremes to climate change. *Nat. Geosci.* **2012**, *5*, 697–700. [CrossRef]
4. O'Gorman, P.A. Precipitation Extremes Under Climate Change. *Curr. Clim. Chang. Rep.* **2015**, *1*, 49–59. [CrossRef]
5. Rossow, W.B.; Mekonnen, A.; Pearl, C.; Goncalves, W. Tropical Precipitation Extremes. *J. Clim.* **2013**, *26*, 1457–1466. [CrossRef]
6. Xie, P.; Arkin, P.A. Analyses of Global Monthly Precipitation Using Gauge Observations, Satellite Estimates, and Numerical Model Predictions. *J. Clim.* **1996**, *9*, 840–858. [CrossRef]
7. Lau, K.M.; Wu, H.T. Detecting trends in tropical rainfall characteristics, 1979–2003. *Int. J. Climatol.* **2007**, *27*, 979–988. [CrossRef]
8. Adler, R.F.; Gu, G.; Wang, J.-J.; Huffman, G.J.; Curtis, S.; Bolvin, D. Relationships between global precipitation and surface temperature on interannual and longer timescales (1979–2006). *J. Geophys. Res.* **2008**, *113*, D2210. [CrossRef]
9. Adler, R.F.; Sapiano, M.R.P.; Huffman, G.J.; Wang, J.-J.; Gu, G.; Bolvin, D.; Chiu, L.; Schneider, U.; Becker, A.; Nelkin, E.; et al. The Global Precipitation Climatology Project (GPCP) Monthly Analysis (New Version 2.3) and a Review of 2017 Global Precipitation. *Atmosphere* **2018**, *9*, 138. [CrossRef]
10. Chou, C.; Neelin, J.D. Mechanisms of Global Warming Impacts on Regional Tropical Precipitation. *J. Clim.* **2004**, *17*, 2688–2701. [CrossRef]
11. Chou, C.; Neelin, J.D.; Chen, C.A.; Tu, J.Y. Evaluating the 'rich-get-richer' mechanism in tropical precipitation change under global warming. *J. Clim.* **2009**, *22*, 1982–2005. [CrossRef]
12. Xie, S.P.; Deser, C.; Vecchi, G.A.; Ma, J.; Teng, H.; Wittenberg, A.T. Global warming pattern formation: Sea surface temperature and rainfall. *J. Clim.* **2010**, *23*, 966–986. [CrossRef]
13. Huang, P.; Xie, S.-P.; Hu, K.; Huang, G.; Huang, R. Patterns of the seasonal response of tropical rainfall to global warming. *Nat. Geosci.* **2013**, *6*, 357–361. [CrossRef]
14. Qian, C.; Chen, G. Warmer-get-wetter or wet-get-wetter? A criterion to classify oceanic precipitation. *J. Ocean Univ. China* **2014**, *13*, 552–560. [CrossRef]
15. Wu, H.-T.J.; Lau, W.K.-M. Detecting climate signals in precipitation extremes from TRMM (1998–2013)—Increasing contrast between wet and dry extremes during the "global warming hiatus". *Geophys. Res. Lett.* **2016**, *43*, 1340–1348. [CrossRef]
16. Bony, S.; Lau, K.-M.; Sud, Y.C. Sea Surface Temperature and Large-Scale Circulation Influences on Tropical Greenhouse Effect and Cloud Radiative Forcing. *J. Clim.* **1997**, *10*, 2055–2077. [CrossRef]
17. Bony, S.; Stevens, B.; Frierson, D.M.W.; Jakob, C.; Kageyama, M.; Pincus, R.; Shepherd, T.G.; Sherwood, S.C.; Siebesma, A.P.; Sobel, A.H.; et al. Clouds, circulation and climate sensitivity. *Nat. Geosci.* **2015**, *8*, 261–268. [CrossRef]
18. Lau, K.-M.; Wu, H.-T.; Bony, S. The Role of Large-Scale Atmospheric Circulation in the Relationship between Tropical Convection and Sea Surface Temperature. *J. Clim.* **1997**, *10*, 381–392. [CrossRef]
19. Lau, W.K.M.; Kim, K.-M.; Chern, J.; Tao, W.-K.; Leung, R. Structural change and variability of the ITCZ induced by radiation-cloud-convection-circulation interactions: Inferences from the Goddard Multi-Model Framework (GMMF) experiments. *Clim. Dym.* **2019**, *54*, 211–229. [CrossRef]
20. Schumacher, R.S. Sensitivity of Precipitation Accumulation in Elevated Convective Systems to Small Changes in Low-Level Moisture. *J. Atmos. Sci.* **2015**, *72*, 2507–2524. [CrossRef]
21. Ahmed, F.; Schumacher, C. Convective and stratiform components of the precipitation-moisture relationship. *Geophys. Res. Lett.* **2015**, *42*, 453-10–462-10. [CrossRef]
22. Lau, W.K.M.; Tao, W. Precipitation–Radiation–Circulation Feedback Processes Associated with Structural Changes of the ITCZ in a Warming Climate during 1980–2014: An Observational Portrayal. *J. Clim.* **2020**, *33*, 8737–8749. [CrossRef]
23. Laing, A.G.; Fritsch, J.M. The global population of mesoscale convective complexes. *Q. J. R. Meteorol. Soc.* **1997**, *123*, 389–405. [CrossRef]
24. Houze, R.A., Jr. Observed structure of mesoscale convective systems and implications for large-scale heating. *Q. J. R. Meteorol. Soc.* **1989**, *115*, 425–461. [CrossRef]
25. Houze, R.A. Mesoscale convective systems. *Rev. Geophys.* **2004**, *42*, RG4003. [CrossRef]
26. Schumacher, R.S.; Johnson, R.H. Organization and Environmental Properties of Extreme-Rain-Producing Mesoscale Convective Systems. *Mon. Weather Rev.* **2005**, *133*, 961–976. [CrossRef]

27. Schumacher, R.S.; Rasmussen, K.L. The formation, character and changing nature of mesoscale convective systems. *Nat. Rev. Earth Environ.* **2020**, *1*, 300–314. [CrossRef]
28. Tao, W.-K.; Lang, S.; Zeng, X.; Shige, S.; Takayabu, Y. Relating Convective and Stratiform Rain to Latent Heating. *J. Clim.* **2010**, *23*, 1874–1893. [CrossRef]
29. Tao, W.-K.; Lang, S.; Zeng, X.; Li, X.; Matsui, T.; Mohr, K.; Posselt, D.; Chern, J.; Peters-Lidard, C.; Norris, P.M.; et al. The Goddard Cumulus Ensemble model (GCE): Improvements and applications for studying precipitation processes. *Atmos. Res.* **2014**, *143*, 392–424. [CrossRef]
30. Soden, B.J.; Broccoli, A.J.; Hemler, R.S. On the Use of Cloud Forcing to Estimate Cloud Feedback. *J. Clim.* **2004**, *17*, 3661–3665. [CrossRef]
31. Stephens, G.L. Cloud Feedbacks in the Climate System: A Critical Review. *J. Clim.* **2005**, *18*, 237–273. [CrossRef]
32. Soden, B.J.; Vecchi, G.A. The vertical distribution of cloud feedback in coupled ocean-atmosphere models. *Geophys. Res. Lett.* **2011**, *38*, L12704. [CrossRef]
33. Bony, S.; Dufresne, J.-L. Marine boundary layer clouds at the heart of tropical cloud feedback uncertainties in climate models. *Geophys. Res. Lett.* **2005**, *32*, L20806. [CrossRef]
34. Klein, S.A.; Hall, A. Emergent Constraints for Cloud Feedbacks. *Curr. Clim. Chang. Rep.* **2015**, *1*, 276–287. [CrossRef]
35. Gettelman, A.; Sherwood, S.C. Processes Responsible for Cloud Feedback. *Curr. Clim. Chang. Rep.* **2016**, *2*, 179–189. [CrossRef]
36. Zelinka, M.D.; Zhou, C.; Klein, S.A. Insights from a refined decomposition of cloud feedbacks. *Geophys. Res. Lett.* **2016**, *43*, 9259–9269. [CrossRef]
37. Ceppi, P.; Brient, F.; Zelinka, M.D.; Hartmann, D.L. Cloud feedback mechanisms and their representation in global climate models. *WIREs Clim. Chang.* **2017**, *8*, e465. [CrossRef]
38. Voigt, A.; Bony, S.; Dufresne, J.-L.; Stevens, B. The radiative impact of clouds on the shift of the Intertropical Convergence Zone. *Geophys. Res. Lett.* **2014**, *41*, 4308–4315. [CrossRef]
39. Voigt, A.; Shaw, T. Circulation response to warming shaped by radiative changes of clouds and water vapour. *Nat. Geosci.* **2015**, *8*, 102–106. [CrossRef]
40. Su, H.; Jiang, J.H.; Zhai, C.; Shen, T.J.; Neelin, J.D.; Stephens, G.L.; Yung, Y.L. Weakening and strengthening structures in the Hadley Circulation change under global warming and implications for cloud response and climate sensitivity. *J. Geophys. Res. Atmos.* **2014**, *119*, 5787–5805. [CrossRef]
41. Su, H.; Jiang, J.H.; Neelin, J.D.; Shen, T.J.; Zhai, C.; Yue, Q.; Wang, Z.; Huang, L.; Choi, Y.-S.; Stephens, G.L.; et al. Tightening of tropical ascent and high clouds key to precipitation change in a warmer climate. *Nat. Commun.* **2017**, *8*, 15771. [CrossRef]
42. Bony, S.; Stevens, B.; Coppin, D.; Becker, T.; Reed, K.A.; Voigt, A.; Medeiros, B. Thermodynamic control of anvil cloud amount. *Proc. Natl. Acad. Sci. USA* **2016**, *113*, 8927–8932. [CrossRef] [PubMed]
43. Lau, W.K.M.; Kim, K.-M. Robust responses of the Hadley circulation and global dryness in CMIP5 model CO_2 warming projections. *Proc. Natl. Acad. Sci. USA* **2015**, *112*, 3630–3635. [CrossRef] [PubMed]
44. Voigt, A.; Albern, N.; Ceppi, P.; Grise, K.; Li, Y.; Medeiros, B. Clouds, radiation, and atmospheric circulation in the present-day climate and under climate change. *WIREs Clim. Chang.* **2021**, *12*, e694. [CrossRef]
45. Romps, D.M. Response of Tropical Precipitation to Global Warming. *J. Atmos. Sci.* **2011**, *68*, 123–138. [CrossRef]
46. Liu, C.; Zipser, E.J. The global distribution of largest, deepest, and most intense precipitation systems. *Geophys. Res. Lett.* **2015**, *42*, 3591–3595. [CrossRef]
47. Tan, J.; Jakob, C.; Rossow, W.B.; Tselioudis, G. Increases in tropical rainfall driven by changes in frequency of organized deep convection. *Nature* **2015**, *519*, 451–454. [CrossRef] [PubMed]
48. Morake, D.M.; Blamey, R.C.; Reason, C.J.C. 2021: Long-Lived Mesoscale Convective Systems over Eastern South Africa. *J. Clim.* **2021**, *34*, 6421–6439. [CrossRef]
49. Roca, R.; Fiolleau, T. Extreme precipitation in the tropics is closely associated with long-lived convective systems. *Comm. Earth Environ.* **2020**, *1*, 18. [CrossRef]
50. Dong, W.; Zhao, M.; Ming, Y.; Ramaswamy, V. Representation of tropical mesoscale convective systems in a general circulation model: Climatology and response to global warming. *J. Clim.* **2021**, *34*, 5657–5671. [CrossRef]
51. Gates, W.L.; Boyle, J.S.; Covey, C.; Dease, C.G.; Doutriaux, C.M.; Drach, R.S.; Fiorino, M.; Gleckler, P.J.; Hnilo, J.J.; Marlais, S.M.; et al. An Overview of the Results of the Atmospheric Model Intercomparison Project (AMIP I). *Bull. Am. Meteorol. Soc.* **1999**, *80*, 29–55. [CrossRef]
52. Leung, L.R.; Bader, D.C.; Taylor, M.A.; McCoy, R.B. An introduction to the E3SM special collection: Goals, science drivers, development, and analysis. *J. Adv. Model. Earth Syst.* **2020**, *12*, e2019MS001821. [CrossRef]
53. Xie, S.; Lin, W.; Rasch, P.J.; Ma, P.-L.; Neale, R.; Larson, V.E.; Qian, Y.; Bogenschutz, P.A.; Caldwell, P.; Cameron-Smith, P.; et al. Understanding cloud and convective characteristics in version 1 of the E3SM atmosphere model. *J. Adv. Model. Earth Syst.* **2018**, *10*, 2618–2644. [CrossRef]
54. Qian, Y.; Wan, H.; Yang, B.; Golaz, J.C.; Harrop, B.; Hou, Z.; Larson, V.E.; Leung, L.R.; Lin, G.; Lin, W.; et al. Parametric sensitivity and uncertainty quantification in the version 1 of E3SM Atmosphere Model based on short Perturbed Parameters Ensemble simulations. *J. Geophys. Res.* **2018**, *123*, 13046–13073. [CrossRef]

55. Golaz, C.; Caldwell, P.M.; Van Roekel, L.P.; Petersen, M.R.; Tang, Q.; Wolfe, J.D.; Abeshu, G.; Anantharaj, V.; Asay-Davis, X.S.; Bader, D.C.; et al. The DOE E3SM coupled model version 1: Overview and evaluation at standard resolution. *J. Adv. Model. Earth Syst.* **2019**, *11*, 2089–2129. [CrossRef]
56. Zhang, G.J.; McFarlane, N.A. Sensitivity of climate simulations to the parameterization of cumulus convection in the Canadian climate centre general circulation model. *Atmos. Ocean* **1995**, *33*, 407–446. [CrossRef]
57. Rasch, P.J.; Xie, S.; Ma, P.-L.; Lin, W.; Wang, H.; Tang, Q.; Burrows, S.M.; Caldwell, P.; Zhang, K.; Easter, R.C.; et al. An overview of the atmospheric component of the Energy Exascale Earth System Model. *J. Adv. Model. Earth Syst.* **2019**, *11*, 2377–2411. [CrossRef]
58. Liu, X.; Ma, P.-L.; Wang, H.; Tilmes, S.; Singh, B.; Easter, R.C.; Ghan, S.J.; Rasch, P.J. Description and evaluation of a new four-mode version of the Modal Aerosol Module (MAM4) within version 5.3 of the Community Atmosphere Model. *Geosci. Model Dev.* **2016**, *9*, 505–522. [CrossRef]
59. Gettelman, A.; Morrison, H.; Santos, S.; Bogenschutz, P.; Caldwell, P.M. Advanced Two-Moment Bulk Microphysics for Global Models. Part II: Global Model Solutions and Aerosol–Cloud Interactions. *J. Clim.* **2015**, *28*, 1288–1307. [CrossRef]
60. Harrop, B.E.; Ma, P.-L.; Rasch, P.J.; Neale, R.B.; Hannay, C. The role of convective gustiness in reducing seasonal precipitation biases in the tropical West Pacific. *J. Adv. Model. Earth Syst.* **2018**, *10*, 961–970. [CrossRef]
61. Wang, J.; Fan, J.; Feng, Z.; Zhang, K.; Roesler, E.; Hillman, B.; Shpund, J.; Lin, W.; Xie, S. Impact of a New Cloud Microphysics Parameterization on the Simulations of Mesoscale Convective Systems in E3SM. *J. Adv. Model. Earth Syst.* **2021**, *13*, e2021MS002628. [CrossRef]
62. Li, H.; Wigmosta, M.S.; Wu, H.; Huang, M.; Ke, Y.; Coleman, A.M.; Leung, L.R. A physically based runoff routing model for land surface and Earth system models. *J. Hydrometeorol.* **2013**, *14*, 808–828. [CrossRef]
63. Li, H.-Y.; Leung, L.R.; Getirana, A.; Huang, M.; Wu, H.; Xu, Y.; Guo, J.; Voisin, N. Evaluating global streamflow simulations by a physically-based routing model coupled with the Community Land Model. *J. Hydrometeorol.* **2015**, *16*, 948–971. [CrossRef]
64. Zhang, X.; Li, H.-Y.; Leung, L.R.; Liu, L.; Hejazi, M.; Forman, B.A.; Yigzaw, W. River regulation alleviates the impacts of climate change on US thermoelectricity production. *J. Geophys. Res.* **2020**, *125*, e2019JD031618. [CrossRef]
65. Houze, R.A. 100 Years of Research on Mesoscale Convective Systems. *Meteor. Monogr.* **2018**, *59*, 17.1–17.54. [CrossRef]
66. Sui, C.; Satoh, M.; Suzuki, K. Precipitation Efficiency and its Role in Cloud-Radiative Feedbacks to Climate Variability. *J. Meteorol. Soc. Japan* **2020**, *98*, 261–282. [CrossRef]
67. Webb, M.J.; Andrews, T.; Bodas-Salcedo, A.; Bony, S.; Bretherton, C.S.; Chadwick, R.; Chepfer, H.; Douville, H.; Good, P.; Kay, J.E.; et al. The Cloud Feedback Model Intercomparison Project (CFMIP) contribution to CMIP6. *Geosci. Model Dev.* **2017**, *10*, 359–384. [CrossRef]
68. Taylor, K.E.; Stouffer, R.J.; Meehl, G.A. An Overview of CMIP5 and the Experiment Design. *Bull. Am. Meteor. Soc.* **2012**, *93*, 485–498. [CrossRef]
69. Lau, K.M.; Wu, H.T. Warm rain processes over tropical oceans and climate implications. *Geophys. Res. Lett.* **2003**, *30*, 2290. [CrossRef]
70. Sui, C.; Li, X.; Yang, M.; Huang, H. Estimation of Oceanic Precipitation Efficiency in Cloud Models. *J. Atmos. Sci.* **2005**, *62*, 4358–4370. [CrossRef]
71. Sui, C.; Li, X.; Yang, M. On the Definition of Precipitation Efficiency. *J. Atmos. Sci.* **2007**, *64*, 4506–4513. [CrossRef]
72. Lutsko, N.J.; Cronin, T.W. Increase in precipitation efficiency with surface warming in radiative-convective equilibrium. *J. Adv. Model. Earth Syst.* **2018**, *10*, 2992–3010. [CrossRef]
73. Narsey, S.; Jakob, C.N.; Singh, M.S.; Bergemann, M.; Louf, V.; Protat, A.; Williams, C. Convective precipitation efficiency observed in the tropics. *Geophys. Res. Lett.* **2019**, *46*, 13574–13583. [CrossRef]
74. Li, R.L.; Studholme, J.H.P.; Fedorov, A.V.; Storelvmo, T. Precipitation efficiency constraint on climate change. *Nat. Clim. Chang.* **2022**, *12*, 642–648. [CrossRef]
75. Allan, R.P.; Barlow, M.; Byrne, M.P.; Cherchi, A.; Douville, H.; Fowler, H.J.; Gan, T.Y.; Pendergrass, A.G.; Rosenfeld, D.; Swann, A.L.S.; et al. Advances in understanding large-scale responses of the water cycle to climate change. *Ann. N. Y. Acad. Sci.* **2020**, *1472*, 49–75. [CrossRef] [PubMed]
76. Wang, Y.; Huang, Y. Understanding the Atmospheric Temperature Adjustment to CO_2 Perturbation at the Process Level. *J. Clim.* **2019**, *33*, 787–803. [CrossRef]
77. Tokinaga, H.; Xie, S.-P.; Deser, C.; Kosaka, Y.; Okumura, Y.M. Slowdown of the Walker circulation driven by tropical Indo-Pacific warming. *Nature* **2012**, *491*, 439–443. [CrossRef] [PubMed]
78. Zhang, L.; Li, T. Relative roles of differential SST warming, uniform SST warming and land surface warming in determining the Walker circulation changes under global warming. *Clim. Dyn.* **2017**, *48*, 987–997. [CrossRef]
79. Byrne, M.P.; O'Gorman, P.A. Understanding Decreases in Land Relative Humidity with Global Warming: Conceptual Model and GCM Simulations. *J. Clim.* **2016**, *29*, 9045–9061. [CrossRef]
80. Berg, A.; Findell, K.; Lintner, B.; Giannini, A.; Seneviratne, S.I.; Hurk, B.v.D.; Lorenz, R.; Pitman, A.; Hagemann, S.; Meier, A.; et al. Land–atmosphere feedbacks amplify aridity increase over land under global warming. *Nat. Clim. Chang.* **2016**, *6*, 869–874. [CrossRef]
81. Chen, J.; Dai, A.; Zhang, Y.; Rasmussen, K.L. Changes in Convective Available Potential Energy and Convective Inhibition under Global Warming. *J. Clim.* **2020**, *33*, 2025–2050. [CrossRef]

82. Neelin, J.D.; Münnich, M.; Su, H.; Meyerson, J.E.; Holloway, C.E. Tropical drying trends in global warming models and observations. *Proc. Natl. Acad. Sci. USA* **2006**, *103*, 6110–6115. [CrossRef] [PubMed]
83. Rio, C.; Hourdin, F.; Grandpeix, J.-Y.; Lafore, J.-P. Shifting the diurnal cycle of parameterized deep convection over land. *Geophys. Res. Lett.* **2009**, *36*, L07809. [CrossRef]
84. Doan, Q.; Chen, F.; Kusaka, H.; Dipankar, A.; Khan, A.; Hamdi, R.; Roth, M.; Niyogi, D. Increased risk of extreme precipitation over an urban agglomeration with future global warming. *Earth Future* **2022**, *10*, e2021EF002563. [CrossRef]
85. Song, F.; Zhang, G.J.; Ramanathan, V.; Leung, L.R. Trends in surface equivalent potential temperature: A more comprehensive metric for global warming and weather extremes. *Proc. Natl. Acad. Sci. USA* **2022**, *119*, e2117832119. [CrossRef] [PubMed]

atmosphere

Communication

Global Warming and Topography Impact the Amplitude of the Synoptic-Scale Surface Temperature Variability across the US

Michael Secor [1], Jie Sun [1], Fucheng Yang [1], Xiaolei Zou [2,*] and Zhaohua Wu [1,*]

1 Department of Earth, Ocean, and Atmospheric Science, and Center for Ocean-Atmospheric Prediction Studies, Florida State University, Tallahassee, FL 32306, USA; msecor@fsu.edu (M.S.); jsun2@fsu.edu (J.S.)
2 Joint Center of Data Assimilation for Research and Application, Nanjing University of Information and Science & Technology, Nanjing 211544, China
* Correspondence: xzou@nuist.edu.cn (X.Z.); zwu@fsu.edu (Z.W.)

Abstract: In recent decades, the United States has experienced changing patterns of extreme temperature. Although much progress has been made, delineating the change in synoptic surface temperature variability (SSTV) and understanding its potential causes remain to be pursued. In this study, we seek to provide a quantitative description of the change in SSTV in the past seven decades across the US and its potential relevant physical factors. To achieve this goal, we develop a spatiotemporally local analysis method based on the ensemble empirical mode decomposition that bypasses the stationary assumption and makes it possible to continuously track the change in SSTV in the spatiotemporal domain. We have found that the change in SSTV across the US is spatially inhomogeneous and temporally non-uniform. The change in the SSTV amplitude ranges from −36% to 39% across the continental United States, Northern Mexico, and surrounding oceans. Higher altitudes and surrounding regions generally see an increase in variability, while elsewhere over land, a reduction is observed, creating a three-band zonal structure across the continental United States. Generally, increases in variability are observed in the subtropics. The shape of the spatiotemporal evolution of SSTV implies that the topography of the United States may play an important role in altering synoptic-scale variability.

Keywords: synoptic surface temperature variability (SSTV); amplitude change in SSTV; global warming; topography

Citation: Secor, M.; Sun, J.; Yang, F.; Zou, X.; Wu, Z. Global Warming and Topography Impact the Amplitude of the Synoptic-Scale Surface Temperature Variability across the US. *Atmosphere* **2023**, *14*, 979. https://doi.org/10.3390/atmos14060979

Academic Editor: Graziano Coppa

Received: 31 March 2023
Revised: 26 May 2023
Accepted: 1 June 2023
Published: 5 June 2023

1. Introduction

With the continuing human-produced greenhouse gas input into the atmosphere, the Earth's climate has been warming at an unprecedented pace and in a spatiotemporally non-uniform fashion [1–3], especially in the high-latitude regions where warming is the most dramatic [4–7]. Associated with this slowly spatiotemporally varying warming, synoptic-scale weather variability at the mid-latitudes has also been changing, and with the reduced meridional temperature gradient caused by polar warming amplification, the synoptic variability generated by meridional advection is anticipated to reduce as the globe continues to warm [8–10]. However, the precondition that led to the theoretical arguments in these studies [8–10] is that the enhanced polar warming is not caused by the meridional heat transport materialized by the synoptic-scale variability. If the latter is the major contributor to the polar warming amplification [5], a valid counterargument may also be that increased synoptic variability is necessary to maintain the polar warming amplification. The conflict between these two arguments remains to be resolved. Adding to this challenge is the fact that the change in the synoptic variability may be spatially non-uniform.

It is indeed these difficulties that led to this study. However, this study is not targeted at resolving the two contradicting arguments; rather, it is focused on quantifying the spatiotemporal change in the synoptic surface temperature variability (SSTV) using a newly developed nonlinear nonstationary method based on multi-dimensional ensemble empirical mode decomposition (MEEMD) [11–13]. Previous studies have chronicled future changes in temperature variability in a multitude of ways, from calculating the standard deviation of monthly temperature anomalies [14] to feature tracking algorithms [15]. These methods generally contain piece-wise stationary assumptions, and the results may be significantly distorted by the multi-decadal variability contained in both the SSTV amplitude variability and the multi-decadal variability of the surface temperature itself. With MEEMD, we can track the variability and change in the amplitude of SSTV and obtain the temporal evolution of the SSTC itself. By applying a nonlinear nonstationary trend-defining method [16], we can extract the spatiotemporal evolution of the amplitude of SSTV. The fine structure of the amplitude evolution of SSTV obtained will help us to determine some physical factors that contribute to the variability and change in the amplitude of SSTV, potentially providing observational evidence to verify the correctness of competing theories.

An additional purpose of this study is to commemorate Prof. Duzheng Ye (also known as Tu-Cheng Yeh), one of the founding fathers of modern meteorological and climate sciences in China and one of the pioneers who initiated the global collaboration in understanding climate change and its impact on future human wellness. Prof. Ye was a student of Carl-Gustaf Rossby and widely considered one of the main figures of the famous Chicago School of Meteorology. His legendary works include, but are not limited to, studies on the development of mid-latitude waves (synoptic variability) and their propagation, the role of topography, especially the Tibetan Plateau in shaping the weather and climate of eastern Asia, and how global climate change has been modulating the weather and climate. All authors of this study can be considered academic descendants of his and have benefitted either directly or indirectly from his guidance. This study, to a significant degree, touches all these aforementioned areas. Other papers in this collection may introduce some other renowned works of Prof. Ye.

2. Data and Method

2.1. Data

This study has two sets of data: surface air temperature (SAT) and surface topography. The temperature data are from the daily averaged NCEP/NCAR Reanalysis 1 project available through NOAA's Physical Sciences Laboratory (PSL) website (https://psl.noaa.gov/data/gridded/data.ncep.reanalysis.html) (accessed on 1 September 2019). The horizontal resolution is 2.5° × 2.5° [17]. The region of interest for this study covers a region bordered by 20° N and 50° N lines in the meridional direction and 60° W and 140° W lines in the zonal direction, encompassing the continental United States as well as parts of the surrounding oceans, Canada, and Mexico. Our analysis uses the data from 1 January 1948 to 31 December 2018, a span of 71 years.

The topography data were obtained from the European Centre for Medium-Range Weather Forecasts (ECMWF) ERA-Interim 1 reanalysis project [18]. The data cover the entire globe, with a resolution of 0.125° × 0.125°. These data were averaged over the 2.5° × 2.5° grid spacing, consistent with the spatial locations of the temperature data.

2.2. Method

A challenging question that remains in our research community is how to quantify systematic change (the monotonic trend) in synoptic variability. Traditionally, we calculate statistical quantities, such as variance or standard deviation, for this purpose. However, this approach leads to a dilemma: to track the non-stationarity of synoptic variability and its potential influence by global climate change, we calculate statistical quantities from either piece-wised sections or running sections of a prescribed length of a time series of synoptic variability. It is well known that even for a stationary time series that contains randomness, the statistical quantities calculated from a limited number of samples can contain uncertainty, as the law of large numbers has stated, let alone nonstationary time series. To some degree, this is a manifestation of Heisenberg's uncertainty principle when the principle is applied to statistical studies. Therefore, when a prescribed piece-wised section of a limited number of samples is applied, the obtained statistical quantities often contain large variability of different timescales caused by the randomness of weather variability and are sensitive to the length of the sections one selects or prescribes, even in the case of stationary processes. The selection of a longer section can reduce the variability and uncertainty of the calculated statistics for stationary processes but leads to the calculated statistical quantities not extending to the two data ends. The longer the data section, the larger the portion of undefined statistical quantities. For non-stationary series, this approach can cause even larger uncertainty. Thus, alternative methods need to be developed.

The method developed here takes advantage of the recent advancements in the adaptive and temporally local analysis of nonlinear nonstationary time series, the empirical mode decomposition (EMD) [11]-based methods [12,13,16]. In this study, EEMD [12] is applied to an SAT time series of a given spatial grid. This time series is decomposed in terms of a set of oscillatory components and a monotonic trend. All these oscillatory components are natural waves of different timescales with amplitude and frequency modulations. Since EEMD is an effective dyadic filter bank, usually, the first component has an averaged period of 3 data points governed by the added white noise in the EEMD algorithm, the second has 6 data points, and so on. Figure 1 presents an example of such a decomposition of the SAT time series near Boston, Massachusetts, USA. The first component, to a significant degree, captures the day-to-day variability [19,20] contained in the SAT time series, while the second, which is highlighted in red and overlapped by a shining blue line, is the synoptic variability with an average period of 6 days. The following four components are sub-seasonal variability, followed by the seasonal cycle, and the lumped inter-annual and longer time scale variability (including the trend).

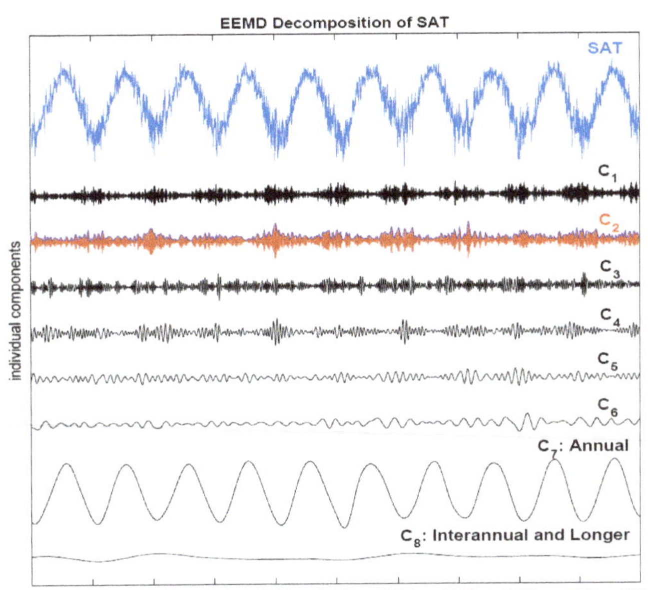

Figure 1. The decomposition of the SAT near Boston, MA, USA. The top navy-blue line is the original SAT time series, and the black and red lines are individual EEMD components of the SAT with their mean periods from about three days and up. In the figure, the amplitude of different timescale variability is in proportion, and each component has a zero mean. The shining blue line that connects the maxima of the second component is the amplitude of the second component.

The EEMD decomposition enables us to isolate typical SSTV, the second component of SAT. Since EMD/EEMD is based on defining upper and lower envelopes that pass-through maxima and minima of a time series, respectively, the varying amplitude of SSTV that encompasses the length of a time series can be well defined. However, this amplitude still contains variability of timescales at least double that of SSTV. Thus, refining needs to be carried out. To achieve this goal, we applied EEMD to this amplitude again, which is shown in Figure 2. After all the oscillatory components of different times scales were removed, we obtained the nonlinear trend [16] that is either monotonic or contains only one internal extremum, as the bold red line for Boston SSTV, which is also shown in the enlarged Figure 3. It can be verified that the mean value of the trend of the amplitude is about 1.13 times the standard deviation of the overall synoptic component, and this proportional number fits the ratios of the amplitude and the standard deviation for individual piece-wised sections well, provided all sections are a few decades long. In this sense, our new method overcomes the dilemma that traditional statistical methods face. It is also noted that the extracted trend in Figure 3 is also visually appealing to the human eye, which serves as the first judgment of the correctness of data analysis.

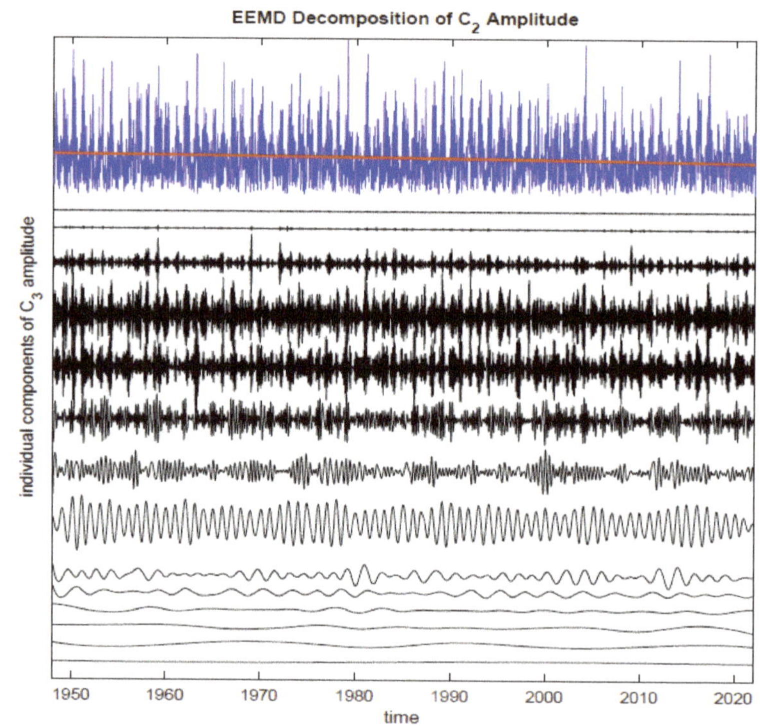

Figure 2. The decomposition of the amplitude of synoptic variability. The shining blue line is the amplitude, the black lines are all EEMD components of the SSTV amplitude, and the red line is the trend.

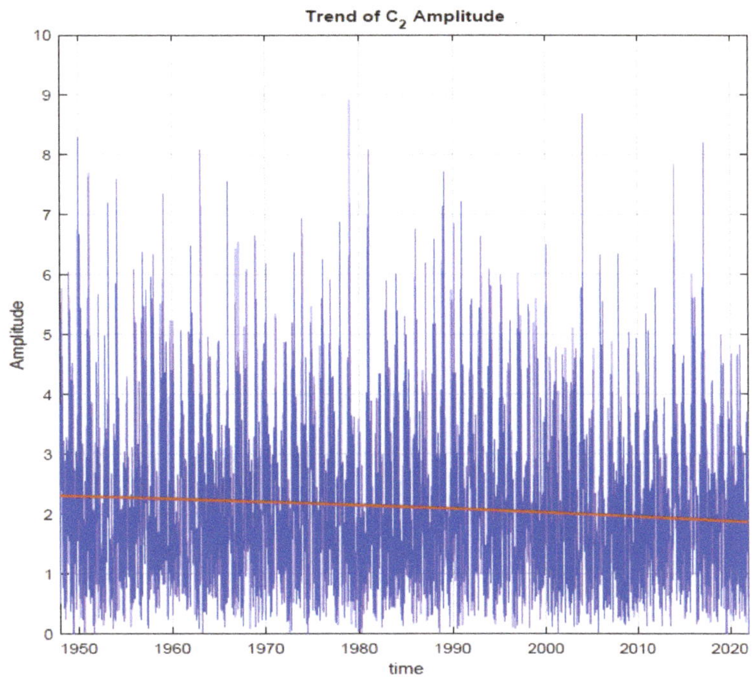

Figure 3. The enlarged top section of Figure 2.

It is noted again that the trend in this study is nonlinear and nonstationary [2,16], not a traditional linear trend. The time-varying nature of this secure trend allows us to derive the accumulated change of any given time with respect to a reference time, as well as the changing rate of the secure trend at any given time. With this isolated trend, denoted as $A(t)$, we can now systematically quantify how synoptic variability has been evolving by defining a standardized amplitude change ratio, i.e.,

$$P(t) = \frac{A(t) - A(t_r)}{A(t_r)} \tag{1}$$

where t_r represents a reference time selected. In this study, this reference time is selected as 1 January 1948, the starting date of the reanalysis data.

The above calculation procedure is applied to all the spatial grid points within the selected domain to obtain the spatiotemporal evolution of the SSTV amplitude evolution.

3. Results

Figure 4 plots the normalized accumulated SSTV amplitude change, $P(t)$, for the decades following the referenced decade, 1948 to 1958. By 1968, we can already see an emerging spatial pattern that persists throughout later decades. In the zonal direction, this pattern is marked by an increase in variability over the ocean off the western coast of the US, a decrease over California and Baja California, an increase over much of the Rocky Mountain region, a decrease in the central longitudinal regions of the US, and an increase again over the southeast US. In the meridional direction, the pattern shows a north–south contrast. In the subtropical region, an increase in SSTV amplitude prevails, while in the mid-latitude, a decrease in SSTV amplitude extends from the Great Plains to the Great Lakes.

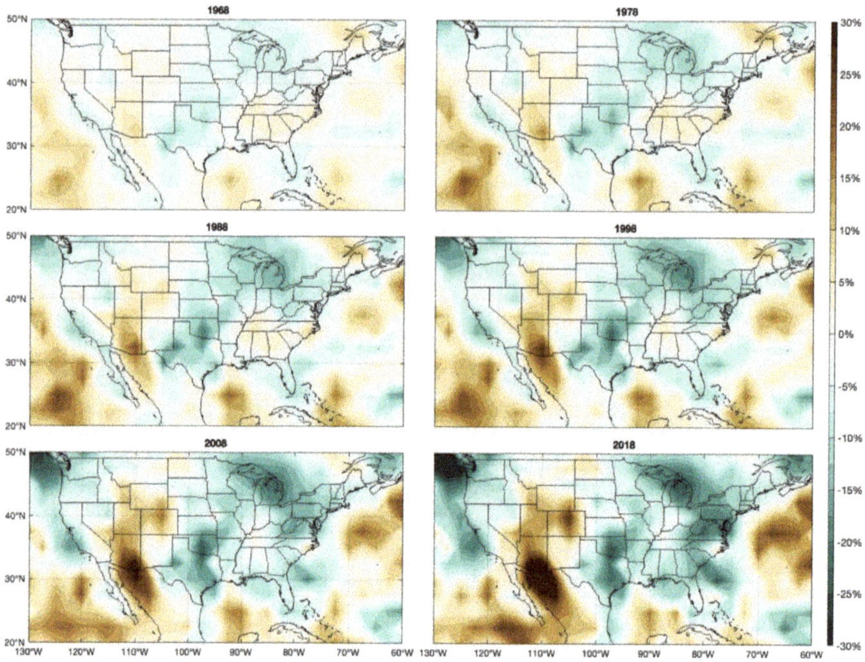

Figure 4. The normalized accumulated SSTV amplitude change, $P(t)$, from the referenced decade ending in 1958. The year label on each panel represents the end of a decade. All panels used the same color scheme of percentage-wise reduction/increment scale.

The pattern continues to develop with time, with small-spatial scale structures fading and those of larger spatial scale becoming more dominant. One of the more notable regions is the southeast US, where the increase in SSTV amplitude in earlier decades transitions to a reduction in SSTV amplitude by 2018. The result of this reversal is that almost all land to the east of the Rocky Mountains is dominated by a decrease in SSTV amplitude, though a notable exception occurs near the Appalachian Mountains, where the early period increase in SSTV amplitude, though reduced compared to early decades, persists.

From 1948 to 2018, there is a significant reduction in SSTV amplitude over much of the central United States, extending northward into Canada, a trend that began in 1968 and has persisted through to 2018. The entire eastern half of the continental United States has seen a reduction in variability over these seven decades. The largest reduction of 36.5% occurs just off the coast of Washington. The reduction in variability is not confined to the eastern US, as reductions are observed along much of the west coast as well. Among the regions with the greatest reductions are the Great Lakes, Northeast, and Southern Plains. Notably, this does not include Mexico, Arizona, or the Rocky Mountain region, where an intensification of a region of increasing variability has occurred. Overall, the total change in variability from 1948 to 2018 over the entire area analyzed is a reduction of 3.33%, due to much of the regional increases/decreases canceling each other out.

The continuous change in SSTV variability, as illustrated by the bold red line in Figure 3, provides an opportunity to quantify the instantaneous change rate of SSTV amplitude, $dP(t)/dt$, by taking the temporal derivative of the normalized amplitude change rate. Since the SSTV amplitude trend at any spatial grid is slowly varying, $dP(t)/dt$ can be well approximated by $[P(t+\Delta t/2) - P(t-\Delta t/2)]/\Delta t$, where Δt is one decade. Figure 5 presents the instantaneous change rate of SSTV for the years 1963, 1973, 1983, 1993, 2003, and 2013. Again, a large-scale spatiotemporal evolution pattern emerges. Of the surrounding oceans of the continental US, the Pacific Ocean experienced a sharp SSTV amplitude increase in the 1960s and a reduced increasing rate in the 1970s and the 1980s, followed by a mild SSTV amplitude reduction and an intensified SSTV amplitude reduction. This change contrasts with the Atlantic Ocean, where initial changes in both directions are observed, gradually transitioning to a large region of increased variability. Over the continental US, the increment in the SSTV amplitude has been generally steady in the Rocky Mountain region through all decades, although the area of increment expands in all directions. The southeast US shares a similar evolution of SSTV amplitude to the Pacific Ocean, but generally with smaller changing rates, apart from the strong reductions in 2013 covering much of South Carolina. In the regions of the continental US, other than the Rocky Mountains and the southeast US, the changing SSTV amplitude changing rates are mostly negative and mild. The maximum decadal change rate of amplitude is about 12% in the Pacific Ocean region off the west coast of the US.

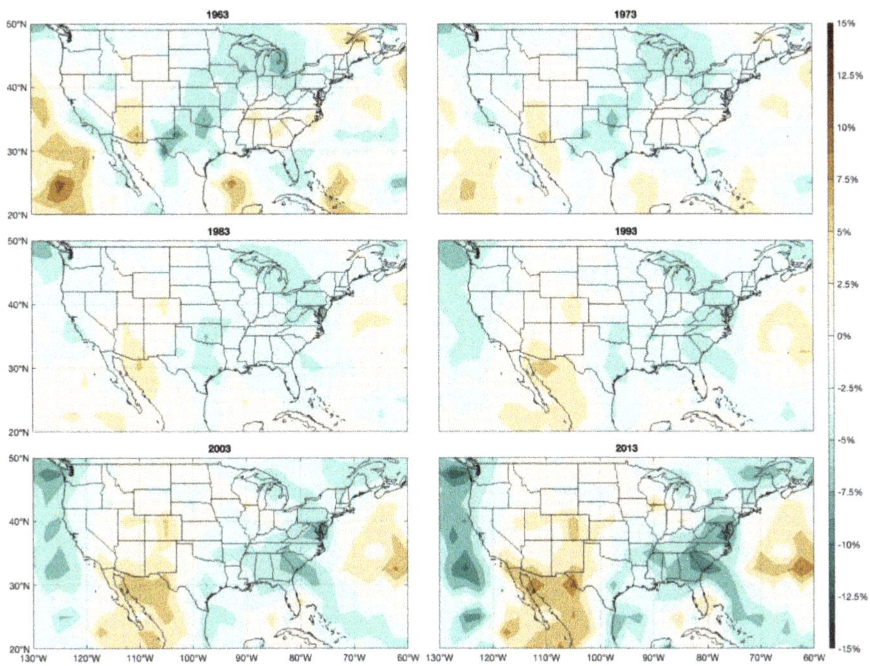

Figure 5. The decade-wise normalized accumulated SSTV amplitude change for the years 1963, 1973, 1983, 1993, 2003, and 2013. All panels used the same color scheme of percentage-wise reduction/increment scale.

The results presented in both Figures 4 and 5 point to the topography of the United States affecting the SSTV amplitude evolution over the past seven decades. To quantify this, we calculated the correlation between the accumulated SSTV amplitude rate of change (the bottom right panel of Figure 4) and the topography of the continental US, which is shown in Figure 6. The correlation coefficient is 0.52. To test whether this correlation coefficient is statistically significant, we designed a Monte Carlo method by repeatedly randomizing either spatial topography data or the normalized accumulated SSTV amplitude changes and calculating the correlation coefficient. The result showed that the correlation coefficient of 0.52 is statistically significant at the 99% confidence level.

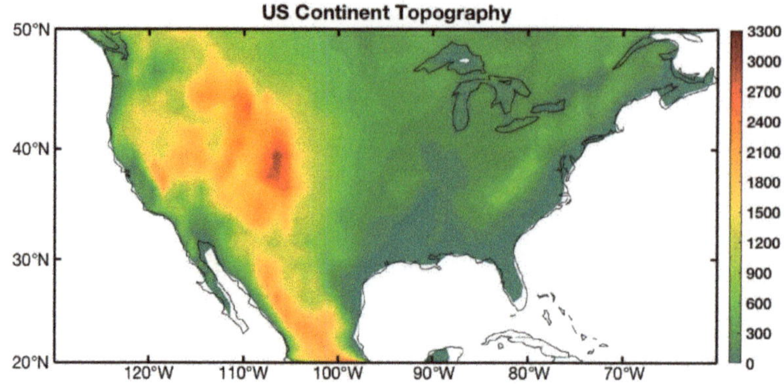

Figure 6. The topography of the continental US. The unit in the color bar is meters above sea level. It is noted that the slight mismatch of the coastal line with the topography display is caused by the grossed horizontal resolution of the topography.

4. Summary and Discussions

In recent years, the world has been experiencing a significant increase in extreme weather events. It is widely suspected that these events are tied closely to global warming in the past century, with a significant number of diagnosis and modeling studies supporting this argument. However, most of these studies used traditional statistical methods that have deficiencies in terms of quantifying the nonlinear nonstationary synoptic variability. An improvement in the methodologies can significantly reduce uncertainty.

In this study, we developed a new analysis method to quantify the nonstationary change in synoptic temperature variability in the last seven decades. The method takes advantage of the most recent progress in nonlinear nonstationary time series analysis, especially the EEMD-based methods for isolating spatiotemporal evolutions of spatiotemporal fields of different spatiotemporal scales. By applying the newly developed method to surface temperature data, we first isolated the nonstationary synoptic-scale surface temperature variability, then tracked its spatiotemporal evolution, and finally extracted important factors that may contribute to the observed SSTV change.

We found that the SSTV change over the US is spatially inhomogeneous and temporally non-uniform. The accumulated maximum changes with respect to the late 1940s reached levels of a 39% increase in the mountainous regions and a 36% reduction in the lowlands. The spatial structure appears to have a zonal three-band structure across the continental United States. In addition to this, increases in variability are observed in the subtropics. We also calculated the correlation between the change in SSTV amplitude and the topography over the USA and found that this relationship is robust. The evolution of the SSTV amplitude is significantly different over land and oceans, with the former regions steadier and the latter region containing a significant change in the changing rate over time.

The most important role of synoptic variability in the mid-latitudes is to provide a carrier to transport heat from low latitudes to high latitudes, and its amplitude is modulated by the meridional temperature gradient. It is also known that the polar region has warmed much more than the lower latitudes in the past seven decades as the globe continues to warm. This inhomogeneous global warming reduces the meridional temperature gradient and infers a reduction in the amplitude of synoptic variability. This result has been confirmed in the lowlands of the continental US but not in the mountainous regions in our study, implying that the evolution of regional synoptic variability in the presence of warming is beyond our current understanding. While our study provides a piece of unambiguous evidence that topography and land–sea contrast are key factors to modify the above dynamical argument, how topography and land–sea contrast affects the SSTV amplitude remains to be investigated.

Author Contributions: Conceptualization, Z.W. and X.Z.; Formal analysis, M.S., J.S., F.Y. and Z.W.; Funding acquisition, Z.W. and X.Z.; Investigation, M.S., F.Y. and Z.W.; Methodology, J.S. and Z.W.; Project administration, Z.W.; Resources, Z.W.; Software, Z.W.; Supervision, Z.W.; Validation, J.S. and F.Y.; Visualization, M.S., J.S. and Z.W.; Writing—original draft, M.S. and Z.W.; Writing—review and editing, X.Z. and Z.W. All authors have read and agreed to the published version of the manuscript.

Funding: M.S., J.S., F.Y. and Z.W. were supported by the US National Science Foundation, grant AGS-1723300. X.Z. was supported by the National Natural Science Foundation of China, grant 91337218.

Data Availability Statement: The temperature data are from the daily averaged NCEP/NCAR Reanalysis 1 project available through NOAA's Physical Sciences Laboratory (PSL) website (https://psl.noaa.gov/data/gridded/data.ncep.reanalysis.html) (accessed on 1 September 2019). The topography data were obtained from the European Centre for Medium-Range Weather Forecasts (ECMWF) ERA-Interim 1 reanalysis project [18].

Conflicts of Interest: The authors declare no conflict of interest.

References

1. Masson-Delmotte, V.; Zhai, P.; Pirani, A.; Connors, S.L.; Péan, C.; Berger, S.; Caud, N.; Chen, Y.; Goldfarb, L.; Gomis, M.I.; et al. Climate change 2021: The physical science basis. In *Contribution of Working Group I to the Sixth Assessment Report of the Intergovernmental Panel on Climate Change*; Cambridge University Press: Cambridge, UK, 2021; Volume 2.
2. Wu, Z.; Huang, N.E.; Wallace, J.M.; Smoliak, B.V.; Chen, X. On the time-varying trend in global-mean surface temperature. *Clim. Dyn.* **2011**, *37*, 759–773. [CrossRef]
3. Ji, F.; Wu, Z.; Huang, J.; Chassignet, E.P. Evolution of land surface air temperature trend. *Nat. Clim. Chang.* **2014**, *4*, 462–466. [CrossRef]
4. Holland, M.M.; Bitz, C.M. Polar amplification of climate change in coupled models. *Clim. Dyn.* **2003**, *21*, 221–232. [CrossRef]
5. Cai, M. Dynamical amplification of polar warming. *Geophys. Res. Lett.* **2005**, *32*, L22710. [CrossRef]
6. Bekryaev, R.V.; Polyakov, I.V.; Alexeev, V.A. Role of polar amplification in long-term surface air temperature variations and modern Arctic warming. *J. Clim.* **2010**, *23*, 3888–3906. [CrossRef]
7. Taylor, P.C.; Cai, M.; Hu, A.; Meehl, J.; Washington, W.; Zhang, G.J. A decomposition of feedback contributions to polar warming amplification. *J. Clim.* **2013**, *26*, 7023–7043. [CrossRef]
8. Hassanzadeh, P.; Kuang, Z.; Farrell, B.F. Responses of midlatitude blocks and wave amplitude to changes in the meridional temperature gradient in an idealized dry GCM. *Geophys. Res. Lett.* **2014**, *41*, 5223–5232. [CrossRef]
9. Schneider, T.; Bischoff, T.; Płotka, H. Physics of changes in synoptic midlatitude temperature variability. *J. Clim.* **2015**, *28*, 2312–2331. [CrossRef]
10. Cattiaux, J.; Peings, Y.; Saint-Martin, D.; Trou-Kechout, N.; Vavrus, S.J. Sinuosity of midlatitude atmospheric flow in a warming world. *Geophys. Res. Lett.* **2016**, *43*, 8259–8268. [CrossRef]
11. Huang, N.E.; Shen, Z.; Long, S.R.; Wu, M.C.; Shih, H.H.; Zheng, Q.; Yen, N.C.; Tung, C.C.; Liu, H.H. The empirical mode decomposition and the Hilbert spectrum for nonlinear and non-stationary time series analysis. *Proc. R. Soc. Lond. Ser. A Math. Phys. Eng. Sci.* **1998**, *454*, 903–995. [CrossRef]
12. Wu, Z.; Huang, N.E. Ensemble empirical mode decomposition: A noise-assisted data analysis method. *Adv. Adapt. Data Anal.* **2009**, *1*, 1–41. [CrossRef]
13. Wu, Z.; Huang, N.E.; Chen, X. The multi-dimensional ensemble empirical mode decomposition method. *Adv. Adapt. Data Anal.* **2009**, *1*, 339–372. [CrossRef]
14. Bathiany, S.; Dakos, V.; Scheffer, M.; Lenton, T.M. Climate models predict increasing temperature variability in poor countries. *Sci. Adv.* **2018**, *4*, eaar5809. [CrossRef] [PubMed]
15. Tamarin-Brodsky, T.; Hodges, K.; Hoskins, B.J.; Shepherd, T.G. Changes in Northern Hemisphere temperature variability shaped by regional warming patterns. *Nat. Geosci.* **2020**, *13*, 414–421. [CrossRef]
16. Wu, Z.; Huang, N.E.; Long, S.R.; Peng, C.K. On the trend, detrending, and variability of nonlinear and nonstationary time series. *Proc. Natl. Acad. Sci. USA* **2007**, *104*, 14889–14894. [CrossRef] [PubMed]
17. Kalnay, E.; Kanamitsu, M.; Kistler, R.; Collins, W.; Deaven, D.; Gandin, L.; Iredell, M.; Saha, S.; White, G.; Woollen, J.; et al. The NCEP/NCAR 40-year reanalysis project. *Bull. Am. Meteorol. Soc.* **1996**, *77*, 437–472. [CrossRef]
18. Dee, D.P.; Uppala, S.M.; Simmons, A.J.; Berrisford, P.; Poli, P.; Kobayashi, S.; Andrae, U.; Balmaseda, M.A.; Balsamo, G.; Bauer, D.P.; et al. The ERA-Interim reanalysis: Configuration and performance of the data assimilation system. *Q. J. R. Meteorol. Soc.* **2011**, *137*, 553–597. [CrossRef]

19. Wu, F.T.; Fu, C.; Qian, Y.; Gao, Y.; Wang, S.Y. High-frequency daily temperature variability in China and its relationship to large-scale circulation. *Int. J. Climatol.* **2017**, *37*, 570–582. [CrossRef]
20. Liu, Q.; Tan, Z.-M.; Sun, J.; Hou, Y.; Fu, C.; Wu, Z. Changing rapid weather variability increases influenza epidemic risk in a warming climate. *Environ. Res. Lett.* **2020**, *15*, 044004. [CrossRef]

atmosphere

MDPI

Article

Diurnal Variation Characteristics of the Surface Sensible Heat Flux over the Tibetan Plateau

Zhu Zhu [1], Meirong Wang [1,*], Jun Wang [2], Xulin Ma [1,*], Jingjia Luo [1] and Xiuping Yao [3]

1 Key Laboratory of Meteorological Disaster, Ministry of Education (KLME)/Collaborative Innovation Center on Forecast and Evaluation of Meteorological Disasters (CIC-FEMD)/Joint International Research Laboratory of Climate and Environment Change (ILCEC)/Joint Center for Data Assimilation Research and Applications, School of Atmospheric Sciences, Nanjing University of Information Science & Technology, Nanjing 210044, China
2 International Institute for Earth System Science, Nanjing University, Nanjing 210033, China
3 China Meteorological Administration Training Center, Beijing 100081, China
* Correspondence: wmr@nuist.edu.cn (M.W.); xulinma@nuist.edu.cn (X.M.)

Abstract: The characteristics of diurnal variation of the surface sensible heat flux (SH) over the Tibetan Plateau (TP) are comprehensively investigated by using the long-term dataset of integrated land–atmosphere interaction observations (2006–2016) on the TP. Results show that the diurnal variation of SH shows obvious seasonal variabilities in terms of amplitude, duration, and peak time. At the Muztagh Ata Westerly Observation and Research Station (MAWORS), the Ngari Desert Observation and Research Station (NADORS), and the Qomolangma Atmospheric and Environmental Observation and Research Station (QOMS), the SH diurnal amplitude is consistently the largest in spring, followed by summer and autumn, and the smallest in winter, with a peak at 15:00. However, for the Southeast Tibet Observation and Research Station (SETORS), the amplitude in winter is rather violent with the peak at 12:00. We find that positive SH at most stations has the longest duration from May to August. Moreover, the peak time fluctuates from month to month, even showing a shift at the QOMS before and after 2015, and the double-peak phenomenon of SH mainly occurs in spring and autumn. Additionally, magnitudes of calculated SH with the conventional heat transfer coefficient (C_{DH}) of 0.004 are about 64% to 100% larger than those of directly observed SH at the QOMS and the Nam Co Monitoring and Research Station (NAMORS). We here additionally recommend a new C_{DH} values of about 2.24×10^{-3} in spring and 2.78×10^{-3} in summer, respectively, to more accurately calculate the TP SH.

Keywords: Tibetan Plateau; surface sensible heat flux; diurnal variation; seasonal variability; heat transfer coefficient

Citation: Zhu, Z.; Wang, M.; Wang, J.; Ma, X.; Luo, J.; Yao, X. Diurnal Variation Characteristics of the Surface Sensible Heat Flux over the Tibetan Plateau. *Atmosphere* **2023**, *14*, 128. https://doi.org/10.3390/atmos14010128

Academic Editor: Jonathan E. Pleim

Received: 1 November 2022
Revised: 26 December 2022
Accepted: 3 January 2023
Published: 6 January 2023

1. Introduction

The Tibetan Plateau (TP) occupies about a quarter of the land area of China, with an average elevation of more than 4000 m. It is well known as "the roof of the world" and "the third pole of the Earth" [1]. Surface sensible heat flux (SH), as one of the significant parameters to characterize the strength of the interaction between surface and atmosphere, accelerates the updraft in spring and summer, which can directly act on the middle troposphere and modulate the atmospheric circulations to affect the Asian summer monsoon variabilities [2–5] and even the development of tropical ENSO and the air–sea interaction in the mid-latitude Pacific [6]. Duan et al. [7] have pointed out that the SH from April to June over the TP can be used as an effective predictor of precipitation in July in the Jianghuai valley of China. Moreover, the abnormal spatial distribution and temporal evolution of SH will also lead to more abnormal characteristics of climate in China [8,9]. Ma et al. [10] demonstrated that the SH over the TP shows obvious change rules in multiple time scales of interannual, interdecadal, seasonal, and diurnal variations, which further jointly act on the weather and climate in local and remote regions [11–14].

In terms of the diurnal variations, Ma et al. [15] suggested that the meadow surface over the northern TP was a strong heating source for the atmosphere in the day and a weak cold source at night. Liu et al. [16] noted that the exchange of the SH between land and atmosphere on the slopes of Mt. Everest mainly occurred in the afternoon. Duan et al. [14] revealed that the SH of dry and wet phases at Shiquanhe and Linzhi station changed with a single peak in the day, but the specific variation characteristics were different. As one of the vital parameters for calculating SH, the heat transfer coefficient C_{DH} also has obvious diurnal variation because it is largely affected by the stability of the atmosphere. Li et al. [17] calculated the C_{DH} at Gaize and Shiquanhe over the western TP by using the profile-flux method, and the C_{DH} showed diurnal variation characteristics with different amplitudes and phases in winter and summer, respectively. Generally, the C_{DH} is large and stable in the day, but small and fluctuating at night [18].

However, due to the lack of observational data over the TP, especially those with high temporal resolutions, more studies are still needed to investigate the diurnal variation of the SH over the TP. Furthermore, an accurate estimate of SH is always a challenge. In this study, based on a long-time dataset with a high temporal resolution (hourly) from 2006 to 2016, the diurnal variation characteristics of the observed SH over the TP are quantitatively analyzed. Furthermore, it is worth noting that numerous studies related to TP SH have been conducted, generally based on the calculated SH, in which the magnitude of the heat transfer coefficient C_{DH} is vital and has not been determined. Wang et al. [19] pointed out that the parameterization of the regional C_{DH} ranged from 2.5×10^{-3} to 5×10^{-3} over the TP based on the GIMMS-NDVI dataset. The C_{DH} estimated by the CHEN-WONG scheme [20] is about 3.6×10^{-3}, depending on the averaged wind speed, which was applied to reveal the TP SH interannual variation and its response to climate change [12]. However, in the majority of previous studies [21–26], they basically recommended to take C_{DH} as 4×10^{-3} when calculating the TP SH. Therefore, we will additionally discuss the differences between the calculated SH and the directly observed SH on a daily scale, and potentially provide a more suitable C_{DH} for improving the accuracy of the SH calculations.

This paper is organized as follows: Section 2 presents the data and methods; Section 3 describes the annual and seasonal mean of the SH diurnal variations over the TP; Section 4 investigates the monthly changes of the SH diurnal variation over the TP; and Section 5 introduces the effect of the C_{DH} on SH diurnal variation. A Conclusions and Discussion are provided in Section 6.

2. Data and methods

2.1. Data

The hourly integrated land-atmosphere interaction observation dataset from the National Tibetan Plateau Science Data Center (2005–2016) (http://data.tpdc.ac.cn/zh-hans/data/b9ab35b2-81fb-4330-925f-4d9860ac47c3, accessed on 15 March 2020) [27,28] is employed as observations in this study, including meteorological gradient data, radiation data, and soil and turbulent flow data. The directly observed SH in this study comes from the turbulent flow data during the period 2006–2016, sampled by the open-path Eddy Covariance (EC) turbulent flux measurement system consisting of an ultrasonic anemometer and an infrared gas analyzer [28]. This dataset contains 6 stations (Figure 1 and Table 1), which are the Muztagh Ata Westerly Observation and Research Station, the Chinese Academy of Sciences (MAWORS), the Ngari Desert Observation and Research Station (NADORS), the Nagqu Station of Plateau Climate and Environment (BJ), the Nam Co Monitoring and Research Station for Multisphere Interactions (NAMORS), the Qomolangma Atmospheric and Environmental Observation and Research Station (QOMS), and the Southeast Tibet Observation and Research Station for the Alpine Environment (SETORS). Additionally, wind speed, air temperature, and surface temperature, derived from the meteorological gradient data in this dataset at these six stations, are also used in this study.

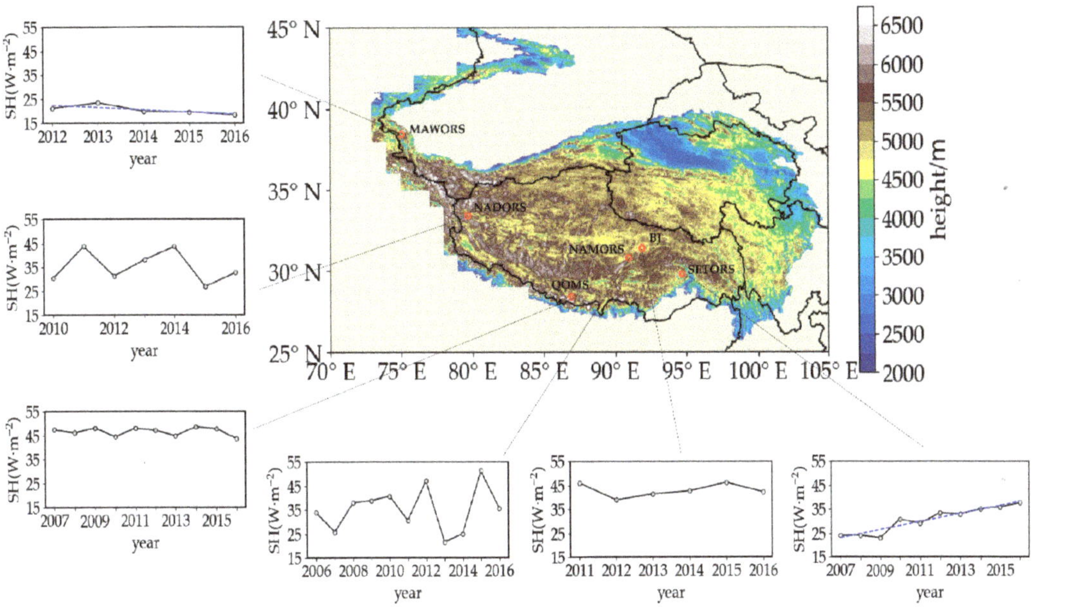

Figure 1. Geographical distributions of the six meteorological stations over the TP and the time series of the annual mean SH (units: W m^{-2}) at these six stations during the period 2006–2016. The blue dotted lines indicate the decreasing and increasing trends at MAWORS and SETORS, respectively.

Table 1. List of the stations in the hourly integrated observational dataset, including the station names, latitudes, longitudes, time periods, and the time difference between local time and Beijing time.

Station	Latitude	Longitude	Time Period	Time Difference between Local Time and Beijing Time (Local Time Is Later than Beijing Time)
BJ	31.37° N	91.90° E	2011–2016	1 h 52 min late
QOMS	28.36° N	86.95° E	2007–2016	2 h 12 min late
SETORS	29.77° N	94.73° E	2007–2016	1 h 41 min late
NAMORS	30.77° N	90.98° E	2006–2016	1 h 56 min late
NADORS	33.39° N	79.70° E	2010–2016	2 h 41 min late
MAWORS	38.41° N	75.05° E	2012–2016	3 h 00 min late

2.2. Methods

For the calculation of the SH, the bulk transfer equation has been widely used [22,29–32] with the formula as:

$$SH = \rho C_p C_{DH} V_{10} (Ts - Ta) \tag{1}$$

where ρ is the density of air, taken as a constant value of 0.8 kg m^{-3} [33]; Cp is the specific heat of dry air under constant pressure, which is 1005 J Kg^{-1} K^{-1}; C_{DH} refers to the heat transfer coefficient, which is usually prescribed as a constant magnitude of 4×10^{-3}; V_{10} is the wind speed at the height of 10 m; *Ts* is the ground surface temperature; and *Ta* refers to the air temperature at the height of 1.5 m. In the following, only the calculated *SH* at QOMS and NAMORS can be obtained by using Equation (1) based on wind speed, ground temperature, and air temperature, respectively, owing to the limitation of data.

Additionally, unitary linear regression analysis is applied in this study to calculate the trend in SH:

$$x_i = a + bt_i \; (i = 1, 2, \cdots, n) \tag{2}$$

where x_i denotes the meteorological variable with the sample size of n, and t_i is the time corresponding to x_i, and the linear regression equation between x_i and t_i can be solved as follows:

$$\begin{cases} b = \frac{\sum_{i=1}^{n} x_i t_i - \frac{1}{n}\left(\sum_{i=1}^{n} x_i\right)\left(\sum_{i=1}^{n} t_i\right)}{\sum_{i=1}^{n} t_i^2 - \frac{1}{n}\left(\sum_{i=1}^{n} t_i\right)} \\ a = \overline{x} - b\overline{t} \end{cases} \quad (3)$$

where a is the regression constant and b is the regression coefficient. The $\overline{x}$ is the average value of the meteorological variable x_i, and $\overline{t}$ is the average value of the time.

The root mean square error (*RMSE*) denotes the extent to which the data deviate from the true value and tends to be applied for assessing the data reliability, which is used in Section 5 to evaluate the accuracy of new *SH* calculated by using new C_{DH}.

$$RMSE = \sqrt{\frac{\sum_{i=1}^{N}(y_i - x_i)^2}{N}} \quad (4)$$

where N is the total number of samples, and x_i and y_i denote the observed samples and the calculated samples, respectively.

3. Annual and Seasonal Mean of the SH Diurnal Variations over the TP

3.1. Annual Mean

Firstly, the annual mean time series of the observed SH at six stations are shown in Figure 1. Although the data length is not uniform for each station (Table 1), we can still see their common features. SH shows obvious interannual variabilities, especially in NADORS and NAMORS, while SH in MAWORS and SETORS displays a clear decreasing and increasing trends with −0.94 and 1.67 W m^{-2} yr^{-1}, respectively. This may be the result of multiscale variabilities in SH, in which the effect of diurnal variation may be important.

The general characteristics of the diurnal variation in the observed SH over the TP can be understood from its climatic state (Figure 2). The magnitudes of annual mean SH at each station are basically negative at night with no clear fluctuations, while they are positive in the day with larger magnitudes and obvious variations, and their peaks appear in the afternoon, which confirms previous results [22,32,34]. Of course, the SH diurnal variations over the TP also show some discrepancies, with different amplitudes and peak timings at those six stations. This results presumably from differences in the underlying surface, altitudes, and climate conditions among these stations. QOMS, with sparse and short surface vegetation, has the largest diurnal amplitude of SH, reaching a peak of 184.35 W m^{-2} at 15:00, while MAWORS, influenced by the westerly wind all year round and being surrounded by large-scale modern glaciers, shows the smallest SH amplitude, with the peak value only reaching 112.49 Wm^{-2} at 15:00. Three other stations (BJ, NADORS, and NAMORS), with altitudes higher than 4000 m, have similar SH amplitudes to the average mean of those six stations, but the peak timing at BJ is earlier, at 14:00. Additionally, SETORS, which is located in a forested valley, close to the southeastern TP, deviates from the average mean, and it has a weak amplitude comparable to that of MAWORS, but its peak occurs much earlier (about 12:00) than other stations. Obviously, six stations distributed along the east–west direction are scattered sparsely across the entire TP, and the time difference of 2–3 h among them cannot be ignored (Table 1). If the Beijing time of all stations is changed to local time, then we will find that the peaks of most stations appear around 12:00 or 13:00 (local time), except SETORS, which is covered by dense vegetation (mainly temperate needleleaf trees and alpine meadows), whose peak timing is around 10:00 (local time) (The following times are Beijing time unless otherwise specified). The conditions of the underlying surface may change the upward radiation flux, affect the peak timing of the diurnal variation in surface net radiation, and then affect the peak timing of SH. Further examination (figure omitted) confirms that the peak timing of the net shortwave radiation flux in SETORS is indeed much earlier (about 13:00) than that of other stations (about 14:00 and 15:00), which is consistent with the peak timing in SH diurnal variation.

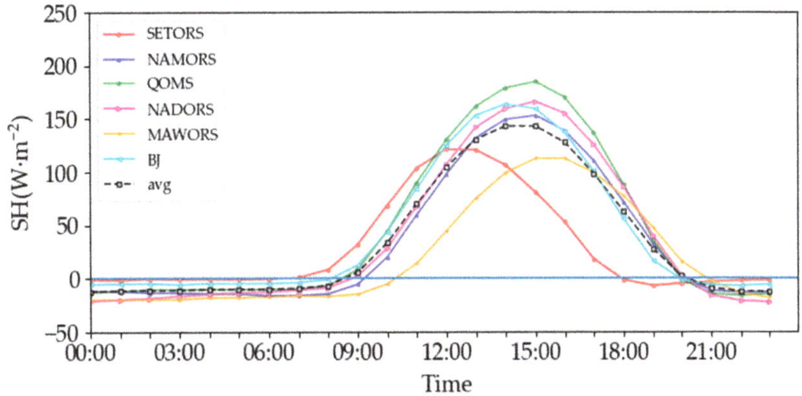

Figure 2. Climatological annual mean of diurnal variations of the observed SH (W m^{-2}) at six stations over the TP during the period 2006–2016. The colored lines represent SH at the six stations, respectively, and the black line indicates their average. The time in x axis is Beijing time.

3.2. Seasonal Mean

Figure 3 shows the seasonal mean of SH diurnal variations at the six stations, which indicates that the diurnal variations in four seasons generally have similar peak timing but different diurnal amplitudes at each station. The peak timing of SH in four seasons is mostly at 15:00 in MAWORS, NADORS, QOMS, and NAMORS, and at 14:00 and 12:00 at BJ and SETORS, respectively, which is consistent with Section 3.1 above.

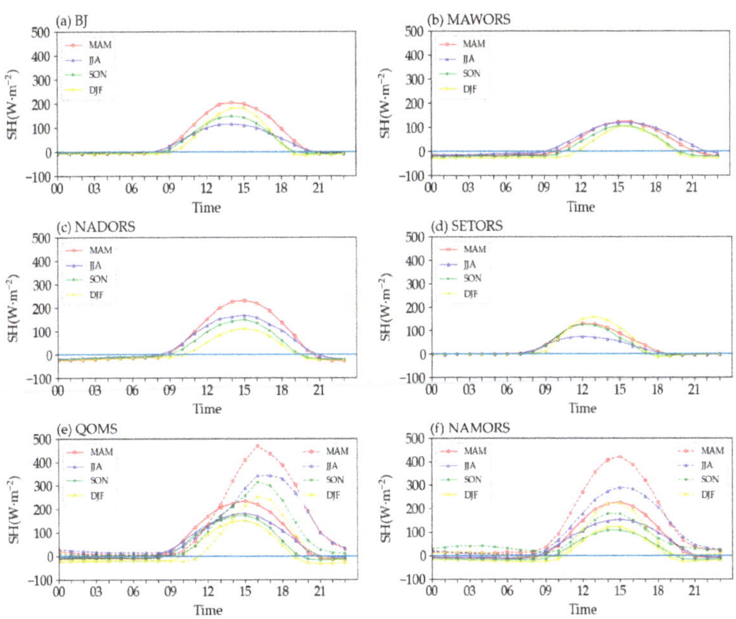

Figure 3. Climatological seasonal mean of the diurnal variation of the observed SH (solid lines, W m^{-2}) at (**a**) BJ, (**b**) MAWORS, (**c**) NADORS, (**d**) SETORS, (**e**) QOMS, and (**f**) NAMORS, and the calculated SH (dashed lines, W m^{-2}) at (**e**) QOMS and (**f**) NAMORS.

The diurnal amplitudes of SH at three stations (MAWORS, NADORS, and QOMS) are consistently largest in spring, followed by summer and autumn, and smallest in winter, which is consistent with previous studies [19,34]. In contrast, winter SH amplitudes at two other stations (BJ and SETORS) are exceptionally strong; the strongest being in SETORS. Both stations are located in the eastern TP and are well covered by vegetation, which allows the surface to capture more net radiation absorption in winter, resulting in stronger SH and latent heat [35]. In spring, the precipitation at SETORS accounts for 25.4% of the whole year, which is 14–18% higher than that of two other stations (QOMS and NAMORS), which measured precipitation by the same method. In a word, more precipitation in spring will weaken the SH to a large extent. Moreover, the amplitudes of BJ and SETORS in summer are the weakest throughout the year; the stronger relative humidity (up to 63.22% and 80.45%) at these two stations and the resulting larger evapotranspiration will probably weaken the SH to a great extent, resulting in an abnormal small amplitude in diurnal variation of the summer SH. For NAMORS, the amplitude of SH diurnal variation in winter is stronger than that in autumn, which may be related to the similar diurnal variations of the difference between surface temperature and air temperature in winter and autumn at this station.

In the past, due to the lack of flux observations over the TP region, most studies on SH were based on the calculated SH by using Equation (1) or reanalysis datasets [36,37], but uncertainties and biases always existed. The diurnal variations of the calculated SH and the observed SH at QOMS and NAMORS are also displayed in Figure 3e,f to detect the bias between them. We find that the diurnal variation of calculated SH shows similar seasonal distribution with the observed one, showing the greatest amplitude in spring and the smallest in winter at QOMS, and an abnormally stronger amplitude in winter at NAMORS. However, significant differences exist in the diurnal variation amplitude between calculated SH and observed SH; that is, the former is about 64–100% larger than the latter, which suggests that the calculated SH over the TP used in previous studies may be largely overestimated. Additionally, the peak timing of the calculated SH also shows different features compared with that of observed SH, mainly in that the latter is at 16:00, which is one hour later than the former at QOMS, suggesting that a significant phase shift occurs between the calculated SH and observed SH, while this phenomenon is not evident at NAMORS.

4. Monthly Changes of the SH Diurnal Variation over the TP

In order to further understand the SH diurnal variation over the TP in detail, we need to shorten the time scale to obtain the monthly mean of the SH diurnal variation.

4.1. Monthly Changes of the Diurnal Variation in Observed SH

Figure 4 shows the monthly mean of the diurnal variation of the observed SH. Obviously, it can be seen that the diurnal amplitudes of the SH among stations are significantly different, with smaller values at MAWORS and SETORS and larger values at the other four stations, which is also consistent with the results in Section 3. Furthermore, the diurnal variation characteristics of the SH differ significantly from month to month for each station. Generally, the duration of positive SH increases from January, with the longest duration in May or June about 13 h, from 8:00 to 21:00, at BJ, NADORS, and NAMORS, while it is about 10 h, from 10:00 to 22:00, at MAWORS, and it then decreases in wintertime. However, at QOMS, which has the higher altitude, the particularly long duration of positive SH in July or August is up to about 19 h from 2:00 to 21:00. The special feature of SETORS is that the changes of the positive SH duration are less distinctive, with more scattered positive values at night.

For the timing of peak values in SH diurnal variation (Figure 5), it also fluctuates month by month and varies from station to station, almost within a time range of 1 h. For example, at MAWORS (Figure 5a), the timing of the SH peak fluctuates between 15:00 and 16:00 for most months; at SETORS and NAMORS (Figure 5c), the peak timing fluctuates during 12:00–13:00 and 14:00–15:00, respectively. It should also be noted that although the peak timing of monthly SH fluctuates steadily within a certain range, there are still interannual variations at some stations. For example, the peak timing at BJ (Figure 5a) varies during 13:00–14:00 before 2013, but during 14:00–15:00 after 2013. Similarly, the peak timing at NADORS (Figure 5b) almost fluctuates between 14:00 and 15:00 before 2014, but between 15:00 and 16:00 after 2014. Especially in QOMS (Figure 5b), this phenomenon is more obvious; that is, before 2015, the peak timing fluctuates between 14:00 and 15:00 for most months in the eight years, while it fluctuates between 15:00 and 16:00 after 2015. Surface radiation flux tends to largely affect the peak timing changes of SH. During the period 2006–2016, the incoming radiation may have changed due to the variations in atmospheric conditions, such as aerosol, cloud cover, and water vapor, and the outgoing radiation may have changed due to the changes in underlying surface conditions, such as vegetation growth and soil moisture. These variations of surface radiation flux will be reflected in the land–air temperature difference, and then result in the SH changes.

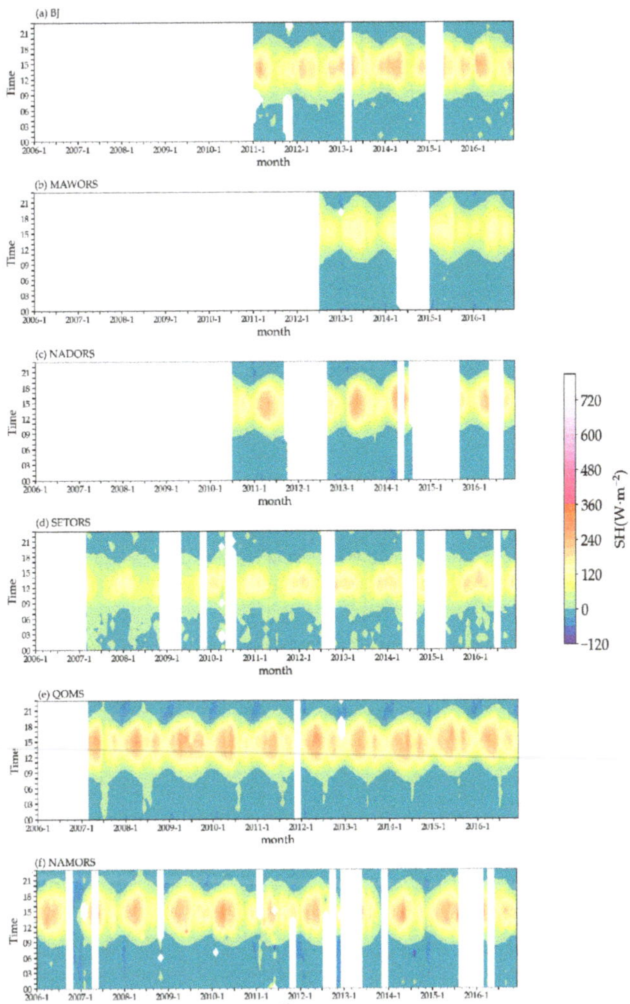

Figure 4. Monthly mean of the diurnal variation in observed SH (shaded, W m^{-2}) at (**a**) BJ, (**b**) MAWORS, (**c**) NADORS, (**d**) SETORS, (**e**) QOMS, and (**f**) NAMORS during the period 2006–2016.

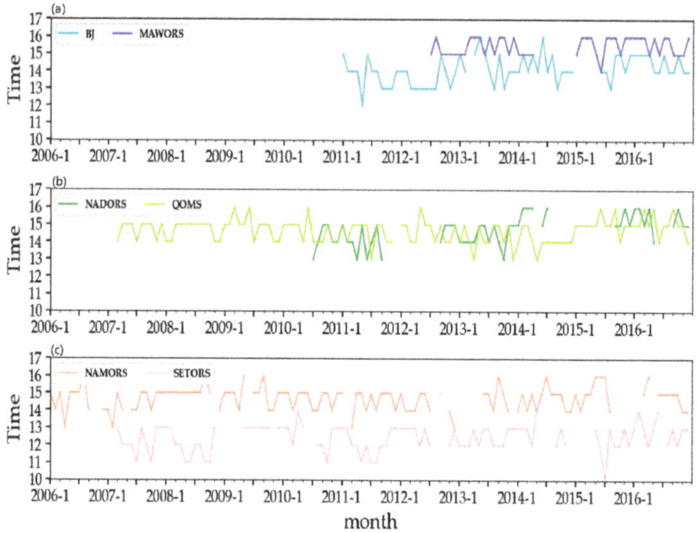

Figure 5. Peak timing (hour) of monthly diurnal variation in SH at (**a**) BJ and MAWORS, (**b**) NADORS and QOMS, and (**c**) NAMORS and SETORS.

Another important characteristic (Figure 4) is that the SH diurnal variations at QOMS and BJ have two distinct centers per year, as detailed in Table 2. In fact, the other four stations (MAWORS, NADORS, SETORS, and NAMORS) have similar characteristics to the two centers, but the second center is inconspicuous, so only the information of the first center of these stations is presented in Table 2. We can clearly see that, in general, the first center of the SH diurnal variations at all stations appears in spring afternoon for most years, except for SETORS, which appears in January to March. However, the second center appears in different months at QOMS and BJ, occurring almost at 14:00–15:00 in September or October for QOMS but at 14:00–15:00 in October to December for BJ. One thing of note is that the first center value is about 27% stronger than the second one on average, which indicates the spring SH is dominant on multiple time scales. Another point worth noting is that the SH diurnal variation in SETORS shows an increasing trend during the period 2007–2016 (Figure 4d), which is obviously manifested by the increasing peak values during the day. This is consistent with the trend in annual mean SH (Figure 1), indicating the internal consistency of different timescales.

Table 2. The month and the specific timing of the occurrence of the two large centers in SH diurnal variation at six stations from 2006 to 2016.

	QOMS		BJ	
	Month1 (Time)	**Month2 (Time)**	**Month1 (Time)**	**Month2 (Time)**
2006				
2007	May (15:00)	October (15:00)		
2008	April (15:00)	September (15:00)		
2009	April (15:00)	October (15:00)		
2010	June (16:00)	October (15:00)		
2011	May (15:00)	September (15:00)	March (14:00)	December (14:00)
2012	May (16:00)	October (14:00)	February (14:00)	October (14:00)
2013	May (15:00)	September (15:00)		December (14:00)
2014	June (14:00)	September (14:00)	May (14:00)	November (14:00)
2015	June (16:00)	September (16:00)		October (15:00)
2016	April (15:00)	October (15:00)	March (15:00)	November (14:00)

Table 2. *Cont.*

	MAWORS Month1 (Time)	SETORS Month1 (Time)	NADORS Month1 (Time)	NAMORS Month1 (Time)
2006				April (13:00)
2007		March (13:00)		
2008		February (13:00)		April (15:00)
2009				April (14:00)
2010		January (13:00)		April (15:00)
2011		February (13:00)	May (15:00)	May (15:00)
2012		February (13:00)		May (15:00)
2013	April (16:00)	February (12:00)	May (15:00)	
2014	March (15:00)			May (14:00)
2015	May (15:00)			May (16:00)
2016	July (16:00)	February (14:00)		

Furthermore, the main factors leading to the obvious centers of SH are discussed by taking QOMS (Figure 4e) as an example. Considering Equation (1) and Figure 6, we can clearly see that two centers in land–air temperature difference are similar to those in SH compared to wind speed, which shows many irregular centers. For the diurnal variation in land–air temperature difference, the first centers always appear in March to May, and the second ones with a smaller value almost occur in September or October, which corresponds well to those in SH (Table 2) each year. This indicates that the land–air temperature difference is the main factor causing the diurnal variation features of the SH at QOMS.

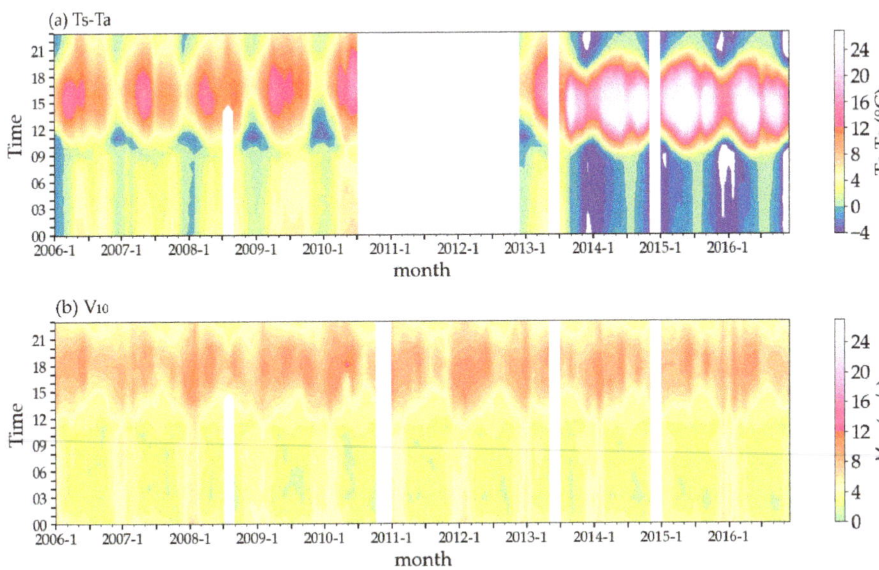

Figure 6. Monthly mean of the diurnal variation in (**a**) land–air temperature difference (Ts-Ta, unit: °C) and (**b**) wind speed at the height of 10 m (V_{10}, unit: m/s) at QOMS during the period 2006–2016.

4.2. Monthly Changes of the Diurnal Variation in Calculated SH

From the above, we have understood that there is a certain difference in the seasonal mean diurnal variation between the calculated SH and the observed SH, so it is necessary to explore the condition for the monthly mean diurnal variation between them, which is of great significance to improve the SH calculation.

Figure 7 shows the monthly diurnal variation of the calculated SH at QOMS and NAMORS. Generally, comparing the diurnal variation characteristics of the observed (Figures 4 and 5) and calculated (Figure 7) SH, we can see that the calculated SH at these two stations have two large-value centers in each year, and the second center is more obvious at QOMS than at NAMORS, similar to the observed characteristics. However, some clear differences still exist between them. The diurnal amplitude in the calculated SH at QOMS and NAMORS can be up to 2.86 and 3.09 times stronger than those in observed SH, respectively. Moreover, the differences between calculated SH and observed SH also exist in the peak timing of the SH diurnal variation and the peak timing fluctuations range. For QOMS, the peak timing shows a clear delay in calculated SH. Before 2011, it is basically stable around 17:00, and then remains at around 16:00 after 2013, which is relatively delayed by 1~3 h compared with the observed one. For NAMORS, the range of the peak timing fluctuations in calculated SH is much greater than that in the observations. Especially before 2009, the peak timing of calculated SH changes during 14:00–17:00 in most months; From 2009 to 2013, it fluctuates within 2 h during 14:00–16:00, and then fluctuates within 1 h during 14:00–15:00 after 2013, but the peak timing of observed SH mainly varies within the range of about one hour.

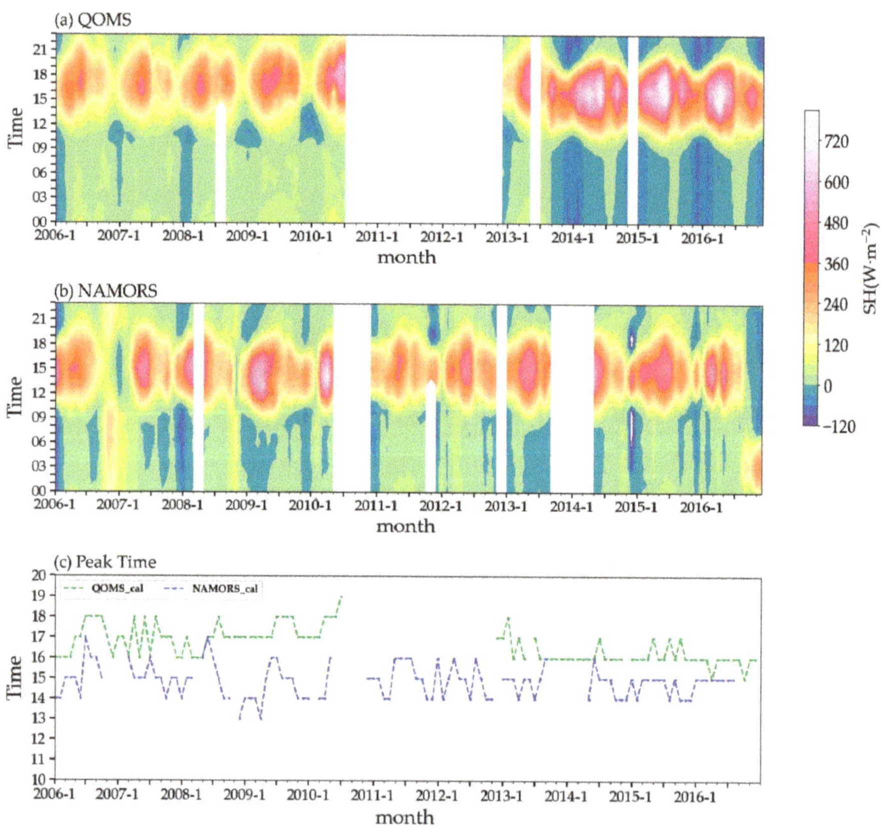

Figure 7. Monthly mean of the diurnal variation in calculated SH (shaded, W m^{-2}) at (**a**) QOMS and (**b**) NAMORS during the period 2006–2016. The curve lines in (**c**) represent the peak timing (hour) of monthly SH diurnal variation at each station, respectively.

Heat transfer coefficient C_{DH} is widely used as a constant value of 4×10^{-3} in the SH calculations over the TP, but the above facts show that there is a significant deviation between the observed SH and the calculated SH by Equation (1) on the diurnal scale, not only in the diurnal amplitude, but also in the peak timing and its fluctuation range. From Equation (1), we can clearly see that SH is calculated by using the heat transfer coefficient (C_{DH}), wind speed (V_{10}), and difference between surface temperature and air temperature (Ts-Ta). Actually, V_{10}, Ts, and Ta are originally from the same dataset and the same stations (QOMS and NAMORS), hence the C_{DH} may be the key factor resulting in the deviation in diurnal variation between the observed and calculated SH at these two stations.

5. Effect of the C_{DH} on SH Diurnal Variation

Previous studies [38] have pointed out that the uncertainty in the estimate of SH over the TP can be strongly attributed to the heat transfer coefficient C_{DH}. C_{DH} is affected by ground roughness and atmospheric stratification stability and has obvious seasonal and diurnal variations [39]. In order to address whether C_{DH} is the dominant factor giving rise to the bias between calculated SH and observed SH, here the diurnal variation characteristics of the heat transfer coefficient ($C_{DH\text{-}O}$ hereafter) derived with the observed SH according to Equation (1) is examined (Figure 8). Due to the limitation of observational data, only the $C_{DH\text{-}O}$ at NAMORS can be obtained.

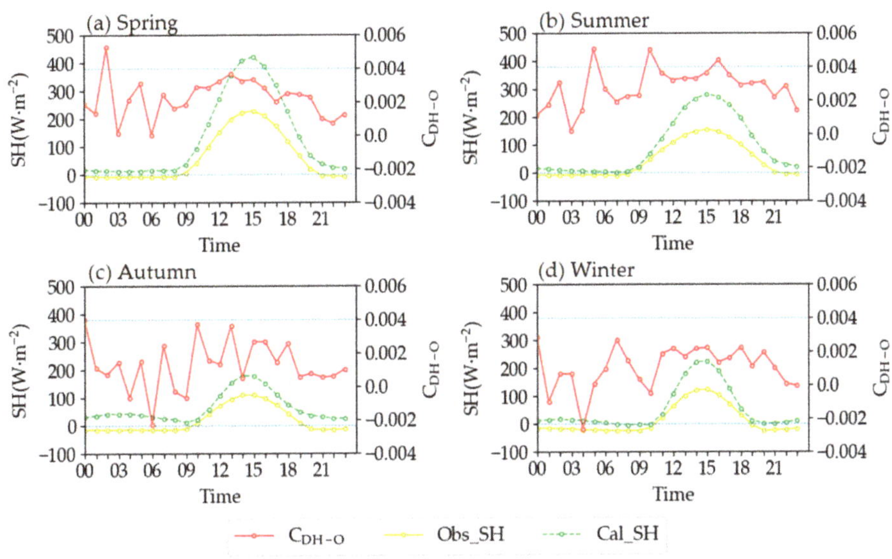

Figure 8. Seasonal mean of the diurnal variation in the derived heat transfer coefficient ($C_{DH\text{-}O}$), observed SH (Obs_SH, unit: W m^{-2}) and calculated SH (Cal_SH, unit: W m^{-2}) at NAMORS in (**a**) spring, (**b**) summer, (**c**) autumn, and (**d**) winter. The two horizontal dashed lines indicate zero SH and 0.004 $C_{DH\text{-}O}$, respectively.

As shown in Figure 8, the value of $C_{DH\text{-}O}$ in summer is the largest, followed by spring and autumn, and the smallest is in winter. This may be due to the unstable atmospheric stratification in summer over the TP and the greater roughness of the underlying surface vegetation, resulting in a larger $C_{DH\text{-}O}$ in summer. In general, the $C_{DH\text{-}O}$ values in all seasons are basically larger and stable in the day but fluctuate significantly at night, especially during the time from 00:00 to 10:00. Because the surface roughness at the fixed station has almost no diurnal variation, the $C_{DH\text{-}O}$ should be mainly considered to be affected by the stability of atmospheric stratification.

For the calculated SH, the C_{DH} is set as 4×10^{-3}, which is the average value often used in previous studies. However, the derived transfer coefficient $C_{DH\text{-}O}$ is much lower than 4×10^{-3} throughout the day (Figure 8), which suggests that the calculation of SH by choosing a fixed value of 4×10^{-3} will be overestimated, and also be biased in the diurnal variation. Clearly, the difference in diurnal variation between the calculated SH and the observed SH is the largest in spring, followed by summer, winter, and autumn, and the deviation can be up to 69.53 W m^{-2} and 194.46 W m^{-2} in autumn and spring, respectively. Therefore, the value of C_{DH} is very important, which will bring a certain uncertainty to the SH calculation and its diurnal variation.

A more reasonable value of the heat transfer coefficient is urgently needed to reduce the uncertainty and obtain a more accurately calculated SH. Usually, the TP is a strong heat source in spring and summer, and most studies on SH also focused on these two seasons. Therefore, by further calculations, the seasonal mean $C_{DH\text{-}O}$ values in spring and summer at NAMORS are about 2.24×10^{-3} and 2.78×10^{-3}, respectively, and the new SH in these two seasons is obtained by using these new C_{DH}. Here, Figure 9 shows the relationships between the originally calculated SH, the new SH, and the observed SH. Obviously, the new SH is much closer to the observations on a diurnal scale, especially in spring, with a maximum deviation of 27.89 W m^{-2}, only being 12.7% of the maximum deviation between the originally calculated SH and the observed SH. Moreover, the new SH shows interannual variations comparable with the observed SH (Figure 10). Of note is that, during the spring of 2006–2012, the change of the new SH is completely consistent with the observed SH, and the relative deviation of the new SH is only 13.1%, while that of the calculated SH is 101.9%. Considering the scarcity of directly observed flux data over the TP region, it is inevitable to calculate SH for the research involving TP SH. Therefore, the new and better heat transfer coefficient (2.24×10^{-3} in spring and 2.78×10^{-3} in summer) presented here is conducive to calculating SH more accurately in the future.

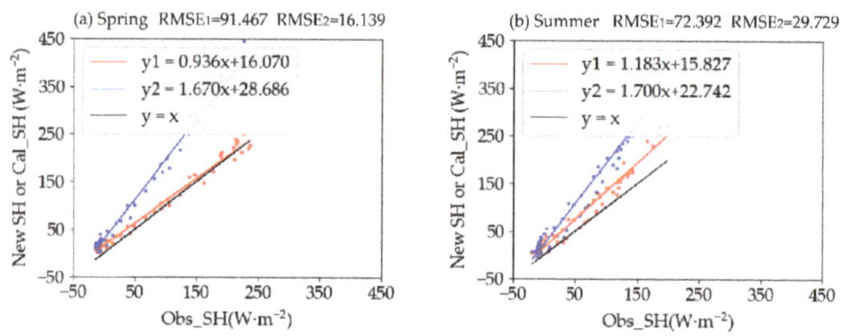

Figure 9. The relationship between the originally calculated SH and observed SH (blue lines, unit: W m^{-2}), and between the new SH and observed SH (red lines, unit: W m^{-2}) in the seasonal mean of the diurnal variation in (**a**) spring and (**b**) summer at NAMORS. $RMSE_1$ and $RMSE_2$ denote the root mean squared error of the originally calculated SH and the new SH, respectively.

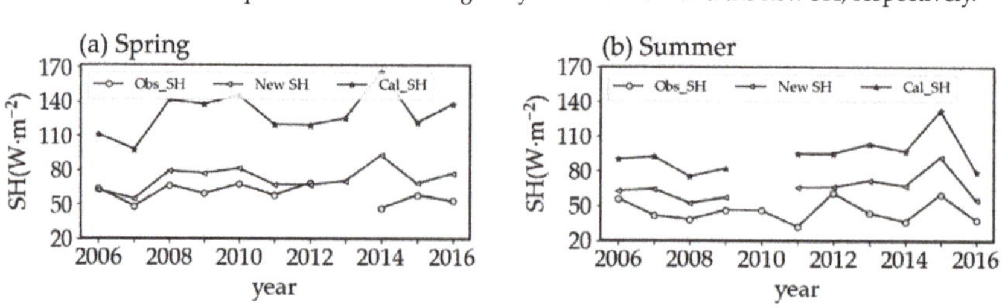

Figure 10. Time series of the annual mean in observed SH (Obs_SH, units: W m^{-2}), originally calculated SH (Cal_SH, units: W m^{-2}), and new SH (New SH, units: W m^{-2}) in (**a**) spring and (**b**) summer at NAMORS during the period 2006–2016.

6. Conclusions and Discussion

Obvious diurnal variation exists in SH over the TP. Here we adopted the hourly observational SH from the Tibetan Plateau Scientific Data Center to deeply understand the characteristics of the SH diurnal variation over the TP. In addition, the differences between the observed and calculated SH are also examined. The main conclusions are as follows:

(1) In general, the magnitude of annual mean SH is negative and stable at night, while it is positive with evident variations in the day, and often reaches its peak at around 12:00 or 13:00 local time, except for at SETORS, whose peak appears at around 10:00 local time.

(2) The SH diurnal variation has obvious seasonal changes, with similar peak timing but different diurnal amplitudes in four seasons at each station. The SH diurnal amplitude is uniformly greatest in spring, followed by summer and autumn, and the smallest in winter at MAWORS, NADORS, and QOMS, while the weakest amplitude in summer and a larger amplitude in winter occur at BJ and SETORS, the strongest amplitude in winter being at SETORS. The peak timing is mostly at 15:00 in four seasons at MAWORS, NADORS, QOMS, and NAMORS, and at 14:00 and 12:00 at BJ and SETORS, respectively.

(3) The SH diurnal variation has significant monthly changes. The positive SH at most stations has the longest duration from May to August. The peak timing of SH fluctuates between 15:00 and 16:00 for most months at MAWORS and fluctuates during 12:00–13:00 and 14:00–15:00 at SETORS and NAMORS, respectively. At other stations, the peak timing even shows a shift; for example, at QOMS the peak timing fluctuates between 14:00 and 15:00 before 2015, while it fluctuates between 15:00 and 16:00 after 2015. Moreover, the double-peak phenomenon of SH diurnal variation mainly occurs in spring and autumn, especially at QOMS, which largely contributes to the similar phenomenon in the land–air temperature difference.

(4) The SH diurnal variations between the observed and calculated SH significantly differ in seasonal and monthly variabilities, including the diurnal amplitude, peak timing, and the range of peak timing fluctuations. For the seasonal mean, the diurnal amplitude of the calculated SH is about 64–100% larger than that of the observed SH. In addition, an obvious phase shift occurs in the peak timing at QOMS, from 15:00 to 16:00. For the monthly changes, the range of the peak timing fluctuations in calculated SH (about 1–3 h) is clearly larger than that in observed SH (about one hour). Furthermore, a new C_{DH} (2.24×10^{-3} in spring and 2.78×10^{-3} in summer) is recommended here for more accurately calculating TP SH, which may provide a valuable implication for future studies on the TP SH.

Research related to turbulent flux has always been the core issue of land–atmosphere interaction [40], and an in-depth understanding of diurnal variation characteristics in SH over the TP can help us to understand the key land surface processes. Due to the uneven distribution of observations in high mountain regions [28], especially over the TP, the model performance is still poor [41], so various parameterization schemes and numerical models for SH are usually developed. An in-depth understanding of the diurnal variation characteristics of SH over the TP can help to improve and calibrate the numerical modes. Moreover, the new C_{DH} obtained by comparing the calculated SH and observed SH on a diurnal scale can boost the accuracy of SH calculations. However, the impact of the diurnal variation in TP SH on the weather has not been mentioned in this study, and it therefore needs further exploration in the future. It is also worth noting that the suggested new C_{DH} is obtained only from NAMORS due to data limitations, so there is still a certain one-sidedness.

Author Contributions: Conceptualization, M.W.; formal analysis, Z.Z.; writing—original draft, Z.Z.; writing—review and editing, M.W., J.W., X.M., J.L. and X.Y. All authors have read and agreed to the published version of the manuscript.

Funding: This work was sponsored by the National Natural Science Foundation of China (42030605), the Second Tibetan Plateau Scientific Expedition and Research (STEP) program (2019QZKK0105), the National Natural Science Foundation of China (41605039, 41807434, 42030611), the Natural Science Foundation of Jiangsu Province, China (grant no. BK20221449), Meteorological soft science of China Meteorological Administration (2022ZDIANXM28), and the open project of Key Laboratory of Meteorological Disaster (KLME), Ministry of Education & Collaborative Innovation Center on Forecast and Evaluation of Meteorological Disasters (CIC-FEMD), Nanjing University of Information Science & Technology (grant no. KLME202203).

Institutional Review Board Statement: Not applicable.

Informed Consent Statement: Not applicable.

Data Availability Statement: The data presented in this study are openly available in [National Tibetan Plateau Data Center] at [https://doi.org/10.11888/Meteoro.tpdc.270910], reference number [27].

Conflicts of Interest: The authors declare no conflict of interest.

References

1. Ma, Y.M.; Hu, Z.Y.; Tian, L.D.; Zhang, F.; Duan, A.M.; Yang, K.; Zhang, Y.L.; Yang, Y.P. Study Progresses of the Tibet Plateau Climate System Change and Mechanism of Its Impact on East Asia. *Adv. Earth Sci.* **2014**, *29*, 207–215.
2. Wu, G.X.; Liu, Y.M.; Liu, X. How the Heating over the Tibetan Plateau Affects the Asian Climate in Summer. *Chin. J. Atmos. Sci.* **2005**, *29*, 47–56.
3. Dai, Y.F. Interannual Variability of Surface Sensible Heat Flux in the Tibetan Plateau and Its Impacts on the Advancing Process of East-Asian Subtropical Summer Monsoon. Master's Thesis, Nanjing University of Information Science and Technology, Nanjing, China, 2016.
4. Zhang, Y.S.; Wu, G.X. Diagnostic Investigations on the Mechanism of the Onset of Asian Summer Monsoon and Abrupt Seasonal Transitions over the Northern Hemisphere Part: II the Role of Surface Sensible Heating over Tibetan Plateau and Surrounding regions. *Acta. Meteor. Sin.* **1999**, *57*, 57–74.
5. Wu, G.X.; Liu, Y.M.; He, B.; Bao, Q.; Wang, Z.Q. Review of the Impact of the Tibetan Plateau Sensible Heat Driven Air-Pump on the Asian Summer Monsoon. *Chin. J. Atmos. Sci.* **2018**, *42*, 488–504.
6. Zhou, X.J.; Zhao, P.; Chen, J.M.; Chen, L.X.; Li, W.L. Impacts of thermodynamic processes over the Tibetan Plateau on the Northern Hemispheric climate. *Sci. China Ser. D-Earth Sci.* **2009**, *39*, 1473–1486. [CrossRef]
7. Duan, A.M.; Liu, Y.M.; Wu, G.X. Heating status of the Tibetan Plateau from April to June and rainfall and atmospheric circulation anomaly over East Asia in midsummer. *Sci. China Ser. D-Earth Sci.* **2005**, *48*, 250–257. [CrossRef]
8. Li, D.L.; Wei, L.; Li, W.J.; Lv, L.Z.; Zhong, H.L.; Ji, G.L. The Effect of Surface Sensible Heat Flux of the Qinghai-Xizang Plateau on General Circulation over the Northern Hemisphere Climatic Anomaly of China. *Clim. Env. Res.* **2003**, *8*, 60–70.
9. Zhang, C.C. The Anomaly of the Surface Sensible Heat of the Tibetan Plateau in the Boreal Spring and Its Influences on the Summertime Rainfall Pattern. Master's Thesis, Nanjing University of Information Science and Technology, Nanjing, China, 2016.
10. Ma, Y.M.; Fan, S.; Ishikawa, H.; Tsukamoto, O.; Yao, T.D.; Koike, T.; Zuo, H.; Hu, Z.Y.; Su, Z.B. Diurnal and inter-monthly variation of land surface heat fluxes over the central Tibetan Plateau area. *Theor. Appl. Climatol.* **2005**, *80*, 259–273. [CrossRef]
11. Wang, H.; Zhang, L.; Shi, X.D.; Li, D.L. Seasonal Differences in the Trend Turning Characteristics of Surface Sensible Heat over the Central and Eastern Tibetan Plateau. *Chin. J. Atmos. Sci.* **2022**, *46*, 133–150.
12. Xie, J.; Liu, C.; Ge, J. Characteristics of Surface Sensible Heat Flux over the Qinghai-Tibetan Plateau and Its Response to Climate Change. *Plateau Meteor.* **2018**, *37*, 28–42.
13. Wang, M.R. Trend in the Atmospheric Heat Source over the Tibetan Plateau and Its Influence on Interdecadal Variation of Summer Precipitation in China during the Past 30 Years. Master's Thesis, Nanjing University of Information Science and Technology, Nanjing, China, 2016.
14. Duan, L.J.; Duan, A.M.; Hu, W.T.; Gong, Y.F. Low Frequency Oscillation of Precipitation and Daily Variation Characteristic of Air–Land Process at Shiquanhe Station and Linzhi Station in Tibetan Plateau in the Summer of 2014. *Chin. J. Atmos. Sci.* **2017**, *41*, 767–783.
15. Ma, Y.M.; Tsukamoto, O.; Wu, X.M.; Tamagawa, I.; Wang, J.M.; Tshikawa, H.; Hu, Z.Y.; Gao, H.C. Characteristics of Energy Transfer and Micrometeorology in the Surface Layer of the Atmosphere above Grassy Marshland of the Tibetan Plateau Area. *Chin. J. Atmos. Sci.* **2000**, *24*, 715–722.
16. Liu, Y.; Zou, H.; Hu, F. Observation Study on Atmospheric Surface Layer in Rongbu Valley in Zumolama Peak Area of Qinghai-Xizang Plateau. *Plateau Meteor.* **2004**, *23*, 512–518.

17. Li, G.P.; Duan, T.Y.; Gong, Y.F. Bulk Transfer Coefficients and Surface Fluxes over the Western Tibetan Plateau. *Chin. Sci. Bull.* **2000**, *45*, 865–869.
18. Li, G.P.; Zhao, B.J.; Lu, J.H. Characteristics of Bulk Transfer Coefficients over the Tibetan Plateau. *Acta. Meteor. Sin.* **2002**, *60*, 60–67.
19. Wang, H.; Hu, Z.Y.; Li, D.L.; Dai, Y.F. Estimation of the surface heat transfer coefficient over the east-central Tibetan Plateau using satellite remote sensing and field observation data. *Theor. Appl. Climatol.* **2019**, *138*, 169–183. [CrossRef]
20. Chen, W.L.; Wong, D.M. A preliminary study on the computational method of 10-day mean sensible heat and latent heat on the Tibetan Plateau. In *Collected Works of the Qinghai-Xizang Plateau Meteorological Experiment*; Series 2; Science Press: Beijing, China, 1984.
21. Li, G.P.; Duan, T.Y.; Wan, J.; Gong, Y.F.; Haginoya, S.; Chen, L.X.; Li, W.L. Determination of the drag coefficient over the Tibetan Plateau. *Adv. Earth Sci.* **1996**, *13*, 511–518.
22. Wang, M.R.; Zhou, S.W.; Duan, A.M. Trend in the atmospheric heat source over the central and eastern Tibetan Plateau during recent decades: Comparison of observations and reanalysis data. *Chin. Sci. Bull.* **2012**, *57*, 178–188. [CrossRef]
23. Li, C.F.; Yanai, M. the Onset and Interannual Variability of the Asian Summer Monsoon in Relation to Land–Sea Thermal Contrast. *J. Clim.* **1996**, *9*, 358–375. [CrossRef]
24. Wang, T.Z.; Zhao, Y. Comparative analysis of five sets of surface sensible heat flux data over the Tibetan Plateau in May. *J. Meteor. Sci.* **2020**, *40*, 819–828.
25. Shan, X.; Zhou, S.W.; Wang, M.R.; Zheng, D.; Wang, C.H. Effects of Spring Sensible Heat in the Tibetan Plateau on Midsummer Precipitation in South China under ENSO. *J. Trop. Meteor.* **2020**, *36*, 60–71.
26. Chen, L.; Pryor, S.C.; Wang, H.; Zhang, R.H. Distribution and Variation of the Surface Sensible Heat Flux over the Central and Eastern Tibetan Plateau: Comparison of Station Observations. *J. Geophys. Res. Atmos.* **2019**, *124*, 6191–6206. [CrossRef]
27. Ma, Y.M. *A Long-Term Dataset of Integrated Landatmosphere Interaction Observations on the Tibetan Plateau (2005–2016)*; National Tibetan Plateau Data Center: Beijing, China, 2020.
28. Ma, Y.M.; Hu, Z.Y.; Xie, Z.P.; Ma, W.Q.; Wang, B.B.; Chen, X.L.; Li, M.S.; Zhong, L.; Sun, F.L.; Gu, L.L.; et al. A long-term (2005–2016) dataset of hourly integrated land–atmosphere interaction observations on the Tibetan Plateau. *Earth Syst. Sci. Data.* **2020**, *12*, 2937–2957. [CrossRef]
29. Duan, A.M.; Li, F.; Wang, M.R.; Wu, G.X. Persistent weakening trend in the spring sensible heat source over the Tibetan Plateau and its impact on the Asian summer monsoon. *J. Clim.* **2011**, *24*, 5671–5682. [CrossRef]
30. Zhu, L.H.; Huang, G.; Fan, G.Z.; Qu, X.; Zhao, G.J.; Hua, W. Evolution of surface sensible heat over the Tibetan Plateau under the recent global warming hiatus. *Adv. Atmos. Sci.* **2017**, *34*, 1249–1262. [CrossRef]
31. Wang, H.; Zhang, L.; Shi, X.D.; Li, D.L. Some New Changes of the Regional Climate on the Tibetan Plateau Since 2000. *Adv. Earth Sci.* **2021**, *36*, 785–796.
32. Duan, A.M.; Wu, G.X. Weakening Trend in the Atmospheric Heat Source over the Tibetan Plateau during Recent Decades. Part I: Observations. *J. Clim.* **2008**, *21*, 3149–3164. [CrossRef]
33. Ye, D.Z.; Gao, Y.X.; Zhou, M.Y. *Qinghai-Xizang Plateau Meteorology*; Science Press: Beijing, China, 1979.
34. Yu, J.H.; Liu, J.M.; Ding, Y.G. Annual and Diurnal Variations of Surface Fluxes in Western Qinghai-Xizang Plateau. *Plateau Meteor.* **2004**, *23*, 353–359.
35. Ji, J.J.; Huang, M. The Estimation of the Surface Energy Fluxes over Tibetan Plateau. *Adv. Earth Sci.* **2006**, *21*, 1268–1272.
36. Jin, R.; Qi, L.; He, J.H. Effect of oceans to spring surface sensible heat flux over Tibetan Plateau and its influence to East China precipitation. *Acta Oceanol. Sin.* **2016**, *38*, 83–95.
37. Wang, X.J.; Yang, M.X.; Wan, G.N. Temporal-Spatial Distribution and Evolution of Surface Sensible Heat Flux over Qinghai-Xizang Plateau during Last 60 Years. *Plateau Meteor.* **2013**, *32*, 1557–1567.
38. Duan, A.M.; Liu, S.F.; Hu, W.T.; Hu, D.; Peng, Y.Z. Long-term daily dataset of surface sensible heat flux and latent heat release over the Tibetan Plateau based on routine meteorological observations. *Big Earth Data* **2022**, *6*, 480–491. [CrossRef]
39. Zhu, Y.X.; Ding, Y.H.; Liu, W.H. Simulation of the Influence of Winter Snow Depth over the Tibetan Plateau on Summer Rainfall in China. *Chin. J. Atmos. Sci.* **2009**, *33*, 903–915.
40. Duan, L.J. Characteristics of Multiscale Air-Land Interaction over Tibetan Plateau in the Summer of 2014. Master's Thesis, Chengdu University of Information Science and Technology, Chengdu, China, 2017.
41. Xie, Z.P.; Hu, Z.Y.; Ma, Y.M.; Sun, G.H.; Gu, L.L.; Liu, S.; Wang, Y.D.; Zheng, H.X.; Ma, H.Q. Modeling Blowing Snow over the Tibetan Plateau with the Community Land Model: Method and Preliminary Evaluation. *J. Geophys. Res.-Atmos.* **2019**, *124*, 9332–9355. [CrossRef]

atmosphere

MDPI

Article

A Parametric Model of Elliptic Orbits for Annual Evolutions of Northern Hemisphere Stratospheric Polar Vortex and Their Interannual Variability

Yueyue Yu [1,2], Jie Sun [3], Michael Secor [3], Ming Cai [3,*] and Xinyue Luo [1,4]

[1] Key Laboratory of Meteorological Disaster, Ministry of Education (KLME)/Joint International Research Laboratory of Climate and Environment Change (ILCEC)/Collaborative Innovation Center on Forecast and Evaluation of Meteorological Disasters (CIC-FEMD), Nanjing University of Information Science & Technology, Nanjing 211544, China; yuyy@nuist.edu.cn (Y.Y.); lh804741@student.reading.ac.uk (X.L.)
[2] State Key Laboratory of Numerical Modeling for Atmospheric Sciences and Geophysical Fluid Dynamics (LASG), Institute of Atmospheric Physics, Chinese Academy of Sciences, Beijing 100029, China
[3] Department of Earth, Ocean & Atmospheric Sciences, Florida State University, Tallahassee, FL 32304, USA; jsun2@fsu.edu (J.S.); msecor@fsu.edu (M.S.)
[4] Reading Academy, Nanjing University of Information Science & Technology, Nanjing 210044, China
* Correspondence: mcai@fsu.edu; Tel.: +1-(850)-644-1551

Abstract: The year-to-year varying annual evolutions of the stratospheric polar vortex (SPV) have an important downward impact on the weather and climate from winter to summer and thus potential implications for seasonal forecasts. This study constructs a parametric elliptic orbit model for capturing the annual evolutions of mass-weighted zonal momentum at 60° N (*MU*) and total air mass above the isentropic surface of 400 K (*M*) over the latitude band of 60–90° N from 1 July 1979 to 30 June 2021. The elliptic orbit model naturally connects two time series of a nonlinear oscillator. As a result, the observed coupling relationship between *MU* and *M* associated with SPV as well as its interannual variations can be well reconstructed by a limited number of parameters of the elliptic orbit model. The findings of this study may pave a new way for short-time climate forecasts of the annual evolutions of SPV, including its temporal evolutions over winter seasons as well as the spring and fall seasons, and timings of the sudden stratospheric warming events by constructing its elliptic orbit in advance.

Keywords: stratospheric polar vortex; annual evolution; interannual variability; parametric model; elliptic orbits

Citation: Yu, Y.; Sun, J.; Secor, M.; Cai, M.; Luo, X. A Parametric Model of Elliptic Orbits for Annual Evolutions of Northern Hemisphere Stratospheric Polar Vortex and Their Interannual Variability. *Atmosphere* **2023**, *14*, 870. https://doi.org/10.3390/atmos14050870

Academic Editor: Yoshihiro Tomikawa

Received: 12 April 2023
Revised: 11 May 2023
Accepted: 12 May 2023
Published: 14 May 2023

1. Introduction

The Northern Hemisphere stratospheric polar vortex (SPV) is characterized by a significant annual evolution or annual cycle [1]. It spins up from autumn and persists through winter into spring, at which point it breaks up and the typical wintertime subpolar westerlies are replaced by easterly summer circulation until the subsequent autumn [2,3]. The transition from a winter to a summer circulation regime is known as the stratospheric final warming (SFW) [4], which is influenced by both solar radiation and dynamics [5].

There are various indices commonly used to describe the strength of the stratospheric polar vortex. These indices include polar cap (60–90° N) height [6], polar cap temperature, potential vorticity on isentropic surfaces [7–9], polar mass [10], zonal-mean zonal wind at subpolar latitudes [11–13] and Northern Annular Mode (NAM) [14–17]. Baldwin and Thompson [18] have made a comprehensive comparison among those indices and found that these indices generally correspond well with each other and are capable of representing the variability in the SPV. The anomalous changes in these indicators always exhibit remarkably larger amplitudes from late fall to early spring each year and are dominated by intra-seasonal variabilities [14,15,19,20].

The dynamic influence of the stratospheric polar vortex during the winter months is known to play a part in shaping the winter circulation patterns of the troposphere such as Arctic Oscillation and blockings [17,20–26]. As a result, the anomalous signals in the stratospheric polar vortex can exert important downward impacts on weather and climate in the extratropical troposphere. For instance, during the 1–2 months after a weaker stratospheric polar vortex event such as sudden stratospheric warming (SSW) and negative stratospheric northern annular mode event, below-normal temperatures tend to occur more frequently over the Northern Hemispheric continents [24,27–38]. In addition, the polar temperature related to the SPV in the late winter months is closely related to the formation of polar stratospheric clouds and thus affects the Arctic ozone depletion [39–42]. The interannual variations in seasonal timing and the features of stratospheric final warming have been found to affect the spring circulation anomalies, which in turn modifies the summer monsoon [4,43], and affect precipitation anomalies over South Asia in May and the temperature anomalies over Central Asia in March [44]. Later, stratospheric final warming and stronger SPV in springtime have also been found to lead to negative ice thickness anomalies in the East Siberian Sea [3]. The strength of SPV can be the sub-seasonal to seasonal forewarning of anomalous atmospheric river frequency [45]. Therefore, identifying and understanding the year-to-year varying annual cycle of SPV has potential implications for winter seasonal forecasts, as the December–February mean behavior may miss important sub-seasonal events.

Hardiman et al. [46] formulate a simple sine wave fit to the observed vacillations in the daily zonal, mean zonal wind anomalies at 10 hPa, which can be used to explain much of the sub-seasonal and interannual variability in the monthly mean vortex strength. Additionally, consistent with wave-mean flow interaction theory, their amplitude correlates positively with the magnitude of winter mean planetary wave driving. They then tested the predictive skill of this simple sine wave model throughout the winter against the persistence of the vortex strength and against a state-of-the-art seasonal forecasting system, and reported improved prediction skills of the vortex strength one month ahead.

Not only the zonal wind surrounding the SPV but also the thermal condition in the stratospheric polar region are important indicators of the SPV intensity changes. The changes in momentum and thermal fields associated with SPV variations are intimately coupled as expected from the thermal wind relationship. According to Zuev and Savelieva [47], the temperature distribution inside the SPV at a specific range of wind speed was approximately in the same temperature range, but the temperature clearly shows year-to-year differences. This study formulates a parametric ellipse orbit model for better capturing the coupled annual evolutions of both the zonal momentum of the polar jet and the total air mass in the polar cap of the stratosphere and their interannual variability.

2. Data and Methods

2.1. Data

Daily mean fields are derived from the hourly ERA5 reanalysis dataset [48], which is on 1.5° × 1.5° grids and covers 42 years (July of the previous year to June of the current year) from 1979/80 to 2020/21. We use variables including the three-dimensional temperature, geopotential height and wind fields at 37 pressure levels from 1000 hPa to 1 hPa, air pressure, 2 m air temperature and wind fields at the surface. A 7-day running mean is applied to minimize synoptic-scale perturbations.

2.2. Methods

In this study, we use the stratospheric polar mass (M) above the isentropic surface of 400 K inside the Arctic polar circle and mass-weighted zonally integrated zonal momentum at 60° N (MU) to represent the polar vortex and polar jet changes. The isentropic level 400 K is about 100 hPa in the subpolar latitudes. This level is chosen in our study because the stratospheric circulation signals above 400 K can actively interact with the upward propagating tropospheric waves [49–56], which in turn exert significant downward impacts on the tropospheric circulation and the weather in winter [19,57,58]. The latitude of 60° N is chosen to better capture the edge of the stratospheric polar vortex and the maximum of the westerly surrounding the vortex in the winter seasons. These two variables are addable and closely linked to the mass budget and angular momentum budget [10,26,59], and thus can be quantitatively attributed to the adiabatic processes mainly driven by wave dynamics and diabatic processes.

Following Pauluis et al. [60,61] and Yu et al. [10,62], we first interpolated the daily potential temperature and wind fields onto 200 equally spaced sigma (σ) levels from 1 to 0. The air mass between two adjacent sigma surfaces per unit area is $m_\sigma = \Delta\sigma/g\ Ps$, where g is the gravitational constant, $\Delta\sigma = 1/200$ and Ps is the surface pressure. According to Yu et al. [10], the *M* can be measured by the integrated mass above 400 K isentropic level north of 60° N (kg), which is derived as

$$M(t) = \int_{\pi/3}^{\pi/2}\int_0^{2\pi}\int_0^1 \mathrm{m}_\sigma(\lambda,\ \phi,\sigma,t)\cdot H(\theta(\lambda,\ \phi,\sigma,t),400\ \mathrm{K})\mathrm{d}\sigma\mathrm{R}\cos\phi\mathrm{d}\lambda\mathrm{R}\mathrm{d}\phi \quad (1)$$

where λ is the longitude and ϕ is the latitude, *R* is the Earth radius; *H* is the Heaviside function, namely $H(x, x_0) = 1$ if $x \geq x_0$, and otherwise $H(x,x_0) = 0$. In (1) and (2) below, $x = \theta(\lambda,\varphi,\sigma,t)$ and $x_0 = 400$ K.

Similarly, the intensity of the polar jet can be measured by the zonally integrated zonal momentum in the stratospheric layer above 400 K across 60° N (*MU*, unit: kg m s^{-1}), which can be obtained according to

$$MU(t) = \int_0^{2\pi}\int_0^1 \mathrm{m}_\sigma(\lambda,\ \phi = 60\ \mathrm{N},\sigma,t)\cdot \mathrm{u}(\lambda,\ \phi = 60\ \mathrm{N},\sigma,t)\cdot H(\theta(\lambda,\ \phi = 60\ \mathrm{N},\sigma,t),400\ \mathrm{K})\mathrm{d}\sigma\mathrm{R}\cos\phi\mathrm{d}\lambda \quad (2)$$

Smaller values of *M* indicate a stronger stratospheric polar vortex, while larger values of *MU* indicate a stronger polar jet and vice versa.

3. Results

3.1. Statistics of the SPV Indices

From the annual evolutions (Figure 1), we see that the stratospheric mass in the polar region (*M*) and the relative zonal momentum around the polar jet (*MU*) vary generally out of phase. Namely, the *MU* increases while the *M* decreases before February and then the *MU* decreases while the *M* increases afterward. This illustrates that the Northern Hemisphere SPV and the polar jet surrounding it persist in the winter months and peak around mid-January. Radiative heating/cooling processes provide overall control of the annual cycle of *M* thermodynamically and adiabatic eddy-driven poleward mass transport processes affect the amplitude and phase of the annual cycle dynamically. The annual cycle of *MU*, besides through the thermal wind balance which is a fast process (less than a day), is controlled by planetary-scale waves that drive the poleward mass transport in the first place. Specifically, a stronger (weaker) poleward transport tends to coincide with the weakening (strengthening) of *MU*, which is about a few days prior to the increase in *M* [10]. In other words, at seasonal scales, the annual cycle of *MU* is driven by the radiative processes via thermal wind balance. However, at sub-seasonal time scales, particularly in the winter seasons, temporal evolutions of *MU* are driven by planetary-scale waves, which in turn drive the temporal evolution of *M* through eddy-driven poleward mass transport. The standard deviations of *MU* and *M* become remarkably larger in the months from December to March compared to the other months throughout the year, consistent with previous studies [14,15,19,20]. Note that the *MU* shows larger interannual variations around its climatology than the *M*. In particular, the maximum standard deviation of *MU* during the winter months is around 60% of its climatological mean, while for the *M*, its maximum standard deviation is only about 10% of its climatological mean. The seasonal changes of the coupled *M* and *MU* also exhibit remarkable interannual variation, as can be seen clearly from Figure 2. The timing of the strongest SPV, which can be represented by the minimum M and maximum MU, ranges from December to April, during the period from 1980 to 2020.

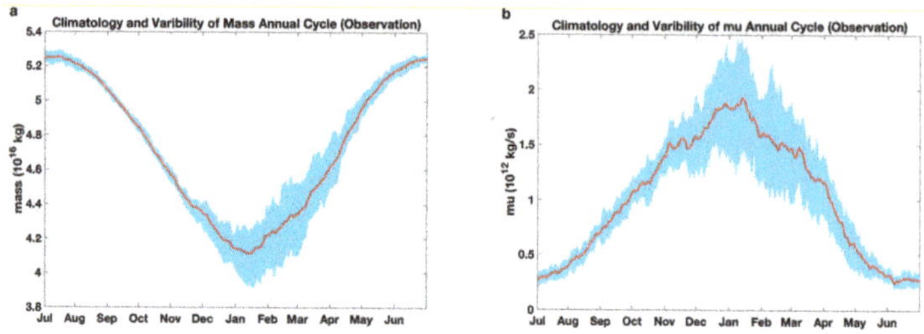

Figure 1. The 42-year climatological annual evolutions (red curves) and their interannual variability (blue shadings) in M, (**a**) units: 10^{16} kg and MU (**b**) units: 10^{12} kg s^{-1}).

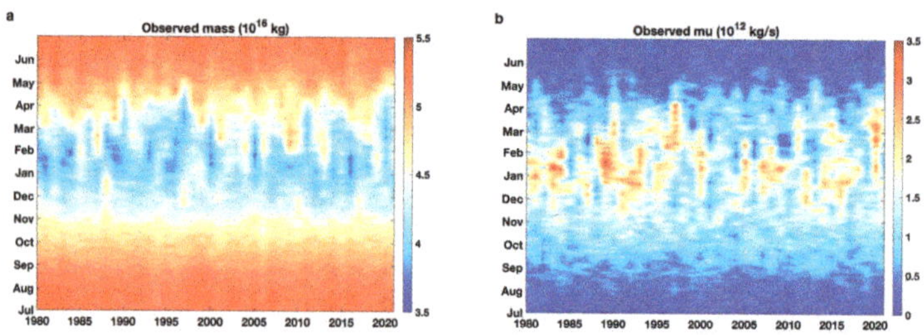

Figure 2. Annual evolutions (the ordinate) of (**a**) M (units: 10^{16} kg) and (**b**) MU (units: 10^{12} kg s^{-1}) in the years from 1980 to 2020 (abscissa).

A closer look at the temporal evolutions of the M and MU in a few selected years (Figure 3) reveals that MU leads M by a few days, even though they are generally negatively correlated. We then investigate the lead–lag correlations of M and MU in each year and the lead days of the MU relative to the M when their maximum negative correlations are reached (Figure 4). It is seen that the maximum negative correlations occur mostly when the MU leads the M. The lead time is mainly in the range of 1–10 days with a climatological mean value of about a week. As discussed above, it is the planetary-scale wave activity that drives the temporal evolutions of MU, which in turn leads to changes in the M field with a lead-time of a few days [10]. Therefore, temporal evolutions of M at sub-seasonal time scales are strongly modulated by eddy-driven poleward energy transport.

In summary, the observed annual evolutions of M and MU, respectively, representing the changes in the thermal and dynamical conditions of the SPV throughout the year, have three main features: (i) the M and MU are dominated by the annual cycle with the largest interannual variability in winter; (ii) the MU has a larger interannual variability than the M; and (iii) the MU varies out of phase with the M but with MU leads to M by 1–10 days.

3.2. The Parametric Elliptic Orbit Model

In this section, we will construct a parametric model of elliptic orbits to capture the coupled changes of the M and MU. To begin with, because the annual cycle of the M or the MU is close to a sine or cosine function, we can first formulate two variables, X, and Y, as

$$\begin{cases} X = X_0 + a\cos\Theta\cos\alpha + b\sin\Theta\sin\alpha \\ Y = Y_0 + a\cos\Theta\sin\alpha - b\sin\Theta\cos\alpha \end{cases} \quad (3)$$

where X_0 and Y_0 are the coordinates of the elliptic central point; a and b are the amplitudes; Θ is the phase angle in the range from 0 to 2π; and α corresponds to the tilting angle of the elliptic obit with the x-axis ranging from $-\pi$ to π. The other form of (3) is the familiar elliptic equation, namely

$$\frac{(X - X_0)^2}{a^2} + \frac{(Y - Y_0)^2}{b^2} = 1 \quad (4)$$

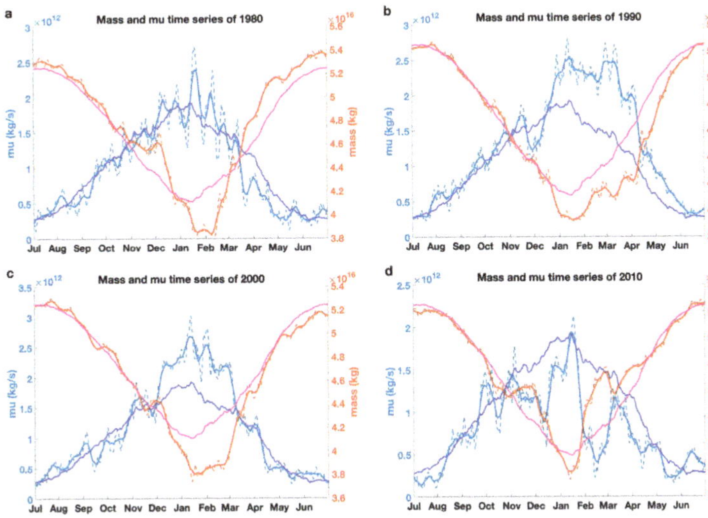

Figure 3. Time series of M (red curves, units: 10^{16} kg) and MU (blue curves, units: 10^{12} kg s^{-1}) in four example years: (**a**) 1980, (**b**) 1990, (**c**) 2000, (**d**) 2010. The solid brown curves and light blue curves are 7-day running means of M (ordinate on the right-hand-side) and MU (ordinate on the left-hand-side), respectively; and dashed curves are original time series of M and MU, respectively. The solid magenta and blue curves in each panel (which are, respectively, identical in the four panels) are the 42-year climatology for M and MU, respectively.

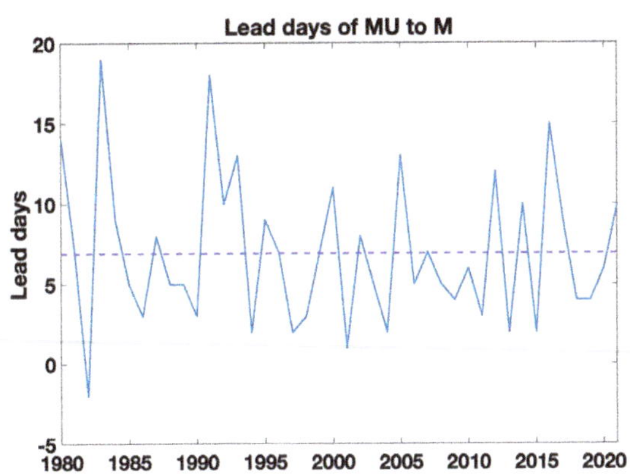

Figure 4. Yearly time series of the lead days of MU with respect to M when their negative correlations reach maximum values in years from 1980 to 2021, derived from observational fields.

Displayed in Figure 5 is an idealized elliptic orbit with $\alpha = -0.25\pi$. It is seen that when the Θ increases from 0 to π, X increases to a maximum while Y decreases to a minimum. Then, from $\Theta = \pi$ to 2π, X decreases to minimum while Y increases to maximum. If we consider MU as the X and M as the Y, such changes that accompany $\Theta = 0$ to 2π are quite consistent with their annual cycle from the previous year, from summer to winter and then to the concurrent summer.

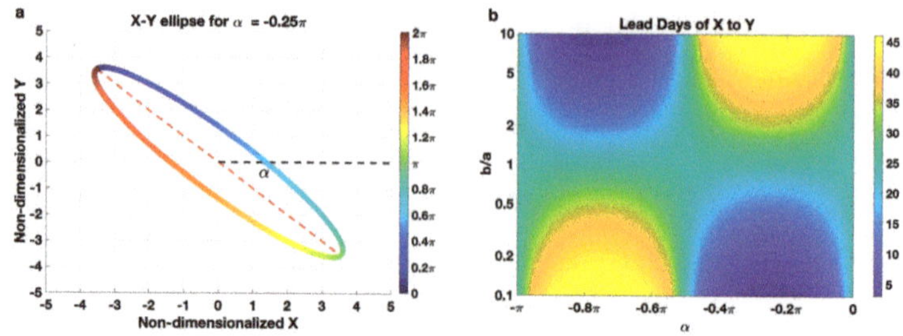

Figure 5. (**a**) Orbits of two variables, X and Y, satisfy the idealized parametric ellipse orbit model, with the tilting angle of the ellipse with the horizontal axis (α) equaling $-0.25\ \pi$. (**b**) Lead days of X with respect to Y when their negative correlations reach maximum values as a function of the ratio of b to a and the tilting angle α.

According to the parametric elliptic orbit model, the maximum negative correlations between X and Y can occur at different lead-lag times, which highly depend on the ratio b/a and the angle α. For the α in the range of $0{\sim}-\frac{\pi}{2}$, a larger b/a would yield a longer lead time for the X relative to Y to achieve their maximum negative correlation. Contrarily, for the α in the range of $-\frac{\pi}{2} \sim -\pi$, a smaller b/a would yield a longer lead time for the X relative to Y.

We apply a MATLAB function, "fit_ellipse" to a pair of daily time series of MU and M from 1 July to 30 June of the next year to determine the parameters X_0, Y_0, a, b and α with $\Theta = \frac{2\pi t}{365}$, where t starts on 1 July as $t = 1$ and ends on June 30 as $t = 365$ (29 February in a leap year is excluded in our analysis). The information from "fit_ellipse" can be found at https://www.mathworks.com/matlabcentral/fileexchange/3215-fit_ellipse (accessed on 15 January 2023). To better capture the observed lead time information of MU with respect to M, we introduce an auxiliary equation when applying the "fit_ellipse" function such that the constructed time series of MU and M from the fitted elliptic orbit meet the same lead time of the observed MU with respect to the observed M. As illustrated in Figure 5, when the value of α is fixed, the lead-lag days of maximum correlation and the ratio b/a have a one-to-one correspondence relation. The climatological mean annual cycles of MU and M and their elliptic orbit are displayed in Figure 6. The close resemblance between the fitted orbit elliptic (dashed magenta) and the scattering plot of the observed MU and M (colored dots) indicates that an elliptic orbit is capable of capturing the annual evolutions of MU and M. In particular, the fitted MU also varies out of phase with the fitted M.

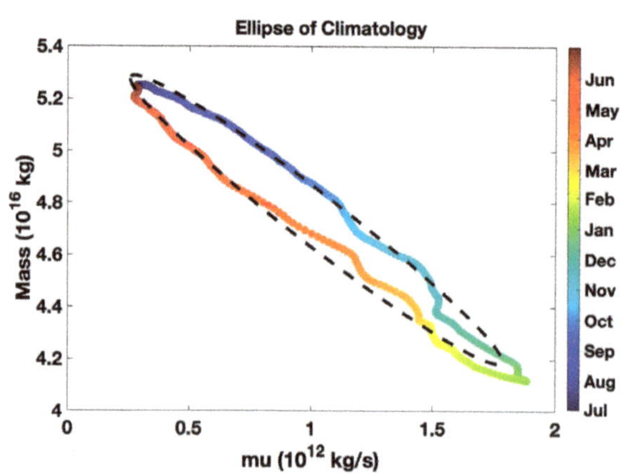

Figure 6. Scatter plot of the climatological mean M versus MU (colored dots) in the period 1980–2021 and its fitted ellipse orbit (dashed black curve).

3.3. Year-to-Year Variations of the Elliptic Orbits for the Annual Evolutions of M and MU

In this section, we will examine the effectiveness of the parametric elliptic orbit model in capturing the inter-annual variations in the annual evolutions of the *MU* and *M*. The yearly time series of the parameters derived from the ellipse orbit fitting function are shown in Figure 7. It is seen that each of these yearly time series exhibits a strong year-to-year variability. In particular, they have pronounced differences to their counterparts derived from the climatological orbit (dashed magenta), indicating a strong year-to-year variation in the elliptic orbits. The generally larger values of X_0 and smaller values of Y_0 than their counterparts for the climatological elliptic orbit are observed, and the general out-of-phase relationship between *MU* and *M* is also present in the year-to-year variation in their annual means. The generally larger values of a and b than their climatological counterparts indicate that the amplitudes of the seasonal cycles of individual years are stronger than those of the climatological annual cycle. The generally larger values of b/a and α than their climatological counterparts indicate that the lead days of *MU* with respect to *M* in individual years are longer than for their climatological counterpart.

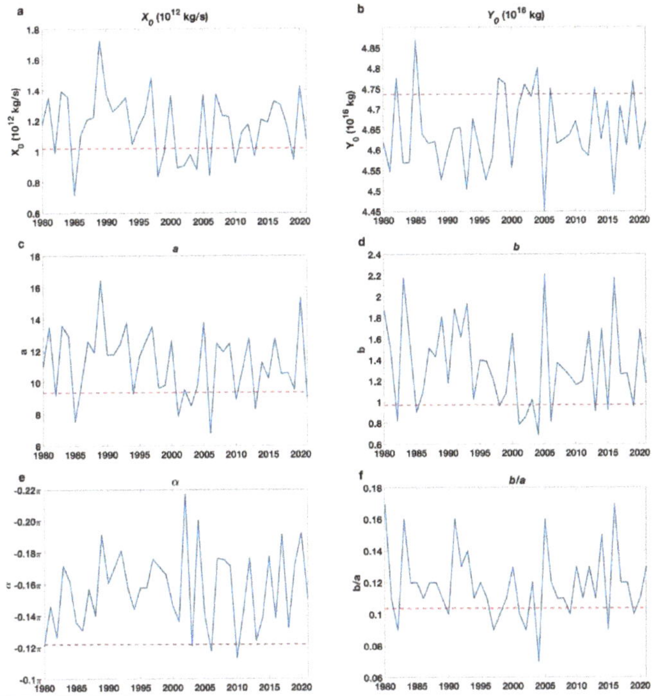

Figure 7. Yearly time series of the parameters of the fitted elliptic orbit model for *M* and *MU*. (**a**) The parameter X_0, (**b**) the parameter Y_0, (**c**) the parameter *a*, (**d**) the parameter *b*, (**e**) the parameter *α*, and (**f**) the ratio of *b* to *a*. The horizontal dashed red lines are the values of the parameters for the fitted orbit of the climatological mean annual cycles of *MU* and *M*.

It is seen from Figure 8 that the fitted elliptic orbits of individual years exhibit similar patterns to the yearly scatter plots of *MU* versus *M* in terms of both timing and intensity. In particular, the fitted orbits capture faithfully the interannual variations in extreme values of both *MU* and *M* in the winter seasons. For example, the fitted orbits capture the weaker SPV in the winters of 1982/83, 2001/02, 2002/03, 2003/04, 2006/07 and 2013/2014, which correspond to the panels in Figure 8 (or the abscissa's tick marks in Figure 9), labeled as 1982, 2001, 2003, 2006 and 2013, respectively. The year-to-year annual evolutions of *MU* and *M* obtained from the fitted elliptic orbits are shown in Figure 9. In comparison with

Figure 2, it can be seen that the fitted elliptic orbits can capture both the annual evolutions and their interannual variability in *MU* and *M*, as well as the generally negative correlation between *MU* and *M* at both intraseasonal and interannual scales.

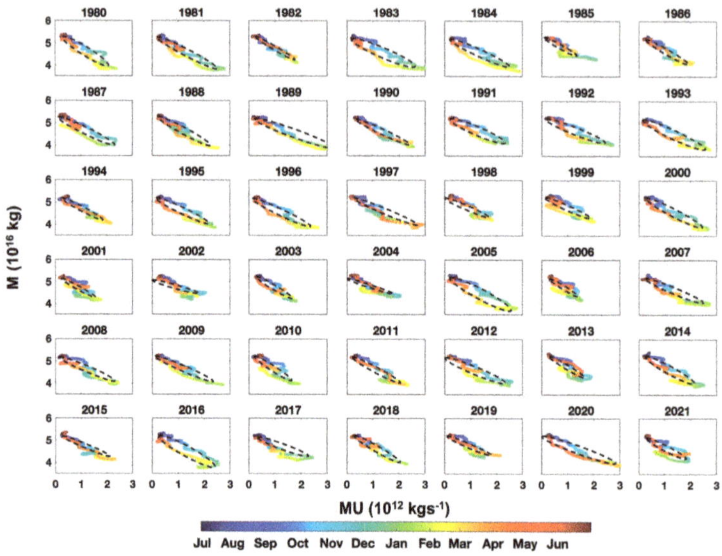

Figure 8. Scatter plot (color dots) of the *M* and *MU* in all years from 1980 to 2021 and the fitted elliptic orbit (black orbit).

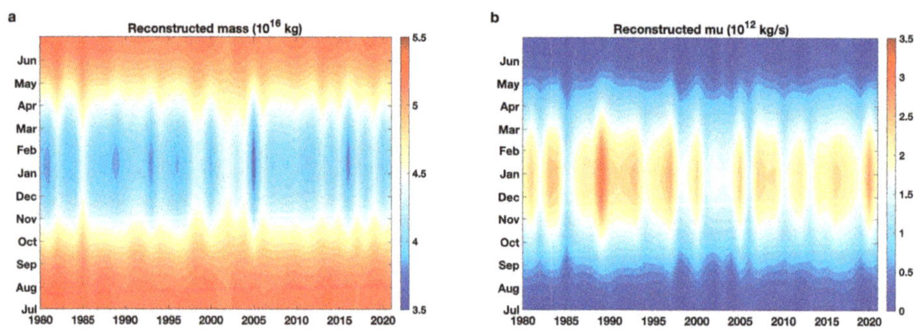

Figure 9. As in Figure 2, but for the (**a**) *M* and (**b**) *MU* derived from the parametric model of ellipse orbit with the parameter *b* calculation scheme adjusted.

Next, we quantitatively compare the extremes of *MU* and *M* in the winter seasons and their annual-cycle amplitudes obtained from the fitted orbits with the original time series of *MU* and *M*. Figure 10a,b shows that the yearly extreme values of both *MU* and *M* in winter seasons obtained from the fitted orbits have a high positive correlation (exceeding 0.94) with their counterparts obtained from the 31-day running mean time series of observed *MU* and *M*. The slightly larger values of the maximum *MU* and minimum *M* indicate that the fitted orbits tend to overestimate the strength of SPV in the winter seasons. The year-to-year variations in the amplitude of the annual cycle of *MU* and *M* (defined as the differences between the maximum and minimum values in the annual evolutions of *MU* and *M*) are displayed in Figure 10c,d, where the orbit-fitted amplitudes are compared with the observed ones. The figures show that the fitted orbits can almost perfectly capture the observed year-to-year variations in the annual cycle amplitudes for both *MU* and *M*, as is evident from the positive correlation exceeding 0.9. Next, we examine if the parametric elliptic orbit model can faithfully reflect the coupling between *MU* and *M* at the interannual time scale, as it does at the annual time scales (e.g., Figure 9 versus Figure 2). Figure 10e shows the scatter plot of the yearly time series of maximum *MU* and minimum *M*, which jointly measure the year-to-year variations in the strength of the SPV in the winter seasons. It is seen that the correlation between the yearly time series of *MU* and *M* derived from the fitted orbits (red circles) is identical to that derived from the observations (blue circles), equaling −0.78. Therefore, the observed negative correlation between the interannual variability in *MU* and that in *M* is well captured by the fitted orbits. Figure 10f shows the scatter plot of the yearly time series of the amplitude of the annual evolutions of *MU* and *M*. The observations show that the annual evolution amplitude of *MU* of individual years is positively and strongly correlated with that of *M* (about 0.69) and so are the fitted orbits (0.72) and the observation (0.69). It can be summarized from Figures 8–10 that both the annual evolutions and their yearly variations in *MU* and *M* are well coupled with significant correlations (about or larger than 0.7), which can be captured with almost identical correlations by the orbit fitting method, with five parameters varying yearly, as illustrated in Figure 7.

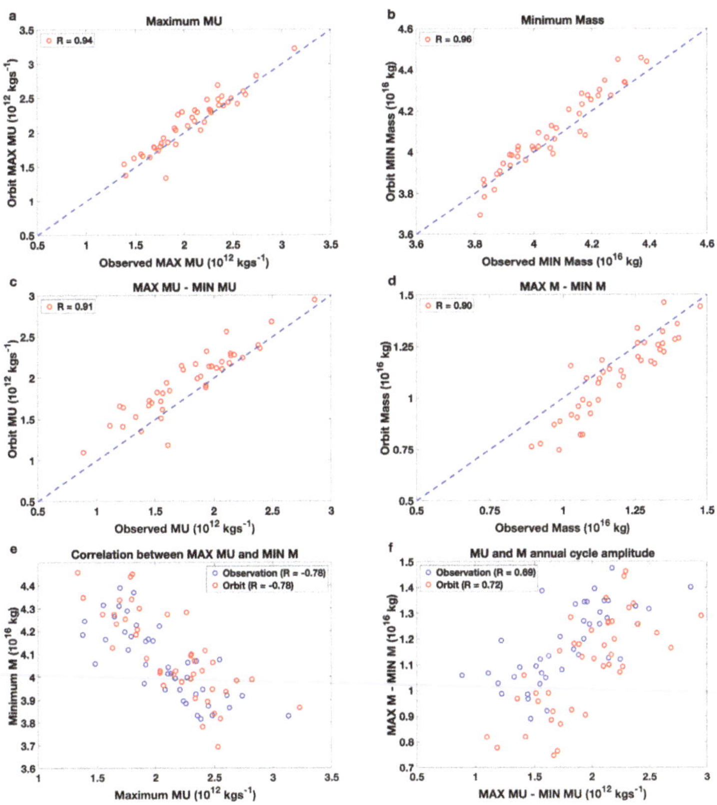

Figure 10. Scatter plot of the orbit-fitted extremes (red circles), and the observed extremes of maximum value of *MU* and minimum value of *M* in each winter season. (**a**) The maximum of *MU* from orbit fitting (ordinate) versus the observation (abscissa), (**b**) the same as (a) but for the minimum of *M*, (**c**) the same as (a) but for the yearly difference between maximum and minimum values of *MU*, (**d**) the same as (a) but for the difference between maximum and minimum values of *M*, (**e**) the maximum of *MU* (abscissa) versus the minimum of *M* (ordinate) for observation (blue circles) and fitted orbits (red circles), and (**f**) the same as (**e**) but for the yearly differences between maximum and minimum values of *MU* and *M*.

4. Conclusions

The year-to-year varying annual evolutions of the stratospheric polar vortex (SPV) have an important downward impact on the weather and climate from winter to summer and thus potential implications for seasonal forecasts. This study considers the necessity of investigating both the thermal and dynamical conditions of SPV by jointly considering the daily time series of the zonally integrated mass-weighted zonal momentum at 60° N (*MU*) and the total air mass above the isentropic surface of 400 K (*M*) over the latitude band of 60–90° N. The annual evolutions of *MU* are characterized by a generally out-of-phase relationship with the annual evolutions of *M*, and the changes in *MU* tend to lead the changes in *M* by 1–10 days. Moreover, their annual evolutions have significant interannual variations, which also exhibit a strong negative correlation between *MU* and *M*.

By constructing a parametric elliptic orbit model to fit the daily time series of *MU* and *M* in each year, we can closely reproduce the year-to-year variations in the annual evolutions of the observed *MU* and *M* jointly from the year-to-year variations in their elliptic orbits, including their amplitude, extremes in winter seasons, as well the strong negative correlation between *MU* and *M* and their amplitudes in the annual cycle at the interannual time scales. The findings of this study may pave a new way for short-time climate forecasts of the annual evolutions of SPV, including its temporal evolutions over individual cold seasons (including the spring and fall seasons). In particular, one would be able to predict timings of the minimum *MU* and maximum *M* in a given year by predicting the five parameters (X_0, Y_0, a, b, and α) of the corresponding yearly elliptic orbit. Because the timings of minimum *MU* and maximum *M* correspond closely to the timings of SSW events, one could in turn predict the timings of the high probability occurrence of SSW events in winters. Such a circulation condition during this timing would provide a favorable background for the break-up of the stratospheric polar vortex, which may yield a higher probability of the occurrence of SSW, which will be a topic of our future studies.

We here wish to add that this study is focused on the seasonal cycle of the stratospheric polar vortex in the Northern Hemisphere. In the Southern Hemisphere, the annual cycle of the stratospheric polar vortex also plays an important role in extreme surface weather conditions and natural hazards, as it may raise the risk of increased rainfall in the latitudinal band of 35–50° S [63–65]. According to previous studies, the Southern Hemisphere stratospheric polar vortex tends to be steadier and varies at a longer period [9,66] and break-up events occur less frequently [65]. It is expected that the annual cycle of the stratospheric polar vortex as well as extreme events such as SSW in the Southern Hemisphere could be also well captured by our parametric elliptic orbit model, which will be verified in our future work.

Author Contributions: Conceptualization, M.C. and Y.Y.; methodology, Y.Y.; validation, J.S., M.S. and Y.Y.; formal analysis, J.S. and M.S.; investigation, M.C.; data curation, J.S.; writing—original draft preparation, Y.Y.; writing—review and editing, M.C., J.S., M.S. and X.L.; visualization, J.S.; supervision, M.C.; project administration, M.C.; funding acquisition, M.C. and Y.Y. All authors have read and agreed to the published version of the manuscript.

Funding: This research was jointly funded in part by the National Science Foundation of China, grant number 42075052, and the Natural Science Foundation of Jiangsu Province, grant number BK20211288, Climate Program Office of National Oceanic and Atmospheric Administration (NA20OAR4310380), and the National Science Foundation (AGS-2032542).

Data Availability Statement: The ERA5 datasets used in this study are freely available on the official website. (https://cds.climate.copernicus.eu/cdsapp#!/dataset/reanalysis-era5-pressure-levels?tab=overview; https://cds.climate.copernicus.eu/cdsapp#!/dataset/reanalysis-era5-single-levels?tab=overview; accessed on 1 November 2022).

Acknowledgments: We would like to thank the three anonymous reviewers for their constructive comments.

Conflicts of Interest: The authors declare no conflict of interest.

References

1. Hu, J.G.; Ren, R.C.; Yu, Y.Y.; Xu, H.M. The boreal spring stratospheric final warming and its interannual and interdecadal variability. *Sci. China Earth Sci.* **2014**, *57*, 710–718. [CrossRef]
2. Schoeberl, M.R.; Newman, P.A. Chapter middle atmosphere: Polar vortex. In *Encyclopedia of Atmospheric Sciences*, 2nd ed.; North, J.G.R., Pyle, F.Z., Eds.; Academic Press: Cambridge, MA, USA, 2015; Volume 4, pp. 12–17.
3. Kelleher, M.E.; Blanca, A.; James, A.S. Interseasonal Connections between the Timing of the Stratospheric Final Warming and Arctic Sea Ice. *J. Clim.* **2020**, *33*, 3079–3092. [CrossRef]
4. Hu, J.G.; Ren, R.C.; Xu, H.M. Occurrence of winter stratospheric sudden warming events and the seasonal timing of spring stratospheric final warming. *J. Atmos. Sci.* **2014**, *71*, 2319–2334. [CrossRef]
5. Wei, K.; Wen, C.M.; Huang, R.H. Dynamical diagnosis of the breakup of the stratospheric polar vortex in the Northern Hemisphere. *Sci. China-Earth Sci.* **2007**, *50*, 1369–1379. [CrossRef]
6. Cohen, J.; Salstein, D.; Saito, K. A dynamical framework to understand and predict the major Northern Hemisphere mode. *Geophys. Res. Lett.* **2002**, *29*, 51. [CrossRef]
7. Cai, M.; Ren, R.C. Meridional and Downward Propagation of Atmospheric Circulation Anomalies. Part I: Northern Hemisphere Cold Season Variability. *J. Atmos. Sci.* **2007**, *64*, 1880–1901. [CrossRef]
8. Ren, R.C.; Cai, M. Meridional and vertical out-of-phase relationships of temperature anomalies associated with the Northern Annular Mode variability. *Geophys. Res. Lett.* **2007**, 34. [CrossRef]
9. Ren, R.C.; Cai, M. Meridional and downward propagation of atmospheric circulation anomalies. Part II: Southern Hemisphere cold season variability. *J. Atmos. Sci.* **2008**, *65*, 2343–2359. [CrossRef]
10. Yu, Y.; Cai, M.; Ren, R.A. Stochastic model with a low-frequency amplification feedback for the stratospheric northern annular mode. *Clim. Dyn.* **2018**, *50*, 3757–3773. [CrossRef]
11. Lorenz, D.J.; Hartmann, D.L. Eddy–zonal flow feedback in the Northern Hemisphere winter. *J. Clim.* **2003**, *16*, 1212–1227. [CrossRef]
12. Christiansen, B. Downward propagation and statistical forecast of the near-surface weather. *J. Geophys. Res.* **2005**, *110*, D14104. [CrossRef]
13. Christiansen, B. Is the atmosphere interesting? A projection pursuit study of the circulation in the Northern Hemisphere winter. *J. Clim.* **2009**, *22*, 1239–1254. [CrossRef]
14. Thompson, D.W.J.; Wallace, J.M. The Arctic Oscillation signature in the wintertime geopotential height and temperature fields. *Geophys. Res. Lett.* **1998**, *25*, 1297–1300. [CrossRef]
15. Thompson, D.W.J.; Wallace, J.M. Annular modes in the extratropical circulation. Part I: Month-to-month variability. *J. Clim.* **2000**, *13*, 1000–1016. [CrossRef]
16. Limpasuvan, V.; Hartmann, D.L. Wave-maintained annular modes of climate variability. *J. Clim.* **2000**, *13*, 4414–4429. [CrossRef]
17. Baldwin, M.P.; Dunkerton, T.J. Stratospheric Harbingers of Anomalous Weather Regimes. *Science* **2001**, *294*, 581–584. [CrossRef]
18. Baldwin, M.P.; Thompson, D.W.J. A critical comparison of stratosphere–troposphere coupling indices. *Q. J. R. Meteorol. Soc.* **2009**, *135*, 1661–1672. [CrossRef]
19. Cai, M.; Yu, Y.; Deng, Y.; van den Dool, H.M.; Ren, R.; Saha, S.; Wu, X.; Huang, J. Feeling the Pulse of the Stratosphere: An Emerging Opportunity for Predicting Continental-Scale Cold-Air Outbreaks 1 Month in Advance. *Bull. Am. Meteorol. Soc.* **2016**, *97*, 1475–1489. [CrossRef]
20. Baldwin, M.P.; Dunkerton, T.J. Propagation of the Arctic Oscillation from the stratosphere to the troposphere. *J. Geophys. Res. Atmos.* **1999**, *104*, 30937–30946. [CrossRef]
21. Christiansen, B. Downward propagation of zonal mean zonal wind anomalies from the stratosphere to the troposphere: Model and reanalysis. *J. Geophys. Res. Atmos.* **2001**, *106*, 27307–27322. [CrossRef]
22. Thompson, D.W.; Baldwin, M.P.; Wallace, J.M. Stratospheric Connection to Northern Hemisphere Wintertime Weather: Implications for Prediction. *J. Clim.* **2002**, *15*, 1421–1428. [CrossRef]
23. Zhou, S.; Miller, A.J.; Wang, J.; Angell, J.K. Downward-Propagating Temperature Anomalies in the Preconditioned Polar Stratosphere. *J. Clim.* **2002**, *15*, 781–792. [CrossRef]
24. Wang, L.; Chen, W. Downward Arctic Oscillation signal associated with moderate weak stratospheric polar vortex and the cold December 2009. *Geophys. Res. Lett.* **2010**, *37*, L09707. [CrossRef]
25. Davini, P.; Cagnazzo, C.; Anstey, J.A. A blocking view of the stratosphere-troposphere coupling. *J. Geophys. Res. Atmos.* **2014**, *119*, 11, 100–111, 115. [CrossRef]
26. Yu, Y.; Ren, R.; Hu, J.; Wu, G. A mass budget analysis on the interannual variability of the polar surface pressure in the winter season. *J. Atmos. Sci.* **2014**, *71*, 3539–3553. [CrossRef]
27. Deng, S.; Chen, Y.; Luo, T.; Bi, Y.; Zhou, H. The possible influence of stratospheric sudden warming on East Asian weather. *Advice Atmos. Sci.* **2008**, *25*, 841–846. [CrossRef]
28. Kenyon, J.; Hegerl, G.C. Influence of Modes of Climate Variability on Global Temperature Extremes. *J. Clim.* **2008**, *21*, 3872–3889. [CrossRef]
29. Scaife, A.A.; Knight, J.R. Ensemble simulations of the cold European winter of 2005–2006. *Q. J. R. Meteorol.* **2008**, *134*, 1647–1659. [CrossRef]

30. Kolstad, E.W.; Breiteig, T.; Scaife, A.A. The association between stratospheric weak polar vortex events and cold air outbreaks in the Northern Hemisphere. *Q. J. R. Meteorol. Soc.* **2010**, *136*, 886–893. [CrossRef]
31. Cohen, J.; Jones, J. Tropospheric Precursors and Stratospheric Warmings. *J. Clim.* **2011**, *24*, 6562–6572. [CrossRef]
32. Mitchell, D.M.; Gray, L.J.; Anstey, J.; Baldwin, M.P.; Charlton-Perez, A.J. The Influence of Stratospheric Vortex Displacements and Splits on Surface Climate. *J. Clim.* **2013**, *26*, 2668–2682. [CrossRef]
33. Kidston, J.; Scaife, A.A.; Hardiman, S.C.; Mitchell, D.M.; Butchart, N.; Baldwin, M.P.; Gray, L.J. Stratospheric influence on tropospheric jet streams, storm tracks and surface weather. *Nat. Geosci.* **2015**, *8*, 433–440. [CrossRef]
34. Domeisen, D.I.V.; Butler, A.H. Stratospheric drivers of extreme events at theEarth's surface. *Commun. Earth Environ.* **2020**, *1*, 59. [CrossRef]
35. Huang, J.; Hitchcock, P.; Maycock, A.C.; McKenna, C.M.; Tian, W. Northern hemisphere cold air outbreaks are more likely to be severe during weak polar vortex conditions. *Commun. Earth Environ.* **2021**, *2*, 147. [CrossRef]
36. Yu, Y.; Li, Y.; Ren, R.; Cai, M.; Guan, Z.; Huang, W. An isentropic mass circulation view on the extreme cold events in the 2020/21 winter. *Adv. Atmos. Sci.* **2022**, *39*, 643–657. [CrossRef]
37. Zhang, X.D.; Fu, Y.F.; Han, Z.; Overland, J.E.; Rinke, A.; Tang, H.; Vihma, T.; Wang, M.Y. Extreme Cold Events from East Asia to North America in Winter 2020/21: Comparisons, Causes, and Future Implications. *Adv. Atmos. Sci.* **2022**, *39*, 553–565. [CrossRef]
38. Zhang, Y.X.; Si, D.; Ding, Y.H.; Jiang, D.B.; Li, Q.Q.; Wang, G.F. Influence of Major Stratospheric Sudden Warming on the Unprecedented Cold Wave in East Asia in January 2021. *Adv. Atmos. Sci.* **2022**, *39*, 576–590. [CrossRef]
39. Arnone, E.; Castelli, E.; Papandrea, E.; Carlotti, M.; Dinelli, B.M. Extreme ozone depletion in the 2010–2011 Arctic winter strato-sphere as observed by MIPAS/ENVISAT using a 2-Dtomographic approach. *Atmos. Chem. Phys.* **2012**, *12*, 9149–9165. [CrossRef]
40. Chipperfield, M.P.; Jones, R.L. Relative influences of atmospheric chemistry and transport on Arctic ozone trends. *Nature* **1999**, *400*, 551–554. [CrossRef]
41. Daniel, J.S.; Solomon, S.; Portmann, R.W.; Garcia, R.R. Stratospheric ozone destruction: The importance of bromine relative to chlorine. *J. Geophys. Res. Atmos.* **1999**, *104*, 23871–23880. [CrossRef]
42. Solomon, S.; Garcia, R.R.; Ravishankara, A.R. On the role of iodine in ozone depletion. *J. Geophys. Res. Atmos.* **1994**, *99*, 20491–20499. [CrossRef]
43. Hauchecorne, A.; Claud, C.; Keckhut, P.; Mariaccia, A. Stratospheric Final Warmings fall into two categories with different evolution over the course of the year. *Commun. Earth Environ.* **2022**, *3*, 4. [CrossRef]
44. Shen, X.; Xu, G.M.; Hu, J.G. Inter-annual variation of the Northern Hemisphere stratospheric polar vortex breakdown and its impact on the precipitation over the South Asia in May. *J. Meteorol. Sci.* **2017**, *37*, 718–726.
45. Lee, S.H.; Lorenzo, M.P.; Guan, B. Modulation of Atmospheric Rivers by the Arctic Stratospheric Polar Vortex. *Geophys. Res. Lett.* **2022**, *49*, e2022GL100381. [CrossRef]
46. Hardiman, S.C.; Scaife, A.A.; Dunstone, N.J.; Wang, L. Subseasonal vacillations in the winter stratosphere. *Geophys. Res. Lett.* **2022**, *47*, e2020GL087766. [CrossRef]
47. Zuev, V.V.; Savelieva, E. Stratospheric polar vortex dynamics according to the vortex delineation method. *J. Earth Syst. Sci.* **2023**, *132*, 39. [CrossRef]
48. Hersbach, H.; Bell, B.; Berrisford, P.; Hirahara, S.; Thépaut, J. The ERA5 global reanalysis. *Q. J. R. Meteorol. Soc.* **2020**, *146*, 1999–2049. [CrossRef]
49. Perlwitz, J.; Graf, H.F. Troposphere-stratosphere dynamic coupling under strong and weak polar vortex conditions. *Geophys. Res. Lett.* **2001**, *28*, 271–274. [CrossRef]
50. Perlwitz, J.; Harnik, N. Observational evidence of a stratospheric influence on the troposphere by planetary wave reflection. *J. Clim.* **2003**, *16*, 3011–3026. [CrossRef]
51. Perlwitz, J.; Harnik, N. Downward coupling between the stratosphere and troposphere: The relative roles of wave and zonal mean processes. *J. Clim.* **2004**, *17*, 4902–4909. [CrossRef]
52. Butler, A.H.; Thompson, D.W.; Heikes, R. The steady-state atmospheric circulation response to climate change-like thermal forcings in a simple general circulation model. *J. Clim.* **2010**, *23*, 3474–3496. [CrossRef]
53. Domeisen, D.I.; Sun, L.; Chen, G. The role of synoptic eddies in the tropospheric response to stratospheric variability. *Geophys. Res. Lett.* **2013**, *40*, 4933–4937. [CrossRef]
54. Hitchcock, P.; Simpson, I.R. The downward influence of stratospheric sudden warmings. *J. Atmos. Sci.* **2014**, *71*, 3856–3876. [CrossRef]
55. Karpechko, A.Y.; Hitchcock, P.; Peters, D.H.; Schneidereit, A. Redictability of downward propagation of major sudden stratospheric warmings. *Q. J. R. Meteorol. Soc.* **2017**, *143*, 1459–1470. [CrossRef]
56. Rupp, P.; Birner, T. Tropospheric eddy feedback to different stratospheric conditions in idealised baroclinic life cycles. *Weather Clim. Dyn.* **2021**, *2*, 111–128. [CrossRef]
57. Yu, Y.; Cai, M.; Ren, R.; Rao, J. A closer look at the relationships between meridional mass circulation pulses in the stratosphere and cold air outbreak patterns in northern hemispheric winter. *Clim. Dyn.* **2018**, *51*, 3125–3143. [CrossRef]
58. Yu, Y.; Cai, M.; Shi, C.; Ren, R. On the linkage among strong stratospheric mass circulation, stratospheric sudden warming, and cold weather events. *Mon. Weather Rev.* **2018**, *146*, 2717–2739. [CrossRef]

59. Cai, M.; Shin, C.S. A Total Flow Perspective of Atmospheric Mass and Angular Momentum Circulations: Boreal Winter Mean State. *J. Atmos. Sci.* **2014**, *71*, 2244–2263. [CrossRef]
60. Pauluis, O.; Czaja, A.; Korty, R. The Global Atmospheric Circulation on Moist Isentropes. *Science* **2008**, *321*, 1075–1078. [CrossRef]
61. Pauluis, O.; Shaw, T.; Laliberte, F. A Statistical Generalization of the Transformed Eulerian-Mean Circulation for an Arbitrary Vertical Coordinate System. *J. Atmos. Sci.* **2011**, *68*, 1766–1783. [CrossRef]
62. Yu, Y.; Cai, M.; Ren, R.; Dool, H.M. Relationship between warm airmass transport into the upper polar atmosphere and cold air outbreaks in winter. *J. Atmos. Sci.* **2015**, *72*, 349–368. [CrossRef]
63. Gillett, N.P.; Kell, T.D.; Jones, P.D. Regional climate impacts of the southern annular mode. *Geophys. Res. Lett.* **2006**, *33*, L23704. [CrossRef]
64. Lim, E.; Hendon, H.H.; Boschat, G.; Hudson, D.; Thompson, D.W.J.; Dowdy, A.J.; Arblaster, J.M. Australian hot and dry extremes induced by weakenings of the stratospheric polar vortex. *Nat. Geosci.* **2019**, *12*, 896–901. [CrossRef]
65. Wang, L.; Hardiman, S.C.; Bett, P.E.; Comer, R.E.; Kent, C.; Scaife, A.A. What chance of a sudden stratospheric warming in the southern hemisphere? *Environ. Res. Lett.* **2020**, *15*, 104038. [CrossRef]
66. Scaife, A.A.; James, I.N. Response of the stratosphere to interannual variability of tropospheric planetary waves. *Q. J. R. Meteorol. Soc.* **2020**, *126*, 275–297. [CrossRef]

atmosphere

MDPI

Communication

A Neural-Network Based MPAS—Shallow Water Model and Its 4D-Var Data Assimilation System

Xiaoxu Tian *, Luke Conibear and Jeffrey Steward

Tomorrow.io, Boston, MA 02210, USA

* Correspondence: xtian15@terpmail.umd.edu

Abstract: The technique of machine learning has been increasingly applied in numerical weather predictions. The aim of this study is to explore the application of a neural network in data assimilation by making use of the convenience in obtaining the tangent linear and adjoint (TL/AD) of a neural network (NN) and formulating a NN-based four-dimensional variational (4D-Var) DA system. A NN-based shallow water (SW) model is developed in this study. The NN model consists of three layers. The weights and biases in the NN-based SW model are trained with 60 years of hourly ERA5 geopotentials and wind field at 500 hPa as initial conditions and the corresponding 12-h forecasts by Model for Prediction Across Scales (MPAS)-SW, in total of 534,697 sets of samples. The 12-h forecasts from independent dates made by NN-based SW prove to closely emulate the simulations by the actual MPAS-SW model. This study further shows that the TL/AD of an NN model can be easily developed and validated. The ease of obtaining the TL/AD makes NN conveniently applicable in various aspects within a data assimilation (DA) system. To demonstrate such, a continuous 4D-Var DA system is also developed with the forward NN and its adjoint. To demonstrate the functionality of the NN-based 4D-Var DA system, the results from a higher resolution simulation will be treated as observations and assimilated to analyze the low resolution initial conditions. The forecasts starting from the analyzed initial conditions will be compared with those without assimilation to demonstrate improvements.

Keywords: machine learning; data assimilation; 4D-Var; neural network; MPAS-shallow water; global modeling

Citation: Tian, X.; Conibear, L.; Steward, J. A Neural-Network Based MPAS—Shallow Water Model and Its 4D-Var Data Assimilation System. *Atmosphere* **2023**, *14*, 157. https://doi.org/10.3390/atmos14010157

Academic Editor: Jimy Dudhia

Received: 1 December 2022
Revised: 27 December 2022
Accepted: 5 January 2023
Published: 10 January 2023

1. Introduction

Forecasts of future atmospheric state has mainly been accomplished by numerical weather predictions (NWP), which is a technology and capability after five decades of development and improvement [1]. Data assimilation (DA) further promoted the capability of NWP to more accurately predict future weather, which is achieved by better capturing initial conditions for NWP and quantifying its uncertainties [2]. In recent years, machine learning (ML), especially neural networks (NN), has been increasingly applied in the atmospheric sciences and has shown great potential [3–6]. Its capability of recognizing patterns in high-dimensional data sets without needing underlying theoretical equations has been appealing and has benefited various research disciplines [7,8].

Various previous studies have applied ML to NWP. One area is postprocessing NWP model outputs to reduce systematic biases. Ref. [9] used logistic regression and random forests to calibrate the probabilistic precipitation forecasts and improve verification statistics. Ref. [10] applied machine learning to postprocess NWP outputs in high-impact weather events to further improve the forecast skill. Ref. [11] trained a nonlinear NN to predict physical variables such as 2 m temperatures and achieved significant improvement compared to conventional postprocessing methods. Ref. [12] applied deep learning in precipitation nowcasting and 1-hour predictions from radar images. In [13], a deep NN was trained with ensemble weather forecasts for postprocssing, which achieved a relative improvement in ensemble forecast skill of over 14%. Ref. [3] developed a global prediction model using a Fourier Forecasting Neural Network that takes the atmospheric state in the initial conditions and predicts a few 2D variable in future times. Ref. [14] proposed a deep neural network in the form of Graph Neural Network (GNN) [15] to make forecasts in six-hour increments and trained with ERA5 dataset. The forecast performance was shown to outperform the global high resolution operational product, HRES, by the European Centre for Medium-Range Weather Forecasts.

Hybrid modeling combining ML and NWP has also been explored in numerous recent studies. Ref. [16] investigated the possibility of replacing the longwave radiative transfer model with a NN-based model and achieved an accuracy comparable to the conventional algorithm in a general circulation model. Ref. [17] emulated the longwave radiation parameterization for the National Center for Atmospheric Research (NCAR) Community Atmospheric Model with a NN and produced almost identical results 50–80 times faster. Ref. [18] trained a deep neural network to resolve atmospheric subgrid processes in climate modeling by learning from a multiscale model with explicit convections. As promising as the results in these studies show, Ref. [19] pointed out that in hybrid modeling, the feedback between the NN and the General Circulation Model (GCM) can cause instability in simulations and make the experiment crash within days. Similarly, hybrid approaches in DA have been explored in a few studies. Ref. [20] emulated the nonorographic gravity wave drag parametrization with a NN and developed the corresponding tangent linear and adjoint components, which were successfully used in a 4D-Var DA system. Ref. [21] formulated a Lorenz96 model emulator with a NN, generated its Jacobians using the emulator, and applied them in the contexts of 4D-Var DA.

In this study, we develop a feedforward NN [22] with an input layer, one hidden layer, and the output layer to emulate the global shallow water (SW) dynamics in the Model for Prediction Across Scales (MPAS) framework. We train the model on fluid heights and winds from real atmospheric states and then developed the tangent linear and adjoint models of the trained neural network to formulate a continuous 4D-Var DA system, which are described in details in Section 2. The performance in analyzing initial conditions of this DA system as well as the forecast improvements are examined and shown in Section 3. Finally, Section 4 summarizes the study.

2. Methodology

2.1. MPAS-SW Dynamics

The SW dynamics under the MPAS spherical centroidal Voronoi tessellation (SCVT) was developed in [23,24]. The forward nonlinear continuous SW dynamics can be written as:

$$\frac{\partial h}{\partial t} + \nabla(h\mathbf{u}) = 0, \tag{1}$$

$$\frac{\partial \mathbf{u}}{\partial t} + (\mathbf{u}\nabla)\mathbf{u} + f\mathbf{k} \times \mathbf{u} = -g\nabla(h+b) \tag{2}$$

where the fluid height h and edge-normal wind $\mathbf{u}$ are the model prognostic variables, f the Coriolis parameter θ being latitudes, and b the bottom height. This dynamical relationship has been widely applied in meteorology and oceanography. In this study, the height and wind fields at 500 hPa from ERA5 (European Centre for Medium-Range Weather Forecasts Reanalysis v5) [25] will first be interpolated into a 1000 km resolution mesh consisting of 611 cells (shown in Figure 1a) and the MPAS-SW model will make forecasts forward in time.

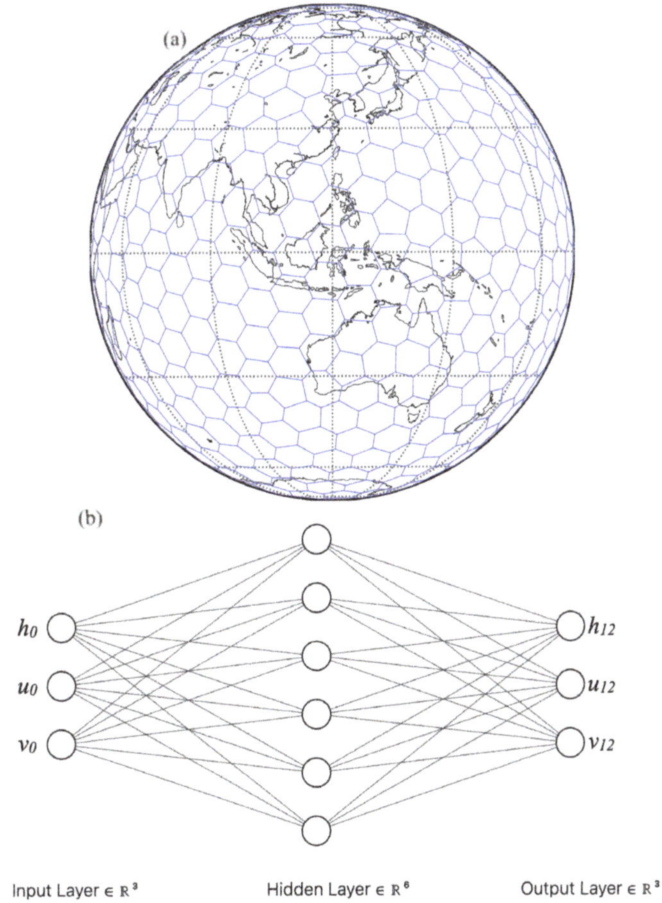

Figure 1. (**a**) Spatial distribution of the Spherical Centroidal Voronoi Tessellation (SCVT) mesh at 1000 km with 611 cells globally. (**b**) The neural network diagram showing the structure of the NN-based MPAS-SW model. The actual number of the neurons for the input and output layers is N = 1833, and N = 3666 for the hidden layer.

2.2. NN Emulator of MPAS-SW

A feedforward NN is first formulated to emulate the SW dynamical behaviors reflected in MPAS-SW simulations. The atmospheric state in MPAS-SW is essentially represented by vectors. The forecasts are also vectors projected from those from a previous time. Similar to the GNN [15], the benefits of such as representation include intrinsical handling of the global spherical structure of the Earth, allowing to resolve the underlying multiscale interactions between cells, and the potential of learning multiresolution models. Thus, densely connected NN layers are chosen in this study. The NN model consists of three layers: an input layer of 1833 values, a hidden layer with 3666 neurons, and an output layer with 1833 neurons. The input layer holds the three variables including height h, zonal wind velocity u, and meridional wind velocity v over the global domain in the initial condition, all three of which are sampledat the 611 cell centers in the mesh shown in Figure 1a. Thus, each layer has 1833 (611 $\times$ 3) neurons. The output layer of the same 1833 dimension holds the same three variables (h, u, and v) of the 12-h forecast. The number of neurons in the hidden layer, 3666, was determined based on heuristics. The hidden layer had an ELU activation function [26] and 10% dropout [27], where ELU can be denoted as

$$y = \begin{cases} \alpha(e^x - 1) & \text{if } x \leq 0 \\ x & \text{if } x > 0 \end{cases} \quad (3)$$

where $\alpha = 1$ in this study, making both ELU and its derivative continuous. The authors also experimented with various other choices of continuous activation functions including identity, tanh, sigmoid, and Sigmoid linear unit. The ELU proves to yield the best performances emulating the shallow water dynamics. The structure of the NN model is illustrated in Figure 1b.

The NN is trained and validated with hourly 500 hPa height h, zonal u, and meridional wind velocities v from the ERA5 dataset over a 60 years (1959–2019) as features and the corresponding 12-h MPAS-SW as targets, totaling 534,697 samples. The training underwent 60 epochs and a learning rate of 3×10^{-4} using the Adam optimizer. Training took 0.5 h on 1 NVIDIA T4 GPU.

The variations of the mean square error as the loss function with respect to the epochs is shown in Figure 2. The model was then independently tested on hourly 500 hPa height and wind fields from 2020 and 2021. The root mean squared error (RMSE) compared with the 12-h MPAS-SW forecasts were 6.32 m and 0.58 m/s in height and wind fields, respectively. Taking the 500 hPa atmospheric state at 00 UTC on 1 January 2021 (shown in Figure 3a) as the initial conditions, the 12-hour forecasts rendered by the NN and the MPAS-SW are given in Figure 3b and Figure 3c, respectively. The distribution of the atmospheric wave patterns from the NN visually resembles the MPAS-SW result to some extent. To further demonstrate the NN emulation of SW dynamics, Figure 4 shows the the differences between the 12-h forecasts and the initial conditions in the case of NN (Figure 4a) and MPAS-SW (Figure 4b). It can be seen that most of the variations in the 12-h forecasts from the initial conditions are captured in the NN results when compared with the actual MPAS-SW simulations.

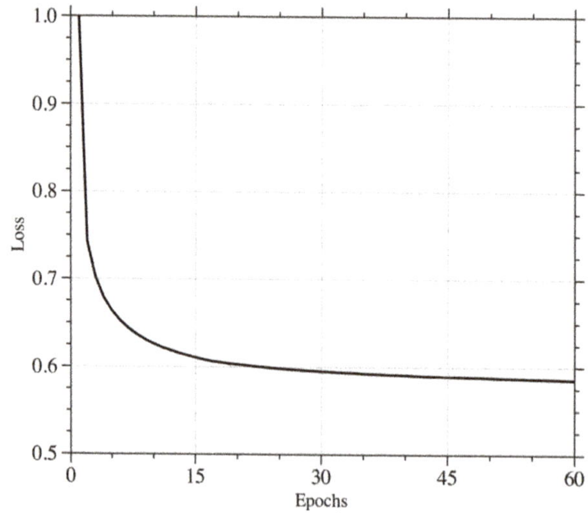

Figure 2. Variations of the loss function with respect to the epochs in the NN training.

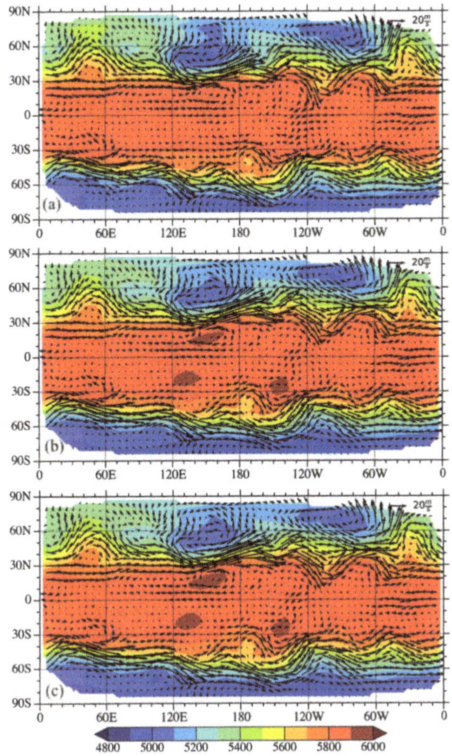

Figure 3. (**a**) The spatial distribution of the fields of height (shaded) and wind (vectors) at 00 UTC 1 January 2021. (**b**,**c**) The 12-h forecasts made by (**b**) the NN-based SW model and (**c**) MPAS-SW model.

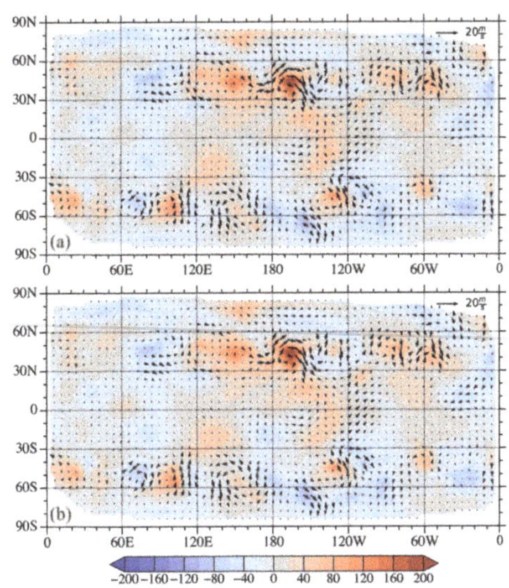

Figure 4. The differences in height (shaded) and wind (vectors) between the 12-h forecasts by (**a**) NN-based SW and (**b**) MPAS-SW with respect to the initial conditions.

2.3. The Tangent Linear and Adjoint Models

A nonlinear forward forecast model can denoted as:

$$\mathbf{x}(t_r) = \mathcal{M}(\mathbf{x}(t_0)) \tag{4}$$

It takes the initial model state $\mathbf{x}(t_0)$ at time t_0 as the initial conditions and predicts the model state $\mathbf{x}(t_r)$ at time t_r. The corresponding tangent linear model is then:

$$\Delta\mathbf{x}(t_r) = \mathbf{M}(\mathbf{x}(t_0))\Delta\mathbf{x}(t_0) = \frac{\partial M(\mathbf{x}(t_0))}{\partial \mathbf{x}}\Delta\mathbf{x}(t_0). \tag{5}$$

The tangent linear model predicts the perturbation distributions forward in time following the nonlinear trajectory given by the nonlinear forward model in Equation (4). The adjoint model is simply the transpose of the tangent linear model as follows [28,29]:

$$\Delta\hat{\mathbf{x}}(t_0) = \mathbf{M}^T(\mathbf{x})\Delta\hat{\mathbf{x}}(t_r) \tag{6}$$

The adjoint model simulates backward in time following the nonlinear trajectory and yields the sensitivity distributions in initial conditions at t_0 to a user-specified response function at time t where $t > t_0$. In the cases where prediction models simulate complex behaviors, the tangent linear and adjoint models are developed at source code levels line-by-line. When the adjoint model is applied in a 4D-Var data assimilation system, the simulation propagated forward in time by the nonlinear forecast model will first be compared with existent observation at the observation time. The discrepency, as a sensitivy or forcing term, will then be taken by the adjoint model and be propagated backward in time in a dynamically consistent manner to the model initial time to inform how the initial condition should be adjusted so the simulation at the observation time can agree closer to the observation. The same forward-backward implementation will be repeated with multiple iterations until an optimal solution is found, which will be discussed further in the next subsection.

In the case of densely connected neural networks, an individual layer in the forward model can be rewritten as:

$$\mathbf{y} = F(\mathbf{x}\mathbf{W} + \mathbf{b}) \tag{7}$$

where $\mathbf{x}$ is the input of the layer, $\mathbf{W}$ and $\mathbf{b}$ the weights and biases resulted from the training, and F denotes activation functions. The tangent linear model in this case becomes:

$$\delta\mathbf{y} = \frac{\partial F}{\partial(\mathbf{x}\mathbf{W} + \mathbf{b})}\delta\mathbf{x}\mathbf{W} \tag{8}$$

The adjoint model is then:

$$\delta\hat{\mathbf{x}} = \delta\hat{\mathbf{y}}\mathbf{W}^T\left[\frac{\partial F}{\partial(\mathbf{x}\mathbf{W} + b)}\right]^T \tag{9}$$

A multilayer or deep NN is simply a repetition of the described above. In this study, the hidden layer has an ELU activation function and the output layer has a linear activation function, Equations (7)–(9) can be further reduced to:

$$\mathbf{y} = \mathbf{x}\mathbf{W} + \mathbf{b} \tag{10}$$

$$\delta\mathbf{y} = \delta\mathbf{x}\mathbf{W} \tag{11}$$

$$\delta\hat{\mathbf{x}} = \delta\hat{\mathbf{y}}\mathbf{W}^T \tag{12}$$

It is to be noted that, when a variational data assimilation system is aimed to be developed, the activation function of choice is preferred to be continuous, such as the ELU defined in Equation (3), as discontinuity (like in the case of Rectified Linear Unit) in activation functions or models in general will make the tangent linear approximation invalid and cause the data assimilation system difficulty or failure to converge when searching for the optimal solution.

2.4. A Continuous 4D-Var DA System

Given the forward and adjoint models, a 4D-Var DA framework can be built to minimize the the following scalar cost function [30–32]:

$$\begin{aligned} J &= J_b + J_o \\ &= \frac{1}{2}(\mathbf{x}_0 - \mathbf{x}_b)^T \mathbf{B}^{-1}(\mathbf{x}_0 - \mathbf{x}_b) \\ &+ \frac{1}{2}\sum_{r=0}^{N}(H_r(\mathbf{x}_r) - \mathbf{y}_r)^T \mathbf{O}_r^{-1}(H_r(\mathbf{x}_r) - \mathbf{y}_r), \end{aligned} \tag{13}$$

where $\mathbf{x}_0$ is the analysis to be solved for, $\mathbf{x}_b$ the first guess, $\mathbf{y}_r$ observations available within the assimilation window, $\mathbf{x}_r$ the model state advanced by model $\mathcal{M}$ from $\mathbf{x}_0$ to observation time t_r, and H_r is the observation operator that maps the model state $\mathbf{x}$ at the observation time t_r to the observation space. The matrices $\mathbf{B}$ and $\mathbf{O}$ are the background and observation error covariance matrices, respectively. Essentially, the term J_b measures the discrepency between the analysis and the model background, weighted by the inverse of the background error covariances, and the term J_o the discrepency between the analysis and the observation weighted by the inverse of observation errors. The solved analysis $\mathbf{x}_0$ with the minimum of J yields the minimum variance estimate. The gradient of the scalar J with respect to $\mathbf{x}_0$ can be obtained following:

$$\nabla_{\mathbf{x}_0} J = \mathbf{B}^{-1}(\mathbf{x}_0 - \mathbf{x}_b) + \sum_{r=0}^{N} \mathbf{M}_r^T \mathbf{H}_r^T \mathbf{O}_r^{-1}(H_r(\mathbf{x}_r) - \mathbf{y}_r), \tag{14}$$

where $\mathbf{M}^T$ denotes the adjoint model and $\mathbf{x}_r$ is obtained by advancing analysis $\mathbf{x}_0$ forward in time to the observation time with the forward model. In the presence of any observations, the correctness of the overall gradient calculation can be validated with the following:

$$\Phi(\alpha) = \frac{J(\mathbf{x}_0 + \alpha\Delta\mathbf{x}) - J(\mathbf{x}_0)}{\alpha\Delta\mathbf{x}^T \nabla J(\mathbf{x}_0)} = 1 + O(\alpha), \tag{15}$$

3. Experiment Design

The forward NN and its adjoint in Section 2.3 are applied into the equations in Section 2.4 describing the 4D-Var DA system. The NN-based 4D-Var DA system is then used to analyze the initial condition. As described in Section 2, the NN-based SW model is trained with MPAS-SW simulation results at 1000 km resolutions. MPAS-SW was run at 250 km resolution initialized with the ERA5 500 hPa atmospheric state at 00 UTC on 1 January 2021. The simulation results at 12 and 24 h will be interpolated into the 1000 km mesh and assimilated as observations to help analyze the initial conditions at the NN native resolutions. The matrices $\mathbf{B}$ and $\mathbf{O}$ are both kept diagonal and assigned with values of RMSE from the test dataset described in Section 2.2. Taking the entire model state (h, u, and v over the globe) as observations, the correctness of the gradient calculation is first checked following Equation (15), the result of which are shown in Figure 5. As the scaling factor decreases in magnitude, the quantity $\Phi(\alpha)$ linearly approaches unity, as expected, proving the accuracy of the calculated gradient of the cost function with respect to the model state vector $\mathbf{x}$. As the accuracy in gradient calculation is an essential and necessary step to ensure that the 4D-Var DA system will perform as expected, the results in Figure 5 is a reassuring signal for a working DA system.

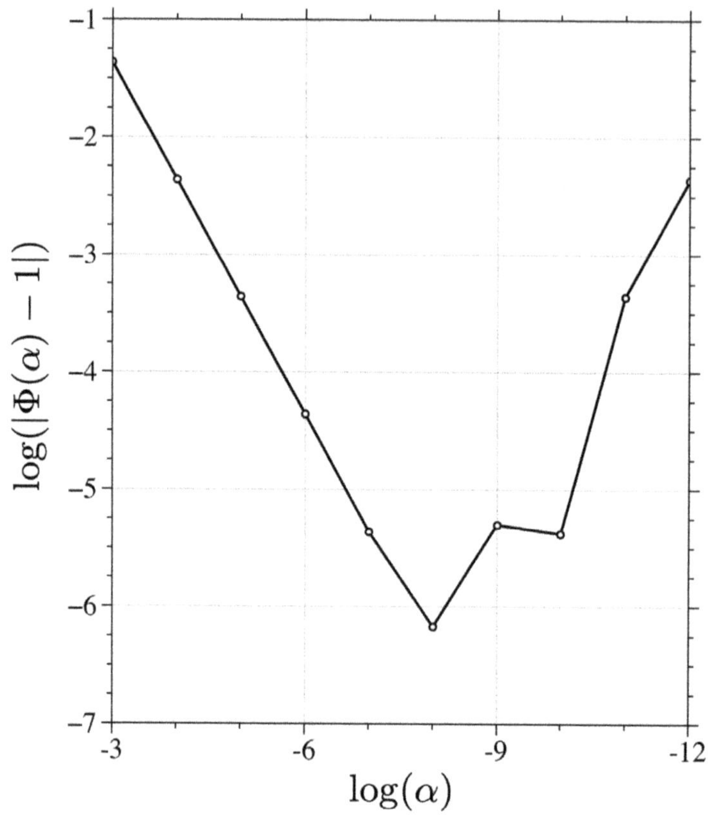

Figure 5. Variations in the gradient-check results $log(|\Phi(\alpha) - 1|)$ as a function of the log of the scaling factor α.

3.1. A Single Observation Experiment

The value of height at a single location of [35.24° N, 164.48° W] from the high-resolution simulations 12 h into the forecast is assimilated. The minimum of the cost function was reached after four iterations as shown in Figure 6a. The norm of the gradient decreased by nearly five orders of magnitude, indicating a local minimum. The analysis increment in both heights and winds are plotted in Figure 6b, with the observation location marked as a white cross. It can be seen that an anticyclonic adjustment was generated near the observation location with some additional adjustments away from the observation location due to the gravity wave mode in the SW dynamics. Notice that only the height at the given location is observed and the background error covariance is kept diagonal in this study, both height and wind adjustments are generated in the analyzed solution, demonstrating the flow dependency in 4D-Var solutions. As simple as this experiment is, a single point observation experiment can be a rather straightforward way of showing the function and feasibility of a DA system.

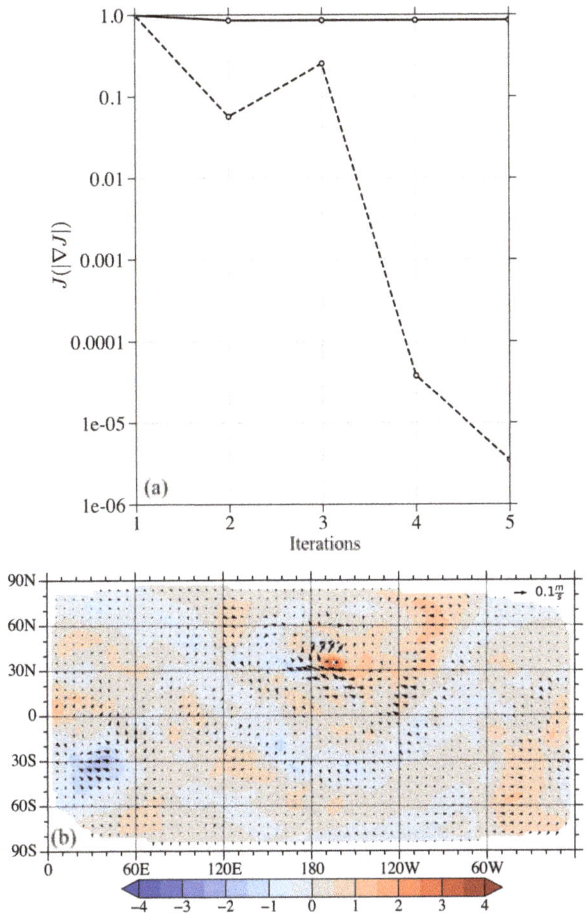

Figure 6. (**a**) Variation of the cost function (solid curve) and the norm of the gradient (dashed curve) with respect to the number of iterations when assimilating a single point observation. (**b**) The analysis increment in height (shaded) and wind (vectors) after assimilating only the height at the location marked with a white cross.

3.2. Full Vector Observation Experiment

The entire atmospheric state (h, u, and v over the global domain) 12 and 24 h after the analysis time simulated by the 250 km resolution MPAS-SW run will be assimilated with the NN-based 4D-Var DA system. The analyzed initial conditions will be used to make forecasts with MPAS-SW at 1000 km resolutions. A control experiment of 1000 km forecast will be made without assimilating any observations, to demonstrate the improvements when observations are assimilated. The predictions with and without DA will be compared against the 250 km simulation results. The evolution of the cost function (solid curve) and the norm of its gradient (dashed curve) are plotted in Figure 7a. After 25 iterations, the value of the gradient norm decreased by more than four orders of magnitude, indicating an extreme point with the solved model state $\mathbf{x}$. The corresponding analysis increment calculated after convergence is given in Figure 7b. Most of the adjustments are located in mid to high latitude regions, especially in the Northern Hemisphere.

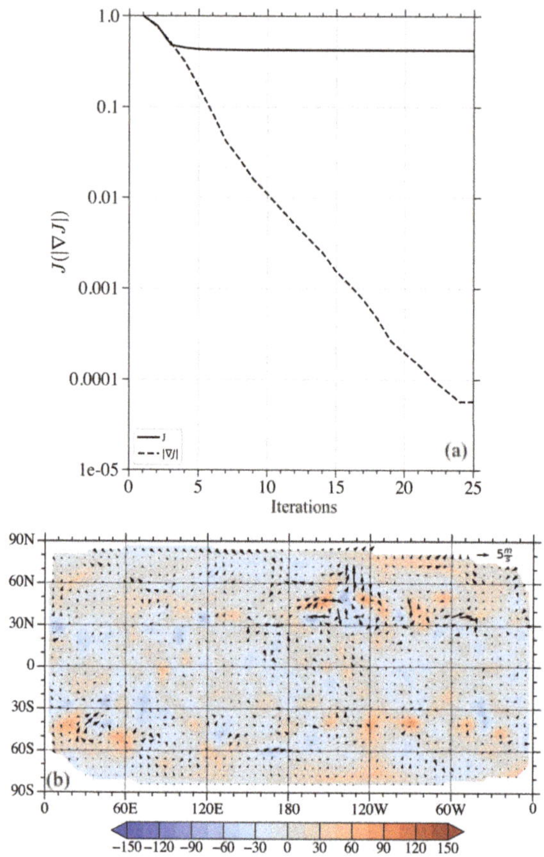

Figure 7. (**a**) Variation of the cost function (solid curve) and the norm of the gradient (dashed curve) with respect to the number of iterations when assimilating the observations of the entire model state. (**b**) The analysis increment in height (shaded) and wind (vectors).

Two MPAS-SW simulations at 1000 km are then initialized with the atmospheric state with and without assimilating observations. As the observations come from the 250 km resolution experiment, the 1000 km forecast results from both simulations are compared against the 250 km forecasts in the form of RMSE. Figure 8 shows the differences in RMSE between the control run and that with DA. It shows that in the first four days, the control outperforms the DA experiment when compared with the 250 km simulation results. However, starting from day four, the DA experiment becomes better than the control and this advantage is maintained over 20 days of forecasts. To compare the forecasts from the two experiments spatially, the differences in forecasts between the control and the 250 km simulation are shown in Figure 9a and those between the DA run and the 250 km simulation are shown in Figure 9b. It can be visually seen that the magnitude of differences in Figure 9b is greater than those in Figure 9a, indicating an inferior performance as also illustrated in Figure 8. The same comparison with the 5-day forecasts are shown in Figure 10. In contrast, the magnitude of differences in both heights and winds in the control run are greater than those in the DA experiment, proving forecasts improvements after assimilating observations with the NN-based DA system.

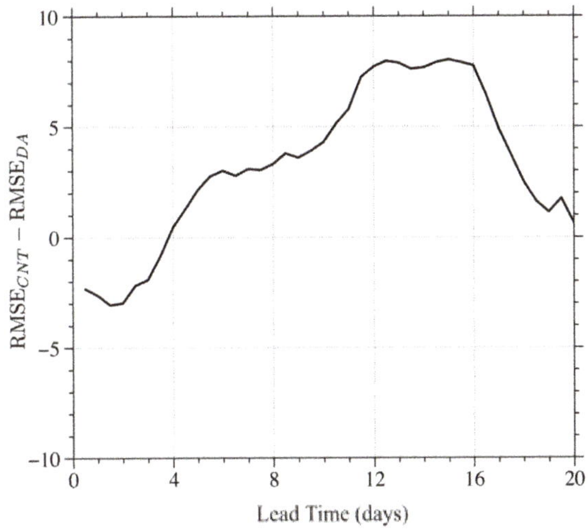

Figure 8. The differences in root mean squared errors (RMSE) between the control and DA experiments with respect to the forecast lead time.

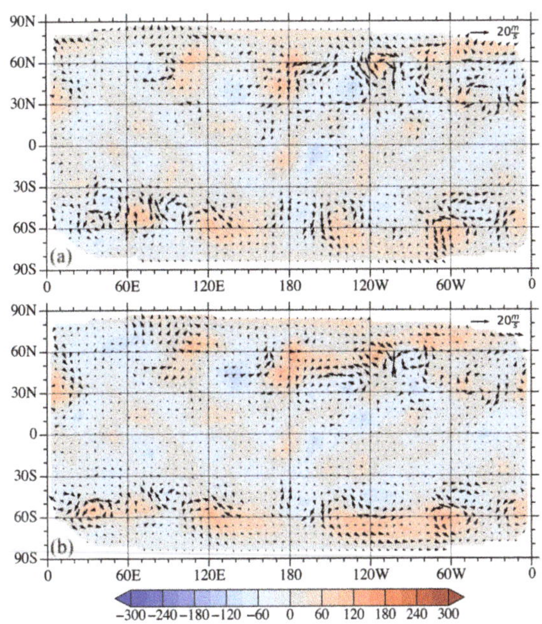

Figure 9. Differences in 2-day forecasts (**a**) between the control experiment and referenced high resolution simulations and (**b**) between the DA experiment and the referenced high-resolution simulations.

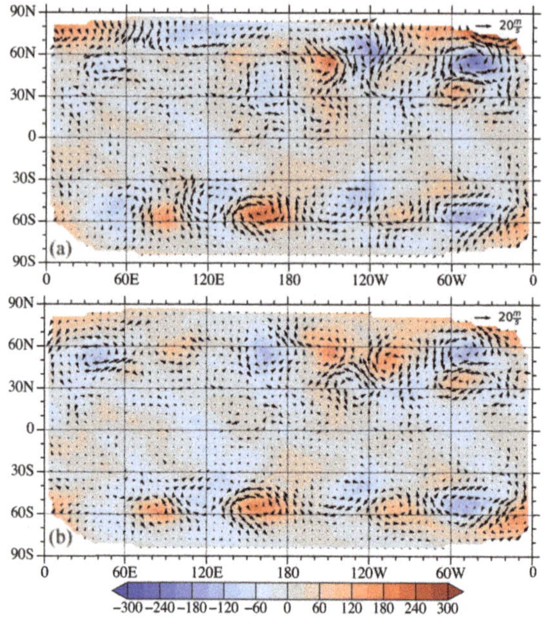

Figure 10. Differences in 5-day forecasts (**a**) between the control experiment and referenced high resolution simulations and (**b**) between the DA experiment and the referenced high-resolution simulations.

3.3. Discussion

In some of the previous studies, the applications of NN techniques in data assimilation has been explored such as replacing physics parameterization components with a NN [20] or a DA system with Lorenz 96 model [21]. Some studies were even exploring the possibility of scaling ML to the entire NWP system like in [3,14]. This study endeavors to extend the application of NN and make use of the convenience in obtaining the TL/AD of an NN with a global shallow water model to formulate a 4D-Var DA system. The convenience of obtaining the TL/AD from a NN is applicable and can potentially benefit various components in a NWP and DA system demonstrated in the aforementioned studies. One example can be to use NN to approximate certain parts of moist physics parameterizations in the nonlinear forward model, which is a process that often involve nonlinear and/or discontinuous calculations and will make the tangent linear approximation invalid and thus make the 4D-Var technique fail. NN can potentially be useful to emulate the physical process while mitigate the nonlinearity/discontinuity. The tangent linear and adjoint of the physics NN can then be conveniently obtained and more reliably incorporated in a variational DA system. The analyses obtained from the purely NN-based 4D-Var DA prove to improve the forecast performances compared with a control experiment, demonstrating the promising prospect of further applications of NN in DA systems.

The potential next steps for this research are numerous. The NN design in this setup was relatively straight forward since it only had a single hidden layer. Additional hidden layers with more sophisticated NN techniques may be experimented in later studies to produce better emulating results. Similar techniques are readily applicable in substituting moise physics parameterizations or observation operators in a DA system. Furthermore, the analyzed results produced in this study may also be compared with a 4D-Var DA system that needs be developed with the traditional MPAS-SW model adjoint. Finally, applying these techniques to larger models such as the MPAS-Atmosphere model will demonstrate whether these techniques are feasible for NWP operations. The recent advances of ML applications in NWP are especially encouraging in this aspect.

4. Conclusions

This study proposes a NN-based SW model that emulates SW dynamics and makes predictions given an initial condition. The NN-based SW model was trained with 60 years (1959 to 2019) of hourly ERA5 atmospheric state and the corresponding 12-h predictions made with MPAS-SW at 1000 km resolution. Taking the ERA5 atmospheric state in 2020 and 2021 and their MPAS-SW predictions as an independent evaluation, the predictions made with the trained NN have an RMSE value of 6.32 m in fluid heights and 0.58 m/s in wind field. An example of the NN-based prediction result is shown to well capture the atmospheric evolution simulated by shallow water dynamics.

The tangent linear and adjoint models of a NN can be conveniently developed, the process of which is described in this study. The NN-based SW model and its adjoint are used to formulate a continuous 4D-Var DA system. Synthetic observations are made with a MPAS-SW experiment at 250 km resolution that is four times higher than the trained NN native 1000 km resolution. In the presence of observations, the calculation of the gradient of the cost function is checked for correctness to ensure that the minimum of the cost function can be found in the 4D-Var DA system.

In a single point observation experiment, the height value at a single point is assimilated as observation. A convergence is achieved rapidly in five iterations. The analysis increment by differing the analyzed initial conditions from the first guess show both local and remote impacts propagated by gravity waves, indicating flow dependency in the solution, even with a simple diagonal background error covariance. In the second DA experiment, the entire model state vectors, i.e., both height and wind fields over the global domain, 12 and 24 h into the forecasts are assimilated. A convergence was achieved with 25 iterations, in which the norm of the cost function gradient decreased by nearly five orders of magnitude. The analysis increments show adjustments throughout the global domain with greater magnitudes in mid and high latitude regions. Forecasts are created with MPAS-SW at 1000 km initialized with the first guess and the analyzed initial conditions. These forecasts are shown to be closer to the 250 km simulations that served as observations. These encouraging results demonstrate the feasibility of the tangent linear and adjoint components obtained from neural networks and the potential value of the proposed DA system.

Funding: This work has received no external funding.

Institutional Review Board Statement: Not applicable.

Informed Consent Statement: Not applicable.

Data Availability Statement: The data analyzed during the current study are available are available at https://xiaoxutian.com/products/, accessed on 30 November 2022.

Acknowledgments: The authors are thank the editor and reviews for their help to make this manuscript better.

Conflicts of Interest: The authors declare no conflict of interest.

Sample Availability: Samples of the compounds are available from the authors.

References

1. Bauer, P.; Thorpe, A.; Brunet, G. The quiet revolution of numerical weather prediction. *Nature* **2015**, *525*, 47–55. [CrossRef] [PubMed]
2. Zhang, Z.; Moore, J.C. *Mathematical and Physical Fundamentals of Climate Change*; Elsevier: Amsterdam, The Netherlands, 2014.
3. Pathak, J.; Subramanian, S.; Harrington, P.; Raja, S.; Chattopadhyay, A.; Mardani, M.; Kurth, T.; Hall, D.; Li, Z.; Azizzadenesheli, K. Fourcastnet: A global data-driven high-resolution weather model using adaptive fourier neural operators. *arXiv* **2022**, arXiv:2202.11214.
4. Schultz, M.G.; Betancourt, C.; Gong, B.; Kleinert, F.; Langguth, M.; Leufen, L.H.; Mozaffari, A.; Stadtler, S. Can deep learning beat numerical weather prediction? *Philos. Trans. R. Soc. A* **2021**, *379*, 20200097. [CrossRef] [PubMed]
5. Irrgang, C.; Boers, N.; Sonnewald, M.; Barnes, E.A.; Kadow, C.; Staneva, J.; Saynisch-Wagner, J. Towards neural Earth system modelling by integrating artificial intelligence in Earth system science. *Nat. Mach. Intell.* **2021**, *3*, 667–674. [CrossRef]

6. Scher, S.; Messori, G. Predicting weather forecast uncertainty with machine learning. *Q. J. R. Meteorol. Soc.* **2018**, *144*, 2830–2841. [CrossRef]
7. Bildirici, M.; Ersin, O.O. Improving forecasts of GARCH family models with the artificial neural networks: An application to the daily returns in Istanbul Stock Exchange. *Expert Syst. Appl.* **2009**, *36*, 7355–7362. [CrossRef]
8. Bildirici, M.; Ersin, O. Modeling Markov switching ARMA-GARCH neural networks models and an application to forecasting stock returns. *Sci. World J.* **2014**, *2014*, 497941. [CrossRef]
9. Gagne, D.J.; McGovern, A.; Xue, M. Machine learning enhancement of storm-scale ensemble probabilistic quantitative precipitation forecasts. *Weather Forecast.* **2014**, *29*, 1024–1043. [CrossRef]
10. McGovern, A.; Elmore, K.L.; Gagne, D.J.; Haupt, S.E.; Karstens, C.D.; Lagerquist, R.; Smith, T.; Williams, J.K. Using artificial intelligence to improve real-time decision-making for high-impact weather. *Bull. Amer. Meteor. Soc.* **2017**, *98*, 2073–2090. [CrossRef]
11. Rasp, S.; Lerch, S. Neural networks for postprocessing ensemble weather forecasts. *Mon. Weather Rev.* **2018**, *146*, 3885–3900. [CrossRef]
12. Agrawal, S.; Barrington, L.; Bromberg, C.; Burge, J.; Gazen, C.; Hickey, J. Machine learning for precipitation nowcasting from radar images. *arXiv* **2019**, arXiv:1912.12132.
13. Grönquist, P.; Yao, C.; Ben-Nun, T.; Dryden, N.; Dueben, P.; Li, S.; Hoefler, T. Deep learning for post-processing ensemble weather forecasts. *Philos. Trans. R. Soc. A* **2021**, *379*, 20200092. [CrossRef] [PubMed]
14. Lam, R.; Sanchez-Gonzalez, A.; Willson, M.; Wirnsberger, P.; Fortunato, M.; Pritzel, A.; Ravuri, S.; Ewalds, T.; Alet, F.; Eaton-Rosen, Z. GraphCast: Learning skillful medium-range global weather forecasting. *arXiv* **2022**, arXiv:2212.12794.
15. Keisler, R. GraphCast: Forecasting Global Weather with Graph Neural Networks. *arXiv* **2022**, arXiv:2202.07575.
16. Chevallier, F.; Chéruy, F.; Scott, N.A.; Chédin, A. A neural network approach for a fast and accurate computation of a longwave radiative budget. *J. Appl. Meteor. Climatol.* **1998**, *37*, 1385–1397. [CrossRef]
17. Krasnopolsky, V.M.; Fox-Rabinovitz, M.S.; Chalikov, D.V. New approach to calculation of atmospheric model physics: Accurate and fast neural network emulation of longwave radiation in a climate model. *Mon. Weather Rev.* **2005**, *133*, 1370–1383. [CrossRef]
18. Rasp, S.; Pritchard, M.S.; Gentine, P. Deep learning to represent subgrid processes in climate models. *Proc. Natl. Acad. Sci. USA* **2018**, *115*, 9684–9689. [CrossRef]
19. Brenowitz, N.D.; Bretherton, C.S. Spatially extended tests of a neural network parametrization trained by coarse-graining. *J. Adv. Model. Earth Syst.* **2019**, *11*, 2728–2744. [CrossRef]
20. Hatfield, S.; Chantry, M.; Dueben, P.; Lopez, P.; Geer, A.; Palmer, T. Building Tangent-Linear and Adjoint Models for Data Assimilation With Neural Networks. *J. Adv. Model. Earth Syst.* **2021**, *13*, e2021MS002521. [CrossRef]
21. Nonnenmacher, M.; Greenberg, D.S. Deep emulators for differentiation, forecasting, and parametrization in Earth science simulators. *J. Adv. Model. Earth Syst.* **2021**, *13*, e2021MS002554. [CrossRef]
22. Scher, S.; Messori, G. Generalization properties of feed-forward neural networks trained on Lorenz systems. *Nonlinear Process. Geophys.* **2019**, *26*, 381–399. [CrossRef]
23. Ringler, T.; Ju, L.; Gunzburger, M. A multiresolution method for climate system modeling: Application of spherical centroidal Voronoi tessellations. *Ocean Dyn.* **2008**, *58*, 475–498. [CrossRef]
24. Ringler, T.D.; Thuburn, J.; Klemp, J.B.; Skamarock, W.C. A unified approach to energy conservation and potential vorticity dynamics for arbitrarily-structured C-grids. *J. Comput. Phys.* **2010**, *229*, 3065–3090. [CrossRef]
25. Hoffmann, L.; Günther, G.; Li, D.; Stein, O.; Wu, X.; Griessbach, S.; Heng, Y.; Konopka, P.; Müller, R.; Vogel, B. From ERA-Interim to ERA5: The considerable impact of ECMWF's next-generation reanalysis on Lagrangian transport simulations. *Atmos. Chem. Phys.* **2019**, *19*, 3097–3124. [CrossRef]
26. Clevert, D.A.; Unterthiner, T.; Hochreiter, S. Fast and Accurate Deep Network Learning by Exponential Linear Units (ELUs). *arXiv*, **2015**, arXiv:1511.07289. https://doi.org/10.48550/ARXIV.1511.07289.
27. Srivastava, N.; Hinton, G.; Krizhevsky, A.; Sutskever, I.; Salakhutdinov, R. Dropout: A Simple Way to Prevent Neural Networks from Overfitting. *J. Mach. Learn. Res.* **2014**, *15*, 1929–1958.
28. Tian, X. Evolutions of Errors in the Global Multiresolution Model for Prediction Across Scales - Shallow Water (MPAS-SW). *Q. J. Royal Meteorol. Soc.* **2020**, *734*, 382–391. [CrossRef]
29. Tian, X.; Ide, K. Hurricane Predictability Analysis with Singular Vectors in the Multiresolution Global Shallow Water Model. *J. Atmos. Sci.* **2021**, *78*, 1259–1273. [CrossRef]
30. Zou, X.; Kuo, Y.H.; Guo, Y.R. Assimilation of atmospheric radio refractivity using a nonhydrostatic adjoint model. *Mon. Weather Rev.* **1995**, *123*, 2229–2250. [CrossRef]
31. Zou, X.; Vandenberghe, F.; Pondeca, M.; Kuo, Y.H. *Introduction to Adjoint Techniques and the MM5 Adjoint Modeling System*; NCAR Technical Note; NCAR: Boulder, CO, USA, 1997.
32. Tian, X.; Zou, X. Validation of a Prototype Global 4D-Var Data Assimilation System for the MPAS-Atmosphere Model. *Mon. Weather Rev.* **2021**, *149*, 2803–2817. [CrossRef]

atmosphere

MDPI

Article

The Effect of Assimilating AMSU-A Radiance Data from Satellites and Large-Scale Flows from GFS on Improving Tropical Cyclone Track Forecast

Zhijuan Lai [1,2,3] and Shiqiu Peng [2,3,4,5,*]

1 South China Sea Marine Forecast Center of State Oceanic Administration, Guangzhou 510000, China
2 State Key Laboratory of Tropical Oceanography, South China Sea Institute of Oceanology, Chinese Academy of Sciences, Guangzhou 510000, China
3 School of Marine, University of Chinese Academy of Sciences, Beijing 100049, China
4 Southern Marine Science and Engineering Guangdong Laboratory, Guangzhou 510000, China
5 Guangxi Key Laboratory of Marine Disaster in the Beibu Gulf (LMDBG), Qinzhou University, Qinzhou 535011, China
* Correspondence: speng@scsio.ac.cn; Tel.: +86-020-8902-1745

Abstract: This study aimed to investigate the effect of assimilating either AMSU-A radiance data from satellites, large-scale flows from the Global Forecast System (GFS), or both together, on improving the track forecast of tropical cyclone (TC). The scale-selective data assimilation (SSDA) approach was employed for the assimilation of large-scale GFS flows, while the conventional 3D variational data assimilation (3DVAR) method was used for that of AMSU-A radiance data. The results show that assimilating either AMSU-A radiance data or large-scale GFS flows has a significant improvement on TC track forecast, but the improvement occurs within the first 72 h and after 72 h, respectively. When assimilating both AMSU-A radiance data and large-scale GFS flows, the forecast can take advantage of both data and thus lead to the smallest 5-day mean errors of the track forecast. These results are instructive to future operational TC track forecasting.

Keywords: tropical cyclone (TC); track forecast; radiance data assimilation; scale-selective data assimilation (SSDA)

Citation: Lai, Z.; Peng, S. The Effect of Assimilating AMSU-A Radiance Data from Satellites and Large-Scale Flows from GFS on Improving Tropical Cyclone Track Forecast. *Atmosphere* **2022**, *13*, 1988. https://doi.org/10.3390/atmos13121988

Academic Editor: Da-Lin Zhang

Received: 11 August 2022
Accepted: 23 November 2022
Published: 28 November 2022

1. Introduction

Tropical cyclones (TCs) occurring over warm tropical oceans are one of the most severely disastrous weather systems. Accurate forecasting of TC tracks, especially landfall locations, is a prerequisite for the formulation of an effective strategy for preventing and mitigating TC-induced disasters. Though continuous improvements have been made over the past several decades [1–6], considerable track forecast errors still exist, especially for forecasting lead times of longer than 24 h, which are far from meeting the requirements of disaster prevention and reduction. Therefore, TC track forecasting remains a major challenge for weather forecasting in the world.

At present, numerical weather forecasting is one of the most widely-used TC forecasting methods, which is a problem of initial condition to a great extent [7]. Thus, in order to improve the forecasting accuracy, many new technologies or methods are being constantly put forward by scientists to optimize the initial fields of a model, among which the use of observations to directly correct the model initial conditions by data assimilation has been proven to be quite effective. In the early days, due to poor observing techniques, only a small amount of conventional observations was available. With the development of satellite remote sensing technology, more and more reliable satellite observations are available with a larger coverage area and higher spatial resolution for atmospheric information; these data have been used in addition to conventional observations to improve a given model's performance in TC forecasting effectively through optimizing the model initial conditions [8,9]. In particular, the polar-orbiting meteorological satellite-borne Advanced Microwave Sounding Unit (AMSU) has the unique ability to penetrate through deep cloud layers, except for precipitation clouds. It can detect vertical profiles of atmospheric temperature and humidity, and its brightness temperature data are very useful for reconstructing the mesoscale structure of a typhoon [10]. Zhang [11] used the Advanced Television and Infrared Observation Satellite (TIROS)-N's Operational Vertical Sounder (ATOVS) microwave data in a three-dimensional variational data assimilation (3DVAR) analysis to study typhoon structures as well as their variations in different periods over the northwest Pacific Ocean; the study showed that ATOVS microwave data assimilation can correctly describe the characteristics and changes of the 3D structure of typhoons over the northwest Pacific Ocean, which cannot be obtained using conventional observations. Since then, ATOVS radiance data have been widely used in the study of TC numerical forecast/simulation, and radiance data assimilation has become a research hotspot in recent years. Previous studies [12–16] have shown that assimilating ATOVS radiance data can improve the initial conditions of a model, including the environmental fields and inner structure of a TC, and is thus able to effectively improve the short-term forecasting accuracy of a TC.

In addition, a method referred to as the scale-selective data assimilation (SSDA) approach was proposed by Peng et al. in 2010 [17], and it has been successfully applied to TC forecasting/simulation in the past few years [18–20]. In principle, the SSDA approach takes into account the multi-scale features in both the observations and model output and only adjusts the model variables on a selective scale through scale separation; technically, the SSDA approach employs a band-pass digital filter and three-dimensional variational data assimilation scheme to correct the large-scale bias by incorporating the large-scale circulations from a global model into a regional model while keeping the regional-scale details unchanged [17]. Studies have showed that both large- and small-scale flows can benefit from the SSDA procedure, resulting in significant improvements in the TC track and intensity forecast/simulation [18–20].

Radiance data assimilation is an effective method to improve the TC forecast accuracy of a model through an optimization of the model initial conditions. However, the effects of radiance data assimilation usually rapidly attenuate with time, or even diminish after 2–3 days, due to the very fast decaying of the effects of the optimized initial conditions [21,22]. Thus, radiance data assimilation might do little to improve the TC forecast accuracy of a model for a longer forecast period. Although the rolling of data assimilation can make up for the disadvantage and be adopted to improve the accuracy of model simulation, it cannot be used in real-time numerical forecasts for the lack of observational data in the future (i.e., forecasting time). The SSDA approach, on the other hand, shows a large advantage in improving TC forecasts with lead times longer than 48 h [18,19], which is accomplished by assimilating large-scale components from global model forecasts instead of real observations at a preset time. Therefore, it is interesting to understand the combining effects of radiance data assimilation and the SSDA approach on TC forecasts, which is the objective of this study.

In this study, we investigated the impacts of assimilating AMSU-A radiance using the 3DVAR technique, assimilating large-scale GFS flows by the SSDA approach, and the combination of both on TC track forecasting through case studies. The rest of this paper is organized as follows. Section 2 briefly describes the model system and methodology. Experiment design is detailed in Section 3. Section 4 presents the results and discussion, followed by a summary in Section 5.

2. Methodology

2.1. The WRF-ARW Model

The weather model used in this study is the Weather Research and Forecasting (WRF) model utilizing the Advanced Research WRF dynamical core (WRF-ARW) [23], which was developed by the National Center for Atmosphere Research (NCAR). There are different dynamical cores and model physics packages accessible in the WRF framework. The WRF-ARW is a non-hydrostatic, fully compressible, primitive equation model that uses a terrain-following hydrostatic pressure coordinate and Arakawa C-grid staggering. In addition, a data assimilation package (WRFDA) is encompassed as a broader component of the WRF system. For more details about the model, readers can refer to Wang et al. [24].

2.2. The Data Assimilation System with SSDA Incorporated

In this study, the SSDA approach proposed by Peng et al. [17] is adopted to incorporate the large-scale circulation from the global model forecasts into the WRF model, driving the model from both the lateral boundaries and the domain interior. The SSDA employs a low-pass filter to separate the large- and small-scale components of the forecasts (or analysis) from both the global and the regional models. After the scale separation, a 3DVAR technique is used to assimilate the large-scale components of the global model forecasts (or analysis) into the regional model to adjust corresponding components periodically (i.e., at the interval of SSDA implementation) as the model integrates forward. The small-scale components of the circulation in the regional model are unconstrained and allowed to freely develop in accordance with the dynamics and physics at a higher resolution. Details about the approach can be referred to in previous studies [17–20].

2.3. Radiative Transfer Model—CRTM

For direct satellite radiance assimilation, a radiative transfer model (RTM), which calculates radiance or brightness temperatures from the input atmospheric state variables, should be incorporated into the WRFDA system. In this study, the Community Radiative Transfer Model (CRTM) [25,26] developed by the Joint Center for Satellite Data Assimilation (JCSDA) in the US is used. The model has already been integrated into the WRFDA system as a part of observation operators, having a flexible and consistent user interface since WRFDA v3.2.1. It includes four major modules, i.e., gaseous absorption model, surface emissivity and reflectivity model, cloud and aerosol optical model, and radiative transfer (RT) solution model [26]. The Optical Path Transmittance (OPTRAN) [27,28] algorithm is used in the gaseous absorption model to calculate the gaseous absorption for the given pressure, temperature, water vapor, and ozone concentration profiles. The surface emissivity and reflectivity model, covering land, ocean, ice, and snow surfaces, is divided into five smaller modules according to the spectral region and surface sub-type. The cloud and aerosol optical model includes six cloud and eight aerosol types, with the optical property pre-computed and stored in a lookup table, which is suitable for multiple cloud/aerosol layers in a vertical column. The advanced doubling-adding method [29] is used in the RT solution model to solve the RT equation.

2.4. AMSU-A Radiance Data and Its Processing

The radiometer AMSU-A is one of the most important components to the Advanced Microwave Sounding Unit (AMSU). It is a cross-track, stepped-line scanning, and total power radiometer with fifteen channels (Table 1) whose primary goal is to measure the temperature profiles of the atmosphere. It observes the Earth with an instantaneous field-of-view of 3.3° at the half-power points and a nominal spatial resolution of 48 km at nadir. Because of the unique ability to penetrate through heavy cloud layers (except for precipitation clouds) and gather information of the inner structure of a tropical cyclone, the AMSU-A is especially useful for tropical cyclone studies.

Table 1. Characteristics of the AMSU-A.

	Channel	Frequency (GHz)	Peak Level	Main Observation
AMSU-A	1	23.8	Surface	Surface condition and the precipitable water
	2	31.4	Surface	As above
	3	50.3	Surface	Surface emissivity
	4	52.8	1000 hPa	Atmospheric temperature
	5	53.59 ± 0.115	700 hPa	Atmospheric temperature
	6	54.4	400 hPa	Atmospheric temperature
	7	59.94	270 hPa	Atmospheric temperature
	8	55.5	180 hPa	Atmospheric temperature
	9	F_{LO} = 57.29	90 hPa	Atmospheric temperature
	10	F_{LO} ± 0.217	50 hPa	Atmospheric temperature
	11	F_{LO} ± 0.322 ± 0.048	25 hPa	Atmospheric temperature
	12	F_{LO} ± 0.322 ± 0.022	12 hPa	Atmospheric temperature
	13	F_{LO} ± 0.322 ± 0.010	5 hPa	Atmospheric temperature
	14	F_{LO} ± 0.322 ± 0.0045	2 hPa	Atmospheric temperature
	15	89	Surface	Surface condition and the precipitable water

It is well-known that some biases related to instrument calibration problems, RTM, and the predictor and zenith angle bias exist. These biases are equivalent to those of the air temperature in the short-term model prediction. Thus, it is essential to correct these biases before radiance data assimilation. The processing includes correcting the relative biases between measurements at different scan angles (Scan bias) with information on the scan angle and a correction for the biases correlated with "air-mass" as sensed by the measurements (air-mass bias). For a detailed description of these correction methods, please refer to relevant publications [30–32]. In WRFDA, there are two schemes available for bias correction. One is carried out using a set of co-efficient files pre-calculated with an off-line statistics package based on the Harris and Kelly [32] method. The other is the variational bias correction (VarBC) [30] scheme, which is of relative simplicity and was thus used in this paper.

In addition, because many factors (such as weather conditions, ground conditions, geographical locations, and so on) can cause large errors in satellite observations [12,15] and the presence of a single data point with large errors can result in a substantial degradation of the analyses and subsequent forecasts [30], quality control is vital and necessary prior to data assimilation. In order to avoid assimilating poor-quality observations, the following quality checks were carried out in this study: (1) performing a location check, which includes removing observations outside the domain, removing observations on both ends of each scan line, rejecting pixels over mixture surface, and rejecting channels with an absolute value of zenith angle >45°; (2) excluding incomplete observations and observations with duplicate locations/times by a thinning procedure to ensure the vertical consistency of upper-air profiles and to keep the radiances relatively uncorrelated; (3) rejecting radiance brightness temperature data outside the range of 150–350 K; (4) removing observations contaminated by precipitation. The presence of precipitation is detected by means of the scattering index (SI) and cloud liquid water (CLW) amount [33,34], respectively. If the SI is >3 K or the CLW is >0.2 mm, the microwave radiances are assumed to be contaminated by precipitation and are rejected; and (5) rejecting channels whose innovation (observation minus background) is 3 times larger than the standard deviation of observation errors, as well as those whose weighting function peak is above the model top or below the surface pressure according to the peak energy contribution level of the sounder channel. Thus, only data on channels 5–9 of AMSU-A are assimilated.

In general, only a small number of observations would be assimilated into the model after the processing of bias correction and quality control. This is beneficial to the minimization procedure in the variational data assimilation system. The number of radiance observations for different channels of different instruments used in the experiments in Sections 3 and 4 is presented in Figure 1.

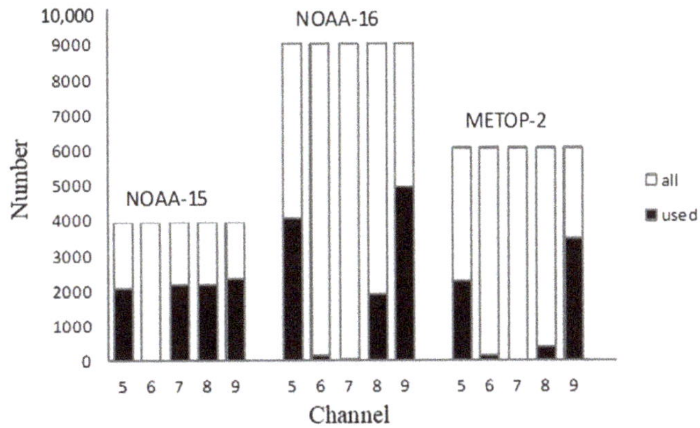

Figure 1. The number of radiance observations for different channels of different instruments used in the experiments in Sections 3 and 4.

3. Experiment Design

Super typhoon Megi (2010) was chosen as a test case for detailed analysis in this study. Megi (2010) was one of the strongest typhoons in the northwestern Pacific since 1979; it generated over the northwestern Pacific far to the east of the Philippines at 1200 UTC 13 October 2010 and reached its peak intensity (with a minimum central pressure of 895 hPa and maximum surface wind speed up to 72 m/s) at 1200 UTC 17 October. After moving westward into the South China Sea, Megi (2010) suddenly made a sharp northward turn around 0000 UTC 20 October and finally landed at the coast of southern Fujian Province in China at about 0455 UTC 23 October. It caused large threats on the safety of life and property in southeastern China, and its sharp northward turn posed major challenges to operational forecasters [35], with an over-prediction of westward motion by nearly all of the official agencies, including the CMA, JTWC, and JMA, which issued five-day track forecasts during the period from 0000 UTC 17 October to 0000 UTC 19 October [4].

To perform TC track forecasts for Megi (2010), the WRF model was configured with 27 sigma levels in the vertical direction, with a model top of 50 hPa and two nested domains with 36 km and 12 km grid spacing, respectively, in a Mercator map projection (see in Figure 2). The parameterization schemes, which are needed for the boundary layer turbulence, cumulus convection, microphysics of the phase transform among ice, water and vapor, and short/long wave radiation, were employed as follows: the Bougeault and Lacarrere (BouLac) TKE PBL scheme [36], the Kain–Fritsch cumulus scheme [37], the Ferrier (new Eta) microphysics scheme [38], and the Dudhia shortwave [39] and rapid radiative transfer model (RRTM) longwave [40] radiation scheme.

The experimental design in this study consisted of four forecasts that are summarized in Table 2. In all of these experiments, the GFS analysis data at the initial time and the afterward 384-h GFS forecasts at 6 h intervals with a $1.0^\circ \times 1.0^\circ$ resolution from NCEP were used for the initial conditions and boundary conditions of the WRF model, and the model was integrated from 0000 UTC 18 October 2010 to 0000 UTC 23 October 2010. The differences of these experiments are as follows. The first experiment was a control run (denoted as CTL) without any data assimilation. In the second experiment (denoted as RAD-DA), AMSU-A radiance data were assimilated into the WRF model using the 3DVAR technique to improve the model initial conditions. In the third experiment (denoted as GFS-DA), the large-scale wind components with wavelengths longer than 2151 m (corresponding to a cutoff wave number of four) above 850 hPa from the GFS global model forecasts were assimilated into the WRF model at an interval of 12 h using the SSDA approach. The fourth experiment (denoted as COM-DA) employed the data assimilation scheme, which combined the AMSU-A radiance data assimilation with the SSDA approach; i.e., AMSU-A radiance data were assimilated into the model using the 3DVAR technique at the initial time, and then large-scale wind fields with wavelengths longer than 2151 m above 850 hPa from the GFS global model forecasts were assimilated into the WRF model after 48 h at an interval of 12 h using the SSDA approach (Figure 3). The JTWC tropical cyclone best track data and NCEP GFS global analyses were used to validate the results of the WRF simulation corresponding to the above four experiments.

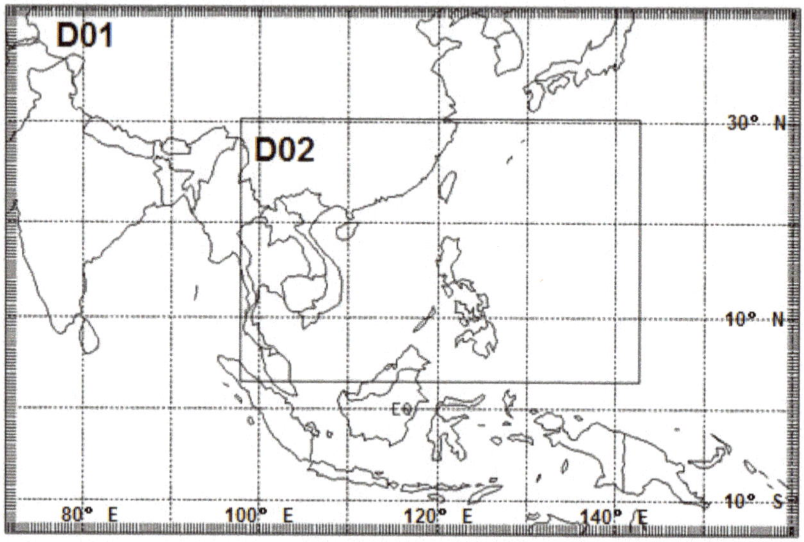

Figure 2. The model domains of the WRF used for all experiments.

Table 2. Experiments designed in the study.

Exp. No.	Exp. Name	Data Assimilated	Assimilation Method Used
1	CTL	No	No
2	RAD-DA	AMSU-A Radiance Data	3 DVAR technique
3	GFS-DA	large-scale GFS flows	Scale-Selective Data Assimilation
4	COM-DA	AMSU-A Radiance Data and large-scale GFS flows	3DVAR technique and Scale-Selective Data Assimilation

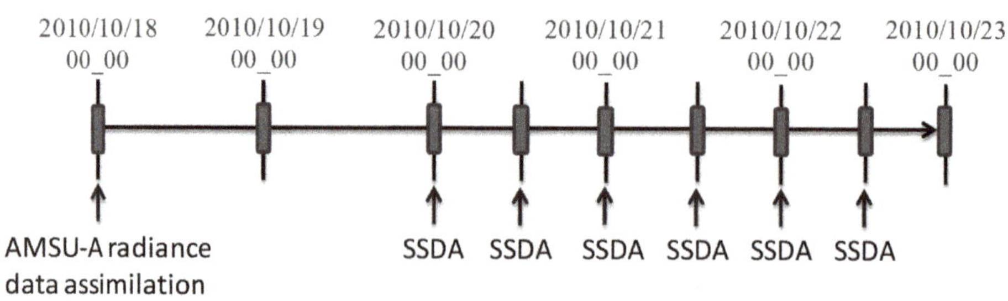

Figure 3. Diagrams schematically illustrating the data assimilation scheme for the COM-DA experiment.

4. Results and Discussion

The TC track position error (TPE, in km) is defined as the great circle distance between the "observed" and the forecast TC center (defined as the location of the minimum sea level pressure), valid at the same time according to the following formulas [41,42]:

$$TPE = 111.11 \cos^{-1}[\sin \varphi_0 \sin \varphi_f + \cos \varphi_0 \cos \varphi_f \cos(\lambda_0 - \lambda_f)],$$

where φ_0 (φ_f) and λ_0 (λ_f) are the latitude and longitude of the "observed" (forecast) TC center, respectively.

4.1. Forecasted Tracks from Different Experiments

Figure 4 shows the observed and forecasted track of Megi (2010) as well as the corresponding errors for all experiments. It is clear that all three data assimilation experiments, including RAD-DA, GFS-DA, and COM-DA, perform better than CTL in terms of track forecast. The forecasted track from CTL deviates to the east of the best track. It curves northward earlier and moves much faster than the best track. After optimizing the initial conditions through assimilating AMSU-A radiance data into the model (RAD-DA), the forecasted track steers west–southwest towards the "observed" track, and the smallest TPE is obtained in the first 3 days. However, the TC then moves much faster to the northeast, similarly to that in CTL. For the GFS-DA experiment, assimilating large-scale flows from the GFS global model forecasts to adjust corresponding components in the regional model periodically at preset intervals not only slows the northeastward track down, but also steers it west–southwest to be closer to the observed track, and thus results in significant improvements in the track forecast of Megi (2010) compared with that from the CTL approach. Though RAD-DA slightly outperforms GFS-DA in the first 72 h, the latter performs much better than the former after 72 h. When assimilating both AMSU-A radiance data and large-scale GFS flows by combining the approaches used in RAD-DA and GFS-DA, the forecasted track from COM-DA steers west–southwestward to be closer to the "observed" one than that from CTL in the first 48 h (which is similar to that from RAD-DA), and then follows that from GFS-DA, leading to the smallest TPEs nearly in all 5 days of forecasts. Detail analyses are presented in the following paragraphs.

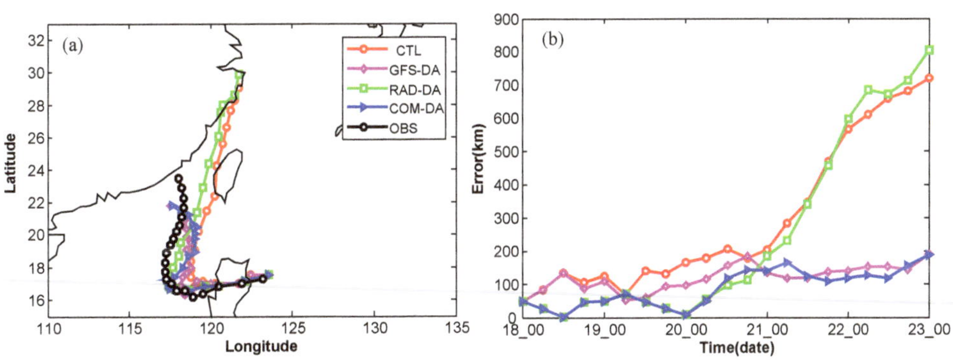

Figure 4. Forecasted (**a**) tracks of typhoon Megi (2010) and (**b**) corresponding errors (unit: km) for experiments; CTL (in red), RAD-DA (in green), GFS-DA (in purple), and COM-DA (in blue). The best-track from the JTWC (OBS, in black) is also given as a reference.

4.2. Analysis and Discussion

Because the same observations have been assimilated into the WRF model in RAD-DA and COM-DA, the initial conditions after the assimilation should be the same and thus only those for RAD-DA are shown here. Figure 5 presents the initial temperature and its increment against CTL at different vertical layers in RAD-DA. Compared with CTL (not shown), the temperature structure at these levels does not appear to change after assimilating the radiance data. However, the increment field (contour line) indicates that the temperature increases (decreases) at the lower and middle (upper) layers of the atmosphere over the western and northwestern areas of the TC center. This may cause temperature stratification that is more unstable in these regions. According to the study by Chen et al. [43], which shows that a TC has the trend to move towards the region with high unstable stratification regarding to temperature and/or humidity, such an adjustment of temperature field after the assimilation of AMSU-A radiance is beneficial for the TC to move towards the west first and then turn to the north. The initial geopotential heights at 500 hPa in Figure 6 indicate that, after the assimilation of AMSU-A radiance, the subtropical high to the north of the TC center in RAD-DA is much stronger than that in CTL. Besides, the mean wind increments at 500–700 hPa for RAD-DA against CTL (Figure 6b) show that there is an anomalously anticyclonic flow, with its southeast branch locating at the TC center forecasted by RAD-DA. All these adjustments of the initial conditions after the assimilation of AMSU-A radiance would inevitably lead to changes of the large-scale environmental fields and the steering flows, which would favor the west–southwestward movement of the TC.

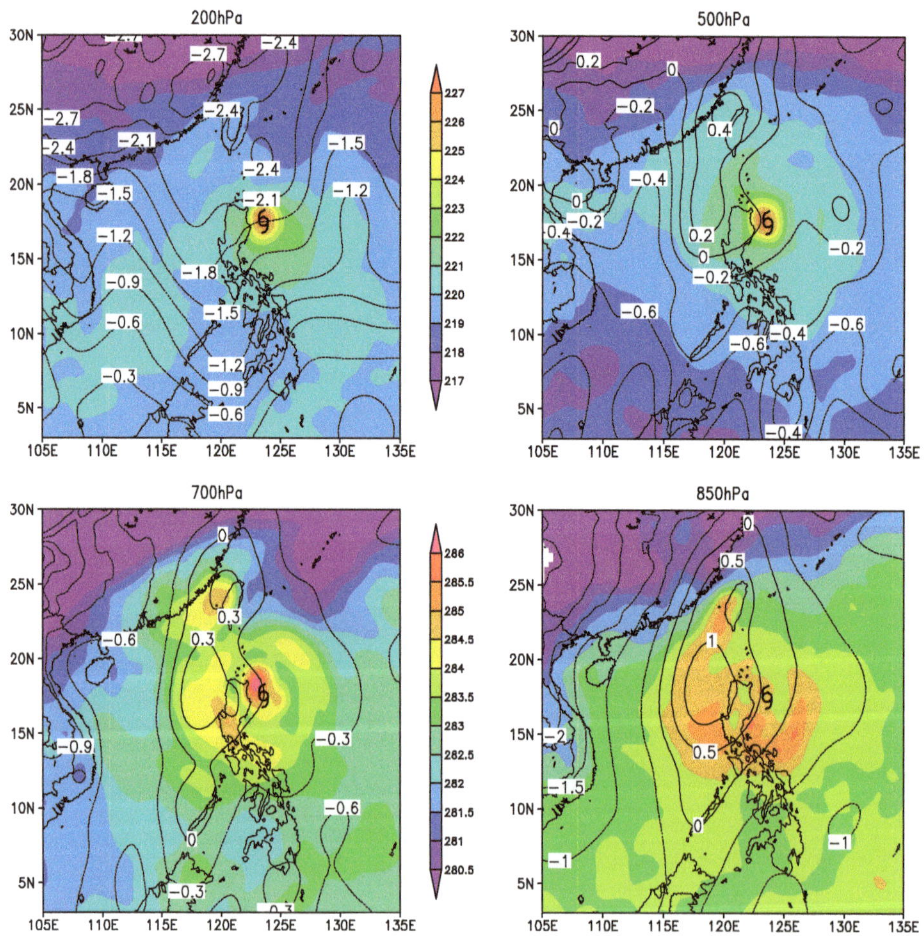

Figure 5. Initial temperature (shaded, unit: K) and temperature increment (RAD-DA minus CTL, contour, unit: K) at 200 hPa, 500 hPa, 700 hPa, and 850 hPa for the RAD-DA experiment.

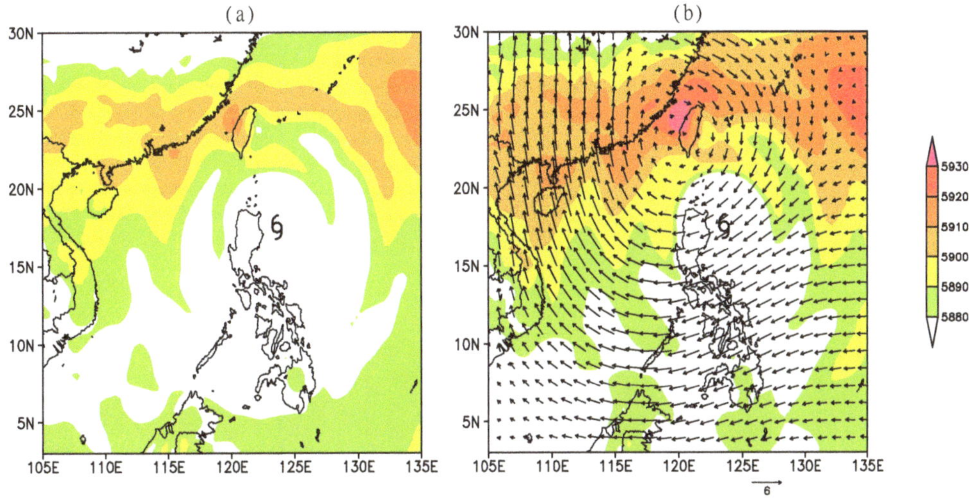

Figure 6. Initial geopotential heights (shaded, unit: gpm) at 500 hPa for the (**a**) CTL and (**b**) RAD-DA experiments, superimposed by the mean wind increments (RAD-DA minus CTL, arrows, unit: m/s) at 500–700 hPa.

In the GFS-DA experiment, the large-scale wind fields forecasted by the GFS global model system were assimilated into the model at an interval of 12 h, starting at the 12th hour of the model integration to adjust the large-scale wind components. As such, the biases of large-scale wind fields for GFS-DA were reduced substantially compared with CTL (Figure 7). Figure 8 displays the root-mean-square errors (RMSEs) of large-scale u and v components for all experiments against the corresponding components of GFS analysis. Compared with CTL, the RMSEs of the large-scale wind components for RAD-DA only slightly decrease at lower layers, while the RMSEs for GFS-DA and COM-DA significantly decrease at nearly all vertical levels, with those for COM-DA being the smallest.

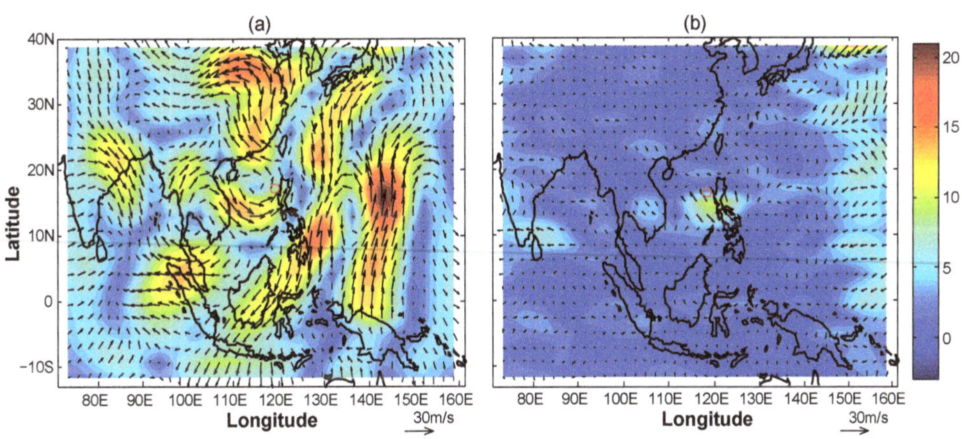

Figure 7. Biases of large-scale wind fields at 200 hPa for experiments (**a**) CTL and (**b**) GFS-DA against those in GFS analysis valid at 1200 UTC 19 Oct 2010 (right after the SSDA implementation; unit: m s^{-1}). TC locations are indicated in red circles.

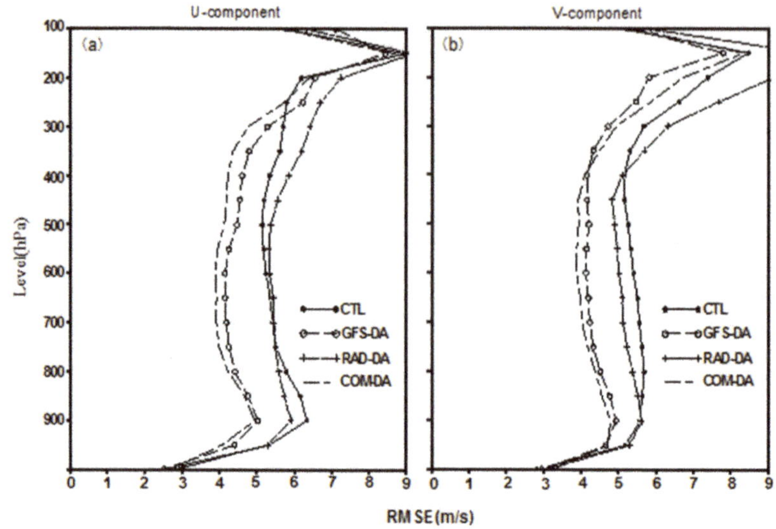

Figure 8. Vertical profile of 5-day mean RMSEs of large-scale (**a**) u and (**b**) v components for each experiment against the corresponding components in the GFS analysis, averaged over all grids in the inner domain. (unit: m s^{-1}).

The geopotential heights at 500 hPa valid at 0000 UTC every day for all experiments are displayed in Figure 9. It is evident that the subtropical high to the north of the TC center in the three data assimilation experiments is much stronger than that in CTL before the TC track turns north, which is beneficial for the TC to move westward. Furthermore, because AMSU-A radiance data have been used to optimize the model initial conditions by the 3DVAR technique, the subtropical high in RAD-DA is stronger than that in GFS-DA, resulting in a better performance of RAD-DA in the forecast of the westward TC track before the big curvature as compared with that of CTL or GFS-DA. After 0000 UTC 20 October, however, the subtropical high in RAD-DA rapidly weakens and steers the TC to quickly move northeastward similar to that in CTL. The subtropical high slowly weakens in the GFS-DA experiment, with a spatial pattern that facilitates the TC to move northward instead of northeastward, resulting in a better performance of GFS-DA in the forecast of the northward TC track after the big curvature as compared with that of CTL or RAD-DA. For the COM-DA experiment, because both the initialization using the AMSU-A radiance data as that in RAD-DA and the large-scale flow adjustment using the SSDA technique as that in GFS-DA are carried out, the subtropical high keeps both the feature found in RAD-DA before the big curvature and that found in GFS-DA after the big curvature, leading to the best performance of the track forecast among all experiments with respect to the entire life cycle of Megi (2010).

As a TC is mainly guided by the large-scale environmental steering flows [44–48], here we calculate the steering flows through averaging the wind vectors in the vertical levels between 700 hPa and 500 hPa along a 5°–7° radial band from the TC center. Figure 10 shows the u and v components of the environmental steering flows valid at different forecast time for all experiments. The steering flows forecasted by CTL and RAD-DA obviously deviate those from GFS analysis after 72 h, which corresponds to the large TPEs in the last 48 h for the two experiments. Enhanced southwestward (negative u and v components) steering flows in the early stage and enhanced northeastward (positive u and v components) steering flows in the later are obtained in the RAD-DA experiment with the assimilation of AMSU-A radiance data. Such an adjustment of steering flows in RAD-DA drives an enhanced southwestward (northeastward) movement of the TC in the early (later) stage, resulting in the smallest TPE in the first 72 h and a larger TPE later. For GFS-DA, the strength of both southerly (positive v component) and westerly (positive u component) of the steering flows in the later stage are significantly reduced as compared with those in CTL and RAD-DA due to the assimilation of large-scale wind components from the GFS global model forecasts, which slows down the northward movement of Typhoon Megi (2010) and steers it west toward the best track. It is worth noting that the adjustment of the steering flows in GFS-DA is relatively small at the beginning; however, it obviously increases after 72 h, which is why larger improvement in the track forecast from GFS-DA is seen after 72 h. It is apparent that the steering flows in COM-DA remain the same as those found in RAD-DA during the first 48 h and become more similar to those found in GFS-DA.

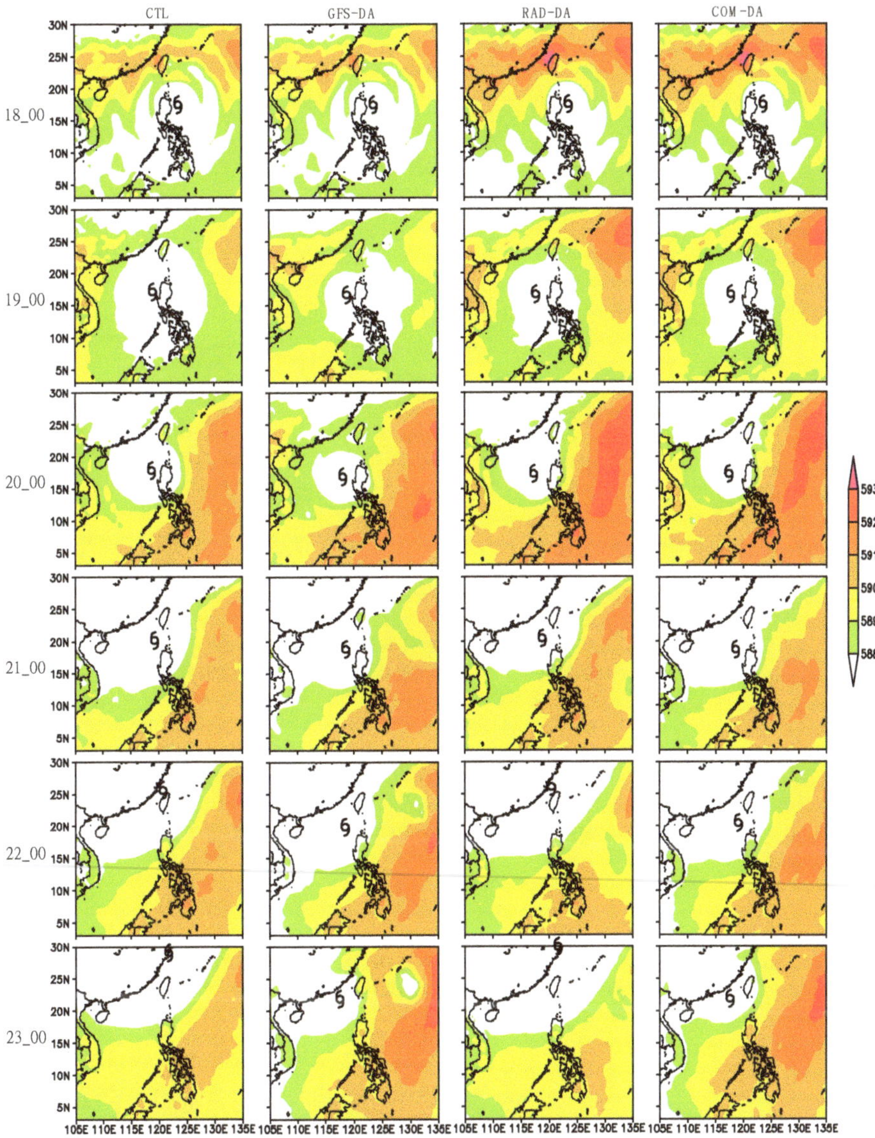

Figure 9. Geopotential heights (unit: gpm) at 500 hPa, valid at 0000 UTC of 18–23 Oct 2010 for experiments CTL, RAD-DA, GFS-DA, and COM-DA.

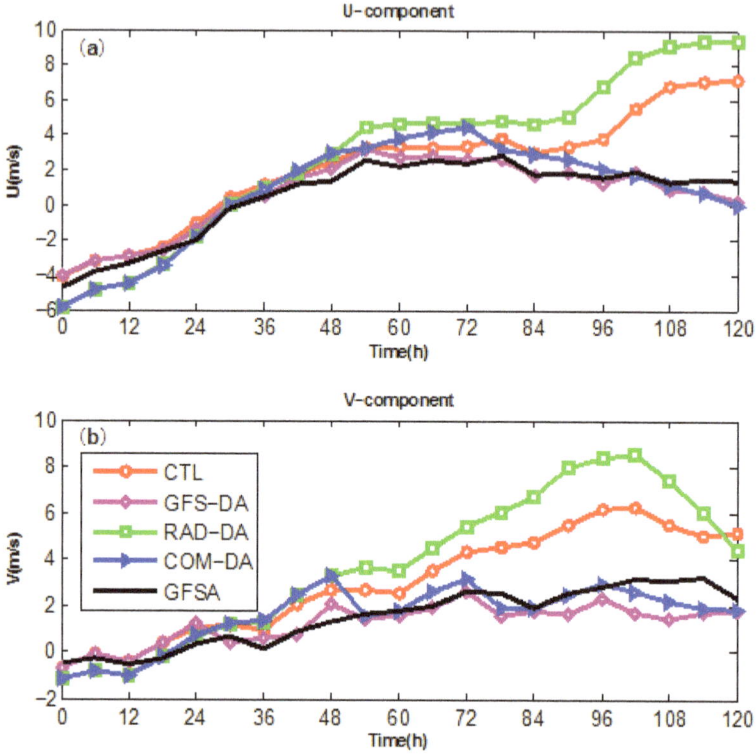

Figure 10. Values of the (**a**) u and (**b**) v components of the steering flows at different forecast times for each experiment (unit: m s^{-1}). The steering flows from the GFS analysis (OBS, in black) are also given as references.

For a further assessment on the effect of assimilating both AMSU-A radiance data and large-scale GFS flows, we carried out the same experiments (TCL, GFS-DA, RAD-DA, and COM-DA) as above with a different initialization time for Megi (2010) as well as for another strong typhoon Nesat (2011), which has completely different track, as depicted in Table 3. The experiments for Megi (2010) and for Nesat (2011) were initialized every 6 h from 0000 UTC 17 Sep 2010 to 1800 UTC 18 Sep 2010 and from 0000 UTC 24 Sep 2011 to 1800 UTC 25 Sep 2011 to create 5-day forecasts, respectively. Thus, there are eight runs for each experiment for both Megi (2010) and Netsat (2011).

Table 3. Experiments included for further assessments in the study.

Exp. Name	Megi	Nesat
CTL RAD-DA GFS-DA COM-DA	8 runs of 5-day forecast initialized every 6 hours from 0000 UTC 17 Sep 2010 to 1800 UTC 18 Sep 2010	8 runs of 5-day forecast initialized every 6 hours from 0000 UTC 24 Sep 2011to 1800 UTC 25 Sep 2011

Table 4 presents the mean track forecast errors of Megi (2010) and Nesat (2011) for different forecast periods for each experiment. Generally, the 5-day mean TPEs from GFS-DA, RAD-DA, and COM-DA are reduced compared with CTL, and COM-DA has the smallest 5-day mean TPEs of 120.4 km and 268.3 km for Megi (2010) and Nesat (2011), respectively. As for different forecast periods, the TPEs from RAD-DA are smaller (larger) than those from GFS-DA before (after) the first 72 h, and COM-DA performs the best for the forecast periods of 12 h, 36 h, 48 h, 60 h, 72 h, 84 h, and 96 h for Megi (2010) and 12 h, 24 h, 36 h, 48 h, and 60 h for Netsat (2011). Therefore, while assimilating AMSU-A radiance data and large-scale GFS flows can achieve the largest improvement of track forecast for forecast periods within 72 h, it could be case dependent for those beyond 72 h.

Table 4. Mean track forecast errors of Typhoon Megi (2010) and Nesat (2011) at different forecast periods for the CTL, GFS-DA, RAD-DA, and COM-DA runs (unit: km).

Forecast Period (h) and (No. of Cases)	TPEs of Megi (2010)				TPEs of Nesat (2011)			
	CTL	GFS-DA	RAD-DA	COM_DA	CTL	GFS-DA	RAD-DA	COM_DA
12(8)	73.4	73.4	58.4	58.4	114.8	111.8	89.3	89.3
24(8)	67.9	69.6	74.1	74.1	134.2	149.1	93.1	93.1
36(8)	82.7	76.3	74.3	74.3	215.7	184.7	142.5	142.5
48(8)	127.5	115.6	98.9	98.9	285.6	215.5	188.4	188.4
60(8)	150.0	130.6	127.2	122.5	353.0	251.8	248.7	229.8
72(8)	147.4	152.9	187.1	126.1	434.6	259.3	311.0	269.4
84(8)	197.6	166.7	292.8	142.3	526.2	307.0	390.2	327.8
96(8)	296.5	163.1	426.7	138.0	617.5	348.1	472.6	375.4
108(8)	378.9	151.0	535.6	166.1	748.8	430.4	595.6	453.5
120(8)	449.9	174.5	652.8	203.5	923.2	480.7	770.4	513.5
Mean errors	197.2	127.3	252.8	120.4	435.4	273.8	330.2	268.3

5. Summaries

Reducing track forecast errors still remains one of top priorities in TC forecasting for forecasters for the sake of improving evacuation planning and disaster mitigation. In order to evaluate the effect of radiance data assimilation by the 3DVAR technique, large-scale GFS flows data assimilation by the SSDA approach and combining the use of both in TC track forecasting, a set of experiments using different data assimilation schemes was performed on Typhoon Megi (2010). The results indicate that AMSU-A radiance data assimilation for the model initialization was effective and better than the SSDA approach in improving the track forecast in the first 3 days. However, the improvements from AMSU-A radiance data assimilation vanish after 3 days because of the rapid decay of the effect of the optimized initial conditions with forecasting time. Assimilating large-scale wind components from GFS global model forecast into the regional model periodically at a preset time by the SSDA approach directly improves the large-scale environmental fields through correcting the large-scale bias of the regional model forecasts and is more effective than the AMSU-A radiance data assimilation in improving TC track forecasts for forecasting periods longer than 72 h. Assimilating both AMSU-A radiance data and large-scale GFS flows inherits the advantages of the both, not only optimizing the model initial conditions but also correcting the large-scale bias of the regional model forecasts, thus leading to the smallest 5-day mean TPEs; however, the improvement could be case dependent for the forecast periods beyond 72 h.

The effect of combining the use of radiance data assimilation and the SSDA approach, however, is still subject to a statistical assessment through a number of TC cases before it can be applied in operational TC track forecasts. Moreover, the setting of some parameters in the SSDA approach, such as the cut-off wave numbers for large scale component and the time interval of SSDA cycle, is also adjustable for different regions. These will be part of our work in the future.

Author Contributions: Data curation, Z.L.; formal analysis, Z.L.; investigation, Z.L.; methodology, Z.L. and S.P.; validation, Z.L.; writing—original draft, Z.L.; writing—review and editing, Z.L. and S.P. All authors have read and agreed to the published version of the manuscript.

Funding: This study is jointly supported by the China Special Fund for Meteorological Research in the Public Interest (Grant No. GYHY201406008), the National Natural Science Foundation of China (Grant No. 41931182, 41521005, and 41676016), the Guangdong Key Project (Grant No. GDME2018B001, 2019BT2H594), the Key Special Project for Introduced Talents Team of Southern Marine Science and Engineering Guangdong Laboratory (Guangzhou) (Grant No. GML2019ZD0303 and GML2019ZD0304), and the Chinese Academy of Sciences (Grant No. ZDRW-XH-2019-2).

Institutional Review Board Statement: Not applicable.

Informed Consent Statement: Not applicable.

Data Availability Statement: The JTWC tropical cyclone best track data were obtained from http://www.usno.navy.mil/NOOC/nmfc-ph/RSS/jtwc/best_tracks/ a few years ago. GFS global model forcast and analyses data were provided by courtesy of NCEP and UCAR (ftp://ftp.ncep.noaa.gov/pub/data/nccf/com/gfs/prod/ and https://rda.ucar.edu/datasets/ds335.0/, accessed on 22 November 2022). AMSU-A Radiance Data are accessible via https://rda.ucar.edu/datasets/ds735.0/, accessed on 22 November 2022.

Acknowledgments: The authors gratefully acknowledge the anonymous reviewers for their excellent comments and efforts, as well as the use of the High Performance Computing Cluster from South China Sea Institute of Oceanology, Chinese Academy of Sciences.

Conflicts of Interest: The authors declare no conflict of interest.

References

1. Goerss, J.S.; Sampson, C.R.; Gross, J.M. A History of Western North Pacific Tropical Cyclone Track Forecast Skill. *Weather Forecast.* **2004**, *19*, 633–638. [CrossRef]
2. Cangialosi, J.P.; Franklin, J.L. *2010 National Hurricane Center Forecast Verfication Report*; NOAA: Silver Spring, MD, USA, 2011; p. 77.
3. Yu, H.; Chan, S.T.; Brown, B.; Kunitsugu, M.; Fukada, E.; Park, S.; Lee, W.; Xu, Y.; Phalla, P.; Sysouphanthavong, B.; et al. Operational tropical cyclone forecast verification practice in the western North Pacific region. *Trop. Cyclone Res. Rev.* **2012**, *1*, 361–372.
4. Qian, C.; Duan, Y.; Ma, S.; Xu, Y. The Current Status and Future Development of China Operational Typhoon Forecasting and Its Key Technologies. *Adv. Meteorol. Sci. Technol.* **2012**, *2*, 6–43.
5. Gao, X. The accuracy of typhoon track forecasts in China hits a new record. *Emerg. Manag. China* **2015**, *11*, 74.
6. Lei, X.; Chen, G.; Zhang, X.; Chen, P.; Yu, H.; Wan, R. Performance of Tropical Cyclone Forecast in Western North Pacific in 2016. In Proceedings of the Forty-ninth Session of ESCAP/WMO Typhoon Committee, Yokohama, Japan, 21–24 February 2017.
7. Bjerknes, V. Dynamic meteorology and hydrographs, Part II. In *Kinematics*; Camegie Institute: New York, NY, USA, 1911.
8. Zhang, S.; Wang, S. Numerical experiments of the prediction of typhoon tracks by using satellite cloud-derived wind. *J. Trop. Meteor.* **1999**, *15*, 347–355.
9. Zhou, X.; Zhu, Y. Numerical study on the effect of asymmetric diabatic heating on tropical cyclone motion. *Q. J. Appl. Meteorol.* **1999**, *10*, 284–292.
10. Kidder, S.Q.; Goldberg, M.D.; Zehr, R.M.; Demaria, M.; Purdom, J.F.W.; Velden, C.S.; Grody, N.C.; Kusselson, S.J. Satellite Analysis of Tropical Cyclones Using the Advanced Microwave Sounding Unit (AMSU). *Bull. Amer. Meteor. Soc.* **2000**, *81*, 1241–1259. [CrossRef]
11. Zhang, H. Chapter 5, the Application of the ATOVS Radiance Microwave Data (I)—The Satellite Observation of the Typhoon Structure in Northwest Pacific, the Direct Assimilation Method and Application Research of the ATOVS Radiance Data. Ph.D. Thesis, Lanzhou University, Lanzhou, China, 2003; pp. 62–86.
12. Cui, L.; Sun, J.; Qi, L.; Lei, T. Application of ATOVS Radiance-Bias Correction to Typhoon Track Prediction with Ensemble Kalman Filter Data Assimilation. *Adv. Atmos. Sci.* **2011**, *28*, 178–186. [CrossRef]
13. Liu, Q.; Weng, F. Radiance assimilation in studying Hurricane Katrina. *Geophys. Res. Lett.* **2006**, *33*, L22811. [CrossRef]
14. Sandeep, S.; Chandrasekar, A.; Singh, D. The impact of assimilation of AMSU data for the prediction of a tropical cyclone over India using amesoscale model. *Int. J. Remote Sens.* **2006**, *27*, 4621–4653. [CrossRef]
15. Zhang, H.; Xue, J.; Zhu, G.; Zhuang, S.; Wu, X.; Zhang, F. Application of Direct Assimilation of ATOVS Microwave Radiances to Typhoon Track Prediction. *Adv. Atmos. Sci.* **2004**, *21*, 283–290. [CrossRef]
16. Zhang, M.; Zupanski, M.; Kim, M.-J. Assimilating AMSU-A Radiances in the TC Core Area with NOAA Operational HWRF (2011) and a Hybrid Data Assimilation System: Danielle (2010). *Mon. Weather Rev.* **2013**, *141*, 3889–3907. [CrossRef]
17. Peng, S.; Xie, L.; Liu, B.; Semazzi, F. Application of Scale-Selective Data Assimilation to Regional Climate Modeling and Prediction. *Mon. Weather Rev.* **2010**, *138*, 1307–1318. [CrossRef]
18. Lai, Z.; Hao, S.; Peng, S.; Liu, B.; Gu, X.; Qian, Y.-K. On Improving Tropical Cyclone Track Forecasts Using a Scale-Selective Data Assimilation Approach: A Case Study. *Nat. Hazards* **2014**, *73*, 1353–1368. [CrossRef]

19. Liu, B.; Xie, L. A Scale-Selective Data Assimilation Approach to Improving Tropical Cyclone Track and Intensity Forecasts in a Limited-Area Model: A Case Study of Hurricane Felix (2007). *Weather. Forecast.* **2012**, *27*, 124–140. [CrossRef]
20. Xie, L.; Liu, B.; Peng, S. Application of scale-selective data assimilation to tropical cyclone track simulation. *J. Geophys. Res.* **2010**, *115*, D17105. [CrossRef]
21. Chou, J. Some properties of operators and the effect of Initial condition. *Acta. Meteorol. Sin.* **1983**, *41*, 385–392.
22. Ding, W.; Wan, Q.; Yan, J.; Huang, Y.; Chen, Z. Impact of The Initialization on Mesoscale Model Predition in South China. *J. Trop. Meteor.* **2006**, *22*, 10–17.
23. Skamarock, W.C.; Klemp, J.B.; Dudhia, J.; Gill, D.O.; Barker, D.M.; Duda, M.G.; Huang, X.Y.; Wang, W.; Powers, J.G. A Description of the Advanced Research WRF Version 3. In *NCAR Tech Note*; Mesoscale and Microscale Meteorology Division, National Center for Atmospheric Research: Boulder, CO, USA, 2008.
24. Wang, W.; Bruyere, C.; Duda, M.; Dudhia, J.; Gill, D.; Kavulich, M.; Keene, K.; Lin, H.-C.; Michalakes, J.; Rizvi, S.; et al. Advanced Research WRF (ARW) Version 3 Modeling system user's guide. In *ARW Tech Note*; Mesoscale and Microscale Meteorology Division, National Center for Atmospheric Research: Boulder, CO, USA, 2014.
25. Han, Y.; Delst, P.V.; Liu, Q.; Weng, F.; Yan, B.; Derber, J. *User's Guider to the JCSDA Community Radiative Transfer Model (Beta Version)*; Joint Center for Satellite Data Assimilation: Camp Springs, MD, USA, 2005.
26. Han, Y.; Delst, P.V.; Liu, Q.; Weng, F.; Yan, B.; Treadon, R.; Derber, J. JCSDA Community Radiative Transfer Model (CRTM), Version 1. In *NOAA Technical Report*; NESDIS: Silver Spring, MD, USA, 2006; Volume 122, p. 33.
27. Kleespies, T.J.; Delst, P.V.; McMillin, L.M.; Derber, J. Atmospheric transmittance of an absorbing gas. 6. An OPTRAN status report and introduction to the NESDIS/NCEP Community Radiative Transfer Model. *Appl. Opt.* **2004**, *43*, 3103–3109. [CrossRef]
28. McMillin, L.M.; Core, L.J.; Kleespies, T.J. Atmospheric transmittance of an absorbing gas. 5: Improvements to the OPTRAN approach. *Appl. Opt.* **1995**, *34*, 8396–8399. [CrossRef]
29. Liu, Q.; Weng, F. Advanced Doubling-Adding Method for Radiative Transfer in Planetary Atmospheres. *J. Atmos. Sci.* **2006**, *63*, 3459–3465. [CrossRef]
30. Derber, J.C.; Wu, W.S. The use of TOVS cloud-cleared radiances in the NCEP SSI analysis system. *Mon. Weather Rev.* **1998**, *126*, 2287–2299. [CrossRef]
31. Eyre, J.R. A bias correction scheme for simulated TOVS brightness temperatures. *Tech. Memo.* **1992**, *176*, 81–109.
32. Harris, B.A.; Kelly, G. A satellite radiance bias correction scheme for data assimilation. *Quart. J. R. Meteor. Soc.* **2001**, *127*, 1453–1468. [CrossRef]
33. Ferraro, R.R.; Weng, F.; Grody, N.C.; Zhao, L. Precipitation characteristics over land from the NOAA-15 AMSU sensor. *Geophys. Res. Lett.* **2000**, *27*, 2669–2672. [CrossRef]
34. Grody, N.C.; Zhao, L.; Ferraro, R.R.; Weng, F.; Boers, R. Determination of precipitable water and cloud liquid water over oceans from the NOAA-15 advanced microwave sounding unit. *J. Geophys Res.* **2001**, *106*, 2943–2953. [CrossRef]
35. Qian, C.; Zhang, F.; Green, B.W. Probabilistic Evaluation of the Dynamics and Prediction of Supertyphoon Megi (2010). *Weather Forecast.* **2013**, *28*, 1562–1577. [CrossRef]
36. Bougeault, P.; Lacarrère, P. Parameterization of orography-induced turbulence in a mesobeta-scale model. *Mon. Weather Rev.* **1989**, *117*, 1872–1890. [CrossRef]
37. Kain, J.S.; Fritsch, J.M. A one-dimensional entraining detraining plume model and its application in convective parameterization. *J. Atmos. Sci.* **1990**, *47*, 2784–2802. [CrossRef]
38. Ferrier, B.S.; Jin, Y.; Lin, Y.; Black, T.; Rogers, E.; DiMego, G. Implementation of a new grid-scale cloud and precipitation scheme in the NCEP Eta model; Preprints. In Proceedings of the 19th Conference on Weather Analysis and Forecasting/15th Conference on Numerical Weather Prediction, San Antonio, TX, USA, 12–16 August 2002; American Meteorological Society: Boston, MA, USA; pp. 280–283.
39. Dudhia, J. Numerical study of convection observed during the Winter Monsoon Experiment using a mesoscale two-dimensional model. *J. Atmos. Sci.* **1989**, *46*, 3077–3107. [CrossRef]
40. Mlawer, E.J.; Taubman, S.J.; Brown, P.D.; Iacono, M.J.; Clough, S.A. Radiative transfer for inhomogeneous atmosphere: RRTM, a validated correlated-k model for the longwave. *J. Geophys. Res.* **1997**, *102*, 16663–16682. [CrossRef]
41. Neumann, C.J.; Pelissier, J.M. An analysis of Atlantic tropical cyclone forecast errors, 1970-1979. *Mon. Weather Rev.* **1981**, *109*, 1248–1266. [CrossRef]
42. Powell, M.D.; Aberson, S.D. Accuracy of United States Tropical Cyclone Landfall Forecasts in the Atlantic Basin (1976-2000). *Bull. Am. Meteor. Soc.* **2001**, *82*, 2749–2767. [CrossRef]
43. Chen, L.; Xu, X.; Xie, Y.; Li, W. The Effect of Tropical Cyclone Asymmetric Thermodynamic Structure on Its Unusual Motion. *Sci. Atmos. Sin.* **1997**, *21*, 83–90.
44. Chan, J.C.L.; Gray, W.M. Tropical cyclone movement and surrounding flow relationships. *Mon. Weather Rev.* **1982**, *110*, 1354–1374. [CrossRef]
45. Deng, G.; Zhou, Y.-S.; Liu, L.-P. Use of a new steering flow method to predict tropical cyclone motion. *J. Trop. Meteor.* **2010**, *16*, 154–159.
46. Dong, K.; Neumann, C.J. On the relative motion of binary tropical cyclones. *Mon. Weather Rev.* **1983**, *111*, 945–953. [CrossRef]

47. Franklin, J. Dropwindsonde observations of the environmental flow of Hurricane Josephine (1984)—Relationships to vortex motion. *Mon. Weather Rev.* **1990**, *118*, 2732–2744. [CrossRef]
48. Wang, B.; Elsberry, R.L.; Wang, Y.; Wu, L. Dynamics in Tropical Cyclone Motion: A Review. *Chin. J. Atmos. Sci.* **1998**, *22*, 416–434.

atmosphere

MDPI

Article

Grid-to-Point Deep-Learning Error Correction for the Surface Weather Forecasts of a Fine-Scale Numerical Weather Prediction System

Yu Qin [1], Yubao Liu [1,*], Xinyu Jiang [2], Li Yang [1], Haixiang Xu [3,*], Yueqin Shi [4] and Zhaoyang Huo [1]

1 Precision Regional Earth Modeling and Information Center, Nanjing University of Information Science and Technology, Nanjing 210044, China
2 Nanjing Xinda Institute of Meteorological Science and Technology, Nanjing 210044, China
3 Jibei Electric Power Corporation of the State Grid Corporation of China, Beijing 100031, China
4 Key Laboratory for Cloud Physics of China Meteorological Administration, Beijing 100081, China
* Correspondence: ybliu@nuist.edu.cn (Y.L.); xhx0617@foxmail.com (H.X.)

Abstract: Forecasts of numerical weather prediction models unavoidably contain errors, and it is a common practice to post-process the model output and correct the error for the proper use of the forecasts. This study develops a grid-to-multipoint (G2N) model output error correction scheme which extracts model spatial features and corrects multistation forecasts simultaneously. The model was tested for an operational high-resolution model system, the precision rapid update forecasting system (PRUFS) model, running for East China at 3 km grid intervals. The variables studied include 2 m temperature, 2 m relative humidity, and 10 m wind speed at 311 standard ground-based weather stations. The dataset for training G2N is a year of historical PRUFS model outputs and the surface observations of the same period and the assessment of the G2N performance are based on the output of two months of real-time G2N. The verification of the real-time results shows that G2N reduced RMSEs of the 2 m temperature, 2 m relative humidity, and 10 m wind speed forecast errors of the PRUFS model by 19%, 24%, and 42%, respectively. Sensitivity analysis reveals that increasing the number of the target stations for simultaneous correction helps to improve the model performance and reduces the computational cost as well indicating that enhancing the loss function with spatial regional meteorological structure is helpful. On the other hand, adequately selecting the size of influencing grid areas of the model input is also important for G2N to incorporate enough spatial features of model forecasts but not to include the information from the grids far from the correcting areas. G2N is a highly efficient and effective tool that can be readily implemented for real-time regional NWP models.

Keywords: deep learning; NWP; post-processing; grid to stations; forecast error correction

Citation: Qin, Y.; Liu, Y.; Jiang, X.; Yang, L.; Xu, H.; Shi, Y.; Huo, Z. Grid-to-Point Deep-Learning Error Correction for the Surface Weather Forecasts of a Fine-Scale Numerical Weather Prediction System. *Atmosphere* **2023**, *14*, 145. https://doi.org/10.3390/atmos14010145

Academic Editors: Xiaolei Zou and Zhemin Tan

Received: 8 December 2022
Revised: 28 December 2022
Accepted: 30 December 2022
Published: 9 January 2023

1. Introduction

Accurate weather forecasting is crucial for the development of society and economy, and human activities and safety. With the rapid development of atmospheric modeling, observation systems, and high-performance computing, numerical weather forecasting capability and accuracy have been improved significantly [1]. Nevertheless, due to the chaotic nature of the weather processes and unavoidable uncertainties in various numerical model components, numerical weather forecasts contain significant errors. Therefore, a correction of model forecast errors is necessary to improve the applications of the model outputs. Several statistical post-processing techniques have been developed for model forecast correction. Among them, model output statistics (MOS) [2] and perfect procedures (PP) [3] are widely used in the current numerical model forecast corrections. While the PP approach achieves a correction by establishing a linear statistical relationship between observations and the NWP model analysis, the MOS method pairs observation data with the output of NWP and then obtains the correction based on linear regression. In addition, the Kalman filter technique has also been applied for bias correction. Unlike MOS, the Kalman filter technique adjusts its filter coefficients in real time [4,5]. An analog ensemble method proposed by [6] showed an improved capability and has been successfully applied for wind and solar energy forecasting [7–9].

In addition to the statistical model output post-processing, several machine learning techniques have been investigated and have demonstrated benefits and great potential [10]. Li et al. [11] proposed a model output machine learning scheme (MOML) that uses multiple linear regression as well as random forest methods to correct the 2 m temperature in the ECMWF model for the Beijing area. Compared with MOS, which works for single-station site correction, MOML incorporates the spatial and temporal structure of the grid data. Cho et al. [12] used machine learning methods including random forests, support vector regression, artificial neural networks, and multimodel ensembles to establish statistical relationships between predictors and predictands to correct the model forecasts of extreme temperatures in urban areas, demonstrating some ability of the machine learning algorithms for modeling nonlinearities of the weather processes.

In the last decade, convolution neural network (CNN)-based deep-learning technology has made significant strides and offers a natural upgrade to the traditional model output post-processing methods. Rasp et al. [13] proposed a neural network-based model for correcting the 2 m temperature model forecasts in Germany. Han et al. [14] proposed a CU-net model to correct the gridded forecasts of four weather variables of the European Centre for Medium-Range Weather Forecast Integrated Forecasting System global model (ECMWF-IFS): 2 m temperature, 2 m relative humidity, 10 m wind speed, and 10 m wind direction. Their approach turned post-processing into an image transformation problem in the context of image processing. Zhang et al. [15] constructed Dense-CUnet and Fuse-CUnet models based on the CU-net model proposed by Han et al. [14]. By introducing a dense convolution module and a variety of meteorological elements and terrain features into the model, they were able to improve the results of Han et al. [14].

In the last decade, fine-grid numerical weather forecasts with grid intervals of 1–3 km became popular. It is very desirable to explore the deep-learning approaches to extract the meso- and small-scale features of weather circulations simulated by high-resolution numerical models and applied them for model forecast error correction. In theory, multi-scale features of weather circulations contain more information about the model forecast errors that the traditional error correction models, which were based on single-point time-sequence data, could not include. In this study, we developed a grid-to-multipoint (G2N) deep-learning model for correcting the 2 m temperature, 2 m relative humidity, and 10 m wind speed forecasts of a rapid-updating high-resolution weather model (named PRUFS: Precision Rapid Updated Forecast System) at multiple weather stations in East China. Sensitivity tests were conducted to study the impact of the scales of the input area (the model grids) and the target area, i.e., the number of target stations. The former helps to determine the scale of the mesoscale circulation features for the optimal error corrections for the target stations and the latter assesses the effect of the number of target stations for simultaneous error correction with multitasking learning.

2. Data and Method

2.1. Data Description

PRUFS is a precision rapid update forecasting system based on the U.S. Weather Research and Forecasting Model (WRF) and four-dimensional data assimilation (FDDA) technology [16]. The model uses the analysis and forecast fields from the NOAA/NCEP Global Forecast System (GFS) to generate boundary conditions and assimilated various observations in the region during the model initialization. PRUFS runs with a four-nested level with a 3 km grid covering east-central China (Figure 1) since the beginning of 2020. The model system runs at hourly cycles and in each cycle, it generates 24 h forecasts, outputting at hourly intervals.

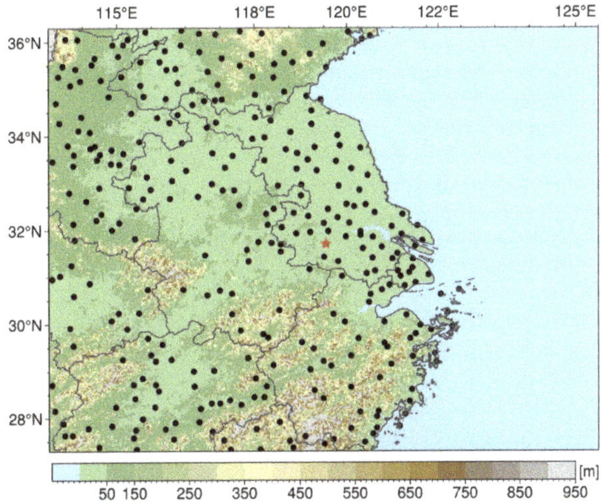

Figure 1. The 3 km horizontal resolution area of PRUFS and the distribution of the 311 automatic weather stations (black dots). The background color shows the height of the terrain, and the red star is the station "Jintan", to be discussed in the later section.

The weather observations were obtained from the state standard ground-based weather stations of the China Meteorological Administration. In this paper, the grid forecast of the PRUFS 3 km domain is corrected by using ground-based meteorological station data. The PRUFS model output and observations are collected for the period from August 2020 to December 2021. The data during the equipment maintenance period September–October 2021 are excluded. The selected computational domain is $(lon \in [113.5° \text{ E}, 125.5° \text{ E}], \text{lat} \in [27.3° \text{ N}, 36.3° \text{ N}])$, with 301×401 grid points. The observation sites and topographic information are shown in Figure 1. The meteorological elements to be corrected are the 24 h hourly forecasts of 2 m temperature (T_2), 2 m relative humidity (RH_2), and 10 m wind speed (W_{10}), generated by PRUFS. With 24 forecast cycles each day, there is a total of 576 model samples (i.e., 24×24) per day.

2.2. G2N Model

Unlike the traditional MOS and PP models that are based on individual station time-sequence weather observations and the model forecast values interpolated at the weather stations (i.e., One-2-One), G2N uses the two-dimensional gridded meteorological information of the model forecasts. Thus, G2N can exploit the two-dimensional gridded meteorological structures (G) of the model forecasts for multiple-site (N sites) weather forecast error correction (i.e., G2N). The gridded model forecast variables can be considered as images in the field of image processing, with the forecast at each grid point corresponding to a pixel. Therefore, the characteristics of meteorological structures can be extracted using image feature engineering technology.

AlexNet is among the simple and effective image-processing deep-learning network models based on a convolutional neural network (CNN). It was first proposed by Krizhevsky et al. [17]. AlexNet consists of five convolutional layers, six pooling layers, and three fully connected layers. The AlexNet model contains a local response normalization layer (LRN) for amplification or suppression of neural activation, and it works together with Dropout methods to prevent the network from overfitting. We construct G2N based on the AlexNet framework.

The change in data distribution due to a change in the network parameters during training is called Internal Covariate Shift [18]. In AlexNet, to avoid the problem of Internal Covariate Shift that slows down convergence and even degrades network performance, a batch normalization [19] was introduced to replace LRN and Dropout, making the network more robust to the changes in the network parameters and activation functions during the training. BN also solves the problem of gradient disappearance and reduces the negative impact of ICS. The BN layer averages the input values of neurons in the layer to redistribute them to a normal distribution with a mean of 0 and a variance of 1. This allows the increasingly distorted distribution to return to the standard distribution. In this study, to adapt AlexNet for weather variables processing, we replaced LRN and Dropout with BN layers to use it as the core of G2N.

The structure of the G2N model is shown in Figure 2a. G2N takes 2D grid model data as input and has 6 convolutional layers, 4 pooling layers, and 3 fully connected layers. The convolutional layer and pooling layer play a role in extracting 2D features of the model forecasts and the fully connected layer integrates the local information and flattens the feature information. Firstly, G2N convolves and pools the input weather forecast data several times to extract the number of multiscale features. Then, three full-connected operations are performed to finally get the correction results at the weather stations. Each convolutional layer in the figure is followed by a BN layer operation, and then the ReLU activation function is connected. In the fully connected layer, ReLU is also used as the activation function.

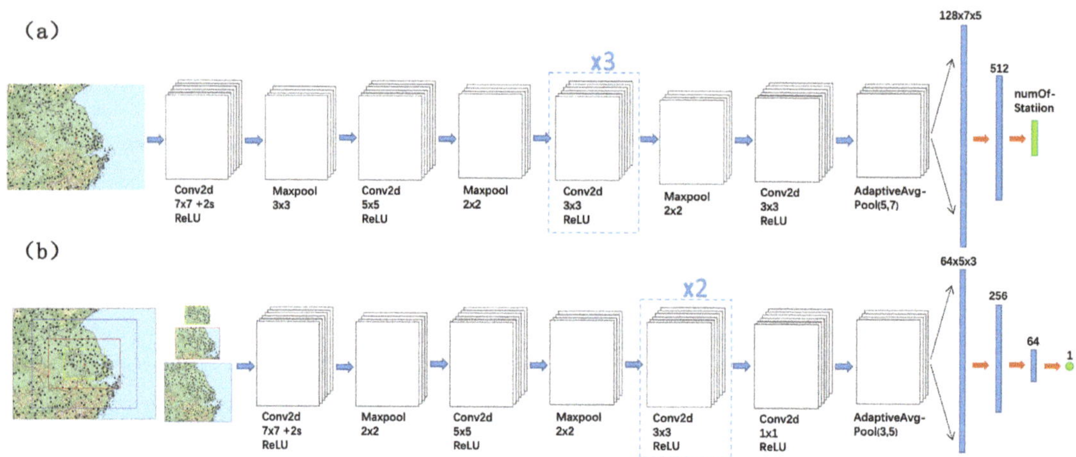

Figure 2. Multisite (G2N, (**a**)) and single-site (G2-One, (**b**)) forecast error correction models for grid forecasting. (**a**) Whole grid area as input for multisite correction of 311 sites. (**b**) Different proportions of grid regions as input for single-point correction experiments.

With the G2N grid-to-station deep-learning architecture, the sizes of model grid forecasts (G) and the numbers of sites (N, weather stations) to be corrected are the two most important model configuration parameters to be considered. Thus, we conducted extensive sensitivity modeling experiments to study the impact of different G and N. Among them, one special case is N = 1, where we let the model work to correct the forecast at only a single site with an input of the model 2D grid data. This is named G2-One (Figure 2b) and is used to study the influence of the G size on the correction results. When changing the size of the input 2D grid data for the G2-One tests, a proper simplification of the G2N model is needed, given in Figure 2b. Adaptive pooling is used in the last layer of the network model, allowing the model to receive flexible sizes of inputs.

In the training process, it is easy to appear over-fitting phenomena with small training errors and large generalization errors. Regularization is a method to solve the overfitting problem in machine learning. The model was trained using L2 regularized weight decay to control overfitting [20]. Adam was chosen as the optimizer for the gradient descent process of the neural network. The loss function is defined as the mean squared error (*MSE*):

$$Loss = MSE = \frac{1}{M}\frac{1}{N}\sum_{m=1}^{M}\sum_{n=1}^{N}(output_n^m - observe_n^m)^2 \quad (1)$$

where m and n represent the sample number and site number and M and N the batch size and site number, respectively. Each epoch traverses the training set's data multiple times, updating the neural network's parameters with each iteration's batch size. The loss functions of the training and validation datasets are computed after each epoch. In the later examples, all calculations are terminated after 200 epochs of training.

2.3. Data Pre-Processing and Dataset Partitioning

The PRUFS model output and surface observations from August 2020 to August 2021 were selected to construct the training and validation set and the real-time operational data of PRUFS and surface observations from November and December 2021 were used as the test dataset to evaluate the G2N model. The input data of G2N are the PRUFS hourly gridded 0–24 h forecasts, and the training labels are the surface weather observations at the corresponding time of each forecast of a given cycle and forecast time. A bilinear interpolation method was used to interpolate the PRUFS model forecast to 311 surface weather stations to match the observations (Figure 3). The sample data with empty or invalid values were rejected. The interpolated PRUFS forecasts, G2N outputs, and observations are used to calculate the loss function during the G2N model training and evaluate the model performances.

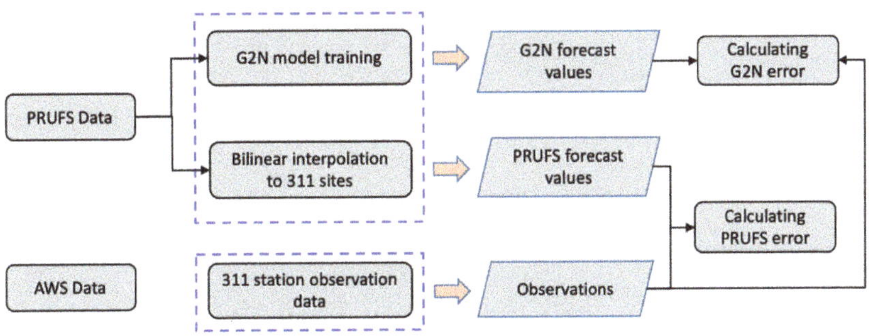

Figure 3. Flow chart of data pre-processing.

After the pre-processing, the number of valid labeled data samples is 167,683,000. The dataset was divided into training and validation sets at an 8:2 ratio. To prevent "information leakage", the 20% validation sets are randomly selected in blocks of continuous 24 forecast cycles, i.e., a whole-day period.

2.4. Model Evaluation Statistics

The model evaluation is conducted by computing the root mean square error (RMSE), systematic bias (BIAS), and the Pearson correlation coefficient (CC), which are the most used metrics for weather forecasting. The root mean square error is calculated as

$$RMSE = \sqrt{MSE} = \sqrt{\frac{1}{M}\frac{1}{N}\sum_{m=1}^{M}\sum_{n=1}^{N}(output_n^m - observe_n^m)^2} \quad (2)$$

The systematic bias (BIAS) formula is

$$BIAS = \frac{1}{M}\frac{1}{N}\sum_{m=1}^{M}\sum_{n=1}^{N}(output_n^m - observe_n^m) \quad (3)$$

The Pearson correlation coefficient (CC) formula is

$$Corr(output, observe) = \frac{\sum(output - \overline{output})\left(observe - \overline{observe}\right)}{\sqrt{\sum(output - \overline{output})^2\left(observe - \overline{observe}\right)^2}} \quad (4)$$

3. Results and Analysis

3.1. Overall Test Results

Model evaluation is based on the datasets collected during real-time G2N applications along with the PRUFS forecast during November and December 2021. The root mean squared error (RMSE) of 2 m temperature, 2 m relative humidity, and 10 m wind speed at 311 stations of the PRUFS forecast and G2N correction for the 0–24 h forecasts were calculated and the results are presented in Table 1. The percentages of improvement (POI) of the corrected forecast accuracy are also given. It can be found that the convolutional-based G2N model corrected the model forecast errors effectively. The RMSE of all three variables is reduced significantly. The POI of the temperature forecast accuracy increased by 19.4%, the relative humidity by 24.5%, and the wind speed forecast has the greatest enhancement rate of 42.8% by the G2N model.

Table 1. The RMSE and improvement percentages (IP) (see Equation (6)) of the 2 m temperature, 2 m relative humidity, and 10 m wind speed 0–24 h forecasts corrected by G2N, averaged at all stations.

Element	Test Dataset		
	PRUFS	**G2N**	**IP**
2 m-T	2.22	1.79	19.4%
2 m-RH	16.25	12.27	24.5%
10 m-WD	1.66	0.95	42.8%

The hour-by-hour errors of the PRUFS forecasts and G2N corrections at each station were computed, and the distribution characteristics of the errors were presented in Figure 4. The error is defined as

$$Error = forecast - observation \quad (5)$$

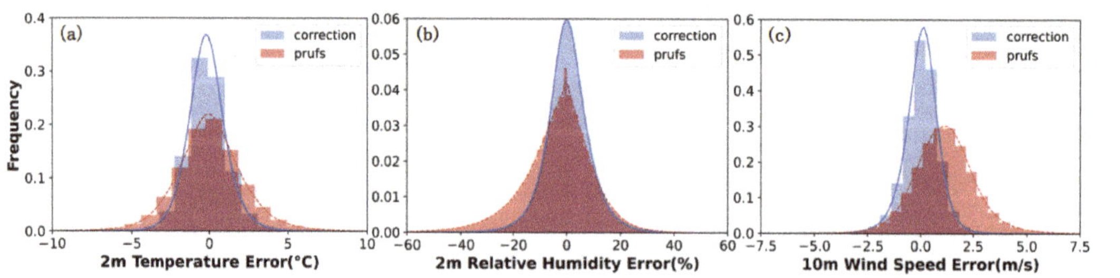

Figure 4. Normalized frequency distributions of the PRUFS forecast errors and the G2N corrected forecast errors. The (**a**) 2 m temperature, (**b**) 2 m relative humidity, and (**c**) 10 m wind speed, with error bins at 1 °C, 1%, and 0.5 m/s, respectively.

Figure 4 shows that the bias of the 2 m temperature, 2 m relative humidity, and 10 m wind speed of the PRUFS forecast are 0.41, −3.43, and 1.15, respectively. After the G2N correction, they are reduced to −0.15, 0.37, and 0.03, respectively. The distribution of the errors of the G2N-corrected 2 m temperature, 2 m relative humidity, and 10 m wind speed are approximately symmetric about and shrunk to the 0-error point, indicating that G2N is effective in eliminating both negative and positive systematic errors. Notably, the distribution of the 10 m wind speed forecast errors by PRUFS shows an overall apparent positive bias. Several previous studies reported similar results [21–23]. G2N effectively corrected such wind speed biases. The overall systematic errors of the 2 m temperature and 2 m humidity forecasts of PRUFS were not as substantial as the wind. Following the G2N correction, the number of samples with larger temperature and humidity errors also significantly decreased.

To further examine the details of the error properties, density scatter plots of the forecast–observation pairs of the PRUFS forecast and the G2N correction are plotted and the results are shown in Figure 5. The samples include all stations and forecast times during the test period.

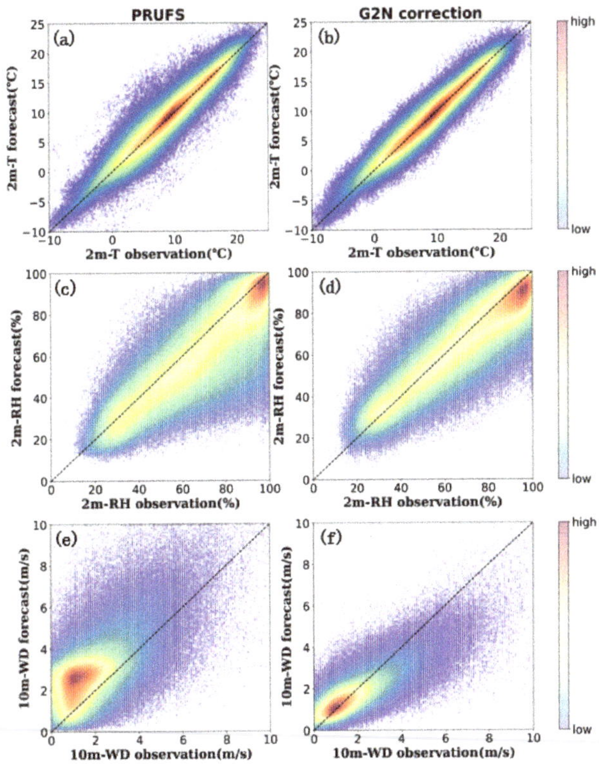

Figure 5. Density scatter plots of the forecast–observation pairs of the PRUFS forecasts (**1st column**) and the G2N correction (**2nd column**). The (**a**,**b**) 2 m temperature, (**c**,**d**) 2 m relative humidity, and (**e**,**f**) 10 m wind speed.

The forecast–observation pairs of all three variables converge more compactly around the black diagonal lines after the correction, i.e., the corrected forecast is closer to the observed value. For example, the variance of the wind speed (Figure 5e,f) is reduced from 2.21 to 0.7. PRUFS underestimated RH (Figure 5c) with the samples skewed to the right of the diagonal and it is removed in the G2N corrected data (Figure 5d), resulting in more centralized and symmetrical error distributions around the diagonal. Similarly, the

10 m wind forecasts (Figure 5e,f) were overall largely overpredicted by PRUFS and G2N dramatically eliminated this bias and the overall errors too.

3.2. Forecast Lead Time and Daily Variation

To analyze the performances of the G2N model for correcting the forecast at different lead times and different times of the day, the samples in the test dataset were grouped according to the forecast lead time and times of day, respectively. After grouping, the error statistics were analyzed for the times in each group. Figure 6a,c,e show the forecast scores of PRUFS and G2N for the 0–24 h lead times. As the forecast lead time increased from 1 to 24 h, the 2 m temperature RMSE increased from approximately 2 to 2.5 °C, the 2 m relative humidity RMSE increased from approximately 12 to 15%, and the 10 m wind speed RMSE increased from 1.8 to 2.0 m/s. After the G2N correction, the RMSE of these three variables is reduced by approximately 1.3 °C, 8%, and 0.8 m/s. Furthermore, G2N can correct the larger errors at longer lead times more effectively for the 24 h forecasts examined here, with the RMSE of the corrected 2 m temperature forecasts increasing only by 0.3 °C, and the corrected 2 m relative humidity and 10 m wind speed errors nearly unchanged with the lead time. This result shows that the G2N model can automatically adjust the magnitude of the error correction according to the error growth for the forecast lead times examined herein.

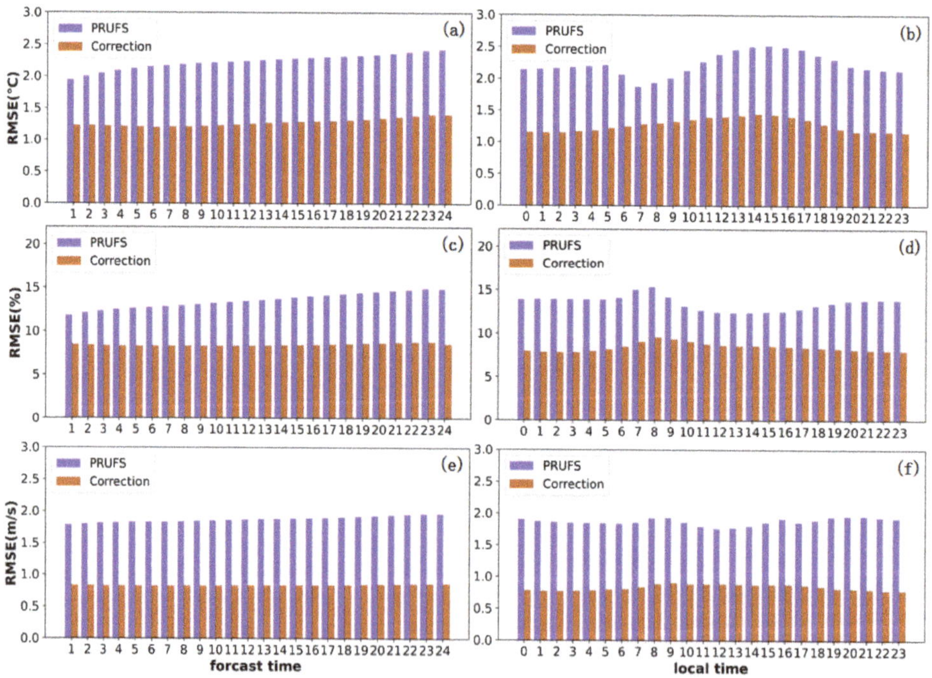

Figure 6. The variation of the RMSE of the PRUFS 0–24 h forecasts and the G2N correction for 2 m temperature (**a**,**b**), 2 m relative humidity (**c**,**d**), and 10 m wind speed (**e**,**f**) with forecast lead time (**left panels**) and diurnal variation (**right panels**); The horizontal coordinate of the left panels is the forecast lead time and that of the right panels is the local time.

The results of the RMSE of the three meteorological variables at different times of day for the PRUFS forecast and G2N correction are shown in Figure 6b,d,f. The errors of the PRUFS forecasts display significant diurnal variations. The RMSE of the 2 m temperature forecasts reached a peak at 15:00 LT, and a minimum at 7:00 LT. The evolution trend of the 2 m relative humidity errors is approximately opposite to the temperature errors, with an error peak at 8:00 LT, and a valley at around noon LT. The error of 10 m wind speed is less fluctuated. After the G2N corrections, the RMSE of the forecasts of all three variables were significantly reduced at all times of day, with a diurnal variation trend generally consistent with the PRUFS forecasts. This suggests that the diurnal variations of physical processes that caused the PRUFS model errors may also lead to some difficulties for the G2N model.

3.3. Spatial Distribution of the G2N Performances

To analyze the horizontal distribution of the G2N performances, the RMSE of 2 m temperature, 2 m relative humidity, and 10 m wind speed were calculated for each station for all samples of the test dataset. The PRUFS forecast RMSE for all three variables was significantly reduced (Figure 7) at all stations by G2N. The PRUFS temperature prediction errors at several stations were higher than 7.55 °C. After the G2N correction, they were reduced to less than 2.0 °C. In general, the central regions of the domain achieved the best correction results, where the overall error of the PRUFS relative humidity forecast is reduced from ~16% to less than 13% by the G2N model, and the wind speed error from ~1.3–1.5 m/s to below approximately 0.5 m/s. Furthermore, the G2N model is more effective for the stations where the PRUFS forecast errors are larger. At most stations, the G2N model gains IP values over 60% for wind speeds. For relative humidity, there are approximately half of the stations yield 60% IP.G2N performed slightly worse at the southern part of the domain and the northwest corner because the peripheral spatial information of the sites at and across the boundary is not included. The correction effect of wind speed is most effective throughout the domain.

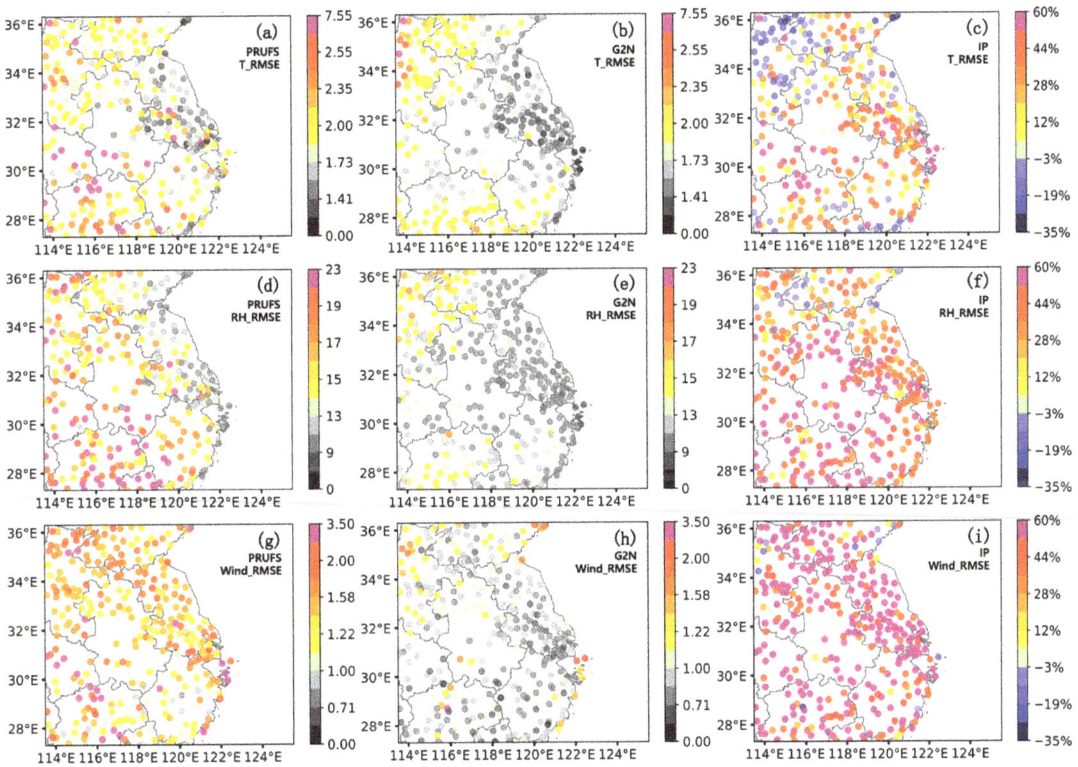

Figure 7. The RMSE of the PRUFS forecasts (**left panels**) and the G2N correction (**middle panels**) and the corresponding G2N improvement percentages (**right panels**, %) of 2 m temperature (**a**–**c**), 2 m relative humidity (**d**–**f**), and 10 m wind speed (**g**–**i**).

To quantitatively compare the G2N effect among the stations, the improvement percentage (IP) of the RMSE of the G2N correction over the PRUFS forecasts was calculated for each site as follows

$$IP = \frac{PRUFS\_forecast_{RMSE} - G2N\_correction_{RMSE}}{PRUFS\_forecast_{RMSE}} \times 100 \tag{6}$$

Figure 7 shows that more than half of the stations gain an IP over 30% for all three meteorological elements although some stations in the northwest marginal area and the southern boundary show a negative effect. A lack of spatial feature information at the edges may impose an unfavorable effect on these sites. Again, the G2N model is most effective for correcting the wind forecast errors, with IPs at most stations larger than 15% and more than a half gained over 50%.

Figures 8 and 9 show the bias and the Pearson correlation coefficients for the forecasts of the three meteorological variables, respectively. The bias of the PRUFS model forecast is significantly reduced by the G2N correction. The PRUFS model temperature forecasts have over 1.3 °C bias at several clustered surface weather stations. They are significantly reduced by the G2N corrections, to less than 0.5 °C. The PRUFS 2 m relative humidity forecast has an overall negative bias (approximately −6.92%) and its wind forecast has a positive bias (approximately 0.95 m/s), and they are reduced to −2.91% and −0.18 m/s, respectively, after the G2N correction.

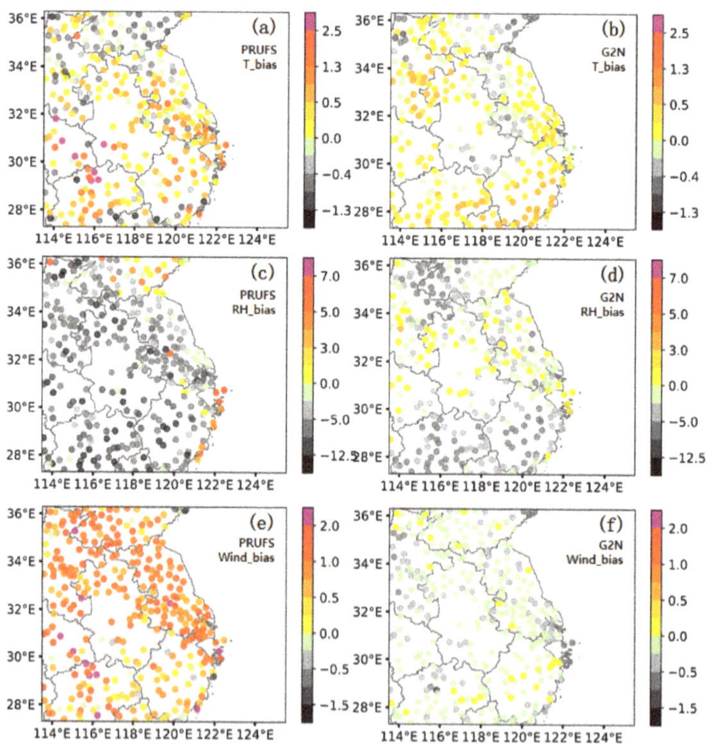

Figure 8. Horizontal distribution of the bias of the PRUFS forecasts (**left panels**) and the G2N correction (**right panels**) of 2 m temperature (**a**,**b**), 2 m relative humidity (**c**,**d**), and 10 m wind speed (**e**,**f**).

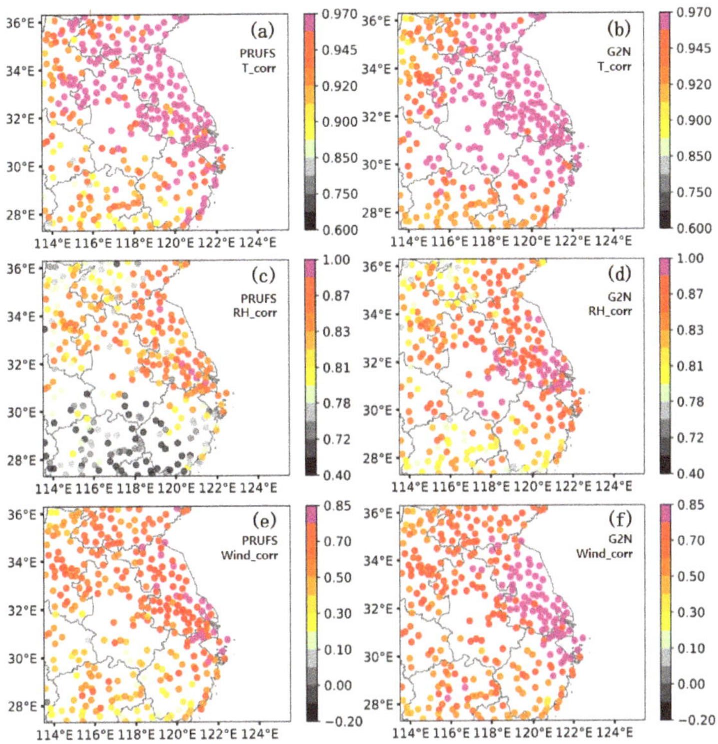

Figure 9. Horizontal distribution of the Pearson correlation coefficients concerning the observations of the PRUFS forecasts (**left panels**) and the G2N correction (**right panels**) of 2 m temperature (**a**,**b**), 2 m relative humidity (**c**,**d**), and 10 m wind speed (**e**,**f**).

In comparison with the PRUFS forecast, the correlation between the G2N corrected forecasts and the observations is also significantly improved for all three meteorological variables (Figure 9). The correlation coefficient (r) can be assessed by the general guidelines proposed by Cohen et al. [24,25], $|r| < 0.3$ is defined as weakly correlated; $0.3 < |r| < 0.6$ as moderately correlated; $0.6 < |r| < 0.8$ as strongly correlated; and $0.8 < |r| < 1$ as extremely strongly correlated.

All station average correlation coefficient for the PRUFS temperature forecast was approximately 0.946 and it reached 0.952 after the G2N correction. For relative humidity, the all-station average correlation coefficient was 0.793 for the PRUFS forecast and ~95% of the stations are strongly correlated. After the G2N correction, the all-station average correlation coefficient was improved to 0.852 and the stations with strong correlation increased to approximately 100%. For the wind, all station average correlation coefficient for the PRUFS forecast was 0.626, the proportion of strong correlation sites is 69%, and the proportion of strong correlation sites was 6.8%. After the G2N correction, all station average correlation coefficient of the corrected sites increased to 0.739, the percentage of strongly correlated sites rose to 92.6%, and the percentage of very strongly correlated sites rose to 33%.

4. Sensitivity Analysis of G2N to the Inputs and Learning Areas

G2N realized the forecast error correction by projecting the PRUFS model grid forecasts to the observation sites. Two natural questions are: what is the optimal size (area) of the gridded input data (G) and what is the proper number of stations for the objective function (loss function)? The size of the input data (G) determines the features of the multiscale characteristics of the PRUFS model forecast that are extracted to infer the information related to the target site. On the other hand, the number of sites (N) of the objective function is a multitask learning problem [26–29], that is, how many adjacent station sites are optimal for simultaneous learning. This section analyzes these two issues by conducting two groups of sensitivity experiments, briefly, G-exp and N-exp.

4.1. Impact of the PRUFS Forecast Input (G-Exp)

A group of G-exps was conducted to investigate the impact of the PRUFS forecast patch sizes, i.e., the areas of G, as the input of G2N, on the G2N correction. For simplicity, the central station "Jintan" (see Figure 1) was selected as a single site for the correction tests, i.e., G2N with N = 1, briefly, G2-One. The structure of G2-One is shown in Figure 2b.

To keep this paper concise, only the 10 m wind speed correction was presented because the results for the other two variables are similar. The experiments were designed by cropping the domain of the input fields with varying area ratios relative to the large default area chosen and discussed in the previous sections (with a dimension of 301 $\times$ 401 grid points). "Jintan" was kept approximately at the centers for all the cropping domains. The area ratio is defined as follows.

$$AreaRatio = Cropped\_dimension / Default\_dimension \tag{7}$$

The G2-One model was trained to correct the 10 m wind speed at the "Jintan" station with AreaRatio = 0.3, 0.4, 0.5, 0.6, 0.7, 0.8, 0.9, and 1.0, respectively. (Figure 2b shows the cases of AreaRatios = 0.3, 0.5, 0.7). The RMSE and improvement percentage of the 10 m wind speed at the station of these eight G2-One experiments corrected on the test dataset were computed and given in Table 2. The table also includes the results of the G2N331 model.

Table 2. RMSE of the 10 m wind speed of the PRUFS forecast and the correction by G2-One at Jintan and Nanjing for different input areas and the corresponding IP. The evaluation was done on all test datasets.

AreaRatios of Input Domain	PRUFS Forecasts (RMSE)-"Jintan"	G2-One Correction (RMSE)-"Jintan"	IP	PRUFS Forecasts (RMSE)-"Nanjing"	G2-One Correction (RMSE)-"Nanjing"	IP
1.0 (G2N331)	1.47	0.89	39.5%	1.36	0.97	28.7%
1.0	1.48	0.98	33.8%	1.37	1.05	23.4%
0.9	1.48	0.96	35.1%	1.37	1.04	24.1%
0.8	1.48	0.96	35.1%	1.37	1.02	25.5%
0.7	1.48	0.90	39.2%	1.37	0.98	28.5%
0.6	1.48	0.97	34.5%	1.37	1.01	26.3%
0.5	1.48	0.97	34.5%	1.37	1.01	26.3%
0.4	1.48	0.98	33.8%	1.37	1.02	25.5%
0.3	1.48	0.98	33.8%	1.37	1.02	25.5%

Table 2 shows again that incorporating information from other surrounding sites in the loss functions improves the error correction at Jintan (i.e., G2N outperforms G2-One). Nevertheless, for clarity and simplicity, G-exps for Jintan only is presented. It can be seen in Table 2 that the G2-One performance is improved as AreaRatios (the sizes of G) increase from 0.3 to 0.7, and thereafter, the performance degrades as the AreaRatios continue to increase. **This indicates that selecting the proper sizes of spatial structures/features is important for G2N.** If it is too small, the model will not be able to take in sufficient information on the spatial features of the PRUFS forecasts. On the other side, if the input domain size is too large, it may introduce unnecessary noises and/or information burdens that hinder the G2N training.

In addition to the Jintan station, we also computed the training at other stations located in the central regions of the domain. The RMSE of the wind speed at Nanjing (Table 3) is smaller than those at Jintan, but the trend of the sensitivity test results with different AreaRatios is consistent with that at Jintan. The results for other stations are similar, but not shown for brevity.

Table 3. Improvement percentages of RMSE for the N-Exps with the G2N model.

N-exps \ Verification	51	101	199	311
(a) 2 m Temperature Statistics Results				
G2N51	16.4%			
G2N101	14.1%	13.8%		
G2N199	19.8%	19.8%	21.8%	
G2N311	20.8%	20.2%	21.8%	18.9%
(b) 2 m Relative Humidity Statistics Results				
G2N51	23.7%			
G2N101	24%	25.5%		
G2N199	25.7%	27.5%	24.6%	
G2N311	27.8%	28.9%	25.7%	24.5%
(c) 10 m Wind Speed Statistics Results				
G2N51	44.5%			
G2N101	46.5%	43.4%		
G2N199	44.5%	41.6%	40.9%	
G2N311	47.4%	44%	43.9%	42.8%

For a possible physical explanation of the optimum size for the G2N model training, we think the mesoscale circulation features are critical. For a given station, the model errors are affected by the mesoscale system over the station and the most important structural features of this mesoscale system should be included for the G2N model input. Thus, the "optimum size" should depend on the size of these most important structural features, which is a few hundred kilometers.

4.2. Impact of the Sites for Multitask Learning(N-Exp)

A set of N-exps is carried out to study the impact of assigning different numbers of surface stations for simultaneous learning, i.e., multitask learning. The stations are selected in regions with "Jintan" approximately at the center (see Figure 1), and the experiments take 51, 101, and 199 sites (Figure 10), respectively.

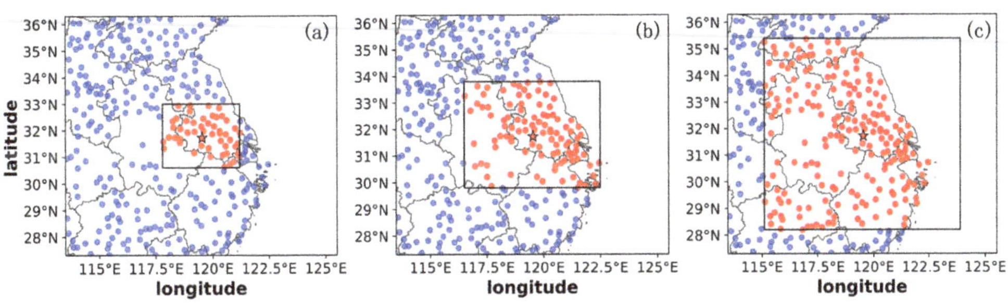

Figure 10. Sub-domains containing 51 (**a**), 101 (**b**), and 199 (**c**) station sites for N-exps. The red star is the station "Jintan".

For N-exp experiments, the G2N model was trained using the same labeled dataset and model forecast input as those discussed in the previous sections, but the loss functions were defined for a varying number of sites (i.e., target domain sizes), i.e., N = 51, 101, 199, and 331, respectively, namely, G2N51, G2N101, G2N199, and G2N311. The performances of these four configurations were assessed based on the outputs of these model runs over the test dataset. The improvement percentages (IP) of RMSE of the G2N outputs, with respect to the PRUFS forecast, were computed for the 51, 101, 199, and 331 site groups, respectively, and presented in Table 3.

The second column of Table 3 labeled as "51" compares the RMSE IP concerning the same 51 sites (shown in red in Figure 10a) corrected by the G2N over the PRUFS forecasts for the four N-exps. The third, fourth, and fifth columns are the same but for the IPs concerning 101, 199, and 311 sites, respectively. It can be seen from Table 3 that for the 51 evaluation sites, as N increases from 51 to 331, the IPs for 2 m T, 2 m RH, and 10 m wind speed gradually grow in general. Similar results can be found for verification statistics computed for 101 and 199 evaluation stations. The learning for all 331 stations achieved the best result. **These results indicate that multistation learning for G2N with more stations is beneficial, not only reducing computing costs dramatically but also increasing the learning skills of the G2N model.**

5. Conclusions

Model output post-processing is a crucial step for correcting the errors of numerical weather prediction. In this study, we established the "grid-to-multipoint" (G2N) convolutional neural network (CNN)-based deep-learning model for correcting the forecast error of an operational high-resolution numerical weather prediction system (PRUFS) running 24 cycles of 0–24 h forecasts, each day over eastern China. G2N corrects model forecast errors by projecting high-resolution weather model gridded forecasts to the surface weather observations. G2N was tested for correcting the forecast of 2 m temperature, 2 m relative humidity, and 10 m wind speed of a high-resolution PRUFS model output. The forecast area contains 311 standard surface weather stations. G2N was trained with one year of data (August 2020 to August 2021) and evaluated by an independent test dataset of the real-time operational PRUFS runs during November and December 2021. The training and testing datasets contain all 24 cycles of 0–24 h forecasts per day. The results show a good performance of G2N for all surface forecast variables corrected and computing efficiency. Furthermore, two groups of sensitivity experiments were conducted to evaluate the impact of changing the input gridded numerical model data sizes and varying the number of stations for multitasking training on the performance of G2N. The main results are as follows.

(1) The G2N model could effectively extract and use the meso- and micro-scale meteorological circulation features, simulated by the high-resolution NWP forecasts, to infer the weather forecast errors at the target stations. The verification of G2N on the test dataset of the 2-month operational runs shows very good improvement percentages of RMSE, 19.0%, 24.5%, and 42.4% for 2 m temperature, 2 m relative humidity, and 10 m wind speed, respectively, in comparison to the PRUFS forecasts.

(2) Sensitivity experiments with selecting mesoscale model forecast (feature) domains show that the size of the input domain has an important impact on the performance of the G2N model. Inputting an excessively small domain will not feed G2N with sufficient spatial features in the PRUFS forecasts that are relevant to the forecast error at the target stations. On the other hand, an excessively large input domain may introduce unnecessary information that hinders the G2N performance.

(3) Sensitivity experiments with multitasking learning strategies (N-exps) show that, for a given input model grid domain, increasing the number of target correction stations within the domain for multitask learning is beneficial to improving the performances of G2N for correcting the errors of all three surface variables. When the three variables (T_2, RH_2, W_{10}) are corrected for the 51 sites, the RMSE improvement percentages of 51 sites

with input threshold are 16.4%, 23.7%, and 44.5%. With the increase in the input threshold, the RMSE improvement percentage of the three variables in 51 sites increased to 20.8%, 27.8%, and 47.4%, with an average increase of approximately 3.8 percentage points.G2N gained the largest error correction when all 311 sites were included in simultaneous learning. This finding indicates that the loss function composed with more target stations could incorporate more relevant spatial loss information and thus increase the G2N model learning abilities.

(4) With its simplicity and high effectiveness, G2N can be readily generalized for post-processing a high-resolution numerical weather prediction system running over other regions. Based on our data and tests, we recommend specifying a patch size of the input model forecast domain with a side dimension of ~600–900 km (200–300 grids) for G2N and including all stations within the domain in the loss function for simultaneous forecast error correction.

The G2N model developed in this paper has been running operationally along with the PRUFS regional numerical weather system to support valuable applications by several customers. For the domain size and stations corrected in this paper, the training time for G2N with one-year samples takes approximately 6 h wall-clock time on a GPU server with Quadro RTX 8000. The G2N real-time run takes only 297 s. Therefore, G2N is a highly efficient and effective tool for post-processing high-resolution NWP forecasts.

Nevertheless, we note that it will be more informative to assess the G2N model performance for a complete year period. Unfortunately, we were not able to access the model data after December 2021. We plan to apply the G2N model for another NWP system in the future and put attention on evaluating the general applicability of the G2N model and its seasonal performance variation characteristics.

We also would like to note that the input for G2N described in this paper only uses a single-element forecast field, e.g., the PRUFS 2 m temperature forecasts for correcting the 2 m temperature at the surface stations. We tested inputting multiple variables, including surface pressure, humidity, and wind, but obtained a degraded performance. Additionally, the results of the present G2N training were obtained without separating the tags of the different forecast lengths, forecast sequences, or forecast cycles of the day. Activating these tags also degraded the G2N performance. Our future work will aim at understanding these limitations and explore more complicated deep-learning models, including refined self-attention algorithms that may amplify the contributions of the key feature in the input and thus gain further improvement on the forecast error correction.

Author Contributions: Conceptualization, Y.L.; methodology, Y.Q., X.J., Y.L. and L.Y.; validation, Y.Q., Y.S. and H.X.; formal analysis, Y.Q., X.J., Y.L. and L.Y.; investigation, Y.Q. and Y.L.; resources, Y.L., Y.S. and H.X.; writing—original draft preparation, Y.Q.; writing—review and editing, Y.L., H.X., Z.H. and L.Y.; visualization, Y.Q., X.J. and Z.H.; funding acquisition, Y.L., H.X. and Y.S.; All authors have read and agreed to the published version of the manuscript.

Funding: This work is supported by the Science and Technology Grant No.520120210003, Jibei Electric Power Company of the State Grid Corporation of China and partially by the National Key R&D Program of China (Grant No. 2018YFC1507901).

Institutional Review Board Statement: Not applicable.

Informed Consent Statement: Not applicable.

Data Availability Statement: Not applicable.

Acknowledgments: The modeling simulations were carried out on the supercomputer provided by the Nanjing University of Information Science and Technology. The authors thank Peng Zhou for providing and processing the model data analyzed in this study and Xing Wang and Daili Qian for valuable discussions.

Conflicts of Interest: The authors declare no conflict of interest.

References

1. Bauer, P.; Thorpe, A.; Brunet, G. The Quiet Revolution of Numerical Weather Prediction. *Nature* **2015**, *525*, 47–55. [CrossRef] [PubMed]
2. Glahn, H.R.; Lowry, D.A. The Use of Model Output Statistics (MOS) in Objective Weather Forecasting. *J. Appl. Meteorol.* **1972**, *11*, 1203–1211. [CrossRef]
3. Klein, W.H.; Lewis, B.M.; Enger, I. Objective prediction of five-day mean temperatures during winter. *J. Meteorol.* **1959**, *16*, 672–682. [CrossRef]
4. Homleid, M. Diurnal Corrections of Short-Term Surface Temperature Forecasts Using the Kalman Filter. *Weather Forecast.* **1995**, *10*, 689–707. [CrossRef]
5. Delle Monache, L.; Nipen, T.; Liu, Y.; Roux, G.; Stull, R. Kalman Filter and Analog Schemes to Postprocess Numerical Weather Predictions. *Mon. Weather Rev.* **2011**, *139*, 3554–3570. [CrossRef]
6. Delle Monache, L.; Eckel, F.A.; Rife, D.L.; Nagarajan, B.; Searight, K. Probabilistic Weather Prediction with an Analog Ensemble. *Mon. Weather Rev.* **2013**, *141*, 3498–3516. [CrossRef]
7. Alessandrini, S.; Davò, F.; Sperati, S.; Benini, M.; Delle Monache, L. Comparison of the Economic Impact of Different Wind Power Forecast Systems for Producers. *Adv. Sci. Res.* **2014**, *11*, 49–53. [CrossRef]
8. Alessandrini, S.; Delle Monache, L.; Sperati, S.; Cervone, G. An Analog Ensemble for Short-Term Probabilistic Solar Power Forecast. *Appl. Energy* **2015**, *157*, 95–110. [CrossRef]
9. Nagarajan, B.; Delle Monache, L.; Hacker, J.P.; Rife, D.L.; Searight, K.; Knievel, J.C.; Nipen, T.N. An Evaluation of Analog-Based Postprocessing Methods across Several Variables and Forecast Models. *Weather Forecast.* **2015**, *30*, 1623–1643. [CrossRef]
10. Whan, K.; Schmeits, M. Comparing Area Probability Forecasts of (Extreme) Local Precipitation Using Parametric and Machine Learning Statistical Postprocessing Methods. *Mon. Weather Rev.* **2018**, *146*, 3651–3673. [CrossRef]
11. Li, H.; Yu, C.; Xia, J.; Wang, Y.; Zhu, J.; Zhang, P. A Model Output Machine Learning Method for Grid Temperature Forecasts in the Beijing Area. *Adv. Atmos. Sci.* **2019**, *36*, 1156–1170. [CrossRef]
12. Cho, D.; Yoo, C.; Im, J.; Cha, D. Comparative Assessment of Various Machine Learning-Based Bias Correction Methods for Numerical Weather Prediction Model Forecasts of Extreme Air Temperatures in Urban Areas. *Earth Space Sci.* **2020**, *7*, e2019EA000740. [CrossRef]
13. Rasp, S.; Lerch, S. Neural Networks for Post-Processing Ensemble Weather Forecasts. *Mon. Weather Rev.* **2018**, *146*, 3885–3900. [CrossRef]
14. Han, L.; Chen, M.; Chen, K.; Chen, H.; Zhang, Y.; Lu, B.; Song, L.; Qin, R. A Deep Learning Method for Bias Correction of ECMWF 24–240 h Forecasts. *Adv. Atmos. Sci.* **2021**, *38*, 1444–1459. [CrossRef]
15. Zhang, Y.; Chen, M.; Han, L.; Song, L.; Yang, L. Multi-element deep learning fusion correction method for numerical weather prediction. *Acta Meteorol. Sin.* **2022**, *80*, 153–167.
16. Skamarock, W.C.; Klemp, J.B.; Dudhia, J.; Gill, D.O.; Barker, D.M.; Duda, M.G.; Huang, X.-Y.; Wang, W.; Powers, J.G. *A Description of the Advanced Research WRF Version 3*; National Center for Atmospheric Research: Boulder, CO, USA, 2008.
17. Krizhevsky, A.; Sutskever, I.; Hinton, G.E. ImageNet Classification with Deep Convolutional Neural Networks. *Commun. ACM* **2017**, *60*, 84–90. [CrossRef]
18. Shimodaira, H. Improving Predictive Inference under Covariate Shift by Weighting the Log-Likelihood Function. *J. Stat. Plan Inference* **2000**, *90*, 227–244. [CrossRef]
19. Ioffe, S.; Szegedy, C. Batch Normalization: Accelerating Deep Network Training by Reducing Internal Covariate Shift. *Proc. Mach. Learn. Res.* **2015**, *37*, 448–456.
20. Rahaman, N.; Baratin, A.; Arpit, D.; Draxler, F.; Lin, M.; Hamprecht, F.A.; Bengio, Y.; Courville, A. On the Spectral Bias of Neural Networks. *Proc. Mach. Learn. Res.* **2018**, *97*, 5301–5310.
21. Pan, L.; Liu, Y.; Roux, G.; Cheng, W.; Liu, Y.; Hu, J.; Jin, S.; Feng, S.; Du, J.; Peng, L. Seasonal Variation of the Surface Wind Forecast Performance of the High-Resolution WRF-RTFDDA System over China. *Atmos. Res.* **2021**, *259*, 105673. [CrossRef]
22. Shi, J.; Liu, Y.; Li, Y.; Liu, Y.; Roux, G.; Shi, L.; Fan, X. Wind Speed Forecasts of a Mesoscale Ensemble for Large-Scale Wind Farms in Northern China: Downscaling Effect of Global Model Forecasts. *Energies* **2022**, *15*, 896. [CrossRef]
23. Zeng, X.-M.; Wang, M.; Wang, N.; Yi, X.; Chen, C.; Zhou, Z.; Wang, G.; Zheng, Y. Assessing Simulated Summer 10-m Wind Speed over China: Influencing Processes and Sensitivities to Land Surface Schemes. *Clim. Dyn.* **2018**, *50*, 4189–4209. [CrossRef]
24. Minton, P.D.; Cohen, J. Statistical Power Analysis for the Behavioral Sciences. *J. Am. Stat. Assoc.* **1971**, *66*, 428. [CrossRef]
25. Cohen, J. *Statistical Power Analysis for the Behavioral*; Routledge: London, UK, 1988.
26. Kumar, A.; Daumé, H., III. Learning Task Grouping and Overlap in Multi-Task Learning. *arXiv* **2012**, arXiv:1206.6417.
27. Gao, Y.; Ma, J.; Zhao, M.; Liu, W.; Yuille, A.L. NDDR-CNN: Layerwise Feature Fusing in Multi-Task CNNs by Neural Discriminative Dimensionality Reduction. In Proceedings of the 2019 IEEE/CVF Conference on Computer Vision and Pattern Recognition (CVPR), Long Beach, CA, USA, 15–20 June 2019; pp. 3200–3209.

28. Crawshaw, M. Multi-Task Learning with Deep Neural Networks: A Survey. *arXiv* **2020**, arXiv:2009.09796.
29. Li, Y.; Lang, J.; Ji, L.; Zhong, J.; Wang, Z.; Guo, Y.; He, S. Weather Forecasting Using Ensemble of Spatial-Temporal Attention Network and Multi-Layer Perceptron. *Asia Pac. J. Atmos. Sci.* **2021**, *57*, 533–546. [CrossRef]

climate

Article

Subseasonal Variation in the Winter ENSO-NAO Relationship and the Modulation of Tropical North Atlantic SST Variability

Wenjun Zhang * and Feng Jiang

CIC-FEMD/ILCEC, Key Laboratory of Meteorological Disaster of Ministry of Education (KLME), Nanjing University of Information Science and Technology, Nanjing 210044, China
* Correspondence: zhangwj@nuist.edu.cn

Abstract: The impact of El Niño–Southern Oscillation (ENSO) on the North Atlantic Oscillation (NAO) has been controversially discussed for several decades, which exhibits prominent seasonality and nonstationarity. During early winter, there appears a positive ENSO-NAO relationship, while this relationship reverses its sign in late winter. Here, we show that this subseasonal variation in the ENSO-NAO relationship could be attributed to the different mechanisms involved in early and late winters. In early winter, the positive linkage between the ENSO and NAO could be simply understood as resulting from the changes in tropical Walker circulation and the associated atmospheric meridional circulation over the North Atlantic. In the following late winter, an opposite NAO-like response appears as the large-scale Pacific–North Atlantic teleconnection pattern fully establishes and evident sea surface temperature anomalies occur over the North Tropical Atlantic (NTA). We further show that the phase shift in NAO during ENSO late winter is largely contributed by the establishment of the ENSO-associated NTA SST anomaly via its excited convection in the subtropical Atlantic. The competing roles of mechanisms explain the subseasonal variation in the ENSO-NAO relationship from early to late winter, providing useful information for seasonal prediction over the North Atlantic–European region.

Keywords: ENSO; NAO; subseasonal change

Citation: Zhang, W.; Jiang, F. Subseasonal Variation in the Winter ENSO-NAO Relationship and the Modulation of Tropical North Atlantic SST Variability. *Climate* **2023**, *11*, 47. https://doi.org/10.3390/cli11020047

Academic Editor: Charles Jones

Received: 10 January 2023
Revised: 9 February 2023
Accepted: 11 February 2023
Published: 14 February 2023

1. Introduction

The North Atlantic Oscillation (NAO), a prominent atmospheric circulation pattern, characterizes a large-scale seesaw of atmosphere mass between the Azores High and the Icelandic Low, which exerts significant climate impacts over the North Atlantic and European sectors [1–4]. Due to large internal mid-latitude atmospheric variability, the skillful prediction of the NAO on seasonal to interannual timescales has been a challenging issue for the climate community [5–8]. As the primary predictability source of the global climate system, the El Niño–Southern Oscillation (ENSO) is suggested to have the capacity in providing the potential seasonal predictability of NAO and, therefore, the associated climate variability in North America and Europe [6,7]. The co-variability of the tropical Pacific signals and the climate anomalies in the extratropical North Atlantic region was first noticed by Sir Gilbert Walker in the 1920s and 1930s [9–11]. The Southern Oscillation in boreal winter is found to be accompanied by the meridional alteration in atmospheric mass over the North Atlantic.

Around half a century later, systematic research started on the relationship between ENSO and NAO and the possible mechanisms involved [12–14]. While the ENSO impacts on the climate variabilities over the North Pacific–American sector are well established [15–18], its influence over the North Atlantic and European sector is still under debate [19–23]. During boreal winter, the ENSO signal can be clearly detected over the North Pacific and North American region via the stationary atmospheric Rossby waves, which is referred to as Pacific–North America (PNA) teleconnection [15,24]. The PNA teleconnection extends downstream and reaches the North Atlantic and leads to changes in the local quasi-stationary wave pattern, which could project on the NAO pattern [13,25,26]. Nevertheless, the atmospheric anomalies associated with the ENSO-induced PNA teleconnection over the North Atlantic are usually shifted eastward in comparison with the classical NAO pattern [21]. Additionally, one recent study argued that no dynamical linkage can be found between the NAO and ENSO-related atmospheric anomalies over the North Atlantic [27]. An alternative mechanism that could account for the ENSO-NAO relationship is that ENSO-related SST anomalies in the Northern Tropical Atlantic (NTA) alter the North Atlantic atmospheric circulation and act as a mediator to connect the ENSO and NAO during winter [28–34]. Prominent SST warming (cooling) is usually observed to lag the Pacific warming (cooling) by a few months [18,35–38], which is conducive to a negative (positive) NAO-like pattern locally [28,29,32–34]. Moreover, stratospheric processes, such as sudden stratospheric warming, have also been proposed to play a part in linking climate variabilities in the tropical Pacific and North Atlantic basins [39–43].

Despite the fact that several mechanisms have been proposed, little consensus has yet been reached on whether a robust ENSO signal can be detected in the North Atlantic–European climate and which physical process is essential for the observed ENSO-NAO linkage. Some studies show that no prominent ENSO signals could be detected over the North Atlantic and adjacent continental European regions [44,45]; however, some other studies suggest that clear climate response to ENSO could be identified in the temperature and precipitation fields [21,40,46,47]. The following studies show that these seemingly contradictory conclusions result from the fact that the ENSO impacts over the North Atlantic regions are associated with highly nonlinear dynamics, which lead to the large uncertainty in the observed relationship between ENSO and NAO [20,21,48]. The ambiguity has also been suggested to be associated with the strong seasonal dependence of the ENSO-NAO relationship [21,43,49]. During early winter, the El Niño events are usually accompanied by a positive NAO response, and La Niña events are usually accompanied by a negative NAO response. In contrast, approximately opposite atmospheric responses could be detected over the North Atlantic sector in the following late winter. The negative relationship between the ENSO and NAO in late winter has been more extensively studied in previous studies, which has been suggested to be related to the delayed stratospheric response to ENSO [40,41,43]. Other studies emphasized the role of the North Atlantic SST variability in the negative ENSO-NAO relationship during late winter [32,34]. So far, the mechanisms underlying the subseasonal variation in the NAO response to ENSO are not clear, which brings considerable difficulties to the seasonal prediction of the NAO and associated climate impacts.

In this study, we demonstrate that the subseasonal variation in the relationship between the ENSO and NAO is closely linked with the constructive or destructive roles of different mechanisms and emphasize the physical process that is dominated in early and late winters, respectively. During El Niño in early winter, the positive NAO response could be understood as the result of changes in the tropical general circulations. In the following late winter, the appearance of warm NTA SST anomalies and the full establishment of PNA teleconnection are both in favor of a negative NAO pattern. Additionally, the anomalous NTA SST is the key factor in the subseasonal phase shift in NAO response to ENSO. Similar mechanisms are also at work during La Niña. The rest of the paper is organized as follows: In Section 2, we show the datasets and methods used in this research. Section 3 presents the observed subseasonal variation in the ENSO-NAO relationship and addresses the possible mechanisms responsible for different NAO responses in early and late winters. Discussions are presented in Section 4, and the conclusion is given in Section 5.

2. Materials and Methods

The monthly SST dataset used in this study is the global sea ice and SST analysis from the Met Office Hadley Centre (HadISST) [50]. The atmospheric circulation, including sea level pressure (SLP) and geopotential height at 500 hPa, were examined based on the

National Center for Environmental Prediction (NCEP)/National Center for Atmospheric Research (NCAR) reanalysis-1 data [51]. The ENSO-related precipitation was investigated using the National Oceanic and Atmospheric Administration's monthly precipitation reconstruction data [52]. The NAO index is based on the normalized SLP of Iceland and Portugal [53]. The PNA index was taken from the Climate Prediction Center (CPC) website. Anomalies for all the variables were measured as the deviation from the monthly climatology in the study period (1960–2020), and all these fields were linearly detrended to exclude potential impacts from global warming. All statistical significance tests were performed using the two-tailed Student's *t*-test. The bootstrap resampling method was used for statistical inference by randomly selecting data with replacements 1000 times from the original data and calculating the average at each time [54]. The statistical significance for the bootstrap ensemble mean of the average in each realization was inferred from the bootstrap probability [55]. For example, the bootstrap ensemble mean is significant at the 95% confidence level when 95% of the bootstrap samples are larger/smaller than zero.

According to the definition of the CPC (see https://origin.cpc.ncep.noaa.gov/products/analysis_monitoring/ensostuff/ONI_v5.php, accessed on 8 January 2023), ENSO events are defined based on a threshold of ±0.5 °C of the 3-month running mean Niño3.4 index (SST anomaly averaged in the region of 5° S–5° N, 120°–170° W) for 5 consecutive months. A total of 21 El Niño events (1963, 1965, 1968, 1969, 1972, 1976, 1977, 1979, 1982, 1986, 1987, 1991, 1994, 1997, 2002, 2004, 2006, 2009, 2014, 2015, and 2018) and 21 La Niña events (1964, 1970, 1971, 1973, 1974, 1975, 1983, 1984, 1985, 1988, 1995, 1998, 1999, 2000, 2005, 2007, 2008, 2010, 2011, 2016, and 2017) were identified. The NTA SST index was defined as the SST anomaly averaged in the region of 0°–20° N, 0°–75° W. We defined the standardized SLP difference between the regions of the equatorial eastern Pacific (5° S–5° N, 80° W–180°) and the equatorial Atlantic (5° S–5° N, 80° W–20° E) as the Walker index in this study to focus on the overturning circulation over the Pacific and Atlantic region. The winter was separated into early (November–December) and late (January–February) periods to investigate the subseasonal variation in the ENSO-NAO relationship.

3. Results

3.1. Subseasonal Variation in the ENSO-NAO Relationship

During the positive phase of NAO, the meridional gradient of atmospheric pressure is strengthened due to the enhanced Icelandic Low and Azores High between the polar and subtropical North Atlantic regions (Figure 1a–d). Most previous studies about the ENSO-NAO relationship focused on the boreal winter season when the ENSO reaches the mature phase and the local air–sea interaction over the North Atlantic is mostly active. However, ENSO impacts on the NAO exhibit a distinctive difference in early and late winters with a prominent subseasonal phase shift in NAO (Figure 1e–h). During November–December of El Niño years, there appear positive SLP anomalies over the subtropical Atlantic and negative SLP anomalies near Iceland, resembling a positive NAO-related atmospheric pattern (Figure 1e,f). In the following January to February, the anomalous atmospheric pattern exhibits a contrasting feature over the North Atlantic when the ENSO-related Aleutian Low is enhanced (Figure 1g,h). An NAO-like atmospheric pattern of the opposite polarity emerges with the spatial correlation between ENSO-related atmospheric pattern and the classical NAO pattern being −0.86 and −0.85 in January and February, respectively. Besides more similarity between the ENSO-related pattern and the classical NAO pattern compared to the early winter, it has also been noted that the active center of the ENSO-related atmospheric response over the North Atlantic in late winter shows an obvious displacement from the east to the west.

To qualitatively measure the subseasonal shift in the ENSO-NAO relationship, we used the commonly adopted NAO index to examine the monthly evolution based on the difference between composite El Niño and La Niña during winter (Figure 2). In El Niño during early winter, the NAO is usually in its positive phase with the strengthened meridional gradient of atmospheric pressure over the North Atlantic. Additionally, in

La Niña during early winter, there usually appears a negative NAO-like pattern with a weakened meridional pressure gradient over the North Atlantic. The NAO of the opposite sign could be observed in later winter. The differences between the NAO index of El Niño and La Niña are significant at the 95% significance level in November and January, while large uncertainty exists for other months.

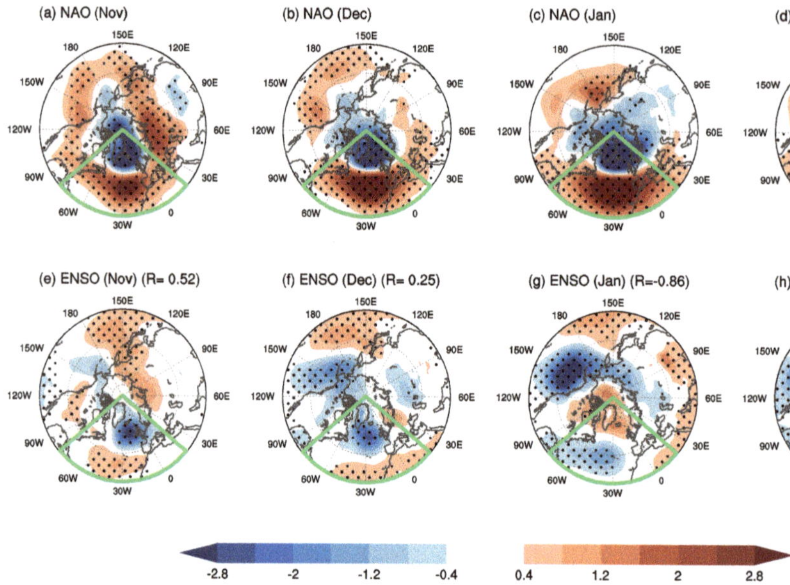

Figure 1. SLP anomalies (hPa) in (**a**) November, (**b**) December, (**c**) January, and (**d**) February regressed upon the simultaneous standardized NAO index. (**e**–**h**) Similar to (**a**–**d**) but for SLP anomalies regressed on the Niño3.4 index. Large (small) dots indicate regression coefficients that are statistically significant at the 95% (90%) confidence level. The green box indicates the North Atlantic region (20°–90° N, 0°–80° W). The pattern correlation is displayed in (**e**–**h**), which is calculated based on the NAO-related SLP anomalies and ENSO-related anomalies over the North Atlantic region for each month, respectively.

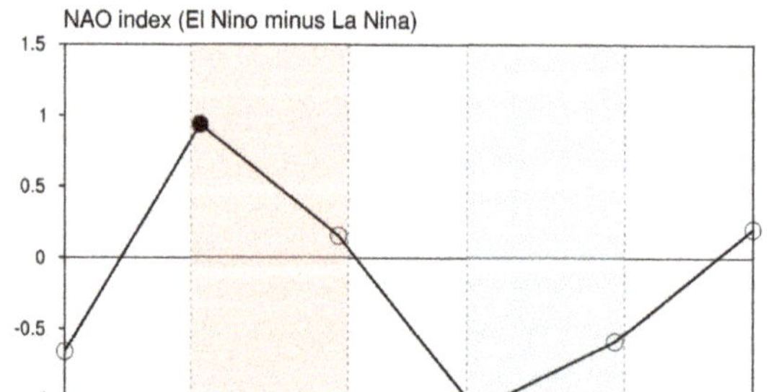

Figure 2. Monthly evolution of the difference of NAO index between composite El Niño and La Niña from October (0) to March (1). Numerals "0" and "1" denote the years in which ENSO events develop and decay, respectively. Solid circles indicate the composite values that are significant at the 95% confidence level. The orange and blue shadings indicate the early and late winter, respectively.

We further used the bootstrap sampling with replacements to generate 1000 realizations of the observed NAO responses during the early and late winters of El Niño and La Niña events, respectively, and show the histograms of the bootstrap samples (Figure 3). The difference in the bootstrapping NAO index in ENSO early and late winter is consistent with the subseasonal phase shift in NAO responses to ENSO, as depicted in Figures 1 and 2. During the El Niño early winter, the bootstrap ensemble mean NAO index is 0.42 (statistically significant at the 95% confidence level), while the ensemble mean NAO index is −0.25 (non-statistically significant at the 95% confidence level) during the following later winter. For La Niña events, the ensemble mean NAO index is −0.29 (non-statistically significant at the 95% confidence level) in early winter and 0.64 (statistically significant at the 95% confidence level) in the late winter. The insignificant values in El Niño late winter and La Niña early winter hint at some asymmetry in the impacts of ENSO on the NAO during both early and late winter.

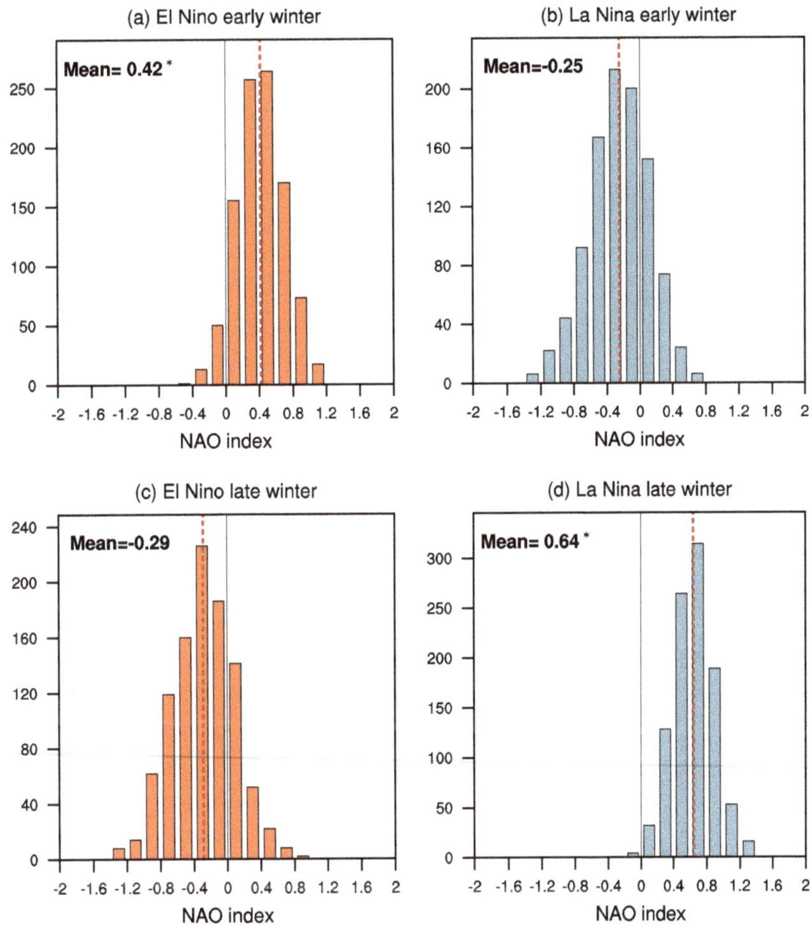

Figure 3. Histograms of the averaged NAO index after 1000 bootstrap resampling during (**a**) El Niño early winter, (**b**) La Niña early winter, (**c**) El Niño late winter, and (**d**) La Niña late winter. The bootstrap ensemble mean is displayed for each case with the asterisk denoting the value that is statistically significant at the 95% confidence level (also indicated by the dotted red line).

3.2. Possible Mechanisms Responsible for Different NAO Responses in Early and Late Winters

We next investigate the possible physical mechanisms responsible for the NAO phase shift during early and late winters of ENSO years. Figure 4 shows anomalous SST, precipitation, SLP, and geopotential height at 500 hPa regressed on the simultaneous Niño3.4 index in early and late winters. During El Niño in early winter, the tropical Pacific SST anomalies feature a horse-shoe pattern with warm SST anomalies in the central to eastern equatorial Pacific and cold SST anomalies in the western Pacific (Figure 4a). Correspondingly, the precipitation anomalies exhibit a dipolar structure, featuring a positive lobe in the central equatorial Pacific and a negative lobe in the western North Pacific (Figure 4c). While the SST anomaly pattern during late winter is quite similar to that during early winter in the tropical Pacific, the precipitation anomalies are obviously enhanced in the central Pacific (Figure 4b,d), which could result from the nonlinear interaction between the ENSO and warm pool SST annual cycle [56–59].

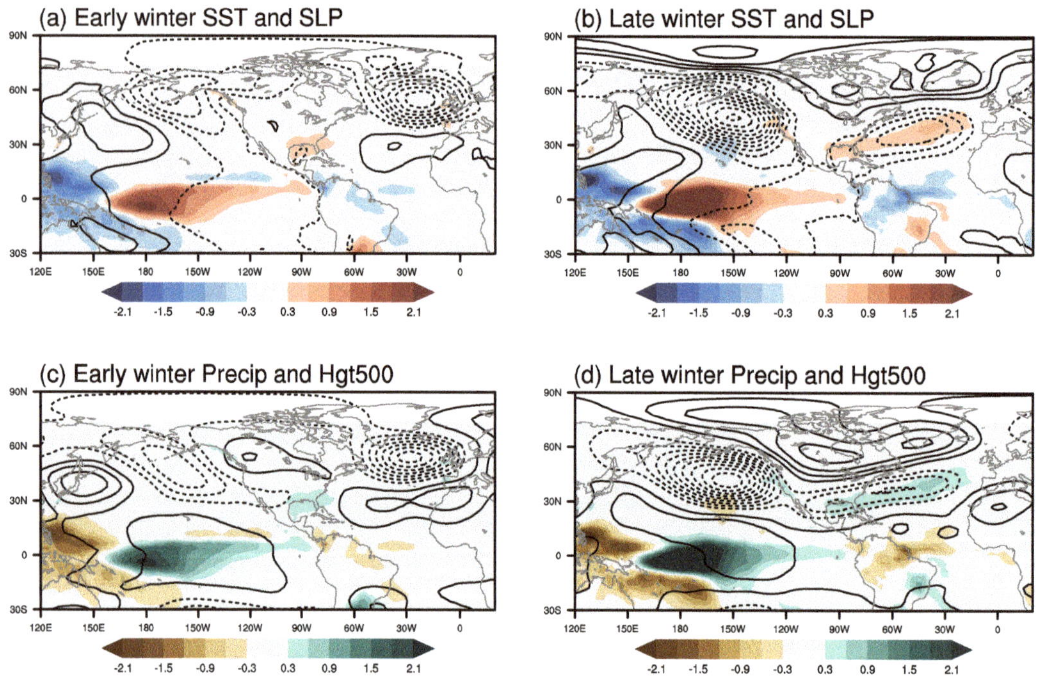

Figure 4. Anomalous SST (shading: °C) and SLP (contour: hPa) anomalies in (**a**) early winter and (**b**) late winter regressed upon the standardized simultaneous Niño3.4 index. Anomalous precipitation (Precip for short; shading, mm/day) and geopotential height at 500 hPa (Hgt500 for short; contour, m) in (**c**) early winter and (**d**) late winter regressed upon the standardized simultaneous Niño3.4 index. Solid lines indicate the positive values and dashed lines indicate the negative values of SLP in (**a**,**b**) and Hgt500 in (**c**,**d**). The contour interval is 0.3 hPa in (**a**,**b**) and 0.3 m in (**c**,**d**).

Accompanied by the enhancement in the tropical Pacific precipitation anomalies, the atmospheric anomalies of the Aleutian Low are evidently enhanced in late winter (Figure 4b,d). Correspondingly, the two downstream centers of the PNA teleconnection pattern are clearly established over North America. These two downstream centers extend from North America to the central North Atlantic ocean with a southwest-to-northeast tilt, resembling the negative NAO pattern. In contrast, the atmospheric response in the Aleutian Low is weak, and the PNA teleconnection pattern is not fully established in the early winter of ENSO years (Figure 4a,c). In early winter, a prominent atmospheric response can be observed over the North Atlantic with a negative center near 50° N and a positive center near 30° N, zonally elongating from the central North Atlantic Ocean to Western Europe. Compared with early winter, the action center of ENSO-related atmospheric response over the North Atlantic is displaced northwestward in late winter.

One open question to be answered is which mechanism drives the positive ENSO-NAO relationship in early winter. As suggested by a previous study [60], the tropical Walker circulation and the North Atlantic meridional circulation can act as a mediator to link the ENSO and climate variability over the North Atlantic region. Figure 5 shows changes in the tropical Walker circulation and North Atlantic meridional circulation associated with ENSO in early and late winters, respectively. During El Niño early winter, the tropical Walker circulation is weakened with an anomalous ascending branch over the central and eastern Pacific and an anomalous descending branch over the Atlantic (Figure 5a). Concomitantly, the meridional circulation over the North Atlantic can be observed to be significantly altered, showing anomalous air subsidence over the tropical Atlantic and anomalous air ascending at 40°–70° N of the North Atlantic (Figure 5c). As a result, the positive NAO pattern emerges along with the enhancement in both the Azores High and Icelandic Low.

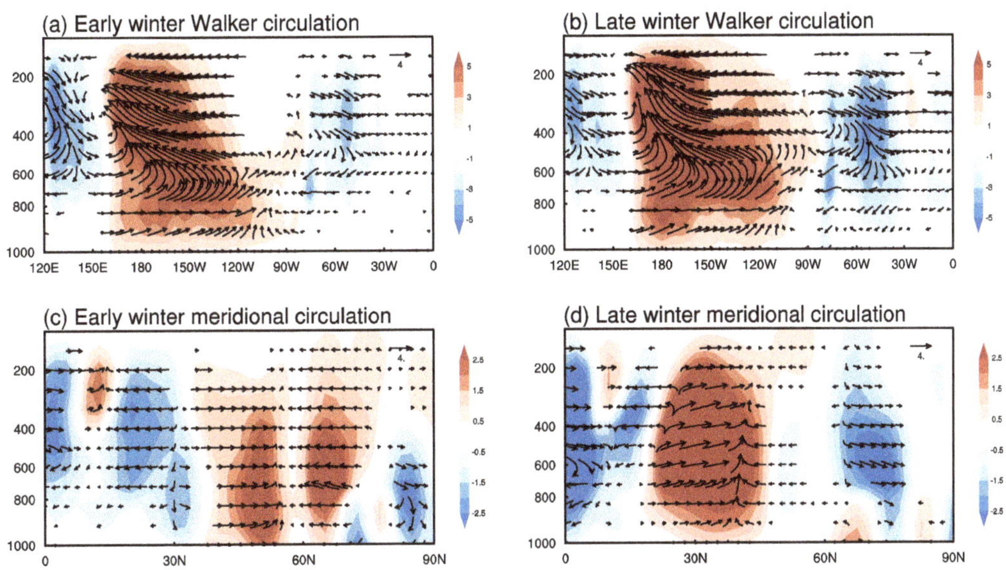

Figure 5. Zonal wind and vertical velocity (omega) anomalies (vector; m/s for zonal wind and Pa/s for omega) averaged meridionally over 5° S–5° N in (**a**) early and (**b**) late winter regressed upon the simultaneous Niño3.4 index. Meridional wind and vertical velocity anomalies (vector; m/s for meridional wind and Pa/s for omega) averaged zonally over 0°–50° W in (**c**) early and (**d**) late winter regressed upon the simultaneous Niño3.4 index. Omega anomalies are multiplied by a factor of −400 in (**a**,**b**) and −800 in (**c**,**d**) for display and are also shown in shading.

During the following late winter, the anomalous tropical Walker circulation exhibits a similar pattern as that of early winter with an increase in amplitude, in particular over the tropical Atlantic (Figure 5b). However, there appears a very different structure of anomalous meridional circulation over the North Atlantic (Figure 5d). A prominent ascending during late winter can be observed near the region of 20°–40° N, which is of the opposite sign with respect to that during early winter. Accordingly, most regions of the subpolar North Atlantic experience the anomalous subsidence. The atmospheric response over the North Atlantic resembles the negative NAO-like pattern during El Niño in late winter. The distinctive meridional circulation anomalies in late winter could be related to the enhanced precipitation near 30° N in the North Atlantic. These precipitation anomalies seem to be excited by the NTA SST warming (Figure 4b). The important role of the NTA SST in impacting the atmospheric circulation over the North Atlantic has been shown in previous studies, as shown in the introduction. The warm NTA SST anomalies could generate negative NAO-like anomalies based on observational analyses and also numerical modeling experiments. El Niño events are usually accompanied by significant SST warming over the NTA via the tropical and extratropical pathways [61]. The NTA SST response is found to lag the ENSO mature phase by about a few months, possibly due to the local adjustment timescale [18,62]. As shown in Figure 4a,b, the warm NTA SST anomalies during El Niño events are evidently strengthened in late winter compared to early winter, which leads to enhanced precipitation over the subtropical Atlantic and acts to explain the distinct meridional circulation structure in late winter. A similar mechanism operates during La Niña late winter.

The respective role of the above-mentioned processes in contributing to the ENSO-NAO linkage is further shown in Figure 6, in which the monthly SLP anomalies are regressed upon the simultaneous Walker index, PNA index, and NTA index. As shown in Figure 6a, the Walker circulation change in the tropics is linked to the alteration in local meridional circulation over the North Atlantic. Accompanied by the strengthening (weakening) of the tropical Walker circulation, there appears a weakened (strengthened) meridional gradient of atmospheric pressure over the North Atlantic, resembling the negative (positive) NAO pattern. During El Niño years, the Walker circulation is weakened throughout the winter season (Figure 7), which tends to favor a positive NAO-like pattern of atmospheric pressure over the North Atlantic. Due to the role of the Walker circulation change, a positive ENSO-NAO relationship can be observed in early winter. However, the opposite relationship between ENSO and NAO emerges in the following late winter, as the PNA fully establishes, and the NTA SST warms up. It seems that the Walker circulation anomalies play a minor role in modulating the atmospheric circulation over the North Atlantic during the late winter of ENSO years. As shown in Figure 6b, the PNA-related atmospheric pattern displays positive anomalies near Greenland and negative anomalies along the eastern coast of North America. The enhancement in PNA during late winter could contribute to the westward displacement of the NAO-related action center. This existence of the linkage between the PNA and NAO has been discussed in many studies; however, the local manifestation of PNA teleconnection can only explain the western edge of the NAO, which might suggest an absence of dynamical linkage between these two phenomena [27].

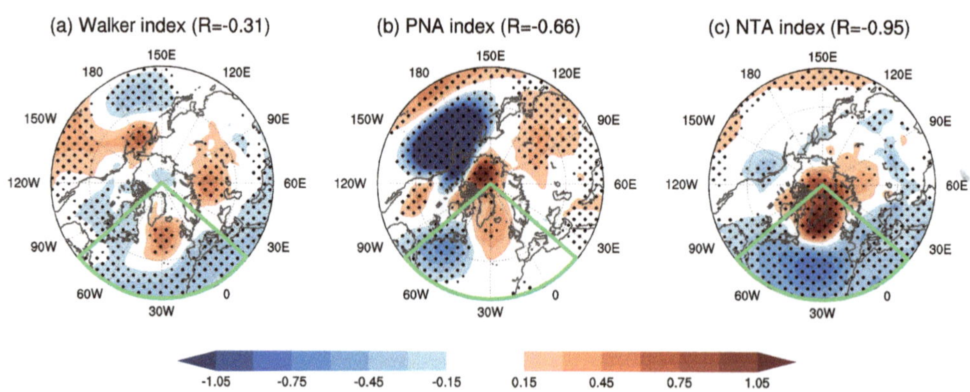

Figure 6. Monthly SLP anomalies (hPa) regressed upon the simultaneous (**a**) Walker index, (**b**) PNA index, and (**c**) NTA index. Dots indicate regression coefficients that are statistically significant at the 95% confidence level. The green box indicates the North Atlantic region (20°–90° N, 0°–80° W).

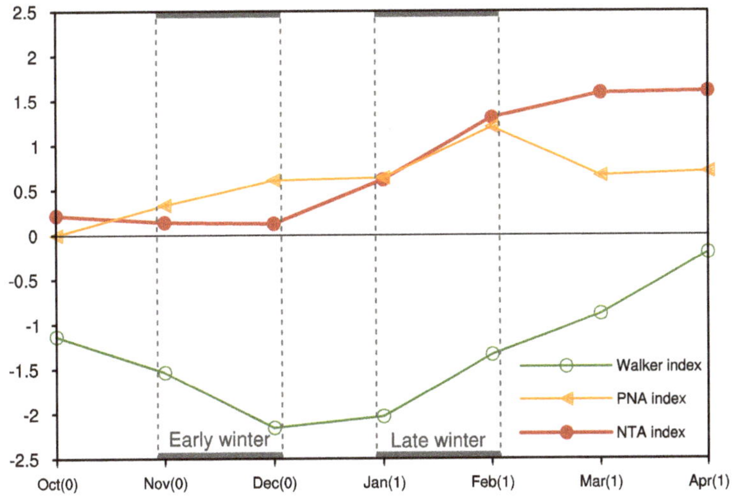

Figure 7. Monthly evolution of difference of the standardized Walker index (green line), PNA index (orange line), and NTA index (red line) between composite El Niño and La Niña.

Figure 6c shows the SLP anomalies regressed on the NTA SST index. The warm NTA SST anomalies tend to be accompanied by weakened Azores High and Icelandic Low, which closely resemble the classical NAO pattern (their spatial correlation being −0.96). During El Niño in late winter, the NTA SST significantly warms (Figure 7) and alters the local atmospheric pattern, which leads to the prominent negative ENSO-NAO relationship. We further examined the role of NTA SST on the NAO phase during ENSO late winter by categorizing the ENSO events into four groups based on late winter NTA SST conditions that are El Niño with a positive NTA index, El Niño with a negative NTA index, La Niña with a positive NTA index, and La Niña with a negative NTA index (Table 1). The El Niño events with warm NTA SST anomalies show a significant negative NAO pattern (Figure 8a), while the El Niño events with cold NTA SST anomalies exhibit no significant NAO-related atmospheric anomaly pattern (Figure 8b). Similarly, the La Niña events with opposite-signed NTA SST anomalies are characterized by a distinct atmospheric response over the North Atlantic. The La Niña events with negative NTA SST anomalies correspond to a positive NAO pattern. However, the La Niña events with warm NTA SST anomalies are accompanied by a weakened meridional gradient of atmospheric pressure, with the centers being shifted eastward relative to the conventional NAO pattern. Therefore, the negative ENSO-NAO relationship can be detected considering that most ENSO years are concurrent with the same-signed NTA SST anomalies. The above results again support that the winter NTA SST variability during ENSO years plays an important role in leading to a remarkable shift in the North Atlantic atmospheric pattern towards an opposite NAO response in late winter compared to that in early winter.

Table 1. El Niño and La Niña years are classified according to late winter NTA SST anomalies.

Categories	Year
El Niño with positive NTA anomalies	1963, 1965, 1968, 1969, 1977, 1979, 1987, 1997, 2002, 2004, 2006, 2009, 2015
El Niño with negative NTA anomalies	1972, 1976, 1982, 1986, 1991, 1994, 2014, 2018
La Niña with positive NTA anomalies	1970, 1995, 2005, 2010, 2016
La Niña with negative NTA anomalies	1964, 1971, 1973, 1974, 1975, 1983, 1984, 1985, 1988, 1998, 1999, 2000, 2007, 2008, 2011, 2017

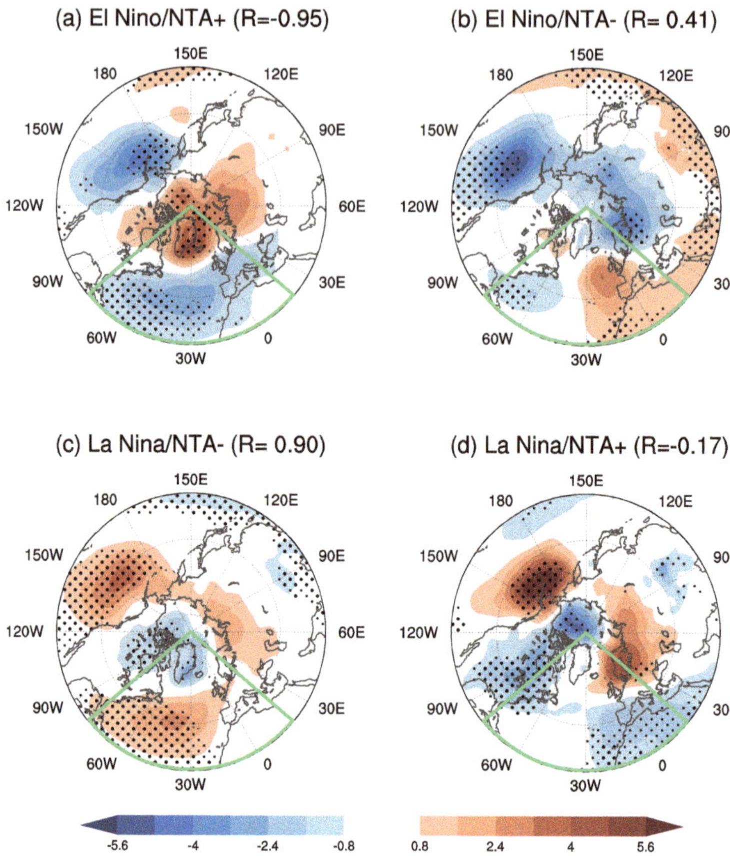

Figure 8. Composite SLP anomalies (hPa) for (**a**) El Niño events concurrent with positive NTA SST anomalies; (**b**) El Niño events concurrent with negative NTA SST anomalies; (**c**) La Niña events concurrent with negative NTA SST anomalies; and (**d**) La Niña events concurrent with positive NTA SST anomalies. Large (small) dots indicate regression coefficients that are statistically significant at the 95% (90%) confidence level. The green box indicates the North Atlantic region (20°–90° N, 0°–80° W).

4. Discussion

Despite the fact that a feasible explanation for the subseasonal variation in the relationship between the ENSO and NAO during winter is provided here, we are still left with considerable levels of uncertainty regarding the impacts of ENSO on NAO in addition to the fact that there is large atmospheric internal variability in the mid–high latitudes [63]. On the one hand, this uncertainty is granted considering that ENSO itself is highly complicated with large inter-event variability in the zonal location and intensity of its associated tropical convection [22,64,65]. On the other hand, the constructive/destructive interference of competing mechanisms operating during different stages of ENSO events could also contribute to the large uncertainty.

At this stage, we cannot determine whether the observed NAO responses to ENSO are simply the superposition of atmospheric patterns associated with different mechanisms or nonlinear interactions between different physical processes are at play. However, even from a linear perspective, the uncertainty of the ENSO-NAO relationship could be partly explained by the time-evolving characteristics of the processes involved. For example, while no significant NTA SST response to ENSO can be detected in early winter, significant positive SST responses appear over the NTA in late winter but with relatively large inter-event variability. We demonstrate that the inter-event difference in NTA SST anomalies during different ENSO events could account for, at least, part of the uncertainty of the ENSO-NAO relationship. Additionally, as noted in a recent study, the ENSO-related precipitation in the Indian Ocean could contribute to the inter-event difference in NAO response during ENSO early winter [66]. Moreover, several studies have argued that stratospheric processes could connect the climate variability between the Pacific and Atlantic basins [43,67]. While the active role of the stratospheric pathway is also emphasized in the ENSO-NAO relationship in late winter, the nonlinearity in the stratospheric response to ENSO largely limits the predictability of the related processes [23,43]. The potential effect of stratospheric processes on the subseasonal variation in the ENSO-NAO relationship is not addressed in this study, which deserves further attention.

5. Conclusions

The present work investigated the different ENSO impacts on the NAO during early and late winter and the possible roles of various mechanisms. Our observational analyses support the finding of the subseasonal variation in the ENSO-NAO linkage and identify the crucial physical processes responsible for the phase shift in NAO during ENSO early and late winters. During El Niño early winter, the North Atlantic experiences a positive NAO-like pattern of atmospheric pressure due to the weakening of the tropical Walker circulation and the associated alteration in the meridional circulation over the North Atlantic. However, the North Atlantic is dominated by a negative NAO-like atmospheric response over the North Atlantic during El Niño late winter. The phase shift in the NAO response to ENSO is closely related to the ENSO-induced NTA SST anomalies in late winter. During El Niño late winter, there appears obvious SST warming over the NTA, which can excite the local convection and thus, generate a negative NAO response. Moreover, the full establishment of the PNA teleconnection in late winter could also make some contribution to the simultaneous negative ENSO-NAO relationship. The strongly enhanced positive PNA teleconnection could partially account for the negative NAO response during El Niño late winter, but its effect is confined to the western edge of the NAO. Similar mechanisms work during La Niña. The understanding of different mechanisms for the subseasonal variation in the ENSO-NAO relationship from early to late winter provides useful information for seasonal prediction over the North Atlantic–European region.

Author Contributions: Conceptualization, W.Z. and F.J.; methodology, W.Z. and F.J.; software, F.J.; validation, F.J.; formal analysis, F.J.; investigation, F.J.; resources, F.J.; data curation, F.J.; writing—original draft preparation, W.Z. and F.J.; writing—review and editing, W.Z.; visualization, F.J.; supervision, W.Z.; project administration, W.Z.; funding acquisition, W.Z. All authors have read and agreed to the published version of the manuscript.

Funding: This research was funded by the National Nature Science Foundation of China (42125501).

Data Availability Statement: Publicly available datasets were analyzed in this study. The datasets are located at https://www.metoffice.gov.uk/hadobs/hadisst/data/download.html (accessed on 10 January 2023), https://psl.noaa.gov/data/gridded/data.ncep.reanalysis.html (accessed on 10 January 2023), https://www.esrl.noaa.gov/psd/data/gridded/data.ncep.reanalysis.html (accessed on 10 January 2023), https://psl.noaa.gov/data/gridded/data.prec.html (accessed on 10 January 2023), https://climatedataguide.ucar.edu/climate-data/hurrell-north-atlantic-oscillation-nao-index-station-based (accessed on 10 January 2023) and https://www.cpc.ncep.noaa.gov/products/precip/CWlink/pna/ (accessed on 10 January 2023).

Conflicts of Interest: The authors declare no conflict of interest. The funders had no role in the design of the study; in the collection, analyses, or interpretation of data; in the writing of the manuscript; or in the decision to publish the results.

References

1. Hurrell, J.W.; Kushnir, Y.; Ottersen, G.; Visbeck, M. An Overview of the North Atlantic Oscillation. In *Geophysical Mono-Graph Series*; Hurrell, J.W., Kushnir, Y., Ottersen, G., Visbeck, M., Eds.; American Geophysical Union: Washington, DC, USA, 2003; Volume 134, pp. 1–35; ISBN 978-0-87590-994-3.
2. Visbeck, M.; Chassignet, E.P.; Curry, R.G.; Delworth, T.L.; Dickson, R.R.; Krahmann, G. The Ocean's Response to North Atlantic Oscillation Variability. In *Geophysical Monograph Series*; Hurrell, J.W., Kushnir, Y., Ottersen, G., Visbeck, M., Eds.; American Geophysical Union: Washington, DC, USA, 2003; Volume 134, pp. 113–145; ISBN 978-0-87590-994-3.
3. Hurrell, J.W.; Kushnir, Y.; Visbeck, M. The North Atlantic Oscillation. *Science* **2001**, *291*, 603–605. [CrossRef] [PubMed]
4. Trigo, R.M.; Pozo-Vázquez, D.; Osborn, T.J.; Castro-Díez, Y.; Gámiz-Fortis, S.; Esteban-Parra, M.J. North Atlantic Oscillation Influence on Precipitation, River Flow and Water Resources in the Iberian Peninsula. *Int. J. Climatol.* **2004**, *24*, 925–944. [CrossRef]
5. Wang, L.; Ting, M.; Kushner, P.J. A Robust Empirical Seasonal Prediction of Winter NAO and Surface Climate. *Sci. Rep.* **2017**, *7*, 279. [CrossRef] [PubMed]
6. Scaife, A.A.; Arribas, A.; Blockley, E.; Brookshaw, A.; Clark, R.T.; Dunstone, N.; Eade, R.; Fereday, D.; Folland, C.K.; Gordon, M.; et al. Skillful Long-range Prediction of European and North American Winters. *Geophys. Res. Lett.* **2014**, *41*, 2514–2519. [CrossRef]
7. Dunstone, N.; Smith, D.; Scaife, A.; Hermanson, L.; Eade, R.; Robinson, N.; Andrews, M.; Knight, J. Skilful Predictions of the Winter North Atlantic Oscillation One Year Ahead. *Nat. Geosci.* **2016**, *9*, 809–814. [CrossRef]
8. Johansson, Å. Prediction Skill of the NAO and PNA from Daily to Seasonal Time Scales. *J. Clim.* **2007**, *20*, 1957–1975. [CrossRef]
9. Walker, G.T. Correlation in Seasonal Variation of Weather VIII: A Preliminary Study of World Weather. *Mem. India Meteorol. Dep.* **1923**, *24*, 75–131.
10. Walker, G.T. Correlation in Seasonal Variation of Weather IX: A Further Study of World Weather. *Mem. India Meteorol. Dep.* **1924**, *24*, 225–232.
11. Walker, G.T.; Bliss, E.W. World Weather V. *Mem. R. Meteorol. Soc.* **1932**, *4*, 53–84. [CrossRef]
12. Rogers, J.C. The Association between the North Atlantic Oscillation and the Southern Oscillation in the Northern Hemisphere. *Mon. Wea. Rev.* **1984**, *112*, 1999–2015. [CrossRef]
13. Huang, J.; Higuchi, K.; Shabbar, A. The Relationship between the North Atlantic Oscillation and El Niño-Southern Oscillation. *Geophys. Res. Lett.* **1998**, *25*, 2707–2710. [CrossRef]
14. Giannini, A.; Cane, M.A.; Kushnir, Y. Interdecadal Changes in the ENSO Teleconnection to the Caribbean Region and the North Atlantic Oscillation. *J. Clim.* **2001**, *14*, 2867–2879. [CrossRef]
15. Hoskins, B.J.; Karoly, D.J. The Steady Linear Response of a Spherical Atmosphere to Thermal and Orographic Forcing. *J. Atmos. Sci.* **1981**, *38*, 1179–1196. [CrossRef]
16. Horel, J.D.; Wallace, J.M. Planetary-Scale Atmospheric Phenomena Associated with the Southern Oscillation. *Mon. Wea. Rev.* **1981**, *109*, 813–829. [CrossRef]
17. Trenberth, K.E.; Branstator, G.W.; Karoly, D.; Kumar, A.; Lau, N.-C.; Ropelewski, C. Progress during TOGA in Understanding and Modeling Global Teleconnections Associated with Tropical Sea Surface Temperatures. *J. Geophys. Res.* **1998**, *103*, 14291–14324. [CrossRef]
18. Alexander, M.A.; Bladé, I.; Newman, M.; Lanzante, J.R.; Lau, N.-C.; Scott, J.D. The Atmospheric Bridge: The Influence of ENSO Teleconnections on Air–Sea Interaction over the Global Oceans. *J. Clim.* **2002**, *15*, 2205–2231. [CrossRef]
19. Pozo-Vázquez, D.; Esteban-Parra, M.J.; Rodrigo, F.S.; Castro-Díez, Y. The Association between ENSO and Winter Atmospheric Circulation and Temperature in the North Atlantic Region. *J. Clim.* **2001**, *14*, 3408–3420. [CrossRef]
20. Mathieu, P.-P.; Sutton, R.T.; Dong, B.; Collins, M. Predictability of Winter Climate over the North Atlantic European Region during ENSO Events. *J. Clim.* **2004**, *17*, 1953–1974. [CrossRef]
21. Brönnimann, S. Impact of El Niño-Southern Oscillation on European Climate: Enso Impact on Europe. *Rev. Geophys.* **2007**, *45*, RG3003. [CrossRef]
22. Toniazzo, T.; Scaife, A.A. The Influence of ENSO on Winter North Atlantic Climate. *Geophys. Res. Lett.* **2006**, *33*, L24704. [CrossRef]
23. Garfinkel, C.I.; Hartmann, D.L. Influence of the Quasi-Biennial Oscillation on the North Pacific and El Niño Teleconnections. *J. Geophys. Res.* **2010**, *115*, D20116. [CrossRef]
24. Wallace, J.M.; Gutzler, D.S. Teleconnections in the Geopotential Height Field during the Northern Hemisphere Winter. *Mon. Wea. Rev.* **1981**, *109*, 784–812. [CrossRef]
25. Pinto, J.G.; Reyers, M.; Ulbrich, U. The Variable Link between PNA and NAO in Observations and in Multi-Century CGCM Simulations. *Clim. Dyn.* **2011**, *36*, 337–354. [CrossRef]
26. Pozo-Vázquez, D.; Gámiz-Fortis, S.R.; Tovar-Pescador, J.; Esteban-Parra, M.J.; Castro-DÍez, Y. North Atlantic Winter SLP Anomalies Based on the Autumn ENSO State. *J. Clim.* **2005**, *18*, 97–103. [CrossRef]
27. Mezzina, B.; García-Serrano, J.; Bladé, I.; Kucharski, F. Dynamics of the ENSO Teleconnection and NAO Variability in the North Atlantic–European Late Winter. *J. Clim.* **2020**, *33*, 907–923. [CrossRef]
28. Watanabe, M.; Kimoto, M. Tropical-Extratropical Connection in the Atlantic Atmosphere-Ocean Variability. *Geophys. Res. Lett.* **1999**, *26*, 2247–2250. [CrossRef]
29. Robertson, A.W.; Mechoso, C.R.; Kim, Y.-J. The Influence of Atlantic Sea Surface Temperature Anomalies on the North Atlan-tic Oscillation. *J. Clim.* **2000**, *13*, 122–138. [CrossRef]

30. Peng, S.; Robinson, W.A.; Li, S. Mechanisms for the NAO Responses to the North Atlantic SST Tripole. *J. Clim.* **2003**, *16*, 1987–2004. [CrossRef]
31. Pan, L.-L. Observed Positive Feedback between the NAO and the North Atlantic SSTA Tripole. *Geophys. Res. Lett.* **2005**, *32*, L06707. [CrossRef]
32. Sung, M.-K.; Ham, Y.-G.; Kug, J.-S.; An, S.-I. An Alterative Effect by the Tropical North Atlantic SST in Intraseasonally Varying El Niño Teleconnection over the North Atlantic. *Tellus A Dyn. Meteorol. Oceanogr.* **2013**, *65*, 19863. [CrossRef]
33. Buchan, J.; Hirschi, J.J.-M.; Blaker, A.T.; Sinha, B. North Atlantic SST Anomalies and the Cold North European Weather Events of Winter 2009/10 and December 2010. *Mon. Weather. Rev.* **2014**, *142*, 922–932. [CrossRef]
34. Ham, Y.-G.; Sung, M.-K.; An, S.-I.; Schubert, S.D.; Kug, J.-S. Role of Tropical Atlantic SST Variability as a Modulator of El Niño Teleconnections. *Asia-Pac. J. Atmos. Sci.* **2014**, *50*, 247–261. [CrossRef]
35. Covey, D.L.; Hastenrath, S. The Pacific El Niño Phenomenon and the Atlantic Circulation. *Mon. Wea. Rev.* **1978**, *106*, 1280–1287. [CrossRef]
36. Enfield, D.B.; Mayer, D.A. Tropical Atlantic Sea Surface Temperature Variability and Its Relation to El Niño-Southern Oscillaion. *J. Geophys. Res.* **1997**, *102*, 929–945. [CrossRef]
37. Nobre, P.; Shukla, J. Variations of Sea Surface Temperature, Wind Stress, and Rainfall over the Tropical Atlantic and South America. *J. Clim.* **1996**, *9*, 2464–2479. [CrossRef]
38. Jiang, F.; Zhang, W. Understanding the Complicated Relationship Between ENSO and Wintertime North Tropical Atlantic SST Variability. *Geophys. Res. Lett.* **2022**, *49*, e2022GL097889. [CrossRef]
39. Castanheira, J.M.; Graf, H.-F. North Pacific–North Atlantic Relationships under Stratospheric Control? *J. Geophys. Res.* **2003**, *108*, 4036. [CrossRef]
40. Ineson, S.; Scaife, A.A. The Role of the Stratosphere in the European Climate Response to El Niño. *Nat. Geosci.* **2009**, *2*, 32–36. [CrossRef]
41. Bell, C.J.; Gray, L.J.; Charlton-Perez, A.J.; Joshi, M.M.; Scaife, A.A. Stratospheric Communication of El Niño Teleconnections to European Winter. *J. Clim.* **2009**, *22*, 4083–4096. [CrossRef]
42. Butler, A.H.; Polvani, L.M.; Deser, C. Separating the Stratospheric and Tropospheric Pathways of El Niño–Southern Oscillation Teleconnections. *Environ. Res. Lett.* **2014**, *9*, 024014. [CrossRef]
43. Jiménez-Esteve, B.; Domeisen, D.I.V. The Tropospheric Pathway of the ENSO–North Atlantic Teleconnection. *J. Clim.* **2018**, *31*, 4563–4584. [CrossRef]
44. Ropelewski, C.F.; Halpert, M.S. Global and Regional Scale Precipitation Patterns Associated with the El Niño/Southern Oscillation. *Mon. Wea. Rev.* **1987**, *115*, 1606–1626. [CrossRef]
45. Halpert, M.S.; Ropelewski, C.F. Surface Temperature Patterns Associated with the Southern Oscillation. *J. Clim.* **1992**, *5*, 577–593. [CrossRef]
46. Fraedrich, K. An ENSO Impact on Europe? *Tellus A* **1994**, *46*, 541–552. [CrossRef]
47. Cassou, C.; Terray, L. Dual Influence of Atlantic and Pacific SST Anomalies on the North Atlantic/Europe Winter Climate. *Geophys. Res. Lett.* **2001**, *28*, 3195–3198. [CrossRef]
48. Zhang, W.; Mei, X.; Geng, X.; Turner, A.G.; Jin, F.-F. A Nonstationary ENSO–NAO Relationship Due to AMO Modulation. *J. Clim.* **2019**, *32*, 33–43. [CrossRef]
49. Moron, V.; Gouirand, I. Seasonal Modulation of the El Niño-Southern Oscillation Relationship with Sea Level Pressure Anomalies over the North Atlantic in October-March 1873-1996: ENSO North Atlantic Sea Level Pressure Relationship. *Int. J. Climatol.* **2003**, *23*, 143–155. [CrossRef]
50. Rayner, N.A. Global Analyses of Sea Surface Temperature, Sea Ice, and Night Marine Air Temperature since the Late Nine-teenth Century. *J. Geophys. Res.* **2003**, *108*, 4407. [CrossRef]
51. Kalnay, E.; Kanamitsu, M.; Kistler, R.; Collins, W.; Deaven, D.; Gandin, L.; Iredell, M.; Saha, S.; White, G.; Woollen, J.; et al. The NCEP/NCAR 40-Year Reanalysis Project. *Bull. Amer. Meteor. Soc.* **1996**, *77*, 437–471. [CrossRef]
52. Chen, M.; Xie, P.; Janowiak, J.E.; Arkin, P.A. Global Land Precipitation: A 50-Yr Monthly Analysis Based on Gauge Observations. *J. Hydrometeorol.* **2002**, *3*, 249–266. [CrossRef]
53. Hurrell, J.W. Decadal Trends in the North Atlantic Oscillation: Regional Temperatures and Precipitation. *Science* **1995**, *269*, 676–679. [CrossRef]
54. Efron, B.; Tibshirani, R.J. *An Introduction to the Bootstrap*; Springer: Boston, MA, USA, 1993; ISBN 978-0-412-04231-7.
55. Yuan, C.; Yamagata, T. Impacts of IOD, ENSO and ENSO Modoki on the Australian Winter Wheat Yields in Recent Decades. *Sci. Rep.* **2015**, *5*, 17252. [CrossRef]
56. Stuecker, M.F.; Timmermann, A.; Jin, F.-F.; McGregor, S.; Ren, H.-L. A Combination Mode of the Annual Cycle and the El Niño/Southern Oscillation. *Nat. Geosci.* **2013**, *6*, 540–544. [CrossRef]
57. Zhang, W.; Li, H.; Jin, F.-F.; Stuecker, M.F.; Turner, A.G.; Klingaman, N.P. The Annual-Cycle Modulation of Meridional Asymetry in ENSO's Atmospheric Response and Its Dependence on ENSO Zonal Structure. *J. Clim.* **2015**, *28*, 5795–5812. [CrossRef]
58. Zhang, W.; Li, H.; Stuecker, M.F.; Jin, F.-F.; Turner, A.G. A New Understanding of El Niño's Impact over East Asia: Dominance of the ENSO Combination Mode. *J. Clim.* **2016**, *29*, 4347–4359. [CrossRef]
59. Stuecker, M.F.; Jin, F.-F.; Timmermann, A.; McGregor, S. Combination Mode Dynamics of the Anomalous Northwest Pacific Anticyclone. *J. Clim.* **2015**, *28*, 1093–1111. [CrossRef]

60. Wang, C. ENSO, Atlantic Climate Variability, and the Walker and Hadley Circulations. In *The Hadley Circulation: Present, Past and Future*; Diaz, H.F., Bradley, R.S., Eds.; Advances in Global Change Research; Springer Netherlands: Dordrecht, The Netherlands, 2004; Volume 21, pp. 173–202; ISBN 978-90-481-6752-4.
61. García-Serrano, J.; Cassou, C.; Douville, H.; Giannini, A.; Doblas-Reyes, F.J. Revisiting the ENSO Teleconnection to the Trop-ical North Atlantic. *J. Clim.* **2017**, *30*, 6945–6957. [CrossRef]
62. Lintner, B.R.; Chiang, J.C.H. Adjustment of the Remote Tropical Climate to El Niño Conditions. *J. Clim.* **2007**, *20*, 2544–2557. [CrossRef]
63. Kumar, A.; Hoerling, M.P. Annual Cycle of Pacific–North American Seasonal Predictability Associated with Different Phases of ENSO. *J. Clim.* **1998**, *11*, 3295–3308. [CrossRef]
64. Graf, H.-F.; Zanchettin, D. Central Pacific El Niño, the "Subtropical Bridge," and Eurasian Climate: Two El Niño Types and Eurasian Climate. *J. Geophys. Res.* **2012**, *117*, D01102. [CrossRef]
65. Zhang, W.; Wang, Z.; Stuecker, M.F.; Turner, A.G.; Jin, F.-F.; Geng, X. Impact of ENSO Longitudinal Position on Teleconnections to the NAO. *Clim. Dyn.* **2019**, *52*, 257–274. [CrossRef]
66. Abid, M.A.; Kucharski, F.; Molteni, F.; Kang, I.-S.; Tompkins, A.M.; Almazroui, M. Separating the Indian and Pacific Ocean Impacts on the Euro-Atlantic Response to ENSO and Its Transition from Early to Late Winter. *J. Clim.* **2021**, *34*, 1531–1548. [CrossRef]
67. Garfinkel, C.I.; Hartmann, D.L. Different ENSO Teleconnections and Their Effects on the Stratospheric Polar Vortex. *J. Geophys. Res.* **2008**, *113*, D18114. [CrossRef]

climate

MDPI

Article

Insight into Asymmetry in the Impact of Different Types of ENSO on the NAO

Peng Zhang and Zhiwei Wu *

Department of Atmospheric and Oceanic Sciences, Institute of Atmospheric Sciences, Shanghai Scientific Frontier Base of Ocean-Atmosphere Interaction, Fudan University, Shanghai 200438, China; zhang_peng@fudan.edu.cn
* Correspondence: zhiweiwu@fudan.edu.cn

Abstract: Understanding the influence of the El Niño–Southern Oscillation (ENSO) on the North Atlantic Oscillation (NAO) is of critical significance for seasonal prediction. The present study found that both Niño3.4 sea surface temperature anomaly (SSTA) intensity and east-west gradient in the mid-low latitude Pacific determine the linkage between ENSO and the NAO. Based on Niño3.4 SSTA intensity and the east-west gradient, ENSO events are classified into three types: strong intensity, weak intensity-strong gradient (WSG), and equatorial ENSOs. Note that the former two types are usually concurrent with a strong zonal SSTA gradient. In contrast, equatorial ENSO is often associated with weak intensity-weak gradient SSTAs confined in the equatorial Pacific. The anomalous circulation patterns in response to the three types of ENSO exhibit asymmetric features over the North Atlantic. The WSG-El Niño associated circulation anomaly resembles a negative NAO-like pattern, yet the strong and equatorial El Niño associated circulation anomalies show a neutral-NAO pattern. For La Niña events, their impact on the NAO mainly depends on the cold SSTA position rather than their intensity. The strong and WSG-La Niña associated negative SSTAs are centered in the equatorial-central Pacific and favor a steady positive NAO-like anomaly. The cold SSTA center of equatorial La Niña shifts to the equatorial-eastern Pacific and cannot profoundly influence the North Atlantic climate. The physical mechanisms are also investigated with a general circulation model.

Keywords: El Niño-southern oscillation; North Atlantic oscillation; asymmetry of ENSO

Citation: Zhang, P.; Wu, Z. Insight into Asymmetry in the Impact of Different Types of ENSO on the NAO. *Climate* **2023**, *11*, 136. https://doi.org/10.3390/cli11070136

Academic Editors: Xiaolei Zou, Guoxiong Wu and Zhemin Tan

Received: 23 May 2023
Revised: 22 June 2023
Accepted: 23 June 2023
Published: 27 June 2023

1. Introduction

As the prominent atmospheric mode occurring in the North Atlantic, the North Atlantic Oscillation (NAO) features an alternation of air mass over Iceland and the Azores [1–3]. Changes in the polarity and intensity of NAO exert profound influences on the surface air temperature and precipitation across the broad areas of North America and Eurasia [4]. In the positive phase, the pressure difference between the Icelandic Low and the Azores High is stronger than average, resulting in stronger westerly winds and storm tracks over the North Atlantic. This can lead to wetter and stormier conditions in Western Europe, while eastern parts may experience milder and drier weather. Conversely, the negative phase of the NAO exhibits almost the opposite characteristics. It is commonly recognized that the formation of NAO arises from the internal stochastic processes of the atmosphere [5,6]. One thing that has become clear is that the forcing stemming from the stratosphere [7–9] and sea surface temperature (SST) anomalies can also generate NAO-like anomalies [10–12].

The El Niño-Southern Oscillation (ENSO) is the dominant interannual air-sea coupled mode in the topics, which exerts salient impacts on global climate via localized forcing or remote teleconnections [13–17]. Whether the ENSO SSTA can excite the NAO-like atmospheric pattern or not, however, is still debated. Early research [18,19] believed that ENSO-related climate variability is hard to observe over the North Atlantic-Eurasia sectors. Despite the strong intrinsic variability of NAO, more and more observational and numerical experimental evidence has proven that a negative NAO-like anomaly sometimes occurs in the El Niño winter; by contrast, a positive NAO-like pattern coincides with the La Niña events [20–22]. The ENSO-NAO relationship, therefore, may provide a potential source for seasonal climate prediction in North America and Eurasia [23]. Nevertheless, many studies also found that the atmospheric signals of ENSO over the North Atlantic exhibit salient inter-event variability [24–28]. This unstable behavior may be attributed to tropical volcanic eruptions, other non-ENSO factors [23,29], the interdecadal variation of ENSO itself [30], and the diversity of ENSO [24,26].

The ENSO diversity involves varying amplitudes and SSTA patterns [31]. The strong ENSOs with SSTAs larger in amplitude usually induced enormous property loss and casualties, grabbing the close attention of the scientific community [32,33]. However, a growing number of studies noticed the impact of moderate ENSOs, emphasizing that atmospheric responses to strong and moderate ENSOs sometimes show patterns with opposite signs rather than reduced amplitudes [27,34,35]. Diverse ENSO patterns have been apparent in recent decades when central Pacific (CP) ENSOs have occurred frequently, exerting impacts that differ from the conventional (eastern Pacific, EP) ENSO in terms of their disparate atmosphere-ocean coupling processes [36–39]. This classification, however, does not distinguish moderate ENSOs from strong ones, although strong El Niños usually exhibit an EP type, making it difficult to explain the climatic impact and evolution of relatively weak ENSOs. Wang et al. [40] classified El Niño events as strong basin-wide, moderate eastern Pacific, moderate central Pacific, and successive events. They revealed that the more frequent occurrence of extreme ENSOs in the past 40 years might be attributed to a background warming in the equatorial western Pacific and associated enhanced zonal SST gradients in the equatorial central Pacific.

The zonal SST gradients reflect the SSTA contrast between the western and eastern Pacific, which accelerates the equatorial zonal wind and increases the tilt of the thermocline, favoring the maintenance of an ENSO event [41]. As an indispensable segment of the Bjerknes positive feedback, zonal SSTA gradient largely furnishes violent tropic deep convections [42] and determines extratropical teleconnections [43–45]. Referencing Wang et al. [40], Zhang P et al. [27] classified ENSO events according to their amplitude and maximum SST, divided the winter La Niña events, based on the ENSO intensity and east-west gradient in the Pacific basin, into three groups: strong intensity La Niñas, weak intensity La Niñas with strong or weak gradients (their features are summarized in Table 1), and those impacted by the three flavors of La Niña on the East Asian winter monsoon. Whether or not El Niño events can be divided into three similar categories can be debated.

This study attempts to determine what types of El Niño and La Niña can profoundly impact NAO. Moreover, owing to the asymmetries that exist in amplitudes [46–48], evolutions [49–51], and effects [52,53], the warm and cold ENSOs, therefore, are not simply mirror images of each other. A systematic contrastive study of the physical process and mechanism for the impact of diverse warm and cold ENSO events on NAO is necessary. To discuss the above questions, Section 2 displays the datasets and model. Section 3 shows the diverse ENSOs and their impacts on NAOs. The possible mechanisms are discussed in Section 4. In Section 5, the major conclusions and discussions are exhibited.

Table 1. The years of different ENSO types.

Events	Years	Common Features
Strong El Niño	1957–1958, **1965–1966**, 1972–1973, 1982–1983, 1991–1992, 1997–1998, 2009–2010, 2015–2016 (8)	The strong warm (cool) SST anomalies in the eastern (western) Pacific with salient Indian Ocean warming
Strong La Niña	**1970–1971**, **1973–1974**, **1975–1976**, **1988–1989**, **1998–1999**, **1999–2000**, **2007–2008**, **2010–2011** (8)	The strong cool (warm) SST anomalies in the eastern (western) Pacific
WSG-El Niño	1958–1959, **1968–1969**, 1976–1977, **1977–1978**, 1979–1980, **1986–1987** (6)	The warm (cool) SST anomalies in the eastern (western) Pacific but amplitude weaker than strong El Nino
WSG-La Niña	1971–1972, **1974–1975**, **2000–2001**, **2008–2009**, **2011–2012** (5)	The same as WSG-El Nino but opposite in sign
Eq-El Niño	**1963–1964**, 1969–1970, 1987–1988, **1994–1995**, **2002–2003**, **2004–2005**, 2006–2007, 2014–2015 (8)	A significant zonal banded warming throughout the tropical oceans without a salient SST cooling in the western Pacific
Eq-La Niña	1964–1965, 1967–1968, 1971–1972, **1983–1984**, 1984–1985, 1985–1986, 1995–1996, 2005–2006 (8)	The same as Eq-El Nino but opposite in sign

Central-Pacific ENSO years are marked with bold fonts.

2. Materials and Methods

The merged Extended Reconstructed SST version 5 (ERSSTv5) [54] and the Hadley Centre Sea Ice and Sea Surface Temperature dataset (HadISST) [55] with the $2° \times 2°$ horizontal resolutions are used in this study. Atmospheric reanalysis datasets include ERA-40 [56] and ERA-Interim [57] data with a horizontal resolution of $1.5° \times 1.5°$. Precipitation Reconstruction (PREC) data were provided by NOAA [58]. The differences between the western Pacific K-shape and eastern Pacific triangle SSTs are defined by the mega-ENSO index [59]. The Niño-3.4 index is obtained from Climate Prediction Center (CPC). The NAO is defined using Hurrell's station-based index [1]. The winter refers to the period from December to the next February. The data used for compositing analysis are de-trended to exclude the potential influence of linear trends.

We normalized the two ENSO indices to contrast them under the same criteria (Figure 1). The present study primarily employs composite analysis; the anomaly refers to the difference between the salient ENSO and ENSO-neutral years. "ENSO-neutral" indicates the normalized Niño3.4 index between -0.5 and 0.5 standard deviations (STD). ENSO events refer to the years with the absolute values of the Niño3.4 index greater than 0.5 STD. Considering that the average intensity of the selected warm and cold ENSO events is around ± 1.1, we classify ENSO events as of weak intensity when the absolute value of the Niño3.4 index falls within 0.5 and 1.1 standard deviations and as strong intensity when the absolute value of the Niño3.4 index exceeds 1.1. To separate strong and weak gradient ENSOs, we set ± 0.7 STD of the mega-ENSO index as the criteria. A strong gradient ENSO needs to satisfy the absolute value of the mega-ENSO index greater than 0.7 STD. If not, we define it as the weak gradient ENSO. Altering the criteria to ± 0.8 STD, the qualitative results will not change.

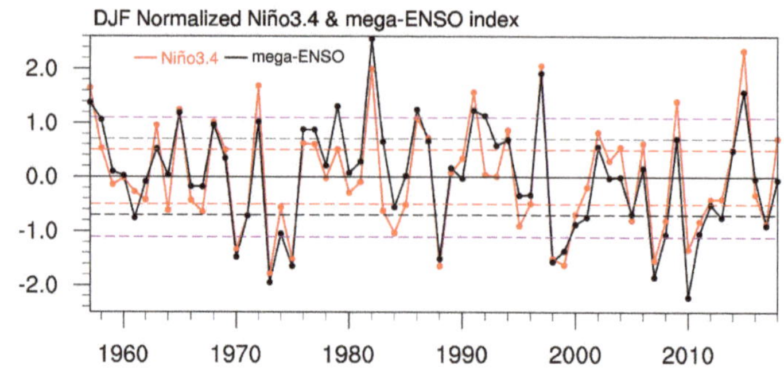

Figure 1. Time series of the normalized December–January–February (DJF) Niño3.4 (red curve) and mega-ENSO (black curve) indices for the period of 1957–2018. The mega-ENSO index is multiplied by -1 for comparison purposes. The red, black, and purple dashed lines denote ± 0.5, ± 0.7, and ± 1.1, which refer to the thresholds for dividing ENSO events, SG/WG ENSO events, and strong/moderate ENSO events, respectively.

The European Center-Hamburg (ECHAM 5.4) [60] model from the Max Planck Institute is applied to illuminate a possible mechanism. The resolution is T63L19 (horizontal grid of 1.875° and 19 vertical levels). Forced by the observational AMIP II SST, the model was integrated from 1950 to 2010. To reduce the potential impact of ENSO, we picked out twenty neutral or weak ENSO winters from 1955 to 2010 as the samples for the control experiments. Next, sensitivity experiments consisted of nine groups of simulation. Each experiment was forced by climatological monthly mean SSTs (same as in the control run) and observed DJF SST anomalies. The initial conditions for the sensitivity simulations were acquired from the control run. The detailed experiment design is shown in Table 2.

Table 2. List of SST perturbation experiments conducted in this study.

Experiments	Description of SST Perturbation
SEN	SSTA associated with strong El Niño events is imposed in the Pacific (40° S–40° N, 120° E–90° W)
WSGEN	SSTA associated with strong gradient weak El Niño events is imposed in the tropical Pacific (40° S–40° N, 120° E–90° W)
EqEN	SSTA associated with weak gradient weak El Niño events is imposed in the tropical Pacific (15° S–15° N, 160° E–90° W)
IO	SSTA associated with strong El Niño events is imposed in the Indian Ocean (20° S–20° N, 40° E–110° E)
SEN_IO	SEN and IO SSTA forcings are added together
EqEN_IO	SSTA associated with weak gradient weak El Niño events is imposed in the tropical Pacific (20° S–20° N, 160° E–90° W) and Indian Ocean (20° S–20° N, 40° E–110° E)
SGLN	SSTA associated with strong gradient (strong and strong gradient weak) La Niña events is imposed in the tropical Pacific (40° S–40° N, 120° E–120° W)
WSGLN	SSTA associated with strong gradient weak La Niña events is imposed in the tropical Pacific (40° S–40° N, 120° E–120° W)
EqLN	SSTA associated with weak gradient weak La Niña events is imposed in the tropical Pacific (15° S–15° N, 150° E–90° W)

3. Results

3.1. Comparison of Different Types of ENSO and Related NAO

3.1.1. SSTA Patterns and Their Associated Tropic Atmospheric Responses

First, since the mega-ENSO index can well represent the strong and weak gradient ENSOs, combined with the Niño3.4 index, the ENSO events with different intensities and gradients might be distinguished. According to the method displayed in Section 2, three types of El Nino and three types of La Nina are classified (Table 1). Interestingly, the strong ENSOs commonly contain a sizeable zonal gradient; only the weak ENSOs can be separated based on the east-west gradient to the weak intensity-strong gradient (WSG) and weak intensity-weak gradient events. Since the weak intensity-weak gradient events are confined in the equatorial region, they are labeled as equatorial (Eq) ENSO. Previous studies subdivided ENSO into the CP and EP groups according to the shift of the maximum SSTA [24,25,61,62]. Therefore, the CP ENSOs are bolded in Table 1, showing that the CP events occur randomly in the three flavors of El Niño; however, they coincide with the strong gradient (SG) La Niña, which consists of the strong and WSG La Niña.

Figure 2 compares the large-scale SST anomalies for the El Niño and La Niña types. Generally speaking, the spatial patterns are well classified. In the strong El Niño (SEN) winters, the obvious warm SSTAs dominate the eastern Pacific and Indian Ocean (IO), while the cold SSTAs appear in the western Pacific (Figure 2a). The strong La Niña (SLN) displays an SSTA distribution opposite to the SEN (Figure 2d). The WSG event also displays a Pacific "seesaw" pattern but a neutral IO SSTA (Figure 2b,e). In contrast, a salient narrow positive (negative) SSTA controls the equatorial central and eastern Pacific for the equatorial EN (equatorial LN) with significant warm (cold) SSTA in the IO (western IO) (Figure 2c,f). In addition, as shown in Table 1, several equatorial (SG) La Niña events coincide with the years of EP (CP) La Niña. Their spatial patterns show the maximum negative SSTA center in the tropical CP for SGLN (Figure 2d,e) but in the tropical EP for equatorial LN (Figure 2f). However, the centers of maximum SST anomalies shifted very little for the three types of El Niño (Figure 2a–c).

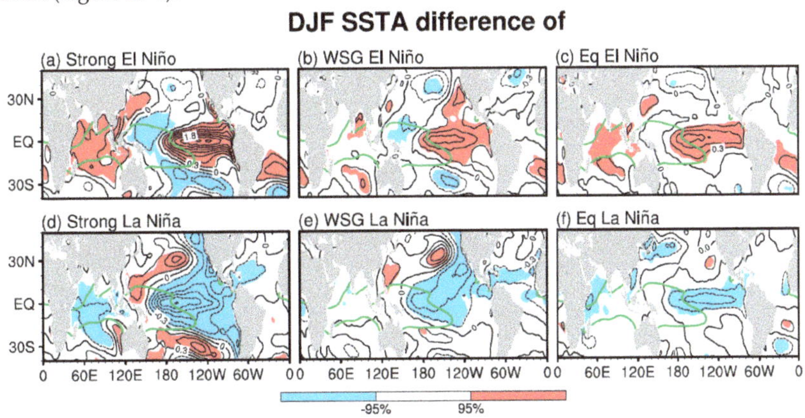

Figure 2. DJF sea surface temperature (SST; K, interval: 0.3 K) composite differences of (**a**) strong El Niño, (**b**) weak intensity-strong gradient (WSG) El Niño, and (**c**) equatorial (Eq) El Niño. (**d–f**) same as (**a–c**), but for La Niña events. The shadings in each panel represent the region with anomalies significant at the 95% confidence level (Student's t-test). The green contour represents 28 °C isotherm.

Due to the diversity in ENSO intensity and spatial patterns, the precipitation anomalies associated with the different ENSOs exhibit distinct features (Figure 3). The tropical precipitation anomalies are stronger in response to intense ENSOs than weak events. For El Niños, the salient positive precipitation anomalies straddle the dateline with the negative anomalies controlling the western Pacific (WP) (Figure 3a–c). Significant discrepancies are found over the IO region with rich tropical rainfall generated by the warm local SSTA during the SEN and equatorial EN winters. However, no salient precipitation anomalies are seen in the WSGEN winters. For the La Niña events (Figure 3d–f), the neutral precipitation anomalies over the IO indicate a relatively weak local air-sea interaction compared to those of the SEN and WSGEN. It also shows the poor precipitation centers in the tropical central Pacific for the two SGLN cases but extends eastward during equatorial LN, which is attributed to the shift of the maximum center of SSTA.

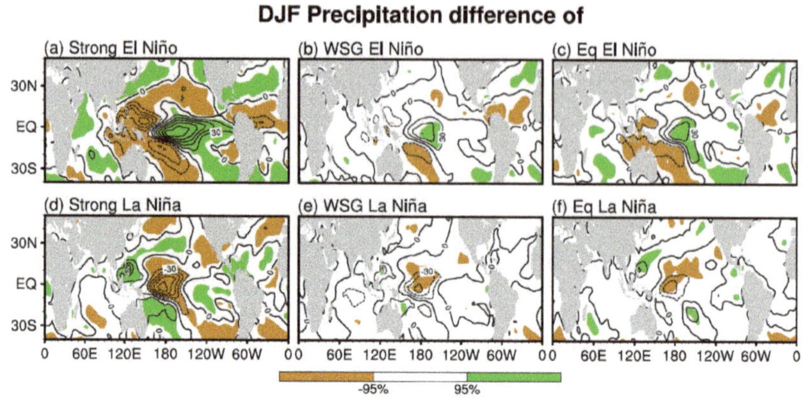

Figure 3. The same as Figure 2, except for tropical precipitation (mm/mon, interval: 30 mm/mon).

3.1.2. Unsteady Relationship of the ENSO with NAO

Next, we display the scatterplot of NAO and Niño3.4 indices in winter (Figure 4), and a complex relationship is detected between ENSO and NAO events. It shows that six WSGENs emerge with a negative NAO; however, the negative or positive NAO occurrence is evenly balanced during the SEN and equatorial EN winters, indicating a steady relationship between the NAO signal and the WSGEN. For the La Niña events, most of the SLN (6 of 8) and WSGLN (4 of 5) events, in other words, 10 of 13 SGLNs, are accompanied by a positive phase of NAO, but such relationships cannot be seen in the equatorial LN cases.

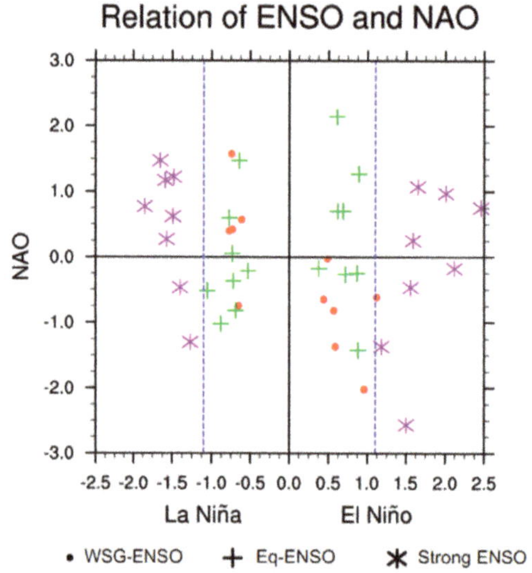

Figure 4. Scatter map of the DJF Niño3.4 and NAO indices for WSG-ENSO (red dots), equatorial ENSO (green crosses), and strong ENSO (purple snowflakes). The blue dashed line denotes ±1.1.

Figure 5 displays the anomalous 850-hPa geopotential height (Z850) and zonal wind at 200-hPa (U200) associated with the different types of ENSO to further inspect the extratropical atmospheric responses. During the EN winters (Figure 5a–c), anomalous positive SST in the tropical central and eastern Pacific induces salient diabatic heating, generating a large-scale Rossby wave train that resembles the PNA teleconnection pattern over the North Pacific (NP) and North America regions. Significant discrepancies are observed over the NP as well as the North Atlantic. Over the NP region, although the apparent negative anomalies, which imply the enhanced Aleutian Low (AL), are seen in the three EN cases, they exhibit different intensities and spatial locations. Generally, the negative Z850 anomalies over NP for SEN (Figure 5a) are much stronger than the weak events (Figure 5b,c). Compared with the SEN and equatorial EN composites (Figure 5a,c), the enhanced AL associated with WSGEN moves eastward to some extent (Figure 5b). Over the North Atlantic, the Z850 anomalies for WSGEN (Figure 5b) exhibit a negative NAO-like anomaly with negative and positive centers over the Azores Island and Iceland, respectively. However, such an anomalous pattern over the North Atlantic cannot be seen in the SEN and equatorial EN cases (Figure 5a,c). For the LN cases, the Z850 anomalies feature a decreased AL over NP in the SGLN winters (Figure 5d). Simultaneously, negative and positive Z850 anomalies are evident in Iceland and Azores Island, respectively, reflecting a positive phase of NAO-like anomalies pattern. In contrast, the equatorial LN-related anomalous Z850 field (Figure 5e) displays the salient negative anomalies over the subtropical North Pacific, representing the southward movement of the AL. Meanwhile, the positive Z850 anomalies, instead of the NAO-like pattern, dominate the mid- to high-latitude North Atlantic.

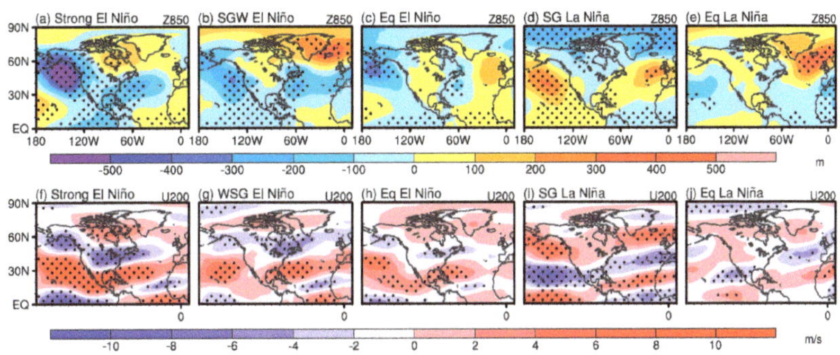

Figure 5. DJF 850-hPa geopotential height (Z850; m) composite differences of (**a**) strong El Niño—neutral, (**b**) WSG-El Niño—neutral, (**c**) Eq-El Niño—neutral, (**d**) SG-La Niña—neutral, and (**e**) Eq-La Niña—neutral. (**f–j**) same as (**a–e**) but for zonal wind at 200-hPa (U200; m/s). The dots in each panel represent the region with anomalies significant at the 90% confidence level (Student's *t*-test).

We observed that only WSGENs, rather than the SEN and equatorial EN, are concurrent with a negative phase of the NAO event, indicating that both the change of intensity and east-west gradient in SST determine the linking of El Niño and NAO. For the La Niña cases, although the SLN and WSGLN display different amplitudes, they both show an obvious east-west gradient in SST. They are accompanied by a positive phase of NAO, which cannot be seen during the equatorial LN winter. Therefore, the influence of LN events on NAO mainly depends on the spatial distribution rather than the intensity of ENSO. We also noticed that compared to the equatorial LN events, the SSTA of SLN and WSGLN show strong east-west gradients, and the maximum negative values center in the equatorial CP. We, therefore, combine the SLN and WSGLN into one category, namely, the SGLN, and discuss the potential mechanisms for the impact of the WSGEN and SGLN on NAO in the next section.

3.1.3. Possible Mechanisms for the Impact of WSGEN and SGLN on NAO

The previous study [25] argued that the jet streams commonly act as an atmospheric bridge, transporting ENSO-induced planetary wave energy to the downstream region and resulting in an oscillation of the remote atmosphere, for instance, the air mass over the North Atlantic. The lower panel of Figure 5 displays the composite U200 anomalies in each ENSO winter.

In the El Niño cases, anomalous U200 displays the tripolar pattern in the NP and extends eastward of a different degree to the North Atlantic (Figure 5f–h). The intensified subtropical jet stream and the weakened mid-latitude westerly jet over North Atlantic tend to elongate farther eastward during the WSGEN winters (Figure 5g) than during the SEN and the equatorial El Niño winters (Figure 5f,h). Such a phenomenon implies that the two branches of westerly jets may serve as the "bridge" connecting the WSGEN and the negative phase of the NAO signal. However, this "bridge" effect is invalid for SEN and equatorial El Niño. For the La Niña events, an opposite situation is observed for the SGLN cases over the North Pacific-North Atlantic sector with the negative and positive U200 anomalies over the subtropical and mid-latitude North Atlantic, indicating the weakened subtropical and the enhanced mid-latitude jets, respectively (Figure 5i). In contrast, the equatorial La Niña-related U200 anomalies show a weakened and northeastward tilted westerly jet from the subtropical North Pacific extent far east to the Barents Sea, but without the salient tripolar structure over the North Pacific region and the dipolar anomalies over the North Atlantic (Figure 5j).

Numerical experiments were performed to identify the potential effects of westerly jets on connecting the tropical SSTAs corresponding to diverse ENSOs and NAO signals. The detailed experiment designs are listed in Table 2. Figure 6 displays the abnormal tropical precipitation, Z850, and U200 response to the SEN_IO, WSGEN, EqEN_IO, SGLN, and EqLN SSTAs forcing versus the CTRL run. In general, the simulated results can well reproduce the observational analysis counterparts. In the EN-type simulations, the prescribed warm SSTA in EP and IO with the cold SSTA in WP for the SEN_IO simulation leads to stronger tropical precipitation responses (Figure 6a) than those of the WSGEN (Figure 6b) and EqEN_IO (Figure 6c) experiments. Additionally, significant positive precipitation anomalies dominate tropical IO in the SEN_IO (Figure 6a) and EqEN_IO (Figure 6c) experiments but disappear in the WSGEN (Figure 6b) experiments. Further inspecting the Z850 responses to each EN-type SSTA forcing can reveal an enhanced AL over the NP in each anomalous Z850 field (Figure 6f–h). Over the North Atlantic region, the WSGEN SSTA forcing (Figure 6g) tends to induce a negative NAO anomaly, which cannot be seen in the SEN_IO (Figure 6f) and EqEN_IO (Figure 6h) simulations. In addition, the anomalous U200 exhibits a tripolar structure over the NP in each simulation (Figure 6k–m), and the salient discrepancies also appear over the North Atlantic region. The WSGEN SSTA forcing triggers the eastward spread of the subtropical and mid-latitude westerly jets (Figure 6l), confirming the observational result. In comparison, the two westerly jets cannot extend that far eastward, indicating the SEN_IO and EqEN_IO-types SSTAs are hard to trigger the NAO-like atmospheric anomaly through the "atmospheric bridge"—the westerly jets (Figure 6k,m).

In the two La Niña-type simulations, considering the weak air-sea interaction over the tropical IO region (Figure 3d–f), we imposed the SSTA associated with the SGLN and the equatorial LN mainly within the Pacific basin. The significant negative precipitation anomalies over the equatorial-CP and EP exhibit a much stronger intensity and extent more westward in response to the SGLN (Figure 6d) than to the EqLN forcing (Figure 6e). This may be due to the maximum SSTA centers straddling the equatorial CP for SGLN but in the equatorial-EP for equatorial LN. The resultant circulation fields for the SGLN simulation are contrary to that of the WSGEN simulation, showing a salient positive phase of NAO-like anomaly (Figure 6i) with the weakened subtropical westerly jet and the intensified mid-latitude jet (Figure 6n) over the North Atlantic. However, the Z850 and U200 responses over the North Atlantic exhibit a neutral pattern in the EqLN simulation (Figure 6j,o).

The above numerical experiments proved that the two branches of westerly jets serve as atmospheric bridges to extend the WSGEN/SGLN signals to the downstream regions, inducing NAO-like anomalies. However, why can the other types of ENSO SSTAs not expand their impact to the downstream North Atlantic?

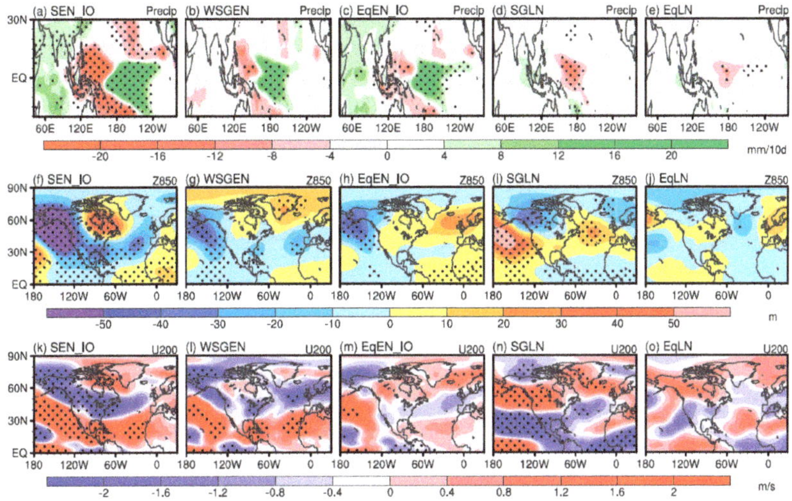

Figure 6. Tropical precipitation responses in the ECHAM5 regarding a difference between (**a**) SEN_IO, (**b**) WSGEN, (**c**) EqEN_IO, (**d**) SGLN, and (**e**) EqLN forcings and the control run (interval: 2 mm/mon). (**f–j**) same as (**a–e**) but for geopotential at 850-hPa (Z850; 10 m). (**k–o**) same as (**a–e**) but for zonal wind at 200-hPa (U200; m/s). The dots represent the anomalies significant at the 90% confidence level (Student's *t*-test).

3.1.4. The Causes for the Abruption of ENSO and NAO Relations

The observational and numerical analysis demonstrated that different El Niño/La Niña flavors exhibit intricate relations with the NAO phase. For the EN cases, we observed the salient positive SSTAs in the tropical IO for the SEN and equatorial EN cases (Figure 2a,c), reflecting the inter-basin coupling of the tropical Pacific and IO, which is the key to comprehending the teleconnection pathway on the globe that is rooted in the tropics [63–66]. Furthermore, the IO is also identified as a principal contributor to the North Atlantic extratropical atmospheric anomalies in winter [10,67,68]. Therefore, it is natural to speculate that the anomalous warm SST in the IO may modulate the linking of El Niño and NAO.

To verify our speculation, an IO warming (IOW) experiment was carried out with the positive SSTAs associated with SEN imposed onto the tropical IO, which induces rich local precipitation (Figure 7a) and a positive phase of NAO anomaly (Figure 7b). This atmosphere experiment confirms the previous studies that the slow variations of IO force the NAO on the seasonal time scale [67,69,70]. Next, the SEN and EqEN experiments were conducted without warm IO SSTA. The dipolar precipitation responses in the Pacific are much weaker in the EqEN simulation (Figure 7b) than in the SEN simulation (Figure 7c); both of them can trigger a negative NAO-like Z850 anomaly (Figure 7e,f). However, the NAO-like Z850 anomalies disappeared when the IO and SEN/EqEN forcings were imposed simultaneously (Figure 6f,h), indicating that the IO SSTA modulates the forcing of El Niño onto the NAO signals.

For the SGLN cases, the SSTA shows a sizeable east-west gradient with the maximum center in the equatorial CP (Figure 2d,e), generating a positive NAO-like anomaly over the North Atlantic (Figures 5d and 6i). By contrast, the equatorial LN features a weak east-west gradient with the maximum SSTA in the equatorial EP (Figure 2f), and the resultant atmospheric anomalies over the North Atlantic exhibit the NAO-neutral pattern (Figures 5e and 6j). However, although the amplitude of WSGLN (Figure 2e) is comparable with equatorial LN (Figure 2f) (their average intensities are −0.72 and −0.82, respectively), their resultant precipitation anomalies exhibit significant discrepancies in intensity (Figure 3e,f). Therefore, the differences in atmospheric anomalies associated with SGLN (Figure 5d) and equatorial LN (Figure 5e) may not depend on the amplitude of SSTA in the Niño3.4 region.

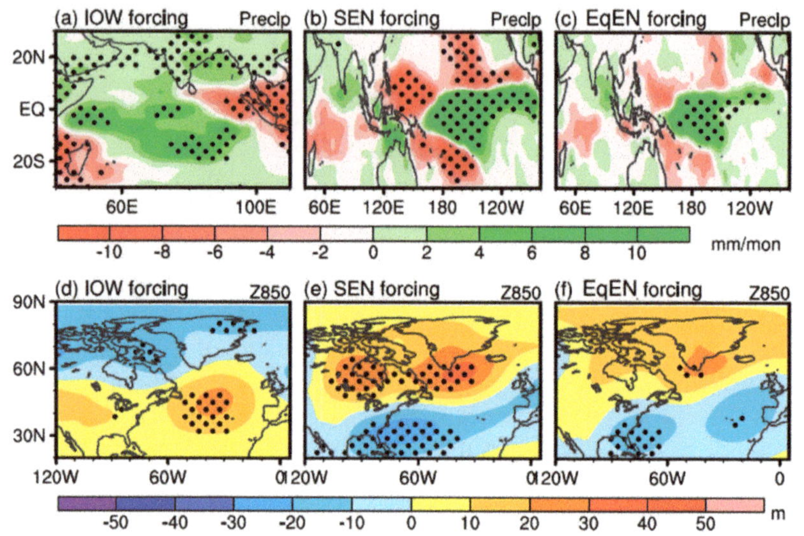

Figure 7. Tropical precipitation responses in the ECHAM5 regarding a difference between (**a**) IOW, (**b**) SEN, (**c**) EqEN forcings, and the control run (interval: 2 mm/mon). (**d–f**) same as (**a–e**) but for geopotential at 850-hPa (Z850; 10 m). The dots represent the anomalies significant at the 90% confidence level (Student's *t*-test).

Regarding the SSTA distributions, the SGLN and equatorial LN correspond to the CP and EP La Niñas, respectively. We know that the climatology SST, to a great extent, determines the occurrence of deep tropical convection; the threshold commonly is about 28 °C [71,72], and the variability of convection and atmospheric responses are primarily dependent on the anomalous tropical SST. Owing to the proximity of the central Pacific to the western Pacific warm pool, the climatology SSTs above 28 °C (green contours in Figure 2), abnormal convective motions, and extra-tropical atmospheric circulation usually present a quasi-linear response to SST anomalies [25]. During the SGLN winter, the negative SSTA in the CP superimposes on the climatology SST, significantly dropping local SST below 28 °C, reducing the deep convection and precipitation over the CP (Figure 3d,e). As a result, a steady forcing of SGLN on the atmospheric anomalies over the North Atlantic is established. In comparison, the low climatological SSTs normally do not reach the threshold to generate convection in the eastern Pacific. Consequently, the negative SSTAs in EP exert a weaker influence on the tropical convective motions and precipitation in the equatorial LN cases (Figure 3f). Therefore, changes in the maximum SSTA center in the tropical Pacific profoundly alter global teleconnections [38,73]. We further executed the WSGLN simulations with the associated SSTA imposed onto the Pacific region. Figure 8a displays the tropical precipitation anomalies in response to the WSGLN forcing versus CTRL run. The negative precipitation anomalies induced by WSGLN are concentrated in the tropical CP with rich rainfall over the WP. The large precipitation differences between WSGLN and EqLN simulations are seen over the tropical CP and EP (Figure 8b). The anomalous dipole mode over the Pacific indicates that WSGLN-type SSTA tends to reduce the tropical convection, weakening the convective motion more pronouncedly over the CP than the EP. The resultant Z850 anomalies for WSGLN forcing also highly resemble a positive phase of the NAO pattern (Figure 8c).

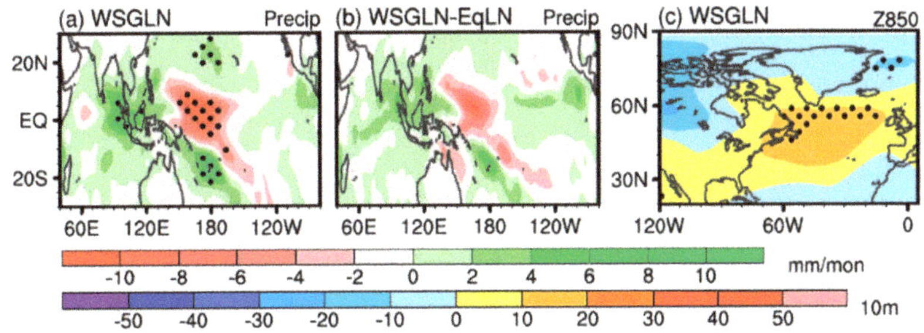

Figure 8. Tropical precipitation responses in the ECHAM5 regarding a difference between (**a**) WSGLN and the control run, (**b**) WSGLN and EqLN (interval: 2 mm/mon), (**c**) same as (**a**) but for geopotential at 850-hPa (Z850; 10 m). The dots represent the anomalies significant at the 90% confidence level (Student's *t*-test).

4. Discussion

In the present work, we separated the different flavors of ENSO, but their physical backgrounds are unclear. Due to the limited space of this article, we give a short discussion. Figure 9 displays the seasonal evolution of sea surface temperatures and surface wind anomalies for the three El Niño types. For SEN and equatorial EN during the springtime (Figure 9a,i), significant westerly wind anomalies prevail in the western Pacific. On the one hand, the westerlies induce the upwelling of the subsurface water in the WP, cooling down the local SST. Conversely, the westerlies blow the warm waters in the western Pacific warm-pool eastward, warming up the equatorial central-eastern Pacific. Such a process can continue to function from summer to winter (upper and lower panels of Figure 9). During the early summer, the cooling in the WP increases the zonal gradient between the North IO (NIO), generating tropical easterlies across the NIO (Figure 9b,j), slowing down the background westerlies and continuing the warm SSTA in the NIO (Figure 9c,k), leading to a positive Indian Ocean Dipole (IOD)-like SSTA pattern in the autumn (Figure 9d,l). However, the IOD decays rapidly thereafter, and SEN and equatorial EN continue to force the IO, inducing an Indian Ocean basin-wide warm SSTA in the ensuing winter [74] (Figure 2a,c).

The SEN and equatorial EN, however, show significant discrepancies in the northwestern Pacific (NWP). For the SEN case, the intense equatorial central-eastern Pacific warming excites a Rossby wave west of the equatorial central Pacific. Anomalous northeasterly flow from North Pacific (NP) down to the tropical WP, speeding up the trades over the western NP (WNP). The resulting wind speed—stronger in the extra-tropical and WNP—induces SSTA changes in the WNP via the Wind-Evaporation-SST (WES) feedback [75]: the structure exhibits a strong east-west gradient (Figure 9d). In the equatorial EN case, the anomalous NWP cooling reaches its maximum in the early summer (Figure 9k) and decays in the autumn (Figure 9i). The weak WNP cold SSTA, going against the local air-sea interaction process proposed by Wang et al. [17], restricts the formation of the WNP anticyclone to accelerate the trade wind and then feedback to the local SST via the WES mechanism.

Nevertheless, the WSGEN exhibits a gradual enhancement of the west-east gradient in SST over the Pacific. The initial SST warming occurs in the equatorial EP, and cooling emerges in the subtropical WP during the early summer, with the westerlies prevailing over the equatorial CP (Figure 9f). The westerly wind anomalies become stronger in the late summer and autumn with the enhancement of the west-east gradient in SST over the Pacific (Figure 9g,h), reflecting a Bjerknes positive feedback process [41]. Moreover, the initial cool SSTA in the South China Sea (SCS) moves eastward from spring to autumn. During the summertime, the deficiency of the salient cool SSTA in SCS restricts the easterlies penetrating NIO and the development of the IOD. We also noticed that the dominant wind and SST anomalies in the extra-tropical NP, which cannot be seen in the equatorial EN cases, may contribute to the cooling of the WP. However, the origin of the circulation and SST anomalies in the NP during the WSGEN autumn call for further investigation.

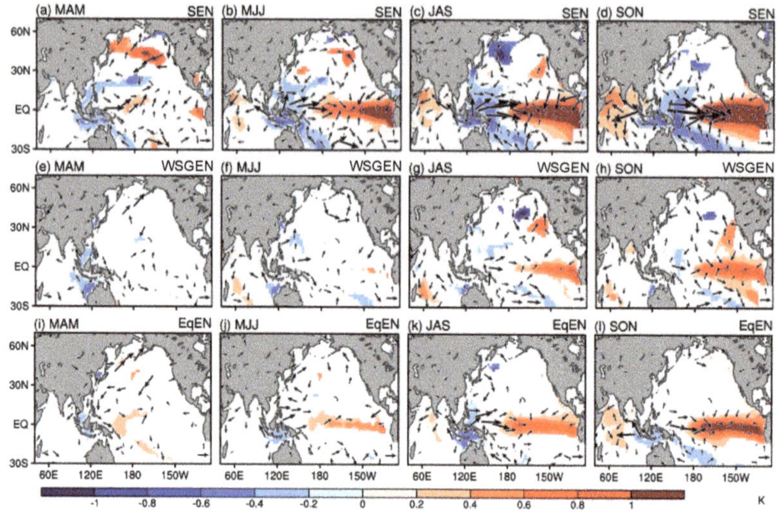

Figure 9. SST (internal: 0.2 K) and horizontal wind at 1000-hPa (UV1000; m/s) composite differences under strong El Niño. (**a**) March–April–May (MAM), (**b**) May–June–July (MJJ), (**c**) July–August–September (JAS), and (**d**) September–October–November (SON). (**e–h**) same as (**a–d**), but for WSG-El Niño. (**i–l**) for equatorial El Niño. The shadings and vectors in each panel represent the region with anomalies significant at the 90% confidence level (Student's *t*-test).

For the La Niña events, the SGLN and equatorial LN coincide with the CP and EP La Niñas, respectively. Prior studies have revealed that the previous winter or spring North Pacific Oscillation (NPO) through seasonal footprint mechanism is largely responsible for triggering the CP ENSO [76–78]. Figure 10 shows the lead-lag composite of UV850 and SSTA for SGLN. An abnormal positive NPO-like pattern appears in the previous spring, driving the North Pacific SST and resulting in a tripolar SSTA pattern in the NP region (Figure 10a). Specifically, the anticyclonic anomalies occur over the subtropical to mid-latitude North Pacific, causing abnormal easterlies in the equatorial CP. The abnormal easterlies in the tropical CP induce convergence and give feedback to maintain the development of negative SSTAs (Figure 10b). Such an air-sea interaction process amplifies the equatorial zonal wind and SST anomalies, driving the evolution of La Niña from summer to autumn (Figure 10c,d). However, such a process cannot be seen for the equatorial LN (lower panel of Figure 10). It starts to develop in the late summer without the salient atmospheric signals. This oceanic process needs to be investigated.

We should also point out that dividing ENSOs based on intensity is still debatable. Alizadeh [79] grouped ENSOs into weak, moderate, strong, and very strong events. This classification is more elaborate but obtains smaller samples than our study does. Then, the unsupervised learning methods, such as clusters, self-organizing maps, and many others, should be examined to determine whether the different types of ENSO can be subdivided more objectively. Furthermore, we only apply the AGCM to mimic the various forcing effects of different types of ENSOs on the extra-tropical atmospheric anomalies. The lack of air-sea interaction process of AGCM makes it difficult to capture the potential effects of extra-tropical SSTAs (e.g., North Pacific), i.e., feedback to ENSO forcing, on the atmospheric anomalies over the North Atlantic. A better method would be to apply the samples of CMIP6 history experiments to verify our hypothesis.

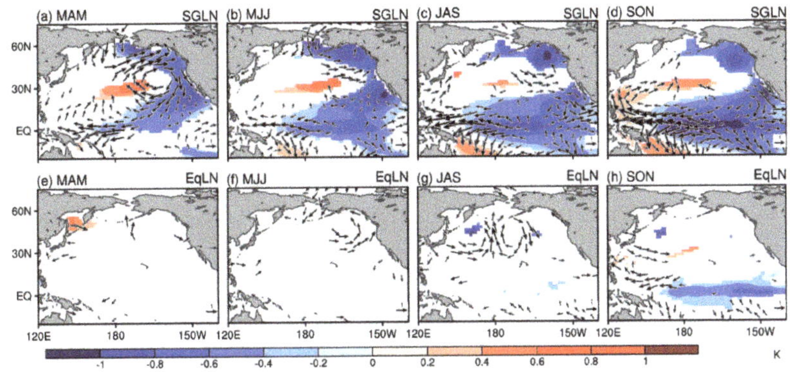

Figure 10. The same as Figure 9, but for SG-La Niña and equatorial La Niña.

5. Conclusions

The previous research pointed out that the combined effect of ENSO and NAO usually results in even more significant impacts than their individual effects [80,81]. It is, therefore, necessary to clarify their potential linkage. In this study, we classified the ENSO events according to their amplitude and east-west gradients in the Pacific. Strong ENSOs commonly contain a sizeable east-west gradient, and only the weak-intensity ENSOs can be separated based on the east-west gradient of the WSG and equatorial events. For the El Niño events, the SEN features a salient east-west gradient in the Pacific with the warming SST in the tropical IO (Figure 2a). The WSGEN also exhibits a strong gradient in the Pacific but without the significant warm SSTA in the IO (Figure 2b). The equatorial EN is characterized by the salient warm SSTA in the equatorial CP and EP, as well as in the tropical IO (Figure 2c). For the La Niña events, their spatial patterns in SST are roughly opposite to their EN counterparts (Figure 2d–f). In addition, the maximum SSTA for SLN and WSGLN centers in the equatorial CP but moves to the EP for the equatorial LN.

Next, we demonstrated that atmospheric anomalies in response to different ENSO types exhibit large discrepancies over the North Atlantic region (Figures 5 and 6). Only the WSGEN and SGLN events can produce NAO-like patterns through the eastward elongation of the two branches of the jet streams. During the WSGEN (SGLN) winter, the salient warm (cool) SSTA in the CP and EP and the opposite anomalies in the WP induce enhanced (weakened) subtropical and weakened (enhanced) mid-latitude westerly jets, leading to a negative (positive) NAO-like anomaly. By contrast, the NAO-like anomaly cannot be detected in other ENSO cases. The further study implied that the warm SSTA in the IO (Figure 7), which triggers abundant precipitation in the tropical IO during the SEN and equatorial EN winters, generates a positive NAO pattern that offsets the effect of El Niño signals in the Pacific region, inducing an NAO-neutral anomalies pattern. However, the IO SSTA cannot modulate the relation of LN and NAO owing to the weak forcing of the cool IO SSTA on local precipitation and remote teleconnections. The SGLN's maximum SSTA is located over the CP, where the climatology SST exceeds the threshold for exciting deep convective motion. The negative SSTA in the CP consistently influences atmospheric circulation, resulting in a response over the North Atlantic that resembles a positive NAO. In comparison, the maximum SSTA of equatorial LN centers in the EP, where the climatology SST is below the threshold for deep convection. The atmospheric response over the North Atlantic shows a neutral pattern.

This study further illustrates the important roles ENSO diversity plays in affecting atmospheric variability over remote regions, especially over the North Atlantic. The atmospheric/coupled general circulation model may offer better analysis for the seasonal prediction of the NAO when considering the ENSO diversity according to its amplitude and the zonal gradient in SST over the Pacific.

Other than the roles of the troposphere, tropical warming excites a poleward-propagating Rossby wave train, which can also extend upward and reach the stratosphere, resulting in a weaker polar vortex that drives a negative NAO anomaly over the NA–Eurasia. In subsequent studies, we would like to examine whether or not SG-ENSO can affect the NAO through stratospheric pathways. Moreover, we will try to classify the summertime ENSO events and inspect their potential forcing effects on extra-tropical atmospheric anomalies.

Author Contributions: Conceptualization, P.Z. and Z.W.; methodology, P.Z.; software, P.Z.; validation, P.Z. and Z.W.; formal analysis, P.Z.; investigation, Z.W.; resources, P.Z.; data curation, P.Z.; writing—original draft preparation, P.Z.; writing—review and editing, Z.W.; visualization, P.Z.; supervision, Z.W.; project administration, Z.W.; funding acquisition, Z.W. All authors have read and agreed to the published version of the manuscript.

Funding: This work is jointly supported by the National Natural Science Foundation of China (NSFC) (Grant Nos. 91937302) and the Ministry of Science and Technology of China (Grant Nos. 2019QZKK0102, 2019YFC1509100 and 2017YFC1502302).

Data Availability Statement: The ERA-40 and ERA-interim datasets identified and investigated in this study are provided by ECMWF websites: https://www.ecmwf.int/en/forecasts/datasets/browse-reanalysis-datasets; The SST data from: https://psl.noaa.gov/data/gridded/data.noaa.ersst.v5.html and https://www.metoffice.gov.uk/hadobs/hadisst/data/download.html; The precipitation data from: https://psl.noaa.gov/data/gridded/data.prec.html. The ERA-40 data is accessed on 1 June 2017, other datasets are accessed on 20 March 2020.

Conflicts of Interest: The authors declare no conflict of interest. The funders had no role in the design of the study; in the collection, analyses, or interpretation of data; in the writing of the manuscript; or in the decision to publish the results.

References

1. Hurrell, J.W. Decadal Trends in the North Atlantic Oscillation: Regional Temperatures and Precipitation. *Science* **1995**, *269*, 676–679. [CrossRef] [PubMed]
2. Thompson, D.W.; Wallace, J.M. The Arctic Oscillation signature in the wintertime geopotential height and temperature fields. *Geophys. Res. Lett.* **1998**, *25*, 1297–1300. [CrossRef]
3. Li, J.; Wang, J. A new North Atlantic Oscillation index and its variability. *Adv. Atmos. Sci.* **2003**, *20*, 661–676.
4. Deser, C.; Hurrell, J.W.; Phillips, A.S. The role of the North Atlantic Oscillation in European climate projections. *Clim. Dyn.* **2017**, *49*, 3141–3157. [CrossRef]
5. Branstator, G.; Selten, F. "Modes of Variability" and Climate Change. *J. Clim.* **2009**, *22*, 2639–2658. [CrossRef]
6. Hurrell, J.W.; Deser, C. North Atlantic climate variability: The role of the north atlantic oscillation. *J. Mar. Syst.* **2009**, *78*, 28–41. [CrossRef]
7. Butler, A.H.; Polvani, L.M.; Deser, C. Separating the stratospheric and tropospheric pathways of El Niño–Southern Oscillation teleconnections. *Environ. Res. Lett.* **2014**, *9*, 024014. [CrossRef]
8. Scaife, A.A.; Arribas, A.; Blockley, E.; Brookshaw, A.; Clark, R.T.; Dunstone, N.; Eade, R.; Fereday, D.; Folland, C.K.; Gordon, M.; et al. Skillful long-range prediction of European and North American winters. *Geophys. Res. Lett.* **2014**, *41*, 2514–2519. [CrossRef]
9. Kidston, J.; Scaife, A.A.; Hardiman, S.C.; Mitchell, D.M.; Butchart, N.; Baldwin, M.P.; Gray, L.J. Stratospheric influence on tropospheric jet streams, storm tracks and surface weather. *Nat. Geosci.* **2015**, *8*, 433–440. [CrossRef]
10. Hoerling, M.P.; Hurrell, J.W.; Xu, T.; Bates, G.T.; Phillips, A.S. Twentieth century north atlantic climate change. Part II. Understanding the effect of indian ocean warming. *Clim. Dyn.* **2004**, *23*, 391–405. [CrossRef]
11. Bracco, A.; Kucharski, F.; Kallummal, R.; Molteni, F. Internal variability, external forcing and climate trends in multi-decadal AGCM ensembles. *Clim. Dyn.* **2004**, *23*, 659–678. [CrossRef]
12. Li, Y.; Lau, N.C. Impact of ENSO in the atmospheric variability over the North Atlantic in late winter—Role of transient eddies. *J. Clim.* **2012**, *25*, 320–342. [CrossRef]
13. Wyrtki, K. El Niño—The Dynamic Response of the Equatorial Pacific Oceanto Atmospheric Forcing. *J. Phys. Oceanogr.* **1975**, *5*, 572–584. [CrossRef]
14. Qadimi, M.; Alizadeh, O.; Irannejad, P. Impacts of the El Niño-Southern Oscillation on the strength and duration of the Indian summer monsoon. *Meteorol. Atmos. Phys.* **2021**, *133*, 553–564. [CrossRef]
15. Held, I.M.; Lyons, S.W.; Nigam, S. Transients and the extratropical response to El Niño. *J. Atmos. Sci.* **1989**, *46*, 163–174. [CrossRef]
16. Alizadeh-Choobari, O.; Adibi, P.; Irannejad, P. Impact of the El Niño–Southern Oscillation on the climate of Iran using ERA-Interim data. *Clim. Dyn.* **2018**, *51*, 2897–2911. [CrossRef]
17. Wang, B.; Wu, R.G.; Fu, X.H. Pacific–East Asia teleconnection: How does ENSO affect East Asian climate? *J. Clim.* **2000**, *13*, 1517–1536. [CrossRef]

18. Ropelewski, C.F.; Halpert, M.S. Global and regional scale precipitation patterns associated with the El Niño/Southern Oscillation. *Mon. Weather Rev.* **1987**, *115*, 1606–1626. [CrossRef]
19. Trenberth, K.E.; Caron, J.M. The Southern Oscillation revisited: Sea level pressure, surface temperatures, and precipitation. *J. Clim.* **2000**, *13*, 4358–4365. [CrossRef]
20. Merkel, U.; Latif, M. A high-resolution AGCM study of the El Niño impact on the North Atlantic/European sector. *Geophys. Res. Lett.* **2002**, *29*, 1291. [CrossRef]
21. Ineson, S.; Scaife, A.A. The role of the stratosphere in the European climate response to El Niño. *Nat. Geosci.* **2009**, *2*, 32–36. [CrossRef]
22. Jiménez-Esteve, B.; Domeisen, D.I.V. The Tropospheric Pathway of the ENSO–North Atlantic Teleconnection. *J. Clim.* **2018**, *31*, 4563–4584. [CrossRef]
23. Brönnimann, S. Impact of El Niño–Southern Oscillation on European climate. *Rev. Geophys.* **2007**, *45*, RG3003. [CrossRef]
24. Zhang, W.; Wang, L.; Xiang, B.Q.; Qi, L.; He, J. Impacts of two types of La Nina on the NAO during boreal winter. *Clim. Dyn.* **2015**, *44*, 1351–1366. [CrossRef]
25. Zhang, W.; Wang, Z.; Stuecker, M.F.; Turner, A.G.; Jin, A.G.; Geng, X. Impact of ENSO longitudinal position on teleconnections to the NAO. *Clim. Dyn.* **2019**, *2*, 257–274. [CrossRef]
26. Zhang, P.; Wang, B.; Wu, Z. Weak El Niño and winter climate in the mid-high latitude Eurasia. *J. Clim.* **2019**, *32*, 402–421. [CrossRef]
27. Zhang, P.; Wu, Z.; Li, J. Reexamining the relationship of La Niña and the east Asian winter monsoon. *Clim. Dyn.* **2019**, *53*, 779–791. [CrossRef]
28. Zhang, P.; Wu, Z. Reexamining the connection of El Niño and North American winter climate. *Int. J. Climatol.* **2021**, *41*, 6133–6144. [CrossRef]
29. Polvani, L.M.; Sun, L.; Butler, A.H.; Richter, J.H.; Deser, C. Distinguishing stratospheric sudden warmings from ENSO as key drivers of wintertime climate variability over the North Atlantic and Eurasia. *J. Clim.* **2017**, *30*, 1959–1969. [CrossRef]
30. Wu, Z.; Zhang, P. Interdecadal variability of the mega-ENSO–NAO synchronization in winter. *Clim. Dyn.* **2015**, *45*, 1117–1128. [CrossRef]
31. Capotondi, A.; Wittenberg, A.T.; Newman, M.; Di Lorenzo, E.; Yu, J.Y.; Braconnot, P.; Cole, J.; Dewitte, B.; Giese, B.; Guilyardi, E.; et al. Understanding ENSO diversity. *Bull. Am. Meteorol. Soc.* **2015**, *96*, 921–938. [CrossRef]
32. McPhaden, M.J. Genesis and evolution of the 1997–1998 El Niño. *Science* **1999**, *283*, 950–954. [CrossRef] [PubMed]
33. Jacox, M.G.; Hazen, E.L.; Zaba, K.D.; Rudnick, D.L.; Edwards, C.A.; Moore, A.M.; Bograd, S.J. Impacts of the 2015–2016 El Niño on the California current system: Early assessment and comparison to past events. *Geophys. Res. Lett.* **2016**, *430*, 7072–7080. [CrossRef]
34. Toniazzo, T.; Scaife, A.A. The influence of ENSO on winter North Atlantic climate. *Geophys. Res. Lett.* **2006**, *33*, L24704. [CrossRef]
35. Xue, F.; Liu, C. The influence of moderate ENSO on summer rainfall in eastern China and its comparison with strong ENSO. *Chin. Sci. Bull.* **2008**, *53*, 791–800. [CrossRef]
36. Alizadeh-Choobari, O.; Najafi, M.S. Climate variability in Iran in response to the diversity of the El Niño-Southern Oscillation. *Int. J. Climatol.* **2018**, *38*, 4239–4250. [CrossRef]
37. Kao, H.; Yu, J. Contrasting Eastern-Pacific and Central-Pacific Types of ENSO. *J. Clim.* **2009**, *22*, 615–632. [CrossRef]
38. Alizadeh-Choobari, O. Contrasting global teleconnection features of the eastern Pacific and central Pacific El Niño events. *Dyn. Atmos. Ocean* **2017**, *80*, 139–154. [CrossRef]
39. Feng, J.; Li, J.P.; Zheng, F.; Xie, F.; Sun, C. Contrasting impacts of developing phases of two types of El Niño on southern China rainfall. *J. Meteorol. Soc. Jpn.* **2016**, *94*, 359–370. [CrossRef]
40. Wang, B.; Luo, X.; Yang, Y.M.; Sun, W.; Cane, M.A.; Cai, W.; Yeh, S.W.; Liu, J. Historical change of El Niño properties sheds light on future changes of extreme El Niño. *Proc. Natl. Acad. Sci. USA* **2019**, *116*, 22512–22517. [CrossRef]
41. Bjerknes, J. Atmospheric teleconnections from the equatorial Pacific. *Mon. Weather Rev.* **1969**, *97*, 163–172. [CrossRef]
42. Hoell, A.; Funk, C. The ENSO-related West Pacific Sea surface temperature gradient. *J. Clim.* **2013**, *26*, 9545–9562. [CrossRef]
43. Liebmann, B.; Hoerling, M.P.; Funk, C.; Bladé, I.; Dole, R.M.; Allured, D.; Quan, X.; Pegion, P.; Eischeid, J.K. Understanding recent eastern horn of Africa rainfall variability and change. *J. Clim.* **2014**, *27*, 8630–8645. [CrossRef]
44. Zinke, J.; Hoell, A.; Lough, J.M.; Feng, M.; Kuret, A.J.; Clarke, H.; Ricca, V.; Rankenburg, K.; McCulloch, M.T. Coral record of Southeast Indian Ocean marine heatwaves with intensified Western Pacific temperature gradient. *Nat. Commun.* **2015**, *6*, 8562. [CrossRef] [PubMed]
45. Zhang, L.; Wu, Z.; Zhou, Y. Different impacts of typical and atypical ENSO on the Indian summer rainfall: ENSO developing phase. *Atmos. Ocean* **2016**, *54*, 440–456. [CrossRef]
46. Meinen, C.S.; McPhaden, M.J. Observations of Warm Water Volume Changes in the Equatorial Pacific and Their Relationship to El Niño and La Niña. *J. Clim.* **2000**, *13*, 3551–3559. [CrossRef]
47. Jin, F.-F.; An, S.; Timmermann, A.; Zhao, J. Strong El Niño Events and Nonlinear Dynamical Heating. *Geophys. Res. Lett.* **2003**, *30*, 1120. [CrossRef]
48. An, S.-I.; Jin, F.-F. Why El Niño is Stronger than La Niña. *J. Clim.* **2004**, *17*, 2399–2412. [CrossRef]
49. Larkin, N.K.; Harrison, D.E. ENSO warm (El Niño) and cold (La Niña) event life cycles: Ocean surface anomaly patterns, their symmetries, asymmetries, and implications. *J. Clim.* **2002**, *15*, 1118–1140. [CrossRef]

50. Okumura, Y.M.; Deser, C. Asymmetry in the duration of El Niño and La Niña. *J. Clim.* **2010**, *23*, 5826–5843. [CrossRef]
51. Chen, M.; Li, T. ENSO evolution asymmetry: EP versus CP El Niño. *Clim. Dyn.* **2021**, *56*, 3569–3579. [CrossRef]
52. Zhang, R.; Li, T.-R.; Wen, M.; Liu, L.-K. Role of intraseasonal oscillation in asymmetric impacts of El Niño and La Niña on the rainfall over southern China in boreal winter. *Clim. Dyn.* **2015**, *45*, 559–567. [CrossRef]
53. Song, X.; Zhang, R.; Rong, X. Influence of intraseasonal oscillation on the asymmetric decays of El Niño and La Niña. *Adv. Atmos. Sci.* **2019**, *36*, 779–792. [CrossRef]
54. Huang, B.; Thorne, P.W.; Banzon, V.F.; Boyer, T.; Chepurin, G.; Lawrimore, J.H.; Menne, M.J.; Smith, T.M.; Vose, R.S.; Zhang, H.M. Extended reconstructed sea surface temperature, version 5 (ERSSTv5): Upgrades, validations, and intercomparisons. *J. Clim.* **2017**, *30*, 8179–8205. [CrossRef]
55. Rayner, N.A.; Parker, D.E.; Horton, E.B.; Folland, C.K.; Alexander, L.V.; Rowell, D.P.; Kent, E.C.; Kaplan, A. Global analyses of sea surface temperature, sea ice, and night marine air temperatures since the late nineteenth century. *J. Geophys. Res.* **2003**, *108*, 4407. [CrossRef]
56. Uppala, S.M.; Kållberg, P.W.; Simmons, A.J.; Andrae, U.; Bechtold VD, C.; Fiorino, M.; Gibson, J.K.; Haseler, J.; Hernandez, A.; Kelly, G.A.; et al. The ERA-40 re-analysis. *Q. J. R. Meteorol. Soc.* **2005**, *131*, 2961–3012. [CrossRef]
57. Dee, D.P.; Uppala, S.M.; Simmons, A.J.; Berrisford, P.; Poli, P.; Kobayashi, S.; Andrae, U.; Balmaseda, M.A.; Balsamo, G.; Bauer, D.P.; et al. The ERA-Interim reanalysis: Configuration and performance of the data assimilation system. *Q. J. R. Meteorol. Soc.* **2011**, *137*, 553–597. [CrossRef]
58. Chen, M.; Xie, P.; Janowiak, J.; Arkin, P.A. Global land precipitation: A 50-yr analysis based on gauge observations. *J. Hydrometeorol.* **2002**, *3*, 249–266. [CrossRef]
59. Wang, B.; Liu, J.; Kim, H.J.; Webster, P.J.; Yim, S.Y.; Xiang, B.Q. Northern Hemisphere summer monsoon intensified by mega-El Niño/southern oscillation and Atlantic multidecadal oscillation. *Proc. Natl. Acad. Sci. USA* **2013**, *110*, 5347–5352. [CrossRef]
60. Roeckner, E.; Bäuml, G.; Bonaventura, L.; Brokopf, R.; Esch, M.; Giorgetta, M.; Hagemann, S.; Kirchner, I.; Kornblueh, L.; Manzini, E.; et al. The atmospheric general circulation model ECHAM5. Part I: Model description. *Max Planck Inst. Rep.* **2003**, *349*, 140.
61. Ashok, K.; Behera, S.; Rao, A.S.; Weng, H.Y.; Yamagata, T. El Niño Modoki and its possible teleconnection. *J. Geophys. Res.* **2007**, *112*, C11007. [CrossRef]
62. Graf, H.-F.; Zanchettin, D. Central Pacific El Niño, the 'subtropical bridge,' and Eurasian climate. *J. Geophys. Res.* **2012**, *117*, D01102. [CrossRef]
63. Saji, N.H.; Yamagata, T. Structure of SST and surface wind variability during Indian Ocean dipole mode events: COADS observations. *J. Clim.* **2003**, *16*, 2735–2751. [CrossRef]
64. Luo, J.-J.; Zhang, R.; Behera, S.K.; Masumoto, Y.; Jin, F.-F.; Lukas, R.; Yamagata, T. Interaction between El Niño and extreme Indian Ocean dipole. *J. Clim.* **2010**, *23*, 726–742. [CrossRef]
65. Cai, W.; van Rensch, P.; Cowan, T.; Hendon, H.H. An asymmetry in the IOD and ENSO teleconnection pathway and its impact on Australian climate. *J. Clim.* **2012**, *25*, 6318–6329. [CrossRef]
66. Ha, K.J.; Chu, J.E.; Lee, J.Y.; Yun, K.S. Interbasin coupling between the tropical Indian and Pacific Ocean on interannual timescale: Observation and CMIP5 reproduction. *Clim. Dyn.* **2017**, *48*, 459–475. [CrossRef]
67. Bader, J.; Latif, M. North Atlantic Oscillation response to anomalous Indian Ocean SST in a coupled GCM. *J. Clim.* **2005**, *18*, 5382–5389. [CrossRef]
68. Lin, H.; Brunet, G.; Derome, J. An observed connection between the North Atlantic Oscillation and the Madden-Julian Oscillation. *J. Clim.* **2009**, *22*, 364–380. [CrossRef]
69. Bader, J.; Latif, M. The impact of decadal-scale Indian Ocean sea surface temperature anomalies on Sahelian rainfall and the North Atlantic Oscillation. *Geophys. Res. Lett.* **2003**, *30*, 2169. [CrossRef]
70. Hurrell, J.W.; Hoerling, M.P.; Phillips, A.; Xu, T. Twentieth century North Atlantic climate change. Part I: Assessing determinism. *Clim. Dyn.* **2004**, *23*, 371–389. [CrossRef]
71. Graham, N.E.; Barnett, T.P. Sea surface temperature, surface wind divergence, and convection over tropical oceans. *Science* **1987**, *238*, 657–659. [CrossRef] [PubMed]
72. Lau, K.-M.; Wu, H.T.; Bony, S. The Role of Large-Scale Atmospheric Circulation in the Relationship between Tropical Convection and Sea Surface Temperature. *J. Clim.* **1997**, *10*, 381–392. [CrossRef]
73. Domeisen, D.I.; Garfinkel, C.I.; Butler, A.H. The teleconnection of El Niño Southern Oscillation to the stratosphere. *Rev. Geophys.* **2019**, *57*, 5–47. [CrossRef]
74. Du, Y.; Xie, S.-P.; Hu, K.; Huang, G. Role of air–sea interaction in the long persistence of El Niño–induced North Indian Ocean warming. *J. Clim.* **2009**, *22*, 2023–2038. [CrossRef]
75. Xie, S.-P. The Shape of Continents, Air-Sea Interaction, and the Rising Branch of the Hadley Circulation. In *The Hadley Circulation: Past, Present and Future*; Diaz, H., Bradley, R.S., Eds.; Kluwer Academic: Amsterdam, The Netherlands, 2004; pp. 121–152.
76. Vimont, D.J.; Alexander, M.A.; Newman, M. Optimal growth of Central and East Pacific ENSO events. *Geophys. Res. Lett.* **2014**, *41*, 4027–4034. [CrossRef]
77. Yu, J.-Y.; Kao, H.Y.; Lee, T. Subtropics-Related Interannual Sea Surface Temperature Variability in the Equatorial Central Pacific. *J. Clim.* **2010**, *23*, 2869–2884. [CrossRef]
78. Yu, J.-Y.; Paek, H. Precursors of ENSO beyond the tropical Pacific. *US CLIVAR Var.* **2015**, *13*, 15–20.

79. Alizadeh, O. Amplitude, duration, variability, and seasonal frequency analysis of the El Niño-Southern Oscillation. *Clim. Chang.* **2022**, *174*, 20. [CrossRef]
80. Wu, Z.; Lin, H. Interdecadal variability of the ENSO-North Atlantic Oscillation connection in boreal summer. *Q. J. R. Meteorol. Soc.* **2012**, *138*, 1668–1675. [CrossRef]
81. Seager, R.; Kushnir, Y.; Nakamura, J.; Ting, M.; Naik, N. Northern Hemisphere winter snow anomalies: ENSO, NAO and the winter of 2009/10. *Geophys. Res. Lett.* **2010**, *37*, L14703. [CrossRef]

remote sensing

MDPI

Technical Note

How Important Is Satellite-Retrieved Aerosol Optical Depth in Deriving Surface $PM_{2.5}$ Using Machine Learning?

Zhongyan Tian [1,†], Jing Wei [2,†] and Zhanqing Li [2,*]

1 Faculty of Geographical Science, Beijing Normal University, Beijing 100875, China; 202021490035@mail.bnu.edu.cn
2 Department of Atmospheric and Oceanic Science, Earth System Science Interdisciplinary Center, University of Maryland, College Park, MD 20740, USA; weijing@umd.edu
* Correspondence: zhanqing@umd.edu
† These authors contributed equally to this work.

Abstract: $PM_{2.5}$ refers to the total mass concentration of tiny particulates in the atmosphere near the surface, obtained by means of in situ observations and satellite remote sensing. Given the highly limited number of ground observation stations of inhomogeneous distribution and an ill-posed remote sensing approach, increasing efforts have been devoted to the application of machine-learning (ML) models to both ground and satellite data. A key satellite-derived parameter, aerosol optical thickness (AOD), has been most commonly used as a proxy of $PM_{2.5}$, although their correlation is fraught with large uncertainties. A critical question that has been overlooked concerns how much AOD helps to improve the retrieval of $PM_{2.5}$ relative to its uncertainty incurred concurrently. The question is addressed here by taking advantage of high-density $PM_{2.5}$ stations in eastern China to evaluate the contributions of AOD, determined as the difference in the accuracy of $PM_{2.5}$ retrievals with and without AOD for varying densities of $PM_{2.5}$ stations, using four popular ML models (i.e., Random Forest, Extra-trees, XGBoost, and LightGBM). Our results reveal that as the density of monitoring stations decreases, both the feature importance and permutation importance of satellite AOD demonstrate a consistent upward trend ($p < 0.05$). Furthermore, the ML models without AOD exhibit faster declines in overall accuracy and predictive ability compared with the models with AOD assessed using the sample-based and station-based (spatial) independent cross-validation approaches. Overall, a 10% reduction in the number of stations results in an increase of 0.7–1.2% and 0.6–1.2% in uncertainty in estimated and predicted accuracies, respectively. These findings attest to the indispensable role of satellite AOD in the $PM_{2.5}$ retrieval process through ML because it can significantly mitigate the negative impact of the sparse distribution of monitoring sites. This role becomes more important as the number of $PM_{2.5}$ stations decreases.

Keywords: machine learning; AOD; $PM_{2.5}$ retrieval; station density; importance assessment

Citation: Tian, Z.; Wei, J.; Li, Z. How Important Is Satellite-Retrieved Aerosol Optical Depth in Deriving Surface $PM_{2.5}$ Using Machine Learning? . *Remote Sens.* **2023**, *15*, 3780. https://doi.org/10.3390/rs15153780

Academic Editor: Stephan Havemann

Received: 30 June 2023
Revised: 26 July 2023
Accepted: 26 July 2023
Published: 29 July 2023

1. Introduction

$PM_{2.5}$ refers to the concentration of airborne particulate matter (PM) with an aerodynamic diameter of less than 2.5 microns. Although small, these particles are abundant and active, and attach easily to toxic and harmful substances. $PM_{2.5}$ can be suspended in the atmosphere for extended periods, ranging from months to even years, which has an important impact on air quality and visibility and also affects human health [1–3]. While $PM_{2.5}$ has been monitored in many parts of the world, observations are still highly limited and very inhomogeneous, with many regions not covered [4–6]. However, satellite remote sensing provides continuous spatial coverage and has been widely used in the estimation of surface $PM_{2.5}$ concentrations [7–9].

Previous studies have made great efforts to infer $PM_{2.5}$ from satellite retrievals of aerosol optical depth (AOD) by virtue of their positive correlation because AOD is much more monitored from both space and the ground. Many factors can influence their relationship, including aerosol vertical distribution, relative humidity, mixed-layer height, and topography, among others [10–12]. The relationship also varies with both location and time scale. Wang and Christopher (2003) [13] used the AOD product retrieved by the Moderate Resolution Imaging Spectroradiometer (MODIS) and in situ measurements of $PM_{2.5}$ at seven ground observation stations in Alabama, USA, finding a sound correlation between them on a monthly time scale. Natunen et al. (2010) [14] explored the relationship at four stations in Helsinki, Finland, on seasonal and monthly time scales and found that time averaging increased the correlation. Likewise, the correlation also varies with spatial resolution [15], indicating that different geographical locations, study area sizes, and spatial resolutions of MODIS AOD products can change the correlation between AOD and $PM_{2.5}$. In general, the relationship varies considerably with location and season [16,17]. Su et al. (2018) [18] studied the relationship extensively across China, one of the most polluted regions of the world, finding that the relationship differs considerably in different parts of China (better in northern than in southern China) and among the four seasons (better in winter than in summer). The relationship can be significantly improved by normalizing against the height of the planetary boundary layer.

Due to the complex relationships between AOD and $PM_{2.5}$, many statistical regression methods have been proposed for estimating $PM_{2.5}$ using satellite AOD retrievals [7,19–21], such as the multiple linear regression model [22], the geographically weighted regression model [23,24], the geographical spatiotemporal weighting regression model [25], and the linear mixed effects model [26]. To a certain extent, these models are capable of estimating surface $PM_{2.5}$ concentrations using satellite AOD data. However, they face challenges when it comes to studying the influences of various factors on $PM_{2.5}$, such as meteorological factors (boundary layer height, relative humidity, etc.) and surface factors (underlying surface types, etc.) [27,28]. Fortunately, machine-learning (ML) models have a strong data-mining capability and can establish robust nonlinear relationships. They allow for the extraction of pertinent information from very large numbers of auxiliary factors to improve the accuracy of $PM_{2.5}$ retrievals. Therefore, various types of ML models have been adopted in $PM_{2.5}$ inversion studies in recent years, e.g., the Random Forest model [29,30], the Extra-trees model [31,32], the XGBoost model [33], and the LightGBM model [34].

AOD has been regarded as an essential input variable in inferring $PM_{2.5}$ from satellites [7,19,27,35]. However, a handful of studies have presented contrasting results [36–39]. Chen et al. (2021) [38], for example, developed a Random Forest model for areas with and without AOD data, finding that the model or areas without AOD can result in better $PM_{2.5}$ retrievals. Yu et al. (2022) [39] developed a deep ensemble ML framework to estimate daily $PM_{2.5}$ concentrations in Italy from 2015 to 2019 and found similar accuracies (cross-validated $R^2 = 0.853$ and 0.857) in comparison with ground observations when including or not including satellite AOD in the model. These conflicting findings pose such critical questions as whether satellite-retrieved AOD plays any significant role in estimating surface $PM_{2.5}$ and what factor, if any, dictates its role in the ML application for estimating $PM_{2.5}$. We attempt to address these questions by taking advantage of rich satellite AOD data and in situ $PM_{2.5}$ measurements in eastern China, together with a large array of other ancillary data, introduced next.

2. Data and Methods

2.1. Study Area

The study area (approximately 1,830,000 km^2) covers 14 provinces in China, including the North China Plain, the Yangtze River Delta, the Pearl River Delta, and parts of central China (Figure 1). As the most populated and advanced in economic development in China, these regions have experienced serious air pollution problems, thus garnering significant public attention. To monitor air pollution, relatively dense $PM_{2.5}$ ground observation stations have been uniformly distributed, enabling us to investigate the effects of satellite AOD on estimating $PM_{2.5}$ concentrations at different levels of ground observation station density.

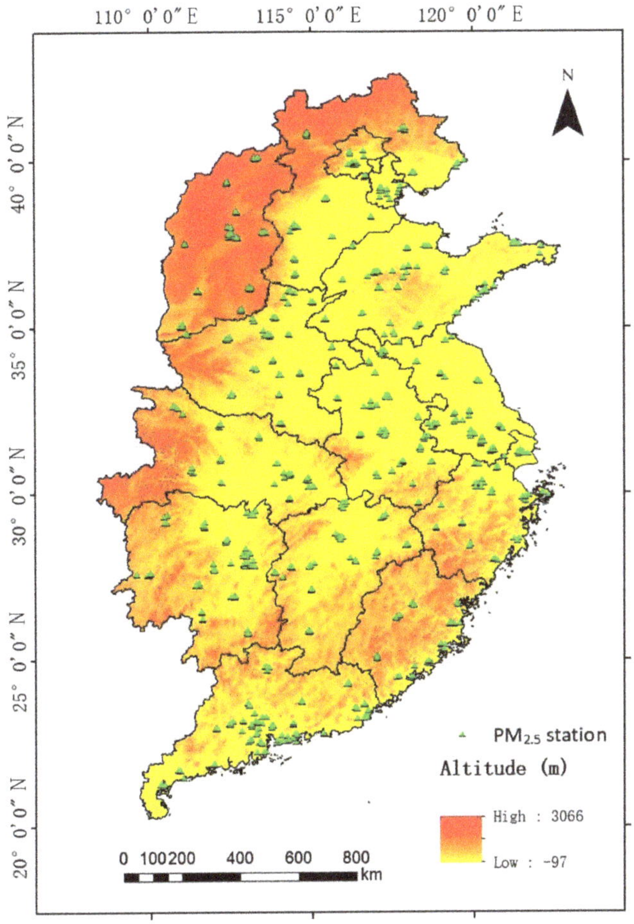

Figure 1. Study area and the distribution of ground stations (green triangles). The colored background shows land elevations (unit: m).

2.2. Data Sources

The datasets used in this study consist of observed $PM_{2.5}$ concentrations, 1-km-resolution MODIS Multi-Angle Implementation of Atmospheric Correction (MAIAC) AOD products, and many auxiliary datasets related to $PM_{2.5}$, such as meteorological and land- and population-related information. The study period spans from 2018 to 2020, ensuring an adequate volume of data for conducting sensitivity analyses.

2.2.1. $PM_{2.5}$ Ground Measurements

The $PM_{2.5}$ observation station data used in this study are real-time ground-measured air quality data (including $PM_{2.5}$) in China from the China National Environment Monitoring Center. In this study, a total of 775 ground observation stations were chosen from the eastern region of China (Figure 1). Daily measurements recorded at each station were subsequently calculated and subjected to rigorous data quality control following Wei et al.

(2019) [30]. These validated ground measurements were then utilized as ground truth for ML-modeling purposes.

2.2.2. MODIS AOD Products

The MODIS AOD product serves as the primary predictor for estimating surface $PM_{2.5}$ in this study. Specifically, Terra and Aqua MCD19A2 AOD products at a spatial resolution of 1 km are employed. This product is retrieved using the MAIAC inversion algorithm over land and incorporates various quality assurance (QA) measures [40,41]. For this study, to ensure the quality of the data, we only employed those MAIAC AOD retrievals passing the recommended QA measures, including cloud screening (QACloudMask = Clear) and adjacency (QAAdjacencyMask = Clear), following the methodology outlined in our previous study [30].

2.2.3. Auxiliary Data

Meteorological reanalysis data used in this paper are collected from the fifth-generation European Reanalysis Interim dataset (ERA5) released by the European Centre for Medium-Range Weather Forecasts. The global hourly dataset has characterized the states of the atmosphere, oceans, and surface since 1979 [42]. Specifically, seven meteorological parameters were employed: boundary layer height (BLH; unit: m), evaporation (ET; unit: mm), relative humidity (RH; unit: %), surface pressure (SP; unit: hPa), 2 m air temperature (TEM; unit: K), and 10 m U and V wind components (unit: m s^{-1}). Copernicus Atmospheric Monitoring Service (CAMS) emission inventories, including the four main precursors of $PM_{2.5}$, i.e., ammonia, nitrogen oxides, sulfur dioxide, and volatile organic compounds, were also considered [43,44]. In addition, parameters related to surface conditions and human activities, including the normalized vegetation index, a digital elevation model, and population density, were involved. In total, 15 predictor variables, including AOD, are utilized for $PM_{2.5}$ modeling through ML.

2.3. *Methodology*

2.3.1. Machine-Learning (ML) Models

ML applies complex statistical theories and algorithms to computer simulations in a somewhat human-learning behavior to acquire new knowledge. It can take advantage of any existing knowledge structure and ample information content to infer a piece of new knowledge. ML has been applied in various fields, including the remote sensing of $PM_{2.5}$ [45–47]. This study compares and analyzes four popular widely used ML algorithms, i.e., Random Forest, Extra-trees, XGBoost, and LightGBM models.

The Random Forest model constructs an ensemble of multiple decision trees, where each tree is generated by bootstrap sampling from the training dataset [48–50]. The Extremely Randomized Trees (Extra-trees) model is also a tree-based ensemble learning method, similar to Random Forest but introduces additional randomness in selecting features and splitting the points from all data samples in the tree-building process [31,32,51]. Both the Random Forest and Extra-trees models are ensemble-learning algorithms based on the bagging technique. In the bagging training process, the base classifiers (decision trees) are trained independently, and there is no strong dependence or correlation between them. This characteristic allows for parallel training of the base classifiers, which can significantly speed up the training process. By training the base classifiers in parallel, these integrated algorithms harness the power of parallel computing, making them efficient and scalable for large datasets.

In contrast, the XGBoost and LightGBM models are based on a boosting ensemble algorithm (Figure 2), where base classifiers are trained sequentially, and each classifier depends on the others. The main goal of boosting is to stack these classifiers on top of each other, with each layer assigning higher weights to samples that were incorrectly classified by the previous layers. However, these two models differ in several ways. XGBoost utilizes a pre-classification algorithm, meaning that all features of a sample are pre-sorted before iterated and repeated operations take place [33,52]. This sorting step significantly reduces the number of calculations required. LightGBM employs a histogram algorithm, which offers advantages such as reduced memory usage and a faster runtime [34,53]. In terms of growth strategy, XGBoost follows a level-wise approach. In this strategy, the child nodes of the same layer are split simultaneously. Conversely, LightGBM adopts a leaf-wise growth strategy, where each layer's child node only needs to find the node with the largest gain (typically the one with the largest data volume) to perform the split.

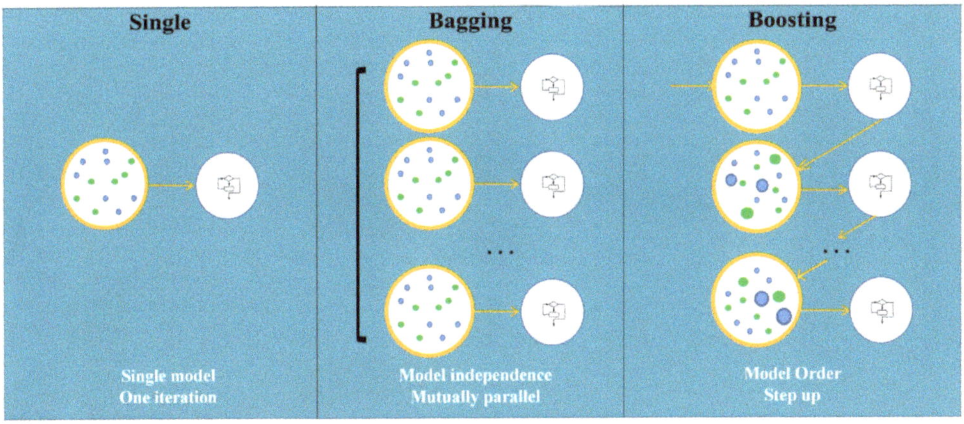

Figure 2. Illustration of the bagging and boosting algorithms.

2.3.2. Importance Assessment Method

(1) Feature importance

The feature importance (FI) score is a common indicator reflecting the importance of input variables that comes with the tree-based ML models. The FI score is calculated via the Gini index based on the mean decrease impurity (MDI) and used to evaluate the importance of each feature by measuring its contribution to splitting in the decision tree [54]. Taking one decision tree as an example, VIM represents variable importance measures, and GI represents the Gini index. Assuming that there are m feature variables, the GI score of a feature (represented by X_i) is calculated (represented by VIM_i^{Gini}), i.e., the average change in the node-splitting impurity of the ith feature in the tree model. The Gini index is calculated as

$$GI_m = \sum_{k=1}^{|K|} \sum_{k' \neq k} p_{mk} p_{mk'} = 1 - \sum_{k=1}^{|K|} p_{mk}^2 \quad (1)$$

where K represents the total number of categories of a feature, and p_{mk} represents the proportion of a category k in node m. The importance of feature X_i in node m, i.e., the GI change before and after the node branch is expressed as

$$VIM_{im}^{(Gini)} = GI_m - GI_l - GI_r \quad (2)$$

where GI_l and GI_r represent the Gini indices of the two new nodes after the branch. The node where feature X_i appears in the jth decision tree is set M. The importance of X_i in the jth tree then is

$$VIM_{ij}^{(Gini)} = \sum_{m \in M} VIM_{im}^{(Gini)} \quad (3)$$

(2) Permutation importance

Although the FI can reflect the characteristic importance of variables, it is more favorable when there are more variable categories. For characteristic variables with multiple correlations, FI may have a bias in describing correlation features, and its assessment could also be overfitted [55]. Therefore, the permutation importance (PI), another method for evaluating the contribution of each feature, is employed. The PI is a model-independent method applicable to almost all types of models, including deep-learning models. Its basic idea is to evaluate the importance of features on a test set by randomly shuffling a feature and then measuring the change in model performance to measure the importance of features. The PI of a feature is calculated as follows (Figure 3). First, the dataset is divided into a training set and a validation set. Second, a baseline metric, defined by scoring, is then evaluated on a (potentially different) dataset defined by the training set. Third, a feature column from the validation set is permutated randomly, and the metric is evaluated again. The PI can be obtained by calculating the difference between the baseline metric and the metric from permutating the feature column.

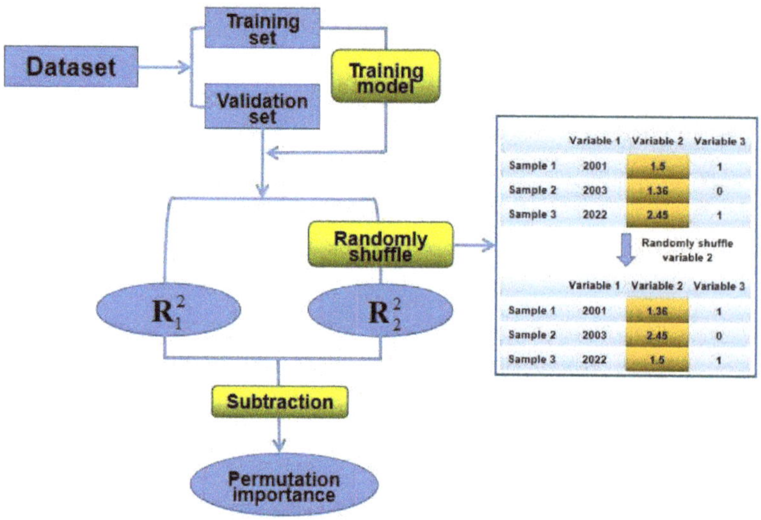

Figure 3. Illustration of the permutation importance (PI).

2.3.3. Model Validation Methods

In this study, two methods are used to evaluate the performance of ML models: sample-based and station-based ten-fold cross-validation (10-CV) [32,56]. In sample-based 10-CV, the sample dataset is randomly divided into ten groups. One group, comprising 10% of the samples, is set aside as the independent validation set, while the remaining nine groups (90%) form the training set. This process is repeated ten times, with each group serving as the validation set once, ensuring that all samples have been tested. The final accuracy is calculated as the average of the results obtained from the ten runs. This approach is commonly used to represent the overall accuracy of ML models in estimating $PM_{2.5}$ levels at locations where ground measurements are available.

The station-based 10-CV is another evaluation method used to assess the predictive ability of ML models in estimating $PM_{2.5}$ concentrations at locations where ground measurements are not available [57]. This method serves as a spatially independent validation technique. Similarly to the sample-based 10-CV, the station-based 10-CV involves dividing the ground observation stations in the study area into ten groups. One group, consisting of 10% of all stations, is designated as the validation set, while the dataset corresponding to the remaining nine groups of stations (90% of all stations) is used as the training set. This approach creates training and validation samples from different locations. This helps isolate spatial autocorrelations among the data samples, making it an effective spatially independent verification method.

2.3.4. Sensitivity Analysis Methods

This study starts by considering the study area as a whole, with 775 ground observation stations. The total number of stations in the study area is then randomly reduced by 10%. The remaining number of stations is again reduced by 10%. This process continues until the number of stations in a group is 31% of the original total number of stations. The end result is 12 groups of stations, each reflecting a certain station density in the study area. The smaller the proportion of remaining stations in a group, the smaller the density of stations in that group. This set of 12 groups of stations and their associated observations is used next to explore the influence of AOD on inversions of $PM_{2.5}$ from different ML models. To assess the importance and contribution of satellite AOD retrievals to ML modeling and quantitatively evaluate its impact on the performance of an ML model, this study incorporates an uncertainty analysis from three key aspects:

(1) The importance scores of satellite AOD were first calculated employing two techniques (FI and PI) for four typical tree-based ML models as the density of ground-based stations in the study area gradually decreased. This analysis offers valuable insights into the role of satellite AOD in the modeling process. It allows for an understanding of the significance of satellite AOD under different station-density conditions.

(2) The accuracies and differences in the estimation of $PM_{2.5}$, with and without satellite AOD as the primary predictor, were calculated using the sample-based 10-CV method. Four typical tree-based ML models were employed, each taking into consideration the decreasing density of ground-based stations in the study area. This analysis allows us to evaluate the importance of satellite AOD in enhancing the overall accuracy of $PM_{2.5}$ estimates for varying station densities.

(3) Similarly, the accuracies and differences in the prediction of $PM_{2.5}$ in regions lacking $PM_{2.5}$ observations, with and without satellite AOD as the main predictor, were calculated using the station-based 10-CV method. Again, four typical tree-based ML models were employed, each taking into consideration the decreasing density of ground-based stations in the study area. This analysis enables us to assess the significance of satellite AOD in improving the predictive ability of $PM_{2.5}$ predictions for varying station densities.

3. Results

3.1. Variations of Satellite AOD Contributions

Figure 4 presents the FI and PI of satellite AOD retrievals in estimating $PM_{2.5}$ obtained by four ML models with the decrease in the density of ground-based monitoring stations. Also shown are the best-fit lines from linear regression and confidence intervals (CIs). CI is a statistical measure used to quantify the uncertainty of an estimate. Here, we used a 95% confidence level (pink-shaded areas in the figure). This indicates that there is a 95% probability that the true value will fall within the specified range, leaving a 5% probability of it falling outside this range. Figure 4a–c show that as the density of the monitoring station decreases, the FI score significantly increases, with regressed correlation of determination (R^2) values of 0.84, 0.81, 0.77, and 0.43 for the Random Forest, Extra-trees, XGBoost, and LightGBM models, respectively. All pass the 99% or 95% confidence ($p < 0.01$ or 0.05) test. The spread of FI in the LightGBM case (Figure 4d) indicates a higher variance compared with other methods, which may be attributed to the use of the specific node-splitting method of the leaf-wise growth strategy. Similar conclusions can be made from the PI score analysis, i.e., the contribution of satellite AOD significantly increases as the density of monitoring stations decreases, with much higher regressed R^2 values (R^2 = 0.94–0.96) for the four ML models (Figure 4e–h). All regressed trends reach the 99% confidence ($p < 0.01$) level. Note that the values of FI and PI are different among the ML methods because these models employ distinct frameworks and operation methods, including sampling, feature selection, and node-splitting techniques. Additionally, the methods of computing FI and PI are also different, e.g., FI relies on the Gini index, which involves calculating the MDI, while PI assesses the change in model performance by randomly shuffling a feature. Results obtained by the two importance verification methods are consistent, suggesting that the two methods can complement and verify each other. These findings reveal that satellite AOD is crucial for $PM_{2.5}$ modeling using ML because it plays a dominant predictive role with the highest importance scores, particularly in regions with a small density of $PM_{2.5}$ ground observation stations.

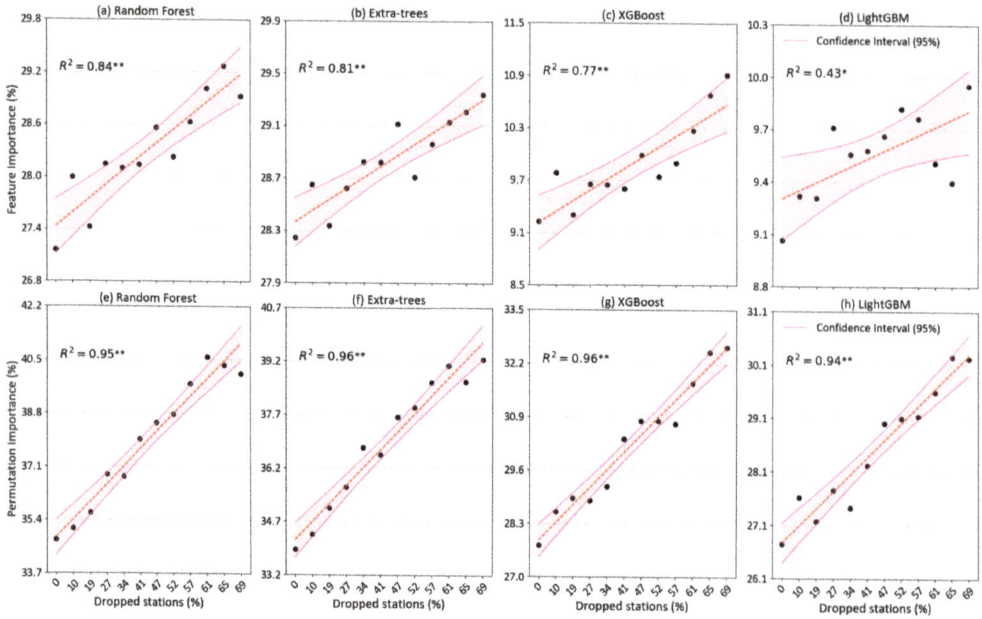

Figure 4. Variations of (**a**–**d**) feature importance (unit: %) and (**e**–**h**) permutation importance (unit: %) of satellite AOD as a function of the decreasing percentage of ground-based monitoring stations for the Random Forest, Extra-trees, XGBoost, and LightGBM models, respectively. Pink-shaded areas are 95% confidence intervals, where * and ** denote the 95% ($p < 0.05$) and 99% ($p < 0.01$) confidence levels of the regressed fits, respectively.

3.2. *Impacts of Satellite AOD on Overall Accuracy*

Figure 5 shows the overall accuracies measured by the CV R^2 (CV-R^2) of the daily estimates of $PM_{2.5}$ with and without AOD as model input as a function of decreasing station density using four ML models. Regarding model results where AOD was included as an input variable (Figure 5, red dots), it is clearly seen that as the station density decreases, the overall accuracy of the $PM_{2.5}$ estimates gradually decreases, showing an average significantly decreasing trend (i.e., change in sample-based CV-R^2 per 1% of discarded stations) of −0.16% ($p < 0.01$), −0.15% ($p < 0.01$), −0.19% ($p < 0.01$), and −0.11% ($p < 0.01$) for the Random Forest, Extra-trees, XGBoost, and LightGBM models, respectively. For model results where AOD was not included as an input variable (Figure 5, blue dots), when the station proportion is 100%, the CV-R^2 is 0.83 with AOD and 0.82 without AOD, a small difference of about 1.48%. However, as the station density decreases, the difference in accuracy between the models with and without AOD gradually increases (black diamonds in the figure). When the station density is 31%, the accuracy of the model with AOD is 0.72, while the accuracy of the model without AOD is 0.65. The relative difference in model accuracy with or without AOD significantly increases to 10.35%. Similarly, Figure 5b–d present comparable results: At 100% station density, the overall accuracies of the Extra-trees, XGBoost, and LightGBM models with AOD are 0.86, 0.84, and 0.85, respectively. Without AOD, the accuracies of these models are 0.86, 0.83, and 0.81, respectively, and the relative differences in model accuracy with or without AOD are 0.10%, 0.58%, and 4.80%, respectively. When the station density drops to 31%, the accuracies of the models with AOD become 0.76, 0.70, and 0.76, respectively, while the accuracies of the models without AOD are 0.71, 0.64, and 0.69, respectively. Consequently, the relative differences in overall accuracy between these three models with and without AOD increase to 6.48%, 9.41%, and 10.44%, respectively. In particular, when the number of monitoring stations decreases, the slopes of the decreased overall accuracy are much steeper for the Random Forest (−0.24%, $p < 0.01$), Extra-trees (−0.22%, $p < 0.01$), XGBoost (−0.27%, $p < 0.01$), and LightGBM (−0.16%, $p < 0.01$) models without using AOD compared to these models using AOD. This is because with the decrease in station density, the sample data volume gradually decreases, as do the accuracies of the models. In general, for every 10% reduction in station proportion, the four ML models without AOD experience a 1.2% ($p < 0.01$), 0.9% ($p < 0.01$), 1.1% ($p < 0.01$), and 0.7% ($p < 0.01$) increase in the uncertainty of the estimated results, respectively, compared with these models using AOD. These findings highlight the indispensable role of satellite AOD in improving the accuracy of estimating $PM_{2.5}$ through ML models, particularly in regions with limited observation stations.

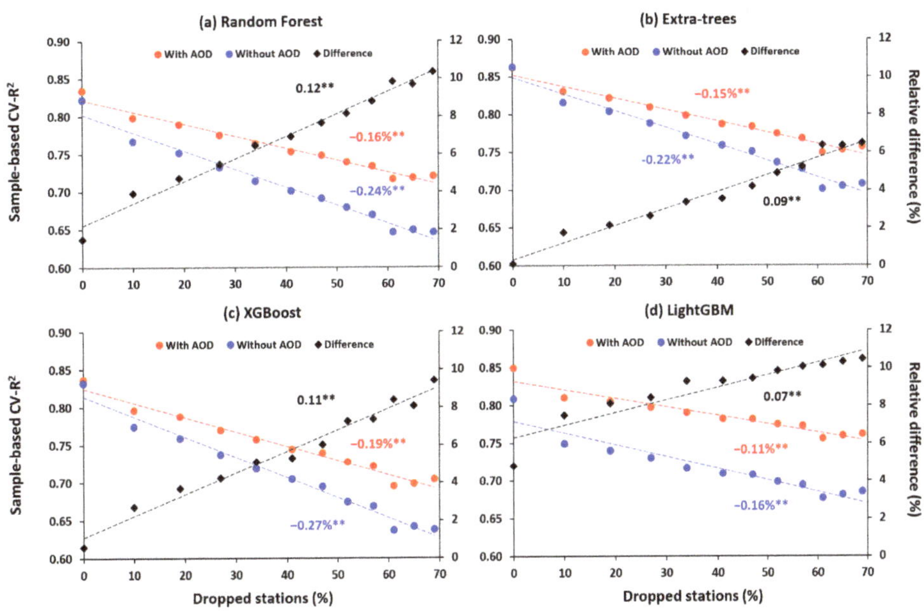

Figure 5. Variation in the overall accuracy (sample-based CV-R^2, left ordinate) and relative difference (%, right ordinate) of daily $PM_{2.5}$ estimates as a function of decreasing station density for four ML models: (**a**) Random Forest, (**b**) Extra-trees, (**c**) XGBoost, and (**d**) LightGBM, with (red dots) and without (blue dots) including satellite AOD as an input predictor. Slopes of the best-fit lines from linear regression for each set of results are given, where ** denotes the 99% confidence ($p < 0.01$) level. Black dashed lines represent the regressed fits of the relative difference between each set of results.

3.3. Impacts of Satellite AOD on Predictive Ability

Here, the model performance in predicting the $PM_{2.5}$ level in areas without surface observations is examined based on results from the station-based 10-CV method. Figure 6 shows the predictive ability of four ML models in retrieving daily $PM_{2.5}$ with (red dots) and without (blue dots) AOD as inputs under different station density conditions. Overall, the trend observed in station-based 10-CV results aligns with those of sample-based 10-CV: When AOD is included as an input variable, with the decreasing number of monitoring stations, the predictive accuracies of the Random Forest, Extra-trees, XGBoost, and Light-

GBM models all show significantly decreasing trends (i.e., change in station-based CV-R^2 per 1% of discarded stations) of −0.14% ($p < 0.01$), −0.13% ($p < 0.01$), −0.18% ($p < 0.01$), and −0.10% ($p < 0.01$), respectively. However, the model's predictive ability experiences a much faster decline, with larger slopes of −0.21%, −0.18%, −0.26%, and −0.13% for the four models without including satellite AOD, respectively. Specifically, when data from all $PM_{2.5}$ stations are used (100%), the Random Forest, Extra-trees, XGBoost, and LightGBM models that include AOD produce slightly better results, with higher station-based CV-R^2 values (0.81, 0.83, 0.80, and 0.79) than those without considering AOD (CV-R^2 = 0.77, 0.81, 0.76, and 0.71, respectively). However, when the proportion of monitoring stations drops to 31%, the predictive accuracies for the same four models decrease to 0.72, 0.75, 0.69, 0.73 (including AOD) and 0.64, 0.69, 0.61, 0.64 (not including AOD), respectively. More importantly, the average relative differences become 2.48, 2.68, 2.19, and 1.33 times larger than the results when all stations are considered. This is because station-based 10-CV uses known stations (regions) to predict unknown stations (regions), reflecting the spatial prediction ability of the model. However, it is difficult for an ML model to make predictions for regions without training samples, inevitably leading to lower prediction accuracies, consistent with existing cognitive rules of an ML model. In general, for every 10% reduction in the station density, the four ML models without and with AOD experience 1.0% ($p < 0.01$), 0.7% ($p < 0.01$), 1.2% ($p < 0.01$), and 0.6% ($p < 0.01$) increases in the uncertainty of the predicted results, respectively. These findings confirm the indispensable role of satellite AOD in predicting $PM_{2.5}$ concentration in areas without observations using ML models, particularly in low-station density situations.

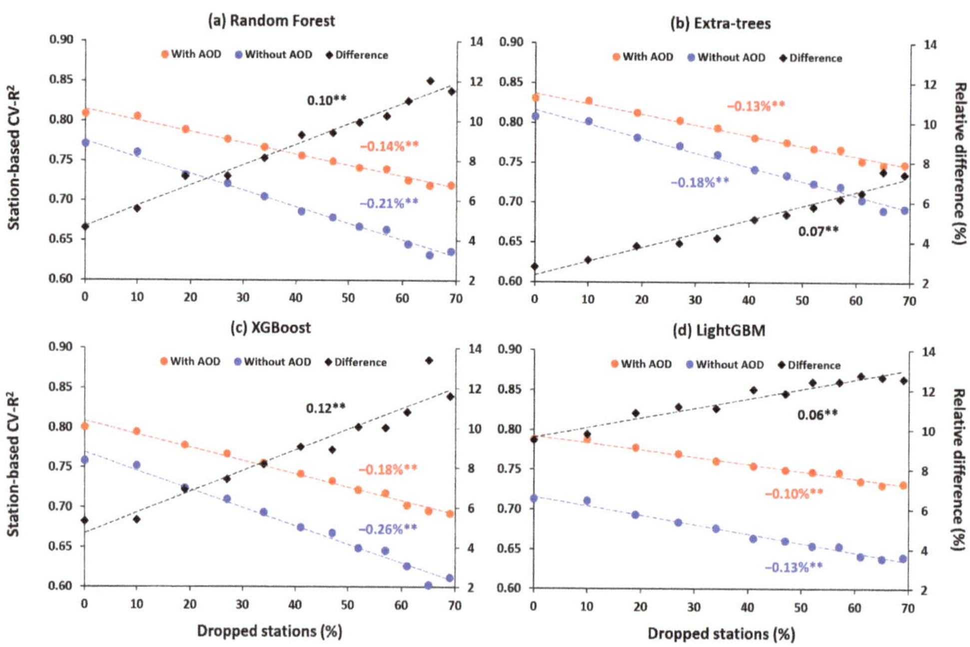

Figure 6. Variation in the predictive ability (station-based CV-R^2, left ordinate) and relative difference (%, right ordinate) of daily $PM_{2.5}$ estimates as a function of decreasing station density for (**a**) Random Forest, (**b**) Extra-trees, (**c**) XGBoost, and (**d**) LightGBM, with (red dots) and without (blue dots) including satellite AOD as an input predictor. Slopes of the best-fit lines from linear regression for each set of results are given, where ** denotes the 99% confidence ($p < 0.01$) level. Black dashed lines represent the regressed fits of the relative difference between each set of results.

The contrasting results with comparable superior accuracy in $PM_{2.5}$ estimation without incorporating AOD input can be attributed to the high density of ground observation stations in the specific study area. In areas with a sufficiently dense network of monitoring sites, excluding AOD can still lead to relatively accurate $PM_{2.5}$ estimations, which largely benefit from the presence of spatial autocorrelation in air pollution, e.g., $PM_{2.5}$ levels and auxiliary factors such as meteorological fields are highly similar in neighboring locations. As a result, $PM_{2.5}$ concentrations in nearby sites can be reasonably estimated based on the established relationships between $PM_{2.5}$ and non-AOD factors from nearby stations. Nevertheless, as the station density decreases, the spatial autocorrelation weakens, and the disparity between natural and human-influenced conditions grows, and consequently, the prediction error rapidly increases. This highlights the critical role of AOD in areas with limited ground monitoring because it significantly enhances the accuracy of $PM_{2.5}$ predictions by providing crucial background pollution information, compensating for the lack of ground observation data.

4. Conclusions

Machine learning (ML) has been used widely to infer ground-level $PM_{2.5}$ using satellite-retrieved aerosol optical depth (AOD) to fill large gaps between $PM_{2.5}$ stations without quantification of its contribution, which is the objective of this study. We rigorously and quantitatively assess the contribution of AOD to the ML-based estimation of $PM_{2.5}$ by applying four common ML models (the Random Forest model, the Extra-trees model, the XGBoost model, and the LightGBM model) to ample measurements from China's high-density $PM_{2.5}$ observation network, the MODIS Multi-Angle Implementation of Atmospheric Correction satellite AOD retrieval product, and many other ancillary meteorological and environmental data from the eastern half of China. Two assessment methods are used, i.e., feature importance (FI) and permutation importance (PI). The contribution of AOD is also assessed by comparing the retrieval results obtained by including and not including AOD. All assessment tests are made for varying numbers of $PM_{2.5}$ stations whose data are sampled by station-based and sample-based 10-CV.

The major findings are summarized as follows: (1) As the station density decreases, the FI and PI of AOD in the four ML models have clear upward trends. This trend indicates the importance and contribution of AOD to improving the accuracy of estimating $PM_{2.5}$, becoming more pronounced in areas with sparse observation stations. (2) As the density of observation stations decreases, the ML models without AOD exhibit a more pronounced decline in overall accuracy compared to the models that incorporate AOD. Additionally, for every 10% reduction in the number of stations, the uncertainty in the estimated accuracy increases by approximately 0.7–1.2%. (3) As the station density decreases, the ML models without AOD exhibit a faster decline in predictive ability compared with these models with AOD. On average, for every 10% reduction in the number of stations, the uncertainty in the predicted accuracy increases by approximately 0.6–1.2%. These findings demonstrate the indispensable role of AOD in any ML model to effectively counteract the negative impact of no or sparse $PM_{2.5}$ stations, resulting in improved accuracy for both estimating and predicting $PM_{2.5}$ levels.

AOD represents the degree of light attenuation caused by the scattering and absorption of atmospheric aerosols in the vertical direction and has served as a crucial indicator in deriving surface particulate matter concentrations. The importance of AOD was confirmed through a sensitivity analysis showing that satellite AOD has the highest FI and PI values in $PM_{2.5}$ modeling using various ML models. As the number of ground stations decreases, the AOD contribution is more apparent, as is the faster drop in ML model performance without AOD. This further underlines how using satellite AOD is essential, providing key background pollution information in areas without ground stations, thereby improving the prediction capability of ground-based $PM_{2.5}$. On the contrary, in this case, relying solely on auxiliary factors such as meteorological fields is far from adequate.

Even though the chosen area in eastern China benefits from a dense and reasonably evenly distributed network of ground observation stations for $PM_{2.5}$, enhancing its representativeness, there still exists the question of the uniformity of the spatial distribution of stations. Further analysis incorporating spatial-block cross-validation is needed to effectively reduce the impact of this issue. This will be undertaken in our future study. Additionally, this approach could be applied to PM_{10}, considering its high similarities with $PM_{2.5}$, i.e., AOD retains its significance as a crucial input variable in PM_{10} predictions [58]. However, additional sensitivity analyses are warranted to confirm this hypothesis. Regarding other pollutants, since they possess entirely different key input variables, further investigations are needed to accurately understand their behaviors [e.g., the importance of satellite tropospheric NO_2 in surface NO_2 modelling [59]].

Author Contributions: Conceptualization: Z.L. and J.W.; Methodology: J.W., Z.T. and Z.L.; Analysis: Z.T. and J.W.; Writing: Z.T., J.W. and Z.L.; Funding Acquisition: Z.L. All authors have read and agreed to the published version of the manuscript. Zhanqing Li's contribution to this publication was not part of his University of Maryland duties or responsibilities.

Funding: This research was funded by the National Natural Science Foundation (42030606).

Data Availability Statement: The $PM_{2.5}$ observation station data used in this study are real-time ground-measured air quality data (including $PM_{2.5}$) in China from the China National Environment Monitoring Center (http://www.cnemc.cn, accessed on 1 January 2023). The NASA data center (https://ladsweb.modaps.eosdis.nasa.gov/search/, accessed on 1 January 2023) provides MODIS MAIAC aerosol products. Meteorological reanalysis data are collected from the fifth-generation European Reanalysis Interim dataset (ERA5, https://www.ecmwf.int/, accessed on 1 January 2023) released by the European Centre for Medium-Range Weather Forecasts.

Acknowledgments: We thank M. Cribb from the University of Maryland for helping in editing the paper.

Conflicts of Interest: The authors declare no conflict of interest.

References

1. IPCC 2021. *Climate Change, 2021: The Physical Science Basis*; Contribution of Working Group I to the Fifth Assessment Report of the Intergovernmental Panel on Climate Change IPCC Working Group I Contribution to AR5Rep.; Cambridge University Press: Cambridge, UK; New York, NY, USA, 2021.
2. Guo, J.; Xia, F.; Zhang, Y.; Liu, H.; Li, J.; Lou, M.; He, J.; Yan, Y.; Wang, F.; Min, M.; et al. Impact of diurnal variability and meteorological factors on the $PM_{2.5}$-AOD relationship: Implications for $PM_{2.5}$ remote sensing. *Environ. Poll.* **2017**, *221*, 94. [CrossRef] [PubMed]
3. Li, Z.; Wang, Y.; Guo, J.; Zhao, C.; Cribb, M.C.; Dong, X.; Fan, J.; Gong, D.; Huang, J.; Jiang, M.; et al. East Asian Study of Tropospheric Aerosols and their Impact on Regional Clouds, Precipitation, and Climate (EAST-AIR_{CPC}). *J. Geophys. Res. Atmos.* **2019**, *124*, 13026–13054. [CrossRef]
4. Duyzer, J.; van den Hout, D.; Zandveld, P.; van Ratingen, S. Representativeness of air quality monitoring networks. *Atmos. Environ.* **2015**, *104*, 88–101. [CrossRef]
5. Alsahli, M.M.; Al-Harbi, M. Allocating optimum sites for air quality monitoring stations using GIS suitability analysis. *Urban Clim.* **2018**, *24*, 875–886. [CrossRef]
6. Chen, N.; Yang, M.; Du, W.; Min, H. $PM_{2.5}$ estimation and spatial-temporal pattern analysis based on the modified support vector regression model and the 1 km resolution MAIAC AOD in Hubei, China. *ISPRS Int. J. Geo-Inf.* **2021**, *10*, 31. [CrossRef]
7. van Donkelaar, A.; Martin, R.V.; Park, R.J. Estimating ground-level $PM_{2.5}$ using aerosol optical depth determined from satellite remote sensing. *J. Geophys. Res. Atmos.* **2006**, *111*, D21201. [CrossRef]
8. Ma, X.; Wang, J.; Yu, F.; Jia, H.; Hu, Y. Can MODIS AOD be employed to derive $PM_{2.5}$ in Beijing-Tianjin-Hebei over China? *Atmos. Res.* **2016**, *181*, 250–256. [CrossRef]
9. Li, S.; Joseph, E.; Min, Q. Remote sensing of ground-level $PM_{2.5}$ combining AOD and backscattering profile. *Remote Sens. Environ.* **2016**, *183*, 120–128. [CrossRef]
10. Li, Z.; Goloub, P.; Devaux, C.; Gu, X.; Qiao, Y.; Zhao, F.; Chen, H. Aerosol polarized phase function and single-scattering albedo retrieved from ground-based measurements. *Atmos. Res.* **2004**, *71*, 233–241. [CrossRef]
11. Gupta, P.; Christopher, S.A.; Wang, J.; Gehrig, R.; Lee, Y.; Kumar, N. Satellite remote sensing of particulate matter and air quality assessment over global cities. *Atmos. Environ.* **2006**, *40*, 5880–5892. [CrossRef]
12. Kumar, N. What can affect AOD–$PM_{2.5}$ association? *Environ. Health Perspect.* **2010**, *118*, A109–A110. [CrossRef] [PubMed]

13. Wang, J.; Christopher, S.A. Intercomparison between satellite-derived aerosol optical thickness and $PM_{2.5}$ mass: Implications for air quality studies. *Geophys. Res. Lett.* **2003**, *30*, 2095. [CrossRef]
14. Natunen, A.; Arola, A.; Mielonen, T.; Huttunen, J.; Lehtinen, K.E.J. A multi-year comparison of $PM_{2.5}$ and AOD for the Helsinki region. *Boreal Environ. Res.* **2010**, *15*, 544–552. [CrossRef]
15. Kloog, I.; Nordio, F.; Coull, B.A.; Schwartz, J. Incorporating local land use regression and satellite aerosol optical depth in a hybrid model of spatiotemporal $PM_{2.5}$ exposures in the Mid-Atlantic states. *Environ. Sci. Technol.* **2012**, *46*, 11913–11921. [CrossRef] [PubMed]
16. Ma, Z.; Hu, X.; Sayer, A.M.; Levy, R.; Zhang, Q.; Xue, Y.; Tong, S.; Bi, J.; Huang, L.; Liu, Y. Satellite-based spatiotem-poral trends in $PM_{2.5}$ concentrations: China 2004-2013. *Environ. Health Perspect.* **2016**, *124*, 184–192. [CrossRef]
17. Qu, W.; Wang, J.; Zhang, X.; Sheng, L.; Wang, W. Opposite seasonality of the aerosol optical depth and the surface particulate matter concentration over the North China Plain. *Atmos. Environ.* **2016**, *127*, 90–99. [CrossRef]
18. Su, T.; Li, Z.; Kahn, R. Relationships between the planetary boundary layer height and surface pollutants derived from lidar observations over China: Regional pattern and influencing factors. *Atmos. Chem. Phys.* **2018**, *18*, 15921–15935. [CrossRef]
19. Koelemeijer, R.B.A.; Homan, C.D.; Matthijsen, J. Comparison of spatial and temporal variations of aerosol optical thickness and particulate matter over Europe. *Atmos. Environ.* **2006**, *40*, 5304–5315. [CrossRef]
20. Brauer, M.; Amann, M.; Burnett, R.T.; Cohen, A.; Dentener, F.; Ezzati, M.; Henderson, S.B.; Krzyzanowski, M.; Martin, R.V.; Van Dingenen, R.; et al. Exposure assessment for estimation of the global burden of disease attributable to outdoor air pollution. *Environ. Sci. Technol.* **2012**, *46*, 652–660. [CrossRef]
21. Chen, J.; de Hoogh, K.; Gulliver, J.; Hoffmann, B.; Hertel, O.; Ketzel, M.; Bauwelinck, M.; van Donkelaar, A.; Hvidtfeldt, U.A.; Katsouyanni, K.; et al. A comparison of linear regression, regularization, and machine learning algorithms to develop Europe-wide spatial models of fine particles and nitrogen dioxide. *Environ. Int.* **2019**, *130*, 104934. [CrossRef]
22. Gupta, P.; Christopher, S. Particulate matter air quality assessment using integrated surface, satellite, and meteorological products: Multiple regression approach. *J. Geophys. Res. Atmos.* **2009**, *114*, D14205. [CrossRef]
23. Ma, Z.; Hu, X.; Huang, L.; Bi, J.; Liu, Y. Estimating ground-level $PM_{2.5}$ in China using satellite remote sensing. *Environ. Sci. Tech.* **2014**, *48*, 7436–7444. [CrossRef] [PubMed]
24. You, W.; Zang, Z.; Zhang, L.; Li, Y.; Wang, W. Estimating national-scale ground-level $PM_{2.5}$ concentration in China using geographically weighted regression based on MODIS and MISR AOD. *Environ. Sci. Pollut. Res.* **2016**, *23*, 8327–8338. [CrossRef]
25. He, Q.; Huang, B. Satellite-based mapping of daily high-resolution ground $PM_{2.5}$ in China via space-time regression modeling. *Remote Sens. Environ.* **2018**, *206*, 72–83. [CrossRef]
26. Xiao, Q.; Wang, Y.; Chang, H.H.; Meng, X.; Liu, Y. Full-coverage high-resolution daily $PM_{2.5}$ estimation using MAIAC AOD in the Yangtze River Delta of China. *Remote Sens. Environ.* **2017**, *199*, 437–446. [CrossRef]
27. Liu, Y.; Sarnat, J.A.; Kilaru, A.; Jacob, D.J.; Koutrakis, P. Estimating ground-level $PM_{2.5}$ in the eastern United States using satellite remote sensing. *Environ. Sci. Technol.* **2005**, *39*, 3269–3278. [CrossRef]
28. Lee, H.J.; Liu, Y.; Coull, B.A.; Schwartz, J.; Koutrakis, P. A novel calibration approach of MODIS AOD data to predict $PM_{2.5}$ concentrations. *Atmos. Chem. Phys.* **2011**, *11*, 7991–8002. [CrossRef]
29. Chen, W.; Ran, H.; Cao, X.; Wang, J.; Zheng, X. Estimating $PM_{2.5}$ with high-resolution 1-km AOD data and an improved machine learning model over Shenzhen, China. *Sci. Total Environ.* **2020**, *746*, 141093. [CrossRef]
30. Wei, J.; Huang, W.; Li, Z.; Xue, W.; Cribb, M. Estimating 1-km-resolution $PM_{2.5}$ concentrations across China using the space-time random forest approach. *Remote Sens. Environ.* **2019**, *231*, 111221. [CrossRef]
31. Wei, J.; Li, Z.; Cribb, M.; Huang, W.; Xue, W.; Sun, L.; Guo, J.; Peng, Y.; Li, J.; Lyapustin, A.; et al. Improved 1-km-resolution $PM_{2.5}$ estimates across China using enhanced space-time extremely randomized trees. *Atmos. Chem. Phys.* **2020**, *20*, 3273–3289. [CrossRef]
32. Wei, J.; Li, Z.; Lyapustin, A.; Sun, L.; Peng, Y.; Xue, W.; Su, T.; Cribb, M. Reconstructing 1-km-resolution high-quality $PM_{2.5}$ data records from 2000 to 2018 in China: Spatiotemporal variations and policy implications. *Remote Sens. Environ.* **2021**, *252*, 112136. [CrossRef]
33. Pan, B. Application of XGBoost algorithm in hourly $PM_{2.5}$ concentration prediction. *IOP Conf. Ser. Earth Environ. Sci.* **2018**, *113*, 012127. [CrossRef]
34. Wei, J.; Li, Z.; Pinker, R.T.; Sun, L.; Li, R. Himawari-8-derived diurnal variations of ground-level $PM_{2.5}$ pollution across China using the fast space-time Light Gradient Boosting Machine (LightGBM). *Atmos. Chem. Phys.* **2021**, *21*, 7863–7880. [CrossRef]
35. Fang, X.; Zou, B.; Liu, X.; Sternberg, T.; Zhai, L. Satellite-based ground $PM_{2.5}$ estimation using timely structure adaptive modeling. *Remote Sens. Environ.* **2016**, *186*, 152–163. [CrossRef]
36. Meng, X.; Fu, Q.; Ma, Z.; Chen, L.; Zou, B.; Zhang, Y.; Xue, W.; Wang, J.; Wang, D.; Han, H. Estimating ground-level PM_{10} in a Chinese city by combining satellite data, meteorological information and a land use regression model. *Environ. Pollut.* **2016**, *208*, 177–184. [CrossRef] [PubMed]
37. Pereira, G.; Lee, H.J.; Bell, M.; Regan, A.; Malacova, E.; Mullins, B.; Knibbs, L.D. Development of a model for particulate matter pollution in Australia with implications for other satellite-based models. *Environ. Res.* **2017**, *159*, 9–15. [CrossRef] [PubMed]
38. Chen, G.; Li, Y.; Zhou, Y.; Shi, C.; Liu, Y. The comparison of AOD-based and non-AOD prediction models for daily $PM_{2.5}$ estimation in Guangdong province, China with poor AOD coverage. *Environ. Res.* **2021**, *195*, 110735. [CrossRef] [PubMed]

39. Yu, W.; Li, S.; Ye, T.; Xu, R.; Song, J.; Guo, Y. Deep ensemble machine learning framework for the estimation of $PM_{2.5}$ concentrations. *Environ. Health Perspect.* **2022**, *130*, 037004. [CrossRef]
40. Lyapustin, A.; Martonchik, J.; Wang, Y.; Laszlo, I.; Korkin, S. Multi-Angle Implementation of Atmospheric Correction (MAIAC): 1. Radiative transfer basis and look-up tables. *J. Geophys. Res. Atmos.* **2011**, *116*, D03210. [CrossRef]
41. Lyapustin, A.; Wang, Y.; Korkin, S.; Huang, D. MODIS Collection 6 MAIAC algorithm. *Atmos. Meas. Tech.* **2018**, *11*, 5741–5765. [CrossRef]
42. Hersbach, H.; Bell, B.; Berrisford, P.; Hirahara, S.; Horányi, A.; Muñoz-Sabater, J.; Nicolas, J.; Peubey, C.; Radu, R.; Schepers, D.; et al. The ERA5 global reanalysis. *Q. J. R. Meteorol. Soc.* **2020**, *146*, 1999–2049. [CrossRef]
43. Peuch, V.H.; Engelen, R.; Ades, M.; Barre, J.; Suttie, M. The use of satellite data in the Copernicus Atmosphere Monitoring Service (CAMS). In *IGARSS 2018–2018 IEEE International Geoscience and Remote Sensing Symposium*; IEEE: Manhattan, NY, USA, 2018.
44. Wei, J.; Li, Z.; Wang, J.; Li, C.; Gupta, P.; Cribb, M. Ground-level gaseous pollutants (NO_2, SO_2, and CO) in China: Daily seamless mapping and spatiotemporal variations. *Atmos. Chem. Phys.* **2023**, *23*, 1511–1532. [CrossRef]
45. Malakar, N.K.; Lary, D.J.; Moore, A.; Gencaga, D.; Roscoe, B.; Albayrak, A.; Petrenko, M.; Wei, J. Estimation and bias correction of aerosol abundance using data-driven machine learning and remote sensing. In Proceedings of the 2012 Conference on Intelligent Data Understanding (CIDU 2012), Boulder, CO, USA, 24–26 October 2012.
46. Lary, D.J.; Faruque, F.S.; Malakar, N.; Moore, A.; Roscoe, B.; Adams, Z.L.; Eggelston, Y. Estimating the global abundance of ground level presence of particulate matter ($PM_{2.5}$). *Geospat. Health* **2014**, *8*, S611–S630. [CrossRef]
47. Reid, C.E.; Jerrett, M.; Petersen, M.L.; Pfister, G.G.; Morefield, P.E.; Tager, I.B.; Raffuse, S.M.; Balmes, J.R. Spatiotemporal prediction of fine particulate matter during the 2008 northern California wildfires using machine learning. *Environ. Sci. Technol.* **2015**, *49*, 3887–3896. [CrossRef] [PubMed]
48. Breiman, L. Random forests. *Mach. Learn.* **2002**, *45*, 5–32. [CrossRef]
49. Chen, G.; Li, S.; Knibbs, L.D.; Hamm, N.A.S.; Cao, W.; Li, T.; Guo, J.; Ren, H.; Abramson, M.J.; Guo, Y. A machine learning method to estimate $PM_{2.5}$ concentrations across China with remote sensing, meteorological and land use information. *Sci. Total Environ.* **2018**, *636*, 52–60. [CrossRef]
50. Hu, X.; Belle, J.H.; Meng, X.; Wildani, A.; Waller, L.A.; Strickland, M.J.; Liu, Y. Estimating $PM_{2.5}$ concentrations in the conterminous United States using the random forest approach. *Environ. Sci. Tech.* **2017**, *51*, 6936–6944. [CrossRef]
51. Geurts, P.; Ernst, D.; Wehenkel, L. Extremely randomized trees. *Mach. Learn.* **2006**, *63*, 3–42. [CrossRef]
52. Chen, T.; Guestrin, C. XGBoost: A scalable tree boosting system. In Proceedings of the 22nd ACM SIGKDD International Conference on Knowledge Discovery and Data Mining, San Francisco, CA, USA, 13–17 August 2016; pp. 785–794.
53. Ke, G.; Meng, Q.; Finley, T.; Wang, T.; Chen, W.; Ma, W.; Ye, Q.; Liu, T. LightGBM: A highly effificient gradient boosting decision tree. In *Advances in Neural Information Processing Systems*; ACM: Long Beach, CA, USA, 2017; pp. 3149–3157. Available online: https://dl.acm.org/doi/10.5555/3294996.3295074 (accessed on 1 January 2020).
54. Loecher, M. Unbiased variable importance for random forests. *Commun. Stat. Theory Methods* **2022**, *51*, 1413–1425. [CrossRef]
55. Kim, H.; Loh, W.-Y. Classification trees with unbiased multiway splits. *J. Am. Stat. Assoc.* **2001**, *96*, 589–604. [CrossRef]
56. Rodriguez, J.; Perez, A.; Lozano, J. Sensitivity analysis of k-fold cross validation in prediction error estimation. *IEEE Trans. Pattern Anal. Mach. Intell.* **2010**, *32*, 569–575. [CrossRef] [PubMed]
57. Wei, J.; Li, Z.; Chen, X.; Li, C.; Sun, Y.; Wang, J.; Lyapustin, A.; Brasseur, G.; Jiang, M.; Sun, L.; et al. Separating daily 1-km $PM_{2.5}$ inorganic chemical composition in China since 2000 via deep learning integrating ground, satellite, and model data. *Environ. Sci. Tech.* **2023**. [CrossRef] [PubMed]
58. Wei, J.; Li, Z.; Xue, W.; Sun, L.; Fan, T.; Liu, L.; Su, T.; Cribb, M. The ChinaHighPM_{10} dataset: Generation, validation, and spatiotemporal variations from 2015 to 2019 across China. *Environ. Int.* **2021**, *146*, 106290. [CrossRef] [PubMed]
59. Wei, J.; Liu, S.; Li, Z.; Liu, C.; Qin, K.; Liu, X.; Pinker, R.; Dickerson, R.; Lin, J.; Boersma, K.; et al. Ground-level NO_2 surveillance from space across China for high resolution using interpretable spatiotemporally weighted artificial intelligence. *Environ. Sci. Tech.* **2022**, *56*, 9988–9998. [CrossRef] [PubMed]

remote sensing

Article

Influence of the Indian Summer Monsoon on Inter-Annual Variability of the Tibetan-Plateau NDVI in Its Main Growing Season

Xin Mao [1,2], Hong-Li Ren [2,3,*], Ge Liu [1,2], Baohuang Su [2] and Yinghan Sang [2]

1 Collaborative Innovation Center on Forecast and Evaluation of Meteorological Disasters (CIC-FEMD), School of Atmosphere Sciences, Nanjing University of Information Science & Technology, Nanjing 210044, China
2 State Key Laboratory of Severe Weather, and Institute of Tibetan Plateau Meteorology, Chinese Academy of Meteorological Sciences, Beijing 100081, China
3 Department of Atmospheric Science, School of Environmental Studies, China University of Geosciences, Wuhan 430074, China
* Correspondence: renhl@cma.gov.cn

Abstract: The vegetation on the Tibetan Plateau (TP), as a major component of the land–atmosphere interaction, affects the TP thermal conditions. And, as a direct climatic factor of vegetation, precipitation over the TP is significantly regulated by the Indian summer monsoon (ISM). Using remote-sensing-based vegetation images, meteorological observations, and reanalysis datasets, this study deeply explored the influence of the ISM on vegetation on the TP in its main growing season, where the vegetation on the TP is indicated by the normalized difference vegetation index (NDVI). The findings reveal that the ISM is a critical external factor impacting the TP vegetation and has a significantly positive correlation with the TP precipitation and NDVI. Corresponding to a strong ISM, the South Asia high moves northwestward toward the TP and Iranian Plateau with an increase in intensity, and the cyclonic circulation develops over the south of the TP in the middle-lower troposphere. This tropospheric circulation structure aids in the transportation of more water vapor to the TP and enhances convection there, which facilitates more precipitation and thus the TP vegetation growth, featuring a uniform NDVI pattern. Since the positive correlation between precipitation over the TP and NDVI is weaker than that between the ISM and NDVI, we suggest that the ISM can influence the TP vegetation growth not only through changing precipitation but also through other local climatic factors. The increased convection and precipitation over the TP induced by the ISM can also affect the surface thermal conditions, featuring an interaction between the TP vegetation and heat sources. The evapotranspiration of vegetation and its coverage affect local latent and sensible heat fluxes, while the TP thermal condition changes affect in return the vegetation growth. In addition, the changes in thermal conditions over the TP caused by the substantial increase in vegetation may have a de-correlation effect on the relationship between the ISM and uniform NDVI pattern after the TP vegetation reaches its maximum coverage.

Keywords: Tibetan Plateau; Indian summer monsoon; vegetation; climatic factor; thermal conditions

Citation: Mao, X.; Ren, H.-L.; Liu, G.; Su, B.; Sang, Y. Influence of the Indian Summer Monsoon on Inter-Annual Variability of the Tibetan-Plateau NDVI in Its Main Growing Season. *Remote Sens.* **2023**, *15*, 3612. https://doi.org/10.3390/rs15143612

Academic Editor: Fernando Camacho

Received: 13 June 2023
Revised: 14 July 2023
Accepted: 17 July 2023
Published: 20 July 2023

1. Introduction

The Tibetan Plateau (TP) is situated in the subtropical area of eastern Eurasia and is referred to as "The Third Pole" and the "Roof of the World", exceeding 4000 m above sea level on average. Its dynamic and thermal forcing can affect the occurrence and development of climate and weather in China and East Asia [1–4]. Land–air hydrothermal exchange processes in the TP (such as surface heat sources, atmospheric heat sources, vegetation cover, and snow cover) regulate the thermodynamic forcings of the TP, which have an important effect on the monsoon, Asian atmospheric circulation, as well as global climate [4–9].

As one of the major elements on the TP land surface, vegetation is critical to land–atmosphere interactions. Local weather and climate change affect vegetation growth [10,11]. Owing to the geography and altitude of the TP, the vegetation on the TP responds to climate change more rapidly than that in other regions at the same latitude. In turn, changes in vegetation can alter the surface properties of the TP, including surface albedo and soil moisture, thus affecting the land–air hydrothermal exchange and carbon cycle over the TP [12,13]. Model experiments showed that the enhanced greening of the TP vegetation induces changes in heat consumption by plant transpiration and surface evaporation and surface heat sources, thus influencing the climate and weather of the TP and its adjacent surroundings [14]. Also, the enhanced vegetation dynamics on the TP may attenuate the local surface warming [15], meaning that thermal conditions over the TP can be affected by changes in the TP vegetation. Therefore, it is essential to understand the characteristics of the TP vegetation in terms of its variability and trends.

Several indices can reflect vegetation activity, such as the normalized difference vegetation index (NDVI), soil-adjusted vegetation index (SAVI), enhanced vegetation index (EVI), leaf area index (LAI), and net primary productivity (NPP). Most of these vegetation indices are derived from remote-sensing images of vegetation; many factors (e.g., atmospheric conditions, soil types, topography, shading effects, and solar angle) may introduce noise into these indices [16]. Of these vegetation indices, the NDVI is calculated as the ratio of the difference between near-infrared (NIR) and red (Red) light to the sum [17]. Due to the ratio of band intensities, the NDVI can eliminate a large proportion of noise caused by instrument calibration, solar angle, topography, cloud shadows, and atmospheric attenuations existing in visible red and infrared bands [18,19], which enhances the response to vegetation and reduces the susceptibleness of illumination conditions [20]. Matsushita et al. [21] also noted that the NDVI may be indirectly affected by topography, which can be somewhat neglected. Moreover, the NDVI seems to have good performance in the TP with complex and fragile ecosystems and vegetation species [22]. Considering the altitude and ecosystem of the study area (i.e., the TP exceeding 3000 m above sea level; Figure 1) in this paper, the NDVI is a suitable vegetation index for indicating the growth activity and cover of vegetation on the TP [23]. Therefore, we chose the NDVI as an indicator to explore the characteristics and trends of vegetation on the TP in the current study.

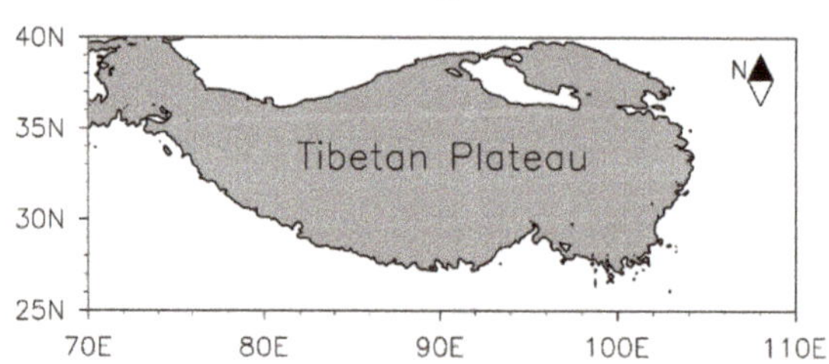

Figure 1. Map showing the location of the study area within the boundary of the Tibetan Plateau (TP) (3000 m above sea level; gray shading).

The NDVI on the entire TP generally exhibits a rising trend under global warming and some human activities [24–27], while certain regions of the TP are suffering from vegetation degradation [28,29]. The apparent inconsistency in the regional and overall NDVI trends on the TP may be due to its relatively distinct climatic characteristics and geographical location. From one perspective, complicated climate change, such as different change trends and intensities of climatic factors, leads to inconsistent vegetation growth trends in different areas of the TP [24,28]. Moreover, when the vegetation on the TP is becoming denser and reaches a certain threshold, the NDVI may no longer increase with the anomalous increase in climatic factors due to the saturation effect of NDVI. From the other perspective, at different altitudes and geographical locations (windward and leeward slopes), vegetation on the TP responds to climatic anomalies in very distinct ways [16,29–31]. For example, the intensity of solar radiation varies greatly between windward and leeward slopes, which can lead to differences in the amount of water lost from vegetation to the atmosphere due to its evapotranspiration, thus affecting the growth of vegetation [16].

The above studies mainly concentrated on the characteristics and trends of longstanding variations in the TP vegetation rather than its inter-annual variations. Considering the inter-annual variability of climatic factors affecting vegetation growth [18,32–34] and the fact that the inter-annual variability of vegetation can also adjust the TP thermal conditions, the inter-annual characteristics of vegetation on the TP in its growing season deserve exploration. Therefore, this study focuses on the growth of TP vegetation on inter-annual scales. Our previous study revealed that several local climatic factors jointly regulate the inter-annual variability of two NDVI patterns dominating the TP in June–September (JJAS, i.e., the main growing season) [35]. However, it is not enough to merely understand the local factors modulating the inter-annual variability of NDVI on the TP, as the variations in local climatic factors over the TP are inseparable from external influences.

Earlier studies have demonstrated that the complex surface environment and anomalously variable ocean, as well as associated atmospheric teleconnections, can alter hydrothermal conditions over the TP. For example, the Indian summer monsoon (ISM) [36–38], North Atlantic Oscillation [39,40], El Niño-Southern Oscillation (ENSO) [41], and Indian Ocean Basin Mode (IOBM) [42] can all modify precipitation over the TP. Sea surface temperature anomalies in several key oceans [42–45], and Indian soil moisture [46], can modulate the TP thermal conditions. Among these external climatic factors, the ISM seems to be a factor more closely related to the TP. The ISM is essential to the variation in summer precipitation over the TP [37,42,47–50]. Precipitation over the TP can be governed by the deep convection in the Indian subcontinent, which is connected to the ISM [37]. The ISM can also affect precipitation over the southern TP by modulating the transportation of water vapor entering the TP [47,48]. In turn, the ISM onset is directly associated with the TP's atmospheric heat source [51–53]. The TP diabatic heating is pivotal in modulating the location and intensity of the ISM [47,54].

The current study intends to explain what role the ISM plays in the inter-annual variability of vegetation on the TP and whether the vegetation can in turn affect the ISM. This study may be of great practical significance to TP ecological environmental protection and the fields of short-term climate prediction. The following section describes the study period and datasets, index definitions, and analysis methods involved in the study. The findings are introduced in Section 3, where Section 3.1 examines relationships between the ISM, precipitation over the TP, and the NDVI on the TP, and Section 3.2 explores the influence of the ISM on vegetation on the TP in its main growing season. Section 4 (i.e., Discussions) and Section 5 (i.e., Conclusions) explore and sum up the relationship between the ISM and vegetation on the TP, respectively.

2. Data and Methods

2.1. Data

2.1.1. Remote-Sensing-Based NDVI Datasets

Remote-sensing images have been widely used in monitoring vegetation dynamics at a regional scale due to their wide coverage and frequent capture of surface information. Based on a comprehensive detailed review related to the use of remote-sensing products (such as LANDSAT, SPOT-Vegetation, Sentinel-2, Himawari-8/9) in the estimation of vegetation growth, we employed two remote-sensing-based NDVI datasets to reflect the variability of the TP vegetation exceeding 3000 m above sea level [35]; the study area is shown in Figure 1. One is GIMMS NDVI3g derived from the National Oceanic and Atmospheric Administration (NOAA), and the other is MCD19A3CMG NDVI of the MODIS products derived from the National Aeronautics and Space Administration (NASA). The former has an 8-km spatial resolution and spans from January 1982 to December 2014, and the latter has a horizontal precision of 0.05° × 0.05° grid and spans from February 2000 to December 2020.

The GIMMS and MODIS NDVIs have been considered as the more commonly used remote-sensing-based NDVI datasets. This is because as the third generation of AVHRR sensor data, GIMMS NDVI3g has been proven to be a better dataset for describing vegetation dynamics in applications [55]. The GIMMS NDVI3g has a longer time scale and has been extensively applied in regional and global-scale studies of vegetation dynamics and degradation [56,57]. In addition, the MCD19A3CMG is a MAIAC BRDF corrected product in the MODIS sensor datasets, which improves spatial resolution (0.05° × 0.05° grid), the accuracy of atmospheric correction, aerosol retrievals, and cloud detection [58,59]. Therefore, they are combined to study vegetation activities due to the linear correlation and compatibility between the two datasets [60–62].

2.1.2. Reanalysis and Meteorological Observation Datasets

The monthly geopotential height, zonal/meridional wind, water vapor, and vertical pressure velocity were derived from the European Centre for Medium-Range Weather Forecasts Reanalysis v5 (ERA5) [63], utilizing a reduced horizontal precision of 2.5° × 2.5° grid. The ERA5 provides a total of 37 vertical pressure levels ranging from 1000 to 1 hPa. In this paper, we used the ERA5 from 1000 to 200 hPa. The monthly outgoing longwave radiation was obtained from NOAA satellite observations [64]. Additionally, the monthly latent and sensible heat flux were derived from the long-term Japanese 55-year Reanalysis (JRA-55) [65], utilizing a horizontal precision of 1.25° × 1.25° grid. These datasets were used to examine how the ISM influences the vegetation on the TP for the period 1982–2020. To explore the relationship between the TP vegetation and ENSO/IBOM, this study also used the Niño 3.4 index obtained from the NOAA CPC and the IOBM index obtained from the National Climate Centre of China Meteorological Administration (NCC/CMA).

The monthly meteorological variables were derived from the station-observed dataset of the National Meteorological Information Center of China, including precipitation, surface air temperature, ground surface temperature, and sunshine duration. Daily meteorological elements at 88 observational stations in the TP (with the average altitude exceeding 3000 m above sea level) were processed into the monthly data on a 0.5° × 0.5° grid by daily accumulation and Cressman spatial interpolation [66]. Besides the Cressman interpolation, we performed the elevation correction of the meteorological variables following the elevation correction equation of He et al. [67], which involves the calculation of the elevation lapse rate. These processed elements were used to illustrate the local effects on the TP vegetation.

2.2. The Study Period and Methods

The ISM is a crucial element of the Asian summer monsoon system [36] and a major source of water vapor for India that is responsible for over 2/3 of the annual precipitation over India [68]. Based on precipitation in Kerala, the southernmost state of the Indian subcontinent, the Indian Meteorological Department defines early June (early October) as the time for the ISM onset (demise) [69]. The main growing season for the TP vegetation is generally from June to September [35]. Considering the overlapped period of the ISM and vegetation growing season, June to September (JJAS) was selected as the main study period.

Following previous studies [25,60,61], we spliced the GIMMS and MCD19A3CMG NDVIs datasets. Before splicing the datasets, the MCD19A3CMG NDVI was downscaled and interpolated to the same resolution as the GIMMS NDVI. Further comparisons revealed that the two NDVIs on the TP have consistent characteristics in the growing seasons and significantly correlate with each other. As such, we can establish linear regression equations between the GIMMS and MCD19A3CMG NDVIs and eventually obtain a longer JJAS TP NDVI dataset by fitting and splicing the two datasets [35]. To capture the varying characteristics of the TP vegetation on inter-annual scales, the rotated empirical orthogonal function (REOF) decomposition and the North test were applied in this paper, ensuring that the leading REOF modes are homogeneous and independent [70].

All data involved in this study were subtracted from the monthly mean climatology, seasonally averaged, and detrended. Several frequently used statistical analysis methods were utilized to examine the influence of the ISM on vegetation on the TP in its growing season, and the specific methods were as follows: (1) REOF analysis and regional average were used to derive the uniform NDVI pattern (UNP) and ISM indices, respectively. (2) Linear (partial) correlation analyses were applied to determine the correlations of the TP vegetation with the ISM, IOBM, ENSO, precipitation on the TP, etc., where partial correlation analysis was used to determine the actual correlations between two of the three related variables. (3) Univariate linear regression analyses were employed to explore the process of ISM influencing the JJAS TP NDVI and the influence of ISM on climatic factors on the TP. (4) The contributions of ISM and four climatic factors to the inter-annual variability of vegetation on the TP in the growing season were also explored through multiple linear regression analysis. (5) The potential feedback between the TP thermal conditions and vegetation growth was discussed through linear correlation and composite analysis. The importance of the findings was determined via Student's *t*-test. Note that all data do not have the same horizontal precision, which we unified in our processing.

2.3. The Definition of the Pattern and Indices

2.3.1. The Uniform NDVI Pattern and Its Index

We found two NDVI patterns dominating the TP in JJAS through the empirical orthogonal function (EOF) decomposition. The first dominant EOF mode features a uniform variation in NDVI anomalies on the TP, which is called the uniform NDVI pattern [35]. In this study, we used the REOF analysis to capture the prevalent and homogeneous patterns of the TP NDVI on inter-annual timescales, since the REOF analysis is more appropriate for the high-resolution and inhomogeneous distribution of the TP NDVI datasets.

The dominant modes obtained based on the EOF and REOF analyses highly resemble each other, and both show a uniform NDVI pattern. Moreover, their corresponding time series are significantly and positively linked. Their distinction is that the uniform pattern based on the REOF analysis does not show larger loadings on the southeast of the TP, which may be due to the high vegetation coverage and insignificant reactions to anomalous climate change over this region. Therefore, the first REOF mode of NDVI anomalies on the TP in JJAS from 1982 to 2020 is called the uniform NDVI pattern (Figure 2a), and its corresponding PC index is called the uniform NDVI pattern index (UNPI).

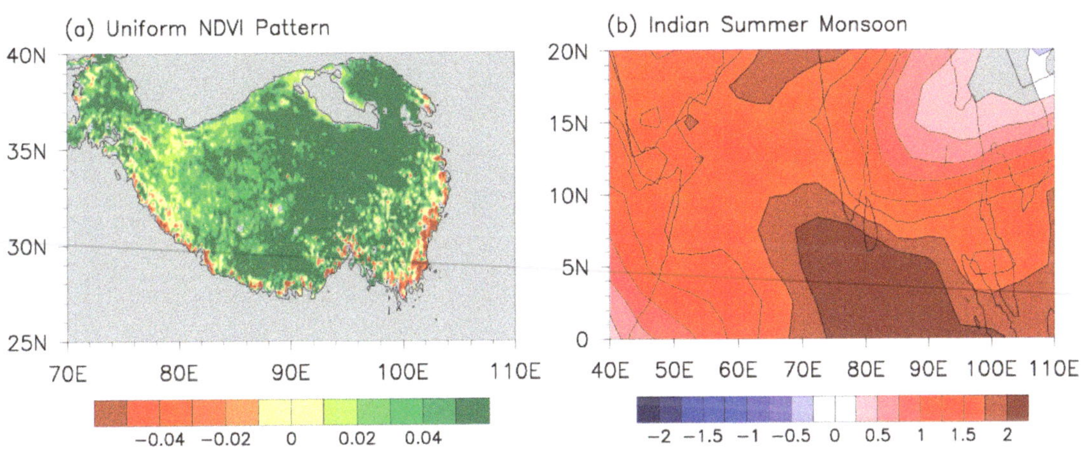

Figure 2. (**a**) The first REOF mode (REOF1) of NDVI anomalies on the TP in JJAS from 1982 to 2020. (**b**) Vertical shear of JJA zonal winds (unit: m/s) over the region (40–110°E, 0–20°N) from 1982 to 2020.

2.3.2. The Indian Summer Monsoon Index

In earlier studies, the Indian summer monsoon (ISM) index was defined as the all-Indian summer monsoon rainfall (AISMR) [71]. With the deepening of research, other indices representing the ISM were also proposed, such as the Webster–Yang index (WYI) [72], the monsoon Hadley circulation index (MHI) [73], the extended Indian monsoon rainfall index (EIMRI) [73], the Indian monsoon index (IMI) [74], and the Indian monsoon trough index (IMTI) [75]. Before exploring the influence of the ISM on the TP vegetation, the definitions and features of the abovementioned ISM indices were compared. Using the singular value decomposition and linear correlation, the WYI was most closely related to the uniform NDVI pattern (figures omitted). Therefore, following Webster and Yang [72], the area-mean vertical shear (U850-U200) of zonal winds over the region (40–110°E, 0–20°N) in June–August (JJA) was referred to as the ISM index (Figure 2b).

3. Results

3.1. Correlations between the ISM and TP Precipitation and Vegetation

Precipitation is one of the main climatic factors affecting TP vegetation in the main growing season [35]. As mentioned in Section 1, the ISM is highly associated with TP precipitation. As such, the ISM should influence the inter-annual variability of vegetation on the TP by modulating precipitation over the TP. Moreover, the ENSO and IOBM can modulate the onset time and intensity of ISM [37,38,48,76,77]. This implies that the ENSO and IOBM may influence the TP precipitation by adjusting the ISM.

Figure 3 reveals the contribution of ENSO, IOBM, and ISM to the TP vegetation on inter-annual scales. A clear and significant positive correlation between the JJA ISM and JJAS TP NDVI appears over most of the TP, manifesting a similarly uniform NDVI pattern (Figure 3a). Moreover, the correlation coefficient between the UNPI and the ISM is 0.45, exceeding the confidence level of 99%. The JJAS TP NDVI is negatively correlated with the spring IBOM/previous winter ENSO, with large loadings (coefficients) roughly distributed on the southwestern TP/northeast–southwest oriented region (Figure 3b,c). These negative correlations reveal that corresponding to positive IOBM/El Niño, the ISM weakens [77]. The correlation coefficient between the UNPI and the spring IBOM is −0.25 and that between the UNPI and the previous winter ENSO is −0.26, which are lower than that between the UNPI and ISM (0.45). After removing the influence of the ISM via the partial correlation, the negative correlation between the spring IBOM (previous winter ENSO) and the JJAS TP NDVI significantly decreased (Figure 3e,f),and the coefficient dropped to −0.09 (−0.23) not reaching the 90% confidence level. To some extent, this implies that the ISM fulfills a "bridge" role linking the influence of ENSO and IOBM with the vegetation growth on the TP. The contribution of ENSO and IOBM to the TP vegetation becomes weaker due to the absence of the bridge effect of ISM.

In contrast, the correlation between the ISM and JJAS TP NDVI slightly decreases after removing the influence of ENSO and IOBM via the partial correlation, but a significantly positive correlation still covers the eastern TP (Figure 3d), showing a closer relationship than the ENSO-UNPI and IOBM-UNPI ones. After removing the influence of ENSO and IOBM, the correlation coefficient between the UNPI and ISM still reaches 0.29, significant at the 95% confidence level. This suggests that the ISM is not only a "bridge" relaying the influence of the ENSO and IOBM on the TP vegetation growth but also has a significant and direct effect on vegetation growth on the TP, albeit in the absence of ENSO and IOBM. Thus, we focus on the relationship between the ISM and the TP precipitation and NDVI in the following study.

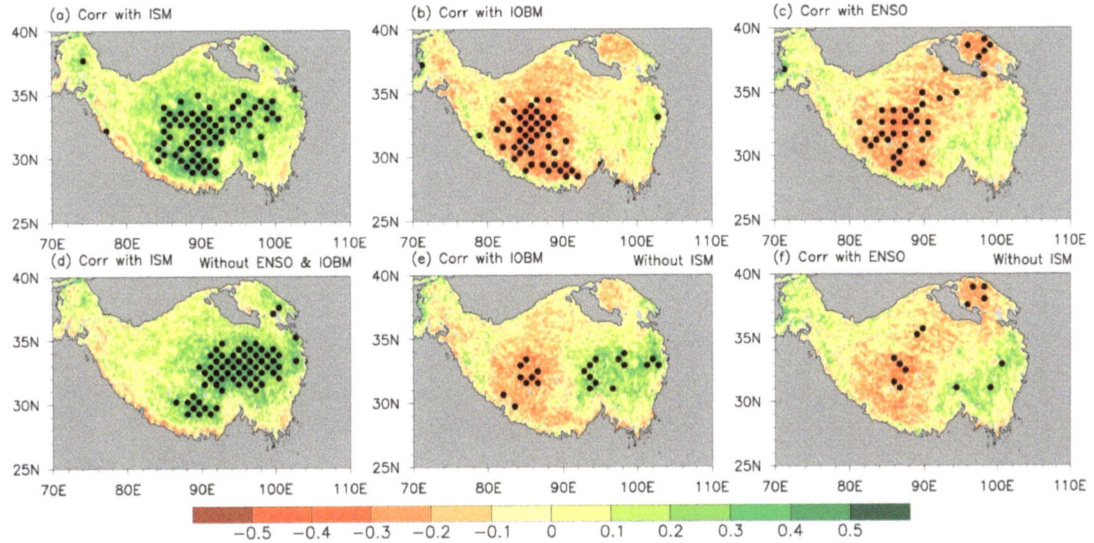

Figure 3. Correlations of the JJA ISM (**a**), MAM IOBM (**b**), and D(−1)JF Niño 3.4 (**c**) indices with the JJAS NDVI anomalies on the TP, respectively. (**d**) Partial correlations of the JJA ISM index with the JJAS NDVI anomalies on the TP, where the influences of the IOBM and ENSO were linearly removed. Partial correlations of the MAM IOBM (**e**) and D(−1)JF Niño 3.4 (**f**) indices with the JJAS NDVI anomalies on the TP, where the influence of the ISM index was linearly removed. Black dots indicate coefficients exceeding the confidence level of 95%. The "MAM" and "D(−1)JF" denote the spring (March–May) and previous winter (December–February), respectively.

Figure 4a presents the correlations of the ISM with the TP precipitation in May–August (MJJA). Note that the periods for the correlations are different from Figure 3c since precipitation has a one-month-lagged effect on the TP NDVI [24,25,29,35]. The ISM is highly positively linked with precipitation over the southwestern TP, exhibiting a central coefficient of 0.52, which reaches the confidence level of 95%. Based on the key region with significant correlations (Figure 4a), we referred to the area-mean precipitation anomaly over the southwest of the TP (80–92°E, 28–35°N) in MJJA as the precipitation index (PRE index). Time series of the ISM index, PRE index, and UNPI are compared in Figure 4b–d. In these figures, we can detect a slightly better correlation between the ISM and precipitation (0.50) than that between the ISM and NDVI (0.43), but both reach the confidence level of 99% (Figure 4b,c). Since precipitation directly affects the TP vegetation growth, the correlation between precipitation and the NDVI was expected to be greater than that between the ISM and NDVI. However, the former exceeds the 95% confidence level by 0.34 (Figure 4d), lower than the latter (Figure 4c). Note that all three indices exhibit a clear upward trend (Figure 4b–d); these correlations could be influenced by global warming. Therefore, their linear trends are removed in the subsequent sections.

After removing their linear trends, the ISM still maintains a significant correlation with the other two indices, which has an approximate coefficient of 0.49 (0.41) with the PRE index (UNPI) exceeding the confidence level of 99%. The PRE index mainly represents the ISM, exhibiting a regional correlation with precipitation over the TP (Figure 4a), while the UNPI represents the variation in vegetation on the overall TP. Thus, the PRE index's correlation with the UNPI decreases from 0.34 to 0.16, which is significantly lower than the correlation between the ISM and UNPI (Figure 4c,d). Such a result implies that the influence of the ISM on the TP vegetation does not depend merely on precipitation, while other ISM-induced climatic factors may contribute to the TP vegetation growth. Clearly, the influence of the ISM on the TP vegetation requires further analysis.

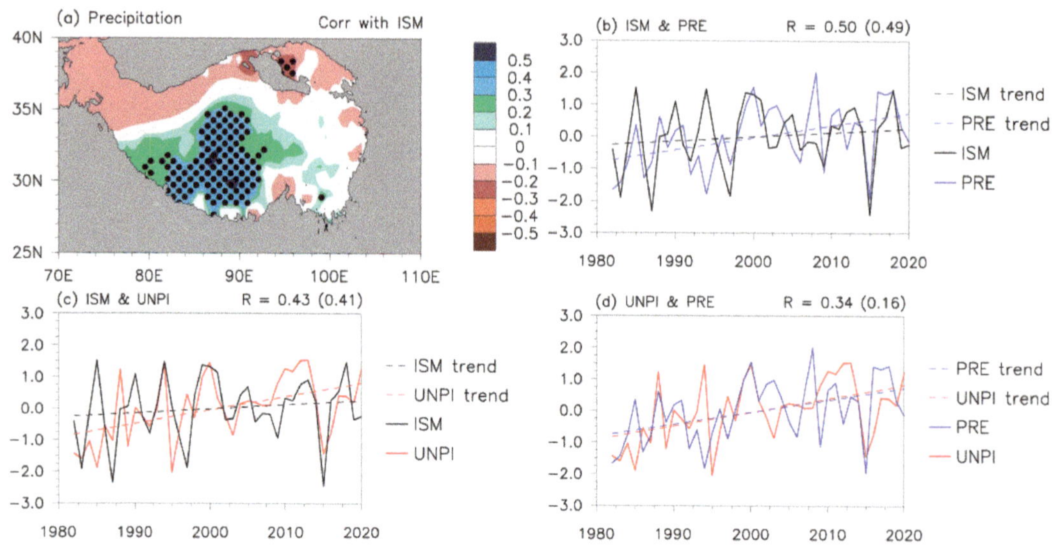

Figure 4. Correlation between the ISM and the MJJA precipitation ((**a**); unit: mm) anomalies. (**b**) Correlation between the ISM index (black lines) and the PRE index (blue lines). The other two figures are as (**b**), but correlations are between the UNPI (red lines) and the ISM index (**c**), and between the UNPI and the PRE index (**d**). Dashed lines indicate the trend of the indices. The R outside the parentheses is the coefficient between these time series of the unremoved trend, while the R in parentheses is removed. Black dots indicate variables exceeding the 95% confidence level.

3.2. Physical Process of the ISM Affecting the TP Vegetation

As an external climatic factor, the ISM plays a role in modifying the vegetation growth on the TP NDVI by stimulating atmospheric circulation to alter the thermal and moisture conditions over the TP. In this section, the process of the ISM influencing the JJAS TP NDVI is explored by linear regression and partial correlation analyses in terms of the atmospheric circulation (Figure 5), water vapor transportation, convection (Figure 6), and thermal conditions over the TP (Figure 7).

In June, the climatological South Asia high (SAH) [78] in the upper troposphere (200 hPa) is situated south of 30°N, with its center around the southern TP. A strong positive geopotential height anomaly appears over the Iranian Plateau, indicating that the SAH significantly strengthens and shifts northwestward to the western TP (Figure 5a). In the middle-lower troposphere (500–850 hPa), two cyclonic convergences occur around the Indian and Indo-China Peninsulas where air pressure significantly decreases (Figure 5e,i). The enhanced SAH induces the ISM onset, which drives a significant increase in the transportation of water vapor from the Indian Ocean to the lower latitudes and accordingly facilitates water vapor converging over the southern TP (Figure 6e). The strengthened upward motion is induced by the low-pressure convergence and high-pressure divergence in the lower and upper troposphere, which occurs over the southern TP to the south of 30°N (Figure 6a). A negative outgoing longwave radiation (OLR) anomaly over the southern TP coincides with the entry of water vapor (Figure 6i).

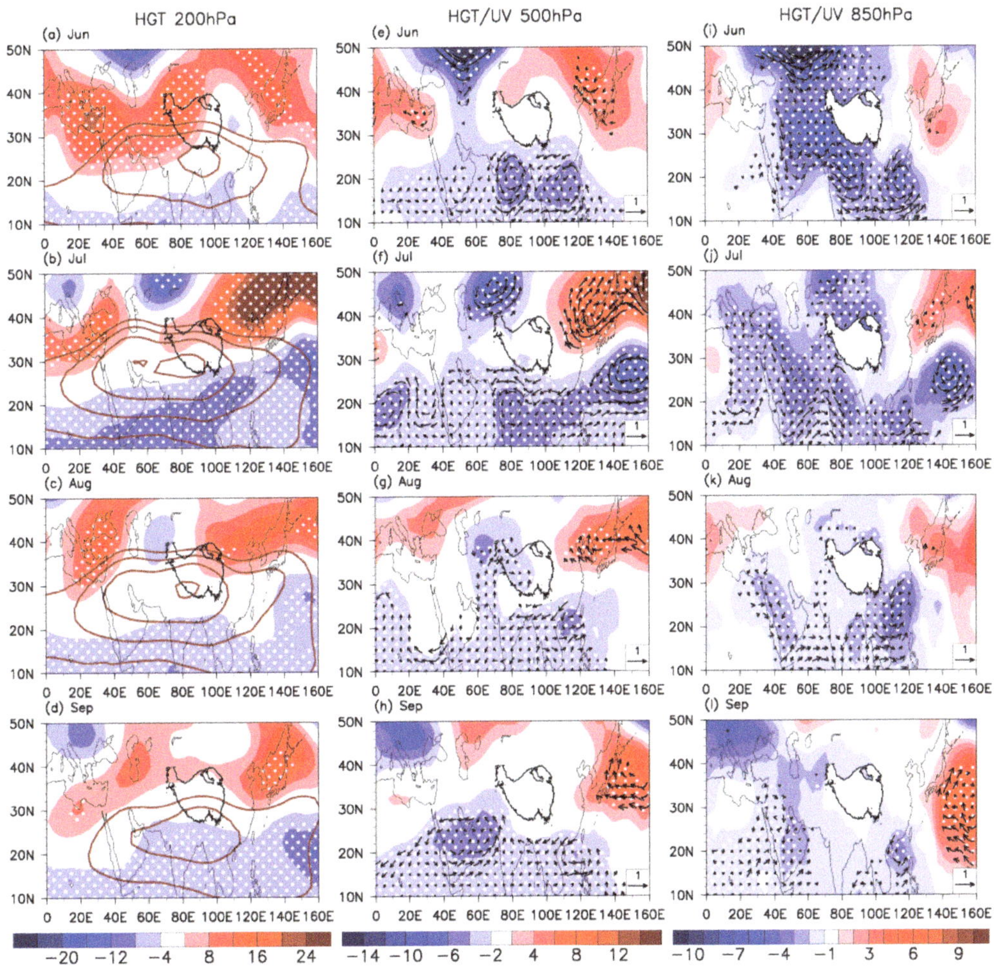

Figure 5. Geopotential height (HGT; unit: gpdm; shading) anomalies at 200 hPa (**a–d**), 500 hPa (**e–h**), and 850 hPa (**i–l**) regressed upon the ISM from June to September, respectively. The brown contours indicate the climatological SAH. Black vectors indicate horizontal wind (UV; unit: m/s) anomalies. White dots indicate variables exceeding the confidence level of 95%.

By July, the location of the SAH in the upper troposphere essentially remains unmoved, but its area and intensity are significantly enhanced (Figure 5b). From June to July, the cyclonic convergence over the Indo-China Peninsula moves eastward to the northwest Pacific in the middle-lower troposphere. During the same period, the low pressure over the Indian Peninsula intensifies and slightly moves eastward (Figure 5f,j). Water vapor fluxes are shifted significantly northward, and then large quantities of water vapor are carried to the lower-latitude TP, where they then converge (Figure 6f). The enhanced upward motion over 25°–35°N and downward motion on either side (Figure 6b) promote convection, featuring a significant negative OLR anomaly over the southwestern TP (Figure 6j). This sufficient water vapor and strengthened convection contribute to the increase in precipitation.

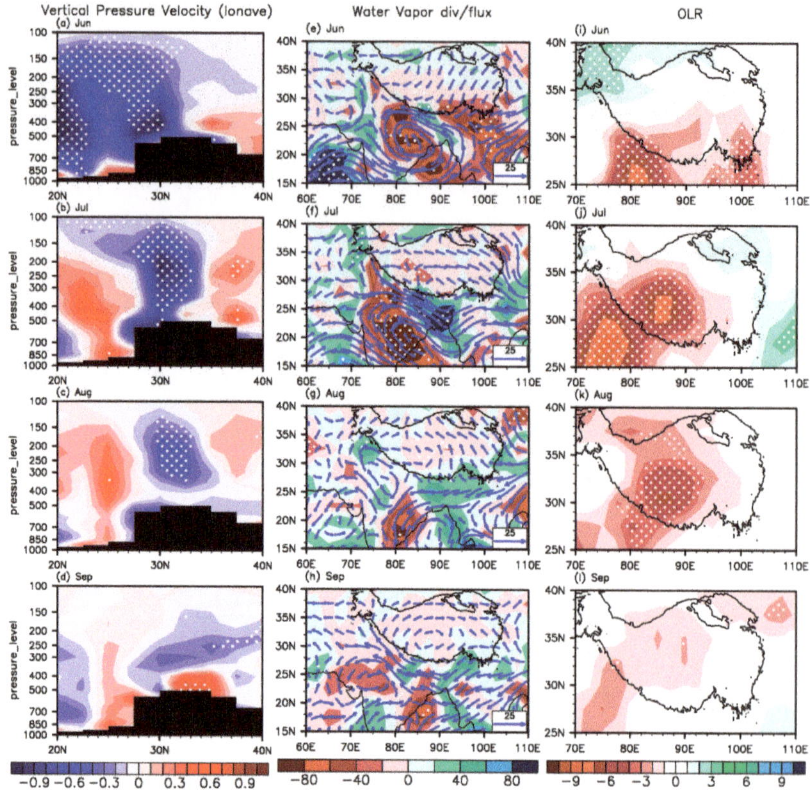

Figure 6. (**a–d**) Vertical pressure velocity (unit: hPa/s) anomalies regressed upon the ISM from June to September in latitude vertical cross section (averaged for 80–90°E), with negative values for upward motion. (**e–h**) Integrated water vapor flux from surface to 300 hPa (unit: kg/(m·s); blue vector) and water vapor flux divergence (unit: 10×10^{-7}kg/(m^2·s·hPa); shading) anomalies regressed upon the ISM from June to September. (**i–l**) Outgoing longwave radiation (OLR; unit: W/m^2) anomalies regressed upon the ISM from June to September. White dots indicate variables exceeding the confidence level of 95%.

As the SAH intensity decreases slightly in August (Figure 5c), the corresponding cyclonic convergences at 500 hPa and 850 hPa move westward to the Indo-China Peninsula, resulting in a weakening low pressure over the Bay of Bengal (Figure 5g,k). Changes in atmospheric circulation reduce the transportation of water vapor, which causes a consequent reduction in water vapor flux convergence to the southern TP (Figure 6g). The significant upward motion still appears over the overall TP (Figure 6c). Correspondingly, the negative OLR anomaly still appears over the central and western TP (Figure 6k), but the intensity begins to decrease, with the numerical value of the maximum OLR anomaly decreasing from approximately 9 W/m^2 to 6 W/m^2.

In September, the SAH intensity weakens substantially (Figure 5d), and a cyclonic circulation develops near the Arabian Sea in the mid-troposphere (Figure 5h). This causes the entire circulation system to move westward, with a significant downward motion (Figure 6d) and reduced convection (Figure 6l) over the TP.

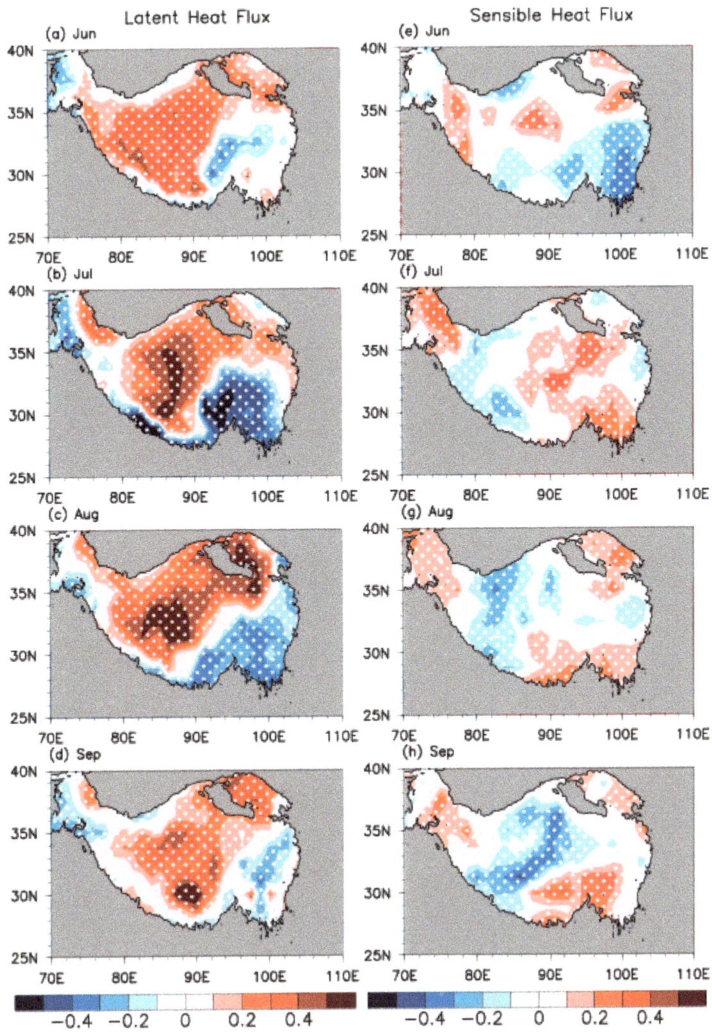

Figure 7. Partial correlations of the ISM with latent heat flux ((**a**–**d**); unit: W/m^2) and sensible heat flux ((**e**–**h**); unit: W/m^2) anomalies over the TP from June to September, where the influence of the PRE index was linearly removed. White dots indicate coefficients exceeding the confidence level of 95%.

As mentioned in Section 3.1, the ISM can influence vegetation on the TP through other climatic factors induced by the ISM besides precipitation, such as thermal factors (e.g., latent and sensible heat fluxes). Vegetation dynamics could largely affect the thermal conditions over the TP [12–15]. For example, the greening of TP vegetation can reduce surface albedo and thus increase sensible heat flux [14,15]. Based on that, we examined the partial correlation of the ISM with surface latent and sensible heat flux over the TP after removing the influence of the PRE index (Figure 7). Due to the annual cycle of vegetation growth on the TP, its coverage gradually increases from June to July. The consequently enhanced evapotranspiration of vegetation leads to substantial heat absorption, which promotes the transportation of latent heat across the TP and cools some regions of the TP. Meanwhile, the decrease in surface albedo induced by the substantial vegetation leads to an increase in sensible heat over most of the TP. Thus, the latent heat flux over the TP exhibits a distinct positive anomaly in JJA (Figure 7a,b), as does the sensible heat flux (Figure 7e,f). When the TP vegetation reaches its maximum coverage in August, the positive latent heat flux anomalies over the TP manifest the greatest magnitude, and the range of negative sensible heat anomalies also increases (Figure 7c,g). With the reduction in the TP vegetation in September, a significant decrease in the positive latent heat anomaly occurs over the TP, whereas the increased albedo leads to a wide range of negative sensible heat flux anomalies (Figure 7d,h).

Based on the above sections, we suggest the ISM may change air and ground temperature over the TP through varying vegetation growth. Additionally, the ISM-induced precipitation can affect the sunshine duration over the TP. Anomalous changes in these ISM-induced climatic factors jointly affect the NDVI inter-annual variability on the TP. Considering the one-month-lagged impact of precipitation on vegetation, the JJA precipitation anomaly, and the JAS surface air temperature, ground surface temperature and sunshine duration anomaly were regressed upon the ISM index (Figure 8).

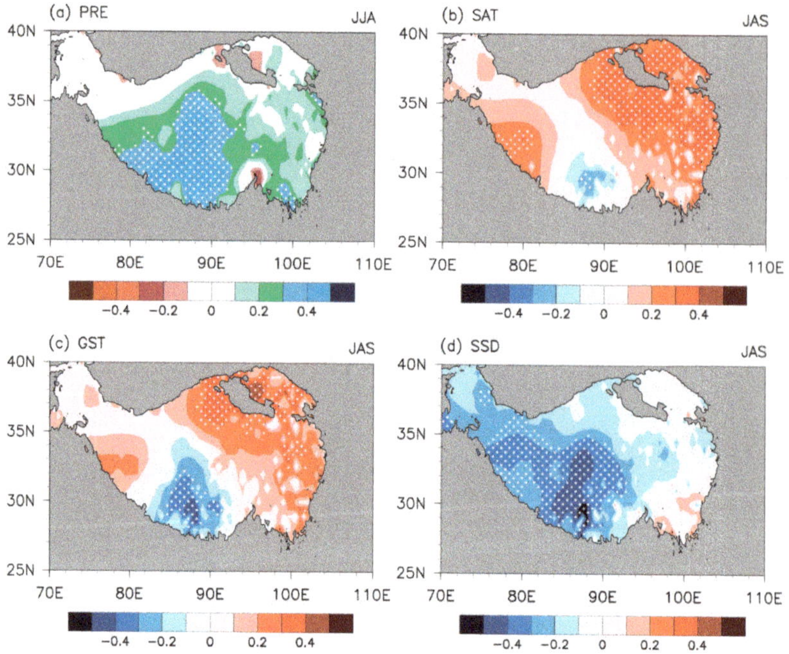

Figure 8. Precipitation (PRE) anomalies in JJA ((**a**); unit: mm) and surface air temperature (SAT) ((**b**); unit: °C), ground surface temperature (GST) ((**c**); unit: °C), and sunshine duration (SSD) ((**d**); unit: hours) anomalies in JAS regressed upon the ISM index. White dots indicate variables exceeding the confidence level of 95%.

Regulated by the ISM, precipitation increases significantly over the southwest of the TP and decreases over the Pamir Plateau and the southeast of the TP (Figure 8a). Surface air and ground temperatures exhibit significant warming over the northeastern TP to the north of 33°N (Figure 8b,c), which can contribute to higher NDVI across almost the entire TP. In fact, sunshine duration over the TP is not affected by the ISM, and the shorter sunshine duration should be attributed to the simultaneously increased precipitation. Precipitation increases abnormally over the southwest of the TP, where the sunshine duration is significantly shortened (Figure 8d). This finding is consistent with Mao et al. [35].

To examine the impact of the ISM-induced climatic factors on the NDVI inter-annual variability on the TP, we defined the corresponding indices in accordance with the significant regions (Figure 8) affected by these climatic factors. The JAS area-mean surface air temperature (ground surface temperature) anomalies over the north of the TP (70–104°E, 34–40°N) are referred to as the SAT (GST) index. The SSD index is determined in the same definition as the PRE index in Section 3.1, but in JAS. Based on the ISM and these indices, the following regression equation was established to estimate the UNPI.

$$UNPI_e = 9.9 \times 10^{-9} - 0.15 * PRE - 0.39 * SAT + 0.69 * GST - 0.12 * SSD + 0.28 * ISM \quad (1)$$

in which the term on the left represents the estimate of UNPI ($UNPI_e$), and the terms on the right represent the effect of these indices (*PRE*, *SAT*, *GST*, *SSD* and *ISM*) with different weights. For ease of calculation, the estimated intercept of 9.9×10^{-9} in Equation (1) is usually negligible.

Figure 9 presents the regression of the monthly NDVI anomaly upon the $UNPI_e$ in JJAS. Positive NDVI anomalies appear on the TP for each month (Figure 9a–d), displaying an approximately uniform NDVI pattern after four-month vegetation accumulation (Figure 9e). This suggests that the inter-annual variability of NDVI on the TP in JJAS can be attributed to the ISM and its induced changes in the local climatic factors in the TP, which can account for more than 52% of the variation in the UNPI.

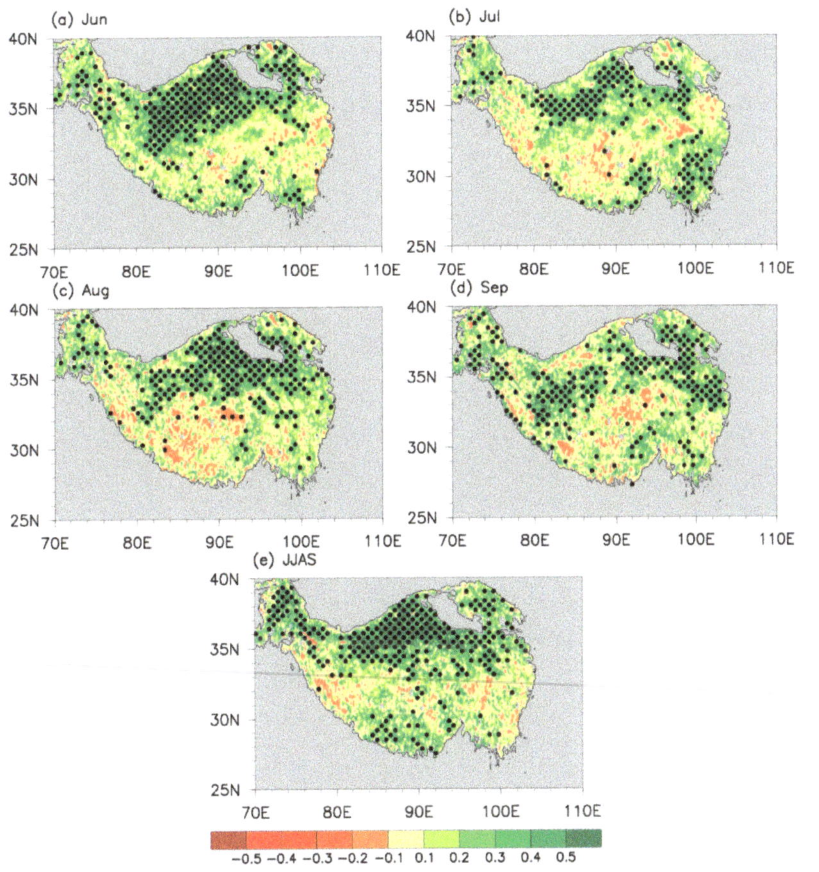

Figure 9. (**a**) NDVI anomalies in June (**a**), July (**b**), August (**c**), September (**d**), and JJAS (**e**), respectively, as regressed upon the $UNPI_e$. Black dots indicate variables exceeding the confidence level of 95%.

4. Discussion

This paper indicates that the ISM is a significant external factor affecting the inter-annual variation in TP vegetation in the growing season, and examined the correlations among the ISM, the TP precipitation, and the TP vegetation. The findings reveal that the correlation between ISM and UNPI is much greater than that between UNPI and ISM-induced precipitation, especially when the linear trends of the three indices were removed (Figure 4b–d). The variations in the uniform NDVI pattern on the TP in JJAS on inter-annual scales are caused by a combination of several local climatic factors [35]. Therefore, instead of precipitation, other ISM-induced climatic factors dominate the inter-annual variations in vegetation on the TP. In addition to regulating the atmospheric circulation and associated precipitation over the TP, the ISM can also influence the inter-annual variability of vegetation by inducing changes in the TP thermal conditions. Figure 10 summarizes the process of the ISM influencing the vegetation growth on the TP.

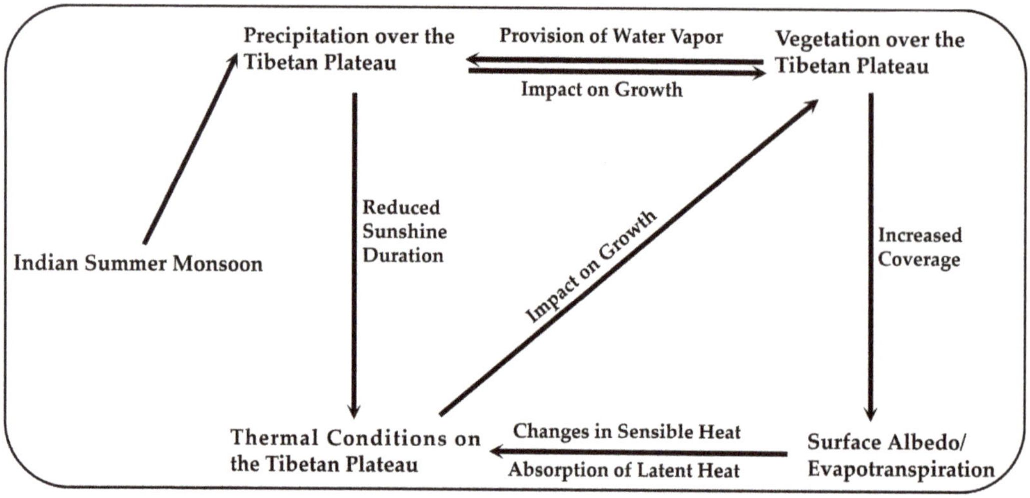

Figure 10. The influence process of the ISM on the inter-annual variability of vegetation on the TP in its main growing season.

In the main growing season (JJAS), changes in the SAH concerning its location and intensity cause higher pressure (positive HGT anomalies) over the TP in the upper troposphere, and lower pressure (negative HGT anomalies) at the lower-latitude TP in the middle-lower troposphere (Figure 5). Such an atmospheric circulation structure, with atmospheric divergence and convergence in the upper and middle-lower troposphere, respectively, enhances the upward motion over the TP (Figure 6a–d). Furthermore, the transport of water vapor through the Indian Ocean entering the lower-latitude TP is facilitated by the strengthened cyclone activity in the lower troposphere (Figure 6e–h). The sufficient water vapor and strengthened convection can increase precipitation over the TP, thus promoting vegetation growth.

Vegetation evapotranspiration modulates thermal conditions over the TP as a result of the physical phase transition of water, as well as altering the local atmospheric water vapor content and associated precipitation. Additionally, the variations in vegetation coverage also can affect surface sensible heat flux by modifying the TP surface elements, such as albedo and roughness. With the gradual increase in vegetation, the enhanced evapotranspiration replenishes the atmospheric water vapor over the TP while absorbing latent heat (Figure 7a–d) and altering surface sensible heat (Figure 7e–h). Due to the interaction between vegetation and temperature, a certain degree of warming promotes vegetation growth. According to earlier studies, changes in thermal conditions over the TP may also affect vegetation growth [23–26], and moderate warming of the northern TP effectively promotes the growth of local vegetation [35]. Additionally, the ISM-induced increase in precipitation can also lead to the lack of sunshine, which interferes with vegetation growth on the TP [25].

We further explored the joint effects of the ISM and its induced changes in local climatic factors (precipitation, air temperature, ground temperature, sunshine duration, etc.) on the uniform NDVI pattern using multiple regression (Equation (1)). These factors jointly modulate more than 52% of the variation in the UNPI. In other words, the ISM not only influences the TP vegetation growth through precipitation, but also can regulate the TP vegetation growth through modulating the variations in thermal factors in the TP. The current findings are based on statistical analyses, which are insufficient to clarify the influence of the ISM on the TP vegetation. We will further verify the results through numerical experiments in the future.

Various factors have an impact on the ISM, such as the TP diabatic heating, land–sea thermal contrast, ENSO, and internal atmospheric processes [37,47–49,53]. Among these factors, the TP atmospheric heat source directly affects the ISM onset, and its diabatic heating also plays a decisive role in the intensity and location of the ISM [52,53]. Additionally, the increased vegetation on the TP can cause changes in local thermal conditions (Figure 7). This raises an interesting question, that is, can the TP vegetation regulate the ISM through the alteration of TP thermal conditions in its main growing season?

Table 1 exhibits the correlation coefficients between the monthly ISM from June to September and the one-month-lagged, simultaneous, and summer (JAS) UNPIs. The ISM and the one-month-lagged (summer) UNPI are positively correlated, with the highest association between the July ISM and August (summer) UNPI. The finding indicates the response of the TP NDVI to the influence of the ISM in its main growing season. The ISM is closely correlated with the simultaneous UNPI in June–July. However, the correlation between the ISM and the simultaneous UNPI is insignificant in August–September. These imply the possible influence of the UNPI on the ISM. The September ISM is hardly relevant with summer UNPI with a coefficient of only −0.03, which suggests the ISM-UNPI correlation has vanished at this time.

Table 1. Correlation of the monthly ISM with different months of the UNPI. The superscript "*" denotes the coefficients that significantly exceed the confidence level of 95%.

	June	July	August	September
With the one-month-lagged UNPI	0.17	0.31 *	0.19	0.30 *
With the simultaneous UNPI	0.27 *	0.39 *	0.21	0.19
With the summer UNPI	0.24	0.39 *	0.33 *	−0.03

The differences in latent and sensible heat fluxes over the TP, and the vertical shear of zonal winds over the region (40–110°E, 0–20°N), are analyzed for typical positive and negative UNPI years (Figure 11), where the changes in the ISM intensity can reflect the correlation between the ISM and UNPI. The typical UNPI years are characterized as having an absolute value of the UNPI larger than a threshold of 0.5 standard deviations (see Table 2). In June–July, the substantial growth and increased coverage of the TP vegetation result in enhanced evapotranspiration and transportation of latent heat (positive anomalies) over the central and northern TP (Figure 11a,b). A clear decrease in sensible heat fluxes (negative anomalies) appears there, while the sensible heat fluxes increase significantly over the northwest corner and southeast of the TP due to the reduced surface albedo (Figure 11e,f). During this period, the relationship between the ISM and UNPI gradually becomes stronger (Figure 11i,j). When the NDVI reaches its maximum coverage in August, positive latent heat flux anomalies keep increasing due to the vegetation evapotranspiration (Figure 11c), and the magnitude of negative sensible heat flux anomalies increases throughout the whole TP except for the southeast (Figure 11g). In contrast, there is no significant increase in the sensible heat fluxes over the southeastern TP because the vegetation coverage (albedo) no longer increases (decreases) (Figure 11g). The overall TP exhibits cooling, while the correlation between the UNPI and ISM is somewhat weakened (Figure 11k). In September, the vegetation still shows strong overall evapotranspiration over the TP despite the decrease in TP vegetation coverage. Most of the TP experiences significantly increased latent heat flux (Figure 11d), and the western TP to the west of 90°E experiences a significant decrease in sensible heat flux (Figure 11h). Meanwhile, the ISM-UNPI correlation seems to disappear (Figure 11l).

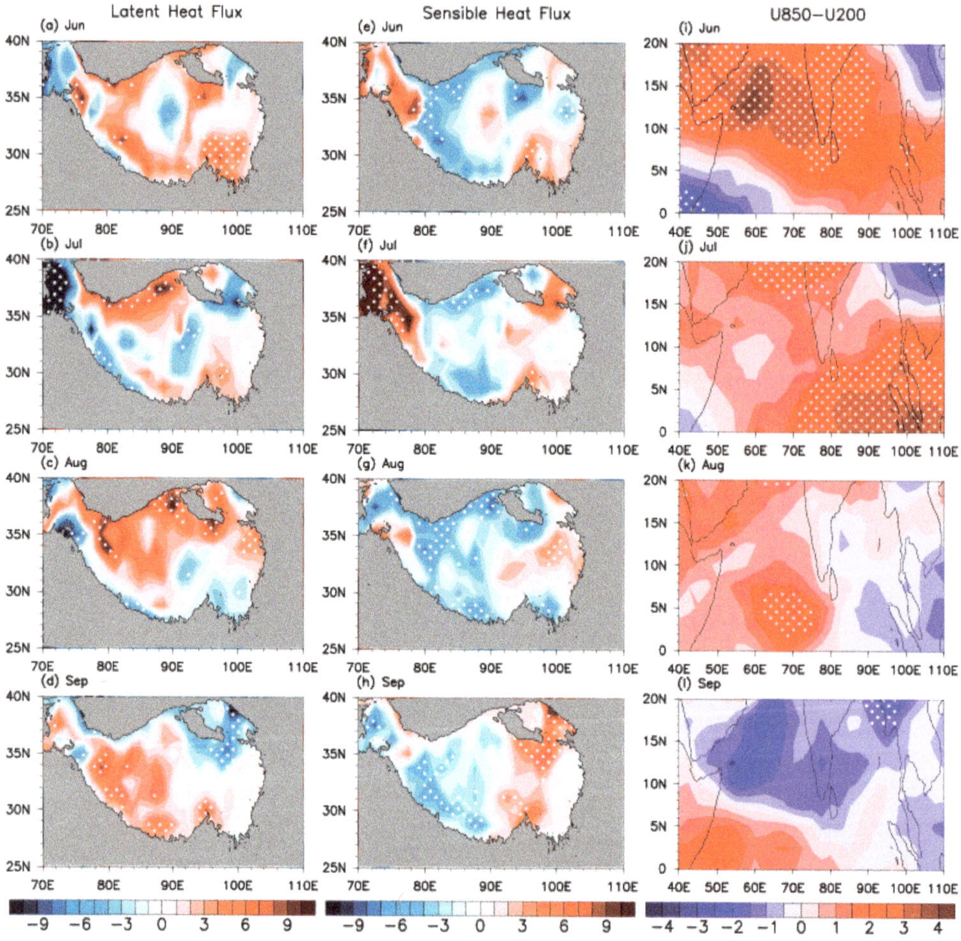

Figure 11. Differences in latent heat flux ((**a**–**d**); unit: W/m^2), sensible heat flux ((**e**–**h**); unit: W/m^2) anomalies over the TP, and vertical shear of zonal wind (U850-U200) anomalies ((**i**–**l**); unit: m/s) over the region (40–110°E, 0–20°N), composited for positive and negative UNPI years. White dots indicate variables exceeding the confidence level of 95%.

Table 2. Typical years of the UNPI in positive and negative phases.

Positive UNPI (Std > 0.5) Years	Negative UNPI (Std < −0.5) Years
1988, 1994, 1997, 1999, 2000, 2010, 2011, 2012, 2013	1982, 1983, 1985, 1989, 1995, 2003, 2015, 2016, 2019

Comparing Table 1 and Figure 11, we speculate that the influence of the NDVI on the TP thermal conditions could play a de-correlation role between the ISM and the TP vegetation in late summer and early autumn. However, our current findings are insufficient to confirm the above opinion. The specific physical mechanism requires further study. The current findings reveal a closer relationship of the TP vegetation with the ISM than that with the IOBM and ENSO on inter-annual scales, explore the influence of the ISM on the TP vegetation, and preliminarily present the speculation of the potential physical process. However, as a "bridge" linking the potential contributions of ENSO and IOBM with vegetation growth on the TP, the effect of the ISM on the growth of the TP vegetation needs to be further quantified. Therefore, a series of numerical experiments should be performed in the future to quantify the specific contributions of several external climatic factors (e.g., the ISM, IOBM, ENSO) to the inter-annual variability of the TP NDVI, especially the role of the ISM. These studies may further deepen our understanding of the relationship between the TP vegetation and regional and global climate change and facilitate future predictions of vegetation activity on the TP on inter-annual scales.

5. Conclusions

Using remote-sensing-based NDVI, several sets of reanalysis, and meteorological station observed data, this paper explored the influence of the ISM on the NDVI inter-annual variability on the TP in JJAS. The findings reveal that the ISM is an external factor affecting the inter-annual variation of TP vegetation in the growing season. Furthermore, the contribution of ISM is more direct and significant than that of ENSO and IOBM. The ISM, TP NDVI, and TP precipitation are all positively correlated with each other. Although precipitation is a direct factor of vegetation growth, the correlation between precipitation and NDVI greatly decreases after removing their linear trends and is much smaller than that between the ISM and NDVI. This implies that the ISM influences the TP vegetation not only by changing precipitation but also by inducing the changes in thermal factors in the TP.

Corresponding to a strong ISM, anticyclonic circulation develops over the TP in the upper troposphere, and significant cyclonic circulation develops over the southern TP in the middle-lower troposphere, which also represents a strengthened SAH. This upper and middle-lower tropospheric circulation structure enhances upward motion over the TP. Moreover, the middle-lower tropospheric cyclones can induce more water vapor to the south of the TP. The sufficient water vapor and strengthened upward motion both facilitate more precipitation over the southwest of the TP, which affects vegetation growth. The ISM-induced increase in precipitation over the TP also affects the TP thermal conditions by modulating sunshine duration. Moreover, vegetation can affect TP thermal conditions through its evapotranspiration and coverage. The increased vegetation causes the TP warming, and the TP warming can in turn promote vegetation growth. Further multiple regression analysis revealed that the ISM and its induced changes in local climatic factors can account for more than 52% of the NDVI inter-annual variability on the TP in JJAS.

Additionally, changes in TP thermal conditions, which are regulated by the NDVI in late summer and early autumn, may influence the relationship between the ISM and TP vegetation. In the early growing season (June–July), the UNPI and ISM are significantly correlated. Composite analysis suggests that the TP NDVI causes the changes in TP thermal conditions and thus affects the ISM intensity. Relative to June–July, the ISM intensity is weaker in August–September. This weakening will be more severe in the case of increased vegetation, which may disturb and weaken the correlation between the ISM and TP vegetation in late summer and early autumn.

Author Contributions: Conceptualization, H.-L.R.; methodology, H.-L.R. and G.L.; formal analysis, X.M.; data curation, X.M.; writing—original draft preparation, X.M., H.-L.R. and G.L.; writing—review and editing, X.M., H.-L.R. and G.L.; funding acquisition, B.S. and Y.S. All authors edited the manuscript. All authors have read and agreed to the published version of the manuscript.

Funding: The National Key R&D Program of China (Grant 2022YFF0801603), the Second Tibetan Plateau Scientific Expedition and Research (STEP) program (Grant 2019QZKK0105), the Strategic Priority Research Program of the Chinese Academy of Sciences (Grant XDA2010030807), and the Basic Research Fund of CAMS (Grant 2021Z007) all provided financial support for this study.

Data Availability Statement: ECOCAST supplied the GIMMS NDVI3g data for this study, which can be obtained from the website at https://ecocast.arc.nasa.gov/data/pub/gimms/ (accessed on 29 November 2020). LAADS DAAC/NASA made the MCD19A3CMG data available on the website at https://ladsweb.modaps.eosdis.nasa.gov/ (accessed on 6 July 2023). The National Meteorological Information Center supplied the daily meteorological dataset of basic meteorological elements of China National Surface Weather Station (V3.0) on the website at http://data.cma.cn/ (accessed on 30 December 2021). The ERA5 reanalysis datasets were publicly available at https://cds.climate.copernicus.eu/cdsapp#!/home/ (accessed on 21 October 2021). The NOAA PSL, Boulder, Colorado, USA, supplied the outgoing longwave radiation (OLR) data from their website at https://psl.noaa.gov (accessed on 21 October 2021). The JRA-55 dataset was publicly available at https://rda.ucar.edu/datasets/ds628.1/ (accessed on 5 October 2022). The Niño 3.4 index was publicly available at https://psl.noaa.gov/data/correlation/nina34.anom.data (accessed on 1 May 2023). The IOBM index was publicly available at http://cmdp.ncc-cma.net/Monitoring/cn_nino_index.php?product=cn_nino_index_iobw (accessed on 1 May 2023).

Acknowledgments: We are grateful to LAADS DAAC/NASA and ECOCAST for supplying the remote sensing datasets. We thank National Meteorological Information Center for supplying all daily station-observed datasets. We thank NOAA PSL for providing the NOAA Interpolated Outgoing Longwave Radiation (OLR) dataset. We thank ECMWF and NCAR UCAR for providing data. We thank NOAA CPC for providing the Niño 3.4 index and NCC/CMA for providing the IOBM index.

Conflicts of Interest: The authors declare no conflict of interest.

References

1. Ye, D.Z. Some characteristics of the summer circulation over the Qinghai-Xizang (Tibet) Plateau and its neighborhood. *Bull. Amer. Meteorl. Soc.* **1981**, *62*, 14–19. [CrossRef]
2. Ye, D.Z.; Wu, G.X. The role of the heat source of the Tibetan Plateau in the general circulation. *Meteorol. Atmos.Phys.* **1998**, *67*, 181–198. [CrossRef]
3. Liu, G.; Zhao, P.; Chen, J. Possible effect of the thermal condition of the Tibetan Plateau on the interannual variability of the summer Asian-Pacific oscillation. *J. Clim.* **2017**, *30*, 9965–19977. [CrossRef]
4. Liu, Y.M.; Lu, M.M.; Yang, H.J.; Duan, A.M.; He, B.; Yang, S.; Wu, G.X. Land–atmosphere–ocean coupling associated with the Tibetan Plateau and its climate impacts. *Natl. Sci. Rev.* **2020**, *7*, 534–552. [CrossRef]
5. Jiang, X.; Li, Y.; Yang, S.; Yang, K.; Chen, J. Interannual Variation of Summer Atmospheric Heat Source over the Tibetan Plateau and the Role of Convection around the Western Maritime Continent. *J. Clim.* **2016**, *29*, 121–138. [CrossRef]
6. Wu, Z.W.; Zhang, P.; Chen, H.; Li, Y. Can the Tibetan Plateau snow cover influence the interannual variations of Eurasian heat wave frequency? *Clim. Dyn.* **2015**, *46*, 3405–3417. [CrossRef]
7. Xu, X.D.; Zhao, T.L.; Shi, X.H.; Lu, C.G. A study of the role of the Tibetan Plateau's thermal forcing in modulating rainband and moisture transport in eastern China. *Acta Meteorol. Sin.* **2015**, *73*, 20–35. (In Chinese) [CrossRef]
8. Wu, G.X.; Liu, Y.M.; He, B.; Bao, Q.; Wang, Z.Q. Review of the impact of the Tibetan Plateau sensible heat driven air-pump on the Asian summer monsoon. *Chin. J. Atmos. Sci.* **2018**, *42*, 488–504. [CrossRef]
9. Cai, Y.; Han, X.; Zhao, H.; Klotzbach, P.J.; Wu, L.; Raga, G.B.; Wang, C. Enhanced Predictability of Rapidly Intensifying Tropical Cyclones over the Western North Pacific Associated with Snow Depth Changes over the Tibetan Plateau. *J. Clim.* **2022**, *35*, 2093–2110. [CrossRef]
10. Chen, B.X.; Zhang, X.Z.; Tao, J.; Wu, J.S.; Wang, J.S.; Shi, P.L.; Zhang, Y.J.; Yu, C.Q. The impact of climate change and anthropogenic activities on alpine grassland over the Qinghai-Tibet Plateau. *Agric. For. Meteorol.* **2014**, *189–190*, 11–18. [CrossRef]
11. Wang, Z.; Niu, B.; He, Y.; Zhang, J.; Wu, J.; Wang, X.; Zhang, Y.; Zhang, X. Weakening summer westerly circulation actuates greening of the Tibetan Plateau. *Glob. Planet. Chang.* **2022**, *221*, 104027. [CrossRef]
12. Yang, K.; Ye, B.S.; Zhou, D.G.; Wu, B.Y.; Foken, T.; Qin, J.; Zhou, Z.Y. Response of hydrological cycle to recent climate changes in the Tibetan Plateau. *Clim. Chang.* **2011**, *109*, 517–534. [CrossRef]
13. Piao, S.; Tan, K.; Nan, H.; Ciais, P.; Fang, J.; Wang, T.; Vuichard, N.; Zhu, B. Impacts of climate and CO2 changes on the vegetation growth and carbon balance of Qinghai–Tibetan grasslands over the past five decades. *Glob. Planet. Chang.* **2012**, *98–99*, 73–80. [CrossRef]
14. Fan, G.Z.; Chen, G.D. Interactions between Physiological Process of the Tibetan Plateau Vegetation and CO2 Concentration and Climate Change. *Chin. J. Atmos. Sci.* **2002**, *26*, 509–518. [CrossRef]

15. Shen, M.G.; Piao, S.L.; Jeong, S.J.; Zhou, L.M.; Zeng, Z.Z.; Ciais, P.; Chen, D.L.; Huang, M.T.; Jin, C.S.; Li, L.; et al. Evaporative cooling over the Tibetan Plateau induced by vegetation growth. *Proc. Natl. Acad. Sci. USA* **2015**, *112*, 9299–9304. [CrossRef]
16. Kumari, N.; Saco, P.M.; Rodriguez, J.F.; Johnstone, S.A.; Srivastava, A.; Chun, K.P.; Yetemen, O. The grass is not always greener on the other side: Seasonal reversal of vegetation greenness in aspect-driven semiarid ecosystems. *Geophys. Res. Lett.* **2020**, *47*, e2020GL088918. [CrossRef]
17. Rouse, J.W.; Haas, R.H.; Schell, J.A.; Deering, D.W. Monitoring the Vernal Advancement and Retrogradation (Green Wave Effect) of Natural Vegetation. Contractor Report; 1973. Available online: https://ntrs.nasa.gov/citations/19750020419 (accessed on 8 July 2023).
18. Huete, A.; Didan, K.; Miura, T.; Rodriguez, E.P.; Gao, X.; Ferreira, L.G. Overview of the radiometric and biophysical performance of the MODIS vegetation indices. *Remote Sens. Environ.* **2002**, *83*, 195–213. [CrossRef]
19. Justice, C.O.; Wharton, S.W.; Holben, B.N. Application of digital terrain data to quantify and reduce the topographic effect on Landsat data. *Int. J. Remote Sens.* **1981**, *2*, 213–230. [CrossRef]
20. Martín-Ortega, P.; García-Montero, L.G.; Sibelet, N. Temporal patterns in illumination conditions and its effect on vegetation indices using Landsat on Google Earth Engine. *Remote Sens.* **2020**, *12*, 211. [CrossRef]
21. Matsushita, B.; Yang, W.; Chen, J.; Onda, Y.; Qiu, G. Sensitivity of the enhanced vegetation index (EVI) and normalized difference vegetation index (NDVI) to topographic effects: A case study in high-density cypress forest. *Sensors* **2007**, *7*, 2636–2651. [CrossRef]
22. Huang, K.; Zhang, Y.; Zhu, J.; Liu, Y.; Zu, J.; Zhang, J. The Influences of Climate Change and Human Activities on Vegetation Dynamics in the Qinghai-Tibet Plateau. *Remote Sens.* **2016**, *8*, 876. [CrossRef]
23. Piao, S.; Mohammat, A.; Fang, J.; Qiang, C.; Feng, J. NDVI- based increase in growth of temperate grasslands and its responses to climate changes in China. *Glob. Environ. Chang.* **2006**, *16*, 340–348. [CrossRef]
24. Cao, X.J.; Ganjurjav, H.; Liang, Y.; Gao, Q.Z.; Zhang, Y.; Li, Y.E.; Wan, Y.F.; Danjiu, L.B. Temporal and spatial distribution of grassland degradation in northern Tibet based on NDVI. *Acta Pratacult. Sin.* **2016**, *25*, 1–8. (In Chinese) [CrossRef]
25. Du, J.Q.; Zhao, C.X.; Shu, J.M.; Jiaerheng, A.; Yuan, X.J.; Yin, J.Q.; Fang, S.F.; He, P. Spatiotemporal changes of vegetation on the Tibetan Plateau and relationship to climatic variables during multiyear periods from 1982–2012. *Environ. Earth Sci.* **2016**, *75*, 77. [CrossRef]
26. Li, W.X.; Xu, J.; Yao, Y.Q.; Zhang, Z.C. Temporal and Spatial Changes in the Vegetation Cover (NDVI) in the Three-River Headwater Region, Tibetan Plateau, China under Global Warming. *Mt. Res.* **2021**, *39*, 473–482. (In Chinese) [CrossRef]
27. Chai, L.F.; Tian, L.; Ao, Y.; Wang, X.Q. Influence of Human Disturbance on the Change of Vegetation Cover in the Tibetan Plateau. *Res. Soil Water Conserv.* **2021**, *28*, 382–388.
28. Liu, S.; Zhang, Y.; Cheng, F.; Hou, X.; Zhao, S. Response of Grassland Degradation to Drought at Different Time-Scales in Qinghai Province: Spatio-Temporal Characteristics, Correlation, and Implications. *Remote Sens.* **2017**, *9*, 1329. [CrossRef]
29. Ding, M.J.; Zhang, Y.L.; Liu, L.S.; Wang, Z.F. Temporal and spatial distribution of grassland coverage change in Tibetan Plateau since 1982. J. *Nat. Resour.* **2010**, *25*, 2114–2122. [CrossRef]
30. Salzer, M.W.; Larson, E.R.; Bunn, A.G.; Hughes, M.K. Changing climate response in near-treeline bristlecone pine with elevation and aspect. *Environ. Res. Lett.* **2014**, *9*, 114007. [CrossRef]
31. Zhang, B.; Zhang, Y.; Wang, Z.; Ding, M.; Liu, L.; Li, L.; Li, S.; Liu, Q.; Paudel, B.; Zhang, H. Factors Driving Changes in Vegetation in Mt. Qomolangma (Everest): Implications for the Management of Protected Areas. *Remote Sens.* **2021**, *13*, 4725. [CrossRef]
32. Weltzin, J.F.; Loik, M.E.; Schwinning, S.; Williams, D.G.; Fay, P.A.; Haddad, B.M.; Harte, J.; Huxman, T.E.; Knapp, A.K.; Lin, G.; et al. Assessing the response of terrestrial ecosystems to potential changes in precipitation. *BioScience* **2003**, *53*, 941–952. [CrossRef]
33. Sarkar, S.; Kafatos, M. Interannual variability of vegetation over the Indian sub-continent and its relation to the different meteorological parameters. *Remote Sens. Environ.* **2004**, *90*, 268–280. [CrossRef]
34. Yu, M.; Wang, G.L.; Parr, D.; Ahmed, K.F. Future changes of the terrestrial ecosystem based on a dynamic vegetation model driven with RCP8.5 climate projections from 19 GCMs. *Clim. Chang.* **2014**, *127*, 257–271. [CrossRef]
35. Mao, X.; Ren, H.-L.; Liu, G. Primary Interannual Variability Patterns of the Growing-Season NDVI over the Tibetan Plateau and Main Climatic Factors. *Remote Sens.* **2022**, *14*, 5183. [CrossRef]
36. Wu, G.X.; Duan, A.M.; Liu, Y.M.; Yan, J.H.; Liu, B.Q.; Ren, S.L.; Zhang, H.Y.; Wang, T.M.; Liang, X.Y.; Guan, Y. Recent Advances in the Study on the Dynamics of the Asian Summer Monsoon Onset. *Chin. J. Atmos. Sci.* **2013**, *37*, 211–228. [CrossRef]
37. Dong, W.; Lin, Y.; Wright, J.S.; Ming, Y.; Xie, Y.; Wang, B.; Luo, Y.; Huang, W.; Huang, J.; Wang, L.; et al. Summer rainfall over the southwestern Tibetan Plateau controlled by deep convection over the Indian subcontinent. *Nat. Commun.* **2016**, *7*, 10925. [CrossRef]
38. Jiang, X.W.; Ting, M. A Dipole Pattern of Summertime Rainfall across the Indian Subcontinent and the Tibetan Plateau. *J. Clim.* **2017**, *30*, 9607–9620. [CrossRef]
39. Liu, X.; Yin, Z.-Y. Spatial and temporal variation of summer precipitation over the eastern Tibetan Plateau and the North Atlantic Oscillation. *J. Clim.* **2001**, *14*, 2896–2909. [CrossRef]
40. Wang, Z.Q.; Yang, S.; Lau, N.C.; Duan, A.M. Teleconnection between summer NAO and East China rainfall variations: A bridge effect of the Tibetan Plateau. *J. Clim.* **2018**, *31*, 6433–6444. [CrossRef]
41. Hu, S.; Zhou, T.; Wu, B. Impact of Developing ENSO on Tibetan Plateau Summer Rainfall. *J. Clim.* **2021**, *34*, 3385–3400. [CrossRef]

42. Chen, X.Y.; You, Q.L. Effect of Indian Ocean SST on Tibetan Plateau precipitation in the early rainy season. *J. Clim.* **2017**, *30*, 8973–8985. [CrossRef]
43. Gao, Y.; Wang, H.; Li, S. Influences of the Atlantic Ocean on the summer precipitation of the southeastern Tibetan Plateau. *J. Geophys. Res. Atmos.* **2013**, *118*, 3534–3544. [CrossRef]
44. Jiang, X.W.; Zhang, T.T.; Tam, C.Y.; Chen, J.W.; Lau, N.C.; Yang, S.; Wang, Z.Y. Impacts of ENSO and IOD on Snow Depth Over the Tibetan Plateau: Roles of Convections Over the Western North Pacific and Indian Ocean. *J. Geophys. Res. Atmos.* **2019**, *124*, 11961–11975. [CrossRef]
45. He, K.; Liu, G.; Zhao, J.; Li, J. Co-variability of the summer NDVIs on the eastern Tibetan Plateau and in the Lake Baikal region: Associated climate factors and atmospheric circulation. *PLoS ONE* **2020**, *15*, e0239465. [CrossRef]
46. Wang, H.; Liu, G.; Wang, S.; He, K. Precursory Signals (SST and Soil Moisture) of Summer Surface Temperature Anomalies over the Tibetan Plateau. *Atmosphere* **2021**, *12*, 146. [CrossRef]
47. Chen, G.; Huang, R. Excitation Mechanisms of the Teleconnection Patterns Affecting the July Precipitation in Northwest China. *J. Clim.* **2012**, *25*, 7834–7851. [CrossRef]
48. Feng, L.; Zhou, T. Water vapor transport for summer precipitation over the Tibetan Plateau: Multidata set analysis. *J. Geophys. Res. Atmos.* **2012**, *117*, 85–99. [CrossRef]
49. Yao, T.; Masson-Delmotte, V.; Gao, J.; Yu, W.S.; Yang, X.X.; Risi, C.; Sturm, C.; Werner, M.; Zhao, H.B.; He, Y.; et al. A review of climatic controls on δ18O in precipitation over the Tibetan Plateau: Observations and simulations. *Rev. Geophys.* **2013**, *51*, 525–548. [CrossRef]
50. Wei, W.; Ren, Q.; Lu, M.; Yang, S. Zonal Extension of the Middle East Jet Stream and Its Influence on the Asian Monsoon. *J. Clim.* **2022**, *35*, 4741–4751. [CrossRef]
51. Yanai, M.; Li, C.F.; Song, Z.S. Seasonal heating of the plateau and its effects on the evolution of the Asian monsoon. *J. Meteor. Soc. Jpn.* **1992**, *70*, 319–351. [CrossRef]
52. Ting, M.F. Maintenance of Northern Summer Stationary Waves in a GCM. *J. Atmos. Sci.* **1994**, *51*, 3286–3308. [CrossRef]
53. Wu, G.X.; Zhang, Y. Tibetan Plateau Forcing and the Timing of the Monsoon Onset over South Asia and the South China Sea. *Mon. Weather Rev.* **1998**, *126*, 913–927. [CrossRef]
54. Wu, R. A mid-latitude Asian circulation anomaly pattern in boreal summer and its connection with the Indian and East Asian summer monsoons. *Int. J. Climatol.* **2002**, *22*, 1879–1895. [CrossRef]
55. Fensholt, R.; Rasmussen, K.; Nielsen, T.T.; Mbow, C. Evaluation of earth observation based long term vegetation trends—Intercomparing NDVI time series trend analysis consistency of Sahel from AVHRR GIMMS, Terra MODIS and SPOT VGT data. *Remote Sens. Environ.* **2009**, *113*, 1886–1898. [CrossRef]
56. Alcaraz-Segura, D.; Liras, E.; Tabik, S.; Paruelo, J.; Cabello, J. Evaluating the consistency of the 1982–1999 NDVI trends in the Iberian Peninsula across four time-series derived from the AVHRR sensor. LTDR, GIMMS, FASIR, and PAL-II. *Sensors* **2010**, *10*, 1291–1314. [CrossRef]
57. Fensholt, R.; Rasmussen, K. Analysis of trends in the Sahelian 'rain-use efficiency' using GIMMS NDVI, RFE and GPCP rainfall data. *Remote Sens. Environ.* **2011**, *115*, 438–451. [CrossRef]
58. Lyapustin, A.; Wang, Y.; Laszlo, I.; Korkin, S. Improved cloud and snow screening in MAIAC aerosol retrievals using spectral and spatial analysis, Atmos. *Meas. Tech.* **2012**, *5*, 843–850. [CrossRef]
59. Lyapustin, A.; Wang, Y.; Korkin, S.; Huang, D. MODIS Collection 6 MAIAC Algorithm. *Atmos. Meas. Tech.* **2018**, *11*, 5741–5765. [CrossRef]
60. Gallo, K.; Li, J.; Reed, B.; Eidenshink, J.; Dwyer, J. Multi-platform comparisons of MODIS and AVHRR normalized difference vegetation index data. *Remote Sens. Environ.* **2005**, *99*, 221–231. [CrossRef]
61. Fensholt, R.; Proud, S.R. Evaluation of Earth Observation based global long term vegetation trends—Comparing GIMMS and MODIS global NDVI time series. *Remote Sens. Environ.* **2012**, *119*, 131–147. [CrossRef]
62. Du, J.Q.; Shu, J.M.; Wang, Y.H.; Li, Y.C.; Zhang, L.B.; Guo, Y. Comparison of GIMMS and MODIS normalized vegetation index composite data for Qinghai-Tibet Plateau. *Chin. J. Appl. Ecol.* **2015**, *25*, 533–544. (In Chinese)
63. Hersbach, H.; Bell, B.; Berrisford, P.; Biavati, G.; Horányi, A.; Muñoz Sabater, J.; Nicolas, J.; Peubey, C.; Radu, R.; Rozum, I.; et al. ERA5 Monthly Averaged Data on Pressure Levels from 1959 to Present. Copernicus Climate Change Service (C3S) Climate Data Store (CDS) 2019. Available online: https://cds.climate.copernicus.eu/cdsapp#!/home/ (accessed on 21 October 2021).
64. Liebmann, B.; Smith, C.A. Description of a complete (interpolated) outgoing longwave radiation dataset. *Bull. Am. Meteor. Soc.* **1996**, *77*, 1275–1277.
65. Kobayashi, S.; Ota, Y.; Harada, Y.; Ebita, A. The JRA-55 Reanalysis: General specifications and basic characteristics. *J. Meteor. Soc. Jpn.* **2015**, *93*, 5–48. [CrossRef]
66. Cressman, G.P. An operational objective analysis system. *Mon. Wea. Rev.* **1959**, *87*, 367–374. [CrossRef]
67. He, Y.; Gao, J.; Yao, T.D.; Ding, Y.J.; Xin, R. Spatial distribution of stable isotope in precipitation upon the Tibetan plateau analyzed with various interpolation methods. *J. Glaciol. Geocryol.* **2015**, *37*, 351–359.
68. Sahana, A.S.; Ghosh, S.; Ganguly, A.; Murtugudde, R. Shift in Indian summer monsoon onset during 1976/1977. *Environ. Res. Lett.* **2015**, *10*, 054006. [CrossRef]
69. Karmakar, N.; Misra, V. The relation of intraseasonal variations with local onset and demise of the Indian summer monsoon. *J. Geophys. Res. Atmos.* **2019**, *124*, 2483–2506. [CrossRef]

70. North, G.R.; Bell, T.L.; Cahalan, R.F.; Moeng, F.J. Sampling errors in the estimation of empirical orthogonal function. *Mon. Weather Rev.* **1982**, *110*, 699–706. [CrossRef]
71. Parthasarathy, B.; Munot, A.A.; Kothawale, D.R. All-India monthly and seasonal rainfall series: 1871-1993. Theor. *Appl. Climatol.* **1994**, *49*, 217–224. [CrossRef]
72. Webster, P.J.; Yang, S. Monsoon and ENSO: Selectively interactive systems. *Q. J. R. Meteorol. Soc.* **1992**, *118*, 877–926. [CrossRef]
73. Goswami, B.N.; Krishnamurthy, V.; Annmalai, H. A broad-scale circulation index for the interannual variability of the Indian summer monsoon. *Q. J. R. Meteorol. Soc.* **1999**, *125*, 611–633. [CrossRef]
74. Wang, B.; Wu, R.G.; Lau, K.-M. Interannual variability of the Asian summer monsoon: Contrasts between the Indian and the western North Pacific–East Asian monsoons. *J. Clim.* **2001**, *14*, 4073–4090. [CrossRef]
75. Zhang, T.T.; Jiang, X.W.; Yang, S.; Chen, J.W.; Li, Z.N. A predictable prospect of the South Asian summer monsoon. *Nat. Commun.* **2022**, *13*, 7080. [CrossRef] [PubMed]
76. Zhang, J.; Chen, H.; Zhao, S. A tripole pattern of summertime rainfall and the teleconnections linking northern China to the Indian subcontinent. *J. Clim.* **2019**, *32*, 3637–3652. [CrossRef]
77. Hu, P.; Chen, W.; Chen, S.; Liu, Y.; Wang, L.; Huang, R. The leading mode and factors for coherent variations among the sub-systems of tropical Asian summer monsoon onset. *J. Clim.* **2021**, *35*, 1597–1612. [CrossRef]
78. Cen, S.; Chen, W.; Chen, S.; Liu, Y.; Ma, T. Potential impact of atmospheric heating over East Europe on the zonal shift in the South Asian high: The role of the Silk Road teleconnection. *Sci. Rep.* **2020**, *10*, 6543. [CrossRef]

remote sensing

MDPI

Communication

Tropical Surface Temperature and Atmospheric Latent Heating: A Whole-Tropics Perspective Based on TRMM and ERA5 Datasets

Yue Gao [1], Xiaolin Liu [1] and Jianhua Lu [1,2,*]

1 School of Atmospheric Sciences, Sun Yat-sen University and Southern Marine Science and Engineering Guangdong Laboratory (Zhuhai), Zhuhai 519082, China; gaoy278@mail2.sysu.edu.cn (Y.G.); liuxlin39@mail.sysu.edu.cn (X.L.)

2 Guangdong Province Key Laboratory for Climate Change and Natural Disaster Studies, Sun Yat-sen University, Guangzhou 510275, China

* Correspondence: lvjianhua@mail.sysu.edu.cn

Abstract: Tropical surface temperature (TST) and its connection with atmospheric heating, including tropical latent heating (TLH), is essential to the interannual variability of tropical atmospheric circulation and global teleconnection. Utilizing seasonally averaged satellite-based TRMM precipitation data as a proxy of TLH and ERA5-based TST data from 1998 to 2018, we reveal some new features in terms of cross-hemispheric connection in the TLH and TST variability by decomposing them into equatorially symmetric and antisymmetric components. We find surprisingly that the spatial patterns of TLH projected upon the first principal components (PC1) of symmetric and antisymmetric TSTs over the whole-tropics, are very similar to each other, seemingly at odds with the classic Mastuno–Gill theory. The similarity in the projected TLH patterns is mainly because the PC1s of symmetric and antisymmetric TSTs co-vary temporally with a very high correlation. We use the spatial pattern of local correlation between symmetric and antisymmetric components, for both TST and TLH to depict geographic dependence of the symmetric–antisymmetric connection. We suggest that a whole-tropics perspective, which takes the different but connected nature of equatorially symmetric and antisymmetric modes across the whole-tropics into consideration, may well be useful in understanding and predicting tropical climate variability because clarifying the puzzle raised in this research from such a perspective about the consistency between the observation and the classic Mastuno–Gill theory is directly related to the fundamental dynamics of tropical systems, such as Walker circulation, monsoons, and their relationship with underlying land and sea conditions.

Keywords: TRMM; tropical latent heating; tropical surface temperature; symmetry; antisymmetry; empirical orthogonal function; singular value decomposition

Citation: Gao, Y.; Liu, X.; Lu, J. Tropical Surface Temperature and Atmospheric Latent Heating: A Whole-Tropics Perspective Based on TRMM and ERA5 Datasets. *Remote Sens.* **2023**, *15*, 2746. https://doi.org/10.3390/rs15112746

Academic Editor: Itamar Lensky

Received: 19 April 2023
Revised: 16 May 2023
Accepted: 22 May 2023
Published: 25 May 2023

1. Introduction

We define equatorial symmetry, antisymmetry, and asymmetry of any tropical climate variable as the exact same, exact opposite, and different states of that variable at the same latitude in the two hemispheres centered at the Equator. Clearly, any equatorially asymmetric variable can be decomposed as the sum of equatorially symmetric and antisymmetric components of that variable.

Matsuno, in his classic paper, obtained a mathematical solution for the eigenmodes of free tropical waves [1]. Based on their equatorially symmetric or antisymmetric nature, these eigenmodes are associated with totally different spatiotemporal structures of wind and pressure. Gill (1980) presented the spatial structure of stationary, forced tropical atmospheric general circulation in response to thermal forcing, which may be either equatorially symmetric or antisymmetric [2]. For instance, in response to an equatorially symmetric forcing with the maximum located at the Equator, stationary Kevin wave response

(Walker-circulation type response) appears at the east side of the forcing, while equatorially symmetric stationary Rossby waves appear at the northwest and southwest sides of the forcing. On the contrary, when the thermal forcing is equatorially antisymmetric, the equatorially antisymmetric stationary mixed Rossby-gravity wave and Rossby wave responses, which may mainly be zonally confined nearby the forcing, lead to a Monsoon-type response with the presence of a cross-equatorial flow and cyclonic response in one hemisphere and anti-cyclonic response in the other hemisphere.

It is well known that tropical sea surface temperature (SST) variability, including the ENSO variability in the Pacific and the modes of SST variability in the tropical Indian Ocean and Atlantic, may affect both the tropical atmospheric and global circulation through atmospheric heating and teleconnections [3–5]. Indeed, various meteorological disasters, such as drought, flooding, cold spells, and heat waves, can be associated with the tropical surface temperature forcing in different tropical ocean basins. While many of the studies have been focused on the modes of thermal forcing in individual ocean basins, some studies have considered the cross-basin interaction in the tropics during recent years [6,7].

Given the importance of atmospheric heating in shaping the response of general atmospheric circulation to tropical SST variability, many studies have paid attention to the linkage between SST and tropical atmospheric convection (or tropical latent heating, TLH). While the SST-convection (also TLH or precipitation) linkage can be local, this local connection is highly nonlinear and depends on the large-scale background of tropical circulation [8–12]. On the other hand, the changes in tropical rainfall patterns as responses to modes of tropical SST variability such as ENSO, Indian Ocean Dipole (IOD), and Atlantic Niño are also extensively investigated [6,7].

Due to the hemispheric asymmetry of land–sea distribution, equatorial asymmetry prevails in both the mean climate and the variability of tropical SST and precipitation/ latent heating. Indeed, various mechanisms responsible for the equatorial asymmetry of tropical SST and ITCZ have been proposed, such as the wind–evaporation-feedback mechanism [13], asymmetric upwelling associated with meridional oceanic heat transport [14,15], and coupled atmosphere–ocean energy balance constraint associated with cross-equatorial ocean transport [16], the asymmetric oceanic transport by north equatorial countercurrent (NECC) [17], among others.

It is interesting to note, however, that the decomposition of an asymmetric SST pattern into the sum of equatorial symmetric and antisymmetric components has rarely been conducted, except in an analysis of the interaction of annual-cycle and interannual variability of SST in the tropical Eastern Pacific [18]. The reason is that the hemispheric asymmetry of land–sea distribution makes it impossible to find an SST pair when the underlying boundary is ocean in one hemisphere but is land in the other. However, this difficulty can be easily overcome by using tropical surface temperature, including both SST and land surface temperature, rather than SST only.

Because of the small heat inertia of the land surface, the variability of tropical land surface temperature (LST), in which there are more short time-scale oscillations, is usually considered as a response to SST forcing. While this is physically reasonable, it should be noted that lower frequency variability of LST also exists, and just as SST, LST renders the atmosphere with underlying boundary conditions of momentum, heat, and moisture. Furthermore, the SST-LST asymmetry itself may well lead to an asymmetric response in atmospheric motions. Therefore, we take a different, whole-tropics (30°S–30°N) perspective, which considers tropical SST and LST together, labeled as tropical surface temperature (TST) hereafter.

In this short note, we decompose the TST and TLH into equatorially symmetric and antisymmetric parts and then investigate the relation between the symmetric and antisymmetric components separately for TST and TLH based on correlation analysis. Furthermore, we investigate the relation between interannual TST variability and interannual variability of TLH across the whole-tropics, rather than with the SST confined in a single tropical ocean basin, by applying standard singular value decomposition (SVD), empirical orthogonal function (EOF), correlation, and regression methods.

The layout of the note is as follows. The data and method are presented in Section 2, while Section 3 presents the analysis of the main results. The summary and discussion are in Section 4.

2. Data and Methods

2.1. Data

The Tropical Rainfall Measuring Mission (TRMM) produces tropical and subtropical estimates of precipitation based on remote observations. The TMPA (TRMM Multisatellite Precipitation Analysis) version 7 dataset [19] obtained from the NASA archive (ftp://disc2.nascom.nasa.gov/ftp/data/s4pa//TRMM_L3/, accessed on 24 April 2021) and aggregated to a high spatial (0.25°) and temporal (daily) resolution are employed in this study, with the time domain ranging from 1998 to 2019.

The single-level atmospheric and oceanic variables, named skin temperature and sea surface temperature, are obtained from the European Centre for Medium-Range Weather Forecasts (ECMWF) Reanalysis 5 (ERA5) datasets. These two variables are merged to obtain the TST dataset. They were downloaded with a spatial resolution of 0.25° × 0.25° and monthly temporal resolution from 1998 to 2019 and then interpolated into a resolution of 1.0° × 1.0° using bilinear interpolation for this study.

The Oceanic Niño Index (ONI) is one of the ENSO indices, based on the SST in the Niño 3.4 region, which is obtained from https://climatedataguide.ucar.edu/climate-data/nino-sst-indices-nino-12-3-34-4-oni-and-tni (accessed on 8 November 2022).

2.2. Methods

The variables, for example, the TST or TLH, are divided into two components: equatorially symmetric and antisymmetric. Assuming the TST over the Northern Hemisphere is A, and the TST over the Southern Hemisphere is B with the latitude reversed. The symmetric component is defined as 1/2 (A + B), and the antisymmetric component is 1/2 (A − B) in the Northern Hemisphere and 1/2 (B − A) in the Southern Hemisphere. Therefore, the original field can be divided into the sum of symmetric and antisymmetric fields. Note such a decomposition on the TST, rather than on SST only, can take the intrinsic asymmetry of land–sea contrast into consideration, and hence it may better reveal the coupling of land, sea, and atmosphere over the tropics.

The singular value decomposition analysis (SVD) method of detecting temporally synchronous spatial patterns is also used in this study. The method is based on a singular value decomposition of the matrix whose elements are covariances between observations made at different grid points in two geophysical fields, for example, TST and TLH, in this study. Here, we briefly describe the SVD method following Wallace et al. (1992) [20] and Hu (1997) [21]. TST and TRMM-based TLH data are denoted as $s(x,t)$ and $z(x,t)$, respectively, where x is space location, and t is time. The SVD analysis is then a linear transformation:

$$s(x,t) \approx \sum_{n=1}^{N} a_n(t)p_n(x) \tag{1}$$

$$z(x,t) \approx \sum_{n=1}^{N} b_n(t)q_n(x) \tag{2}$$

in which the pairs of coupled spatial patterns, $p_n(x)$ and $q_n(x)$ (also called left and right SVD spatial patterns, respectively), and their temporal expansion coefficients, $a_n(t)$ and $b_n(t)$, are identified. Here N is the number of SVD modes. As described by Zhang et al. (2018) [22], the SVD analysis was conducted as follows: First, the cross-covariance matrix of $s(x,t)$ and $z(x,t)$, C_{sz}, was calculated. Secondly, the eigenvalues (also called singular values) σ_n of the matrix were obtained by solving $|C_{sz} - \sigma I| = 0$ where I is the identity matrix. Next, the eigenvectors (that is, the SVD patterns, $p_n(x)$ and $q_n(x)$) corresponding to each eigenvalue were obtained. $p_n(x)$, $q_n(x)$, $a_n(t)$, and $b_n(t)$ are the components of an SVD mode. All modes are arranged so that their σ_n appear in descending order. The first pair of singular patterns describes the largest fraction of the squire covariance between the two fields, and each succeeding pair describes a maximum fraction of square covariance that is unexplained by the previous pairs. The contribution of the nth mode to the total covariance of the two fields is measured by squared covariance fraction:

$$SCF_n = \sigma_n^2 / \sum_m^N \sigma_n^2 \quad (3)$$

To verify the results obtained from the SVD analysis, we further apply the standard empirical orthogonal function (EOF) decomposition to the symmetric and antisymmetric TST fields and then apply the linear regression method onto the TLH field, i.e., by regressing the monthly TLH field upon the principal components (PCs) of the first EOF (EOF1) of symmetric and antisymmetric TST fields. The independent EOF and regression analyses are necessary to verify the robustness of the conclusion obtained from the SVD analysis.

We also calculate the correlation coefficients between symmetric and antisymmetric components of TST (or TLH) locally at each grid point over the tropics. Understandably when the variable at a given location is equatorially asymmetric, the correlation is positive in one hemisphere and negative in another hemisphere; but when the variable at a given location is mainly equatorially symmetric or antisymmetric, then the correlation is weak because one of the symmetric or antisymmetric components is close to zero in this case. As such, the spatial pattern of the correlation may reveal the meridional structure of the variation of TST or TLH.

Figure 1 summarizes the flowchart of methods adopted in this study.

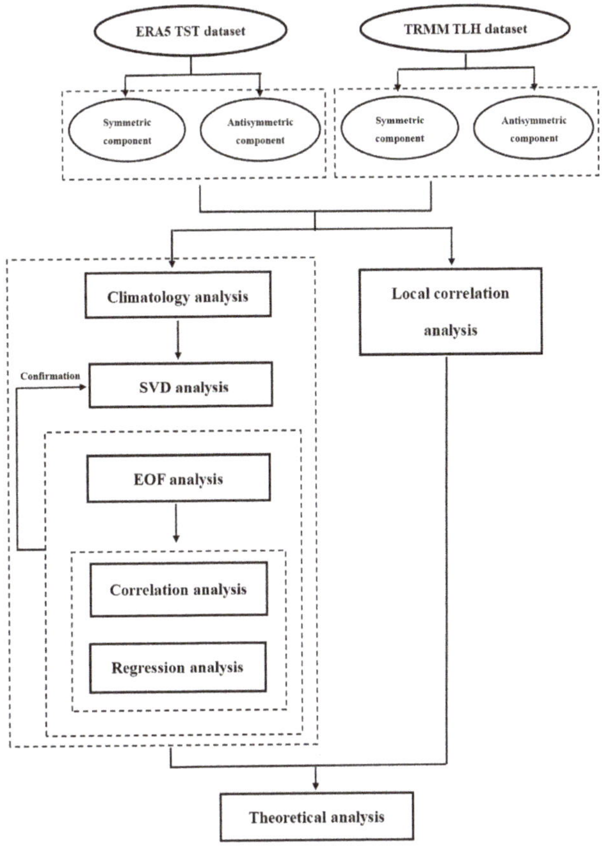

Figure 1. Flowchart of the methods used in this study.

3. Results

Climatology of Annual TLH, TST, Their Symmetric and Antisymmetric Parts

Figure 2 shows the climatology of the annual means of TLH and TST and their symmetric and antisymmetric parts, respectively. The annual mean TLH in Figure 2a shows a zonal band of TLH maxima located between 20°S and 20°N over the whole-tropics, with the maxima value reaching over 150 W m^{-2}, corresponding to the ITCZ and SPCZ over the Pacific and the ITCZ over the equatorial Atlantic and the monsoonal rainfall over the vast region from Africa to Asia and the Western Pacific, and the rainfall maxima over the Amazon Basin. The symmetric component of the TLH (Figure 2b) exhibits three centers: the largest is located at the Equatorial Indian Ocean to the Western Pacific Ocean, which corresponds well to the largest warm pool of SST (Figure 2); the second is located over the northern part of South America, which corresponds to a relatively low (land) surface temperature, but is also nearby warm Equatorial Atlantic SST at its east flank; the third and smallest center is located over Equatorial Africa, which again corresponds to a relatively low land surface temperature. The antisymmetric component of the TLH (Figure 2c) reveals that: (1) The positive atmospheric latent heating regions in the Northern Hemisphere are associated with the ITCZ over the Pacific and Atlantic and monsoonal precipitation over Asia and maritime continent; (2) The positive atmospheric latent heating regions in the Southern Hemisphere are associated with the SPCZ over the Pacific and ITCZ over the Indian Ocean, and also the two broad northwest–southeast-oriented regions over Southern Africa to the Western Indian Ocean and over South America to West Atlantic. By comparing the corresponding annual-mean TST pattern (Figure 2d) and its symmetric (Figure 2e) and antisymmetric (Figure 2f) components with the TLH pattern, we find that the high TLH (rainfall) zones over the sea surface correspond to high SST, but the high TLH zones over the land surface are associated with relatively lower LST. Indeed, the high LST regions are usually desert areas with little rainfall, but the low LST regions are usually associated with tropical forests. This asymmetry in the TLH-TST relation between the land and sea surfaces may have important implications for tropical atmospheric dynamics.

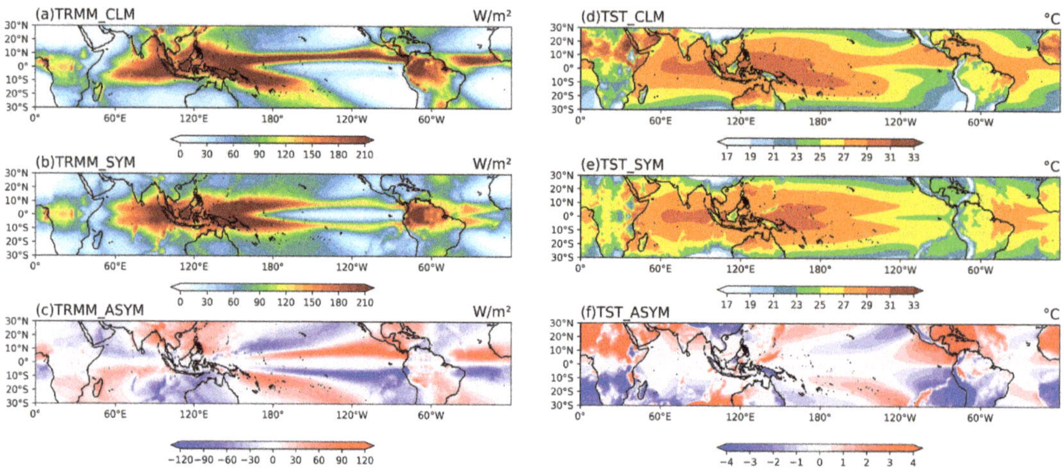

Figure 2. The left panel: (**a**) climatology of the annual-mean TRMM-based tropical latent heating (TLH), and its (**b**) symmetric and (**c**) antisymmetric parts. Unit: W/m^2. The right panel: (**d**) climatology of the annual-mean tropical surface temperature (TST) and its (**e**) symmetric and (**f**) antisymmetric parts. Units: °C.

To further investigate the relationship between TST and TLH, SVD analysis is employed for the original seasonal TST anomaly and the original seasonal TLH anomaly. Note the anomalies in both fields mean that the mean seasonal cycles of TLH and TST have been removed. It can be observed in Figure 3a,b that the leading SVD mode of the TST and TLH is related to El Niño–Southern Oscillation (ENSO), with anomalously warmer SST over the Equatorial Central and Eastern Pacific, cooler SST over the Western Pacific, and also anomalously warmer SST over the Indian Ocean. Correspondingly, anomalously larger TLH is associated with more precipitation over the Central Pacific (near 180°E) and the Equatorial Indian Ocean; also, equatorially shifted ITCZ and SPCZ are found over the Pacific, while anomalously smaller TLH dominates over the Western Pacific (Figure 3b).

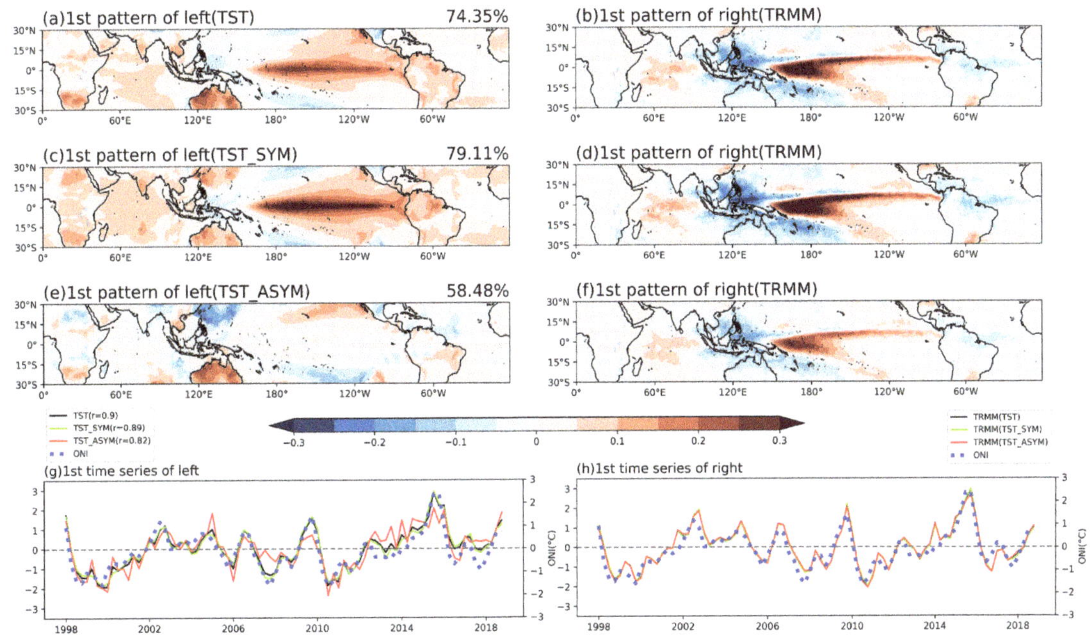

Figure 3. The left (**a**) and right (**b**) spatial pattern of the first mode of SVD using TST and TRMM-based TLH. The squared covariance fraction of the first mode, expressed as a percent, is printed on the upper right-hand corner of each map. (**c**,**d**) and (**e**,**f**): same as (**a**,**b**), but using symmetric and antisymmetric components of TST and TLH, respectively. (**g**,**h**): The time series of the expansion coefficient of the three left patterns and three right patterns, respectively, the blue dotted line is the seasonal-averaged Nino 3.4 index.

To compare the difference between the symmetric and antisymmetric TSTs' connection with the original TLH, SVD analyses are also applied to the symmetric and antisymmetric components of TST paired with the original TRMM-based TLH as shown in Figure 3c,d for the first SVD mode of symmetric TST and original TLH, and Figure 3e,f for the first SVD mode of antisymmetric TST and original TLH. We surprisingly find that the first SVD modes of original TLH corresponding to the first SVD modes of symmetric and antisymmetric components of TST show very high similarities (Figure 3d,f), which resemble the pattern in Figure 3b. This seems at odds with the classic Mastuno–Gill theory, which implies a one-to-one correspondence between the (anti) symmetric forcing and (anti)symmetric atmospheric response. To verify the above results, we check the time series of the expansion coefficient of the three left (TST) patterns (Figure 3g) and three right (TLH) patterns (Figure 3h). The high correlation coefficients between each other indicate that the leading symmetric and antisymmetric SVD modes co-vary temporally (Table 1) and are highly consistent with the evolution of El Niño events in the tropical Pacific (Table 2).

Table 1. The correlation coefficients between the time series of the expansion coefficient of three left and right SVD patterns.

	r	Significance Level
TST vs. TLH (original)	0.9	<0.01
Symmetric TST (TST_SYM) vs. original TLH	0.89	<0.01
Antisymmetric TST (TST_ASYM) vs. original TLH	0.82	<0.01

Table 2. The correlation coefficients between the time series of the expansion coefficient of three left and right SVD patterns and the Oceanic Niño Index.

	r	Significance Level
TST vs. ONI	0.915	<0.01
TLH (TST) vs. ONI	0.933	<0.01
TST_SYM vs. ONI	0.928	<0.01
TLH (TST_SYM) vs. ONI	0.935	<0.01
TST_ASYM vs. ONI	0.763	<0.01
TLH (TST_ASYM) vs. ONI	0.925	<0.01

To further prove the results obtained from the SVD analysis, we independently employed the EOF analysis for the original seasonal TST anomaly and its symmetric and antisymmetric components (Figure 4). We find the first EOF (EOF1) mode of the seasonal TST anomaly resembles the first left (TST) SVD model in Figure 3a, both indicating the pattern of El Niño events. Furthermore, the EOF1 mode of the symmetric TST in Figure 4b shows a very high similarity with the EOF1 of the original TST (Figure 4a), with the pattern correlation being 0.99, with a significance level less than 0.01. However, the EOF1 mode of the antisymmetric TST (Figure 4c and Table 3) indicates there is also a weaker but non-negligible antisymmetric TST component across the whole-tropics. In addition, the correlation between the corresponding principle components (PCs) of the EOF1 of symmetric and antisymmetric TSTs is 0.82 (Figure 4c and Table 3), with a significance level of less than 0.01. In short, it can be concluded that while the principal EOF mode of TST interannual variability is dominated by its equatorially symmetric component, there is a non-negligible equatorially antisymmetric component that well co-varies with the symmetric part over most parts of the tropical land and ocean areas. Figure 4d (see also the correlation coefficients in Table 4) shows that the PC1s of original, symmetric, and antisymmetric TST fields are consistent with the evolution of El Niño conditions in the tropical Pacific.

Then we further reveal the link between the symmetric or antisymmetric components of TST and the TRMM-based TLH by calculating the correlation (Figure 5a,e) and regression (Figure 5b,f) of the TLH with/onto the corresponding PCs of the EOF1 of symmetric and antisymmetric TSTs. The regression patterns of TLH in Figure 5b,f are also further separated into equatorially symmetric and antisymmetric components in Figure 5c,d (regressed onto PC1 of the symmetric TST) and in Figure 5g,h (regressed onto PC1 of antisymmetric TST), respectively. It is obvious that the correlation or regression patterns of the TLH associated with the PC1 of symmetric TST (the left panel of Figure 5) are very similar to that associated with the PC1 of antisymmetric TST (the right panel of Figure 5). As such, we confirm the results obtained from the SVD analysis are right.

Table 3. The correlation coefficients between the principle components of the EOF1 of TST, TST_SYM, and TST_ASYM.

	r	Significance Level
TST_PC1& TST_SYM_PC1	0.99	<0.01
TST_PC1&TST_ASYM_PC1	0.87	<0.01
TST_SYM_PC1&TST_ASYM_PC1	0.82	<0.01

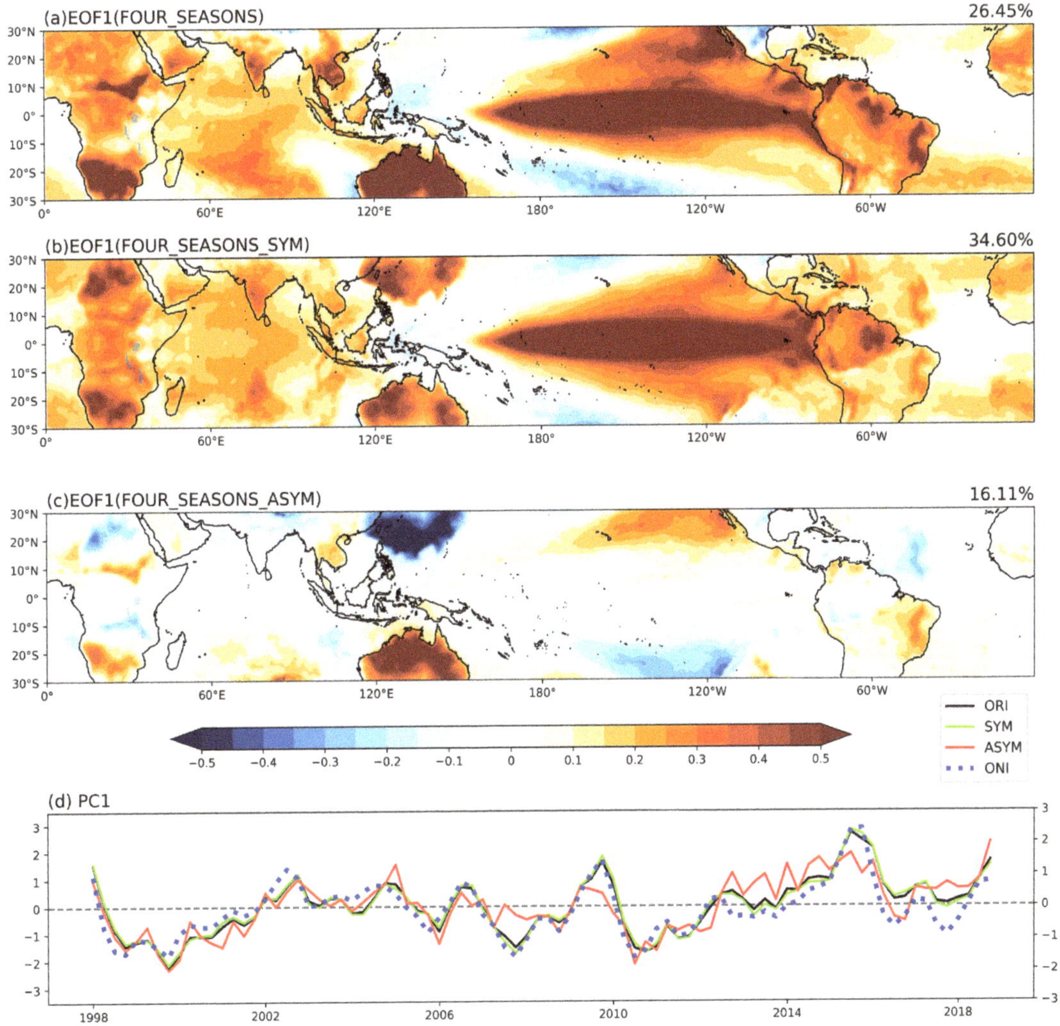

Figure 4. The first EOF (EOF1) mode of the seasonal-mean: (**a**) original, (**b**) the symmetric part and (**c**) antisymmetric components of TST over the whole-tropics, and (**d**) the corresponding principle components for (**a**–**c**).

Table 4. The correlation coefficients between principle components of the EOF1 of TST, TST_SYM, TST_ASYM, and the Oceanic Niño Index.

	R	Significance Level
TST_PC1& ONI	0.897	<0.01
TST_SYM_PC1&ONI	0.907	<0.01
TST_ASYM_PC1&ONI	0.721	<0.01

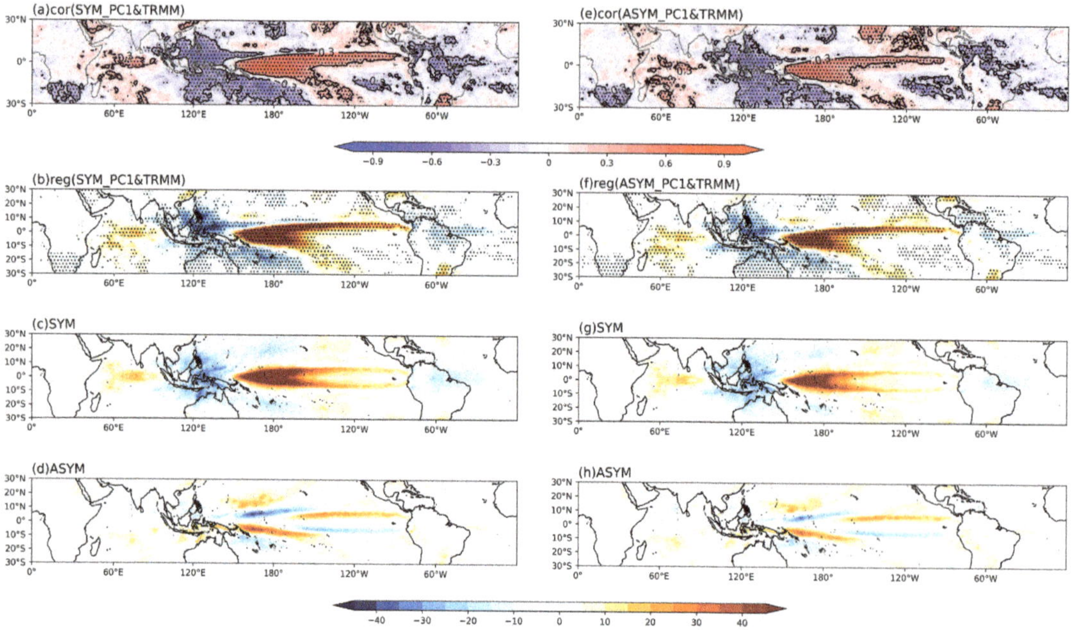

Figure 5. (**a**) The correlation and (**b**) the regression of the TRMM-based TLH with/onto the PC1 of equatorially symmetric TST. (**c**) and (**d**) are the symmetric and antisymmetric parts of (**b**), respectively. (**e–h**) are the same as (**a–d**), but associated with the PC1 of equatorially antisymmetric TST. The areas with a significance level of less than 0.05 are dotted.

To depict the geographic dependence of the symmetric–antisymmetric connection, the local correlations between the symmetric and antisymmetric components of both TST and TLH are calculated in Figure 6a,b. By definition, the high correlation in Figure 6 implies the co-variation between symmetric and antisymmetric components, in which the regions with positive correlation further indicate the dominant regions in the symmetric–antisymmetric connection. Indeed, this corresponds to the condition of equatorially asymmetric variation, i.e., temporal variation is strong in one hemisphere only while very weak in the other hemisphere. On the other hand, low correlations mean that one of the equatorially symmetric and antisymmetric components is dominant, and the symmetric and antisymmetric components vary independently (i.e., being mathematically orthogonal with each other). The low correlation region for TST mainly extends from the Equatorial West Pacific to the Middle Equatorial Pacific (Figure 6a), consistent with the equatorially symmetric SST variability over this region, while the high positive correlations (Figure 6a) in the tropics are located over the Southeastern Pacific, Northern Equatorial Atlantic, North Africa, Australia, South Asia, Southern South America, clearly being consistent with asymmetric TST variability associated with the equatorially asymmetric land–sea distribution. On the other hand, the low correlation regions for TLH (Figure 6b) are mainly limited in the narrow regions of equatorial Africa, the Indian Ocean, western to middle equatorial Pacific, and equatorial South America, but the regions with a high positive correlation of TLH are consistent with the monsoonal rainfall over Africa, Asia, and North America, and also the ITCZs over the Pacific and Atlantic in the Northern Hemisphere, with the SPCZ over the Pacific, ITCZ over the Indian Ocean, and rainfall region over Southern Africa and Southern America in the Southern Hemisphere.

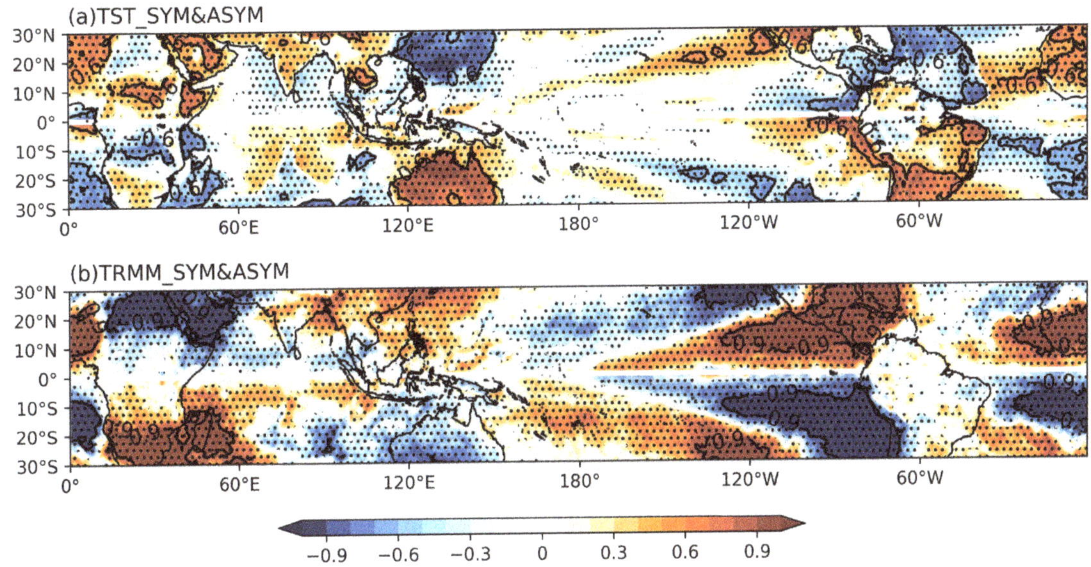

Figure 6. The local correlation between equatorially symmetric and antisymmetric components of (**a**) TST and (**b**) TRMM-based TLH. The areas with a significance level of less than 0.05 are dotted.

Figure 6 indicates that the symmetric and antisymmetric components of both TST and TLH co-vary over vast regions of the tropics, and hence the corresponding Matsuno–Gill modes are intrinsically coupled with each other in the tropical ocean-atmosphere–land system. Untangling their relationship through model simulations and mechanistic analysis is needed for a better theoretical understanding of tropical dynamics.

4. Summary and Discussion

By utilizing seasonally averaged satellite-based TRMM precipitation data as a proxy of tropical latent heating (TLH) and ERA5-based tropical surface temperature (TST) data from 1998 to 2018, we investigate the cross-hemispheric connection in the TLH and TST variability and their co-variability. The interannual variability of both the TST and the TLH is equatorially asymmetric and can be decomposed as the sum of equatorially symmetric and antisymmetric components. Based on the decomposition, we reveal some new features of variability and co-variability of the TST and TLH. The main results are summarized as follows:

(1) While the principal mode of TST interannual variability is dominated by its equatorially symmetric component, there is a non-negligible equatorially antisymmetric component that well co-varies with the symmetric part over most parts of the tropical land and ocean areas (Figures 4 and 6a);

(2) For the principal mode of TLH (and tropical rainfall) interannual variability, we find that equatorial symmetric components dominate only over a zonal band over from the near-Equator Indian Ocean to western and middle equatorial Pacific, salient equatorially antisymmetric variability exists over the middle and Eastern Pacific (Figure 5). In general, the symmetric and antisymmetric TLH exhibits high co-variability over most areas of the tropics (Figure 6b).

(3) We find surprisingly that the spatial patterns of TLH projected upon the first principal components (PC1) of symmetric and antisymmetric TSTs over the whole-tropics, are very similar to each other, seemingly at odds with the classic Mastuno–Gill theory. The similarity in the projected TLH patterns is mainly due to the fact that the symmetric

and antisymmetric PCs of TST are both nearly coincident with the ENSO index during the 21 years of 1998–2018.

While these results are obtained for all four seasons with the mean seasonal cycle being removed, we note they basically hold individually for each season. Results on individual season-based analyses will be reported later in more detail.

The above results raise some interesting puzzles in the theoretical understanding of tropical atmospheric dynamics. First is that if they are really at odds with the classic Matsuno–Gill theory [1,2] or not. We might expect from the linear Matsuno–Gill theory a one-to-one correspondence between equatorially (anti) symmetric TST forcing and equatorially (anti) symmetric TLH pattern, but due to the intrinsic nonlinearity in tropical dynamics, say, related to convective precipitation, the one-to-one correspondence may at least partially be broken up. Because of the strong co-variability in the observed TST-TLH relation, we may not obtain mechanistic understanding directly from statistical analysis of the observations. Well-designed modeling experiments and theoretical analysis are needed for a decisive solution to the puzzle.

The second puzzle is to what extent are the equatorially symmetric and antisymmetric components of the joint TST-TLH variability interactive with each other? What is the underlying mechanism responsible for the interaction? Indeed, it is quite reasonable to assume that the Matsuno–Gill theory still holds to some degree for the interannual joint TST-TLH variability, but the departure from the theory due to nonlinear interactions may be essential to better understand and predict the tropical variability.

While the PCs of joint TST-TLH variability are clearly associated with the ENSO cycle in the tropical Pacific, the co-variability over the Indian Ocean, tropical Atlantic, and tropical land areas should not be neglected from our analysis. Recent studies have suggested the importance of pantropical interaction or cross-basin interaction in tropical dynamics [6,7]. We further suggest that a whole-tropics perspective that takes the different but connected nature of equatorially symmetric and antisymmetric modes across the whole-tropics into consideration may well be useful in understanding and predicting tropical climate variability.

5. Historical Note

After graduating from the University of Chicago with his famous thesis on energy dispersion in the atmosphere, T.C. Yeh stayed there and became a member of the team on tropical dynamics, second to H. Riehl, the head of the team. The other two members were J. Malkus (J. Simpson) and N. LaSeur. Before his return to China, he published two classic papers on the intensity of the Hadley cell [23] and on trade inversion [24], among others. Although later on he did not focus on tropical dynamics in his lifelong career, his works on tropical dynamics also clearly reflect his style and character in research, i.e., being thorough, systematic, and insightful. This short note is devoted to Prof. Yeh's contribution to the field of tropical dynamics—J. Lu.

Author Contributions: Conceptualization, J.L.; methodology, J.L.; validation, Y.G., X.L. and J.L.; formal analysis, Y.G. and X.L.; resources, J.L.; data curation, Y.G.; writing—original draft preparation, J.L. and X.L.; writing—review and editing, J.L.; visualization, Y.G.; supervision, J.L.; project administration, J.L.; funding acquisition, J.L. All authors have read and agreed to the published version of the manuscript.

Funding: This research was funded by National Natural Science Foundation of China, grant number 42175070.

Data Availability Statement: All of data are available by acquiring from the authors.

Conflicts of Interest: The authors declare no conflict of interest.

References

1. Matsuno, T. Quasi-Geostrophic Motions in the Equatorial Area. *J. Meteorol. Soc. Jpn. Ser. II* **1966**, *44*, 25–43. [CrossRef]
2. Gill, A.E. Some Simple Solutions for Heat-Induced Tropical Circulation. *Q. J. R. Meteorol. Soc.* **1980**, *106*, 447–462. [CrossRef]
3. Hoerling, M.P.; Kumar, A. Atmospheric Response Patterns Associated with Tropical Forcing. *J. Clim.* **2002**, *15*, 2184–2203. [CrossRef]
4. Qu, X.; Huang, G. An Enhanced Influence of Tropical Indian Ocean on the South Asia High after the Late 1970s. *J. Clim.* **2012**, *25*, 6930–6941. [CrossRef]
5. Hogikyan, A.; Resplandy, L.; Fueglistaler, S. Cause of the Intense Tropics-Wide Tropospheric Warming in Response to El Niño. *J. Clim.* **2022**, *35*, 2933–2944. [CrossRef]
6. Cai, W.; Wu, L.; Lengaigne, M.; Li, T.; McGregor, S.; Kug, J.-S.; Yu, J.-Y.; Stuecker, M.F.; Santoso, A.; Li, X.; et al. Pantropical Climate Interactions. *Science* **2019**, *363*, eaav4236. [CrossRef] [PubMed]
7. Wang, C. Three-Ocean Interactions and Climate Variability: A Review and Perspective. *Clim. Dyn.* **2019**, *53*, 5119–5136. [CrossRef]
8. Zhang, C. Large-Scale Variability of Atmospheric Deep Convection in Relation to Sea Surface Temperature in the Tropics. *J. Clim.* **1993**, *6*, 1898–1913. [CrossRef]
9. Lau, K.-M.; Wu, H.-T.; Bony, S. The Role of Large-Scale Atmospheric Circulation in the Relationship between Tropical Convection and Sea Surface Temperature. *J. Clim.* **1997**, *10*, 381–392. [CrossRef]
10. Lu, R.; Lu, S. Local and Remote Factors Affecting the SST–Precipitation Relationship over the Western North Pacific during Summer. *J. Clim.* **2014**, *27*, 5132–5147. [CrossRef]
11. Sabin, T.P.; Babu, C.A.; Joseph, P.V. SST–Convection Relation over Tropical Oceans. *Int. J. Climatol.* **2013**, *33*, 1424–1435. [CrossRef]
12. Williams, A.I.L.; Jeevanjee, N.; Bloch-Johnson, J. Circus Tents, Convective Thresholds, and the Non-Linear Climate Response to Tropical SSTs. *Geophys. Res. Lett.* **2023**, *50*, e2022GL101499. [CrossRef]
13. Xie, S.-P. Oceanic Response to the Wind Forcing Associated with the Intertropical Convergence Zone in the Northern Hemisphere. *J. Geophys. Res. Ocean.* **1994**, *99*, 20393–20402. [CrossRef]
14. Li, T.; Philander, S.G.H. On the Annual Cycle of the Eastern Equatorial Pacific. *J. Clim.* **1996**, *9*, 2986–2998. [CrossRef]
15. Masunaga, H.; L'Ecuyer, T.S. Equatorial Asymmetry of the East Pacific ITCZ: Observational Constraints on the Underlying Processes. *J. Clim.* **2011**, *24*, 1784–1800. [CrossRef]
16. Schneider, T.; Bischoff, T.; Haug, G.H. Migrations and Dynamics of the Intertropical Convergence Zone. *Nature* **2014**, *513*, 45–53. [CrossRef]
17. Sun, Z.; Lu, J. The North Equatorial Countercurrent and the Zonality of the Intertropical Convergence Zone. *Geophys. Res. Lett.* **2021**, *48*, e2021GL095657. [CrossRef]
18. An, S.-I.; Ham, Y.-G.; Kug, J.-S.; Timmermann, A.; Choi, J.; Kang, I.-S. The Inverse Effect of Annual-Mean State and Annual-Cycle Changes on ENSO. *J. Clim.* **2010**, *23*, 1095–1110. [CrossRef]
19. Huffman, G.J.; Bolvin, D.T.; Nelkin, E.J.; Wolff, D.B.; Adler, R.F.; Gu, G.; Hong, Y.; Bowman, K.P.; Stocker, E.F. The TRMM Multisatellite Precipitation Analysis (TMPA): Quasi-Global, Multiyear, Combined-Sensor Precipitation Estimates at Fine Scales. *J. Hydrometeorol.* **2007**, *8*, 38–55. [CrossRef]
20. Wallace, J.M.; Smith, C.; Bretherton, C.S. Singular Value Decomposition of Wintertime Sea Surface Temperature and 500-Mb Height Anomalies. *J. Clim.* **1992**, *5*, 561–576. [CrossRef]
21. Hu, Q. On the Uniqueness of the Singular Value Decomposition in Meteorological Applications. *J. Clim.* **1997**, *10*, 1762–1766. [CrossRef]
22. Zhang, L.; Liu, Y.; Zhao, F. Singular Value Decomposition Analysis of Spatial Relationships between Monthly Weather and Air Pollution Index in China. *Stoch. Environ. Res. Risk Assess.* **2018**, *32*, 733–748. [CrossRef]
23. Riehl, H.; Yeh, T.C. The Intensity of the Net Meridional Circulation. *Q. J. R. Meteorol. Soc.* **1950**, *76*, 182–186. [CrossRef]
24. Riehl, H.; Yeh, T.C.; Malkus, J.S.; la Seur, N.E. The North-East Trade of the Pacific Ocean. *Q. J. R. Meteorol. Soc.* **1951**, *77*, 598–626. [CrossRef]

remote sensing

MDPI

Article

Turbulence: A Significant Role in Clear-Air Echoes of CINRAD/SA at Night

Yupeng Teng [1,*], Tianyan Li [1,2], Shuqing Ma [3] and Hongbin Chen [1,2]

[1] Institute of Atmospheric Physics, Chinese Academy of Sciences, Beijing 100029, China
[2] College of Earth Sciences, University of Chinese Academy of Sciences, Beijing 100049, China
[3] Meteorological Observation Centre, China Meteorological Administration, Beijing 100089, China
* Correspondence: tengyp@mail.iap.ac.cn

Abstract: It is commonly believed that clear-air echoes detected by weather radars are caused by atmobios migration. However, clear-air echoes are sometimes inconsistently related to the activity of living creatures. In some cases, the characteristics of radar products seem to conform to biological scattering, but the movement of echoes cannot be observed. For these reasons, we sought to expand the cause of clear-air echoes from a Chinese Doppler S-band Weather Radar (CINRAD/SA) in Beijing. Some contradictions were discovered in a case which diverged from previous conclusions. It was found that the progression and movement of clear-air echoes do not conform to the rules of biological activities. The frequency distribution of dual-wavelength ratio peaks is 21.5 dB, which is in accordance with Villars–Weisskopf's turbulence theory. From 1 May to 20 May, the 58% dual-wavelength ratio between the S-band and the X-band was distributed between 18 dB and 24 dB. These results show that more than half of the clear-air echoes of CINRAD/SA at night were caused by turbulence in Beijing. A new model of troposcatter propagation, the reflecting-layers model, was then introduced to explain the radar observations. According to the reflecting-layers model, the echoes' diurnal variation and reflectivity characteristics are attributed to the effects of turbulent mixing. Excessive turbulent mixing affects the generation of the reflective layer, thereby weakening the echo signal. It is necessary to re-examine the position of turbulence in clear-air echoes.

Keywords: weather radar; clear-air echo; turbulence; troposcatter propagation; aeroecology

Citation: Teng, Y.; Li, T.; Ma, S.; Chen, H. Turbulence: A Significant Role in Clear-Air Echoes of CINRAD/SA at Night. *Remote Sens.* **2023**, *15*, 1781. https://doi.org/10.3390/rs15071781

Academic Editor: Kenji Nakamura

Received: 14 February 2023
Revised: 10 March 2023
Accepted: 23 March 2023
Published: 27 March 2023

1. Introduction

Billions of atmobios cross the sky each year in search of food, partners, and habitats. With the imminent threat to the ecological state driven by human activity, broad-scale biological monitoring may prove crucial to successful conservation efforts. However, efforts to monitor atmobios are hampered by the unpredictability of their movements [1]. Traditional methods, such as visual and auditory observations, laboratory research, trapping, and ringing studies, have left a wide gap caused by limitations of space, time, and labor [2]. Since weather radars, which were originally used to observe clouds and rain, have been found to be able to observe the bio-scattering of atmobios after World War II, many researchers deemed that atmobios are responsible for clear-air echoes. For instance, Wilson et al. compared the reflectivity at different wavelengths to conclude that particulate scattering dominates in the boundary layer [3]. Martin et al. used data from the Weather Surveillance Radar-1988 Doppler (WSR88D) and X- and W-band research radars and deemed that the targets of nocturnal clear-air echoes are almost insects [4]. Further, Broeke found that biological scatterers, consisting of birds and insects, may become trapped near the circulation center of tropical cyclones [5]. Westbrook et al. used a WSR-88D radar to detect corn earworm moth migration [6]. Now, radars are believed to be a practical tool and an important data source for monitoring atmobios activity.

Although two mechanisms, turbulence-scattering mechanisms caused by turbulent inhomogeneities and biological scatterers exemplified by insects and birds, can dominate the scattering process of clear-air echoes [7], it is recognized that most of the echoes at centimeter wavelengths are primarily caused by insects and birds. This viewpoint is based on the theory of locally homogeneous isotropic turbulence developed by Kolmogorov [8,9].

The Kolmogorov–Obukhov theory holds that turbulent motion is homogeneous and isotropic, and its average properties are uniquely determined by the average rate of dissipation of the turbulent kinetic energy per unit mass of fluid within a subrange or regime of turbulent eddy sizes [9]. According to this theory, Ottersten clarified the relationship of turbulence scattering (frequently referred to as Bragg scattering) from refractive index irregularities to the atmospheric structure [8]. In light of this relationship, Wilson compared the reflectivity at different wavelengths and examined the differential reflectivity at the S-band, concluding that biological scatterers dominate clear-air echoes because a smaller reflectivity difference and a nonzero value of the differential reflectivity are not consistent with Bragg scattering [3].

However, Landau's query (1957) caused Kolmogorov and Obukhov (1961) to introduce important modifications to Kolmogorov's theory. They took into account spatial fluctuations in the turbulent energy dissipation and chose a specific form (lognormal) for the probability density as a third hypothesis [10]. However, Mandelbrot (1976) pointed out that the lognormal assumption is only a special, probably physically unrealistic, case of weighted curdling [11]. Moreover, many experimental cases for the fine-structure intermittency of turbulence showed the inhomogeneity of physical quantities in space and time [12–21]. Batchelor and Towsend firstly observed that the turbulence and its energy dissipation were very unevenly distributed over space, and the velocity gradients became increasingly intermittent as the Reynolds number increased [13]. Siggia made a numerical simulation and found that 95% of the energy dissipation is concentrated in a tiny region of space [15]. Turbulent mixing of a passive scalar is an extremely intermittent phenomenon [21]. Intermittency has been shown to be one of the fundamental mechanisms of turbulence. Conclusions regarding clear-air echoes are overshadowed by theoretical defects.

Although it is essentially appropriate to use Kolmogorov's theory in atmospheric science, it is a fact that non-Kolmogorov turbulence is widely present in the boundary layer [22–27]. Experiments show that the Kolmogorov theory is sometimes incomplete to describe atmospheric statistics properly, in particular, in portions of the troposphere and stratosphere [25]. The power-law exponent for the inverse spatial frequency dependence has been observed experimentally to be both larger and smaller than the value of 11/3 that derives from Kolmogorov's model [24]. Consortini, Ronchi, and Stefanutti illustrated in the laboratory that the statistics of laser beam wander differed for horizontal and vertical orientations, implying the presence of turbulent anisotropy [23]. Other observations also have shown that the atmospheric structures of refractive index irregularities often differ between the horizontal and vertical planes at the same height and distance [27]. This means that a nonzero value of differential reflectivity is also consistent with turbulent scattering. Additionally, skepticism towards scattering mechanisms has been exacerbated by these contradictions.

With the continued decline of global species biodiversity, China wants to shoulder more environmental responsibilities as a global economic power and implements the strictest possible systems for environmental protection. Ecological monitoring is an important part of environmental protection. Similar to the Next-Generation Weather Radar (NEXRAD) network in the US, the China Meteorological Administration (CMA) deployed the China Next-Generation Weather Radar (CINRAD) network. The CMA wants to use the radar network for monitoring the movement and abundance of animals in the airspace. However, a few irrationalities were found in the monitoring. The characteristics of clear-air echoes are not completely consistent with the law of seasonal biological activities in China. The source of the contradictions seems to be the influence of turbulence, which can be disregarded in clear-air echoes. Therefore, a rethinking of scattering mechanisms regarding clear-air echoes is needed for an expanded set of causes on the cause of clear-air echoes.

If Bragg scattering was the cause of echoes, the dependence of the echo strength on the radar wavelength would be expected. Since 2015, the Beijing Meteorological Service has built multiple X-POL radars (wavelength λ~3 cm) in the observation coverage area of the CINRAD/SA radar (λ~10 cm) in the Daxing district. Moreover, the CINRAD/SA radar has completed a polarization upgrade in April 2021. The feasible conditions for studying the cause of clear-air echoes in Beijing have been met. Thus, this study is focused on determining the cause of clear-air echoes and their scattering mechanism in Beijing to clarify the mechanism of clear-air echo causes and help the quantitative observation of biology. Observations of the multi-time and dual-wavelength characteristics of clear-air echoes were used to confirm the diagnosis of clear-air echoes. A troposcatter mechanism was introduced to provide some explanations of the observed phenomena.

Section 2 introduces some basic concepts and theories. The data and methods are described in Section 3. The clear-air echo characteristics are analyzed in Section 4. In Section 5, a theory in communication engineering is used to explain the phenomenon of clear-air echoes. Section 6 presents the conclusion.

2. Concepts and Theory

2.1. Dual-Polarization Radar Products

A dual-polarization weather radar is an advanced sensor with a high observation accuracy and many products. It can provide multiple products at the same time. The reflectivity factor (denoted by Z in dBZ) is the most conventional product in radar meteorology, which is a more meteorologically meaningful way of expressing the radar reflectivity η. The radar's property is independent of the radar's wavelength in the case of small scatterers such as precipitation. The reflectivity factor Z is expressed as

$$Z = \frac{\eta\lambda^4}{\pi^5 K^2} \tag{1}$$

where λ is the wavelength; K^2 is a dielectric constant and is often taken to be 0.9 for water and 0.2 for ice, respectively.

Another conventional product of the weather radar is the Doppler velocity (or called the radial velocity) which is the velocity vector of a scattering object along the radial direction as observed by the Doppler radar. In general, the Doppler velocity is positive for object motion away from the radar.

Prior to the dual-polarization upgrade, the weather radar transmitted and received in a purely horizontal polarization and could not measure any polarization-dependent attributes of a target. When the upgrade was finished, the radar could transmit and receiving two orthogonal signals. The ratio of the received wave powers between horizontal and vertical polarization is called the differential reflectivity (Z_{DR} in units of dB), which helps to infer the shape of the scatterer. Z_{DR} can be defined as the difference between the measured radar reflectivity factor in the horizontal polarization (Z_H in dBZ) and the vertical polarization (Z_V in dBZ):

$$Z_{DR} = Z_H - Z_V \tag{2}$$

The correlation coefficient is another useful polarimetric product which depends on the similarity of the received signals at the horizontal and vertical polarizations across multiple pulses. It is a measure of the variety of hydrometeor shapes in a pulse volume.

There are also some other radar products that have not been introduced, and their definitions can be found in the Glossary of Meteorology, American Meteorological Society (http://glossary.ametsoc.org/wiki/climatology, accessed on 27 January 2023).

2.2. Turbulence

During the 1940s, Kolmogorov developed a model to illustrate how energy is transported from large-scale turbulent eddies to small-scale turbulent eddies [9,10]. Kolmogorov's model provides a spatial power spectrum for index of refraction fluctuations.

Kolmogorov proposed that turbulence in the inertial subrange would reach a statistical equilibrium which is called the "homogeneous isotropic turbulence". Kolmogorov introduced a structure function defined as the squared mean of the difference between the meteorological elements from two independent points in space. For Kolmogorov's turbulence, the structure function of the index of refraction is

$$D_n(r) = \langle [n(r_1 + r) - n(r_1)] \rangle^2 = C_n^2 r^{2/3}, \ l_0 < r < L_0 \tag{3}$$

where $n(r_1)$ is the index of refraction at point r_1; $r = |r|$; C_n^2 is the refractive index structure constant; and l_0 and L_0 are the inner and outer scales, respectively.

Tatarskii applied Kolmogorov's model and concluded that the structure constant of the refractive index C_n^2 is related to meteorological parameters, as shown in Equation (4) [28]:

$$C_n^2 = a^2 {L_0}^{4/3} \left(-\frac{79 \times 10^{-5} P}{T^2} \frac{\partial \theta}{\partial h} \right)^2 \tag{4}$$

where a^2 is a constant with the laboratory value being 2.8; L_0 is the outer scale of turbulence; T is the air temperature; P is the atmosphere pressure; θ is the potential temperature; and h is the altitude. The conversion relationship between θ and T is expressed in the form of Equation (5):

$$\theta = T \left(\frac{1000}{P} \right)^{0.286} \tag{5}$$

In Kolmogorov's model, there is no preferred direction through the turbulence, and the turbulent fluctuations of the refractive index are homogeneous, statistically stationary, and isotropic. However, some experimental results have shown that atmospheric turbulence may not always obey Kolmogorov's law [24–26]. The power spectrum exponent has been observed experimentally to be both larger and smaller than the value from Kolmogorov's model. Since Batchelor and Townsend's first observations of intermittency in 1949 [13], which found that turbulence and its energy dissipation are not space-filling but are intermittent in space, various turbulence theories have attempted to account for and reduce the intermittency geometry of turbulent dissipation [11,12,15,16,19–21].

2.3. Bragg Scattering

Bragg scattering is caused by turbulent inhomogeneities with sizes around one-half of the transmitted wavelength of a radar [29]. Technically, the atmospheric turbulence with refractivity gradients is essentially a dipole and causes scattering. Ottersten provides the radar reflectivity's relationship with the atmospheric structure constant of the refractive index C_n^2 and the radar wavelength λ [8]. The radar reflectivity η (or radar cross-section per unit volume) is given in Equation (6):

$$\eta = 0.38 C_n^2 \lambda^{-1/3} \tag{6}$$

According to Equation (4), the reflectivity factor Z is given in Equation (7):

$$Z = \frac{0.38 C_n^2 \lambda^{11/3}}{\pi^5 K^2} \tag{7}$$

Therefore, the differential reflectivity Z_{DR} can be equated to C_n^2 as

$$Z_{DR} = \frac{Z_H}{Z_V} = \frac{C_{nH}^2}{C_{nV}^2} \tag{8}$$

Based on isotropic turbulence, C_n^2 is equal in the horizontal and vertical directions, and the value of Z_{DR} will be zero in conventional dB form. Additionally, the ratio of the Z values for two radar wavelengths (also called dual-wavelength ratio, DWR) is

$$\frac{Z_1}{Z_2} = \left(\frac{\lambda_1}{\lambda_2}\right)^{11/3} \tag{9}$$

Wilson used Equation (9) to study clear-air echoes over Florida and Colorado and concluded that the clear-air echoes over these areas were caused by living creatures [3].

However, Equation (6) is not the only relation between the radar reflectivity and the radar wavelength. The value of the radar reflectivity is variable based on the different applied turbulence theories, and Equation (4) is based on the Kolmogorov–Obukhov theory. According to the Villars–Weisskopf theory, the radar reflectivity's relation is given in Equation (10) [30]:

$$\eta = C\lambda^{1/3} \tag{10}$$

where C is constant. Researchers hold different views regarding the relationship, which changes from $\eta \propto \lambda^{-1/3}$ to $\eta \propto \lambda$, and the ratio of the Z values for the two radar wavelengths changes according to the researcher's view.

2.4. Biological Scattering

Unlike raindrops, biological scatterers have complex shapes that result in highly aspect-dependent scattering characteristics. As a result, radar cross-sections (RCS) are normally modeled by prolate spheroids of equivalent mass [7,31]. It is apparent that insects and birds, which are highly non-spherical, would exhibit large Z_{DR} signals and low copular correlation coefficient values.

However, there are polarimetric differences between birds and insects. Insects often have a high Z_{DR} (up to 10 dB) and a relatively low differential phase, while birds may have a lower Z_{DR} (1 to 3 dB) and a much larger differential phase [5]. Moreover, for both types of echoes, the cross-correlation coefficient is between 0.3 and 0.5, which is lower than the hydrometeorological signal. Polarimetry becomes a technical standard of the application of a dual-polarization radar to delineate meteorological and non-meteorological signals in the areas of aeroecology.

Based on fuzzy logic, some researchers have differentiated bio-scatterers by using typical values of polarimetric variables [32]. Kilambi proposed an estimate of the depolarization ratio for separate types of echoes [33]. Overall, the polarimetric characteristics of bio-scatterers are the primary means by which to solve this problem.

For radars that cannot measure the polarization characteristics, the difference derived from the Doppler velocity is another indicator of the scatterer type. Insects reasonably represent actual winds, whereas birds have a much larger independent velocity component [34]. Therefore, the radial velocity standard deviation obtained from the velocity–azimuth display retrieval is an indicator of migration. For high migration densities, the raw weather radar wind vectors could be up to 15 m/s for birds and 6 m/s for strong insects [7]. However, for micro-insects, which are weak flyers, their motions are primarily wind-driven and behave as quasi-passive wind tracers. Thus, migrating and wandering birds are the main targets of Doppler wind measurement.

Biological scattering and Bragg scattering are always rivals. Compared with Bragg scattering discussed above, the ratio of the Z values for two radar wavelengths is more complicated. When the physical diameter of the spherical particles is considerably smaller than the radar wavelength (approximately 6.25 mm for the S-band), Rayleigh scattering can be used. On the other hand, for larger particles, Mie scattering occurs. The different scattering makes the relation between their size and Z values far from straightforward. Provided that Mie scattering is occurring at one or both of the wavelengths, the ratio of the Z values with a spherical diameter is nonlinear [3].

For example, using prolate spheroids of a spinal cord dielectric, a new model that is closer to real insects [35], the RCS could be simulated by the method of moments, as shown in Table 1, according to biological datasets provided by the Chinese Academy of Agricultural Sciences. Additionally, based on the simulation, the DWRs were 1.9 dB, 13.8 dB, and 17.0 dB between the wavelengths of the S-band and X-band. Further, all birds and many common insects are above the 2 mm size threshold for Mie scattering at the X-band. Measurements of volumes containing multiple scatterers are likely often biased toward the characteristics of the largest scatterers [31]. Thus, comparing the value of the DWR is a valid way to be certain of the dominant mechanism of scattering.

Table 1. Parameters of several insects and their RCSs at different wavelengths. The biometric data were provided by the Chinese Academy of Agricultural Sciences from captured insects in North China. The RCSs were simulated by FEKO simulation software using the prolate spheroid model of a spinal cord dielectric.

Species	Average Weight (mg)	Average Length (mm)	Average Width (mm)	RCS of S-Band (dBsm)	RCS of X-Band (dBsm)
Conogethes punctiferalis, Hawaiian beet webworm, Athetis lepigone	22.1	13.0	3.2	−52.5	−25.0
Cotton bollworms, Plusia agnata	114.8	16.7	5.4	−39.8	−34.2
Armyworms, Black cutworms, Sprodoptera litura	145.4	19.0	5.8	−36.2	−33.8

3. Instruments and Data

3.1. Instruments

The data of clear-air echoes used in this study were collected by China's New Generation Weather Radar (CINRAD/SA radar) and three X-band dual-polarization (X-POL) radars with the same technical parameters [36]. CINRAD/SA was developed from the American WSR-88D (NEXRAD) through a joint agreement between the two countries [37]. The CINRAD/SA radar of Beijing is located in the Daxing district of Beijing and was fully upgraded with polarimetric capabilities in April 2021. The products of CINRAD/SA have a radial distance resolution of 250 m and an azimuthal resolution of 1 degree. The volume coverage pattern 21 (VCP21) scan mode was selected which sweeps 9 elevation angles of 0.5, 1.5, 2.4, 3.4, 4.3, 6.0, 9.9, 14.6, and 19.5 degrees in 6 min.

The X-POL radars were built by the Beijing Meteorologic Service to improve radar usage in weather monitoring. The stations of the X-POL radars used in this study are in the Fangshan, Shunyi, and Tongzhou districts of Beijing (the BJXFS, BJXSY, and BJXTZ sites, respectively), located around the CINRAD/SA radar sites. The positions of the four radars and terrain are shown in Figure 1. The scan strategy for the X-Pol radars is the same as the CINRAD/SA radar. The detailed system characteristics of the CINRAD/SA radar and the X-POL radars are shown in Table 2.

The operations of all these radars are under the supervision of the CMA. The CINRAD/SA and the X-POLs are also calibrated weekly and monthly according to the technical standard of the CMA, which includes the system internal calibration, the receiving link calibration, the rotary joint calibration, and others.

Some meteorological parameters were used in this study. The profiles of the air temperature, the relative humidity, and the absolute humidity were collected using a microwave radiometer (RPG-HATPRO-G5, Meckenheim, Germany). The range resolution of the microwave radiometer was 50 m for heights below 1 km, 100 m for heights between 1 km and 2 km, and 250 m for heights from 2 km to 10 km, and the temporal resolution was 1 s. The wind vertical velocity and the wind shear were retrieved using a 3D Doppler wind lidar (Windcube 100 s, Leosphere, Saclay, France). The wind measurement products had a spatial resolution of 25 m with a temporal resolution of 20 s, and the products were not used when the carrier-to-noise ratio was less than −30 dB. A radar wind profiler (RWP) CFL-06, which was manufactured by the 23rd Institute of the China Aerospace Science and Industry Corporation, was also used to detect and process the profiles of the horizontal wind speed and direction. The radar wind profiler operates in the L-band (1290 MHz) and provides data with a vertical resolution of 120 m for heights between 150 m and 2.9 km. The microwave radiometer, the wind lidar, and the wind profile radar were located several tens of meters from the S-band weather radar in the same observation field.

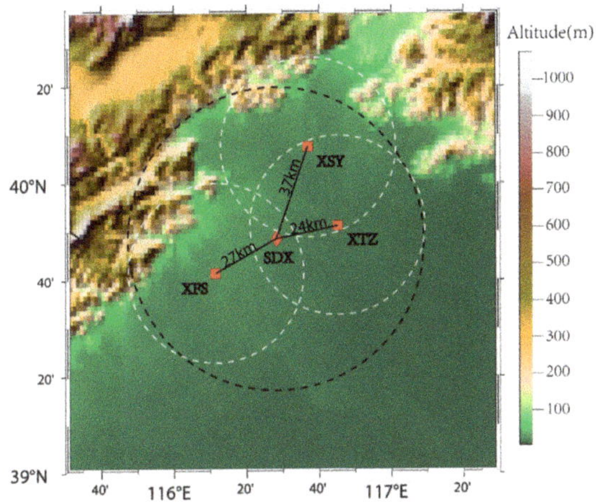

Figure 1. Distribution of radars (square signs and diamonds) and topography (coloring) of Beijing and its vicinity. The three square signs indicate the locations of the three X-POL radar sites (XFS, XSY, and XTZ). The diamond shows the location of the SDX site. The distance of each X-POL radar relative to the SDX site is labeled. The black dotted circle is the distance circle of the S-band with a 58 km radius, and the radius of the white dotted circles of the X-POL radar sites are 34 km, respectively. These circles show the detection zones of the CINRAD/SA and X-POLs where their minimum detectable reflectivity is less than −5 dBZ.

Table 2. System characteristics of the CINRAD/SA radar and the X-POL radars.

Parameter	CINRAD/SA Radar	X-POL Radars
Frequency	2700–3000 MHz	9300–9500 MHz
Antenna cover diameter	11.9 m	≥4 m
Polarization	Linear H and V	Linear H and V
Volume coverage patterns	VCP 21	VCP 21
Time of VCP 21	6 min	3 min
Range resolution	250 m	75 m
Minimum detectable reflectivity	−7.5 dBZ @ 50 km	5 dBZ @ 60 km

3.2. Preprocessing

Preprocessing was needed before data comparison between multiple radars because of the differences temporally and spatially. It was necessary to preprocess and use multiple actual sounding data to create a time–height cross-section for the comparison about the DWR. The time–height cross-section is composed of vertical profiles of continuous time. Thus, the calculation of the vertical profile is introduced in the following sections.

3.2.1. Threshold

It is commonly believed that the signal of clear-air echoes is generally weak. A distorted signal would hamper the estimation of echoes. Hence, first, the signal-to-noise ratio (SNR) is set as the standard of thresholding. The minimum value of the SNR thresholding is 6 dB.

Secondly, the setting of the threshold needs to consider the limits of the radar's minimum detectable reflectivity. If the echoes are out of the range of the radar system's designed criteria, weak echoes may be distorted. The distortion values of the echoes then introduce errors into the statistics and the comparison. Meanwhile, the different minimums of the reflectivity values on different radar systems may affect the statistics as well. On the other hand, an excessive detection range would reduce the effectiveness of the comparison with the vertical observation system. Consequently, the radial distance from the radar to the objective is limited.

As shown in Table 2, the minimum detectable reflectivity of the CINRAD/SA and X-POLs was less than −7.5 dBZ at 50 km and 5 dBZ at 60 km. Therefore, balancing the minimum detectable reflectivity and the volume of the data, the range threshold of the data was 59 km for the S-band and 17 km for the X-band. The range threshold was set according to the radar equation to ensure that the radars with different wavelengths had the same minimum detectable reflectivity, which was −6 dBZ.

3.2.2. Vertical Profiles

The vertical profile of the reflectivity factor is useful to estimate the rainfall intensity because of the complexity of the vertical structure of radar echoes [38–41]. From radar data recorded at multiple elevation angles, the mean value of the reflectivity at each altitude can be calculated. Thus, the mean vertical profile of reflectivity (MVPR) can be easily extracted from volume-scan data. Unlike the MVPR, which focuses on the precision of the radar precipitation estimation, a method for determining the state of clear-air echoes needs to be presented.

For weak clear-air echoes, an extreme value can cause fluctuations in the mean value. The limited number of antenna elevation angles also introduces a discretization of the sampling of echoes and lessens the accuracy of the profile. Therefore, a minor alteration is being made to the MVPR.

Assuming the value of the reflectivity factor in a certain altitude range obeys the Gaussian distribution, the expected value of the reflectivity factor can be used to accurately estimate the state of the clear-air echoes at the sampling altitude. By applying Gauss curve fitting to the frequency distribution function of the reflectivity factor, the expected value can be obtained. The fitting uses the bi-square method for robustness, and the adjusted R-square of the effective fitting needs to be larger than 0.95.

Of special note is that a multiplicative adjustment factor is applied to the frequency distribution function because of the characteristics of the radar beam that make the volume of echoes different. The adjustment factor (called the weight factor) is the ratio of the single-sampling volume to the sum of the volume of the scan in the altitude range, or simply denoted by the sampling volume of the single echo.

It is also noticed that the fit coefficients are affected by the number of samples, so the sample size of the reflectivity factor needs to be checked at the sampling altitude. The minimum ratio of the sample size is set to 10%, which means the ratio of the valid sample to the total of the scan volume at the sampling altitude needs to be greater than one-tenth experientially in this study.

3.2.3. Dual-Wavelength Ratio

Unlike the differential reflectivity measured by means of two orthogonal signals that differ in polarization, the DWR describes the ratio of the radar reflectivity measured with two signals of different wavelengths. In Section 2, it has been found that biological scattering and turbulent scattering have different characteristics of the DWR. The DWR of Bragg scattering is dependent on the radar wavelength, as shown in Equation (9). Moreover, the DWR change in biological scattering is unpredictable because of the biologically complex shape. Hence, the predominant scattering mechanism can be determined by the DWR.

However, it is also known that the value of Z from biological scatterers has an asymmetric pattern which depends on the angle between the radar beam and the main orientation of the biological scatterers. Moreover, values of the DWR may be more complicated and confusing because of the asymmetric pattern. Thus, it needs to examine the effect of the asymmetric pattern.

For the examination, the RCS of biological scatterers is simulated by the computer utilizing FEKO simulation software in this study [35]. The simulative value of RCS is dependent on many factors. One of the factors is the incident angle of the radar beam. Although biological scatterers appear randomly at each azimuth angle of the radar, the incident angle of the simulation is only decided by the angle between the incident beam and the biological scatterer. By changing the coordinate origin of the simulation from the location of the scatterer to the location of the radar, the incident angle can be converted to the azimuth angle when the directions of the biological scatterers are same.

Other parameters of the simulation are obtained from Table 1 based on the model of prolate spheroids of the spinal cord dielectric. Further, supposing that insects fly horizontally in the east–west direction and the density is one per cubic meter, the Z value can be calculated by the simulated RCS shown in Figure 2. Figure 2 exemplifies the asymmetric pattern of the Z values and is almost consistent with previous studies [42]. It is found that the asymmetric pattern may lead to confusion about the conclusions of Section 2.

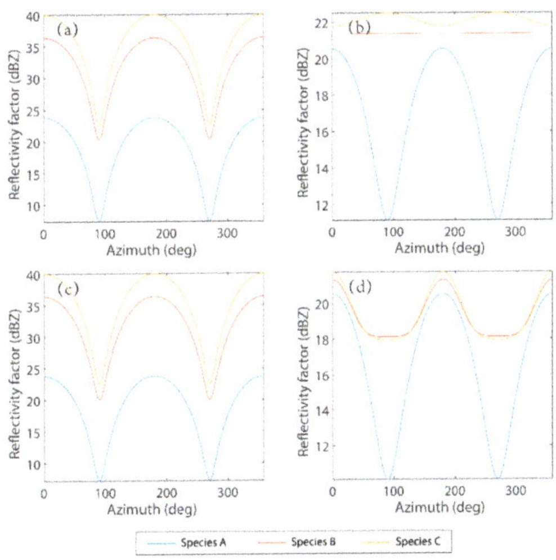

Figure 2. Azimuthal dependencies of the reflectivity factor for three species at the S-band (**a**,**c**) and X-band (**b**,**d**). As in (**a**,**b**), the elevation of the radar beam is 0.5°, which is the minimum elevation of VCP 21, and in (**c**,**d**), the elevation is 19.5°, which is the maximum elevation of VCP 21.

However, the asymmetric pattern does not affect the results of the method in Section 3.2.3. The vertical profiles of the DWR between the S- and X-band are calculated and shown in Figure 3. The computation is deduced in terms of the scan mode of VCP21. Figure 3 shows that the vertical profiles of the DWR are same on each level of height, and the values are close to the simulation values of the insects' body sizes. This is because the RCS of the insect size changes more slowly with angle. However, it is also found that the Z difference values of large insects and the values of turbulence are close. Thus, the echo of turbulence is likely confused with that of large insects in the DWR.

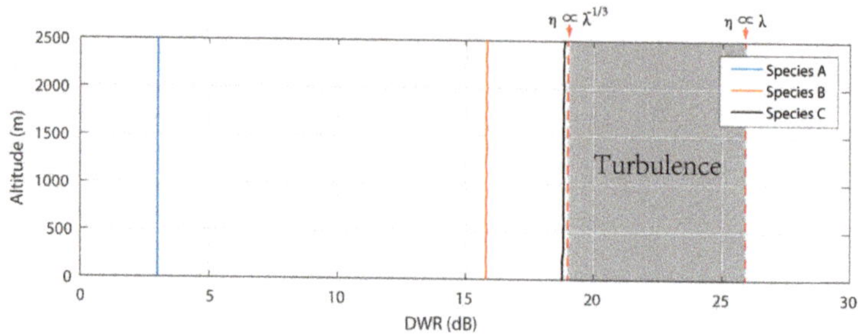

Figure 3. The simulated vertical profiles of the DWR between the S-band and X-band with the vertical height. The vertical profiles of turbulence fall in the grey-shaded area because of the different turbulence theories. The parameters of the three species are referenced from Table 1. All species are major agricultural pests in North China.

Fortunately, large insects have a much stronger flight ability, and flight speed is correlated with body size in animals [43]. Research on the migratory behavior of armyworms expresses that the flight speed of armyworms during migration is greater than 4 m/s [44]. Thus, large insects can be easily distinguished by the velocity azimuth display (VAD). Analysis of the VAD is necessary to exclude confusion with large insects.

4. Results

4.1. Plan Position Indicator

For the sake of discussion, the radar data from May 2 were taken as the case for further study. The focus of this case was from 11:30 (UTC), 2 May 2021, when the clear-air echo first appeared in the radar display, until it began to disappear at 20:30, when the echo became less distinct. The products in Figure 4 show that the echo was characterized by a low Z, a Z_{DR} higher than that typically observed in meteorological echoes, and a correlation coefficient ρ_{HV} lower than that observed in meteorological echoes. Generally, similar echoes have been observed in bird scattering comprising many species flying over Southern Kansas from Wichita, KS, USA [5].

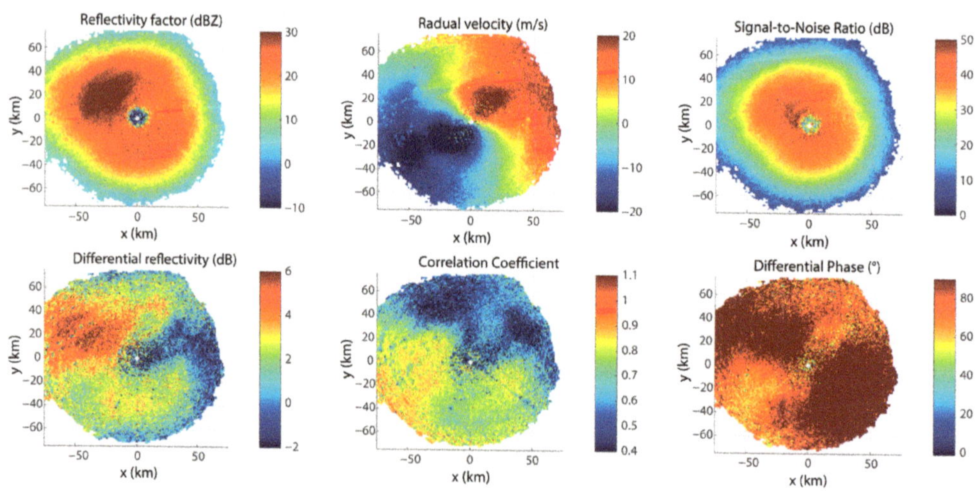

Figure 4. Radar products from the CINRAD/SA radar (SDX), Daxing, Beijing, on 2 May 2021 at 13:00 UTC for an elevation angle of 2.4°. The mapped domain is 75 km by 75 km.

Some researchers have indicated that the ρ_{HV} varies with scatterer orientation, with birds flying away producing low ρ_{HV} values and birds flying toward producing higher ρ_{HV} values. However, the velocity–azimuth display (VAD) confirmed that there was no intense bird migration. Analysis of the VAD is used to determine spatially averaged kinematic properties of the velocity field. According to the function of the radar azimuth angle, the VAD of the wind field resembles a sine function. Moreover, during intense bird migrations, a huge difference in scattering of the observed radial velocity around the modeled radial velocities will be evident.

Figure 5 shows a high-quality wind VAD, which was the same passage of a cold front as observed by an operational weather radar in De Bilt [45], Netherlands. Only a small part of the residual error exhibited signs of bird activity. Since the nonzero value of the Doppler velocity due to bio-scatterers is representative of biological target movement, the temporal change in Z is shown in Figure 6 to check for movement.

Figure 6 shows a continuous change in Z; the echoes began to increase rapidly at 11:30 and then remained basically unchanged after 12:06. It is strange that the echoes changed with the radar station as the center but not the "habitat", and the echoes with a larger Z also maintained their appearance. Whereas the value of the Doppler velocity was rhythmical, shown in Figure 7, the characteristics of the spatial distribution were also unchanged after 48 min, indicating that the scatterers did not move.

To confirm the state of the scatterers' movement, in Figure 8, the range–height cross-section of Z is displayed in the azimuth of the wind direction. As shown in Figure 8, the signal was enhanced along the height at each range and did not change following wind direction. The phenomena discovered by Adriaan were not observed [46]. Paradoxically, the scatterer remained in the air, but the Doppler velocity was a nonzero value.

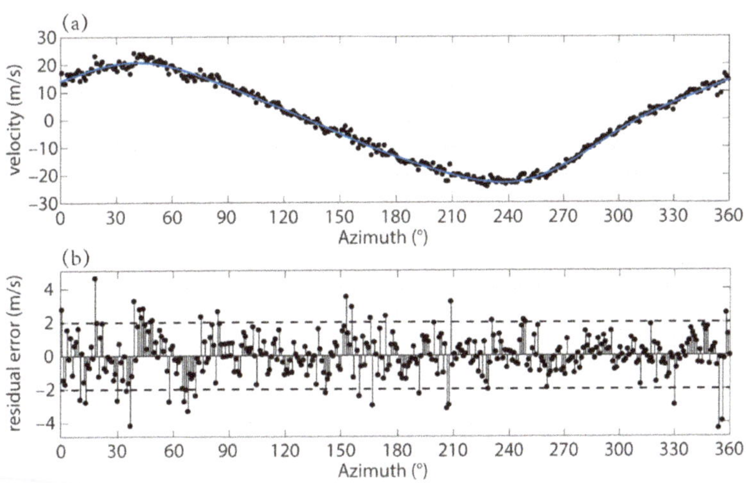

Figure 5. (**a**) An example of VAD data from the CINRAD/SA in Beijing. The line is the modeled radial velocities as a sine function of the azimuth, and the dots are the data of the Doppler velocity. The data samples are in the range of 30 km and the elevation of 2.4 deg. (**b**) The residual error (the dot) is essentially less than 2 m/s. The root mean squared error is 1.278, and the adjusted R-square is 0.9924.

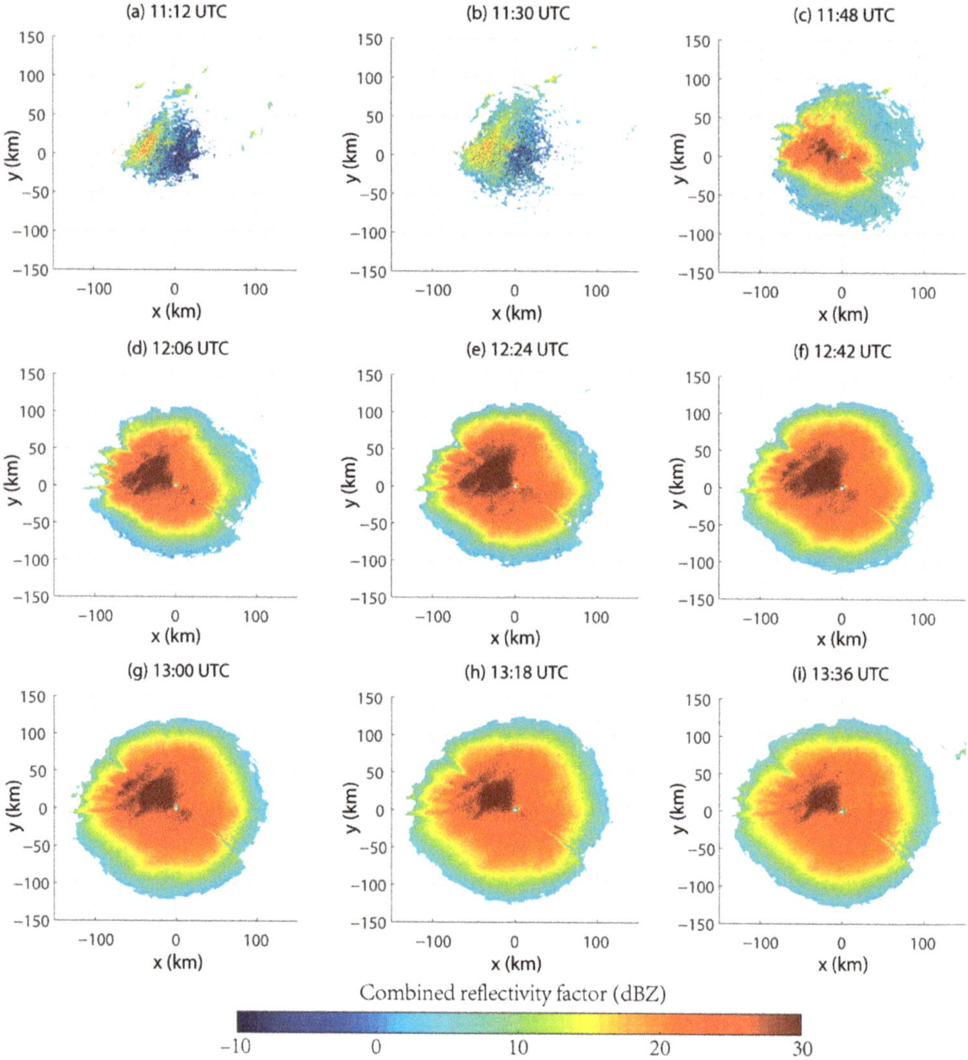

Figure 6. Combined reflectivity factor for 2.4 h from the CINRAD/SA, 2 May 2021. (**a**–**i**) Continuous observations from the CINRAD/SA at 12 min intervals from 11:12 to 13:36. The horizontal and vertical coordinates are, respectively, the ranges (km) in the west–east direction and the south–north direction.

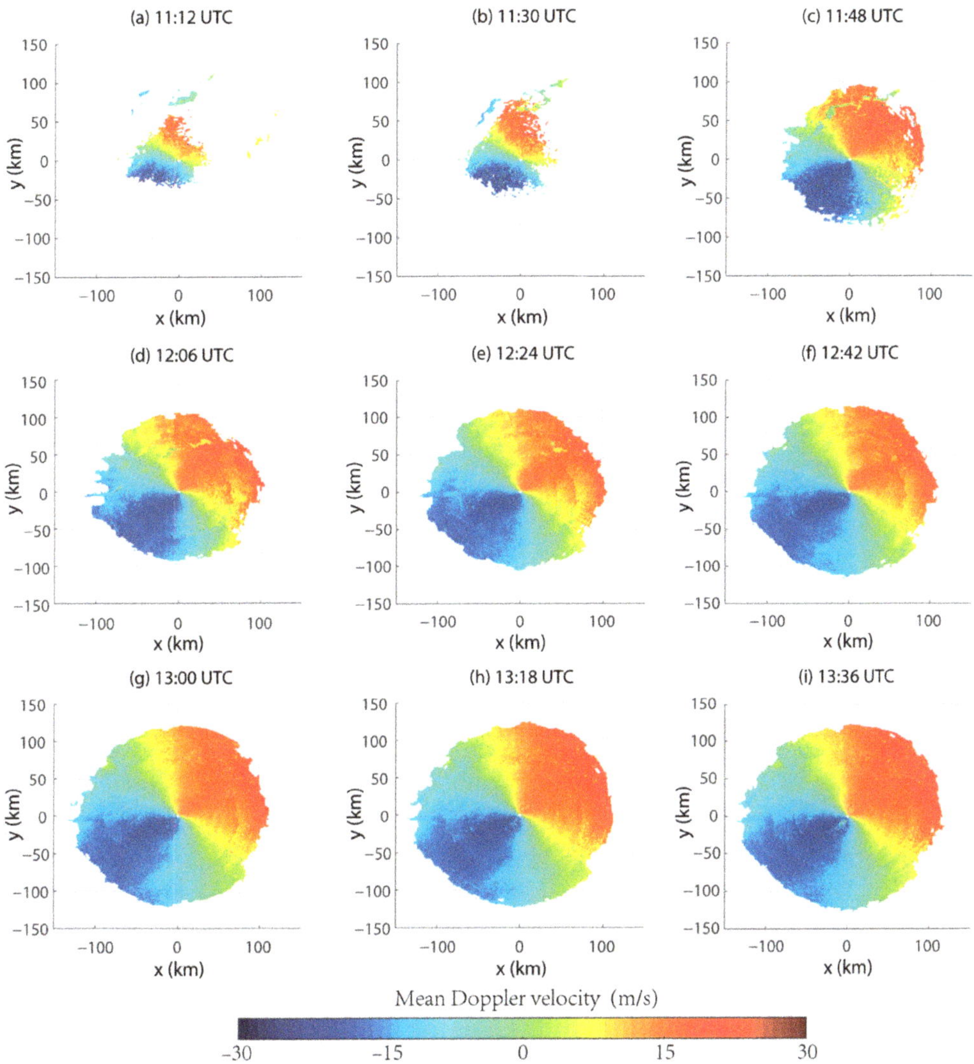

Figure 7. Mean Doppler velocity for 2.4 h from the CINRAD/SA, 2 May 2021. (**a**–**i**) Continuous observations from the CINRAD/SA at 12 min intervals from 11:12 to 13:36. The horizontal and vertical coordinates are, respectively, the ranges (km) in the west–east direction and the south–north direction.

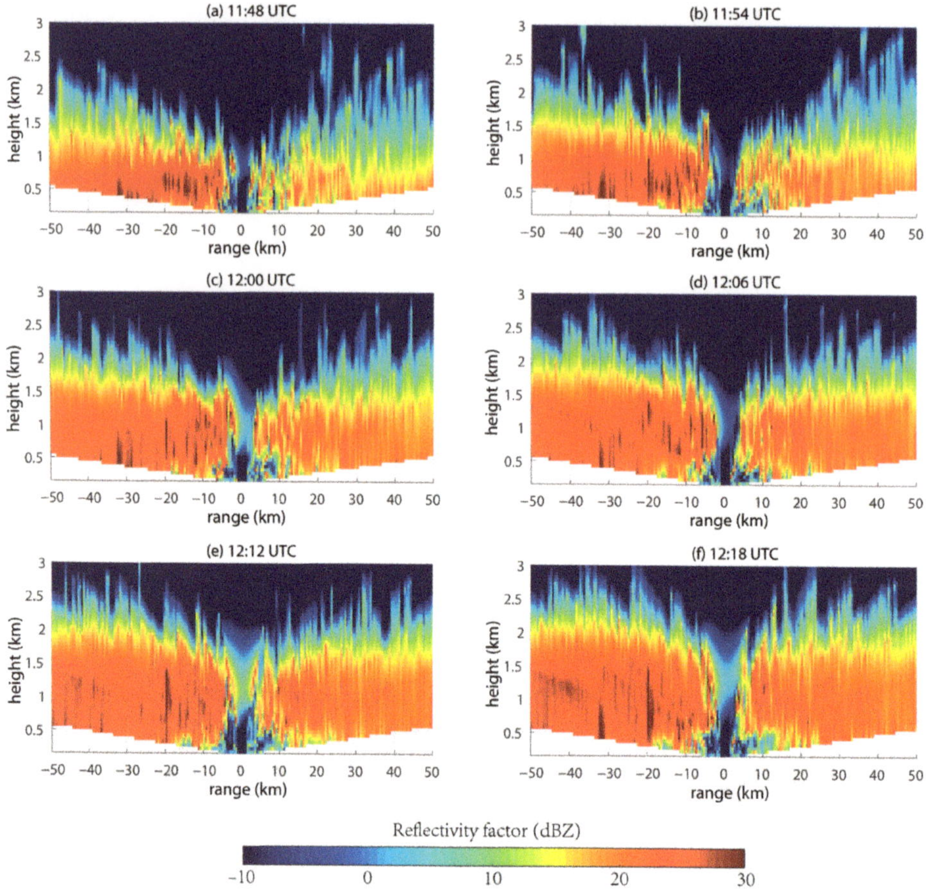

Figure 8. Range–height cross-section of Z following the azimuthal direction (azimuth: 225°). (**a**–**f**) Continuous observations from the CINRAD/SA at 6 min intervals from 11:48 to 12:28, 2 May 2021. The horizontal axis is the range (km) along the wind direction, and the vertical axis is the height (km).

4.2. Time–Height Cross-Section

Movement is the most important feature of bio-scatterers. To reveal the causes of clear-air echoes, the radar data were displayed as a time–height cross-section. The signal of the clear-air echoes showed significant diurnal variations, as shown in Figure 9. It was thought that the activities of nocturnal creatures caused greater echoes. However, the rapid growth in the signal in the time–height cross-section caught our attention. Surprisingly, the value of Z grew fast. The 23 dBZ echo only took 13 min to increase from 500 m to 1000 m, and the increasing velocity was 0.64 m/s. When using the wind lidar, as Figure 10 shows, the vertical velocity of the wind was below 0.5 m/s and was smaller than the velocity of the echoes. This indicates that a small bio-scatterer, which does not fly well, could not be the scatterer. Despite the impractical state of motion, perhaps these clear-air echoes were still caused by large bio-scatterers. However, the vertical profiles of Z_{DR} shown in Figure 9 deny this view.

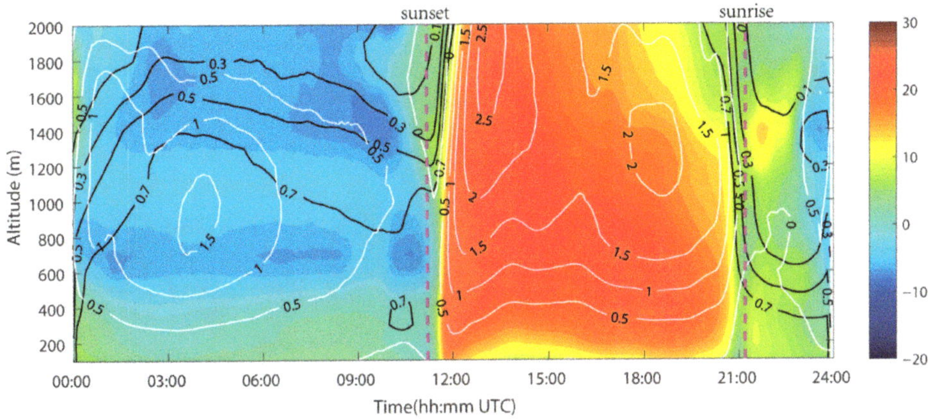

Figure 9. Time–height cross-section of radar products for 24 h from the CINRAD/SA at Daxing, Beijing, 2 May 2021. The fill color is Z (unit: dBZ); the black isopleths are valid data proportions; and the white isopleths are Z_{DR} (unit: dB). The times of sunset and sunrise were 11:08 and 21:13, UTC, respectively.

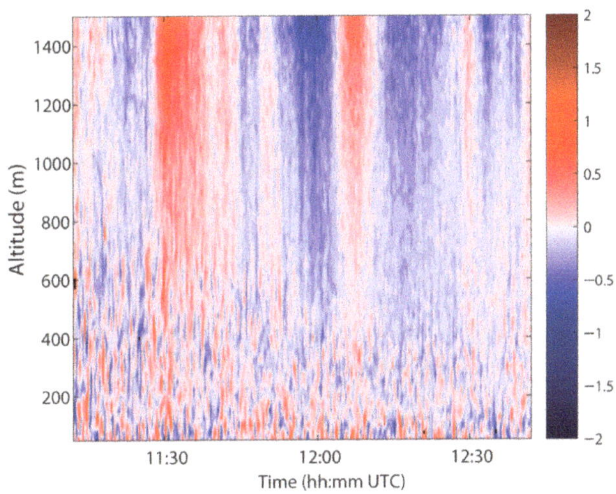

Figure 10. Vertical wind profiles measured using the Windcube 100 s on 2 May 2021. The positive and negative values represent the vertical upward and downward wind speeds (unit: m/s), respectively.

It is known that Z_{DR} is the radar reflectivity difference between the horizontal and vertical polarization and represents the dimension of the scatterer. If the measurements of the volumes contain multiple scatterers, Z_{DR} will be biased toward the characteristics of the largest scatterers. In Figure 9, the profiles of Z_{DR} increased with height and were temporarily greater than 2.5 dB above an altitude of 1200 m. Scatterers with a 2.5 dB Z_{DR} do not climb from the ground and do not land. This implies that the echoes did not float from the ground into the air, but they suddenly appeared in the middle of the air without immigration.

4.3. Velocity Analysis

Although we have preliminarily analyzed the VAD in Section 4.1, to avoid confusion with large insects and birds, the Doppler velocity and the results of the VAD analysis are displayed in Figure 11 again. In Figure 11, the velocity fields, which were obtained from the VAD analysis, are similar to the radar Doppler velocity. The residual values of the velocity fields are shown in Figure 12a. in total, 86.4% of the residual values are less than 4 m/s, and 65.3% of the values are less than 2 m/s. The echoes were unlikely caused by large insects and birds mainly because large insects and birds have quite a great flight speed and would make the speed field messy and produce great deviations.

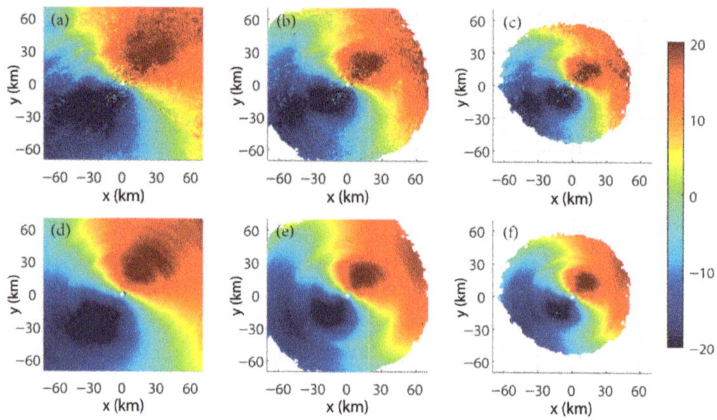

Figure 11. Doppler velocity fields (**a**–**c**) and the fields which were obtained from the VAD analysis (**d**–**f**). The elevation angles of (**a**–**c**) are 0.5°, 1.5°, and 2.4°, respectively, the same as (**d**–**f**). Except for some error points and point targets, the velocity fields (**d**–**f**) which were obtained from the VAD analysis are similar to the Doppler velocity (**a**–**c**).

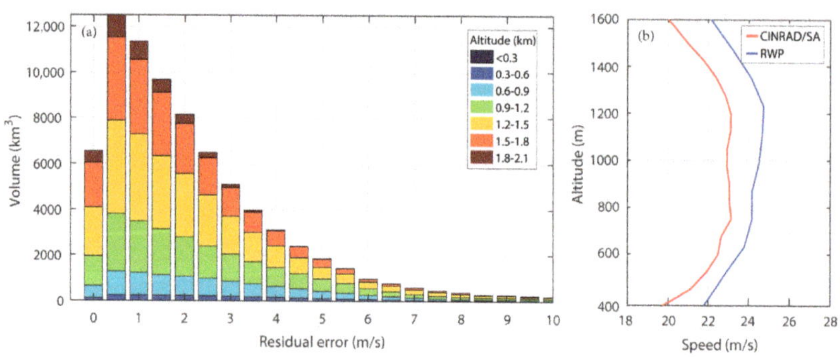

Figure 12. (**a**) Histogram of the residual error of the VAD analysis at different altitudes. The volume of the radar beam is used to calculate the ordinate. (**b**) The profiles of the mean wind speed during 12:30 to 13:30 (UTC) on 2 May 2021.

The comparison of the speed profiles with the RWP and the CINRAD/SA confirms the conclusion. The profiles of the horizontal wind speed which were sensed by the RWP are not different from the speed profiles calculated by the CINRAD/SA, as shown in Figure 12b. Despite the system deviation in Figure 12b, which is common in the results of the VAD wind profile [47], the changes in the speed profiles are consistent, generally. The profiles of the CINRAD/SA indicate that clear-air echoes are moving with the wind speed. However, the speed of biological autonomous movement is not detected by the weather radar. This means that these echoes are unlikely to be caused by creatures with a strong flight ability such as birds and large insects.

4.4. Comparison of the S-Band and X-Band

In Sections 4.1 and 4.2, some contradictions of bio-scatterers were revealed, and a quantitative analysis of the echoes was thus needed to convincingly demonstrate the cause of the clear-air echoes. A comparison of the reflectivity factors of the two bands is another method that can be used to determine the cause of clear-air echoes. Since the DWR of the turbulence echoes in different bands was more regular than biological, the time–height cross-section of Z using the data from BJXFS is shown in Figure 13, and the DWR between the S-band and the X-band is displayed in Figure 14.

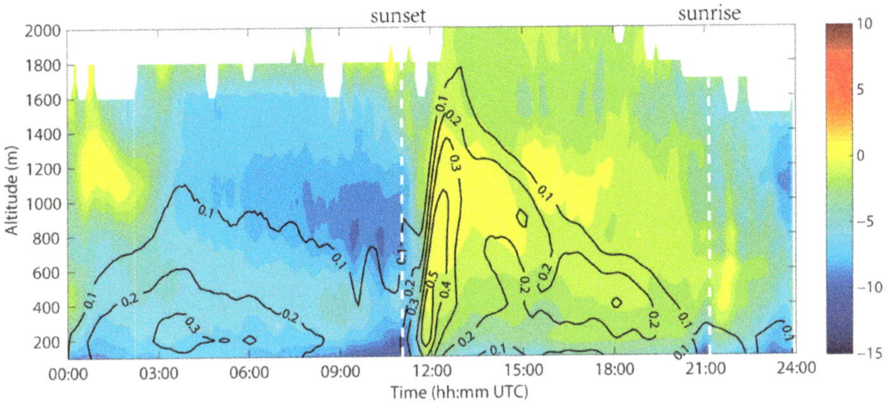

Figure 13. Time–height cross-section of Z (the fill color) and the valid data proportions (the black isopleths) for 24 h from the X-band radar at FS, Beijing, 2 May 2021.

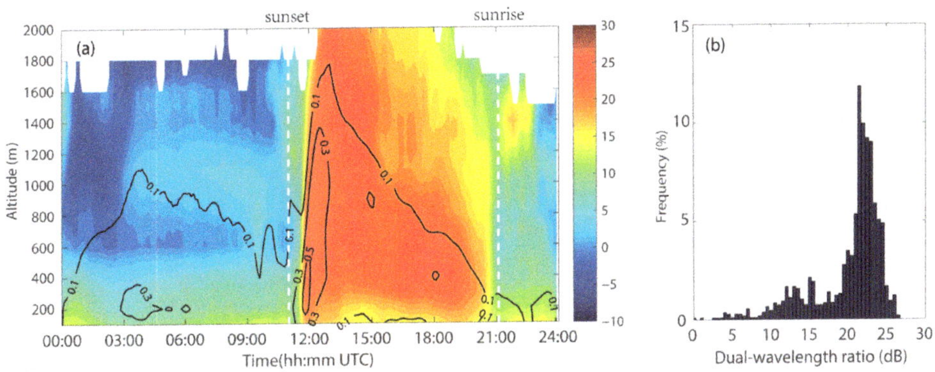

Figure 14. The DWR between the values of Figures 8 and 12. The time–height cross-section and the histogram of the DWR at nighttime are exhibited in (**a**) and (**b**), respectively. The black isopleth in (**a**) is the proportion of the X-band valid data. The bar chart (**b**) represents the normalized frequency distribution of the DWR from sunset to sunrise, and the data in the histogram from 19 dB to 25 dB occupy 80%.

In Figure 13, intuitively, the signal of the X-band differed from the S-band signal in the value of Z; the value of the X-band was much smaller than that of the S-band. The signal of the X-band also did not have an as pronounced diurnal variation as that of the S-band echoes. Although the signal of the X-band echoes appeared as fast as those of the S-band signal at dusk, the signal did not last long and slowly vanished from the top to the bottom after 13:00. Meanwhile, the value of Z in Figure 13 is low at lower altitudes and increases with height, which is different in Figure 9. This is due to the minimum scale of turbulence which increases with height. When the minimum scale exceeds the half-wavelength of the X-band, which is still much smaller than the wavelength of the S-band, the scattering vanishes before the refractivity gradient weakens. Hence, the Z value of the X-band seems to increase progressively without reducing.

Figure 14a shows the time–height cross-section of the DWR between the Z values of the CINRAD/SA and X-POLs, and Figure 14b shows the normalized frequency histograms for the radar fields in Figure 14a. The histogram of the DWR shows the characteristic values near 21.5 dB. Moreover, the Z value distributions of the S-band and X-band also show that the peaks of the Z values are at intervals of about 22 dB (in Figure 15), whereas the Z_{DR} distribution does not peak at 0 dB. In Figure 15, the frequency distribution of the Z value resembles the Gaussian distribution and slightly changes with height, which is in line with the condition of the hypothesis in Section 3.2. However, it is not expected that the correlation or other products should follow the Gaussian distribution. The distributions of turbulence variables deviate from normality because of the intermittence [48–50].

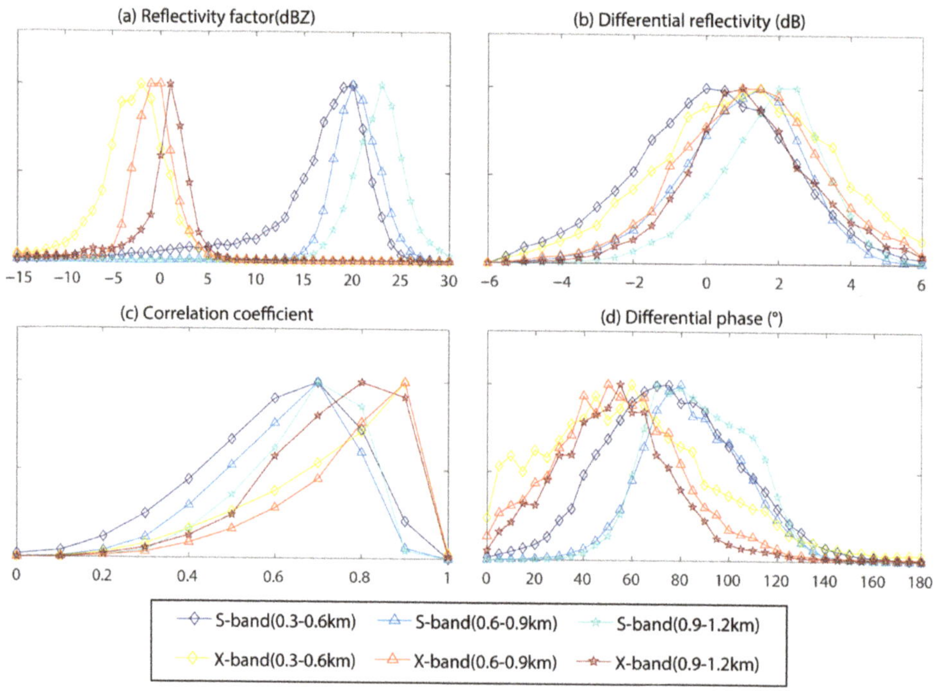

Figure 15. Histograms of the radar products of the CINRAD/SA (S-band) and X-POL (X-band), respectively, on 2 May 2021 at 13:00 UTC for a height ranging between 300 m and 1.2 km.

The large widths of these distributions match with the characteristics of bird echoes seemingly. However, the echo is unlikely to be mainly caused by birds for the following reasons. First, the analysis of the VAD has already denied the dominance of birds. Second, for north China, May is almost the end of the spring migration and is close to the period of preincubation, yet the scatterers of the clear-air echo still travel to the north until late July [51,52]. Third, there are two international airports located at the azimuth angles of 22° and 188° and the distances of 31 km and 34 km, respectively. The large numbers of birds migrating would obviously be a threat to the safety of the flights if the clear-air echo was caused by birds. Therefore, the suspicion of birds and large insects has been ruled out.

It is worth noting that the CINRAD/SA and X-POL are located along the same migration path. It means that the two radars should observe the same group of scatterers with the same scattering characteristic, moving from upstream (XFS) to downstream (SDX) the whole night, continuously. Although Z values from atmospheric insects may have asymmetric patterns which depend on the angle between the radar beam and the main orientation of the biological scatterers, the time–height cross-section of the Z values will not be affected because the cross-sections are counted by the whole coverage volume of the radar observation and are not dependent on the azimuth. Thus, the asymmetric patterns cannot stain the ratio between the Z values of the CINRAD/SA and X-POL.

It is already known that, according to Equation (6) and Equation (10), the Z difference in the turbulence of the S-band (2870 MHz, 10.45 cm) and the X-band (9455 MHz, 3.17 cm) is about 19.0 dB and 21.4 dB, respectively. Since many creatures are within the size threshold for Mie scattering and have different RCSs at different wavelengths, the reflectivity relationships to the wavelength are far from straightforward. Meanwhile, wide discrepancies are found between the characteristics of clear-air echoes and the law of biological activities. Therefore, this proves that the Beijing CINRAD/SA echoes are mainly caused by turbulence.

The DWR statistics for the region of Beijing reveal that, from 1 May 2021 to 20 May 2021, 58% of the DWR was distributed between 18dB and 24dB, which is the characteristic interval of turbulence. Moreover, the clear-air echoes maintain the features of the variations in the Z values and the VAD analysis. It means that nearly 58% of the echoes were probably caused by turbulence and proves that the influence of turbulence is erroneously ignored in clear-air echoes. Almost all these echoes did not show the classic features of Bragg scattering, such as a low Z and zero Z_{DR}. These echoes could wrongly be identified as bio-scatterers and cause a misestimation of biomass. The number of flying creatures observed by the weather radar was much smaller than previously thought.

5. Discussion

Many scholars deny or reject the domination of turbulence in clear-air echoes. First, they propose that a nonzero Z_{DR} indicates that the scattering is due to creatures rather than turbulence because, according to the statement of Kolmogorov's theory that turbulence is homogeneous and isotropic, the Z_{DR} of turbulent echoes should be zero. However, a problem with this viewpoint is that non-Kolmogorov turbulence widely exists in the atmospheric boundary layer. Some observations have already shown that the atmospheric structure constant of the refractive index differs between the horizontal direction and the vertical direction. Therefore, the Z_{DR} cannot support their rejections. We agree that biological scatterers cause large values of Z_{DR}, but the contribution from the effect of turbulence still needs to be investigated.

Second, some scholars believe that because the C_n^2 calculated from echoes is larger than the high end of the observed values, clear-air echoes should not be attributed to turbulence. For example, clear-air echoes of a 10 to 20 dB Z require C_n^2 to be greater than $10^{-11}\ m^{-2/3}$, which is far above the observed C_n^2. The phenomenon of clear-air echoes cannot be explained by Bragg scattering. Therefore, it is assumed that the domination of clear-air echoes is not turbulence. However, scholars have ignored the effect of another scattering mechanism. We found another reasonable scattering mechanism to explain clear-air echoes in communication and wireless areas.

In communication, tropospheric scattering (also known as troposcatter) is admitted as an efficient propagation method by the Radiocommunication Sector of the International Telecommunication Union, which describes the mechanism by which microwave radio systems inadvertently achieve beyond-the-horizon communications [53]. One of the three models of the troposcatter is the reflecting-layers model. The other two models are scattering from turbulence and reflections from an exponential atmosphere. Tropospheric scattering is the result of the combination of these three models. Zhang points out that the reflectivity of the troposcatter can be written as follows [54]:

$$\eta = B\overline{\left(\frac{d\varepsilon_r}{dh}\right)^2 \lambda^n (2\varphi)^{-m}} \tag{11}$$

where B, n, and m are constants and are measured by experiments; ε_r is the dielectric constant; h is the height; φ is the glancing angle; the bar is the sign of the mean; and the relationship between ε_r and h can refer to the assumption by K. Bullington [55]. Interestingly, the scattering from turbulence is essentially the same as Bragg scattering, but few studies have highlighted the role of reflecting layers in clear-air echoes.

To demonstrate how reflecting layers affect clear-air echoes, we calculated the gradient Richardson number and the intensity of the turbulence (expressed by the ratio of the standard deviation of the wind speed to the mean wind speed) using a microwave radiometer and a wind lidar [56,57]. Moreover, the atmospheric profiles were collected by a microwave radiometer. Although the geometrical structure of the reflecting layers remains obscure, it can be inferred that turbulent mixing is detrimental to the reflecting layers because a reflecting layer in the atmosphere is formed by relatively sharp gradients of the refractive index, but turbulent mixing makes the temperature and humidity homogeneous and reduces the gradient of the refractive index. Thus, the Richardson number and the turbulence intensity, which indicate turbulent mixing, are shown in Figure 16 to give the relationship between clear-air echoes and turbulent mixing.

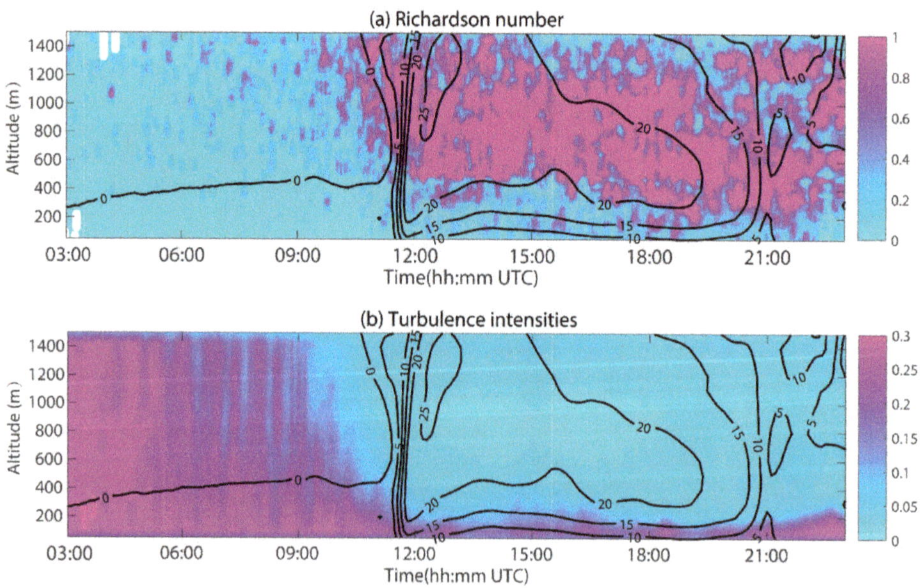

Figure 16. Turbulence intensities (**a**) and Richardson number (**b**) for 20 h at Daxing, Beijing, 2 May 2021. The black isopleth is the time–height cross-section of Z (unit: dBZ), which is the same as the fill color in Figure 6.

In Figure 16, large values of Z correspond to large Richardson numbers and turbulent mixing. When the sun sets in the west, the turbulent mixing rapidly weakens. Additionally, the unmixed air masses with relatively sharp gradients become the reflecting layer and generate the scattering signal of the clear-air echoes. In contrast, in the daytime, the strong turbulent mixing breaks the structure of the reflecting layer, and only the turbulent scattering remains. Thus, turbulence has a greater impact on the echoes at night than in the daytime.

The same diurnal variation has been observed in the field-strength variation in the short radio wave propagation in Arizona [58]. Other studies have found that troposcatter propagation only occurs when the turbulence scale is larger than the wavelength of the radio signal, and the signal level with a wavelength of 9 cm is much larger than that with a wavelength of 3 cm [58,59]. These phenomena have the same characteristics as clear-air echoes, with the value of the reflectivity factor at the S-band being bigger than that at the X-band. Thus, we propose that the reflecting layers cause a diurnal variation in clear-air echoes and enhance the signal of clear-air echoes at night. Therefore, the coexistence of the reflecting layers and Bragg scattering is the reason why the value of C_n^2 calculated from the echoes is larger than the value from the theoretical calculations. Further, the reflecting layers explain some of the characteristics of Z_{DR}. For example, the vertical signal of clear-air echoes is weaker than the horizontal signal because turbulent mixing is stronger in the vertical direction, and the vertical reflecting layers cannot easily survive. Thus, the values of Z_{DR} are generally greater than zero.

An interesting note is that the high reflectivity area in the upper-left region of Figure 4 analyzes the characteristics of a low correlation coefficient, and in this area, Z_{DR} presents a mixture of high and low values. This characteristic is ascribed to the atmospheric response to the underlying surface, which belongs to a mountainous region and is quite different from the others in Figure 1.

The rough terrain of mountains makes the turbulence more chaotic and intense, which produces a stronger refractive index and scattering. A thin layer of shear turbulence excited on the shear plane makes the Z_{DR} larger, but the layer is not stable, bringing a mixture of high and low values to the Z_{DR}. However, it is noted that, by upthrust, too high mountains may affect the turbulent scales, which are too large to match up to the radar wavelength for scattering. The relationship between the underlying surface and clear-air echoes is worth further exploration in the future.

Another interesting characteristic of the Z_{DR} is that its value grows weak and becomes close to zero in the day–night shift scenes. We speculate that the turbulent mixing reaches a quasi-equilibrium state, and this will be further investigated in future studies.

Yet despite all this, some scholars still express a slightly different point of view. They claim that the reflection from turbulent air at the S-band has been studied by some studies [29,60–63], and it has been well established with polarization radars that turbulent air has Z_{DR} values close to 0 dB at the S-band. However, these echoes were commonly observed at the top of the convective boundary layer (also called the entrainment layer). The entrainment layer is essentially static in stability. Due to the effects of penetrative convection and entrainment, thermals reciprocate in the layer [64]. The reciprocation brings turbulence generation on the one hand; on the other hand, the stable air brings the turbulence closer to a locally homogeneous isotropic state. Thus, the observed echoes can be regarded as special cases and cannot represent the whole characteristic of clear-air turbulent echoes. Figure 17 shows the vertical profiles of the Z values and rawinsonde data. It is observed that there is a layer of enhanced reflectivity at the entrainment layer, which is consistent with the known results [63].

Some scholars also conject that dust and other particulate matter can be the cause of the clear-air echoes. At the beginning of this study, we also inferred that nocturnal pollutant accumulation is the reason for the echo diurnal variation because of the uncleanliness of the atmosphere of the megacities. Yet, it is hard to explain the variation in the echo signal.

One piece of evidence is based on seasonal variation. In winter, the air quality is the worst because of heating, which uses fossil fuels, but the clear-air echo is hardly observed by radars during winter. The annual and monthly variation in the sand and dust also showed that the sand and dust weather is most frequent in spring, whereas the signal of the spring clear-air echo was generally weaker than summer and autumn.

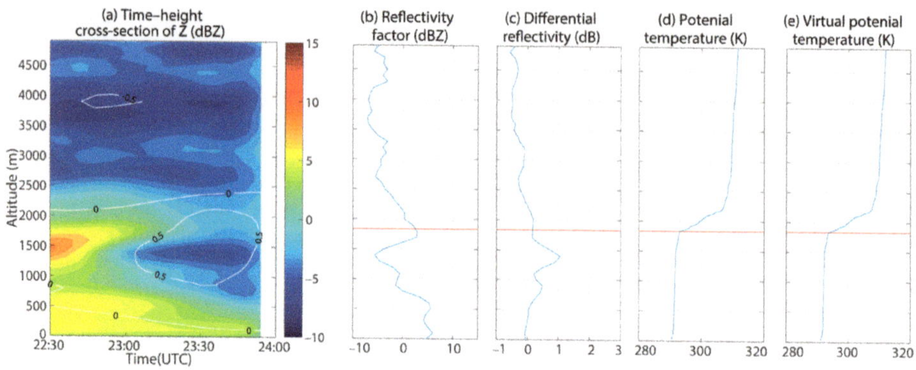

Figure 17. Time–height cross-section of the Z values (**a**) from 22:30 to 24:00, and the vertical profiles of the radar products (**b**,**c**) and rawinsonde products (**d**,**e**) at 23:15 UTC on 2 May 2021. The profiles of (**d**,**e**) are measured by rawinsondes. The estimated entrainment layer is based on the maximum vertical gradient in each variable (orange line).

The other piece of evidence comes from the daily change in pollutants. The concentration of pollutants has two peaks because of traffic congestion in the metropolis, which is different from the Z value. The wind, which is related to pollutant diffusion and dust, also does not show a significant correlation with the clear-air echoes. Although it is known that fire plumes can cause clear-air echoes [65,66], it is unrealistic to detect plumes of wild fires in an urban area. Thus, dust and the pollutants are unlikely to be the main causes of clear-air echoes.

The issue of the influence of meteorological factors on troposcatter communication remains unknown and requires future examination. In troposcatter, the value of the signal level depends on the refractive index and its gradient, which are affected by the intensity of the turbulence fluctuation and atmospheric stratification. Thus, Gaoming Zhang proposed that the inversion of temperature leads to the diurnal variation in the signal [54]. However, few studies focus on the effect of turbulent mixing on signals level. Thus, we plan to provide a more in-depth explanation of reflecting layers forming and the structure of the nocturnal boundary layer in future studies. Meanwhile, Guifu Zhang pointed that bistatic radars have an advantage in the sensitivity of clear echoes [67]. Technically, experiments using bistatic radars are closer to the principle of troposcatter propagation. Moreover, experiments with bistatic radar data would provide more information for the analysis in future research.

6. Conclusions

In this study, clear-air echoes detected by CINRAD were analyzed to find their causes. Some observations diverge from the previous conclusion that bio-scatterers are the main reason for clear-air echoes.

Echoes with a larger Z are not deformed in the air, even if the scatterers are moving. The change in echoes in the vertical direction is also closer to the switching of the physical mode rather than biological flying. The analysis results of the DWR and the VAD support that turbulence plays an important role in clear-air echoes. In the case of May 2, the frequency distribution of the DWR peaks at 21.5 dB, which is consistent with the theory of turbulence. From 1 May to 20 May, 58% of the DWR between the S-band and the X-band is distributed between 18 dB and 24 dB, which means that more than half of the echoes at night were caused by turbulence. It was confirmed that the influence of turbulence is erroneously ignored in clear-air echoes. The reflecting-layers model of troposcatter propagation is the cause of the clear-air echoes, and this model can explain the main phenomena of the radar observations.

This study provides initial evidence for a case study in Beijing that the model based on Kolmogorov's theory may not be tenable for all clear-air echoes, highlighting the need for an expanded set of causes of clear-air echoes. The reflecting-layers model, which is one of the three models of tropospheric scattering, cause a diurnal variation in clear-air echoes and enhances the signal of clear-air echoes at night. Unmixed air masses with relatively sharp gradients become the reflecting layer and generate the scattering signal of clear-air echoes.

With the help of the theory of troposcatter propagation, rapid progress will be made in the ecological monitoring method of weather radars. A more objective and comprehensive study of clear-air echoes can effectively make weather radars acceptable in biological research instead of ignoring irrationalities. With the help of weather radars, ecology will be able to develop strongly and continuously in the near future.

Author Contributions: Conceptualization, Y.T. and S.M.; methodology, Y.T.; software, Y.T.; validation, Y.T.; formal analysis, Y.T.; investigation, Y.T.; resources, S.M. and H.C.; data curation, S.M. and H.C.; writing—original draft preparation, Y.T. and T.L.; writing—review and editing, Y.T. and T.L.; visualization, Y.T. and T.L.; supervision, S.M. and H.C.; project administration, Y.T., S.M. and H.C.; funding acquisition, Y.T., S.M. and H.C. All authors have read and agreed to the published version of the manuscript.

Funding: This research was funded by the National Natural Science Foundation of China, grant number 42205145 and 31727901.

Data Availability Statement: Not applicable.

Acknowledgments: The authors thank Yunjie Xia, Beijing Meteorological Observation Center, China, for providing the RPG-HATPRO-G5 and the Windcube 100 s data.

Conflicts of Interest: The authors declare no conflict of interest. The funders had no role in the design of this study; in the collection, analyses, or interpretation of data; in the writing of the manuscript; or in the decision to publish the results.

References

1. Van Doren, B.M.; Horton, K.G. A continental system for forecasting bird migration. *Science* **2018**, *361*, 1115–1117. [CrossRef] [PubMed]
2. Bruderer, B. The study of bird migration by radar part 1: The technical basis. *Naturwissenschaften* **1997**, *84*, 1–8. [CrossRef]
3. Wilson, J.W.; Weckwerth, T.M.; Vivekanandan, J.; Wakimoto, R.M.; Russell, R.W. Boundary Layer Clear-Air Radar Echoes: Origin of Echoes and Accuracy of Derived Winds. *J. Atmos. Ocean. Technol.* **1994**, *11*, 1184–1206. [CrossRef]
4. Martin, W.J.; Shapiro, A. Discrimination of bird and insect radar echoes in clear air using high-resolution radars. *J. Atmos. Ocean. Technol.* **2007**, *24*, 1215–1230. [CrossRef]
5. Van den Broeke, M.S. Polarimetric Radar Observations of Biological Scatterers in Hurricanes Irene (2011) and Sandy (2012). *J. Atmos. Ocean. Technol.* **2013**, *30*, 2754–2767. [CrossRef]
6. Westbrook, J.K.; Eyster, R.S.; Wolf, W.W. WSR-88D doppler radar detection of corn earworm moth migration. *Int. J. Biometeorol.* **2014**, *58*, 931–940. [CrossRef] [PubMed]
7. Zrnic, D.S.; Ryzhkov, A.V. Observations of insects and birds with a polarimetric radar. *IEEE Trans. Geosci. Remote Sens.* **1998**, *36*, 661–668. [CrossRef]
8. Ottersten, H. Atmospheric Structure and Radar Backscattering in Clear Air. *Radio Sci.* **1969**, *4*, 1179–1193. [CrossRef]
9. Kolmogorov, A.N.; Levin, V.; Hunt, J.C.R.; Phillips, O.M.; Williams, D. The local structure of turbulence in incompressible viscous fluid for very large Reynolds numbers. *Rep. AS USSR* **1941**, *434*, 9–13. [CrossRef]
10. Kolmogorov, A.N. A refinement of previous hypotheses concerning the local structure of turbulence in a viscous incompressible fluid at high Reynolds number. *J. Fluid Mech.* **1962**, *13*, 82–85. [CrossRef]
11. Mandelbrot, B.B. Intermittent turbulence and fractal dimension: Kurtosis and the spectral exponent 5/3 + B. In *Multifractals and 1/f Noise: Wild Self-Affinity in Physics (1963–1976)*; Mandelbrot, B.B., Ed.; Springer: New York, NY, USA, 1976; pp. 389–415.
12. Ringuet, E.; Rozé, C.; Gouesbet, G. Experimental observation of type-II intermittency in a hydrodynamic system. *Phys. Rev. E* **1993**, *47*, 1405–1407. [CrossRef] [PubMed]
13. Batchelor, G.K.; Townsend, A.A.; Jeffreys, H. The nature of turbulent motion at large wave-numbers. *Proc. R. Soc. Lond. Ser. A Math. Phys. Sci.* **1949**, *199*, 238–255. [CrossRef]
14. Pomeau, Y.; Manneville, P. Intermittent transition to turbulence in dissipative dynamical systems. *Commun. Math. Phys.* **1980**, *74*, 189–197. [CrossRef]
15. Siggia, E.D. Numerical study of small-scale intermittency in three-dimensional turbulence. *J. Fluid Mech.* **1981**, *107*, 375–406. [CrossRef]

16. Paladin, G.; Vulpiani, A. Anomalous scaling laws in multifractal objects. *Phys. Rep.* **1987**, *156*, 147–225. [CrossRef]
17. Huang, Y.N.; Huang, Y.D. On the transition to turbulence in pipe flow. *Phys. D Nonlinear Phenom.* **1989**, *37*, 153–159. [CrossRef]
18. Meneveau, C.; Sreenivasan, K.R. Interface dimension in intermittent turbulence. *Phys. Rev. A* **1990**, *41*, 2246–2248. [CrossRef]
19. Vassilicos, J.C. Turbulence and intermittency. *Nature* **1995**, *374*, 408–409. [CrossRef]
20. Benzi, R.; Biferale, L. Intermittency in Turbulence. In *Theories of Turbulence*; Oberlack, M., Busse, F.H., Eds.; Springer: Vienna, Austria, 2002; pp. 1–76.
21. Jiménez, J. Intermittency in Turbulence. In *Encyclopedia of Mathematical Physics*; Françoise, J.-P., Naber, G.L., Tsun, T.S., Eds.; Academic Press: Oxford, UK, 2006; pp. 144–151.
22. Belen'kii, M.S. Effect of the stratosphere on star image motion. *Opt. Lett.* **1995**, *20*, 1359–1361. [CrossRef]
23. Korotkova, O.; Toselli, I. Non-Classic Atmospheric Optical Turbulence: Review. *Appl. Sci.* **2021**, *11*, 8487. [CrossRef]
24. Rao, C.; Jiang, W.; Ling, N. Spatial and temporal characterization of phase fluctuations in non-Kolmogorov atmospheric turbulence. *J. Mod. Opt.* **2000**, *47*, 1111–1126. [CrossRef]
25. Andrews, L.C. Free-space optical system performance for laser beam propagation through non-Kolmogorov turbulence. *Opt. Eng.* **2008**, *47*, 026003. [CrossRef]
26. Li, Y.; Zhu, W.; Wu, X.; Rao, R. Equivalent refractive-index structure constant of non-Kolmogorov turbulence. *Opt. Express* **2015**, *23*, 23004–23012. [CrossRef]
27. Ruizhong, R.; Yujie, L. Light Propagation through Non-Kolmogorov-Type Atmospheric Turbulence and Its Effects on Optical Engineering. *Acta Opt. Sin.* **2015**, *35*, 0501003. [CrossRef]
28. Yang, H.; Fang, Z.; Li, C.; Deng, X.; Xing, K.; Xie, C. Atmospheric Optical Turbulence Profile Measurement and Model Improvement over Arid and Semi-arid regions. *Atmos. Meas. Tech. Discuss.* **2021**, *2021*, 1–14. [CrossRef]
29. Richardson, L.M.; Cunningham, J.G.; Zittel, W.D.; Lee, R.R.; Ice, R.L.; Melnikov, V.M.; Hoban, N.P.; Gebauer, J.G. Bragg Scatter Detection by the WSR-88D. Part I: Algorithm Development. *J. Atmos. Ocean. Technol.* **2017**, *34*, 465–478. [CrossRef]
30. Villars, F.; Weisskopf, V.F. The scattering of electromagnetic waves by turbulent atmospheric fluctuations. *Phys. Rev.* **1954**, *94*, 232–240. [CrossRef]
31. Stepanian, P.M.; Horton, K.G.; Melnikov, V.M.; Zrnic, D.S.; Gauthreaux, S.A. Dual-polarization radar products for biological applications. *Ecosphere* **2016**, *7*, 27. [CrossRef]
32. Park, H.S.; Ryzhkov, A.V.; Zrnić, D.S.; Kim, K.-E. The Hydrometeor Classification Algorithm for the Polarimetric WSR-88D: Description and Application to an MCS. *Weather Forecast.* **2009**, *24*, 730–748. [CrossRef]
33. Kilambi, A.; Fabry, F.; Meunier, V. A Simple and Effective Method for Separating Meteorological from Nonmeteorological Targets Using Dual-Polarization Data. *J. Atmos. Ocean. Technol.* **2018**, *35*, 1415–1424. [CrossRef]
34. Koistinen, J. Bird migration patterns on weather radars. *Phys. Chem. Earth Pt B-Hydrol. Ocean. Atmos.* **2000**, *25*, 1185–1193. [CrossRef]
35. Hu, C.; Fang, L.; Wang, R.; Zhou, C.; Li, W.; Zhang, F.; Lang, T.; Long, T. Analysis of Insect RCS Characteristics. *J. Electron. Inf. Technol.* **2020**, *42*, 140–153.
36. Wang, C.; Wu, C.; Liu, L.; Liu, X.; Chen, C. Integrated Correction Algorithm for X Band Dual-Polarization Radar Reflectivity Based on CINRAD/SA Radar. *Atmosphere* **2020**, *11*, 119. [CrossRef]
37. Chen, Y.; Zou, Q.; Han, J.; Cluckie, I. Cinrad data quality control and precipitation estimation. *Proc. Inst. Civ. Eng.—Water Manag.* **2009**, *162*, 95–105. [CrossRef]
38. Vignal, B.; Andrieu, H.; Creutin, J.D. Identification of Vertical Profiles of Reflectivity from Volume Scan Radar Data. *J. Appl. Meteorol.* **1999**, *38*, 1214–1228. [CrossRef]
39. Joss, J.; Lee, R. The Application of Radar Gauge Comparisons to Operational Precipitation Profile Corrections. *J. Appl. Meteorol.* **1995**, *34*, 2612–2630. [CrossRef]
40. Joss, J.; Waldvogel, A.; Collier, C.G. Precipitation Measurement and Hydrology. In *Radar in Meteorology: Battan Memorial and 40th Anniversary Radar Meteorology Conference*; Atlas, D., Ed.; American Meteorological Society: Boston, MA, USA, 1990; pp. 577–606.
41. Cuihong, W.; Yufa, W.; Tao, W.; Hongxiang, J. Vertical Profile of Radar Echo and Its Deteermination Methods. *J. Appl. Meteorol. Sci.* **2006**, *17*, 232–239.
42. Melnikov, V.M.; Istok, M.J.; Westbrook, J.K. Asymmetric Radar Echo Patterns from Insects. *J. Atmos. Ocean. Technol.* **2015**, *32*, 659–674. [CrossRef]
43. Farisenkov, S.E.; Kolomenskiy, D.; Petrov, P.N.; Engels, T.; Lapina, N.A.; Lehmann, F.O.; Onishi, R.; Liu, H.; Polilov, A.A. Novel flight style and light wings boost flight performance of tiny beetles. *Nature* **2022**, *602*, 96–100. [CrossRef]
44. Xingfu, J. The Physiological and Genetic Characteristics of Migratory Behavior and Genetic Diversity, as Determined by AFLP in the Oriental Armyworm, Mythimna Separata (Walker). Ph.D. Thesis, Chinese Academy of Agricultural Sciences, Beijing, China, 2004.
45. Holleman, I.; van Gasteren, H.; Bouten, W. Quality Assessment of Weather Radar Wind Profiles during Bird Migration. *J. Atmos. Ocean. Technol.* **2008**, *25*, 2188–2198. [CrossRef]
46. Dokter, A.M.; Liechti, F.; Stark, H.; Delobbe, L.; Tabary, P.; Holleman, I. Bird migration flight altitudes studied by a network of operational weather radars. *J. R. Soc. Interface* **2011**, *8*, 30–43. [CrossRef] [PubMed]
47. Pei, L.; Qiu, C. The assessment of velocity azimuth display technique of doppler weather radar. *J. Trop. Meteorol.* **2013**, *29*, 597–606.
48. Benedict, L.H.; Gould, R.D. Towards better uncertainty estimates for turbulence statistics. *Exp. Fluids* **1996**, *22*, 129–136. [CrossRef]

49. Moraghan, A.; Kim, J.; Yoon, S.-J. Density distributions of outflow-driven turbulence. *Mon. Not. R. Astron. Soc. Lett.* **2013**, *432*, L80–L84. [CrossRef]
50. Cael, B.B.; Mashayek, A. Log-Skew-Normality of Ocean Turbulence. *Phys. Rev. Lett.* **2021**, *126*, 224502. [CrossRef]
51. Zhao, Y.; Zhao, X.; Wu, L.; Mu, T.; Yu, F.; Kearsley, L.; Liang, X.; Fu, J.; Hou, X.; Peng, P.; et al. A 30,000-km journey by Apus apus pekinensis tracks arid lands between northern China and south-western Africa. *Mov. Ecol.* **2022**, *10*, 29. [CrossRef]
52. Huang, X.; Zhao, Y.; Liu, Y. Using light-level geolocations to monitor incubation behaviour of a cavity-nesting bird Apus apus pekinensis. *Avian Res.* **2021**, *12*, 9. [CrossRef]
53. Li, L.; Wu, Z.S.; Lin, L.K.; Zhang, R.; Zhao, Z.W. Study on the Prediction of Troposcatter Transmission Loss. *IEEE Trans. Antennas Propag.* **2016**, *64*, 1071–1079. [CrossRef]
54. Zhang, M.G. *Tropospheric Scatter Propagation*; Publishing House of Electronics Industry: Beijing, China, 2004; Volume 10.
55. Bullington, K. Reflections from an exponential atmosphere. *Bell Syst. Tech. J.* **1963**, *42*, 2849–2867. [CrossRef]
56. Zoumakis, N.M. On the relationship between the gradient and the bulk Richardson number for the atmospheric surface layer. *Il Nuovo Cim. C* **1992**, *15*, 111–114. [CrossRef]
57. Ren, G.; Liu, J.; Wan, J.; Li, F.; Guo, Y.; Yu, D. The analysis of turbulence intensity based on wind speed data in onshore wind farms. *Renew. Energy* **2018**, *123*, 756–766. [CrossRef]
58. Day, J.P.; Trolese, L.G. Propagation of Short Radio Waves over Desert Terrain. *Proc. IRE* **1950**, *38*, 165–175. [CrossRef]
59. Katzin, M.; Bauchman, R.W.; Binnian, W. 3- and 9-Centimeter Propagation in Low Ocean Ducts. *Proc. IRE* **1947**, *35*, 891–905. [CrossRef]
60. Melnikov, V.; Zrnić, D.S. Observations of Convective Thermals with Weather Radar. *J. Atmos. Ocean. Technol.* **2017**, *34*, 1585–1590. [CrossRef]
61. Melnikov, V.M.; Doviak, R.J.; Zrnić, D.S.; Stensrud, D.J. Structures of Bragg Scatter Observed with the Polarimetric WSR-88D. *J. Atmos. Ocean. Technol.* **2013**, *30*, 1253–1258. [CrossRef]
62. Richardson, L.M.; Zittel, W.D.; Lee, R.R.; Melnikov, V.M.; Ice, R.L.; Cunningham, J.G. Bragg Scatter Detection by the WSR-88D. Part II: Assessment of Z(DR) Bias Estimation. *J. Atmos. Ocean. Technol.* **2017**, *34*, 479–493. [CrossRef]
63. Banghoff, J.R.; Stensrud, D.J.; Kumjian, M.R. Convective Boundary Layer Depth Estimation from S-Band Dual-Polarization Radar. *J. Atmos. Ocean. Technol.* **2018**, *35*, 1723–1733. [CrossRef]
64. Stull, R.B. *An Introduction to Boundary Layer Meteorology*; Kluwer Academic: Dordrecht, The Netherlands, 1988.
65. Hufford, G.L.; Kelley, H.L.; Sparkman, W.; Moore, R.K. Use of Real-Time Multisatellite and Radar Data to Support Forest Fire Management. *Weather Forecast.* **1998**, *13*, 592–605. [CrossRef]
66. Melnikov, V.M.; Zrnic, D.S.; Rabin, R.M.; Zhang, P. Radar polarimetric signatures of fire plumes in Oklahoma. *Geophys. Res. Lett.* **2008**, *35*, L14815. [CrossRef]
67. Zhang, G.; Doviak, R.; Palmer, R. Bistatic interferometry to measure clear air wind. In Proceedings of the 32nd Conference on Radar Meteorology, Albuquerque, NM, USA, 24–29 October 2005.

remote sensing

MDPI

Technical Note

Radar Echo Reconstruction in Oceanic Area via Deep Learning of Satellite Data

Xiaoqi Yu [1,2], Xiao Lou [1,2], Yan Yan [3], Zhongwei Yan [1,2], Wencong Cheng [4], Zhibin Wang [5], Deming Zhao [1,2] and Jiangjiang Xia [1,2,*]

1 Key Laboratory of Regional Climate-Environment for Temperate East Asia & Center for Artificial Intelligence in Atmospheric Science, Institute of Atmospheric Physics, Chinese Academy of Sciences, Beijing 100029, China; yuxq@tea.ac.cn (X.Y.); louxiao@tea.ac.cn (X.L.); yzw@tea.ac.cn (Z.Y.); zhaodm@tea.ac.cn (D.Z.)
2 College of Earth and Planetary Sciences, University of Chinese Academy of Sciences, Beijing 100049, China
3 93110 Troops, People's Liberation Army of China, Beijing 100843, China; yanyanflykite@163.com
4 Beijing Aviation Meteorological Institute, Beijing 100085, China; emailtocheng@163.com
5 DAMO Academy, Alibaba Group, Beijing 100102, China; zbwang16@126.com
* Correspondence: xiajj@tea.ac.cn

Abstract: A conventional way to monitor severe convective weather is using the composite reflectivity of radar as an indicator. For oceanic areas without radar deployment, reconstruction from satellite data is useful. However, those reconstruction models built on a land dataset are not directly applicable to the ocean due to different underlying surfaces. In this study, we built reconstruction models based on U-Net (named STR-UNet) for different underlying surfaces (land, coast, offshore, and sea), and evaluated their applicability to the ocean. Our results suggest that the comprehensive use of land, coast, and offshore datasets should be more suitable for reconstruction in the ocean than using the sea dataset. The comprehensive performances (in terms of RMSE, MAE, POD, CSI, FAR, and BIAS) of the Land-Model, Coast-Model, and Offshore-Model in the ocean are superior to those of the Sea-Model, e.g., with RMSE being 5.61, 6.08, 5.06, and 7.73 in the oceanic area (Region B), respectively. We then analyzed the importance of different types of features on different underlying surfaces for reconstruction by using interpretability methods combined with physical meaning. Overall, satellite cloud-related features are most important, followed by satellite water-related features and satellite temperature-related features. For the transition of the model from land to coast, then offshore, the importance of satellite water-related features gradually increases, while the importance of satellite cloud-related features and satellite temperature-related features gradually decreases. It is worth mentioning that in the offshore region, the importance of satellite water-related features slightly exceeds the importance of satellite cloud-related features. Finally, based on the performance of the case, the results show that the STR-UNet reconstruction models we established can accurately reconstruct the shape, location, intensity, and range of the convective center, achieving the goal of detecting severe convective weather where a radar is not present.

Keywords: radar; U-Net; Himawari-8; CREF; DeepLIFT

Citation: Yu, X.; Lou, X.; Yan, Y.; Yan, Z.; Cheng, W.; Wang, Z.; Zhao, D.; Xia, J. Radar Echo Reconstruction in Oceanic Area via Deep Learning of Satellite Data. *Remote Sens.* **2023**, *15*, 3065. https://doi.org/10.3390/rs15123065

Academic Editor: Silas Michaelides

Received: 7 April 2023
Revised: 15 May 2023
Accepted: 26 May 2023
Published: 12 June 2023

1. Introduction

Severe convective weather refers to convective weather accompanied by thunderstorm, gale, hail, tornado, local heavy precipitation, and other severe weather phenomena. It is a typical small to medium scale disastrous weather event that seriously threatens the safety of aviation, ship navigation, and occurs frequently in the sea [1–3]. The accurate monitoring and forecasting of severe convective weather are difficult and significant [4]. At present, one of the main means to monitor severe convective weather is achieved by monitoring radar echoes. Radar composite reflectivity >35 dBZ is generally considered as an indicator of the occurrence of severe convective weather [5]. However, in some regions, such as oceans, radar cannot be deployed.

It has been shown that radar echoes (e.g., composite reflectivity, vertically integrated liquid), precipitation, and other data can be inverted to monitor severe convective weather based on satellite data with wide coverage [6]. For example, some scholars proposed the Geostationary Operational Environmental Satellites (GOES) Precipitation Index (GPI) by using the physical properties of cold and warm clouds to establish the relationship between cloud top infrared temperature and rainfall probability and intensity [7–9]. Then, in order to improve the inversion accuracy, the results obtained by the GPI method are accumulated over a longer time scale [10,11]. Further, researchers introduced more characteristic variables, such as relative humidity and precipitable water, and developed the Hydro-Estimator algorithm [12]. On this basis, some scholars used exponential functions and quadratic curves to estimate the rainfall intensity, and improved the satellite inversion precipitation algorithm using humidity correction factors and cloud growth rate correction factors [9,13]. Traditional satellite inversion methods are usually based on the understanding of physical processes, and rely on parametric relationships between cloud properties and rainfall and convective processes [14].

With the development of artificial intelligence science and technology, machine learning algorithms have been gradually introduced into the field of atmospheric science in the context of meteorological big data. Machine learning has nonlinear mapping capability and is good at finding patterns in input and output signals, which can better solve nonlinear problems compared with traditional statistical regression methods [15]. Several studies have shown that models based on deep learning network structures outperform traditional methods in experiments to invert severe convective weather [16]. For example, some scholars have conducted preliminary research on precipitation reconstruction based on artificial neural networks (ANN), and the results show that the performance of ANN satellite inversion algorithms is superior to traditional linear methods [17,18]. Later, with the emergence of convolutional neural networks (CNN) [19], more and more scholars have used CNN to invert precipitation and vertically integrated liquid [20,21], demonstrating the effectiveness of CNN in fusing spatial data under different underlying surfaces, and combining data with the physical multichannel inputs in infrared spectroscopy precipitation estimations. On this basis, for the improvement and development of CNN, U-Net is widely used in the field of image segmentation [22]. The U-Net-based reconstruction algorithm is also used to reconstruct radar reflectivity fields to improve short-term convective-scale forecasts of high impact weather hazards and to identify the location, shape, and intensity of convective systems [23–27].

However, most studies that use satellite information to reconstruct data such as radar echoes and precipitation to monitor severe convective weather are based on data from the land area in order to construct the reconstruction model. These studies have defaulted that the models can be applied directly to the oceanic area, whereas there is no assessment of the applicability of the reconstruction model to the ocean. Due to the existence of underlying surface differences, differences in climate situations, lightning, and storm characteristics can occur. This indicates that it is not rigorous to apply the satellite reconstruction model constructed by datasets in non-maritime regions to the oceans directly. However, there are also many problems if radar data from ocean are directly used for data reconstruction. Given that the radar base stations are located on land, with the increase in offshore distance, the elevation of radar detection radiation is too high, and the composite reflectivity of the area far from the radar base station only contains a small amount of the basic reflectivity factors of elevation, which is biased from the real data [28]. The accuracy of radar data in oceanic surface is affected. Therefore, it is urgent to find a data reconstruction method suitable for the ocean.

In addition, with the rapid development of deep learning, it is difficult for us to understand the deep learning model and fully trust it. Therefore, the interpretability of the model has also been highly valued by scholars. In 2004, the academic community proposed the concept of interpretable artificial intelligence [29]. After that, methods of interpretable research such as Local Interpretable Model-Agnostic Explanations (LIME), Layer-wise Relevance Propagation (LRP), Shapley Additive Explanation (SHAP), saliency map, attention mechanism, and DeepLIFT were proposed [30–36]. In previous reconstruction studies, few studies have focused on the differences in the feature importance of models when underlying surface conditions change. Due to the differences between land and ocean, it is extremely necessary to conduct interpretable research on deep learning models generated on different underlying surfaces.

In this study, we build deep learning models for the reconstruction of composite reflectivity from satellite bright temperature data using U-Net with jump connections under different underlying surfaces (land, coast, offshore, and sea). The accuracy is compared to derive a deep learning reconstruction method that is relatively more applicable to the ocean. Then, the importance of the features on different underlying surfaces is analyzed to obtain an interpretable reconstruction model. The model achieves more accurate and credible monitoring of severe convective weather in remote areas without radar deployment.

2. Materials and Methods

2.1. Materials

This study uses Himawari-8 satellite data as a model input to reconstruct the radar composite reflectivity (CREF, unit dBZ), which is the maximum reflectivity from any of the reflectivity angles of the weather radar. Usually, when the CREF > 35 dBZ, a severe convective weather (SWC) can be considered to occur.

2.1.1. Himawari-8 Satellite Data

The Himawari-8 satellite data can be downloaded from http://www.eorc.jaxa.jp/ptree/index.html, accessed on 1 June 2020, which includes visible bands (central wavelength ranges from 0.47 to 0.64 μm), near-infrared bands (central wavelength ranges from 0.86 to 2.3 μm), and infrared bands (central wavelength ranges from 3.9 to 13.3 μm), with a total of 16 bands and by collecting data on the distribution of clouds, air temperature, wind, precipitation, and aerosols. In order to produce a generalized model that can be used during both daytime and nighttime, only infrared bands were chosen in this study. Band 12 is abandoned because it characterizes O_3 content.

In addition, the brightness temperature differences (BTDs) between bands can also characterize cloud property information and facilitate the capture of severe convective regions [37]. Therefore, according to previous studies [37–39], 17 bands in total are chosen or calculated as the model input, including 9 single infrared bands, and 8 BTD bands, as shown in Table 1.

Table 1. 17 satellite bands selected or calculated in this study, along with physical meaning of each band. '-' indicates minus. E.g., tbb08-tbb10 (Band 08 minus Band10) indicates the BTDs between band 08 and band 10.

Band	Central Wavelength	Physical Meaning	Type
Band 07	3.9	Shortwave infrared window, low clouds, fog	Cloud
Band 08	6.2	Mid and high level water vapor	Water
Band 09	6.9	Middle level water vapor	Water
Band 10	7.3	Middle and low level water vapor	Water
Band 11	8.6	Water vapor, Cloud phase state	Water, Cloud
Band 13	10.4	Cloud imaging	Cloud
Band 14	11.2	Surface temperature	Temperature
Band 15	12.4	Surface temperature	Temperature
Band 16	13.3	Temperature, Cloud top height	Temperature, Cloud
Band 08-14	6.2–11.2	Temperature, Cloud top height	Temperature, Cloud
Band 10-15	7.3−12.4	Temperature, Cloud top height	Temperature, Cloud
Band 08-10	6.2–7.3	Water vapor detection above cloud top	Water

Table 1. *Cont.*

Band	Central Wavelength	Physical Meaning	Type
Band 08-13	6.2−10.4	Water vapor detection above cloud top	Water
Band 11-14	8.6−11.2	Cloud phase state	Cloud
Band 14-15	11.2−12.4	Cloud phase state	Cloud
Band 13-15	10.4–12.4	Detection of ice cloud	Cloud
Band 13-16	10.4–13.3	Detection of ice cloud	Cloud

The temporal resolution is 10 min and the spatial resolution is 2 km. The latitude and longitude ranges of 20°N–40°N, and 110°E–130°E are used in this study.

2.1.2. Composite Reflectivity (CREF)

The output variable used in the reconstruction model in this study is the composite reflectivity (CREF), which is obtained from the China Meteorological Administration. The CREF data have a 10 min time interval before June 2016, 6 min time interval in July 2016 and beyond, and the spatial resolution is 1 km. The latitude and longitude ranges of the study area are consistent with the selected range of Himawari-8 satellite data, specifically 20°N–40°N, and 110°E–130°E.

The data (both the satellite data and the CREF) from May to October for the period 2016–2018 are used in this study.

2.1.3. GPM Precipitation Data

The Global Precipitation Measurement (GPM) is the next generation of the Global Satellite Precipitation Measurement Program carried out in collaboration with NASA and JAXA. The precipitation and radar CREF have a certain correlation, although the data of GPM are difficult to fully quantify in regard to the effectiveness of CREF reconstruction, the data can qualitatively verify the effectiveness of the models in areas without radar coverage, which can be used as supplementary information to indicate the area of severe radar echoes [26].

2.1.4. Data Preprocessing

Spatial and Temporal Matching

The Himawari-8 satellite data are matched with the spatial and temporal resolution of CREF data as features and labels of the model, respectively. At the temporal level, satellite data and radar data that do not match are discarded to ensure that they are consistent in time (the time difference is less than 5 min). Spatially, the CREF data are sampled onto a network with a spatial resolution of 2 km, maintaining the same spatial resolution as the Himawari-8 satellite data.

Normalization

Data standardization processing can increase the learning ability of the model, improve the speed of the convergence, and avoid the difficulty of the model's training due to the non-uniformity of magnitudes.

In this study, both the satellite data and the CREF data are normalized by z-score normalization. The formula is as follows:

$$x^* = \frac{x - \mu}{\sigma} \quad (1)$$

where μ and σ are the mean and variance of the original data, respectively. x denotes the original data, and x^* denotes the result after z-score normalization.

2.2. Method

2.2.1. Satellite to Radar U-Net

U-Net has been shown to demonstrate good performance in the reconstruction of the radar data in previous studies [23–27]. We use the U-Net architecture to construct a CREF reconstruction model, namely satellite-to-radar U-Net (STR-UNet, Figure 1).

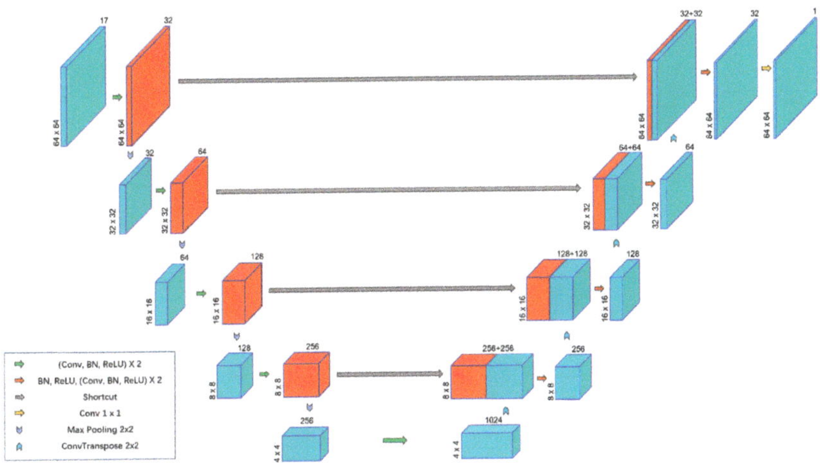

Figure 1. The STR-UNet architecture.

Overall, the STRU-Net designed in this study is of an encoder–decoder structure [27]. The left side of the network is often referred to as the contracting path and the right side is referred to as the expanding path. The shortcut in the middle is called the jump connection layer, which is also known as the feature splicing layer.

The left half of the model (contracting path) is used for feature extraction, which is repeatedly composed of convolution blocks and 2 × 2 pooling layers, and each convolution block contains the 3 × 3 convolution layer, batch normalization layer, and ReLU activation function. The input size of the model is a 64 × 64 × 17 satellite image, where 64 × 64 represents the length and width of the satellite image after padding, and 17 represents the number of input channels (i.e., bands). Then, after each convolution block and pooling, the number of feature maps is doubled and the length and width are halved, respectively.

The right half of the model (expanding path) performs the up-sampling operation, which is composed of several transposed convolution layers, feature splicing layers, and convolution blocks repeatedly, and the convolution block also encapsulates the batch normalization layer, 3 × 3 convolution layer, and ReLU activation function. In the expanding path, first we perform transposed convolution on the feature map obtained on the contracting path; next, the obtained feature map is spliced on the channel with the feature map at the corresponding position on the contracting path; then, the convolution operation is performed on the feature map after splicing, and so on and so forth. After each transposed convolution and convolution block, the number of feature maps is halved, and the length and width are doubled. In the last layer of the model, a 1 × 1 convolution layer maps the tensor of 32 channels to 1 channel, which in turn yields a target image of size 64 × 64 × 1. For this study, the reconstruction of CREF data is completed.

STRU-Net combines the low-resolution information in down-sampling process and high-resolution information in up-sampling process, and applies long-range jump connection combined with the feature details from the shallow convolution layer at the bottom of the satellite images, which can effectively compensate for the lack of spatial information of satellite images during the down-sampling process, and help the network to achieve more accurate localization. It is very important for reconstructing accurate radar data and boundary information.

2.2.2. Research Scheme of the CREF Reconstruction

This paper aims to construct a satellite reconstruction model with satellite data that is suitable for monitoring severe convective weather in the ocean without deploying a radar. In order to achieve this objective, we designed the following research scheme, as shown in Figure 2.

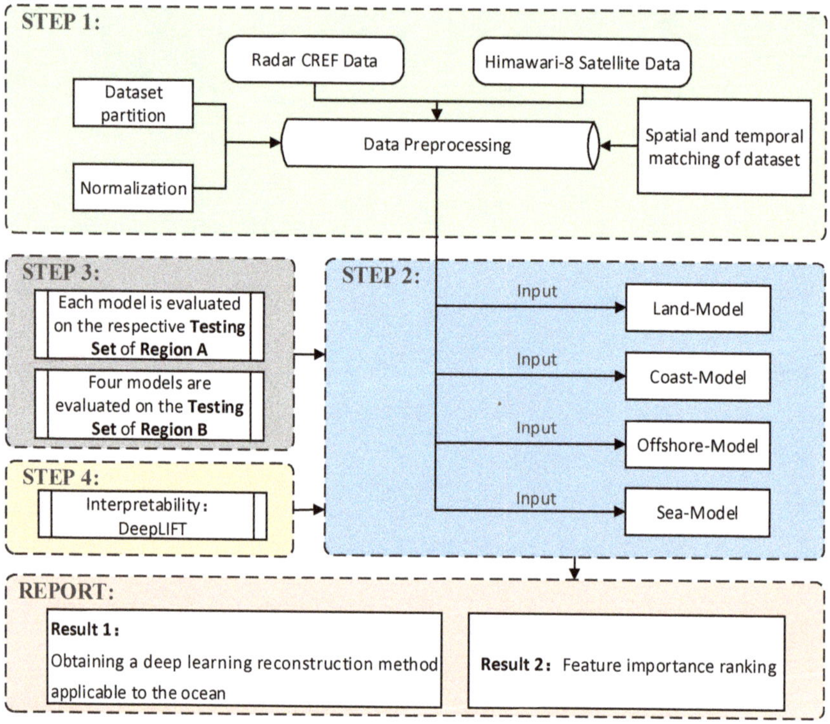

Figure 2. Research scheme of the CERF reconstruction.

Step 1. Preprocess the dataset, as described in Section 2.1.4.

Step 2. Build four STR-UNet models with different underlying surfaces. As shown in Figure 3 (Left), Region A includes four different underlying surfaces: land, coast, offshore, and sea. Four STR-UNet models, namely Land-Model, Coast-Model, Offshore-Model, and Sea-Model, are constructed.

Step 3. Train and test the STR-UNet models. The first 24 days of each month (May to October) in 2016 and 2017 are used as the training set, and the remaining days of these months are used as the validation set. The data in 2018 are used as the test set. It is worth noting that each model is trained and tested on its own underlying surface data. For example, the Coast-Model is trained and tested by using the data from the coastal area.

Then, the performances of the four STR-UNet models on the oceanic areas are assessed. The orange box regions, as shown in Figure 3 (Right), are defined as oceanic areas that are not overlapping with the "offshore" areas shown in Figure 3 (Left). It is difficult for us to obtain radar CREF from the ocean, and at this time, the offshore radar CREF has relatively high accuracy and its data comes from the ocean, which means it has data features of oceanic underlying surface. Based on this, we assume that Region B can represent the "ocean" underlying surface. The performance of the four models will be evaluated based on the test set (2018) in this area.

Step 4. Perform interpretability study of STR-UNet. See Section 2.2.4 for more details.

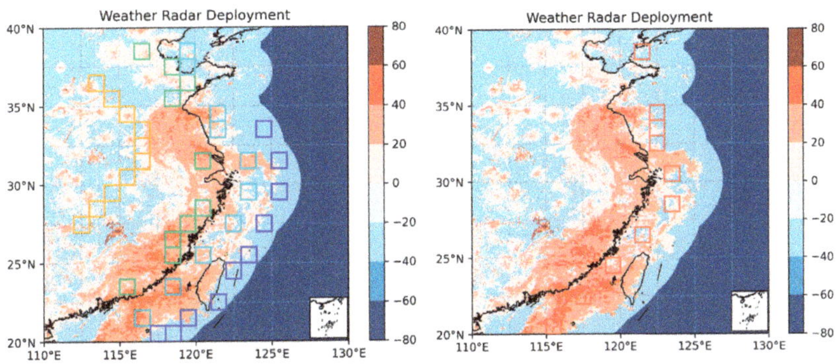

Figure 3. Schematic diagram of Region A (**left**) and Region B (**right**). Region A includes four different underlying surfaces: land (yellow box), coast (green box), offshore (cyan box), and sea (blue box). Region B only includes one underlying surface: offshore, and it does not coincide with the four underlying surfaces of Region A.

2.2.3. Evaluation Metrics

In this study, root mean square error (RMSE) and mean absolute error (MAE) are used to quantitatively verify the performance of the four STR-UNet models built in this paper. RMSE and MAE can measure the deviation between the reconstructed CREF and the radar CREF. The equations are as follows:

$$\text{RMSE} = \frac{1}{n}\sqrt{\sum_{i=1}^{n}\left(y_i' - y_i\right)^2} \tag{2}$$

$$\text{MAE} = \frac{1}{n}\sum_{i=1}^{n}\left|y_i' - y_i\right| \tag{3}$$

where n represents the number of samples, y_i' represents the reconstructed CREF value, and y_i represents the radar CREF value.

The classification metrics used in this study include probability of detection (POD), false alarm ratio (FAR), and critical success index (CSI), BIAS, as shown in Table 2. The model's ability to reconstruct for CREF above 35 dBZ, a critical issue for many industries, including aviation and ship navigation, was evaluated using the classification criteria.

$$\text{POD} = \frac{Hits}{Hits + Misses} \tag{4}$$

$$\text{CSI} = \frac{Hits}{Hits + Misses + False\ alarms} \tag{5}$$

$$\text{CSI} = \frac{Hits}{Hits + Misses + False\ alarms} \tag{6}$$

$$\text{BIAS} = \frac{Hits + False\ alarms}{Hits + Misses} \tag{7}$$

Table 2. Contingency table of the classification score parameters.

	Reconstructed CREF (<35 dBZ)	Reconstructed CREF (≥35 dBZ)
True CREF (<35 dBZ)	Correct negatives	False alarms
True CREF (≥35 dBZ)	Misses	Hits

2.2.4. Interpretability

In recent years, with the rapid development of deep learning, the interpretability of models has received more and more attention from scholars at home and abroad. Most deep learning models are "black box" models [40]. Their "black box" nature makes it difficult for scholars to understand the decision logic of the models in many cases, and thus they cannot fully trust the deep learning models. In order to improve the interpretability and transparency of deep learning models, this study investigates the interpretability of models.

For this study, a total of 17 features of satellite data were used as the input, after obtaining a model in different underlying surfaces, respectively, besides being interested in the effect of the model, it was essential to determine which features played an important role in the reconstruction.

In this paper, the DeepLIFT algorithm is used to conduct interpretability research on the above models. The DeepLIFT [36] method allocates the prediction results of the neural network to each dimension of the input. Its working principle is to compare the activation of each neuron with its "reference" activation, and back propagate the importance signal in order to assign a contribution score based on the difference. In essence, this is a method of tracing the internal feature selection of the algorithm, which uses the input differences of some "reference" inputs to explain the output differences of some "reference" outputs.

In this study, for each band feature, first, we conducted the normalization on Region B, and the vector with the "reference" value of all zeros was set to calculate the attribution of each feature, which is the contribution of each input feature to the results. Finally, the absolute value of attribution was taken, and the ratio of the absolute value of each feature attribution to the sum of the absolute values of all feature attributions was expressed as the importance of the feature.

Based on this method, several more important features on different underlying surfaces can be selected, and whether surface information affects the band selection of the model can be analyzed. Finally, we explored the relationship between the importance of bands and underlying surfaces, as well as the reasons why different underlying surfaces differed in importance.

3. Results

3.1. Performances of the Four STR-UNet Models

The performances of the four STR-UNet models are shown in Table 3.

First, in the process of advancing from land to coast, to offshore, and then to ocean, it can be found that the RMSE and MAE are getting smaller on the test set in Region A. However, it does not indicate that the model's performance is getting better. The main reason is that the CREF refers to the ratio of the radar waves reflected from clouds of different heights within a certain range received by the meteorological radar. With the increasing distance from the coastline to the ocean, the radiation elevation of the radar is also increasing. It means that at a distance from the radar, the CREF is only calculated based on a small number of basic reflectivity factors of higher elevations. As a result, the proportion of CREF larger than 35 dBZ decreases significantly as one moves from land to coast, offshore, and finally to the ocean, as depicted in Figure 4. Especially for the sea areas in Region A, the proportion of CREF larger than 35 dBZ is only a few tenths of that of the other three areas.

Table 3. Performances of the four STR-UNet models.

Model	Metric	Region A Test (on Each of the Four Underlying Surfaces, Respectively)	Region B Test (Ocean)
Land-Model	RMSE	7.4392	5.6120
	MAE	3.2353	1.8542
	POD (35 dBZ)	0.1478	0.1815
	CSI (35 dBZ)	0.1274	0.1484
	FAR (35 dBZ)	0.5195	0.5509
	BIAS (35 dBZ)	0.3076	0.4042
Coast-Model	RMSE	7.1517	6.0755
	MAE	3.0315	2.1929
	POD (35 dBZ)	0.2663	0.2954
	CSI (35 dBZ)	0.2177	0.1958
	FAR (35 dBZ)	0.4560	0.6327
	BIAS (35 dBZ)	0.4895	0.8042
Offshore-Model	RMSE	5.0824	5.0591
	MAE	1.4646	1.4444
	POD (35 dBZ)	0.2107	0.2144
	CSI (35 dBZ)	0.1755	0.1703
	FAR (35 dBZ)	0.4879	0.5469
	BIAS (35 dBZ)	0.4115	0.4732
Sea-Model	RMSE	4.1744	7.7300
	MAE	0.7019	2.1525
	POD (35 dBZ)	0.0000	0.0000
	CSI (35 dBZ)	0.0000	0.0000
	FAR (35 dBZ)	0.0000	0.0000
	BIAS (35 dBZ)	0.0000	0.0000

Then, we observe the performance of the four models on the test set in Region B. For all the metrics, it can be seen that when the models are evaluated in Region B, the performances of the Land-Model, Coast-Model, and Offshore-Model are significantly better than those of Sea-Model. Four metrics (POD, CSI, BIAS and FAR) of the Sea-Model are 0, and the RMSE and MAE are the largest among all these models. It indicates that the Sea-Model does not have the ability to reconstruct the CREF.

Since the Offshore-Model's data selection resembles that of Region B (without overlap), it can serve as a proxy for the highest level of precision that the reconstruction model is capable of. Compared to the Offshore-Model, the Land-Model's performances are a little worse on the test set in Region B for all the evaluation metrics.

It is a little complicated to evaluate the performance of the Coast-Model compared to the Offshore-Model. The RMSE, MAE, and FAR of the Coast-Model are a little larger (worse) than those of Offshore-Model. However, the Coast-Model has better POD, CSI, and BIAS than the Offshore-Model. This is due to the fact that, as shown in Figure 4, the coastal area has the highest fraction of CREF larger than 35 dBZ compared to the other three areas. It denotes a complicated meteorological situation affected by the complex underlying surface of the coast [41,42]. Thus, compared to the Offshore-Model, the prediction of Coast-Model is bolder.

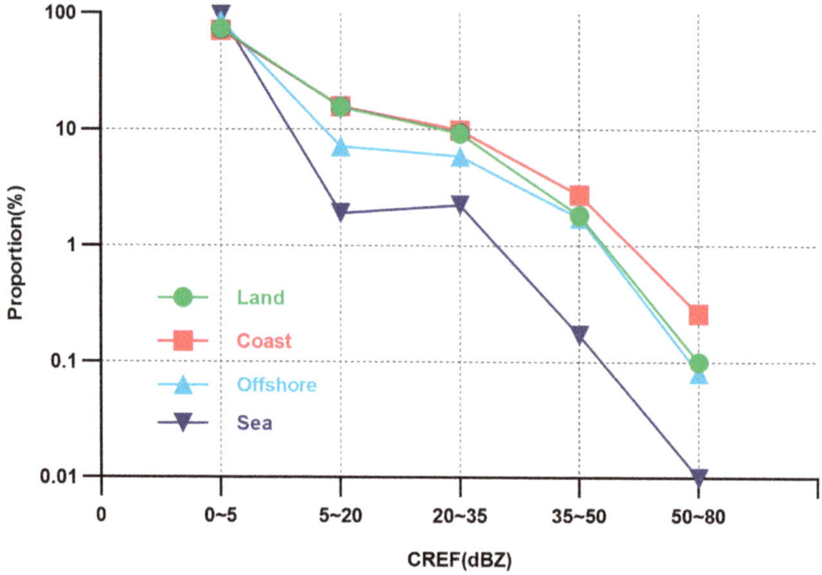

Figure 4. Probability statistics of CREF data in different regions. The horizontal axis represents the range of CREF values, while the vertical axis represents the corresponding proportion. The different colors on the figure correspond to different underlying surfaces.

In conclusion, a mix of land-dataset, coast-dataset, and offshore-dataset can be taken into consideration when employing satellite data to reconstruct radar data. The Offshore-Model can give the medium results, while the Coast-Model can give bolder results. Due to the abundance of data on land area, the Land-Model can provide a good baseline for reconstructing the radar data, despite being slightly inferior to the Offshore-Model on the test set in Region B.

3.2. Case Study

In order to demonstrate the actual combat effect of the models, we selected several severe convective weather cases from the test set (UTC) to visually show the reconstruction effect of the models.

Typhoon is one of the important disastrous weather systems that affects the safety of people's lives and property. It often brings rainstorm, strong wind, and secondary disasters [43]. Typhoon "Yagi" was generated on 7 August 2018 (Beijing time, the same below) with the intensity of a tropical depression. The intensity of "Yagi" increased to a tropical storm on 8 August, moving towards the north by east, turning to the northwest at night on 9 August, and entering the eastern region of the East China Sea at night on 11 August [44]. During the influence of "Yagi", extreme precipitation and a large-scale rainstorm had been brought to the cities along the way, resulting in heavy economic losses. Figure 5 shows the GPM precipitation and radar echo distribution of severe convective events that occurred in the research area in the test set. It can be seen that the models can more accurately reconstruct the shape, location, intensity, and range of the convective center, whether it is a slightly lower intensity convective event or a severe convective event such as a typhoon. In addition, for areas beyond the radar coverage, the radar echoes can be also reconstructed, and the distribution of reconstructed CREF is quite consistent with the pattern of the GPM precipitation.

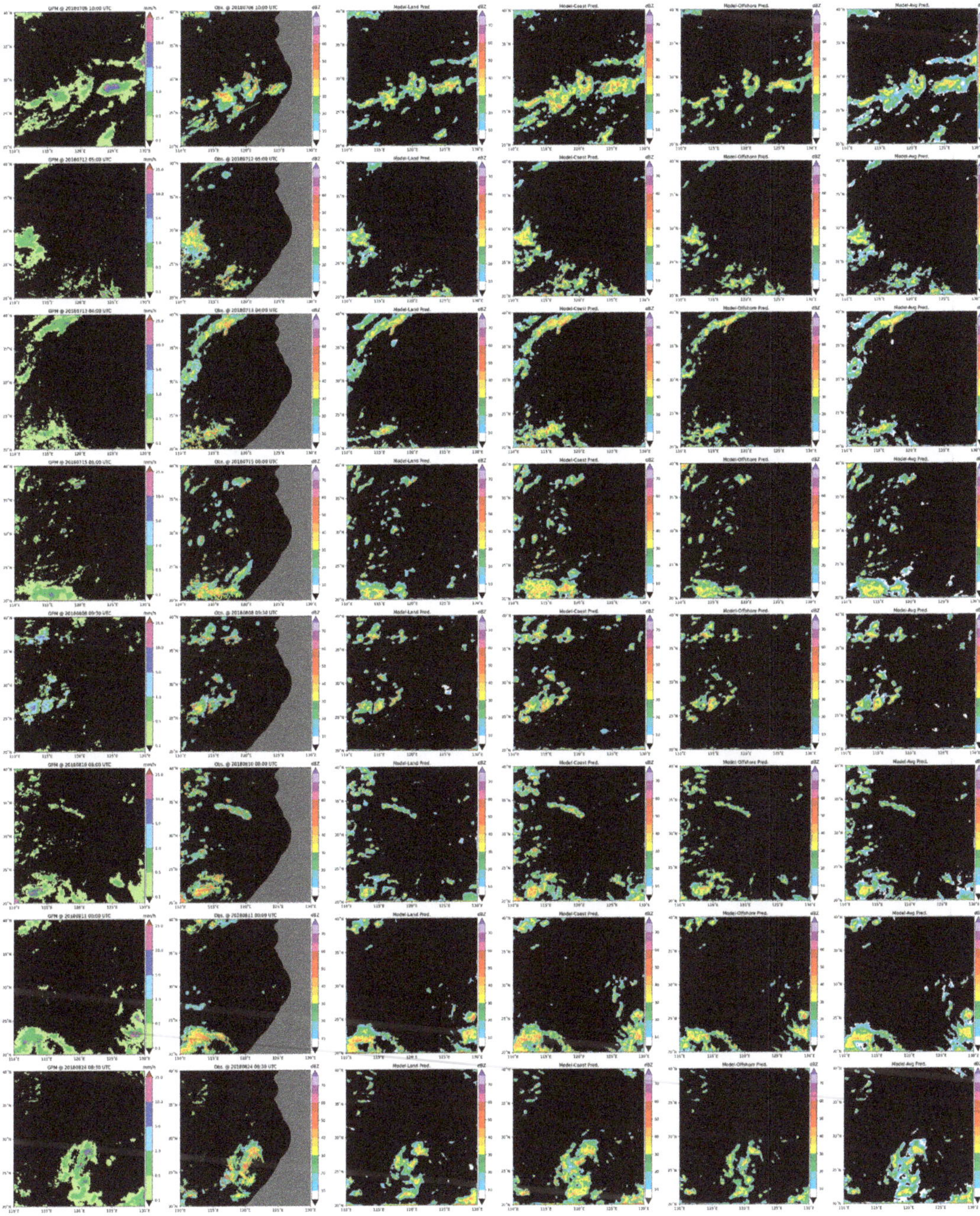

Figure 5. Radar echo map: comparative study of observation and reconstruction. The first column shows GPM precipitation distribution at different times; the second column shows the radar echo observed; the gray areas on the map represent areas outside the radar deployment range. The third, fourth, and fifth columns show the reconstructions of the Land-Model, Coast-Model, and Offshore-Model, respectively, and the last column represents the average of the three reconstructed models mentioned above.

3.3. Results of Interpretability

For the models selected above: Land-Model, Coast-Model and Offshore-Model, the DeepLIFT method is adopted to analyze the differences in feature importance under different underlying surfaces (land, coast, and offshore). The results are shown in Figure 6.

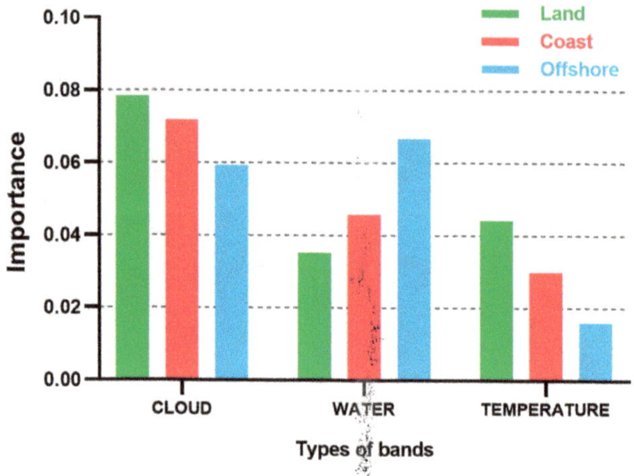

Figure 6. The importance of each type of band (cloud, water, and temperature) under different underlying surfaces (land, coast, offshore).

According to the previous description, and their physical meaning (Table 1), the 17 input features of satellite bands can be classified as satellite cloud-related features, satellite water-related features, and satellite temperature-related features. It is worth mentioning that for bands with more than one type of physical meaning, such as Band 11, the center wavelength is 8.6 μm, which can both measure water vapor and cloud phase state. When classifying the input features, Band 11 is classified as both satellite water-related features and satellite cloud-related features. It means that after calculating the importance of Band 11 using the DeepLIFT method, we will calculate the average importance of satellite water-related features along with other bands that measure water vapor. At the same time, the importance of Band 11 will be used along with bands that measure the cloud phase state to calculate the average importance of satellite cloud-related features. Similarly, we calculated the average value of the importance of the bands under each type. For the land's underlying surface, it can be intuitively seen that satellite cloud-related features are more important to the reconstruction, far outweighing the importance of satellite water-related features and satellite temperature-related features.

Overall, satellite cloud-related features are the most important, followed by satellite water-related features, and satellite temperature-related features are the least important. When the underlying surface changes to the coast, then to the offshore, the importance of satellite cloud-related features gradually decreases, but they still play an important role in reconstruction, while the importance of satellite water-related features gradually increases, which is also important for reconstruction. When the underlying surface is located on the ocean, it is clear that satellite water-related features are more important than satellite cloud-related features. The importance of satellite temperature-related features gradually decreases as the model changes to the ocean, compared with the former two, they are relatively unimportant in reconstruction.

In summary, during the transition of the model from the land to the ocean, for all the underlying surface cases, clouds have a great impact on the amount of solar radiation reaching the Earth and play a crucial role in the water cycle of the climate system [45,46]; the cloud phase state can also reflect the temperature and humidity state, and dynamic characteristics of the atmosphere to a certain extent [47]. In addition, water vapor has a strong correlation with severe convective weather; the increase in water vapor content is conducive to the development of convective weather and it can easily cause the rapid growth of convective weather. Therefore, the satellite features characterizing cloud amount, cloud phase state, and water vapor play an important role in reconstruction. Secondly, for the underlying surface of the ocean, because the ocean has the characteristics of high heat capacity and high thermal inertia, it means that it needs more energy to make its temperature change greatly. Therefore, compared with the land's underlying surface, satellite temperature-related features have a lower significance in the reconstruction of severe convective weather.

4. Conclusions

In this study, we, respectively, sampled land, coast, offshore, and sea areas in the eastern area (20°N–40°N, 110°E–130°E), built four deep learning models using U-Net, and compared their accuracy. The results show that a mix of land-dataset, coast-dataset, and offshore-dataset can be taken into consideration when deploying satellite data to reconstruct radar data. This allows for more accurate reconstruction and monitoring of severe convective weather in the ocean without radar deployment.

In addition, in previous studies, there was a lack of research on the interpretability of the models. In this paper, the DeepLIFT method was used to obtain the feature importance ranking and the differences in different underlying surfaces. Overall, satellite cloud-related features are most important, followed by satellite water-related features, and satellite temperature-related features are the least important. The importance of satellite water-related features gradually increases, and the importance of satellite cloud-related features and satellite temperature-related features gradually decreases as the model changes from land to ocean. Then, the reasons for this phenomenon are briefly analyzed in combination with physical meaning. It is beneficial to the research in the oceanic area, which is of great significance for aviation, navigation, and the maintenance of people's lives and property safety.

In addition to the research tasks outlined in this study, future research will be conducted from the following aspects.

Firstly, the data used in this paper are infrared band data, while the data used in previous studies include lightning and other data. In subsequent studies, we will also increase the data types in order to further reduce the error and improve the reconstruction effect. Secondly, we used the DeepLIFT method to preliminarily analyze the differences in feature importance caused by the differences in the underlying surfaces of the model. However, using only one method to study the interpretability means the results lack credibility [48]. In the future, we will use more interpretable methods and optimize them to obtain more convincing interpretable conclusions. We hope the reconstruction method we have proposed will spur new developments in the deep learning and meteorological fields.

Author Contributions: Conceptualization, X.Y., J.X. and D.Z.; methodology, X.L., X.Y., Y.Y., J.X. and Z.W.; software, X.L., W.C., X.Y. and Y.Y.; validation, X.Y. and X.L.; formal analysis, X.Y.; resources, J.X. and Z.Y.; data curation, X.L. and X.Y.; writing—original draft preparation, X.Y. and J.X.; writing—review and editing, Z.Y., Y.Y., W.C., Z.W. and D.Z.; visualization, X.Y. and X.L. All authors have read and agreed to the published version of the manuscript.

Funding: This study was funded by the National Natural Science Foundation of China (Grant No. 42275158).

Data Availability Statement: Not applicable.

Acknowledgments: We would like to express our gratitude to the Japan Meteorological Agency (JMA) for freely providing the Himawari-8 satellite data used in this research.

Conflicts of Interest: The authors declare no conflict of interest.

References

1. Maddox, R.A. Mesoscale convective complexes. *Bull. Am. Meteorol. Soc.* **1980**, *61*, 1374–1387. [CrossRef]
2. Brimelow, J.C.; Hanesiak, J.M.; Burrows, W.R. On the Surface-Convection Feedback during Drought Periods on the Canadian Prairies. *Earth Interact.* **2011**, *15*, 1–26. [CrossRef]
3. Zheng, Y.; Tian, F.; Meng, Z.; Xue, M.; Yao, D.; Bai, L.; Zhou, X.; Mao, X.; Wang, M. Survey and Multi-Scale Characteristics of Wind Samage Caused by Convective Storms in the Surrounding Area of the Capsizing Accident of Cruise Ship "Dongfangzhixing". *Meteorol. Mon.* **2016**, *42*, 1–13. (In Chinese)
4. Zheng, Y.; Zhou, K.; Sheng, J.; Lin, Y.; Tian, F.; Tang, W.; Lan, Y.; Zhu, W. Advances in Techniques of Monitoring, Forecasting and Warning of Severe Convective Weather. *J. Appl. Meteorol. Sci.* **2015**, *26*, 641–657. (In Chinese)
5. Roberts, R.D.; Rutledge, S. Nowcasting storm initiation and growth using GOES-8 and WSR-88D data. *Weather Forecast.* **2003**, *18*, 562–584. [CrossRef]
6. Stampoulis, D.; Anagnostou, E.N. Evaluation of global satellite rainfall products over continental Europe. *J. Hydrometeorol.* **2012**, *13*, 588–603. [CrossRef]
7. Arkin, P.A.; Meisner, B.N. The relationship between large-scale convective rainfall and cold cloud over the Western Hemisphere during 1982–1984. *Mon. Weather Rev.* **1987**, *115*, 51–74. [CrossRef]
8. Arkin, P.A.; Joyce, R.; Janowiak, J.E. The estimation of global monthly mean rainfall using infrared satellite data: The GOES Precipitation Index (GPI). *Remote Sens. Rev.* **1994**, *11*, 107–124. [CrossRef]
9. Liu, Y.; Fu, Q.; Song, P. Satellite retrieval of precipitation: An overview. *Adv. Atmos. Sci.* **2011**, *26*, 1162–1172. (In Chinese)
10. Bastiaanssen, W.G.M.; Pelgrum, H.; Wang, J.; Ma, Y.; Moreno, J.F.; Roerink, G.J.; van der Wal, T. A remote sensing surface energy balance algorithm for land (SEBAL): Part 2: Validation. *J. Hydrol.* **1998**, *212*, 213–229. [CrossRef]
11. Liang, L.; Liu, C.; Xu, Y.Q.; Guo, B.; Shum, H.Y. Real-time texture synthesis by patch-based sampling. *ACM Trans. Graph.* **2001**, *20*, 127–150. [CrossRef]
12. Scofield, R.A.; Kuligowski, R.J. Status and Outlook of Operational Satellite Precipitation Algorithms for Extreme-Precipitation Events. *Weather Forecast.* **2003**, *18*, 1037–1051. [CrossRef]
13. Ba, M.B.; Gruber, A. GOES Multispectral Rainfall Algorithm (GMSRA). *J. Appl. Meteor. Climatol.* **2001**, *40*, 1500–1514. [CrossRef]
14. Zhang, Y.; Wu, K.; Zhang, J.; Zhang, F.; Xiao, H.; Wang, F.; Zhou, J.; Song, Y.; Peng, L. Estimating Rainfall with Multi-Resource Data over East Asia Based on Machine Learning. *Remote Sens.* **2021**, *13*, 3332. [CrossRef]
15. Chang, G.W.; Lu, H.J.; Chang, Y.R.; Lee, Y.D. An improved neural network-based approach for short-term wind speed and power forecast. *Renew. Energ.* **2017**, *105*, 301–311. [CrossRef]
16. Beusch, L.; Foresti, L.; Gabella, M.; Hamann, U. Satellite-Based Rainfall Retrieval: From Generalized Linear Models to Artificial Neural Networks. *Remote Sens.* **2018**, *10*, 939. [CrossRef]
17. Hsu, K.; Gao, X.; Sorooshian, S.; Gupta, H.V. Precipitation Estimation from Remotely Sensed Information Using Artificial Neural Networks. *J. Appl. Meteor. Climatol.* **1997**, *36*, 1176–1190. [CrossRef]
18. Hong, Y.; Hsu, K.L.; Sorooshian, S.; Gao, X. Precipitation estimation from remotely sensed imagery using an artificial neural network cloud classification system. *J. Appl. Meteorol.* **2004**, *43*, 1834–1853. [CrossRef]
19. Krizhevsky, A.; Sutskever, I.; Hinton, G.E. ImageNet classification with deep convolutional neural networks. *Commun. ACM.* **2012**, *60*, 84–90. [CrossRef]
20. Veillette, M.S.; Hassey, E.P.; Mattioli, C.J.; Iskenderian, H.; Lamey, P.M. Creating Synthetic Radar Imagery Using Convolutional Neural Networks. *J. Atmos. Ocean. Technol.* **2018**, *35*, 2323–2338. [CrossRef]
21. Wang, C.; Xu, J.; Tang, G.; Yang, Y.; Hong, Y. Infrared Precipitation Estimation Using Convolutional Neural Network. IEEE Trans. Geosci. *Remote Sens.* **2020**, *58*, 8612–8625. [CrossRef]
22. Ronneberger, O.; Fischer, P.; Brox, T. 2015: U-Net: Convolutional Networks for Biomedical Image Segmentation. In Proceedings of the 18th International Conference on Medical Image Computing and Computer-Assisted Intervention-MICCAI, Munich, Germany, 5–19 November 2015; Springer: Cham, Switzerland, 2015; pp. 234–241.
23. Hilburn, K.A.; Ebert-Uphoff, I.; Miller, S.D. Development and Interpretation of a Neural Network-Based Synthetic Radar Reflectivity Estimator Using GOES-R Satellite Observations. *J. Appl. Meteor. Climatol.* **2020**, *60*, 1–21. [CrossRef]
24. Duan, M.; Xia, J.; Yan, Z.; Han, L.; Zhang, L.; Xia, H.; Yu, S. Reconstruction of the Radar Reflectivity of Convective Storms Based on Deep Learning and Himawari-8 Observations. *Remote Sens.* **2021**, *13*, 3330. [CrossRef]
25. Sun, F.; Li, B.; Min, M.; Qin, D. Deep Learning-Based Radar Composite Reflectivity Factor Estimations from Fengyun-4A Geostationary Satellite Observations. *Remote Sens.* **2021**, *13*, 2229. [CrossRef]
26. Yang, L.; Zhao, Q.; Xue, Y.; Sun, F.; Li, J.; Zhen, X.; Lu, T. Radar Composite Reflectivity Reconstruction Based on FY-4A Using Deep Learning. *Sensors.* **2023**, *23*, 81. [CrossRef]
27. Veillette, M.; Samsi, S.; Mattioli, C. Sevir: A storm event imagery dataset for deep learning applications in radar and satellite meteorology. *Adv. Neural Inf. Process. Syst.* **2020**, *33*, 22009–22019.
28. Zhang, P.; Du, B.; Dai, T. *Radar Meteorology*, 2nd ed.; China Meteorological Press: Beijing, China, 2010.
29. Van Lent, M.; Fisher, W.; Mancuso, M. An explainable artificial intelligence system for small-unit tactical behavior. In Proceedings of the National Conference on Artificial Intelligence, San Jose, CA, USA, 25–29 July 2004; AAAI Press: Menlo Park, CA, USA; MIT Press: Cambridge, MA, USA; London, UK, 1999; pp. 900–907.

30. Ribeiro, M.T.; Singh, S.; Guestrin, C. "Why Should I Trust You?" Explaining the Predictions of Any Classifier. In Proceedings of the 22nd ACM SIGKDD International Conference on Knowledge Discovery and Data Mining, New York, NY, USA, 13–17 August 2016; pp. 1135–1144.
31. Bach, S.; Binder, A.; Montavon, G.; Klauschen, F.; Müller, K.-R.; Samek, W. On Pixel-Wise Explanations for Non-Linear Classifier Decisions by Layer-Wise Relevance Propagation. *PLoS ONE* **2015**, *10*, e0130140. [CrossRef]
32. Lundberg, S.M.; Lee, S.-I. A unified approach to interpreting model predictions. *Adv. Neural Inf. Process. Syst.* **2017**, *30*, 4768–4777, ArXiv:1705.07874.
33. Selvaraju, R.R.; Cogswell, M.; Das, A.; Vedantam, R.; Parikh, D.; Batra, D. Grad-CAM: Visual Explanations from Deep Networks via Gradient-based Localization. *Int. J. Comput. Vis.* **2020**, *128*, 336–359. [CrossRef]
34. Cho, K.; van Merriënboer, B.; Gulcehre, C.; Bahdanau, D.; Bougares, F.; Schwenk, H.; Bengio, Y. Learning phrase representations using RNN encoder-decoder for statistical machine translation. In Proceedings of the 2014 Conference on Empirical Methods in Natural Language Processing (EMNLP), Doha, Qatar, 26–28 October 2014; pp. 1724–1734.
35. Zhou, Y.; Wang, H.; Zhao, J.; Chen, Y.; Yao, R.; Chen, S. Interpretable attention part model for person re-identification. *Acta Autom. Sin.* **2020**, *41*, 116. (In Chinese)
36. Shrikumar, A.; Greenside, P.; Kundaje, A. Learning important features through propagating activation differences. In Proceedings of the 34th International Conference on Machine Learning. PMLR, Sydney, Australia, 6–11 August 2017; pp. 3145–3153.
37. Yasuhiko, S.; Hiroshi, S.; Takahito, I.; Akira, S. Convective Cloud Information derived from Himawari-8 data. In *Meteorological Satellite Center Technical Note*; Meteorological Satellite Center (MSC): Kiyose, Tokyo, 2017; p. 22.
38. Sun, S.; Li, W.; Huang, Y. Retrieval of Precipitation by Using Himawari-8 Infrared Images. *Acta Sci. Nat. Univ. Pekinensis* **2019**, *55*, 215–226. (In Chinese)
39. Sadeghi, M.; Asanjan, A.A.; Faridzad, M.; Nguyen, P.; Hsu, K.; Sorooshian, S.; Braithwaite, D. PERSIANN-CNN: Precipitation Estimation from Remotely Sensed Information Using Artificial Neural Networks–Convolutional Neural Networks. *J. Hydrometeor.* **2019**, *20*, 2273–2289. [CrossRef]
40. Bathaee, Y. The Artificial Intelligence Black Box and the Failure of Intent and Causation. *Harvard J. Law Technol.* **2018**, *31*, 889.
41. Zhai, G.; Ding, H.; Gao, K. A Numerical Experiment of the Meso-scale Influence of Underlying Surface on a Cyclonic Precipitation Process. *J. Hangzhou Univ. (Nat. Sci.)* **1995**, *22*, 185–190. (In Chinese)
42. Tian, C.; Zhou, W.; Miao, J. Review of lmpact of Land Surface Characteristics on Severe Convective Weather in China. *Meteorol. Sci. Technol.* **2012**, *40*, 207–212. (In Chinese)
43. Lyu, X.; Xu, Y.; Dong, L.; Gao, S. Analysis of characteristics and forecast difficulties of TCs over Northwestern Pacific in 2018. *Meteor. Mon.* **2021**, *47*, 359–372. (In Chinese)
44. Sun, S.; Chen, B.; Sun, J.; Sun, Y.; Diao, X.; Wang, Q. Periodic Characteristics and Cause Analysis of Continuous Heavy Rainfall Induced by Typhoon Yagi (1814) in Shandong. *Plateau Meteorol.* **2022**, 1–15. (In Chinese) [CrossRef]
45. Zhang, Q.; Li, Y.; Yang, Y. Research Progress on the Cloudage and Its Relation with Precipitation in China. *Plateau Mt. Meteorol. Res.* **2011**, *31*, 79–83. (In Chinese)
46. Zou, Y.; Wang, Y.; Wang, S. Characteristics of lighting activity during severe convective weather in Dalian area based on satellite data. *J. Meteorol. Environ.* **2021**, *37*, 128–133. (In Chinese)
47. Cao, Z.; Wang, X. Cloud Characteristics and Synoptic Background Associated with Severe Convective Storms. *J. Appl. Meteorol. Sci.* **2013**, *24*, 365–372. (In Chinese)
48. McGovern, A.; Lagerquist, R.; Gagne, D.J.; Jergensen, G.E.; Elmore, K.L.; Homeyer, C.R.; Smith, T. Making the black box more transparent: Understanding the physical implications of machine learning. *Nat. Mach. Intell.* **2019**, *100*, 2175–2199. [CrossRef]

remote sensing MDPI

Article

A Preliminary Analysis of Typical Structures and Microphysical Characteristics of Precipitation in Northeastern China Cold Vortexes

Jingshi Wang [1,2], Xiaoyong Zhuge [2,*], Fengjiao Chen [2,3,*], Xu Chen [4] and Yuan Wang [1]

1 Key Laboratory of Mesoscale Severe Weather, Ministry of Education, School of Atmospheric Sciences, Nanjing University, Nanjing 210023, China
2 Key Laboratory of Transportation Meteorology of China Meteorological Administration, Nanjing Joint Institute for Atmospheric Sciences, Nanjing 210041, China
3 Anhui Province Key Laboratory of Atmospheric Science and Satellite Remote Sensing, Institute of Meteorological Sciences, Hefei 230031, China
4 Suzhou Meteorological Bureau, Suzhou 215131, China
* Correspondence: zhugexy@cma.gov.cn (X.Z.); cfj@mail.ustc.edu.cn (F.C.)

Abstract: The northeastern China cold vortex (NCCV) is the main weather system affecting Northeast China. Based on the precipitation products from the dual-frequency precipitation radar (DPR) onboard the Global Precipitation Measurement core observatory (GPM) satellite, the precipitation structures and microphysical properties for different rain types in 6432 NCCVs from 2014 to 2019 were studied using dynamic composite analysis. Our results show that the precipitation in NCCVs is dominated by stratiform precipitation. Regions with high stratiform and convective precipitation frequency have a comma shape. The growth mechanism of precipitation particles changes at ~4 km in altitude, the lower particles grow through collision (more pronounced in convective precipitation), and the upper hydrometeors grow through the Bergeron process. Additionally, the precipitation structures and microphysical properties exhibit great regional variations in NCCVs. The rainfall for all rain types is the strongest in the southeast region within an NCCV, mainly characterized by higher near-surface droplet concentration, while precipitation events occur more frequently in the southeast region for all rain types. There are active rimming growth processes above the melting layer for convective precipitation in the western region of an NCCV. In the southeast region of an NCCV, the collision growth of droplets in both types of precipitation is the most obvious.

Keywords: northeastern China cold vortex; precipitation structure; GPM; microphysical properties

Citation: Wang, J.; Zhuge, X.; Chen, F.; Chen, X.; Wang, Y. A Preliminary Analysis of Typical Structures and Microphysical Characteristics of Precipitation in Northeastern China Cold Vortexes. *Remote Sens.* **2023**, *15*, 3399. https://doi.org/10.3390/rs15133399

Academic Editor: Seon Ki Park

Received: 10 May 2023
Revised: 28 June 2023
Accepted: 30 June 2023
Published: 4 July 2023

1. Introduction

The cutoff low over the northeastern China–Siberian section of the northwestern Pacific coast is usually called the northeastern China cold vortex (NCCV) [1]. At present, the widely used definition of an NCCV is as follows: (1) At 500 hPa, there is a low-pressure system with an evident cold trough or a cold core, and at least one closed geopotential height contour (4 dagpm interval). (2) The system appears in the region (35–60°N, 115–145°E). (3) The system in the defined area must last for at least three days [2–4]. The NCCV can be accompanied by strong convective weather such as rainstorms, tornadoes, and hail during its formation, development, persistence, and dissipation, which brings economic loss and human casualties to Northeast China. It is also worth noting that an NCCV can provide favorable conditions for the initiation of mesoscale convective systems [5], which cause asymmetries of precipitation in their interior. Obviously, it is particularly important and urgent to study the precipitation structures of the NCCV in the background of the frequent occurrence and serious impact of NCCVs on society in recent years.

During the past two decades, progress has been made in the study of macrocharacteristics and favorable conditions of precipitation inside the NCCV [6–13]. The NCCV precipitation has obvious asymmetry distribution, i.e., the east side of the NCCV is the main area of precipitation [6,7], and the mesoscale weather generally occurs in the southern half of the NCCV [8]. During different life stages of the NCCV, the precipitation characteristics are different. Rainstorms generally occur in the development stage of the NCCV [9]. At this stage, the cloud system at its head develops vigorously [10], and most of these systems are characterized by large-scale mixed precipitation on the south side of the NCCV. At the mature stage, the precipitation in the NCCV is mostly isolated convective precipitation [11], and the precipitation center is located closer to the NCCV center [12]. In the dissipation stage of the NCCV, the comma-shaped cloud system is not obvious, and the main form of precipitation is isolated convective precipitation [10,11], with less frequent heavy rainfall events [9]. The variations of environmental characteristics in the NCCV are responsible for the different precipitation distributions. The positive vorticity advection is located in the east side of the NCCV, and its forcing produces an updraft, which is favorable for precipitation formation [13]. The convergence of water vapor occurs in the head and tail of the NCCV comma-shaped cloud system, associated with the low-level jet and high-energy wet tongue in the tail of the NCCV, making the convection develop violently [10]. During the development period of the NCCV, the main factors affecting the precipitation area are convective available potential energy and water vapor flux. During the mature period, the convection is basically in the area with high convective available potential energy and humidity. During the dissipation period, the convective precipitation is also related to the low-level convergence line [11]. These previous studies are mostly limited to individual cases or typical periods such as summer and flood seasons. The comprehensive statistical analysis from multiple factors for different rain types in the NCCV is relatively rare.

It is worth noting that the microphysical structures and processes inside the NCCV play an important role on precipitation and clouds [14]. Studying the microphysical characteristics of the NCCV's precipitation is crucial for understanding the NCCV's structures, providing the reference for cloud and precipitation modules in NCCV modeling, and improving the accuracy of numerical predictions. Using observations from an aircraft, Qi et al. [15] found that there was a high-concentration area of ice particles in the upper part of the convective cloud band, and that ice particles increased rapidly in the areas with high supercooled water content, which plays an important role on precipitation. Zhao and Lei [16] studied the precipitation microphysics of the NCCV from the late-mature stage to the dissipation stage using aircraft observations. They found that the particle concentration in the warm layer of the clouds was larger, while in the supercooled water layer, the particle concentration was much smaller. The high concentration of ice particles in the layer between −3 and −6 °C is caused by the Hallett–Mossop ice-crystal multiplication process. Zhong et al. [17] analyzed an NCCV in July based on CloudSat satellite data. In the development stage of the NCCV, the convective clouds in the warm front on the east side of the NCCV were mainly composed of ice water, corresponding to the strong echo band, and the liquid water was mainly in the southeast quadrant of the NCCV. In the mature stage, the convective systems with more ice water content were mostly located on the north side of the NCCV, and the liquid water was mainly distributed below the 0 °C layer of the NCCV center. At present, due to the high cost of aircraft observation and the incomplete information of cloud and precipitation obtained by ground-based radar [18], our understanding of the precipitation structure and microphysical characteristics in NCCVs is still insufficient. Hence, it is necessary to introduce more refined data to study the internal microphysical structure of the NCCV systematically and comprehensively.

The Global Precipitation Measurement core observatory (GPM) satellite can detect precipitation activity in the range of 65°S–65°N, which provides us with an excellent opportunity to study the precipitation characteristics of the NCCV in mid- and high-latitude regions [19]. The dual-frequency precipitation radar (DPR) onboard the GPM satellite has the ability to detect microphysical characteristics, which have been widely used in the study of tropical and extratropical cyclones [20–24]. Therefore, the three-dimensional structures and microphysical properties and process of precipitation in NCCVs will be analyzed using GPM DPR from 2014 to 2019 in the present study.

2. Materials and Methods

2.1. GPM Satellite and Precipitation Data

GPM DPR was designed by the Japanese Aerospace Exploration Agency (JAXA) and the National Institute of Communication Technology (NICT). It consists of a Ku-band precipitation radar (KuPR, 13.6 GHz) and a Ka-band precipitation radar (KaPR, 35.5 GHz). The detection area can cover 65°S–65°N on the Earth [24–26]. The KuPR has only one scanning mode (normal scan, NS), and its minimum detectable reflectivity is 14.5 dBZ. The KaPR has two scanning modes, matched scan (MS) and high-sensitivity scan (HS). The minimum detectable reflectivity is 16.7 dBZ and 10.2 dBZ, respectively [27]. The droplet-size parameters of hydrometeors can be inverted by different responses of the KuPR and the KaPR to hydrometeors [28]. Using ground-based observations, scholars have verified the reliability of the droplet-size distribution retrieved by DPR [29]. We used the GPM dual-frequency precipitation product 2ADPR from 2014 to 2019, including the rain type, near-surface rain rate, storm-top height, and droplet-size distributions (DSDs). Above the melting layer, for mixed-phase and solid-phase hydrometeors, the liquid-equivalent DSDs were retrieved from the official 2ADPR dataset [30].

2.2. GPM Products for Rain Types

The variable *typePrecip* in the Classification Module (CSF Module) of GPM data provides rain-type classification made by various methods: the single-frequency horizontal pattern method (H-method), vertical profiling method (V-method) and measured dual-frequency ratio method (DFR_m-method) [19]. In the V-method, the melting-layer bright band (BB) is first detected. The detection of the BB is to determine whether the vertical profile of radar reflectivity satisfies certain conditions which are typical for the profile of the radar reflectivity factor when a BB exits by examining the vertical profile of the radar reflectivity factor. When a BB is detected, if the reflectivity factor in the rain area does not exceed the convection threshold (46 dBZ), then the rain type is stratiform. When no BB is detected and the reflectivity factor exceeds the conventional convective threshold (40 dBZ), the rain type is convective. If rain type is neither stratiform nor convective, the rain type is other. In the H-method, the horizontal distribution of representative radar reflectivity factors, i.e., the maximum value of the reflectivity factor along the considered radar beam in the rain region, is detected. Rain-type classification adopts a modified University of Washington convective/stratiform separation method, which is divided into three categories: stratiform, convective, and other. Detection of convective precipitation is made first. The rain type is stratiform if it is not convective, unless the reflectivity factor is very small and has almost identical noise. If rain type is neither stratiform nor convective, it is other.

For the same precipitation profile, due to the different rain-type classification methods, the results may be different. Therefore, the variable *typePrecip* unifies rain types by the above classification methods, that is, the single-frequency method and the dual-frequency method are used to unify the rain types. In the dual-frequency method, the DFR_m-method is used to detect the BB, which classifies rain into three types: stratiform, convective, and transition. The dual-frequency method merges the rain type of the DFR_m-method with the single-frequency Ku-band rain type, and it outputs a unified rain type: stratiform, convective, and other. The features for the unification of the dual-frequency method are as follows: When the DFR_m rain type is convective or stratiform, if no BB is detected, the unified rain type follows the DFR_m rain type. If a BB is detected, however, the unified rain type is stratiform. When the DFR_m rain type is transition or the DFR_m processing is skipped, the single-frequency Ku-band rain type is the unified rain type. If heavy ice precipitation (HIP) or winter convection is detected, some stratiform type is changed into a convective type.

Since the rain-type identification by the DFR_m-method considers observations from two bands, the rain types are probably more reliable. Therefore, the rain types in the present study are retrieved from the dual-frequency method. Figure 1 shows the spatial distribution of 10.4 µm brightness temperature from the Japanese Himawari-7 meteorological satellite, as well as the rain type, storm-top height, and radar echo at 6 km height from GPM DPR for the NCCV at 1200 UTC on 12 June 2014. The brightness temperature of the cloud system in the region of 24–28°N and 136–140°E is low (Figure 1a), and the radar echo intensity at 6 km is higher than 20 dBZ (Figure 1d). However, this region is classified as stratiform by GPM in Figure 1b. From the vertical profile of the radar echo of the NCCV (Figure 2), there is a clear BB inside the cloud system in this region, and the echo intensity in the BB is larger than that in the upper and lower layers. The BB is the main feature of the stratiform precipitation echo, indicating that the airflow in the stratiform precipitation is stable and there is no obvious convective activity. For the convective precipitation on the southeastern edge of the stratiform region, the storm-top height is less than 5 km. It can be seen in Figure 2d that there is no BB in the precipitation system, and there is a strong echo higher than 40 dBZ in the radar reflectivity factor. Therefore, the rain types from the dual-frequency method in GPM have relatively high reliability from the vertical and horizontal distribution of radar echoes.

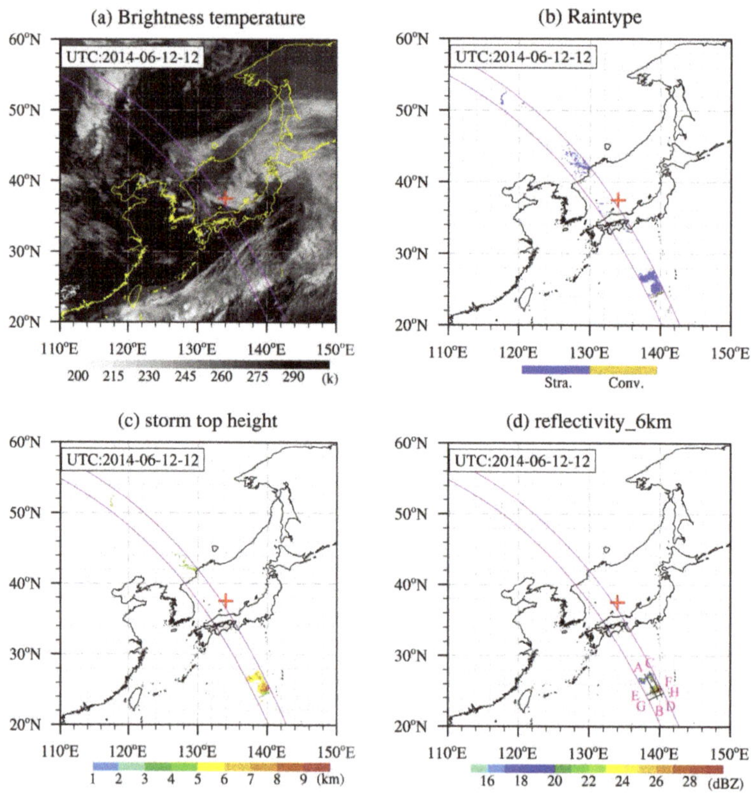

Figure 1. (**a**) The 10.4 µm infrared brightness temperature of the Himawari-7 satellite (shading, k), (**b**) the rain type identified by GPM DPR (shading, unitless), with yellow (blue) representing convection (stratiform), (**c**) the storm-top height (shading, km), and (**d**) the reflectivity at 6 km height (shading, dBZ) within 2000 km distance of the NCCV center at 1200 UTC on 12 June 2014 with GPM orbit No.001631. (The purple lines represent the swath of GPM DPR; the '+' represents the NCCV center; and lines AB, CD, EF, and GH represent the section position in Figure 2).

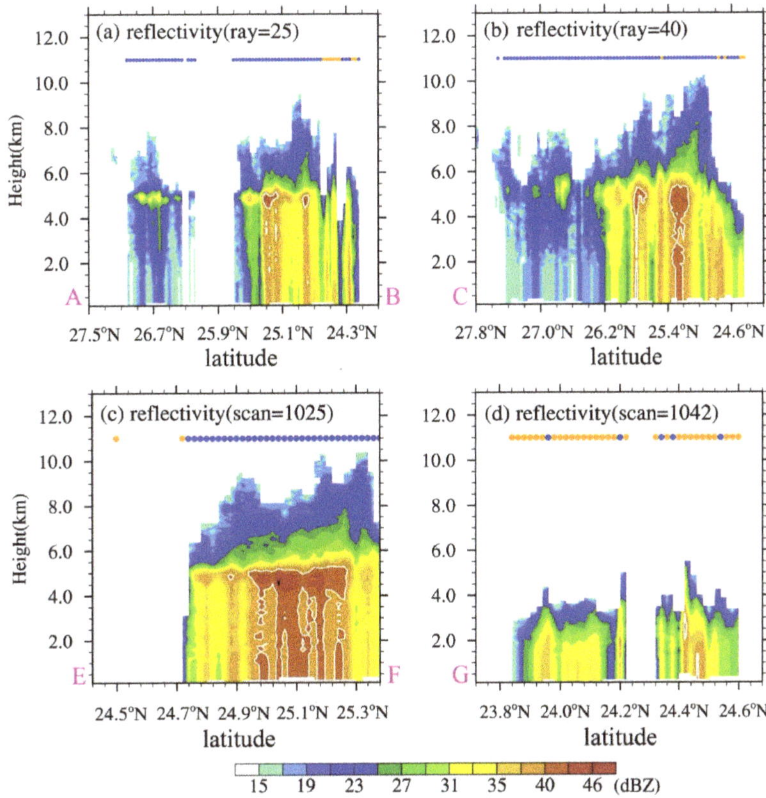

Figure 2. Vertical sections of the radar echo along lines (**a**) AB, (**b**) CD, (**c**) EF, and (**d**) GH in Figure 1d. The *X* axis represents the latitude along the cross-section direction. The yellow and blue dots at the 11 km altitude represent the convective and stratiform precipitation retrieved from the dual-frequency method in GPM, respectively.

2.3. Reanalysis Data

The 500 hPa geopotential height fields are extracted from the fifth-generation European Centre for Medium-Range Weather Forecasts (ECMWF) atmospheric reanalysis (ERA5) data with a temporal resolution of 1 h and a spatial resolution of $0.25° \times 0.25°$ [31]. They were used to correct the initial 6 h NCCV centers provided by Chen et al. [1] which were identified from a reanalysis dataset with a temporal resolution of 6 h and a spatial resolution of $1° \times 1°$. Firstly, NCCV centers at hourly resolution are obtained from the initial 6 h ones by a linear interpolation method. Secondly, the 1° resolution NCCV centers are refined to 0.25° resolution using the ERA5 500 hPa geopotential height field. The NCCV center is corrected by the lowest geopotential height within 1° from the original NCCV center. In this way, we can obtain the NCCV centers at hourly resolution, and match the NCCV centers with GPM data at the same time.

2.4. NCCV Coordinate System and Dynamic Composite Analysis

For the convenience of statistics, an NCCV coordinate system is defined in this study. In this coordinate system, the origin represents the center of the NCCV, the *x* axis indicates the east–west direction, and the *y* axis indicates the north–south direction.

The near-surface rain rate and other physical variables for precipitation were analyzed using dynamic composite analysis [7,32] in the NCCV coordinate system. The formula of dynamic composite analysis is

$$\overline{S}(x,y) = \frac{1}{N}\sum_{t=1}^{N} S_t(x,y), \tag{1}$$

where $S_t(x,y)$ are the physical variables at time t, and (x,y) are the coordinates in the composite area, which moves with the position of the NCCV center. N is the total number of samples, and $\overline{S}(x,y)$ is the average value of the samples obtained after the dynamic composite analysis. In the composite analysis, the NCCV center is used as the composite center; we took 2000 km in each direction around the NCCV center [24] to perform dynamic composite analysis in a square area with a side length of 4000 km.

In total, 6432 NCCV cases from 2014 to 2019 are used in the dynamic composite analysis. These things considered, the samples with near-surface rain rate greater than 0.5 mm h^{-1} are defined as precipitation samples.

3. Results

3.1. Subsection Horizontal Distributions of Stratiform and Convective Precipitation in the NCCV

3.1.1. Precipitation Characteristics

In order to understand the internal precipitation structure of the NCCV, the horizontal distribution of near-surface rain rate (the total rain rate of the precipitation samples in the grid divided by the number of precipitation samples), precipitation frequency (the number of precipitation samples in the grid divided by the number of total observational samples), and storm-top height of total, stratiform, and convective precipitation in the NCCV coordinate system are given, respectively (Figure 3). There are obvious regional differences in the distribution of near-surface rain rate for two types of precipitation. The rain rate is high in the south of the NCCV, especially in the southeast region, while it is the smallest in the northeast (Figure 3a–c), which may be related to the more abundant water vapor transport in the south [1]. The rain rate of stratiform in the southeast quadrant (Figure 3b) is generally higher than 2 mm h^{-1}, while it is generally lower than 2 mm h^{-1} in other regions. Compared to stratiform precipitation, the convective rain rate (Figure 3c) is generally larger in the NCCV, and the rain rate in the south of the NCCV is higher than 3.5 mm h^{-1}, while the rain rate in other regions is also higher than 2 mm h^{-1}. The horizontal distribution of the total rain rate (Figure 3a) is similar to that of stratiform, which is related to the high proportion of stratiform precipitation in the NCCV. The rain rate of stratiform in each region is smaller than that of convective precipitation, while the precipitation frequency is significantly higher, indicating that the precipitation in the NCCV is more composed of stratiform precipitation with weaker precipitation intensity.

There is an obvious comma-shaped rain band for the precipitation frequency distribution of the NCCV, which is consistent with the research results of Chen et al. [1] (Figure 3d–f). For total (Figure 3d) and stratiform (Figure 3e) precipitation, the regions with precipitation frequency higher than 4% are mainly located over the small area near the center in the northeast of the NCCV and a large area in the southeast of the NCCV. This is in good agreement with previous research results that the east side of the NCCV is the main area of precipitation [6,7], while in most of the other areas of the NCCV, precipitation frequency is generally lower than 1.5%. With the increase in the distance from the center of the NCCV, the region with high precipitation frequency shifts from the northeast to the southeast quadrant. Compared to stratiform precipitation, the frequency of convective precipitation is lower in all regions. The precipitation frequency in the comma-shaped rain band is mainly in the range of 0.5–1.5%, while it is basically lower than 0.5% in other regions (Figure 3f). As the distance to the center of the NCCV increases, the high convective precipitation frequency occurs, and the frequency of convective precipitation increases in the southeast quadrant, which is probably related to the higher instability and more moisture in this region (figure omitted).

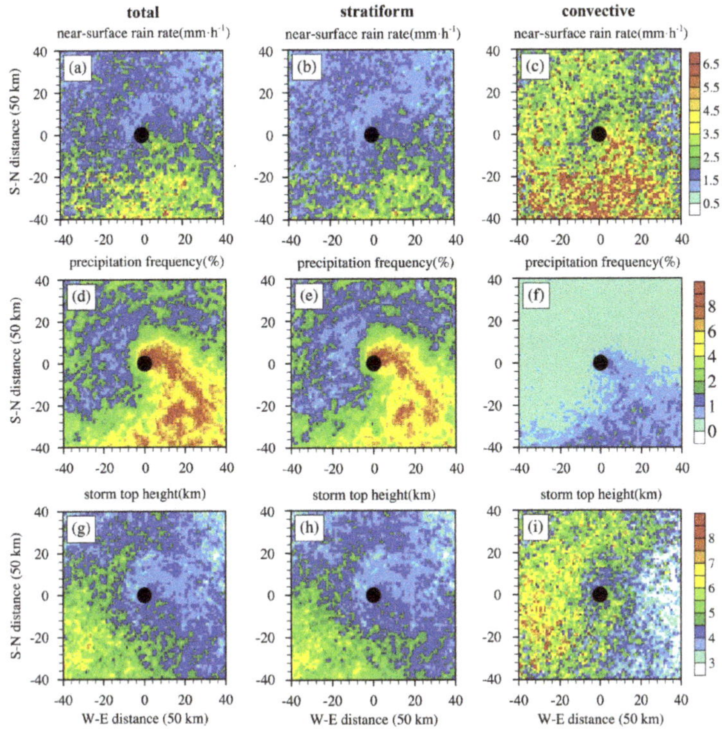

Figure 3. The distribution of (**a–c**) near-surface rain rate (shading, mm h^{-1}), (**d–f**) precipitation frequency (shading, %), and (**g–i**) storm-top height (shading, km) for total, stratiform, and convective precipitation at each 50 km × 50 km grid in the NCCV coordinate system, derived from GPM DPR for 2014–2019. (The black dots represent the NCCV center; 6432 NCCVs are included in the dynamic composite analysis).

The horizontal distribution of storm-top height in the NCCV also has an obvious asymmetric structure (Figure 3g–i). For total (Figure 3g) and stratiform (Figure 3h) precipitation, the storm-top height is highest (4.5–6.5 km) in the southwest quadrant and lowest in the northeast quadrant. Although the precipitation cloud develops deeply in the southwest regions, the rain rate is small. Conversely, the storm-top height in the northern half of the NCCV is basically lower than 4.5 km, indicating shallower precipitation clouds, and the rain rate is also small. The storm-top height in the southeast region of the NCCV is generally lower than 5 km, but the rain rate is the largest in the NCCV, which may be related to the more abundant water vapor in this area [1]. The convective precipitation cloud in the west of the NCCV develops higher (Figure 3i), and the average storm-top height is higher than 5 km, even up to 8 km in some areas which also have a higher rain rate. The storm-top height in the southeast quadrant of the NCCV is basically lower than 4.5 km, but the rain rate is the highest in the NCCV. One possible explanation is that these areas are mostly located over the ocean, where shallow convection with low storm-top height prevails, according to a previous study [22]. The sufficient water vapor over the ocean may provide favorable moisture conditions for the formation of shallow convection.

To better understand the contribution of stratiform and convective precipitation within the NCCV and the distribution of NCCV precipitation, the horizontal distribution of precipitation amount contribution (the total rain rate of stratiform/convective precipitation samples divided by the total rain rate of precipitation samples), and precipitation frequency contribution (the number of stratiform/convective precipitation samples in the grid divided by the number of precipitation samples) for stratiform and convective precipitation within 2000 km of the NCCV center are given (Figure 4a–d). Overall, the precipitation in NCCVs is dominated by stratiform precipitation, and the precipitation frequency (mostly more than 70%) and amount contribution (mostly more than 60%) of stratiform precipitation are greater than that of convection precipitation. The difference between convective and stratiform precipitation amount (frequency) contributions is the largest in the northeast quadrant of the NCCV. The latter is significantly higher than the former. For example, in the northeast of the NCCV, the contribution of stratiform precipitation amount (frequency) can reach 80% (90%), while convective precipitation is less than 20% (10%). It is worth noting that in the southwest of the NCCV, although the contribution of convective precipitation frequency is smaller than that of stratiform precipitation (basically less than 30%), the contribution of precipitation amount is higher than 60% locally, exceeding stratiform precipitation. For stratiform and convective precipitation, the distribution of precipitation frequency contribution of the NCCV (Figure 4c,d) is very similar to that of precipitation amount contribution (Figure 4a,b). Generally, the regions with high (low) precipitation frequency contribution also exhibit a high (low) precipitation amount contribution. However, since the rain rate of stratiform is generally lower than that of convective, the contribution of the rain amount is smaller than the frequency for stratiform precipitation. On the contrary, the contribution of the convective precipitation amount increases compared to the frequency contribution.

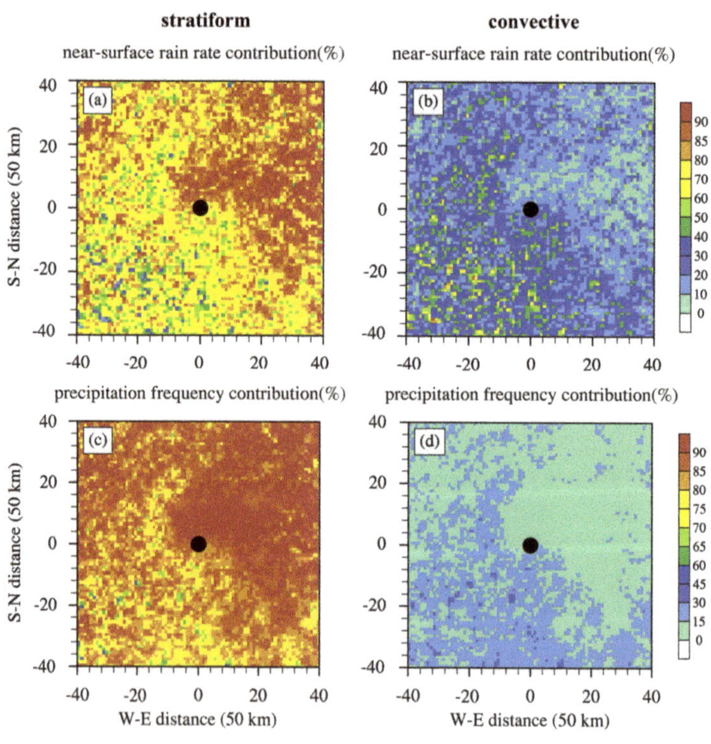

Figure 4. The distribution of (**a**,**b**) precipitation contribution (shading, %), and (**c**,**d**) precipitation frequency contribution (shading, %) for total, stratiform, and convective precipitation at each 50 km × 50 km grid in the NCCV coordinate system, derived from GPM DPR for 2014–2019. (The black dots represent the NCCV center; 6432 NCCVs are included in the dynamic composite analysis).

3.1.2. Characteristics of Near-Surface Microphysics

Rain rate is controlled by both raindrop concentration and raindrop size [28]. The microphysical structure and process inside the NCCV also play a very important role in precipitation [14]. The characteristics of the DSD can reflect the microphysical processes of precipitation. The DSD parameters provided by GPM 2ADPR products are used, including D_m and dBN_w, where D_m represents the size of the particles and N_w represents the particle concentration. In order to reduce the influence of DSD retrieval uncertainties caused by ground clutters, the D_m and dBN_w at 2.5 km height are chosen to represent the near-surface DSDs. In order to facilitate the display, dBN_w is used to represent the particle concentration parameter (dBN_w = $10log_{10}$ (N_w)).

The near-surface particle diameter (Figure 5a–c), concentration (Figure 5d–f), and radar reflectivity (Figure 5g–i) are very similar horizontal distributions for stratiform and convective precipitation. The D_m and reflectivity factor in the southeast quadrant of the NCCV are small and the dBN_w is high, while the dBN_w and reflectivity factor in the northwest quadrant are low and the D_m is large. For stratiform precipitation (Figure 5b), the D_m in the northwest quadrant is generally higher than 1.3 mm and the dBN_w is lower than 33, while the D_m in the southeast is mostly higher than 1.25 mm and the dBN_w is about 34. For convective precipitation, the D_m in the northwest is generally higher than 1.45 mm and the dBN_w is lower than 32. The D_m in the southeast is generally below 1.25 mm and the dBN_w is generally higher than 36. For convective and stratiform precipitation, the radar reflectivity is very similar to the D_m distribution, which is mainly because the near-surface echo intensity is more affected by particle size. For convective and stratiform precipitation, the near-surface rain rate is close to the horizontal distribution of dBN_w. For example, the high rain rate of different types of precipitation in the southeast quadrant of the NCCV (Figure 3a–c) corresponds to the high concentration of particles (smaller size), while the weak precipitation in the northwest quadrant corresponds to the low concentration of hydrometeors (larger size). This shows that the rain rate in the NCCV is closely related to the raindrop concentration. The heavy precipitation in the southeast is mainly contributed to by higher raindrop concentration.

Compared to stratiform precipitation, convective precipitation generally has stronger near-surface echoes, smaller particle concentration, and larger particle size in the western and northwestern regions of the NCCV, while the convective precipitation in the eastern and southeastern quadrants generally has weaker near-surface radar echo, higher particle concentration, and smaller particle size. This shows that there are differences in the microphysical processes of precipitation in different types of precipitation clouds and different quadrants. The following section will further study the variation of precipitation characteristics in different regions in the NCCV.

3.2. Azimuthal Distributions of Stratiform and Convective Precipitation in NCCV

3.2.1. Precipitation Characteristics

The above studies show that there is a significant asymmetric structure in the horizontal distribution characteristics of stratiform and convective precipitation in the NCCV. In order to understand the distribution characteristics and differences of precipitation in different directions within the NCCV, the azimuth distribution of precipitation frequency, near-surface rain rate, precipitation frequency contribution, precipitation amount contribution, and storm-top height within 2000 km of the NCCV center are given in Figure 6.

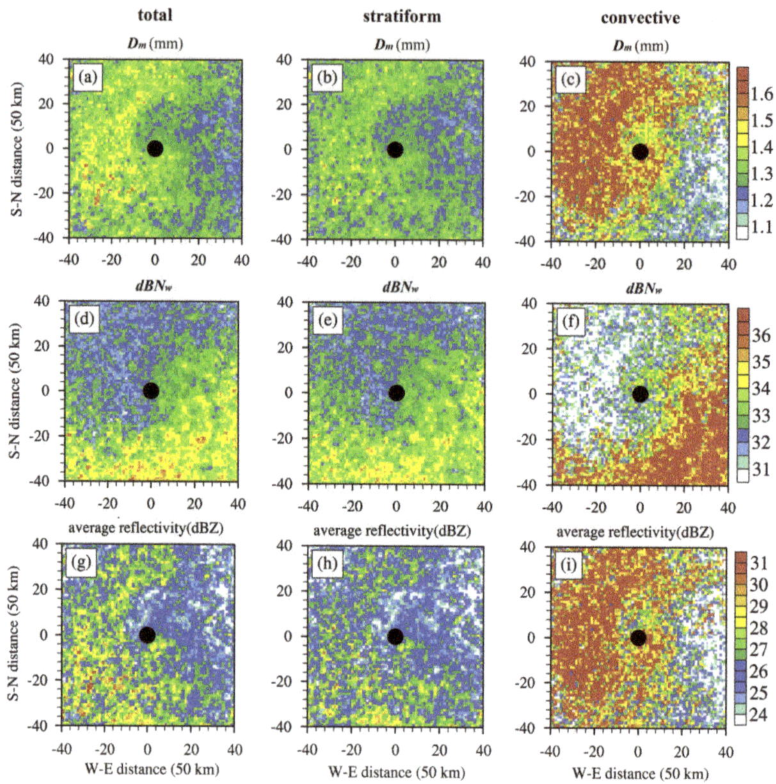

Figure 5. The distribution of near-surface (**a–c**) droplet mass-weighted mean diameter D_m (shading, mm), (**d–f**) particle concentration parameter dBN_w (shading, no unit), and (**g–i**) average reflectivity (shading, dBZ) for total, stratiform, and convective precipitation at each 50 km × 50 km grid in the NCCV coordinate system, derived from GPM DPR for 2014–2019. (The black dots represent the NCCV center; 6432 NCCVs are included in the dynamic composite analysis).

The peak value of precipitation frequency is in the southeast quadrant of the NCCV for different rain types (Figure 6a). In each direction of the NCCV, the frequency of stratiform precipitation is greater than that of convective precipitation. This indicates that the NCCV precipitation is dominated by stratiform precipitation, which is consistent with the horizontal distribution (Figure 3e,f). The convective precipitation frequency changes slightly in the northern half of the NCCV, while it gradually increases from west to southeast, and then gradually decreases to the east. The peak of the precipitation frequency is about 1% in the southeast of the NCCV. The frequency of stratiform precipitation varies greatly in different directions of the NCCV. From the east side of the NCCV to the southwest, the frequency of stratiform precipitation gradually decreases and reaches the minimum in the southwest of the NCCV, about 1.3%. Then to the southeast, the frequency of precipitation increases rapidly and reaches the peak in the southeast of the NCCV, about 4%. The azimuthal distribution of the total precipitation frequency is basically consistent with that of stratiform precipitation but is larger than that of stratiform precipitation in each direction.

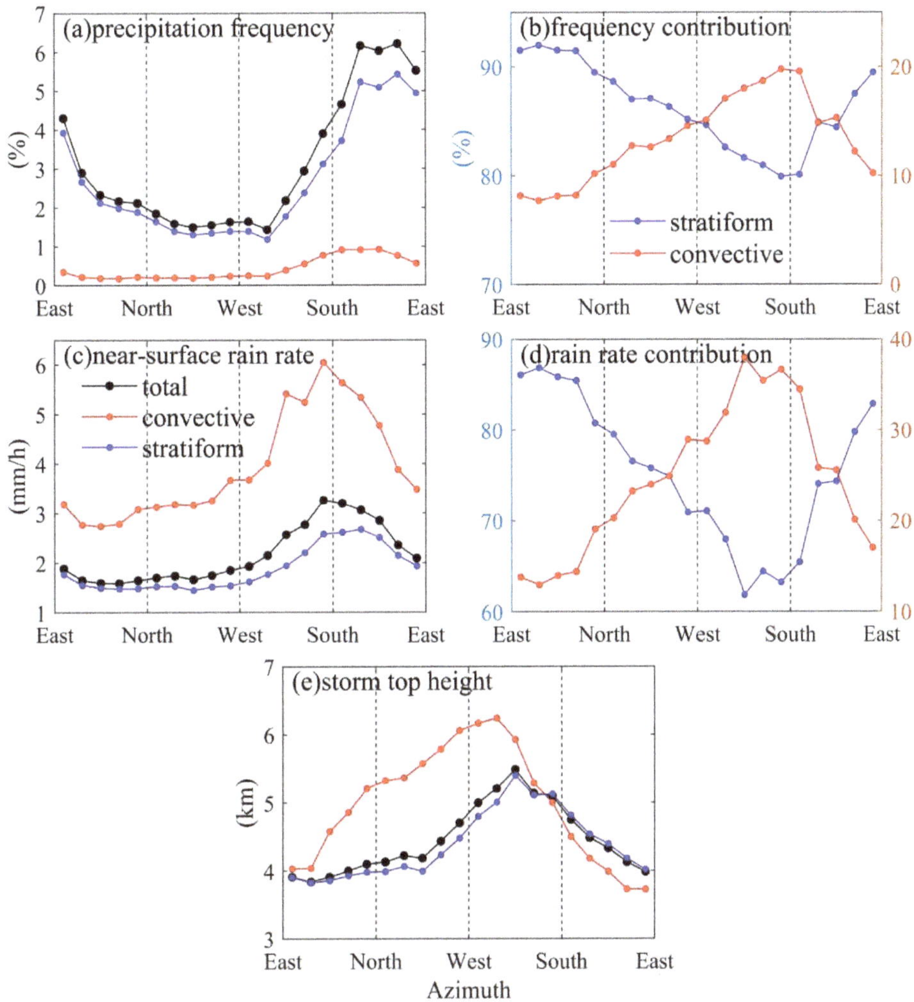

Figure 6. The azimuthal distributions of (**a**) precipitation frequency, (**b**) near-surface rain rate, (**c**) precipitation frequency contribution, (**d**) precipitation contribution, and (**e**) storm-top height for total (black solid lines), stratiform (blue solid lines), and convective (red solid lines) precipitation in the NCCV coordinate system, derived from GPM DPR for 2014–2019. (A total of 6432 NCCVs are included in the dynamic composite analysis; samples should be within 2000 km distance of the NCCV center).

The peak of average rain rate in the NCCV is in the south of the NCCV (Figure 6b) for convective and stratiform precipitation, which is more westward than the peak of precipitation frequency. In all directions in the NCCV, the convective rain rate is higher than that of stratiform precipitation. The convective rain rate varies greatly in different directions of the NCCV. It gradually increases from the east side to the south side in the NCCV and reaches the peak in the southwest (about 6 mm h^{-1}), and then gradually decreases to the east side. The stratiform rain rate changes little in the northern region of the NCCV and gradually increases from the west to the south. The peak is in the south of the NCCV, about 2.67 mm h^{-1}. From south to east, the stratiform rain rate gradually decreases. Due to the high proportion of stratiform precipitation, the azimuth distribution characteristics of the total rain rate are basically the same as the stratiform rain rate, which is not described here.

In all directions within the NCCV, the precipitation frequency and amount contribution for stratiform precipitation is greater than those of convective precipitation; they are generally between 80% and 90%, and 60% and 90%, respectively. The precipitation amount and frequency contribution for stratiform precipitation are the lowest in the southwest quadrant and the highest in the northeast quadrant of the NCCV (Figure 6c,d). On the contrary, the contribution of convective precipitation is the highest in the southwest region and the lowest in the northeast region. The precipitation frequency and amount contribution for stratiform precipitation increase slightly from the east to the south of the NCCV, reaching the peak in the northeast quadrant, about 92% and 87%, respectively. After that, they gradually decrease and reach the valley in the southwest, at about 80% and 62%, respectively. On the contrary, the peak value of convective precipitation contribution is in the southwest of the NCCV center; the contribution of precipitation frequency and the amount is about 19.8% and 37%, respectively. The valley value is in the northeast, where the frequency contribution is about 7.7% and the precipitation amount contribution is about 13%.

The peak of average storm-top height for all types of precipitation in the NCCV is in the southwest quadrant of the NCCV, while the peak of convection is more westward. The valley of the storm-top height is in the east (Figure 6e). The storm-top height of convective precipitation in the southeast of the NCCV is smaller than that of stratiform, while it is opposite in other quadrants. The storm-top height of convective precipitation varies greatly in different directions. From the east side to the southwest side, it gradually increases, reaching a peak in the southwest, about 6.24 km, and then to the east side, the storm-top height gradually decreases, reaching the valley value in the east. The storm-top height peak value of stratiform precipitation is about 5.4 km in the southwest, and the valley value is about 3.83 km in the east of the NCCV. The azimuth distribution of storm-top height for total precipitation is basically consistent with that for stratiform precipitation, which is related to the large proportion of stratiform precipitation in the NCCV.

3.2.2. Characteristics of Near-Surface Microphysics

Figure 7a,b shows the azimuthal distribution of near-surface D_m and dBN_w of the total, stratiform, and convective precipitation in the NCCV. The D_m for different types of precipitation shows obvious asymmetry, and the azimuth distribution characteristics are similar (Figure 7a). The peak values are in the west of the NCCV, and the valley values are in the east. Except for the southeast quadrant, the D_m of convective precipitation is generally larger than that of stratiform in all directions. The D_m of convective precipitation changes the most in different directions within the NCCV (the asymmetry is more obvious). From the east side of the NCCV to the west side, the D_m gradually increases and continues to the east side, where the D_m gradually decreases. The peak value in the west is about 1.66 mm, and the valley value is in the southeast, about 1.24 mm. For stratiform precipitation, the average azimuthal variation of D_m is small, with a peak of about 1.32 mm in the west and a valley of about 1.25 mm in the east. The D_m distribution of total precipitation is similar to that of stratiform precipitation.

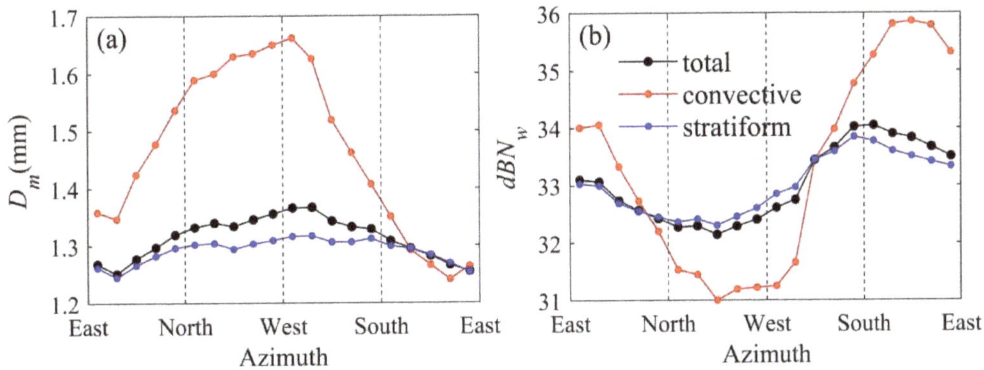

Figure 7. The azimuthal distributions of near-surface (**a**) droplet mass-weighted mean diameter D_m, and (**b**) particle concentration parameter dBN_w for total (black solid lines), stratiform (blue solid lines), and convective (red solid lines) precipitation in the NCCV coordinate system, derived from GPM DPR for 2014–2019. (A total of 6432 NCCVs are included in the dynamic composite analysis; samples should be within 2000 km distance of the NCCV center).

The azimuth distributions of dBN_w for different types of precipitation are also obviously similar (Figure 7b). In the southeast or south of the NCCV, the concentration of precipitation particles is the highest (~36); the lowest is in the northwest (~31) for both convective and stratiform precipitation. The variation of convective precipitation in different directions is larger than that of stratiform clouds. From the east side of the NCCV to the northwest side, the particle concentration gradually decreases, and then to the southeast side, the particle concentration gradually increases. The peak value in the southeast is about 35.9, and the valley value in the northwest is about 31. For stratiform precipitation, the peak value is about 33.8 in the south, and the valley value is about 32.3 in the northwest. The particle concentration distribution of total precipitation is basically consistent with that of stratiform. In general, the concentration of convective precipitation systems will have lower concentration and larger hydrometeors, while in Figure 7b, except in the southeast and northeast quadrants of the NCCV, the concentration of droplets in convective precipitation is relatively smaller, which proves that the microphysical distribution varies with weather systems [21,23,24,33]. However, for a specific type of precipitation, the peak and valley values of particle concentration are opposite to that of D_m, which is related to the characteristics of the droplet-size distribution (high-concentration small particles or low-concentration large particles). This is consistent with the research results of Qi et al. [15] that particle concentration and diameter showed a negative correlation. In addition, compared to D_m, near-surface rain rates for the different types in the NCCV (Figure 6b) are closer to the azimuth distribution of the dBN_w. Especially for stratiform precipitation, the peak of heavy rain rate in the south corresponds well to the peak of high concentrations of hydrometeors in the south.

The above results reveal the different characteristics of near-surface DSDs of the NCCV precipitation in different directions. In order to further analyze the near-surface microphysics at different positions in the NCCV and study the corresponding relationship between particle concentration and particle size, Figure 8 shows the two-dimensional frequency distribution of dBN_w and D_m at 2.5 km in four quadrants for convective and stratiform precipitation in the NCCV coordinate system. The stratiform and convection in each quadrant generally have high concentrations of small particles and low concentrations of large particles. The DSDs of stratiform precipitation are more concentrated, while the DSDs of convective precipitation are much wider, with higher concentrations and larger hydrometeors.

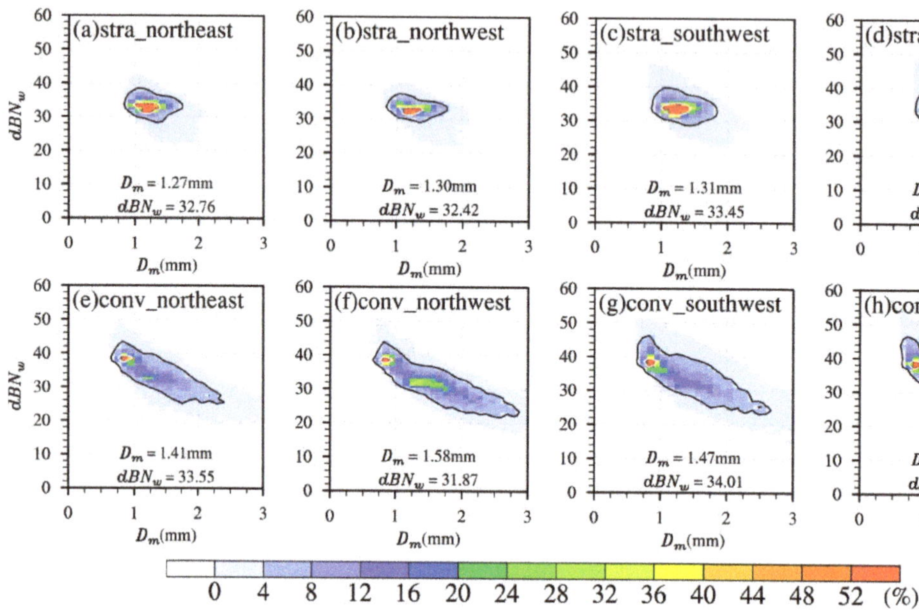

Figure 8. The frequency pattern (shading, %) in two-dimensional space of D_m and dBN_w at 2.5 km in the (**a**,**e**) northeast, (**b**,**f**) northwest, (**c**,**g**) southwest, and (**d**,**h**) southeast regions within 2000 km distance of the NCCV center for stratiform and convective precipitation. (The black (white) solid line represents the frequency higher than 4% (36%)).

For stratiform precipitation (Figure 8a–d), the particle D_m is mainly concentrated between 0.8 and 2.4 mm, and the dBN_w is mainly between 22 and 46. There are differences in DSDs in different quadrants. The DSDs in the southeast quadrant are the widest, followed by the southwest quadrant, and the average particle concentration is higher and the diameter is larger in the two quadrants, corresponding to a high rain rate. For convective precipitation (Figure 8e–h), the D_m is concentrated between 0.6 and 3.0 mm, and the dBN_w is between 18 and 52. The difference of DSDs in different quadrants is more significant than that in stratiform clouds. There are two peak centers of DSDs, located in the northeast and northwest quadrants of the NCCV. One peak center is concentrated on D_m in the range of 0.8–1 mm, and the dBN_w is concentrated in the range of 38–40, which is composed of a high concentration of small particles. Another peak center is concentrated when D_m ranges from 1.25 to 1.75 mm, and the dBN_w is concentrated from 30 to 34, which indicates a low concentration of large particles probably resulted from deep convection. However, there is only the former DSD peak center in the southwest and southeast quadrants, which may be caused by a higher proportion of shallow convective precipitation in these regions. For the southeast and southwest quadrants with higher convective rain rates, the strong convective precipitation on the southeast side of the NCCV is mainly contributed to by near-surface high-concentration hydrometeor particles. The average diameter of the hydrometeor particles is the smallest, 1.27 mm, while the dBN_w is the largest, 35.82. The DSDs are also more concentrated on large dBN_w values and small D_m values (high particle concentration and small particle diameter, indicated by white solid lines), with few particles with D_m higher than 2 mm. The stronger convective rain rate in the southwest quadrant is contributed to by higher particle concentration (dBN_w = 34.01) and larger particle size (D_m = 1.47 mm).

3.3. Vertical Structure Characteristics of Stratiform and Convective Precipitation in NCCV

3.3.1. Precipitation Characteristics

The vertical distribution structure of the radar echo of precipitation can effectively reflect the vertical distributions of solid, liquid, and solid–liquid mixed particles in the NCCV precipitation cloud. Contoured frequency by altitude diagram (CFAD) is helpful to clearly show the vertical structure characteristics of the NCCV precipitation. Figure 9 shows the CFAD distribution of radar reflectivity of total, stratiform, and convective precipitation within 2000 km in NCCV coordinate systems from 2014 to 2019. Above 4 km, the radar reflectivity factors for all types of precipitation increase rapidly with the decrease in height, and the hydrometeors such as supercooled water and ice crystals in this layer continue to freeze and grow through the Bergeron process. For total precipitation (Figure 9a) and stratiform precipitation (Figure 9b), radar echoes are mainly distributed between 20 dBZ and 29 dBZ below 4 km, and as the height decreases, the radar reflectivity factor can reach a larger value, showing a wider radar reflectivity factor spectrum. The frequency of 40 dBZ can reach 5%, reflecting the collision-growth process of particles. For convective precipitation below 4 km, the increase in radar reflectivity factor is more obvious with the decrease in height, reflecting the more obvious collision-growth process. Compared to stratiform clouds, the radar reflectivity factor spectrum of convective precipitation below 4 km is wider, which is distributed from 17 dBZ to 50 dBZ. There is a shallow precipitation characteristic area, the echo is concentrated below 3 km, and the echo is between 17 dBZ and 28 dBZ. Another echo center is below 3 km, and the echo is in the range of 32–37 dBZ, indicating deep convection. These two echo centers corresponded well to the two frequency centers of near-surface DSDs (Figure 8). Particularly, there is a clear BB feature for stratiform precipitation, showing a sudden increase in echo impacted by the melted ice particles at the height of approximately 3~4 km in altitude.

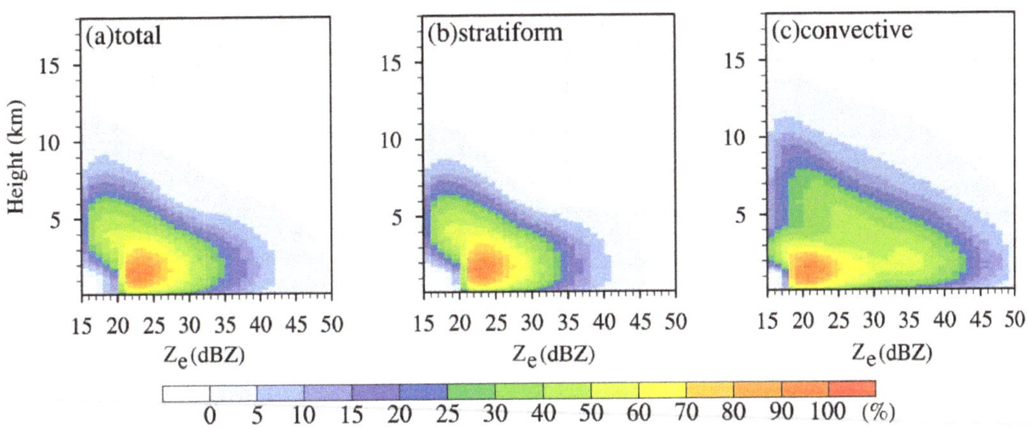

Figure 9. The CFADs (shading, %) of the Ku-band reflectivity for (**a**) total, (**b**) stratiform, and (**c**) convective precipitation within 2000 km distance of the NCCV center, derived from GPM DPR for 2014–2019.

In order to further explore the vertical variation characteristics and differences of precipitation in different quadrants within the NCCV, Figure 10 shows the CFADs of radar reflectivity for total, stratiform, and convective precipitation in each quadrant within 2000 km in the NCCV coordinate system from 2014 to 2019. For stratiform precipitation (Figure 10a–d), the echo top in the southwest quadrant is the highest and the echo is the strongest, followed by the southeast quadrant of the NCCV. The echo top of stratiform precipitation in the southwest quadrant of the NCCV is about 12 km, and the echo is mainly concentrated below 5 km, distributed between 20 dBZ and 30 dBZ. The echo top of the southeast quadrant is about 11 km, and the echo is concentrated below 4 km, distributed in 20–30 dBZ. The storm-top heights of the northwest and northeast quadrants are 10 and 9 km, respectively, and the echoes are concentrated below 4 km, distributed from 20 dBZ to 28 dBZ. Significant BB characteristic areas can be seen in each quadrant. The BB is higher in the southwest and southeast quadrants, indicating that the melting layer is also higher. Below 5 km, as the height decreases, the radar reflectivity factor value in the southeast quadrant can reach a greater value, the spectral width is wider, and the particle collision-growth process is the most obvious. This may be due to the relatively stronger upward movement in the southeast quadrant, and the raindrops fall against the airflow, which is conducive to rapid collision growth.

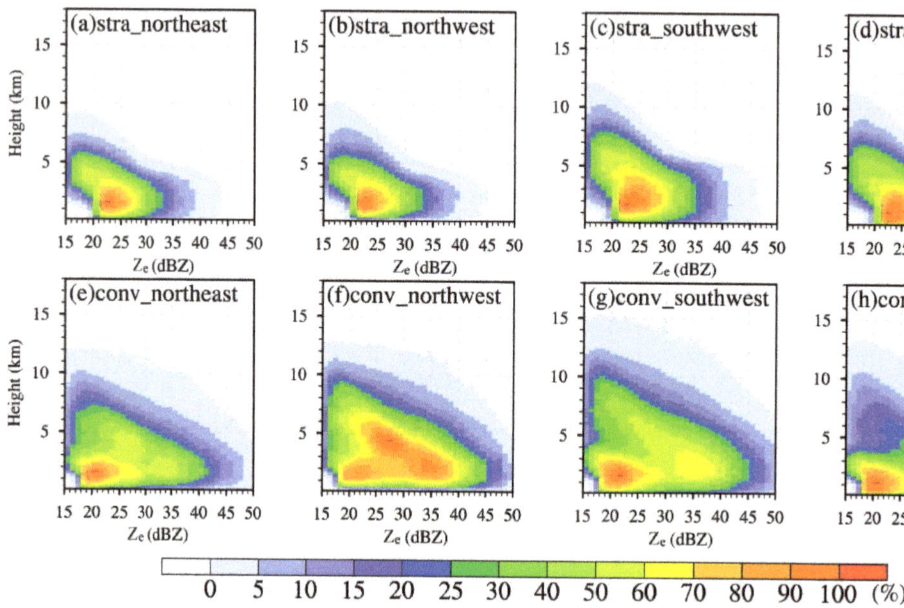

Figure 10. The CFADs (shading, %) of the Ku-band reflectivity in the (**a**,**e**) northeast, (**b**,**f**) northwest, (**c**,**g**) southwest, and (**d**,**h**) southeast regions within 2000 km distance of the NCCV center for stratiform and convective precipitation, derived from GPM DPR for 2014–2019.

For convective precipitation (Figure 10e–h), the echo height and intensity in each quadrant are greater than those of stratiform precipitation, and the echo spectrum width is much wider. As the height decreases, the echo spectrum width increases more significantly, indicating that the collision growth of particles is more obvious. The CFAD in each quadrant is significantly different. In the southeast and northeast quadrants, the width of the echo spectrum increases more significantly as the altitude decreases, indicating that the collision growth of particles is more obvious. The shallow convection characteristic area can be seen in each quadrant, and the echo is concentrated below 3 km, distributed in 18–27 dBZ. This characteristic is particularly significant in the southeast quadrant of the NCCV. Probably because this area is mostly located in the ocean, shallow convection occurs easily on the sea surface, resulting in a high proportion of shallow convection and an obvious characteristic area. In the other three quadrants, a deep convective feature area can be seen, and the echo is generally concentrated in 32–39 dBZ, which is particularly significant in the northwest quadrant of the NCCV. In the northwest and southwest quadrants of convective precipitation (Figure 10f,g), the storm-top height is higher, and the radar echo above 4 km altitude is stronger, which increases rapidly with the decrease in height, reflecting the microphysical processes such as ice-crystal and rime growth. Below 4 km in altitude, the radar echo also increases rapidly with the decrease in height, which reflects the obvious collision-growth process in this area, which is consistent with the discovery that the near-surface particles have larger diameters and smaller concentrations (Figure 5).

In order to further explore the precipitation structures in the NCCV at different directions and heights, the azimuthal average profiles of radar reflectivity for total, stratiform, and convective precipitation within 2000 km in the NCCV coordinate system from 2014 to 2019 are presented in Figure 11. For total precipitation (Figure 11a) and stratiform precipitation (Figure 11b) below 10 km, the reflectivity factor increases with the decrease in height. This phenomenon is the most significant in the southeast quadrant, indicating collision growth of particles, corresponding well to the high rain rate in the quadrant. For convective precipitation (Figure 11c), the echo intensity of each position in the NCCV is greater than that of the stratiform. The reflectivity factor in the southwest and northwest quadrants of the NCCV increase with the decrease in height, indicating that the particles continue to collide and grow during the falling process. The height of the convective echo top in some areas within the southwest of the NCCV is higher than 14 km, indicating that there is penetrating convection in the area [34]. In the southeast and northeast, the echo is smaller than that in the west side of the NCCV, which is different from the results of the individual case [17] that the east of the NCCV corresponds to a strong echo. In the southeast quadrant of the NCCV, the storm heights in some areas can be as high as 14 km, but the frequency of occurrence of these deep clouds over this region is relatively low. The reflectivity of the upper layer is the smallest, part of which is lower than 21 dBZ, and the reflectivity near the height of 4 km is up to 28 dBZ. The reflectivity factor increases rapidly with the decrease in height, indicating that the particle collision growth is more obvious. However, within the height of 2–3 km, there is a relatively weak echo area, which may be due to the breakup of particles in the lower layer, showing that the particle diameter decreases significantly and the concentration increases.

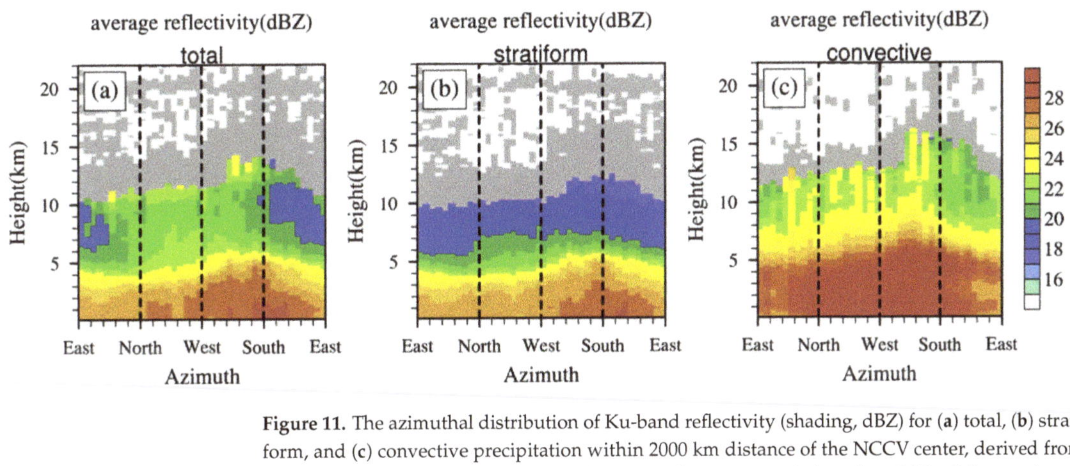

Figure 11. The azimuthal distribution of Ku-band reflectivity (shading, dBZ) for (**a**) total, (**b**) stratiform, and (**c**) convective precipitation within 2000 km distance of the NCCV center, derived from GPM DPR for 2014–2019. (The sample size in the gray area is less than 0.1% of the maximum sample size).

3.3.2. Microphysical Structures

The intensity of precipitation echo is affected by both particle concentration and particle size. Figure 12 shows the CFAD of D_m and dBN_w of total, stratiform, and convective precipitation within 2000 km in the NCCV coordinate system from 2014 to 2019. The dBN_w of convective precipitation is larger (smaller) than that of stratiform below (above) 5 km in altitude. The D_m of convective precipitation is larger than stratiform precipitation above 5 km in altitude. The stronger ascending motion within convective clouds may bring hydrometers to higher altitude and increases the chances of collision, which eventually leads to large-sized hydrometeors and low concentrations. The D_m of total precipitation (Figure 12a) is concentrated between 0.6 and 2.8 mm at each height layer, and the dBN_w is almost concentrated between 20 and 48 (Figure 12d); the D_m of stratiform precipitation (Figure 12b) is mainly concentrated between 0.8 and 2.4 mm, and the dBN_w is almost concentrated between 20 and 46 (Figure 12e). The D_m of convective precipitation (Figure 12c) is mainly concentrated between 0.6 and 3 mm, and the distribution of dBN_w is mainly concentrated between 18 and 52. As the height decreases, the particle diameter and concentration of convective and stratiform precipitation increase, and the spectral width increases, which corresponds to the increase in reflectivity. In convective precipitation, there are two high-frequency regions for particle concentration and size. Particles with a D_m of 0.8 to 1 mm and a dBN_w of 38 to 40 correspond to shallow convection and mainly high concentrations of small particles. Another high-frequency characteristic region corresponds to deep convection. The D_m is within 1.2 to 1.6 mm, and the dBN_w is 30 to 36.

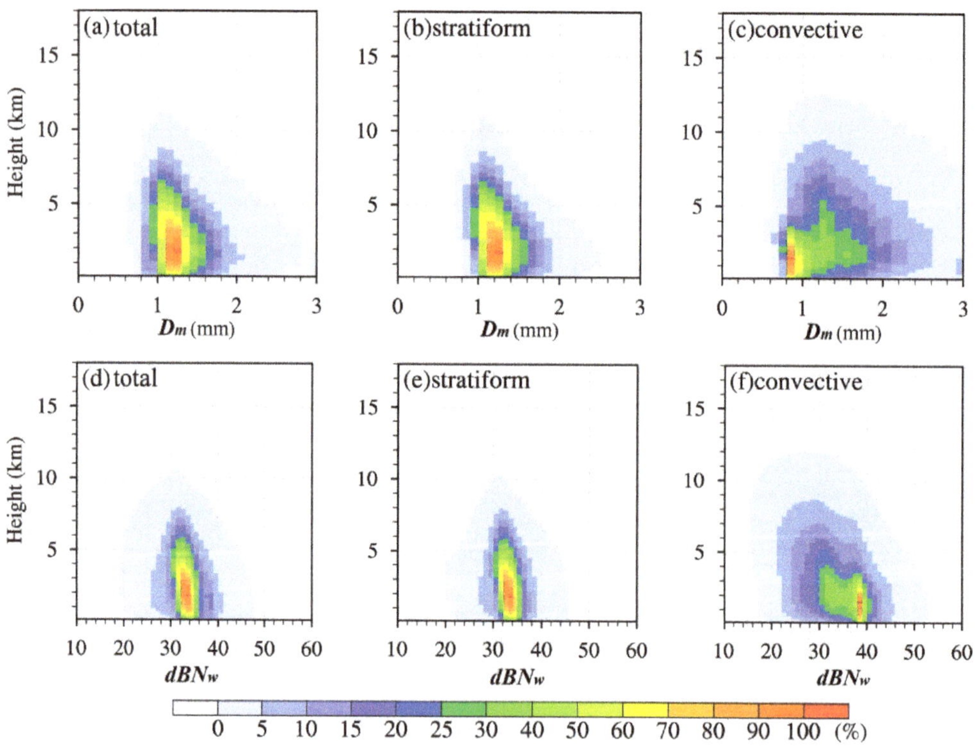

Figure 12. The contoured frequency (shading, %) by altitude diagram of D_m and dBN_w for (**a**,**d**) total, (**b**,**e**) stratiform, and (**c**,**f**) convective precipitation within 2000 km distance of the NCCV center, derived from GPM DPR for 2014–2019.

4. Discussion

It is important to know the limitations of the present study. To obtain the NCCV center, ERA5 data at a temporal resolution of 1 h and a spatial resolution of 0.25 degree are used in this study. Despite the high spatial–temporal resolution, it is still coarser than that of the DPR observations. The uncertainties during the processes in the identification of the NCCV center may introduce uncertainties to the results. However, the relatively large number of samples (6432 NCCVs) included in the synthetic analysis may partially cancel out the uncertainties. These things considered, since the GPM cannot continuously observe the precipitation systems, the analysis of the particle growth and precipitation process formation can only be derived by deduction, based on the dynamic composite maps.

Here, we try to provide some explanations on the distributions of NCCV precipitation frequency and intensity. First, the frequency of convective systems is relatively higher in the southern part of the NCCV. This may be because the NCCV provides a favorable circulation background to guide the cold air southward. If there are heating conditions at the lower level, an unstable stratification with high-level dry, cold air and low-level warm and humid air will be formed, which is conducive to the triggering of the convective system [5,35,36]. The warm and cold air usually intersect in the southern half of the NCCV, which is also the rear part of the warm and humid tongue [8]. At the same time, the coupling of high- and low-level jets under the background of NCCV causes a large-scale upward movement on the southeast side of the system, which invigorates deep convection and heavy rain [37]. In addition, the average storm-top height for all types of precipitation is highest in the southwest quadrant, while average rain rate in the NCCV is largest in the south of the NCCV. The mechanism of the obvious phase differences of storm-top height and rain rate is unknown, and a study of this will be carried out in our subsequent research.

5. Conclusions

To reveal the microphysical characteristics of NCCV precipitation, an NCCV coordinate system is firstly introduced in the present study. Under the coordinate system, the horizontal distribution, azimuth distribution, and vertical structures of stratiform and convective precipitation in the NCCV are hereafter explored during 2014–2019. The main findings are listed as follow.

The near-surface rain rate for convective precipitation in the NCCV is stronger than that for stratiform precipitation, while the convective precipitation frequency is lower than stratiform precipitation. The contribution of precipitation frequency and precipitation amount for stratiform precipitation are both larger than those for convection precipitation, which are generally higher than 70% and 60%, respectively. The regions with high convective and stratiform precipitation frequency have a comma-shaped distribution. With the increase in the distance from the NCCV center, the region with frequent stratiform precipitation occurrence shifts from the northeast quadrant to the southeast quadrant, while it is mainly located in the southeast quadrant for convective precipitation. The near-surface droplet sizes of the strong stratiform and convective rain rates inside the NCCV are not larger than those of smaller rain rates, while the droplet concentration is much higher. This indicates the great contribution of high droplet concentration to intense rain rate in the NCCV. The echo top of convective precipitation is higher than that for stratiform precipitation in the NCCV. Below 4 km, the radar reflectivity increases as the altitude decreases for both convective and stratiform precipitation, but with a much more obvious increase for convective precipitation, indicating more efficient collision-growth processes. Above 4 km, hydrometeor particles such as supercooled water and ice crystals in convective and stratiform precipitation grow through the Bergeron process. The stronger updraft in convective precipitation clouds may provide favorable environmental conditions for the growth of precipitation particles. As a result, the droplet concentration of convective precipitation is greater than that for stratiform below 5 km, but above 5 km, the concentration is smaller than stratiform and the droplet diameter is larger. There are shallow and deep convections in the convective precipitation inside the NCCV. Compared to shallow convection, deep convection has a larger droplet diameter, lower concentration, and stronger echo near the surface.

The precipitation and microphysical structures vary in different regions of the NCCV for stratiform and convective precipitation. Convective and stratiform precipitation mostly occurred in the south part of the NCCV, and the near-surface rain rates are also the largest in this region, especially in the southeast quadrant of the NCCV. The peak convective and stratiform rain rates are 6 mm h^{-1} and 2.67 mm h^{-1} in the southeast quadrant, respectively. In addition, the precipitation frequency in the southeast quadrant of the NCCV is also the largest, with the azimuthal averages of convective and stratiform precipitation frequencies reaching 1% and 4%, respectively. The frequency and precipitation contribution of stratiform precipitation are the lowest in the southwest of the NCCV and the highest in the northeast quadrant (reaching 92% and 87%, respectively). On the contrary, the contribution peaks of convective precipitation frequency and amount in the southwest quadrant of the NCCV are 19.8% and 37%, respectively. Convective and stratiform precipitation have peak storm-top heights in the southwest quadrant of the NCCV. The peak values of droplet concentration and diameter are in the southeast and west of the NCCV, respectively. In the northwest and southwest quadrants of the NCCV, convective clouds develop deeply, and the radar echo above the melting layer is stronger than those in other quadrants, which increases rapidly as the height decreases, indicating the microphysical processes such as collision-growth and rimming processes of ice crystals and other hydrometeors. Below the melting layer, as the height decreases, the radar echo increases rapidly, indicating collision-growth processes, leading to the prevalence of a low concentration of large-sized droplets at the near-surface, while due to the relatively insufficient water vapor in this quadrant, the near-surface rain rate is relatively low. In the southeast quadrant of the NCCV, the storm-top heights are low for convective and stratiform precipitation. The collision growth of droplets is more significant than that in other quadrants. However, due to the fragmentation of droplets during the falling process, the rain hydrometeors near the surface are mainly composed of high-concentration and small-sized droplets. The high concentrations of hydrometeors together with enough water vapor supply provide favorable conditions for heavy rain rate in this quadrant.

Previous studies have revealed the difference of DSDs in different seasons and regions and different precipitation systems [21,23,24,33]. As a continuation of this work, future work on the microphysical structures of NCCV precipitation in different seasons, and their differences among different precipitation systems (such as Meiyu precipitation) are worthy of further investigation using the joint observations from ground-based and satellite-based instruments, which will help us gain a deeper knowledge of NCCV precipitation.

Author Contributions: Conceptualization, X.Z. and F.C.; methodology, X.Z., F.C. and X.C.; software, F.C. and X.C.; validation, X.Z. and F.C.; formal analysis, J.W. and F.C.; investigation, X.Z. and F.C.; resources, X.Z.; data curation, F.C.; writing—original draft preparation, J.W.; writing—review and editing, X.Z. and F.C.; visualization, J.W., X.Z. and F.C.; supervision, X.Z., F.C. and Y.W.; project administration, X.Z; funding acquisition, X.Z. All authors have read and agreed to the published version of the manuscript.

Funding: This work was financially supported by the National Natural Science Foundation of China (42175006), the Fengyun Application Pioneering Project (FY-APP-2021.0101), Jiangsu Youth Talent Promotion Project (2021-084), the Basic Research Fund of CAMS (2020R002), and the National Natural Science Foundation of China (41805023).

Data Availability Statement: The ERA5 data are publicly available at https://cds.climate.copernicus.eu/cdsapp#!/dataset/reanalysis-era5-pressure-levels?tab=overview, accessed on 5 September 2022. The GPM DPR dataset can be obtained from https://gpm.nasa.gov/data/directory, accessed on 4 September 2022. The lists of 20-year NCCVs are available for free at Zenodo via https://doi.org/10.5281/zenodo.5571340, accessed on 1 September 2022. The Himawari-7 data used in our study can be downloaded from http://weather.is.kochi-u.ac.jp/sat/, accessed on 21 October 2022.

Acknowledgments: The authors thank the editors and three reviewers for their helpful comments and valuable suggestions, which improved the manuscript.

Conflicts of Interest: The authors declare no conflict of interest.

References

1. Chen, X.; Zhuge, X.; Zhang, X.; Wang, Y.; Xue, D. Objective Identification and Climatic Characteristics of Heavy-Precipitation Northeastern China Cold Vortexes. *Adv. Atmos. Sci.* **2023**, 40, 305–316. [CrossRef]
2. Wang, W.; Li, J.; Hu, C.; Li, J.; Jiao, M. A review of definition, identification and quantitative investigation on Northeast cold vortex. *Sci. Meteor. Sin.* **2017**, 37, 394–402. (In Chinese)

3. Sun, L.; Zheng, X.; Wang, Q. The climatological characteristics of northeast cold vortex in China. *J. Appl. Meteor Sci.* **1994**, *5*, 297–303. (In Chinese)
4. Zheng, X.; Zhang, T.; Bai, R. *Rainstorm in Northeast China*, 6th ed.; Meteorological Press: Beijing, China, 1992; pp. 129–151.
5. Li, S.; Ding, Z.; Dai, P.; Liu, Y.; Han, Y. Recent Advances in Research on Northeast China Cold Vortex. *J. Arid. Meteor.* **2016**, *34*, 13–19. (In Chinese) [CrossRef]
6. Sun, L.; An, G.; Gao, Z.; Tang, X.; Ding, L.; Shen, B. A composite diagnostic study of heavy rain caused by the northeast cold vortex over Songhuajiang-Nenjiang River Basin in summer of 1998. *J. Appl. Meteor. Sci.* **2002**, *13*, 156–162. (In Chinese)
7. Shen, X.; Zhang, C.; Gao, H.; Wang, L.; Li, X. Classification and dynamic composite analysis of three kinds of high altitude cold vortex. *Torrential Rain Disasters* **2020**, *39*, 1–9. (In Chinese) [CrossRef]
8. Bai, R.; Sun, Y. The Background Analysis Study of Mesoscale Weather of the Cold Vortex in Northeast China. *Heilongjiang Meteor.* **1997**, *3*, 7–8. (In Chinese)
9. Zhang, Y.; Lei, H.; Qian, Z. Analyses of Formation Mechanisms of a Rainstorm during the Declining Phase of a Northeast Cold Vortex. *Chin. J. Atmos. Sci.* **2008**, *32*, 481–498. (In Chinese)
10. Liu, Y.; Wang, D.; Zhang, Z.; Zhong, S. A comprehensive analysis of the structure of a northeast China-cold-vortex and its characteristics of evolution. *Acta Meteor. Sin.* **2012**, *70*, 354–370. (In Chinese) [CrossRef]
11. Chen, L.; Zhang, L.; Zhou, X. Characteristic of lnstable Energy Distribution in Cold Vortex over Northeastern China and lts Relation to Precipitation Area. *Plateau Meteor.* **2008**, *27*, 339–348. (In Chinese)
12. Qi, D.; Yuan, M.; Zhou, Y.; Han, B. Analysis of the Relationship between Structures of a Cold Vortex Process and Rainfall over the Northeast China. *Plateau Meteor.* **2020**, *39*, 808–818. (In Chinese) [CrossRef]
13. Hsieh, Y.-P. An Investigation of A Selected Cold Vortex over North America. *J. Meteor.* **1949**, *6*, 401–410. [CrossRef]
14. Pruppacher, H.R.; Klett, J.D.; Wang, P.K. Microphysics of Clouds and Precipitation. *Aerosol Sci. Tech.* **1998**, *28*, 381–382. [CrossRef]
15. Qi, Y.; Guo, X.; Jin, D. An Observational Study of Macro/Microphysical Structures of Convective Rainbands of a Cold Vortex over Northeast China. *Chin. J. Atmos. Sci.* **2007**, *31*, 621–634. (In Chinese) [CrossRef]
16. Zhao, Z.; Lei, H. Observed Microphysical Structure of Nimbostratus in Northeast Cold Vortex over China. *Atmos. Res.* **2014**, *142*, 91–99. [CrossRef]
17. Zhong, S.; Wang, D.; Zhang, R.; Liu, Y. Vertical Structure of Convective Cloud in a Cold Vortex over Northeastern China Using CloudSat Data. *J. Appl. Meteor. Sci.* **2011**, *22*, 257–264. (In Chinese) [CrossRef]
18. Li, S.; Li, Y.; Sun, G.; Song, W. Cloud microphysical characteristics in the development of stratocumulus clouds over Eastern China. *Chin. J. Geophys.* **2019**, *62*, 4513–4526.
19. Iguchi, T.; Seto, S.; Meneghini, R.; Yoshida, N.; Awaka, J.; Le, M.; Chandrasekar, V.; Brodzik, S.; Kubota, T.; Takahashi, N. GPM/DPR Level-2 Algorithm Theoretical Basis Document. JAXA–NASA Tech. Rep. 2021; 238p. Available online: https://gpm.nasa.gov/sites/default/files/2022-06/ATBD_DPR_V07A.pdf (accessed on 19 April 2023).
20. Chen, F.; Huang, H. Comparisons of Gauge, TMPA and IMERG Products for Monsoon and Tropical Cyclone Precipitation in Southern China. *Pure Appl. Geophys.* **2019**, *176*, 1767–1784. [CrossRef]
21. Chen, F.; Fu, Y.; Yang, Y. Regional Variability of Precipitation in Tropical Cyclones over the Western North Pacific Revealed by the GPM Dual-Frequency Precipitation Radar and Microwave Imager. *J. Geophys. Res. Atmos.* **2019**, *124*, 11281–11296. [CrossRef]
22. Chen, F.; Fu, Y.; Liu, P.; Yang, Y. Seasonal Variability of Storm Top Altitudes in the Tropics and Subtropics Observed by TRMM PR. *Atmos. Res.* **2016**, *169*, 113–126. [CrossRef]
23. Zhang, A.; Chen, Y.; Pan, X.; Hu, Y.; Chen, S.; Li, W. Precipitation Microphysics of Tropical Cyclones over Northeast China in 2020. *Remote Sens.* **2022**, *14*, 2188. [CrossRef]
24. Zhang, A.; Chen, Y.; Zhang, X.; Zhang, Q.; Fu, Y. Structure of Cyclonic Precipitation in the Northern Pacific Storm Track Measured by GPM DPR. *J. Hydrometeorol.* **2020**, *21*, 227–240. [CrossRef]
25. Hou, A.Y.; Kakar, R.K.; Neeck, S.; Azarbarzin, A.A.; Kummerow, C.D.; Kojima, M.; Oki, R.; Nakamura, K.; Iguchi, T. The Global Precipitation Measurement Mission. *Bull. Amer. Meteor. Soc.* **2014**, *95*, 701–722. [CrossRef]
26. Zhang, A.; Fu, Y. Life Cycle Effects on the Vertical Structure of Precipitation in East China Measured by Himawari-8 and GPM DPR. *Mon. Weather Rev.* **2018**, *146*, 2183–2199. [CrossRef]
27. Hamada, A.; Takayabu, Y.N. Improvements in Detection of Light Precipitation with the Global Precipitation Measurement Dual-Frequency Precipitation Radar (GPM DPR). *J. Atmos. Ocean. Tech.* **2016**, *33*, 653–667. [CrossRef]
28. Iguchi, T.; Seto, S.; Meneghini, R.; Yoshida, N.; Awaka, J.; Kubota, T.; Kozu, T.; Chandra, V.; Le, M.; Liao, L.; et al. *An Overview of the Precipitation Retrieval Algorithm for the Dual-Frequency Precipitation Radar (DPR) on the Global Precipitation Measurement (GPM) Mission's Core Satellite. Kyoto, Japan, 9 November 2012*; Shimoda, H., Xiong, X., Cao, C., Gu, X., Kim, C., Kiran Kumar, A.S., Eds.; SPIE: Bellingham, WA, USA, 85281C. [CrossRef]
29. Huang, H.; Zhao, K.; Fu, P.; Chen, H.; Chen, G.; Zhang, Y. Validation of Precipitation Measurements from the Dual-Frequency Precipitation Radar Onboard the GPM Core Observatory Using a Polarimetric Radar in South China. *IEEE Trans. Geosci. Remote Sens.* **2022**, *60*, 1–16. [CrossRef]
30. Seto, S.; Iguchi, T.; Oki, T. The Basic Performance of a Precipitation Retrieval Algorithm for the Global Precipitation Measurement Mission's Single/Dual-Frequency Radar Measurements. *IEEE T. Geosci. Remote* **2013**, *51*, 5239–5251. [CrossRef]
31. Hersbach, H.; Bell, B.; Berrisford, P.; Hirahara, S.; Horányi, A.; Muñoz Sabater, J.; Nicolas, J.; Peubey, C.; Radu, R.; Schepers, D.; et al. The ERA5 global reanalysis. *Q. J. Roy. Meteor. Soc.* **2020**, *146*, 1999–2049. [CrossRef]

32. Li, S.; Ding, Z.; Liu, Y.; Tian, L. Statistical and composite analysis on short-time heavy rainfall of Liaoning province under Northeast China cold vortex. *Sci. Meteor. Sin.* **2017**, *37*, 70–77. (In Chinese) [CrossRef]
33. Radhakrishna, B.; Satheesh, S.K.; Rao, T.R.; Saikranthi, K.; Sunilkumar, K. Assessment of DSDs of GPM-DPR with ground-based disdrometer at seasonal scale over Gadanki, India. *J. Geophys. Res. Atmos.* **2016**, *121*, 11792–11802. [CrossRef]
34. Xian, T.; Fu, Y. Characteristics of Tropopause-Penetrating Convection Determined by TRMM and COSMIC GPS Radio Occultation Measurements: Tropopause-Penetrating Convection. *J. Geophys. Res. Atmos.* **2015**, *120*, 7006–7024. [CrossRef]
35. Knippertz, P.; Martin, J.E. Tropical plumes and extreme precipitation in subtropical and tropical West Africa. *Q. J. Roy. Meteor. Soc.* **2010**, *131*, 2337–2365. [CrossRef]
36. Zhong, S.; Wang, D.; Zhang, R.; Liu, Y. Study of Mesoscale Convective System in Heavy Rainstorm Process at a Cold Vortex Development Stage. *Plateau Meteor.* **2013**, *32*, 435–445. (In Chinese) [CrossRef]
37. Wang, Z.; Li, J.; Wang, F.; Lin, C. Asymmetric Characteristics of the Northeast Cold Vortex and Its Effect on Heavy Rain. *Plateau Meteor.* **2015**, *34*, 1721–1731. (In Chinese) [CrossRef]

remote sensing

MDPI

Article

Assimilating AMSU-A Radiance Observations with an Ensemble Four-Dimensional Variational (En4DVar) Hybrid Data Assimilation System

Shujun Zhu [1,2], Bin Wang [1,2,3,4,*], Lin Zhang [5,6], Juanjuan Liu [1,4], Yongzhu Liu [5,6], Jiandong Gong [5,6], Shiming Xu [2], Yong Wang [2], Wenyu Huang [2], Li Liu [2], Yujun He [1], Xiangjun Wu [5,6], Bin Zhao [5,6] and Fajing Chen [5,6]

1 State Key Laboratory of Numerical Modelling for Atmospheric Sciences and Geophysical Fluid Dynamics, Institute of Atmospheric Physics, Chinese Academy of Sciences, Beijing 100029, China
2 Department of Earth System Science, Tsinghua University, Beijing 100084, China
3 Southern Marine Science and Engineering Guangdong Laboratory, Zhuhai 519082, China
4 College of Ocean, University of Chinese Academy of Sciences, Qingdao 266400, China
5 CMA Earth System Modeling and Prediction Centre, China Meteorological Administration, Beijing 100081, China
6 State Key Laboratory of Severe Weather, China Meteorological Administration, Beijing 100081, China
* Correspondence: wab@lasg.iap.ac.cn

Abstract: Many ensemble-based data assimilation (DA) methods use observation space localization to mitigate the sampling errors due to the insufficient ensemble members. Observation space localization is simpler and more timesaving than model space localization in implementation, but more difficult to directly assimilate satellite radiance observations, a kind of non-local observations. The vertical locations of radiance observations are undetermined and the transmission of observational information is thereby obstructed. To determine the vertical coordinates of radiance observations, a weighted average hypsometry is proposed. Using this hypsometry, AMSU-A radiance observations are directly assimilated with an ensemble four-dimensional variational (En4DVar) DA system. It consists of a four-dimensional ensemble-variational (4DEnVar) system providing ensemble covariance and a 4DVar system. Observing system simulation experiments show that the hypsometry alleviates the degradations in the late period of medium-range forecast in the Northern Extratropics that occur in the traditional peak-based hypsometry. It obviously improves the analysis qualities and forecast skills of the En4DVar system and its two components, especially in the Southern Extratropics, when incorporating AMSU-A radiance observations. The improvement in the En4DVar-initialized forecast is comparable to that in the 4DVar-initialized forecast in the Southern Extratropics and Tropics. It indicates that a proper hypsometry enables efficient extraction of useful information from AMSU-A radiance observations by 4DEnVar with observation space localization. Therefore, the 4DEnVar provides high-quality ensemble covariances for En4DVar.

Keywords: AMSU-A radiance observation; ensemble four-dimensional variational data assimilation; observation space localization; weighted average hypsometry

Citation: Zhu, S.; Wang, B.; Zhang, L.; Liu, J.; Liu, Y.; Gong, J.; Xu, S.; Wang, Y.; Huang, W.; Liu, L.; et al. Assimilating AMSU-A Radiance Observations with an Ensemble Four-Dimensional Variational (En4DVar) Hybrid Data Assimilation System. *Remote Sens.* **2023**, *15*, 3476. https://doi.org/10.3390/rs15143476

Academic Editors: Xiaolei Zou, Guoxiong Wu and Zhemin Tan

Received: 30 May 2023
Revised: 2 July 2023
Accepted: 3 July 2023
Published: 10 July 2023

1. Introduction

The ensemble four-dimensional variational (En4DVar) hybrid data assimilation (DA) approach incorporates the advantage of flow-dependent characteristic of ensemble Kalman filter (EnKF) into the 4DVar DA approach. It has become popular in major operational centers of the world and shown great potential for further improving numerical weather prediction (NWP) skills [1–5]. The En4DVar approach typically uses flow-dependent information extracted from the ensemble forecasts to help estimate the background error covariance (BEC) for 4DVar. When applying this approach to global NWPs, the ensemble size is much smaller than the dimensionality of the state variables of the prediction model due to the limitations of computational resource. The limited ensemble size may easily result in spurious correlations between two grids that are far apart in the BEC matrix (B-matrix). Localization techniques [6–10] can effectively mitigate such spurious correlations and thus improve the analysis quality and forecast skill.

The main idea of localization is to restrict the analysis at a specific grid point to be influenced only by observations within its surrounding local region. Houtekamer and Mitchell [6] performed observation selection by setting a cutoff radius, thus excluding the influence of observations outside the cutoff radius on the analysis at the specific grid point. Houtekamer et al. [11] further proposed a localization scheme using a compactly supported GC function [12] that decreases monotonically with distance, which is realized as a Schür product between the ensemble B-matrix and the localization correlation matrix. Localization can be achieved by assimilating observations one by one to update the analyses within the local region, or implicitly by simultaneously assimilating all observations within the local region around the analysis at a specific grid point [13]. However, for ensemble-based non-sequential DA approaches, using the localized covariance directly in model space can easily lead to large computational costs if one wants to solve directly in the global space. Implementing localization in observation space may be a more economical choice. Approaches that use orthogonal functions (e.g., empirical orthogonal function and sine function) to decompose the localization correlation matrix can also further reduce the localization cost in observation space [8,10,14–17].

The elements of the localization correlation matrix in observation space are dependent on the observation coordinates, which are easy to be calculated for conventional observations with well-defined positions. Following the development of satellite technology, the rapidly increasing radiance observations have significantly improved the medium-range forecasts and have greatly reduced the gap of forecast skills between the northern and southern hemispheres [18]. It is noted that radiance observations are non-local due to the sampling of multiple atmospheric layers, and different satellite channels are sensitive to different atmospheric layers. Therefore, defining the vertical coordinates of radiance observations is an unavoidable challenge in the use of observation space localization. Houtekamer et al. [11] used the pressure at the peak of the weighting function to define the vertical coordinates of radiance observations from the Advanced Microwave Sounding Unit-A (AMSU-A) instruments in the EnKF system. Fertig et al. [19] selected radiance observations within the local region to be assimilated into the LETKF system if a weight above the cutoff value is signed to any model state in the local region. This cutoff-based selection method is significantly different from the abovementioned peak-based selection method of Houtekamer et al. [11] when the cutoff value is small and the weighting function is broad. In addition, it allows a wider range of influence for non-local observations than for local observations. Miyoshi et al. [20] used the normalized weighting functions of satellite channels to provide weights for error covariance localization in the LETKF system.

Based on these studies, the EnKF class approaches have significantly benefited from assimilating radiance observations. In particular, the effective assimilation of AMSU-A radiance observations in the DA system plays important role in improving forecast skills. The 4DVar and En4DVar approaches usually use model space localization method for radiance observations and their DA schemes are globally solved. In contrast, the EnKF class approaches generally adopt the observation space localization method that uses vertical coordinates defined by the pressure at peak weight to achieve the effective assimilation of radiance observations and is solved in local regions. Moreover, Campbell et al. [21] pointed out that when the number of satellite channels is large enough and the observation error is very small, the observation space localization is difficult to recover the true state. Thus, investigating the differences between the effects of model space and observation space localization techniques on the assimilation and forecast performances when incorporating radiance observations is beneficial for efficiently using radiance observations in ensemble-based data assimilations.

The purpose of this study is to use the weighted average pressure to define the vertical coordinates of AMSU-A radiance observations in the En4DVar system so as to investigate the contributions of these observations to the improvements of analysis quality and forecast skill. In Section 2, a brief description of the DA methods used in this study, including En4DVar and its 4DVar and 4DEnVar components, the observation space localization scheme and the vertical positioning method for radiance observations, are presented. Section 2 also displays the DA configurations, experimental details and observations, followed by the analysis and forecast results in Section 3. Finally, the conclusions and discussions are provided in the last section.

2. Materials and Methods

2.1. A Brief Description of DA Methods

The 4DVar system used here is based on the incremental 4DVar scheme [22], which obtains the optimal analysis of atmospheric state on a low-resolution grid by combining forecast and observation information. It adopts a highly parameterized climatological B-matrix: $\mathbf{B}_c = \mathbf{U}\mathbf{U}^{\mathrm{T}}$ [4,23], and relies on the adjoint model (ADM) in the minimization process of solving the optimal analysis.

The ensemble covariance for the En4DVar system is provided by the 4DEnVar system [24]. The 4DEnVar system is established using the dimension-reduced projection four-dimensional variational (DRP-4DVar) method [25]. This method uses the ensemble samples to project the minimization problem of 4DVar in the original model space onto the reduced-dimensional subspace, and to avoid using the ADM in the minimization process.

The En4DVar system used in this study consists of two components including the abovementioned 4DVar and 4DEnVar systems [26]. It uses a hybrid BEC ($\mathbf{B} = \gamma_c\mathbf{B}_c + \gamma_e\mathbf{B}_e$) achieved through the extended control variable approach [27]. Here, $\mathbf{B}_e$ is the ensemble covariance produced by the 4DEnVar component and $\mathbf{B}_c$ is the climatological covariance in the 4DVar component. The variables γ_c and γ_e represent the scalar weights of the climatological and ensemble covariances, respectively. Unlike other variants of 4DEnVar, the hybrid BEC used here consists of a three-dimensional (3D) climatological covariance from 4DVar and a 4D ensemble covariance from 4DEnVar. In addition, calculating the gradient of cost function in the ensemble component does not contain the ADM but uses the same statistical relationship as in the 4DEnVar system. More details about the En4DVar system can be found in Zhu et al. [26].

2.2. Localization

2.2.1. Observation Space Localization

In the En4DVar system, the localization for the climatological covariance is contained in its square root $\mathbf{U}$, while the localization for the ensemble covariance is the same as in the 4DEnVar system. The traditional localization scheme based on the Schür product between the high-dimensionality ensemble B-matrix $\mathbf{B}_e = \mathbf{p}_\mathbf{x}\mathbf{p}_\mathbf{x}^{\mathrm{T}}$ and the high-dimensionality correlation matrix $\mathbf{C}$ may lead to large computational costs. Approximately decomposing the correlation matrix [8,10] and ignoring the time-variation of localization, the localization can be economically achieved in observation space by the Schür products between a finite number of observational perturbation samples and localization leading eigenvectors:

$$\begin{cases} \mathbf{Ep}_\mathbf{x} = \left[(\mathbf{p}_{\mathbf{x},1} \circ \boldsymbol{\rho}_{\mathbf{x},1}, \cdots, \mathbf{p}_{\mathbf{x},1} \circ \boldsymbol{\rho}_{\mathbf{x},L}), \cdots, (\mathbf{p}_{\mathbf{x},N} \circ \boldsymbol{\rho}_{\mathbf{x},1}, \cdots, \mathbf{p}_{\mathbf{x},N} \circ \boldsymbol{\rho}_{\mathbf{x},L})\right] \\ \mathbf{Ep}_\mathbf{y} = \left[\left(\mathbf{p}_{\mathbf{y},1} \circ \boldsymbol{\rho}_{\mathbf{y},1}, \cdots, \mathbf{p}_{\mathbf{y},1} \circ \boldsymbol{\rho}_{\mathbf{y},L}\right), \cdots, \left(\mathbf{p}_{\mathbf{y},N} \circ \boldsymbol{\rho}_{\mathbf{y},1}, \cdots, \mathbf{p}_{\mathbf{y},N} \circ \boldsymbol{\rho}_{\mathbf{y},L}\right)\right] \end{cases}. \quad (1)$$

Here, $\mathbf{p}_\mathbf{x}(i = 1, 2, \cdots, N)$ and $\boldsymbol{\rho}_{\mathbf{x},j}(j = 1, 2, \cdots, L)$ denote the initial perturbation samples and the leading eigenvectors in model space, and $\mathbf{p}_\mathbf{y}(i = 1, 2, \cdots, N)$ and $\boldsymbol{\rho}_{\mathbf{y},j}(j = 1, 2, \cdots, L)$ denote the corresponding observational perturbation samples and leading eigenvectors. To further reduce the cost, the leading eigenvectors are selected based on the cumulative contribution of variance, and each leading eigenvector has three components in zonal, meridional and vertical directions, respectively. The empirical orthogonal function [8] and the sine function [10] are used for the zonal and vertical components, and the meridional component, respectively. Thus, the ensemble component of the En4DVar system is solved in the subspace consisting of the extended perturbation samples that are generated by the abovementioned Schür products. For more details refer to Zhu et al. [24].

For local observations, the approximation in the extended perturbation samples introduces little error. In contrast, it is more complicated for non-local observations, such as radiance observations, which does not have explicit vertical coordinates. In order to implement the vertical localization, we need to properly define the vertical coordinates of radiance observations.

2.2.2. Vertical Positioning of AMSU-A Radiance Observation

As mentioned earlier, each radiance observation depends on the atmospheric states at multiple vertical layers. Therefore, its vertical coordinate cannot be given explicitly as conventional observations. The weighting function of the radiance observation at a specific horizontal position reflects the contribution of the observation to the atmospheric state at different vertical layer [20]. The weighting function is typically calculated by the vertical difference of the transmittance of the satellite channel, which is dependent not only on the satellite channel but also on the atmospheric profile. In this study, we proposed the weighted average hypsometry to define the vertical coordinates of AMSU-A radiance observations:

$$P_{jpr,lch} = \frac{\sum_{k=1}^{K} w_{jpr,lch,k} \times p_{jpr,k}}{\sum_{k=1}^{K} w_{jpr,lch,k}} \tag{2}$$

Here, the subscripts jpr, lch and k denote the atmospheric profile, the satellite channel and the atmospheric vertical layer; K is the number of atmospheric vertical layers; P denotes the vertical coordinate of radiance; w and p represent the weight of the satellite channel and the pressure of the atmospheric profile.

In practice, when the radiance observations are assimilated, the Radiative Transfer for TOVS (RTTOV) model calculates the transmittance for each satellite channel. Therefore, the transmittance can be obtained directly from the RTTOV model for the calculation of weighting functions.

2.3. *DA Configurations, Experimental Details and Observations*

2.3.1. DA Configurations

The model used here is the operational global forecast system of China Meteorological Administration (CMA-GFS), whose original name was the Global/Regional Assimilation and Prediction System (GRAPES-GFS) [28]. The DA systems used in this study include the 4DVar system [23], and the recently developed 4DEnVar system and En4DVar system [24,26].

A dual-resolution framework with 1.0° for the inner loop and 0.5° for the outer loop, and 87 vertical layers are adopted by all DA systems. In the first assimilation window, the 4DEnVar system utilizes random perturbation samples with balanced constraints generated using the "randomcv" method [29]. This method can generate reasonable initial condition (IC) samples with balanced constraints by using the variational variable transform. Then, the 4DEnVar system updates the flow-dependent perturbation samples every 6 h in subsequent assimilation windows by assimilating perturbed observations. An extended-ensemble-sample-based localization method mentioned in Section 2.2.1 is applied in the 4DEnVar system. To alleviate the filter divergence problem, inflation, observation perturbation and SST perturbation approaches are applied [24].

The En4DVar system constructs the hybrid BEC by incorporating the ensemble covariance estimated by 60 ensemble members from the 4DEnVar system into the climatological BEC of the 4DVar system. The scalar weights of the climatological and ensemble covariances for the hybrid BEC in the En4DVar system are 0.25 and 0.8. Moreover, the ensemble covariance of the En4DVar system utilizes the same localization scheme as in the 4DEnVar system.

2.3.2. Experimental Details

Observing system simulation experiment (OSSE) allows an objective assessment of the assimilation and forecast performances of a DA system when the "truth" state is known [25,30–32]. In this study, cycled assimilation experiments and corresponding initialized forecast experiments were performed based on the OSSE.

The design of OSSE is similar to Zhu et al. [24,26]. The background field in the first assimilation window and the "truth" state (or "truth") were generated by the low-resolution and high-resolution versions of the CMA-GFS model, which were initialized from the ERA-Interim 6 h forecast field and the ERA-5 reanalysis field, respectively. For a fair comparison, all assimilation experiments used the same background field in the first assimilation window.

Based on previous experiments assimilating conventional observations [24,26], further experiments adding AMSU-A radiance observations were carried out. The 1-week sensitivity experiments on the basis of the pressure at peak weight and the weighted average pressure were initially conducted to determine the vertical positioning method for the observation space localization. The most favorable vertical positioning method for the forecast performance was adopted. Then, three classes of experiments were designed, which covers a period of about 1 month starting from 0900 UTC 11 September 2016 and taking the first 2 days for spin-up. The first class includes the ensemble DA experiments using the 4DEnVar system and their initialized forecast experiments, and the second one contains the hybrid DA and forecast experiments with the En4DVar system that require the flow-dependent data produced by the first one. Additionally, the third class of experiments, i.e., the standard 4DVar DA and forecast experiments, were conducted for comparisons.

Each class includes two sets of DA experiments, respectively, incorporating only conventional observations [24,26] and both conventional and AMSU-A radiance observations (simply all types of observations, hereinafter), and two sets of corresponding initialized forecast experiments. Totally twelve sets of experiments were conducted to investigate the effects of adding AMSU-A radiance observations on the assimilation and forecast performances of the En4DVar system and its 4DVar and 4DEnVar component systems. The analyses of assimilating only conventional observations and assimilating all types of observations and their initialized forecasts were compared to assess the contributions of AMSU-A radiance observations. The experiments upon the standard 4DVar and 4DEnVar systems were conducted to provide the references for evaluating the performance of the En4DVar system when adding AMSU-A radiance observations.

2.3.3. Observations

The "observations" were extracted from the "truth" state by using the transformations of observation operators and superimposing observation errors. The conventional observations used in this study were obtained from sounding and cloud-derived wind, and more details are presented in Zhu et al. [24]. Additionally, radiance from AMSU-A instruments of NOAA 15, 18, 19, NPP, and Metop A, B were also utilized. Sounding observations are sampled every 6 h, while both cloud-derived wind observations and AMSU-A radiance observations are sampled every 30 min. The radiance observations are assimilated using Version 12 of the RTTOV model [33] as the observation operator. To avoid the negative impacts of ground albedo and interpolation errors at upper layers, only channels 5–14 of the AMSU-A radiance observations were assimilated. Conventional observations cover most of the Northern Hemisphere, with a lower sampling density in the Southern Hemisphere. In contrast, radiance observations have a wider sampling range, which especially compensates for the low coverage of conventional observations in the Southern Hemisphere (Figure 1).

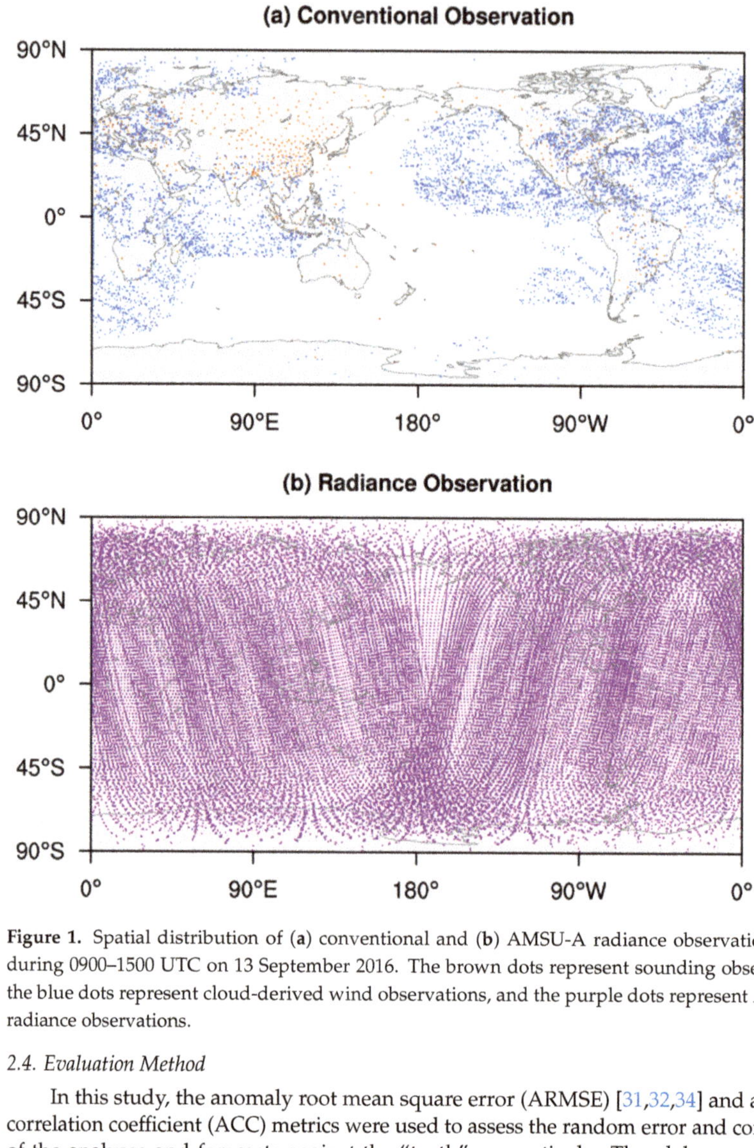

Figure 1. Spatial distribution of (**a**) conventional and (**b**) AMSU-A radiance observations valid during 0900–1500 UTC on 13 September 2016. The brown dots represent sounding observations, the blue dots represent cloud-derived wind observations, and the purple dots represent AMSU-A radiance observations.

2.4. Evaluation Method

In this study, the anomaly root mean square error (ARMSE) [31,32,34] and anomaly correlation coefficient (ACC) metrics were used to assess the random error and correlation of the analyses and forecasts against the "truth", respectively. The globe was divided into Northern Extratropics (20°N~90°N), Southern Extratropics (20°S~90°S) and Tropics (20°S~20°N) for calculating the statistical results of these metrics. Moreover, a score card, which is marked with the significance of performance difference, was used to conveniently exhibit the performance difference between two forecasts initialized from different analyses in terms of ARMSE and ACC. Note that the analyses and forecasts from the 4DEnVar system are its ensemble mean analyses and deterministic 10-day forecasts initialized from these ensemble mean analyses. For more details about the evaluation methods refer to Zhu et al. [24,26].

3. Results

3.1. Vertical Positioning Method

In this subsection, the vertical positioning method was determined by a set of sensitivity experiments. The purpose of these experiments is to investigate the effects of two vertical coordinate definitions of AMSU-A radiance observations including the pressure at peak weight and the weighted average pressure on the forecast skill of the 4DEnVar system.

Figure 2a shows the scorecard of the 4DEnVar-initialized forecasts assimilating all types of observations with the pressure at peak weight as the vertical coordinates of AMSU-A radiance observations against those assimilating only conventional observations in terms of ACC and ARMSE. Encouragingly, the addition of AMSU-A radiance observations leads to significant improvements of the forecasts, especially in the Southern Extratropics and Tropics, except the degradation in the late period of the medium range over the Northern Extratropics (Figure 2a). Meanwhile, similar impacts of the AMSU-A radiance observations on the forecasts can be observed when the weighted average pressure is used as their vertical coordinates, but the degradation shown in Figure 2a is alleviated (Figure 2b). Therefore, the weighted average pressure was finally chosen to define the vertical coordinates of AMSU-A radiance observations in this study.

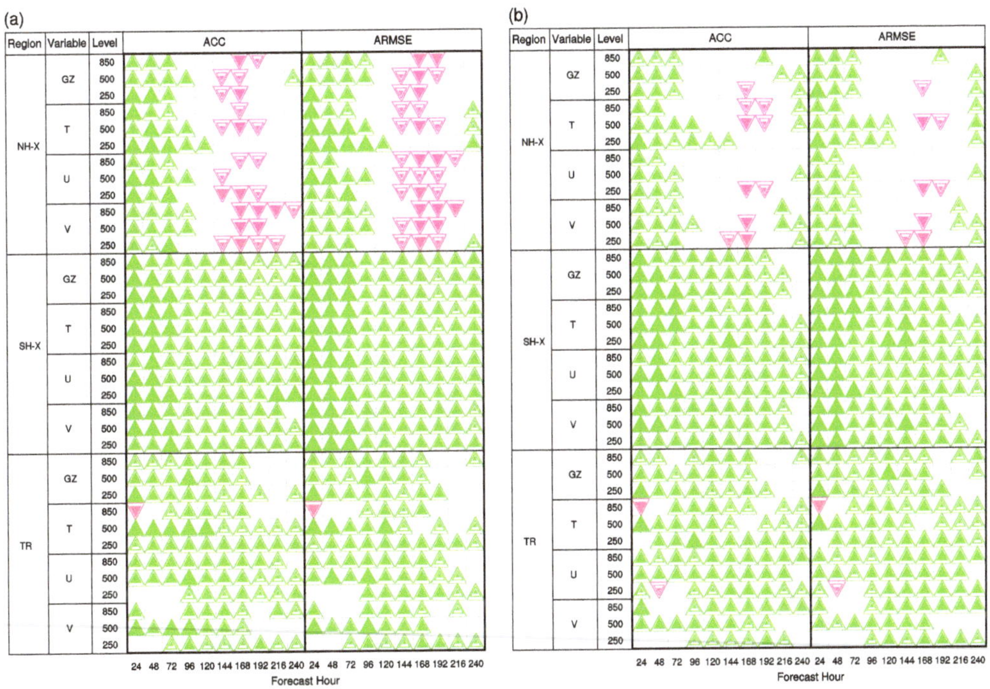

Figure 2. The scorecards of the 4DEnVar-initialized geopotential height (GZ), temperature (T), zonal wind (U) and meridional wind (V) forecasts assimilating all types of observations with (**a**) the pressure at peak weight and (**b**) the weighted average pressure as the vertical coordinates of AMSU-A radiance observations against those assimilating only conventional observations. The filling size of the triangle shows the difference significance of anomaly correlation coefficient (ACC) or anomaly root mean square error (ARMSE) between the evaluated and reference forecasts. The largest filling size represents very significant difference, and the other two decreasing filling sizes represent significant and insignificant differences. The green upward-pointing (purple downward-pointing) triangles are plotted if the evaluated forecast is better (worse) than the reference forecast. No triangles indicate equivalent.

3.2. Effects of AMSU-A Radiance Observations on Analysis Quality

After the vertical positioning method was determined, the effects of AMSU-A radiance observations on the analysis qualities of the DA systems were evaluated. Based on the En4DVar system and its two components, the analysis errors of assimilating all types of observations were compared with those of assimilating only conventional observations so as to investigate whether the AMSU-A radiance observations benefit the analysis quality in different DA systems. The results of the 4DVar and 4DEnVar component systems were used as the references to assess the effectiveness of the En4DVar system on assimilating AMSU-A radiance observations.

Figure 3 shows the contributions of AMSU-A radiance observations to the decreases in analysis error in the En4DVar system and its two components. It is found that all three DA systems reduced the ARMSE on all vertical layers except very few layers over Northern Extratropics and Tropic when the AMSU-A radiance observations joined the analyses. In particular, the decreases in ARMSE in all basic variables except specific humidity are most significant in the Southern Extratropics, especially in the stratosphere where conventional observations are sparsely distributed (Figure 3, column 2). However, the most significant improvement in the specific humidity analysis is located in the Tropics (Figure 3, row 4) where the water vapor content is high. As for the comparisons among three DA systems, they have different performances on different variables in different regions. The improvement in the 4DEnVar ensemble mean analysis is more (less) significant than in the 4DVar analysis on geopotential height (specific humidity), and comparable on zonal wind and temperature. It is more obvious on temperature (zonal wind and temperature) at the middle (upper) layers in the Tropics (Northern and Southern Extratropics), but less obvious on temperature at the lower layers, and on zonal wind at the upper (middle and lower) layers in the Tropics (Northern and Southern Extratropics). The improvement in the En4DVar analysis is generally between those in the analyses from its two component systems. There is larger improvement in zonal wind (temperature) at the middle layers in the Southern Extratropics and on the layers below 100 hPa in the Tropics (at the upper layers in the Tropics). Smaller improvement in temperature is in the middle and upper troposphere in the Northern Extratropics (Figure 3, rows 2 and 3).

The effects of adding AMSU-A observations on the error structures of the En4DVar and 4DVar analyses and the 4DEnVar ensemble mean analyses were shown in Figure 4. First, the analysis errors of all three DA systems are significantly reduced in most regions, indicating that the AMSU-A radiance observations have an overall positive effect on the analysis quality. Second, it is found that AMSU-A radiance observations most significantly reduces the analysis errors of the geopotential height, zonal wind and temperature in the Southern Extratropics, especially near 60°S, where conventional observations are sparsely distributed. In addition, the analysis errors of the specific humidity are significantly reduced not only in the Southern Extratropics, but also in the Tropics (Figure 4, row 4). Finally, the improvement of analysis by the En4DVar is generally between those by the 4DVar and 4DEnVar. These results are consistent with the findings in Figure 3.

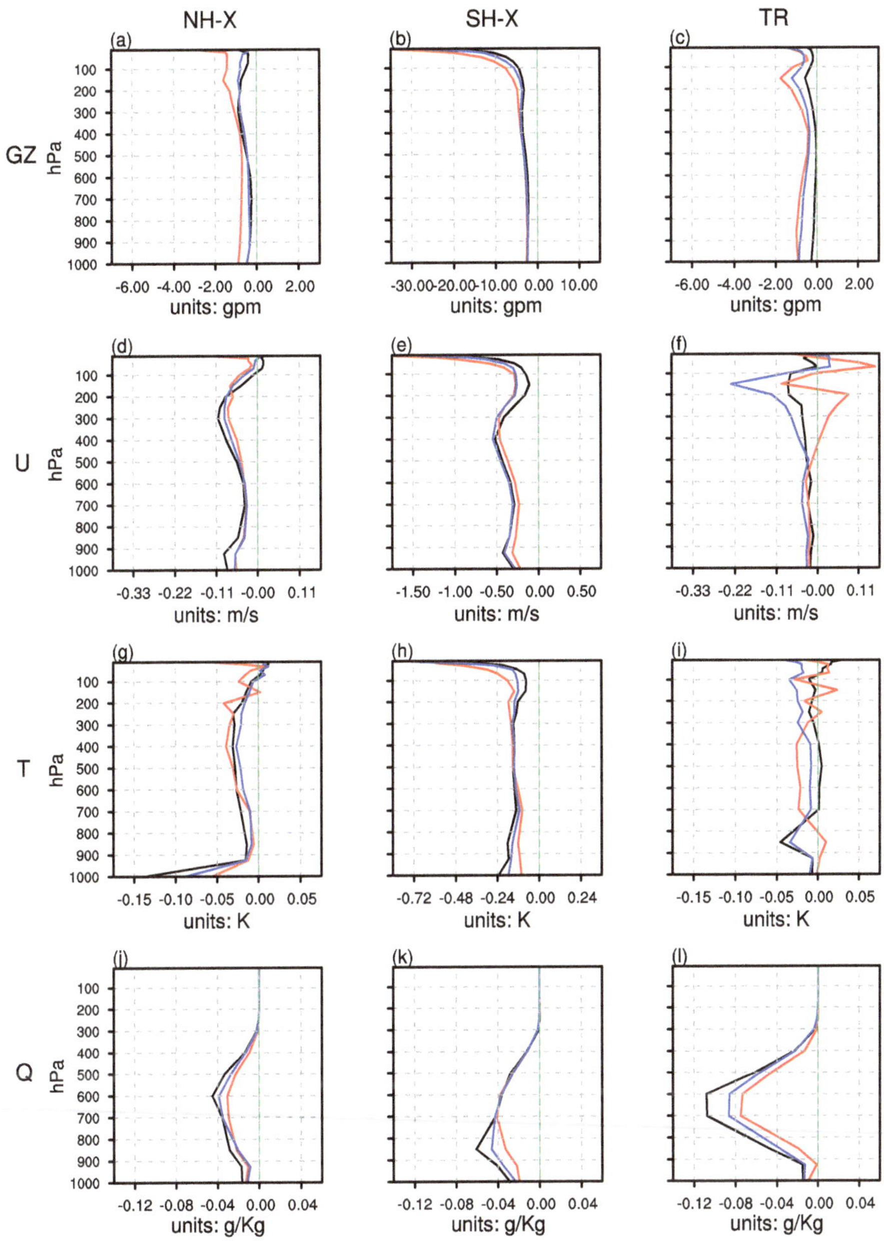

Figure 3. The anomaly root mean square error (ARMSE) differences between the analyses of assimilating all types of observations and those of assimilating only conventional observations by the 4DVar (black), 4DEnVar (red) and En4DVar (blue) systems in the Northern Extratropics (**left** column), Southern Extratropics (**middle** column) and Tropics (**right** column). The results of geopotential height (GZ; units: gpm), zonal wind (U; units: m/s), temperature (T; units: K) and specific humidity (Q; units: g/Kg) are ploted in rows 1–4, respectively. The green line denotes zero.

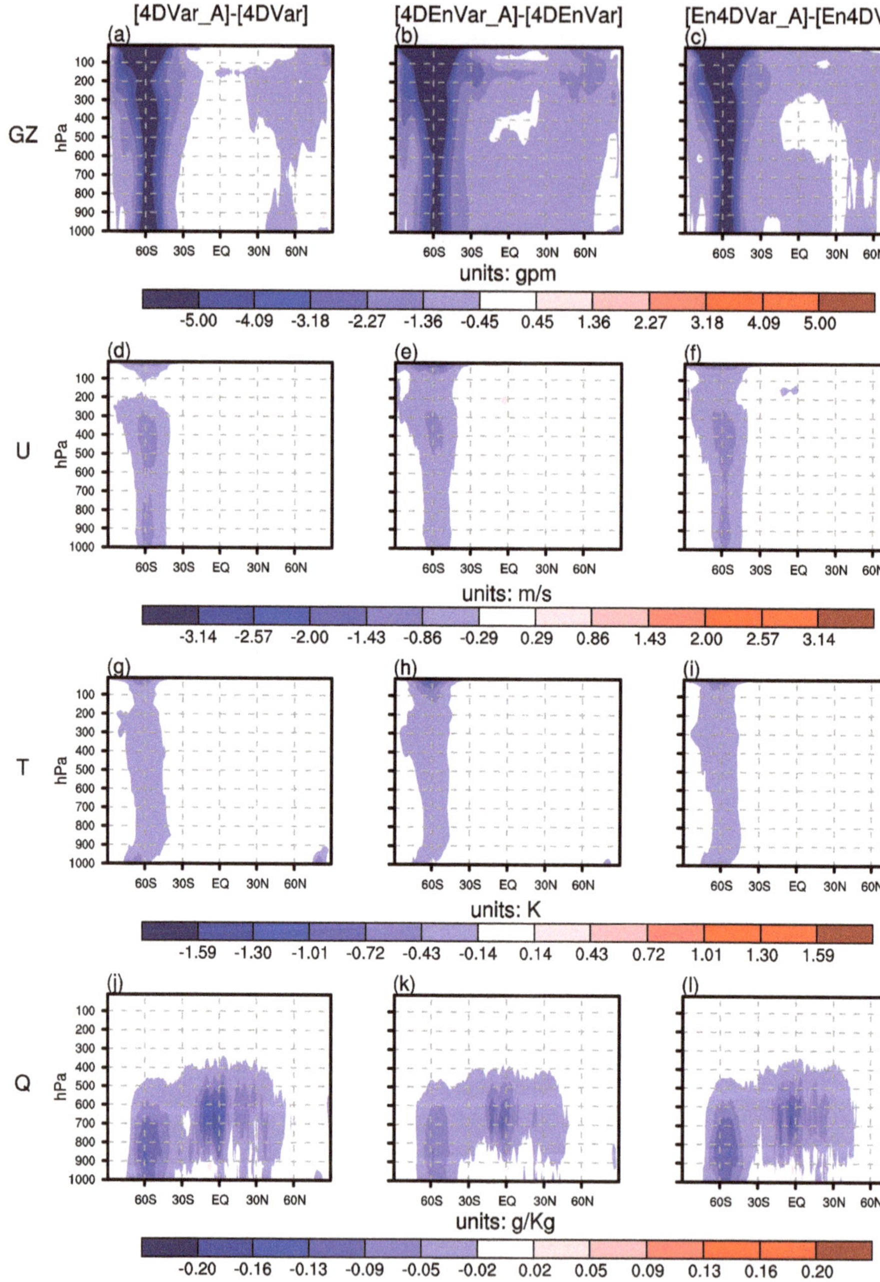

Figure 4. The zonally averaged anomaly root mean square error (ARMSE) differences between assimilating all types of observations and assimilating only conventional observations for the 4DVar (**left** column), 4DEnVar (**middle** column) and En4DVar (**left** column) analyses. The results of geopotential height (GZ; units: gpm), zonal wind (U; units: m/s), temperature (T; units: K) and specific humidity (Q; units: g/Kg) are ploted in rows 1–4, respectively.

3.3. Effects of AMSU-A Radiance Observations on Forecast Skill

Given that the analysis errors of the DA systems are significantly reduced by adding AMSU-A radiance observations, we next focus on whether the improved analysis could benefit the forecast skill as well.

From the comparisons between the geopotential height forecasts initialized from the analyses with and without including AMSU-A radiance observations in all three DA systems, it can be found that AMSU-A radiance observations can generally reduce the geopotential height forecast errors (Figure 5). The largest improvements are mainly located at the middle and upper layers in the Southern Extratropics, followed by the Northern Extratropics and Tropics, which is consistent with the analysis error distributions (Figure 3). In addition, the improvement of the 4DVar-initialized forecast is more obvious than those of the 4DEnVar- and En4DVar-initialized forecasts in the Northern Extratropics, but comparable in the Southern Extratropics and Tropics.

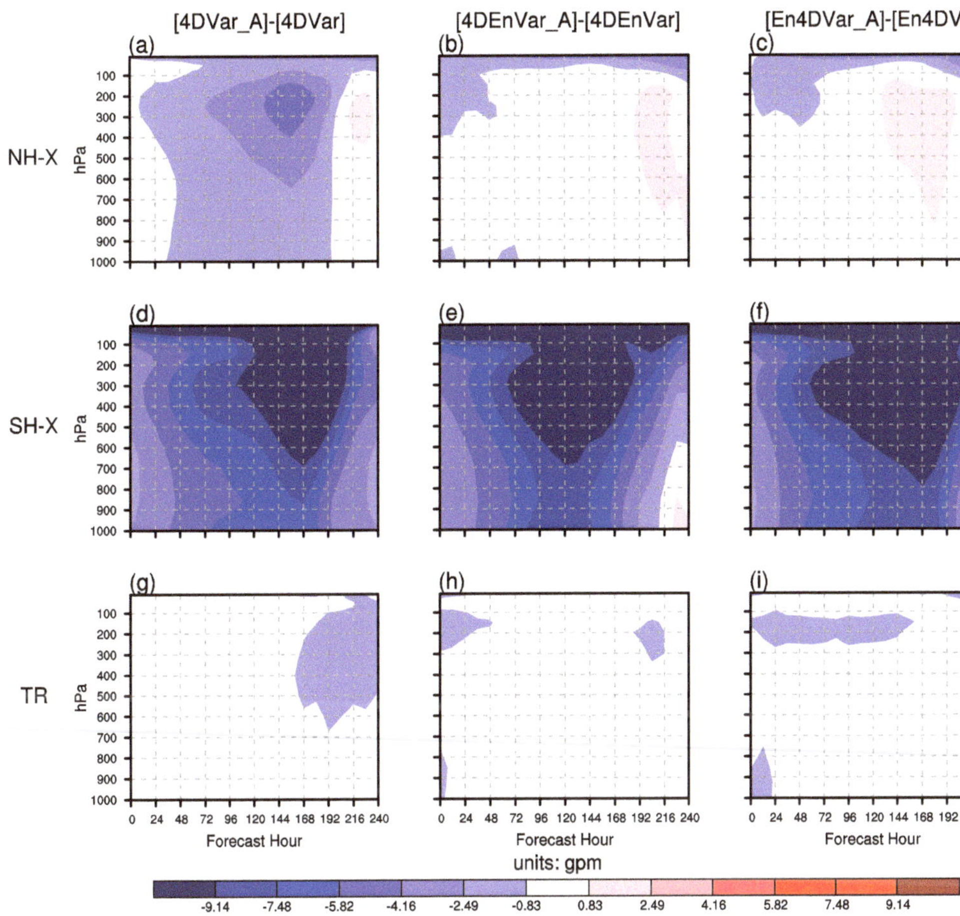

Figure 5. The time-variation of the anomaly root mean square error (ARMSE) differences between assimilating all types of observations and assimilating only conventional observations for the geopotential height forecasts (units: gpm) initialized by the 4DVar (**left** column), 4DEnVar (**middle** column), and En4DVar (**right** column) systems. The results in the Northern Extratropics, Southern Extratropics and Tropics are ploted in rows 1–3, respectively.

Figure 6 shows the effects of AMSU-A radiance observations in the En4DVar, 4DVar, and 4DEnVar systems on the zonal wind forecast errors. AMSU-A radiance observations in all these DA systems generally reduces the zonal wind forecast errors. The locations where the 4DVar- and 4DEnVar-initialized zonal wind forecasts are improved or degraded are generally consistent with the geopotential height. However, inconsistently, the En4DVar-initialized zonal wind forecast shows an improvement at the late period in the Northern Extratropics.

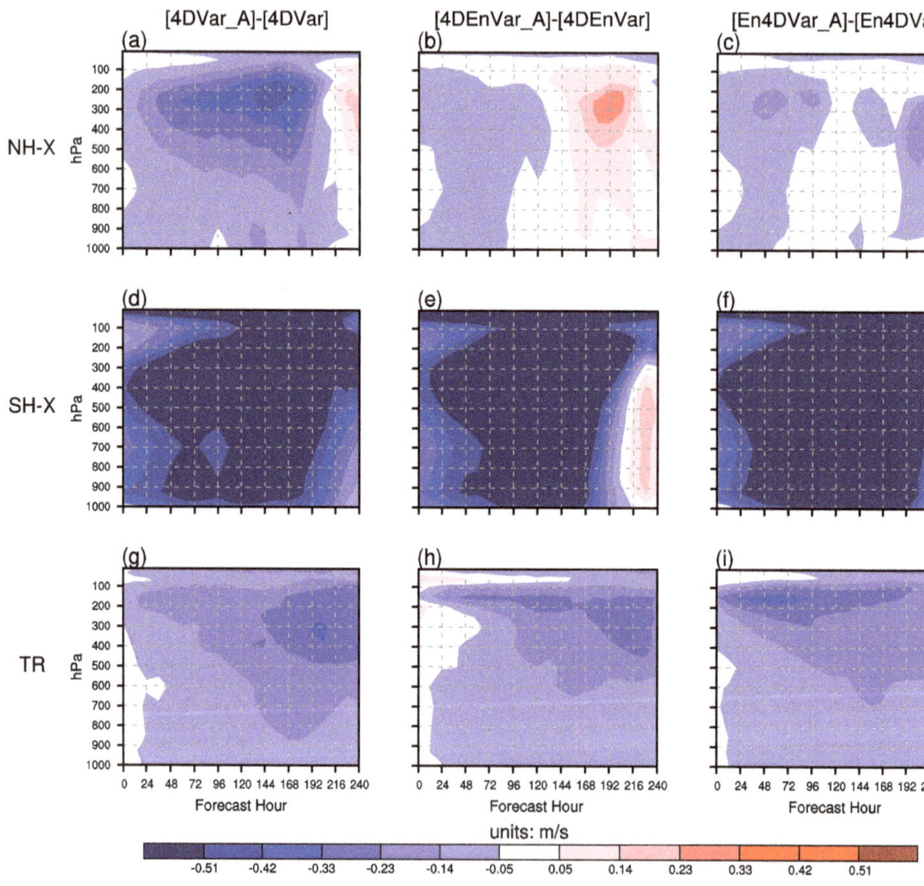

Figure 6. Same as Figure 5, except the zonal wind forecasts (units: m/s).

Adding AMSU-A radiance observations to all three DA systems also reduces most of the temperature forecast errors, with the largest improvement in the Southern Extratropics (Figure 7). Quite different from the geopotential height and zonal wind, the largest improvements in the temperature forecasts are located in the stratosphere, and middle and lower troposphere in the Southern Extratropics (Figure 7, row 2), consistent with the reduced analysis errors (Figure 3h). In addition, while the 4DVar-initialized forecast shows a persistent improvement in the Northern Extratropics, the 4DEnVar- and En4DVar-initialized forecasts performs neutrally. In contrast, the 4DEnVar and En4DVar systems show larger improvements than 4DVar in the Southern Extratropics and the Tropics. In particular, the En4DVar system shows the largest improvement.

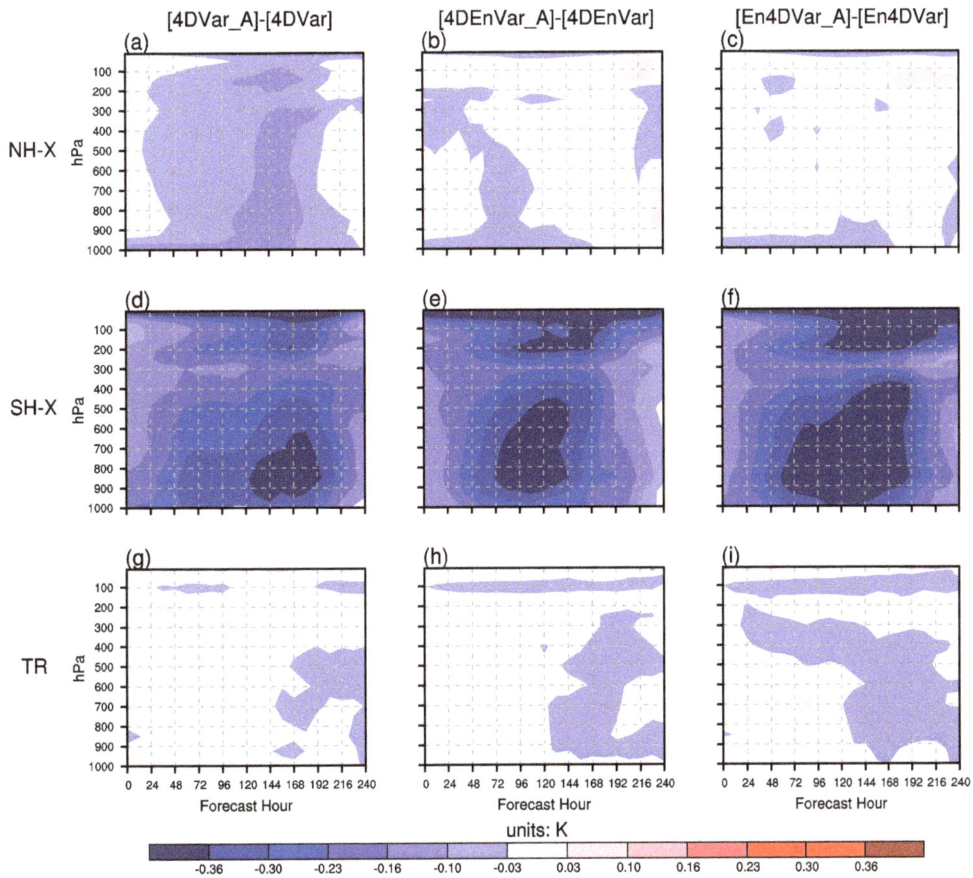

Figure 7. Same as Figure 5, except the temperature forecasts (units: K).

Figure 8 shows the effects of AMSU-A radiance observations in all DA systems on the specific humidity forecasts. Similar to other variables, adding AMSU-A radiance observations steadily improves the specific humidity forecasts of all DA systems except for very few lead days in the Northern Extratropics. The improvement of the 4DVar-initialized forecast is also more significant than those of the 4DEnVar- and En4DVar-initialized forecasts in the Northern Extratropics (Figure 8, row 1). However, different from other variables, the largest improvement in the specific humidity forecasts is mainly distributed in the lower troposphere of the Southern Extratropics (Figure 8, row 2). In addition, there are significant improvements on the first few lead days in the Tropics (Figure 8, row 3), consistent with the regions where analysis errors are significantly reduced (Figure 3l).

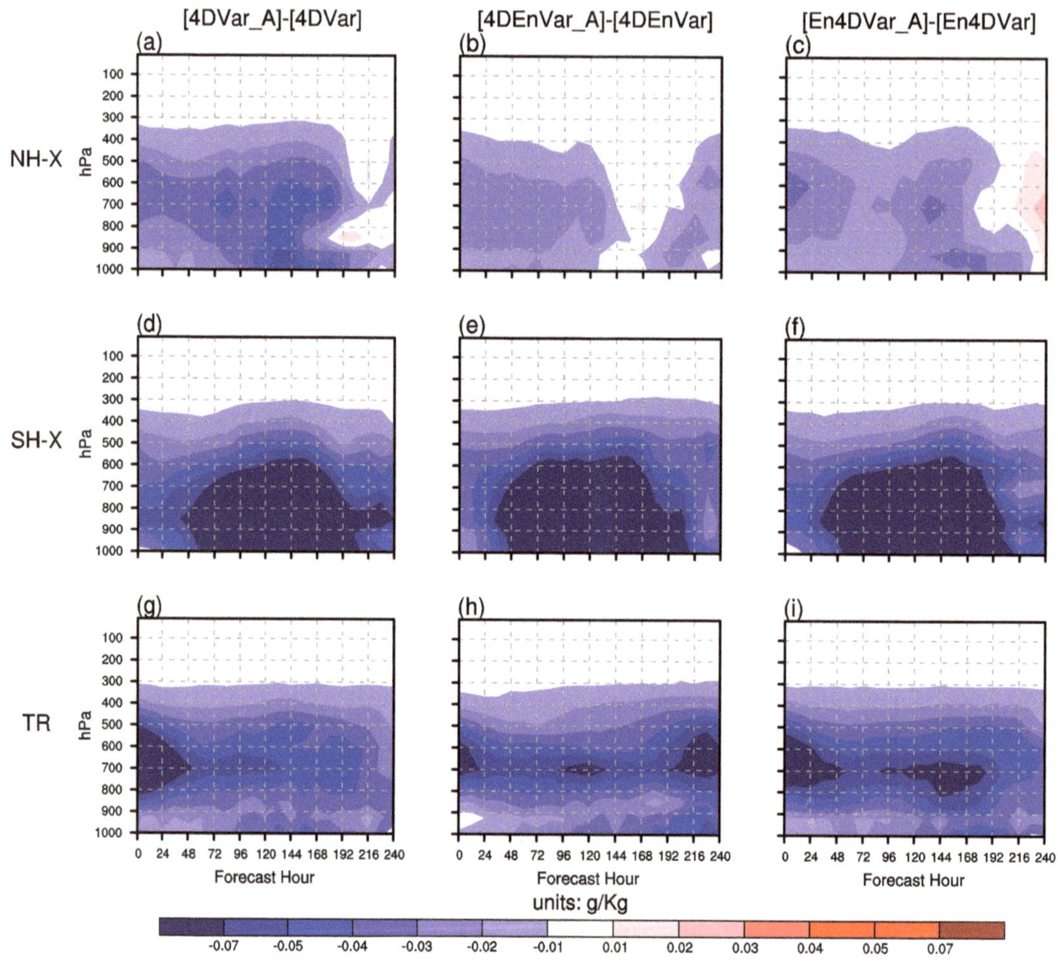

Figure 8. Same as Figure 5, except the specific humidity forecasts (units: g/kg).

Overall, the differences of the 4DVar-initialized forecast performances between assimilating all types of observations and assimilating only conventional observations are statistically significant for almost all lead days in the Southern Extratropics and Tropics and the first few lead days in the Northern Extratropics (Figure 9a). It is encouraging to note that adding AMSU-A radiance observations to the 4DEnVar and En4DVar systems with the weighted average pressure as the vertical coordinates in the observation space localization also has significant positive effects on forecasts. While there are similar improvements in the Southern Extratropics and Tropics for the 4DEnVar- and En4DVar-initialized forecasts, the improvements are less statistically significant than those of the 4DVar-initialized forecast at the last few lead days. In addition, the impacts of AMSU-A observations in the 4DEnVar and En4DVar systems on the medium-range forecasts in the Northern Extratropics are neutral to slightly worse (Figure 9). It is reasonable considering that the 4DVar system uses model space localization, which can simulate close to the true atmospheric state [21,35]. In contrast, the observation space localization may hinder the transfer of some information from the radiance observations.

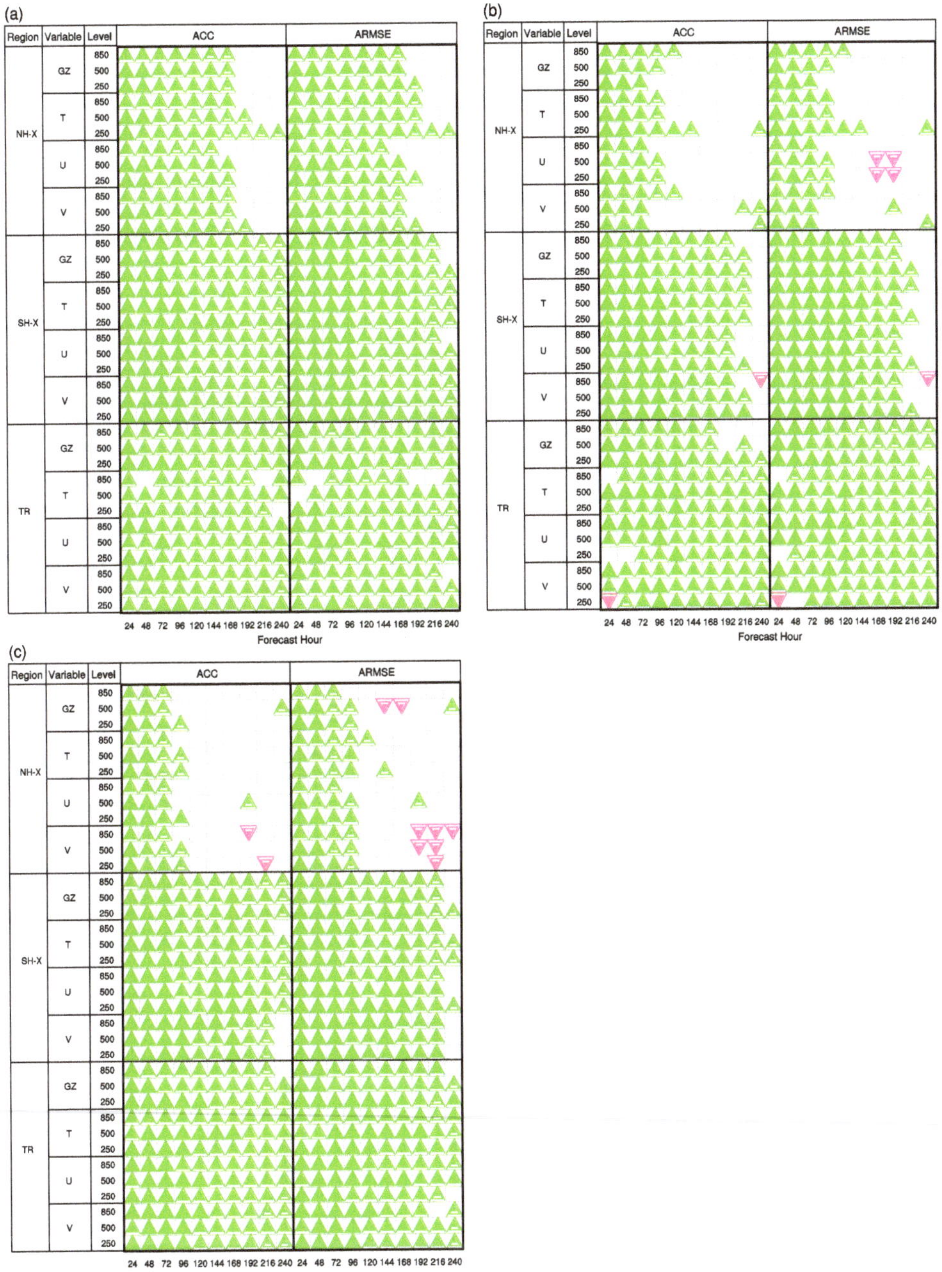

Figure 9. Same as Figure 2, except the forecasts assimilating all types of observations against assimilating only conventional observations initialized by the (**a**) 4DVar, (**b**) 4DEnVar and (**c**) En4DVar systems, respectively.

4. Discussion

This study investigated the effects of incorporating AMSU-A radiance observations on the En4DVar system. Unlike most En4DVar systems that utilize the ensemble covariance produced by the locally solved EnKF class or the ensemble of globally solved 4DVars, this system introduces the ensemble covariance provided by the globally solved 4DEnVar system using an economical observation space localization [26]. To take into account the information of AMSU-A radiance observations at other vertical layers, a weighted average hypsometry was proposed to define the vertical coordinates of radiance observations. The sensitivity experiments indicates that the new hypsometry approach has a wider range of positive effects on the 4DEnVar deterministic forecasts than the traditional peak-based approach.

The impacts of adding AMSU-A radiance observations on the assimilation and forecast performances of the En4DVar system were systematically assessed through 1-month OSSE-based assimilation experiments and its corresponding initialized forecast experiments. The results of the 4DVar and 4DEnVar components are also given as the references for more systematic evaluation of the En4DVar system in assimilating radiance observations. The analyses of all three DA systems benefit from AMSU-A observations, especially in the Southern Extratropics, where conventional observations are sparsely distributed. It is encouraging that the 4DEnVar system using observation space localization improved the analyses on the upper layers of the Northern and Southern Extratropics more significantly than the 4DVar system using model space localization. The improvement in the En4DVar analyses is generally between those of the standalone 4DVar and 4DEnVar components. In terms of ACC and ARMSE, three DA systems further improved the forecasts when adding AMSU-A radiance observations to the ICs. There is a steady improvement in the Southern Extratropics and Tropics, but the impact on the later lead days in the Northern Extratropics is neutral or even slightly negative. In the Northern Extratropics, the improvement of forecast by 4DVar is more significant than by 4DEnVar and En4DVar.

Future improvements in the assimilation of radiance observations based on the En4DVar system will focus on increasing the types of observations and adjusting the filtering radius of localization. In order to further improve the analysis quality, the En4DVar system needs to continue adding more radiance observations with complex multi-peak distribution weighting functions such as those from AMSU-B instruments. In addition, the broad satellite channel weighting function has a significant influence on the filtering radius of localization, and too larger or too small filtering radius will limit the assimilation performance. More flexible and adaptive localization techniques need to be developed for satellite DA with localization in observation space.

Moreover, although encouraging results were obtained using observation space localization method in assimilating AMSU-A observations with a single-peak distribution of weighting function, model space localization has proven to be more beneficial for assimilating radiance observations [21,35]. Therefore, future attempts will also be made to develop efficient model space localization method for the ensemble component of the En4DVar system, in order to obtain better results when assimilating radiance observations with complex multi-peak distribution weighting functions.

Author Contributions: Conceptualization, S.Z. and B.W.; writing—original draft preparation, S.Z.; writing—review and editing, S.Z., B.W., L.Z., J.L., Y.L., J.G., S.X., Y.W., W.H., L.L., Y.H., X.W., B.Z. and F.C. All authors have read and agreed to the published version of the manuscript.

Funding: This research was funded by the National Key Research and Development Program of China (2018YFC1506703) and the Innovation Group Project of Southern Marine Science and Engineering Guangdong Laboratory (Zhuhai) (No. 311022006).

Data Availability Statement: The observations were supported by the Global Telecommunications System (https://public.wmo.int/en/programmes/global-telecommunication-system accessed on 1 February 2021). The ERA-5 reanalysis and the ERA-Interim 6 h forecast are available at https://apps.ecmwf.int/data-catalogues/era5/?class=ea&stream=oper&expver=1&type=an accessed on 1 February 2021 and https://apps.ecmwf.int/datasets/data/interim-full-daily/levtype=sfc/ accessed on 1 February 2021. All data used in this study are available from the authors upon request.

Acknowledgments: The assimilation and forecast experiments were performed on the high-performance computer PI-SUGON of the China Meteorological Administration.

Conflicts of Interest: The authors declare no conflict of interest.

References

1. Raynaud, L.; Berre, L.; Desroziers, G. An extended specification of flow-dependent background error variances in the Météo-France global 4D-Var system. *Q. J. R. Meteorol. Soc.* **2011**, *137*, 607–619. [CrossRef]
2. Bonavita, M.; Isaksen, L.; Hólm, E. On the use of EDA background error variances in the ECMWF 4D-Var. *Q. J. R. Meteorol. Soc.* **2012**, *138*, 1540–1559. [CrossRef]
3. Bonavita, M.; Hólm, E.; Isaksen, L.; Fisher, M. The evolution of the ECMWF hybrid data assimilation system. *Q. J. R. Meteorol. Soc.* **2016**, *142*, 287–303. [CrossRef]
4. Clayton, A.M.; Lorenc, A.C.; Barker, D.M. Operational implementation of a hybrid ensemble/4D-Var global data assimilation system at the Met Office. *Q. J. R. Meteorol. Soc.* **2013**, *139*, 1445–1461. [CrossRef]
5. Lorenc, A.C.; Bowler, N.E.; Clayton, A.M.; Pring, S.R.; Fairbairn, D. Comparison of hybrid-4DEnVar and hybrid-4DVar data assimilation methods for global NWP. *Mon. Weather Rev.* **2015**, *143*, 212–229. [CrossRef]
6. Houtekamer, P.L.; Mitchell, H.L. Data assimilation using an ensemble Kalman filter technique. *Mon. Weather Rev.* **1998**, *126*, 796–811. [CrossRef]
7. Anderson, J.L. Exploring the need for localization in ensemble data assimilation using a hierarchical ensemble filter. *Phys. D Nonlinear Phenom.* **2007**, *230*, 99–111. [CrossRef]
8. Liu, C.; Xiao, Q.; Wang, B. An ensemble-based four-dimensional variational data assimilation scheme. Part II: Observing system simulation experiments with Advanced Research WRF (ARW). *Mon. Weather Rev.* **2009**, *137*, 1687–1704. [CrossRef]
9. Hamill, T.M.; Whitaker, J.S.; Snyder, C. Distance-dependent filtering of background error covariance estimates in an ensemble Kalman filter. *Mon. Weather Rev.* **2001**, *129*, 2776–2790. [CrossRef]
10. Wang, B.; Liu, J.; Liu, L.; Xu, S.; Huang, W. An approach to localization for ensemble-based data assimilation. *PLoS ONE* **2018**, *13*, e0191088. [CrossRef]
11. Houtekamer, P.L.; Mitchell, H.L.; Pellerin, G.; Buehner, M.; Charron, M.; Spacek, L.; Hansen, B. Atmospheric data assimilation with an ensemble Kalman filter: Results with real observations. *Mon. Weather Rev.* **2005**, *133*, 604–620. [CrossRef]
12. Gaspari, G.; Cohn, S.E. Construction of correlation functions in two and three dimensions. *Q. J. R. Meteorol. Soc.* **1999**, *125*, 723–757. [CrossRef]
13. Ott, E.; Hunt, B.R.; Szunyogh, I.; Zimin, A.V.; Kostelich, E.J.; Corazza, M.; Kalnay, E.; Patil, D.J.; Yorke, J.A. A local ensemble Kalman filter for atmospheric data assimilation. *Tellus A Dyn. Meteorol. Oceanogr.* **2004**, *56*, 415–428. [CrossRef]
14. Buehner, M.; Houtekamer, P.L.; Charette, C.; Mitchell, H.L.; He, B. Intercomparison of variational data assimilation and the ensemble Kalman filter for global deterministic NWP. Part I: Description and single-observation experiments. *Mon. Weather Rev.* **2010**, *138*, 1550–1566. [CrossRef]
15. Buehner, M.; Houtekamer, P.L.; Charette, C.; Mitchell, H.L.; He, B. Intercomparison of variational data assimilation and the ensemble Kalman filter for global deterministic NWP. Part II: One-month experiments with real observations. *Mon. Weather Rev.* **2010**, *138*, 1567–1586. [CrossRef]
16. Bishop, C.H.; Hodyss, D.; Steinle, P.; Sims, H.; Clayton, A.M.; Lorenc, A.C.; Barker, D.M.; Buehner, M. Efficient ensemble covariance localization in variational data assimilation. *Mon. Weather Rev.* **2011**, *139*, 573–580. [CrossRef]
17. Kuhl, D.D.; Rosmond, T.E.; Bishop, C.H.; McLay, J.; Baker, N.L. Comparison of hybrid ensemble/4DVar and 4DVar within the NAVDAS-AR data assimilation framework. *Mon. Weather Rev.* **2013**, *141*, 2740–2758. [CrossRef]
18. Simmons, A.J.; Hollingsworth, A. Some aspects of the improvement in skill of numerical weather prediction. *Q. J. R. Meteorol. Soc. A J. Atmos. Sci. Appl. Meteorol. Phys. Oceanogr.* **2002**, *128*, 647–677. [CrossRef]
19. Fertig, E.J.; Hunt, B.R.; Ott, E.; Szunyogh, I. Assimilating non-local observations with a local ensemble Kalman filter. *Tellus A Dyn. Meteorol. Oceanogr.* **2007**, *59*, 719–730. [CrossRef]
20. Miyoshi, T.; Sato, Y. Assimilating satellite radiances with a local ensemble transform Kalman filter (LETKF) applied to the JMA global model (GSM). *Sola* **2007**, *3*, 37–40. [CrossRef]
21. Campbell, W.F.; Bishop, C.H.; Hodyss, D. Vertical covariance localization for satellite radiances in ensemble Kalman filters. *Mon. Weather Rev.* **2010**, *138*, 282–290. [CrossRef]
22. Courtier, P.; Thépaut, J.N.; Hollingsworth, A. A strategy for operational implementation of 4D-Var, using an incremental approach. *Q. J. R. Meteorol. Soc.* **1994**, *120*, 1367–1387. [CrossRef]
23. Zhang, L.; Liu, Y.; Liu, Y.; Gong, J.; Lu, H.; Jin, Z.; Tian, W.; Liu, G.; Zhou, B.; Zhao, B. The operational global four-dimensional variational data assimilation system at the China Meteorological Administration. *Q. J. R. Meteorol. Soc.* **2019**, *145*, 1882–1896. [CrossRef]
24. Zhu, S.; Wang, B.; Zhang, L.; Liu, J.; Liu, Y.; Gong, J.; Xu, S.; Wang, Y.; Huang, W.; Liu, L.; et al. A Four-Dimensional Ensemble-Variational (4DEnVar) Data Assimilation System Based on GRAPES-GFS: System Description and Primary Tests. *J. Adv. Model. Earth Syst.* **2022**, *14*, e2021MS002737. [CrossRef]

25. Wang, B.; Liu, J.; Wang, S.; Cheng, W.; Juan, L.; Liu, C.; Xiao, Q.; Kuo, Y.-H. An economical approach to four-dimensional variational data assimilation. *Adv. Atmos. Sci.* **2010**, *27*, 715–727. [CrossRef]
26. Zhu, S.; Wang, B.; Zhang, L.; Liu, J.; Liu, Y.; Gong, J.; Xu, S.; Wang, Y.; Huang, W.; Liu, L.; et al. A 4DEnVar-Based Ensemble Four-Dimensional Variational (En4DVar) Hybrid Data Assimilation System for Global NWP: System Description and Primary Tests. *J. Adv. Model. Earth Syst.* **2022**, *14*, e2022MS003023. [CrossRef]
27. Lorenc, A.C. The potential of the ensemble Kalman filter for NWP—A comparison with 4D-Var. *Q. J. R. Meteorol. Soc. A J. Atmos. Sci. Appl. Meteorol. Phys. Oceanogr.* **2003**, *129*, 3183–3203. [CrossRef]
28. Su, Y.; Shen, X.S.; Zhang, H.L.; Liu, Y.Z. A study on the three-dimensional reference atmosphere in GRAPES_GFS: Constructive reference state and real forecast experiment. *Acta Meteorol. Sin.* **2020**, *78*, 962–971.
29. Barker, D.M. Southern high-latitude ensemble data assimilation in the Antarctic Mesoscale Prediction System. *Mon. Weather Rev.* **2005**, *133*, 3431–3449. [CrossRef]
30. Wang, X.; Barker, D.M.; Snyder, C.; Hamill, T.M. A hybrid ETKF–3DVAR data assimilation scheme for the WRF model. Part I: Observing system simulation experiment. *Mon. Weather Rev.* **2008**, *136*, 5116–5131. [CrossRef]
31. Kleist, D.T.; Ide, K. An OSSE-based evaluation of hybrid variational–ensemble data assimilation for the NCEP GFS. Part I: System description and 3D-hybrid results. *Mon. Weather Rev.* **2015**, *143*, 433–451. [CrossRef]
32. Kleist, D.T.; Ide, K. An OSSE-based evaluation of hybrid variational–ensemble data assimilation for the NCEP GFS. Part II: 4DEnVar and hybrid variants. *Mon. Weather Rev.* **2015**, *143*, 452–470. [CrossRef]
33. Saunders, R.; Hocking, J.; Turner, E.; Rayer, P.; Rundle, D.; Brunel, P.; Vidot, J.; Roquet, P.; Matricardi, M.; Geer, A.; et al. An update on the RTTOV fast radiative transfer model (currently at version 12). *Geosci. Model Dev.* **2018**, *11*, 2717–2737. [CrossRef]
34. He, Y.; Wang, B.; Liu, L.; Huang, W.; Xu, S.; Liu, J.; Wang, Y.; Li, L.; Huang, X.; Peng, Y.; et al. A DRP-4DVar-based coupled data assimilation system with a simplified off-line localization technique for decadal predictions. *J. Adv. Model. Earth Syst.* **2020**, *12*, e2019MS001768. [CrossRef]
35. Lei, L.; Whitaker, J.S.; Bishop, C. Improving assimilation of radiance observations by implementing model space localization in an ensemble Kalman filter. *J. Adv. Model. Earth Syst.* **2018**, *10*, 3221–3232. [CrossRef]

remote sensing

MDPI

Article

Direct Assimilation of Chinese FY-3E Microwave Temperature Sounder-3 Radiances in the CMA-GFS: An Initial Study

Juan Li [1,2,*], Xiaoli Qian [3], Zhengkun Qin [3] and Guiqing Liu [1,2]

1 Earth System Modeling and Prediction Centre, China Meteorological Administration, Beijing 100081, China
2 State Key Laboratory of Severe Weather, Institute of Meteorological, China Meteorological Administration, Beijing 100081, China
3 Joint Center for Data Assimilation for Research and Application, Nanjing University of Information Science and Technology, Nanjing 210044, China
* Correspondence: lj@cma.gov.cn

Abstract: FengYun-3E (FY-3E), the fifth satellite in China's second-generation polar-orbiting satellite FY-3 series, was launched on 5 July 2021. FY-3E carries a third-generation microwave temperature sounder (MWTS-3). For the first time, this study demonstrates that MWTS-3 radiances data assimilation can improve the China Meteorological Administration Global Forecast System (CMA-GFS). By establishing a cloud detection module based on the retrieval results of the new channels of MWTS-3, a quality control module according to the error characteristics of MWTS-3 data, and a bias correction module considering the scanning position of satellite and weather systems, the effective assimilation of MWTS-3 data in the CMA-GFS has been realized. Through one-month cycling experiments of assimilation and forecasts, the error characteristics and assimilation effects of MWTS-3 data are carefully evaluated. The results show that the observation errors in MWTS-3 data are similar to those in advanced technology microwave sounder (ATMS) data within the same frequency channel, are slightly larger than those in the advanced microwave-sounding unit-A (AMSU-A) data, and are much better than those in the MWTS-2 data. The validation of the assimilation and prediction results demonstrate the positive contribution of MWTS-3 data assimilation, which can remarkably reduce the analysis errors in the Northern and Southern Hemispheres. Specifically, the error growth on the upper layer of the model is obviously suppressed. When all other operational satellite observations are included, the assimilation of MWTS-3 data has a neutral or slightly positive contribution to the analysis and forecast results, and the improvement is mainly found in the Southern Hemisphere. The relevant evaluation results indicate that the MWTS-3 data assimilation has good application prospects for operation.

Keywords: FY-3E; early-morning-orbit satellite; data assimilation; microwave-sounding unit

Citation: Li, J.; Qian, X.; Qin, Z.; Liu, G. Direct Assimilation of Chinese FY-3E Microwave Temperature Sounder-3 Radiances in the CMA-GFS: An Initial Study. *Remote Sens.* **2022**, *14*, 5943. https://doi.org/10.3390/rs14235943

Academic Editor: Stephan Havemann

Received: 15 October 2022
Accepted: 21 November 2022
Published: 24 November 2022

1. Introduction

In recent years, satellite observations have become a key component of the global operational numerical weather prediction (NWP) system due to their high spatial-temporal resolution and wide spatial coverage. Many studies have shown that direct assimilation of microwave-sounding data can remarkably improve the initial conditions of numerical models so as to improve the prediction levels of global and regional models [1–5]. Most NWP centers have reported a substantial reduction in the root mean square error (RMSE) in forecasts by effectively assimilating the data from the Advanced Television and Infrared Observation Satellite (TIROS) operational vertical sounder (ATOVS) onboard the National Oceanic and Atmospheric Administration satellites (NOAA-15, -16, -17, -18 -19, and -20), the Meteorological Operational satellite-A/B (MetOp-A/B) and the Aqua earth observing system. Adjoint sensitivity experiments [6] have proven that microwave temperature-sounding data has become the most influential observation in almost all operational forecasting systems [7–10].

Recently, China's polar-orbiting meteorological satellites have become an important part of the global polar-orbiting satellite observing system. Since the successful launch of China's new generation polar-orbiting satellite Fengyun-3A (FY-3A) on 26 May 2008 [11,12], Fengyun-3B/C/D (FY-3B/C/D) satellites have been launched successively. The performance of microwave sounders onboard these satellites is similar to those of the advanced microwave-sounding unit-A (AMSU-A) onboard the NOAA and MetOp satellites [11,12]. FY-3A/B are equipped with the first-generation microwave temperature sounder (MWTS-1), which has four channels with frequencies comparable to channels 3, 5, 7, and 9 of the AMSU-A [13]. FY-3C/D are equipped with the second-generation microwave temperature sounder (MWTS-2). MWTS-2 has 13 channels, and the channels located in the oxygen absorption band (50–60 GHz) are identical to those of the AMSU-A. Various studies on data evaluation [14,15] and assimilation have been carried out for the MWTS, and many of them have indicated that the assimilation of MWTS-1 and MWTS-2 data has positive impacts on NWP results [15–20].

FY-3E satellite was successfully launched on 5 July 2021, which is the world's first meteorological satellite sent into the early-morning orbit for civil use [21]. It has a local equatorial crossing time of about 5:40 am. This satellite carries a third-generation microwave temperature sounder (MWTS-3). A systematic evaluation study [22] has demonstrated that the performance of MWTS-3 is remarkably better than the previous two generations of instruments, with more observational information and well-suppressed observational noises.

The purpose of this study is to evaluate the impacts of the direct assimilation of the MWTS-3 radiance data on the China Meteorological Administration global forecast system (CMA-GFS) for the first time. By establishing the quality control (QC) and bias correction modules suitable for the MWTS-3 data, the effective assimilation of MWTS-3 data in the CMA-GFS is realized. The influence of MWTS-3 data assimilation on the CMA-GFS is evaluated based on the results of one-month assimilation and forecasting. It should be noted that the original name of the operational numerical prediction system in China was the Global and Regional Estimation and PrEdiction System (GRAPES) [23–25]. After September 2021, it was renamed CMA-GFS.

The remainder of this paper is organized as follows. Section 2 introduces the CMA-GFS four-dimensional variational assimilation (4D-Var) system. The general details of the FY-3E MWTS-3 radiance data is described here. Section 2 also provides the QC and bias correction scheme for the MWTS-3 radiance data, and the initial assessments of MWTS-3 data. Section 3 presents the analysis of the numerical results of the FY-3E MWTS-3 radiance data assimilation experiments. The discussion and conclusion are given in Sections 4 and 5.

2. Materials and Methods

2.1. CMA-GFS 4D-Var System

The main components of CMA-GFS include: four-dimensional variational (4D-Var) data assimilation; fully compressible non-hydrostatical model core with semi-implicit and semi-Lagrangian discretization scheme; modularized model physics package, and global and regional assimilation and prediction systems [23].

The CMA-GFS 4D-Var system is an analysis system designed for operational application [26]. This assimilation system adopts an incremental analysis method, and the assimilation process is divided into outer circulation and inner circulation. In order to reduce the amount of computation, the horizontal resolution of the nonlinear model in the outer circulation of the assimilation is 0.25 degrees, the horizontal resolution of the tangent linear model and the adjoint model in the inner circulation is 1.0 degrees, and only the simplified physical process is applied. The model has 87 vertical layers, with the top being approximately 0.1 hPa. The 4D-Var data assimilation system applies the incremental analysis scheme proposed by Courtier et al. (1998) [2]. By using the observations distributed within a time interval (t0, tn) in the assimilation, the cost function can be defined as follows:

$$J(x(t_0)) = \frac{1}{2}\left(x(t_0) - x^b(t_0)\right)^T \mathbf{B}^{-1}\left(x(t_0) - x^b(t_0)\right) + \frac{1}{2}\sum_{i=0}^{N}(H(x_i) - y_i^o)^T \mathbf{R}_i^{-1}(H(x_i) - y_i^o) + J_c$$

where $x(t_0)$ is a state vector composed of atmospheric and surface variables; $x^b(t_0)$ is a background estimate of the state vector provided by a 6 h forecast, and y_i^o is a vector of all the observations; H is the observation operator that transforms the state vector $\boldsymbol{x}$ into observation space; $\boldsymbol{R}_i$ is the estimated error covariance of the observations at time i; J_c is a constraint term added to control various noises and errors generated in variational analysis. For the CMA-GFS data assimilation system, J_c is the weak constraints of the digital filtering. $\mathbf{B}$ is the error covariance matrix of x^b. In order to solve the problem that the inverse of the background error covariance matrix ($\mathbf{B}^{-1}$) is too large to be computed, the background term is preconditioned, which improves the convergence in the minimization process and avoids calculating $\mathbf{B}^{-1}$ directly. In the CMA-GFS 4D-Var system, the limited-memory Broyden-Fletcher-Goldfarb-Shanno (L-BFGS) algorithm [27] is used to perform the minimization.

Currently, the CMA-GFS can directly assimilate radiosonde data, surface synoptic observations (SYNOPs), ship reports, aircraft reports (Airep), atmospheric motion vectors (AMVs), the AMSU-A and the microwave humidity sounder (MHS) data of NOAA-15/18/19, the AMSU-A, MHS and infrared atmospheric sounding interferometer (IASI) data of MetOp-A/B, Suomi National Polar-orbiting Partnership (NPP) ATMS data, the MWHS-2, micro-wave radiation imager (MWRI) and Global Navigation Satellite System (GNSS) radio occultation sounder (GNOS) data of FY-3C/D, the MWTS-2 and hyperspectral infrared atmospheric sounder-2 (HIRAS) radiance data of FY-3D, the Constellation Observing System for Meteorology, Ionosphere and Climate radio occultation (COSMIC RO) data, etc.

The radiative transfer for TIROS operational vertical sounder-12 (RTTOV-12) [28] is used as the observation operator for the direct assimilation of satellite radiance data in the CMA-GFS 4D-Var system. The transmittance coefficients applicable to the RTTOV-12 for FY-3E MWTS-3 simulation are provided by the National Satellite Meteorological Center of CMA.

2.2. FY-3E MWTS-3 Observations

The MWTS-3 radiance data in L1b format from September to October 2021 were used in this study. Channel characteristics of FY-3E MWTS-3 are shown in Table 1. Compared with the MWTS-2, MWTS-3 has improved its detection capability and performance indicators. The number of detection channels of MWTS-3 is 17, which is 4 more than that of MWTS-2. Channels 1 and 2 are horizontally polarized, while the other channels are vertically polarized. The noise equivalent differential temperature (NEDT) of channels 1–11 is about 0.3–0.35 K. The NEDT of channels 12–17 is slightly larger, around 0.6–2.1 K. The swath width of MWTS-3 is 2700 km, which is much larger than that of the MWTS-2 (2250 km) and is also wider than that of similar instruments in the world, such as the AMSU-A (2300 km) and ATMS (2500 km). For MWTS-3, the number of fields of view (FOV) in a single scan line also increases to 98 from 90 for the MWTS-2, which is larger than that of AMSU-A (30) and ATMS (96).

In terms of channel settings, two channels that can detect cloud water content were added into MWTS-3 for the first time, with the detection frequencies being 23.8 GHz and 31.4 GHz, respectively. As a result, the existing mature scheme can be used to identify the microwave data in cloudy areas [22] based on cloud liquid water path (CLWP) retrieval [29–31]. In addition, MWTS-3 also has two detection channels at the oxygen absorption band 50–60 GHz, which can be used to detect the atmospheric temperature information at central altitudes of about 500 and 700 hPa.

Table 1. Channel characteristics of FY-3E MWTS-3.

Channel Number	Center Frequency (GHz)	Bandwidth (MHz)	Polarization	NEΔT (K)
1	23.8	270	QH	0.3
2	31.4	180	QH	0.35
3	50.3	180	QV	0.35
4	51.76	400	QV	0.3
5	52.8	400	QV	0.3
6	53.246 ± 0.08	2 × 140	QV	0.35
7	53.596 ± 0.115	2 × 170	QV	0.3
8	53.948 ± 0.081	2 × 142	QV	0.35
9	54.40	400	QV	0.3
10	54.94	400	QV	0.3
11	55.50	330	QV	0.3
12	57.290	330	QV	0.6
13	57.290 ± 0.217	2 × 78	QV	0.7
14	57.290 ± 0.3222 ± 0.048	4 × 36	QV	0.8
15	57.290 ± 0.3222 ± 0.022	4 × 16	QV	1.0
16	57.290 ± 0.3222 ± 0.010	4 × 8	QV	1.2
17	57.290 ± 0.3222 ± 0.0045	4 × 3	QV	2.1

The weighting function of MWTS-3 is shown in Figure 1, which is calculated using the RTTOV-12 based on the American standard atmosphere profile. The MWTS-3 can detect atmospheric temperature information from the troposphere to the stratosphere. The peaks of the weighting function of channels 1–4 are mainly located on the ground, and those of channels 5–17 are uniformly distributed in the vertical direction, which allows the MWTS-3 to detect the atmospheric temperature information at different heights. The weighting function of channel 17 has the highest peak at about 2 hPa.

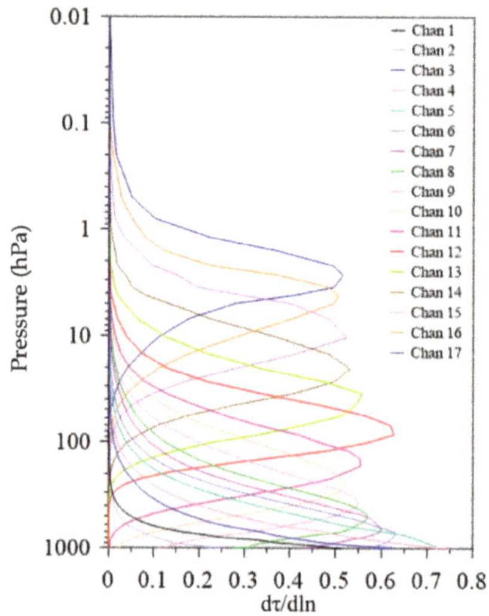

Figure 1. Weighting Functions of FY-3E MWTS-3 calculated by RTTOV based on US standard atmosphere profile.

2.3. Cloud Detection

The MWTS-3 instrument observes the Earth from outer space, which is inevitably affected by clouds. Although the long wavelength allows microwave radiation to penetrate most nonprecipitation clouds, it is inevitably influenced by cloud absorption, large-particle scattering, etc. At present, the assimilation of radiance data in cloudy areas is very challenging due to the lack of reliable information about clouds in the input atmospheric profiles and the inability to accurately involve the cloud impact in the fast radiative transfer model. Many schemes have been developed to assimilate the cloud-influenced observations of microwave-sounding data [32–34]. However, in order to ensure the stability of the operational NWP system, the CMA-GFS is still assimilating the clear sky data of microwave temperature-sounding. Hence, it is necessary to perform cloud detection on the MWTS-3 data in this study.

The microwave sounders onboard the satellites (from FY-3A to FY-3D) lack channels that are sensitive to cloud absorption and scattering, which makes it difficult to perform cloud detection in MWTS-1/2 data assimilation. In the early stage, cloud products of the visible and infrared radiometer (VIRR) mounted on the same platform were used to assist in cloud detection [15,16]. In order to meet the needs of cloud detection, the MWTS-3 onboard FY-3E has included the channels of 23.8 GHz and 31.4 GHz for the first time. Previous studies have developed a mature CLWP retrieval method over the ocean area based on the brightness temperatures observed at these two frequencies [29], which provides an effective way for cloud detection in MWTS-3.

Figure 2 shows the distribution of FY-3E MWTS-3 observed brightness temperatures at channels 1–2 and the retrieved CLWP during 0300–1500 Universal Time (UTC) on 1 July 2014. Note that only the CLWP over the ocean area is retrieved (areas covered by sea ice are also excluded), which ranges from 0.01 to 2.0 g kg^{-1}.

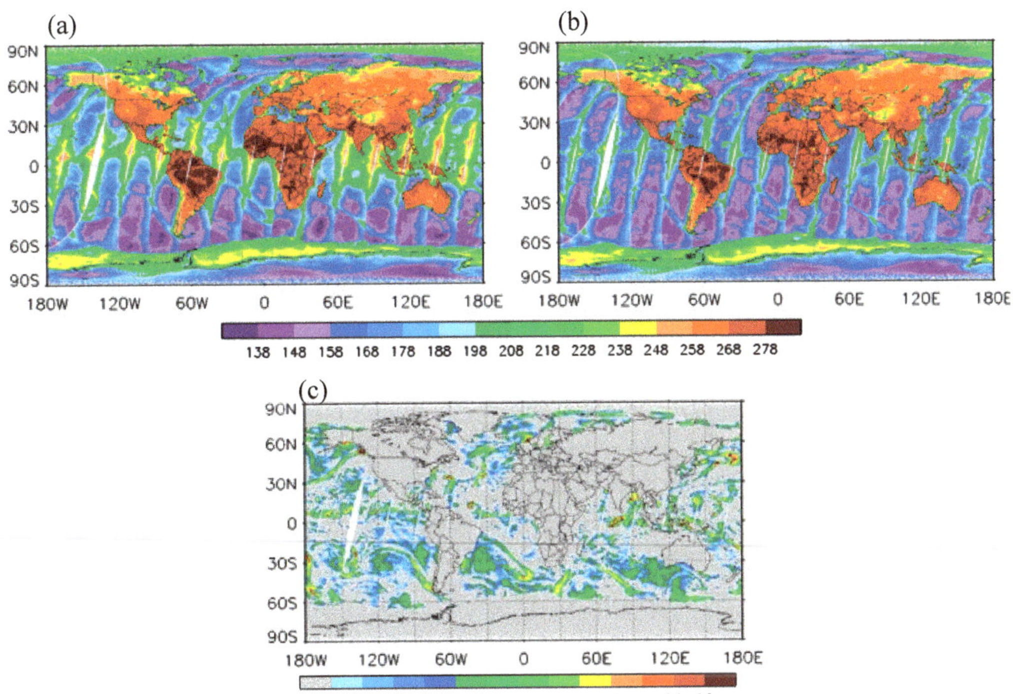

Figure 2. Spatial distribution of observed brightness temperature of FY-3E MWTS-3 channel 1 (**a**), channel 2 (**b**) and retrieved cloud LWP (**c**) for descending orbit data on 24 September 2021.

The accuracy of the retrieval product is assessed by comparing it with the brightness temperature of a 12 μm-channel (channel 7) in the medium resolution spectral imager with a low light level (MERSI-II) [21] onboard the same platform. Figure 3 shows the distribution of the retrieved CLWP and MERSI channel 7 brightness temperature during

0300–1500 UTC on 24 September 2021. As shown in Figure 3, there is a tropical cyclone over the north Pacific with an obvious high brightness temperature center, which has a good spatial correspondence with the large-value area of the retrieved CLWP. A larger CLWP indicates thicker clouds.

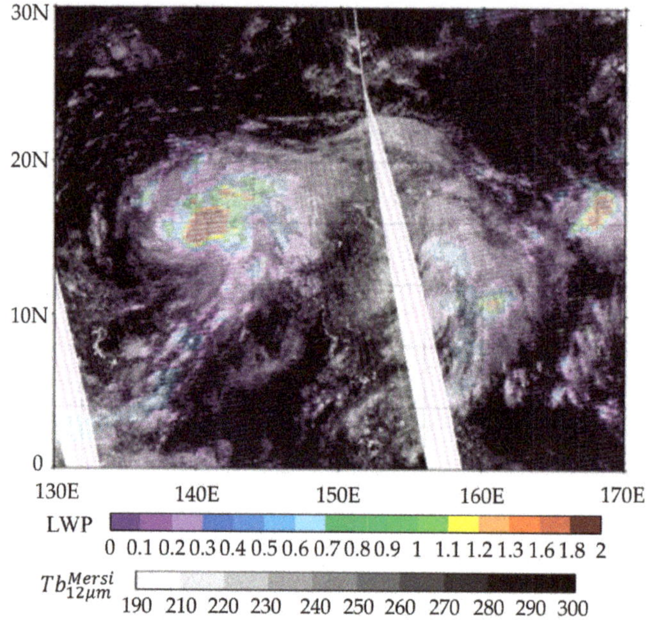

Figure 3. Spatial distribution of retrieved cloud LWP from FY-3E MWTS-3 channel 1 and 2, and brightness temperature of MERSI channel 7 during 0300–1500 UTC 24 September 2021.

For the land area, the differences between the observed and simulated brightness temperature (O-B) on window channel 3 of MWTS-3 is used for cloudy data identification. When the O-B exceeds 1.5 K, this FOV is determined to be the data over cloud and will be rejected.

2.4. *The Initial Evaluation of Observation Bias and Error*

The accurate estimation of observation bias and error is an important prerequisite for the effective assimilation of satellite data. Observation from 10–23 September 2021 was selected for the evaluation of MWTS-3 data before assimilation. The RTTOV-12 was used to simulate the brightness temperature during the same period based on the ERA-5 reanalysis data released by the European Centre for Medium-Range Weather Forecasts (ECMWF). On this basis, the observation bias and error of MWTS-3 were estimated by analyzing the difference between the observed (O) and simulated (B) brightness temperature. In order to avoid the influence of the uncertainty of land surface emissivity, only clear-sky observations over the ocean were selected for the estimation. The means and standard deviations (STDs) of the calculated O-B of the MWTS-3 data are shown in Figure 4.

It can be seen that the biases and errors have great channel differences. For all channels, the biases are basically between ±2.0 K. Specifically, the biases of channels 1, 4, 7, 8, 10, and 11 are negative, whereas the biases of channels 4, 8, and 10 reach about −2.0 K. Channels 2, 3, 5, and 9 and the four channels in the upper stratosphere, i.e., channels 14–17, all have positive biases, which are basically around 1.5 K. While the biases of rest channels are close to 0.

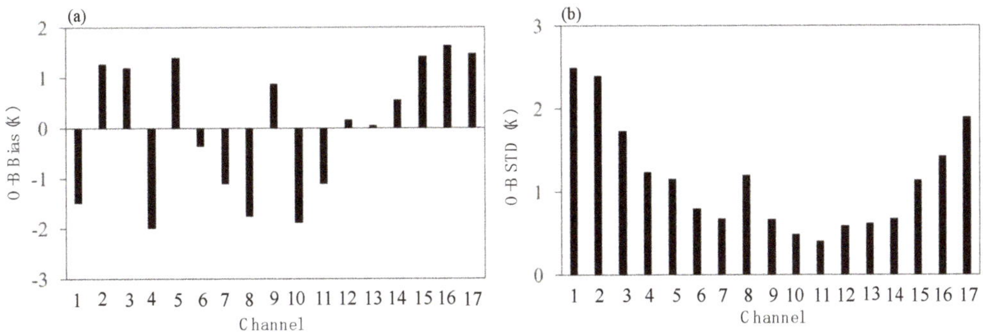

Figure 4. Bias (**a**) and standard deviation (STD) (**b**) of the differences between the brightness temperature observations and ERA simulations for FY-3E MWTS-3 channels during 10–23 September 2021.

The O-B STD (Figure 4b) gradually decreases and then increases with the increasing height from the ground to higher altitudes. The O-B STDs of channels 1–5 are sensitive to clouds and are also greatly affected by weather systems. Influenced by the relatively larger error of the lower-layer background, the O-B STDs of these channels are the largest. The peak values of the weighting function of channels 6–14 are mainly between 20–700 hPa, and the overall STD is within 1 K (except channel 8). The STD of channel 8 is about 1.2 K, obviously higher than those of adjacent channels. The peak values of the weighting function of channels 15–17 appear in the upper stratosphere, where remarkable increases in the observation errors are found in these channels, which may be due to the large error of the upper-level temperature profile in the background field and large NEDTs of these channels.

In general, the observation errors of MWTS-3 are within the normal range; only the noise of channel 8 is greater than expected. In addition, the biases of channels 6–7 also exceed those of similar channels of the same-type instruments, such as the ATMS (personal communication with Prof. Wen F. Z.).

2.5. Channel Selection

As shown in Figure 1, it is found that the maximums of the weighting function of MWTS-3 channels 1–5 are close to the ground and are sensitive to the underlying surface. Due to those inaccurate surface physical variables, such as the surface temperature and surface emissivity, these near-ground channels were not included in the data assimilation. When considering the bias problems of channels 6–8 in the preliminary evaluation in Section 2.4, these three channels were also excluded. For the upper tropospheric or stratospheric channels 11–17, since the error of the CMA-GFS is relatively larger near the model top (10 hPa to 0.1 hPa), the two high-level channels of 16 and 17 were excluded. As a preliminary study, MWTS-3 channels 9–15 were directly assimilated in the CMA-GFS.

2.6. Quality Control Based on Scan and Surface Characteristics

In addition to cloud detection, some extra QC procedures were applied to eliminate the observation data with abnormal O-B values caused by complex underlying surfaces and large terrain height.

Extra QC procedures were carried out in the following order. (I) The observations of channels 9 and 10 in the cloudy area are removed. (II) All FOVs covering the coastline are removed. A land mask database with longitudinal and latitudinal resolutions of 0.1° is used for land/ocean/coast identification. (III) The 10 outermost FOVs on each side of a scan line are not used; (IV) The observations of channel 9 over the sea ice or the land are not used. Sea ice surface is identified by the criteria that the sea surface temperature is lower than 271.45 K. (V) If the terrain height is greater than 500 m, the data of channel 10 is rejected. This threshold is based on previous experience. In the CMA-GFS, the QC of AMSU-A, ATMS, MWTS-1/2 data all adopt this threshold. Lastly, the MWTS-3 data, which passes all the above QC procedures, is thinned to a spatial resolution of 120 km according to the distance between the observations and the nearest model grid.

2.7. Bias Correction

The CMA-GFS 4D-VAR system was used in this study. The basic theory of variational data assimilation is the Bayesian conditional probability theorem [35]. This theory assumes that the background error and observation error satisfy the Gaussian distribution, and that there is no systematic bias. However, in practical application, systematic biases generally exist in the background and are mainly caused by the continuous forward integration of the numerical model. Meanwhile, there are inevitably systematic errors in the radiation transfer model simulations. These also lead to a certain degree of systematic biases in the O-B, meaning effective bias correction is necessary.

The importance of bias correction for satellite radiances in data assimilation has been realized by many meteorologists, and a lot of studies have been conducted to develop effective bias correction methods. It is found that systematic bias mainly consists of the bias caused by the scanning position difference and the bias depending on the air-mass property. Harris and Kelly (2001) developed a static bias correction scheme [36]. After a lot of practical applications, it has proven to be an effective correction scheme and is widely used in operational NWP centers around the world [3,4]. In 2007, Liu et al. (2007) added this scheme to the CMA-MESO model and also achieved good results [37]. However, considering the limitation of the static bias correction scheme in estimating the bias caused by the change in the weather system, an air-mass bias correction method has been proposed to take into account the impact of weather systems on systematic biases [3]. In addition, a variational bias correction scheme has also been established, which considers the variation of biases in combination with the minimization process of the assimilation system [38]. At present, this scheme has been applied in many national operational forecast centers, such as the National Centers for Environmental Prediction (NCEP) and the ECMWF [3,39,40]. After selecting the appropriate forecast factors, the variational bias correction scheme statistically updates the correction coefficients in the minimization process of the cost function. This scheme has also been tested in the CMA-GFS, and it is expected to achieve operational application in 2023. However, only the scan bias correction and the air-mass bias correction are involved in this study.

2.7.1. Scan Bias Correction

Since the scan angle bias obviously changes with the latitude, the statistics of scan bias also need to be conducted in different latitude bands. The whole hemisphere was divided into 18 latitudinal bands using 10° intervals. For each latitude band, the O-B difference between each scanning position and the nadir point in each scan line was calculated, and then the average value of all the O-B differences in the same latitude band was obtained as the systematic bias in this latitude band. A linear smoothing method was also applied to avoid discontinuous correction between the two adjacent latitudinal bands.

2.7.2. Air-Mass Bias Correction

In this study, two predictors were selected for the air-mass bias correction, namely, the thicknesses between 300–1000 hPa and 50–200 hPa of the background. Using the two-week thickness data, a linear regression equation was established for each channel, and the coefficients, a_{jo} and a_{ji}, in the regression equation were obtained for the channel j data with a scan angle of θ. The regression equation is as follows:

$$\text{Bias}_j(\theta) = a_{j0} + \sum_{i=1}^{2} a_{ji}(\theta)X_{ji}(\theta)$$

Here $Bias_j$ is the O-B bias, and X_{ji} is for the thickness. a_{jo} and a_{ji} represent the linear relationship between the O-B bias and the two thickness data. Using these coefficients, the O-B bias was calculated and subtracted from each observation in the assimilation process.

After the bias correction, the QC module also removes the observation data with large O-B values, and the pixels with O-B values greater than two times that of the observation error are rejected. According to the analysis results in Figure 4, the observation errors were set to 0.55 K for MWTS-3 channels 9, 10, and 12–14, 0.4 K for channel 11, and 1.1 K for channel 15 in this study.

3. Results

3.1. Experimental Design

Four experiments were conducted to demonstrate the impact of MWTS-3 data on the CMA-GFS during the period from 24 September to 25 October 2021. Table 2 shows the specific experimental designs. Experiment 1 assimilated only the conventional observations, called CTRL1. The conventional observations contain a global set of surface and upper-air reports, including radiosondes, SYNOP, ship reports, Aireps, and AMVs from the Global Telecommunications System (GTS). Experiment 2 assimilated the conventional observations: NOAA-15/18/19 AMSU-A, NOAA-18/19 MHS, MetOp-A/B AMSU-A, MHS and IASI, NPP ATMS, FY-3C/D MWHS-2 and MWRI, FY-3D MWTS-2 and HIRAS radiance data, FY-3C/D GNOS, COSMIC RO data, etc., called CTRL2. The setup of the two sensitive experiments (TEST1 and TEST2) is identical to the control experiments (CTRL1 and CTRL2), except that the FY-3E MWTS-3 radiance data were added in TEST1 and TEST2.

Table 2. Experiment design for the four cycle experiments.

EXP	Observation Data
CTL1	Conventional data
CTL2	Conventional data+ NOAA-15/18/19 AMSU-A+ NOAA-18/19 MHS+ MetOp-A/B AMSU-A/MHS/IASI+ NPP ATMS + FY-3C/D MWHS-2/MWRI + FY-3D MWTS-2/HIRAS + FY-3C/D GNOS + COSMIC RO, etc
TEST1	CTL1+FY-3E MWTS-3
TEST2	CTL2+FY-3E MWTS-3

Notes: conventional data consists of radiosondes, SYNOP, ship, Airep, and AMVs.

3.2. Analysis and Forecast of the Cycling Experiments

3.2.1. Characteristics of Data after Quality Control and Bias Correction

Figure 5 shows scatter plots of the observed and simulated brightness temperature of MWTS-3 channels 11 and 14 before and after QC during September 24–30, 2021. It can be seen that the differences between the O and B of channel 11 are larger before QC, which are scatter distributed, especially in the range of 210–220 K. Besides, the scatter plots obviously deviate from the diagonal. After QC, only the clear-sky observations over the ocean are retrained, which makes the distributions of O and B closer to each other, and the differences between them are from −3 K to −5 K. Figure 5b is for channel 14, where the scatter plots are already close to the diagonal before QC, only the plots with a brightness temperature higher than 230 K slightly deviate from the diagonal. The QC removes those abnormal observations effectively and makes the plots closer to the diagonal after QC.

Figure 6 shows the probability density functions of O-B for channels 11 and 14 before and after bias correction. Before bias correction, the biases of channels 11 and 14 are about −0.8 K and 0.8 K, and the STDs are 0.31 K and 0.55 K, respectively. The biases after correction are within ±0.1 K, and the STDs are slightly reduced to 0.29 K and 0.48 K, respectively. This indicates that the systematic biases of O-B have been corrected.

3.2.2. Comparisons of Observation Biases and Errors between MWTS-3 and Other Microwave Temperature Sounders

In order to further clarify the performance of MWTS-3, the bias and error characteristics of various microwave temperature-sounding data assimilated in TEST2 are given in this subsection, where the microwave temperature sounders include FY-3E MWTS-3, FY-3D MWTS-2, AMSU-A onboard NOAA-15/18/19, MetOp-A/B, and NPP ATMs. Figure 7 shows the biases and STDs of O-B from MWTS-3 and AMSU-A before and after bias correction in TEST2 during the period from 24 September to 25 October 2021. Among them, the frequencies of AMSU-A channels 5 and 6–14 are the same as those of MWTS-3 channels 7 and 9–17.

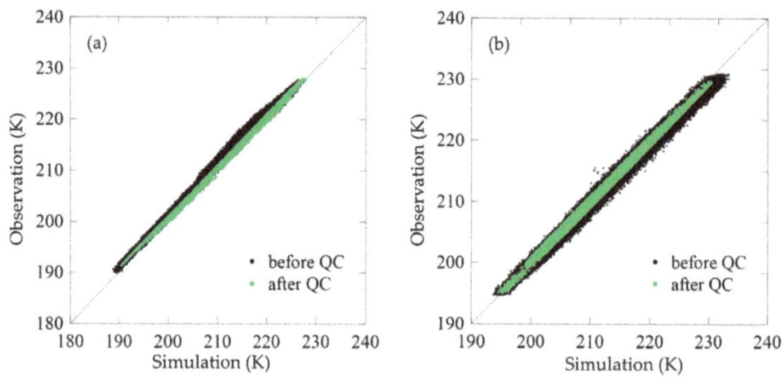

Figure 5. Scatterplots of observed (y-axis) and simulated (x-axis) brightness temperature for MWTS-3 channels 11 (**a**) and 14 (**b**) before (black dots) and after (green dots) quality control during 24–30 September 2021.

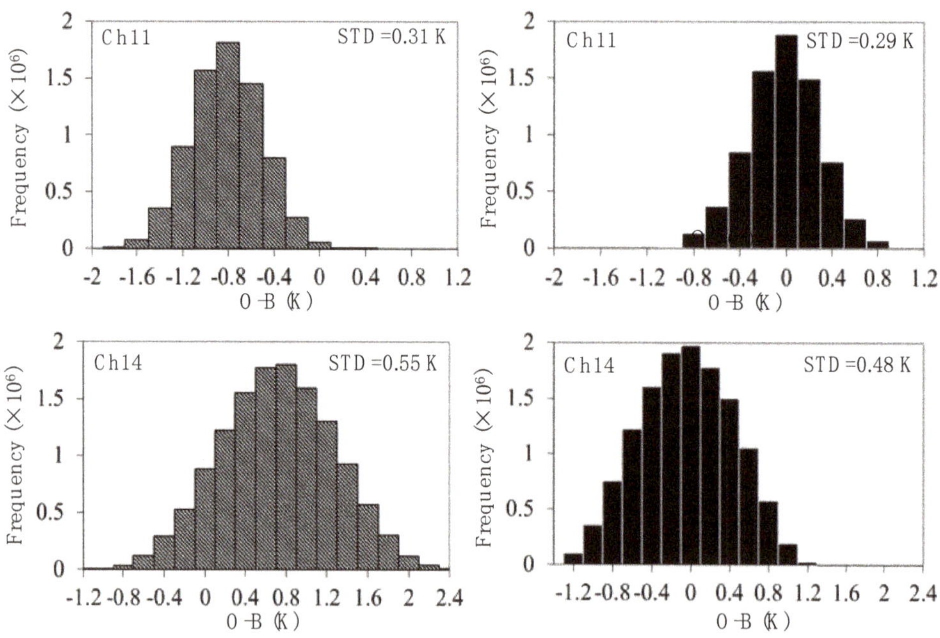

Figure 6. Frequency distributions of O-B differences for channels 11 (**top**) and 14 (**bottom**) before (hatched bars) and after (solid bars) bias correction for MWTS-3 channels 11 (upper panels) and 14 (down panels) during 24 September–3 October 2021.

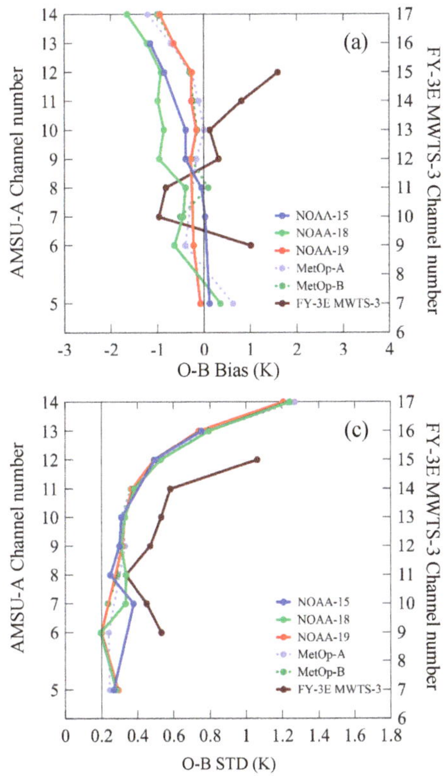

Figure 7. Bias (upper panels) and STD (lower panels) of the O-B for FY-3E MWTS-3, NOAA-15/18/19, and MetOp-A/B AMSU-A channels before (**a**,**c**) and after (**b**,**d**) bias correction calculated from the analysis results of the TEST2 experiment during 24 September–25 October 2021.

As shown in Figure 7, the overall bias of AMSU-A mid- and low-level channels is generally small before bias correction, where most channels show negative biases. Among them, the negative bias of AMSU-A onboard NOAA-18 is the most obvious. The bias of MWTS-3 is slightly larger than that of the same frequency channel of other instruments, where channel 9 shows a positive bias, while channel 6 of AMSU-A with the same frequency shows a slightly smaller negative bias. The biases of MWTS-3 channels 10–11 are twice those of the AMSU-A channels with the same frequency. Channels 12–17 (of MWTS-3) show positive biases that are opposite to those of AMSU-A. The upper-level channels of MWTS-3 and AMSU-A both exhibit large biases, which may be related to the large temperature errors in the upper-level of the background. Figure 7c shows that the O-B STD of MWTS-3 is also larger than that of AMSU-A before the bias correction, which may be related to the fact that the MWTS-3 has more pixels per scan line and a shorter sampling residence time. After the bias correction, the biases of all instruments are close to 0 (Figure 7b), indicating that the bias correction method for the CMA-GFS data assimilation system has a good correction effect. Besides, the STDs of all instruments also obviously decrease after the correction (Figure 7d).

Figure 8 shows the biases and STDs of FY-3D MWTS-2 and NPP ATMS for the same period. It can be found that, before bias correction, the bias of ATMS is also smaller than that of MWTS-3, which is comparable to that of AMSU-A, but the STD is larger than that of AMSU-A and is only slightly smaller than that of MWTS-3. The magnitudes of the bias and STD of MWTS-2 are comparable to those of MWTS-3. After the bias correction, the biases of all instruments are close to 0, and the STDs are also remarkably reduced. However, the STD of MWTS-3 is smaller than that of FY-3D MWTS-2 but is more similar to the STD features of the ATMS channels with the same frequency.

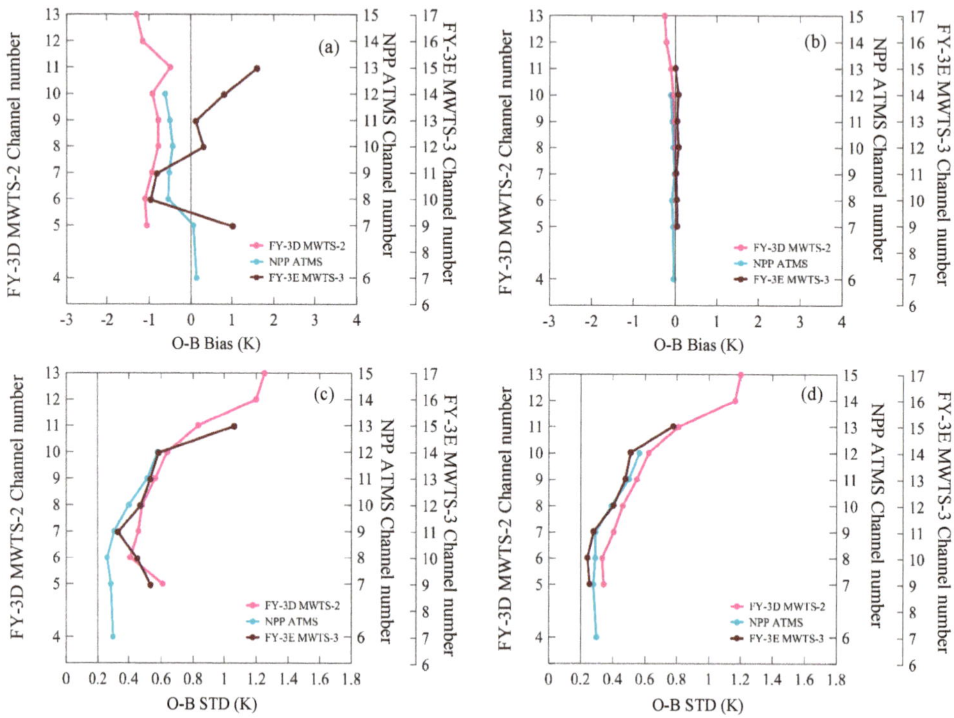

Figure 8. Bias (upper panels) and STD (lower panels) of the O-B for FY-3E MWTS-3, FY-3D MWTS-2 and NPP ATMS channels before (**a**,**c**) and after (**b**,**d**) bias correction calculated from the analysis results of the TEST2 experiment during 24 September–25 October 2021.

As indicated above, the comparisons among the observation errors of these microwave-sounding data before and after bias correction reveal that the error of AMSU-A is the smallest, followed by that of MWTS-3 and ATMS, and the observation error of MWTS-2 is the largest.

3.2.3. Analysis and Forecast

After investigating the characteristics of the MWTS-3 data, the assimilation effect of MWTS-3 data was further evaluated. The effect of adding the MWTS-3 data to the conventional data assimilation was explored first. Figure 9 shows that the RMSE of the geopotential height and the potential temperature differences between the analysis field and ERA-5 reanalysis data in the southern and Northern Hemispheres are reduced remarkably during the period from 24 September to 25 October 2021. Due to the lack of conventional observations in the Southern Hemisphere, the RMSE reduction in the Southern Hemisphere is most pronounced by adding the MWTS-3 data. Since only channels 9–15 of MWTS-3 are assimilated, and the peak heights of the weighting functions are located in the range of 10–400 hPa, the variables in the middle and high layers of the model are improved the most. Because there are a large number of conventional observations in the middle and lower layers of the Northern Hemisphere, the influence of MWTS-3 data assimilation over these regions is very small. However, as there are few conventional data above the height of

10 hPa, the improvement of adding MWTS-3 data on the geopotential height and potential temperature above 10 hPa is more obvious.

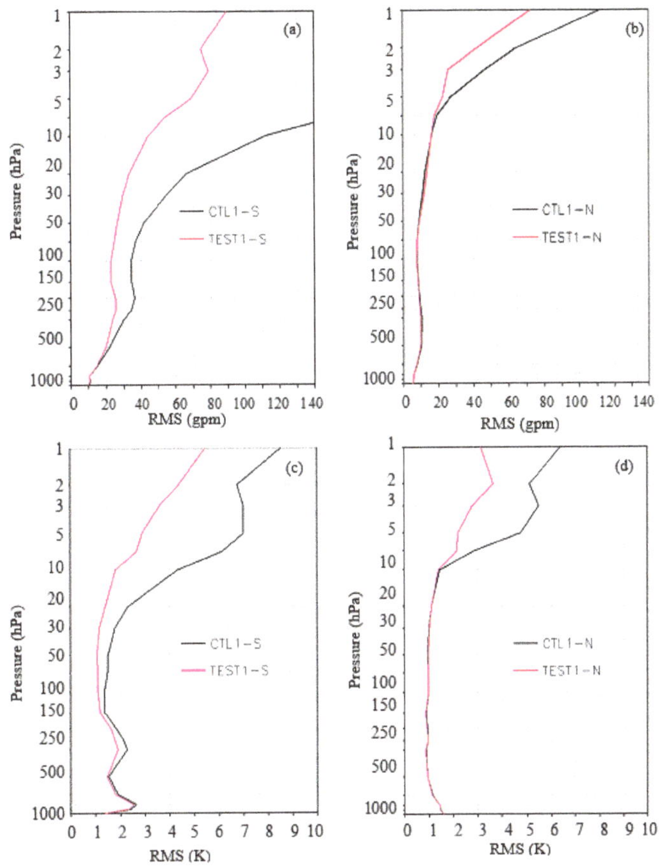

Figure 9. RMS of geopotential height from the analysis field difference between CTL1 and ERA (black) and TEST1 and ERA (red) in the (**a**) Southern Hemisphere and (**b**) Northern Hemisphere from 24 September–25 October 2021. (**c**,**d**) are similar to (**a**,**b**) but for the potential temperature.

Figure 10 shows the daily RMSE of the geopotential height differences between the analysis results and the ERA-5 reanalysis data for CTRL1 and TEST1 in the Southern Hemisphere at 10 hPa during the period from 24 September to 25 October 2021. It can be seen that, in the CTRL1, due to the lack of observation data above the altitude of 10 hPa, the model error increases rapidly with time. On the other hand, although only the MWTS-3 data is added in TEST1, the growth of the model error above 10 hPa is obviously suppressed, and the RMSE is greatly reduced in the first 15 days and then stably maintained within 60 gpm.

For operational assimilation applications, the impact of assimilating the FY-3E MWTS-3 data on the operational NWP system using all observation data needs to be paid more attention. CTRL2 assimilates all observation data used in the operations, including conventional and various satellite data, while TEST2 assimilates the MWTS-3 data additionally. The comparison shows that, after adding the MWTS-3 data, the errors of geopotential height, potential temperature, the U and V wind exhibit little change compared with the CTRL2 results at almost all altitudes. As shown in Figure 11, below an altitude of 2 hPa, the errors of the TEST2 results are slightly lower than those of the CTRL2 results. However, near the model top of above 2 hPa, the analysis results are slightly worse. This may be related to the imperfection of the bias correction scheme for the upper-level satellite data. Figure 12 shows that the error of wind below 5 hPa is slightly reduced. Overall, the effects of MWTS-3 data assimilation are neutral or slightly positive.

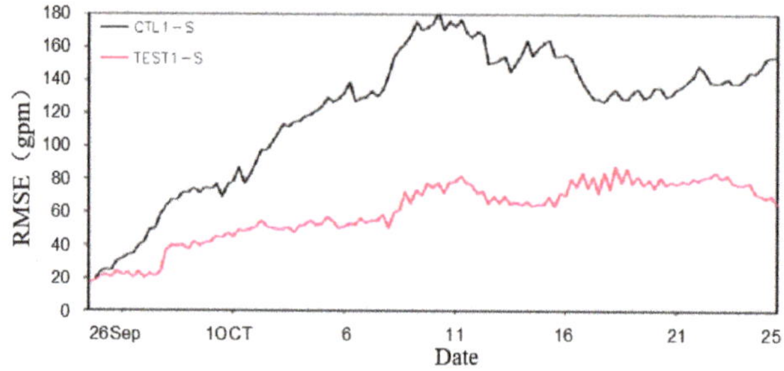

Figure 10. The daily RMS of geopotential height for the analysis field difference between CTL1 and ERA (black) and TEST1 and ERA (red) at 10 hPa in the Southern Hemisphere from 24 September–25 October 2021.

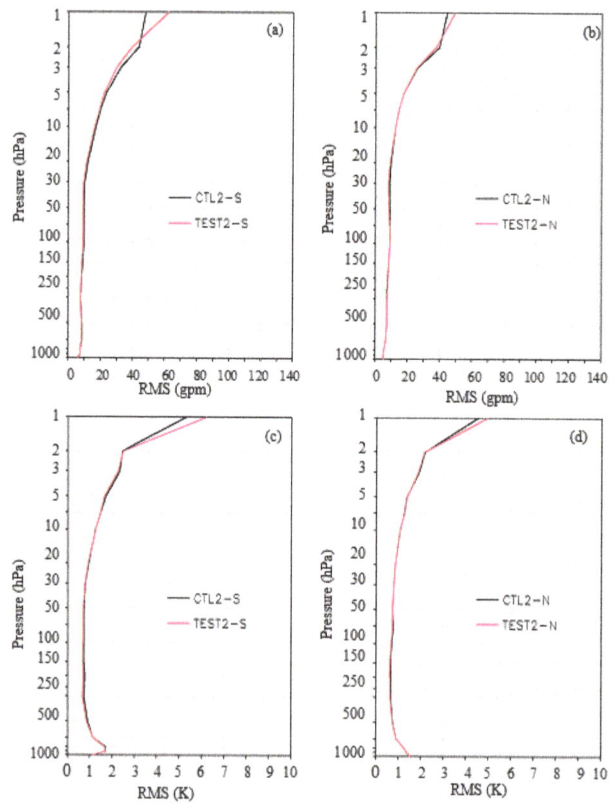

Figure 11. RMS of geopotential height for the analysis field difference between CTL2 and ERA (black) and TEST2 and ERA (red) in the (**a**) Southern Hemisphere and (**b**) Northern Hemisphere from 24 September–25 October 2021. (**c**,**d**) are similar to (**a**,**b**) but for the potential temperature.

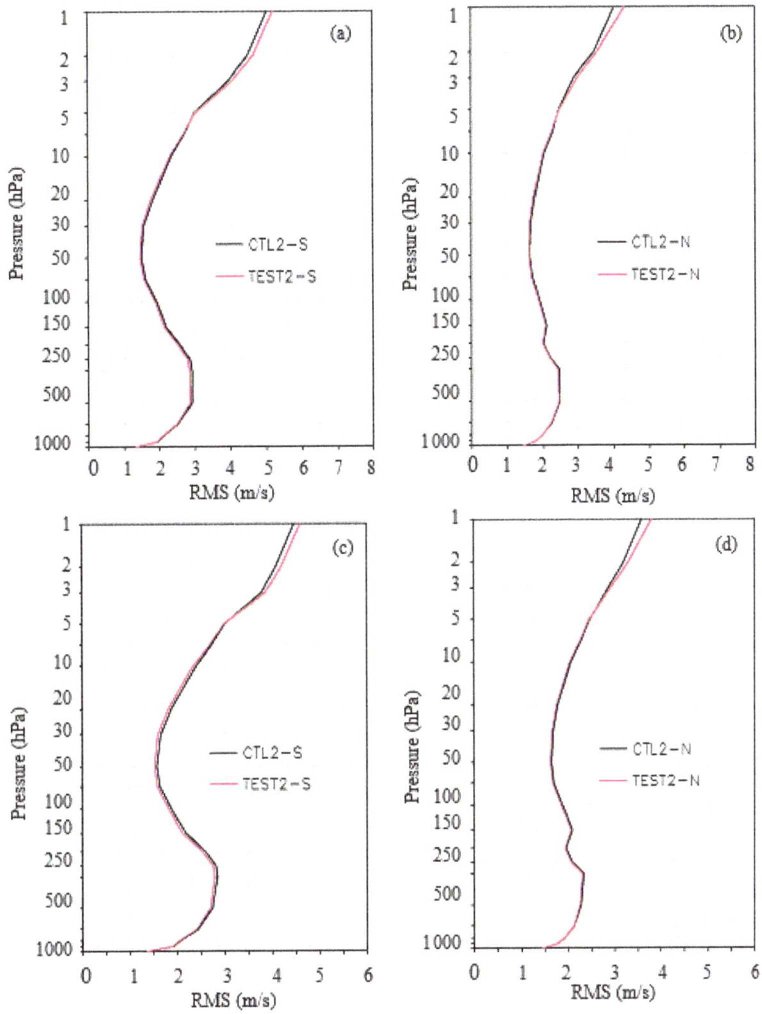

Figure 12. RMS of U wind for the analysis field difference between CTL2 and ERA (black), TEST2 and ERA (red) in the (**a**,**c**) Southern Hemisphere and (**b**,**d**) Northern Hemisphere.

Using the analysis results of CTRL2 and TEST2 (at 1200 UTC of each day) as the initial conditions, a 10-day prediction was achieved. The comprehensive scorecard for the evaluation of the forecast results shows the abnormal correlation coefficients (ACCs) and RMSEs of various variables at different levels and in different regions (Figure 13). It can be seen that the assimilation of MWTS-3 data has a positive contribution to the 10-day forecasts in the Northern and Southern Hemispheres, especially to the first two-day forecasts of the Southern Hemisphere. The overall impact in East Asia is neutral. In tropical areas, the impact on the ACCs is also generally neutral, but the RMSEs have increased, especially for the errors of geopotential height and potential temperature, which need to be further investigated in the future.

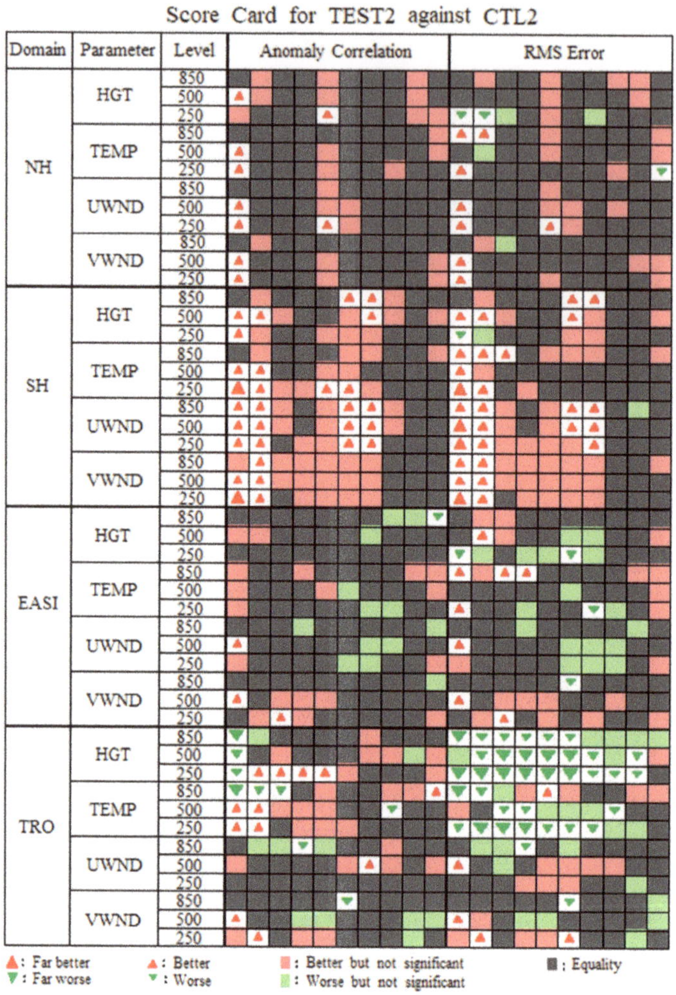

Figure 13. The score card for TEST2 against CTL2.

4. Discussion

This study demonstrates the impact of the FY-3E MWTS-3 radiances data assimilation for the first time. In the follow-up study, a further comparison of the impacts of data assimilation between MWTS-3 and MWTS-2 will be conducted to provide a reference for the improvement of the Fengyun satellite microwave temperature sounder. For the two newly added channels (6 (53.246 ± 0.08 GHz) and 8 (53.948 ± 0.081 GHz)) of MWTS-3, further assessment and application research for their assimilation is needed. In addition, as the new generation of early-morning orbiting satellites, the FY-3E, NOAA series, and the MetOp series polar-orbiting satellites have formed a complete three-orbit observation system. The supplementary effect of MWTS-3 is worth further exploration.

5. Conclusions

FY-3E is the fifth polar-orbiting satellite in the FY-3 series launched in July of 2021 in China, which carries the new-generation microwave temperature-sounding instruments. Compared with its predecessor satellites, two channels capable of retrieving the CLWP have been designed for the first time, which is of great help for cloud detection in data assimilation. In addition, two detection channels have been added with their peak weighting functions near 700 hPa and 500 hPa, and the ability to detect atmospheric temperature has been improved compared with the previous generation MWTS-2.

After the effective QC, bias correction, and accurate error specification of the MWTS-3 data, the direct assimilation of MWTS-3 radiance data has been realized in the CMA-GFS. The near one-month cycling experiments have indicated that the errors of analysis results can be remarkably reduced by adding the MWTS-3 data to the conventional data, especially for the variables on the upper layer of the model, where there is a lack of sufficient conventional observations. When all the observations in operation are included, the MWTS-3 data assimilation has a neutral contribution to the forecasts in the Northern Hemisphere and a slightly positive contribution in the Southern Hemisphere. However, in the tropics, the forecast errors of geopotential height and potential temperature have increased after adding the MWTS-3 data, which needs further investigation.

Author Contributions: Conceptualization, J.L.; methodology, J.L.; software, J.L., X.Q. and G.L.; validation, J.L., X.Q. and G.L.; formal analysis, J.L. and X.Q.; investigation, J.L. and G.L.; data curation, G.L.; writing—original draft preparation, J.L. and G.L.; writing—review and editing, J.L. and Z.Q. All authors have read and agreed to the published version of the manuscript.

Funding: This research was jointly funded by the Fengyun Application Pioneering Project (FY-APP-2021.0201) and FY-3 Meteorological Satellite Ground Application System Project (FY-3 (03)-AS-11.08).

Data Availability Statement: All data used in this paper are available from the authors upon request (lj@cma.gov.cn).

Acknowledgments: The authors would like to acknowledge the National Satellite Meteorological Centre of CMA for providing the satellite data.

Conflicts of Interest: The authors declare no conflict of interest.

References

1. Andersson, E.; Pailleux, J.; Thepaut, J.N.; Eyre, J.R.; McNally, A.P.; Kelly, G.A.; Courtier, P. Use of cloud-cleared radiances in three-four-dimensional variational data assimilation. *Q. J. R. Meteorol. Soc.* **1994**, *120*, 627–653. [CrossRef]
2. Courtier, P.; Andersson, E.; Heckley, W.; Vasiljevic, D.; Hamrud, M.; Hollingsworth, A.; Rabier, F.; Fisher, M.; Pailleux, J. The ECMWF implementation of three-dimensional variational assimilation (3D-Var). I. Formulation. *Q. J. R. Meteorol. Soc.* **1998**, *124*, 1783–1807. [CrossRef]
3. Derber, J.C.; Wu, W.S. The use of TOVS cloud-cleared radiances in the NCEP SSI analysis system. *Mon. Weather Rev.* **1998**, *126*, 2287–2299. [CrossRef]
4. McNally, A.P.; Derber, J.C.; Wu, W.; Katz, B.B. The use of TOVS level-1b radiances in the NCEP SSI analysis system. *Q. J. R. Meteorol. Soc.* **2000**, *126*, 689–724. [CrossRef]
5. Okamoto, K.; Kazumori, M.; Owada, H. The assimilation of ATOVS radiances in the JMA global analysis system. *J. Meteor. Soc.* **2005**, *83*, 201–217. [CrossRef]
6. Baker, N.L.; Daley, R. Observation and background adjoint sensitivity in the adaptive observation-targeting problem. *Q. J. R. Meteorol. Soc.* **2000**, *126*, 1431–1454. [CrossRef]
7. Fourrié, N.; Doerenbecher, A.; Bergot, T.; Joly, A. Adjoint sensitivity of the forecast to TOVS observations. *Q. J. R. Meteorol. Soc.* **2002**, *128*, 2759–2777. [CrossRef]
8. Langland, R.H.; Baker, A.L. Estimation of observation impact using the NRL atmospheric variational data assimilation adjoint system. *Tellus A* **2004**, *56*, 189–201. [CrossRef]
9. Cardinali, C. Monitoring observation impact on short-range forecast. *Q. J. R. Meteorol. Soc.* **2009**, *135*, 239–250. [CrossRef]
10. Gelaro, R.; Langland, R.H.; Pellerin, S.; Todling, R. The THORPEX observation impact intercomparison experiment. *Mon. Weather Rev.* **2010**, *138*, 4009–4025. [CrossRef]
11. Dong, C.H.; Yang, J.; Yang, Z.D.; Lu, N.M.; Shi, J.M.; Zhang, P.; Liu, Y.J.; Cai, B.; Zhang, W. An overview of a new Chinese weather satellite FY-3A. *Bull. Am. Meteorol. Soc.* **2009**, *90*, 1531–1544. [CrossRef]
12. Zhang, P.; Yang, J.; Dong, C.H.; Lu, N.M.; Yang, Z.D.; Shi, J.M. General introduction on payloads, ground segment and data application of Fengyun 3A. *Front. Earth. Sci.* **2009**, *3*, 367–373.
13. You, R.; Gu, S.; Guo, Y.; Chen, W.; Yang, H. Long-term calibration and accuracy assessment of the FengYun-3 Microwave Temperature Sounder radiance measurements. *IEEE Trans. Geosci. Remote Sens.* **2012**, *50*, 4854–4859. [CrossRef]
14. Zou, X.; Wang, X.; Weng, F.; Guan, L. Assessments of Chinese FengYun Microwave Temperature Sounder (MWTS) measurements for weather and climate applications. *J. Atmos. Ocean. Technol.* **2011**, *28*, 1206–1227. [CrossRef]

15. Li, J.; Zou, X. A quality control procedure for FY-3A MWTS measurements with emphasis on cloud detection using VIRR cloud fraction. *J. Atmos. Ocean. Technol.* **2013**, *30*, 1704–1715. [CrossRef]
16. Li, J.; Zou, X. Impact of FY-3A MWTS radiances on prediction in GRAPES with comparison of two quality control schemes. *Front. Earth Sci.* **2014**, *8*, 251–263. [CrossRef]
17. Li, J.Z.Q.; Liu, G. A new generation of Chinese FY-3C microwave sounding measurements and the initial assessments of its observations. *Int. J. Remote Sens.* **2016**, *37*, 4035–4058. [CrossRef]
18. Li, J.; Liu, G. Assimilation of Chinese FengYun 3B Microwave Temperature Sounder radiances into Global GRAPES system with an improved cloud detection threshold. *Front. Earth Sci.* **2016**, *10*, 145–158. [CrossRef]
19. Lu, Q.; Bell, W.; Bauer, P.; Bormann, N.; Peubey, C. *An Initial Evaluation of FY-3A Satellite Data*; ECMWF Technical Memoranda Number 631; ECMWF: Reading, UK, 2010.
20. Lu, Q.; Bell, W. *Evaluation of FY-3B Data and an Assessment of Passband Shifts in AMSU-A and MSU during the Period 1978–2012*; Interim Report of Visiting Scientist mission NWP_11_05, Document NWPSAF-EC-VS-023, Version 0.1, 28; Met. Office: Exeter, UK, 2012.
21. Zhang, P.; Hu, X.Q.; Lu, Q.F.; Zhu, A.J.; Lin, M.Y.; Sun, L.; Chen, L.; Xu, N. FY-3E: The first operational meteorological satellite mission in an early morning orbit. *Adv. Atmos. Sci.* **2021**, *39*, 1–8. [CrossRef]
22. Qian, X.; Qin, Z.; Li, J.; Han, Y.; Liu, G. Preliminary Evaluation of FY-3E Microwave Temperature Sounder Performance Based on Observation Minus Simulation. *Remote Sens.* **2022**, *14*, 2250. [CrossRef]
23. Chen, D.H.; Xue, J.S.; Yang, X.S.; Zhang, H.L.; Shen, X.S.; Hu, J.L.; Wang, Y.; Ji, L.R.; Chen, J.B. New generation of multi-scale NWP system (GRAPES): General scientific design. *Chin. Sci. Bull.* **2008**, *53*, 3433–3445. [CrossRef]
24. Xue, J.S.; Chen, D.H. *Numerical Prediction System Design and Application of Science GRAPES*; Science Press: Beijing, China, 2008.
25. Xue, J.S.; Zhuang, S.Y.; Zhu, G.F.; Zhang, H.; Liu, Z.Q.; Liu, Y.; Zhuang, Z.R. Scientific design and preliminary results of three-dimensional variational data assimilation system of GRAPES. *Chin. Sci. Bull.* **2008**, *53*, 3446–3457. [CrossRef]
26. Zhang, L.; Liu, Y.; Liu, Y.; Gong, J.; Lu, H.; Jin, Z.; Tian, W.; Liu, G.; Zhou, B.; Zhao, B. The operational global four-dimensional variational data assimilation system at the China Meteorological Administration. *Q. J. R. Meteorol. Soc.* **2019**, *145*, 1882–1896. [CrossRef]
27. Navon, I.M.; Legler, D.M. Conjugate gradient methods for large scale minimization in meteorology. *Mon. Weather Rev.* **1987**, *115*, 1479–1502. [CrossRef]
28. Saunders, R.W.; Matricardi, M.; Brunel, P. An Improved Fast Radiative Transfer Model for Assimilation of Satellite Radiance Observations. *Q. J. R. Meteorol. Soc.* **1999**, *125*, 1407–1425. [CrossRef]
29. Weng, F.; Zhao, L.; Ferraro, R.R.; Poe, G.; Li, X.; Grody, N.C. Advanced microwave sounding unit cloud and precipitation algorithms. *Radio Sci.* **2003**, *38*, 8068. [CrossRef]
30. Grody, N.; Zhao, J.; Ferraro, R.; Weng, F.; Boers, R. Determination of precipitable water and cloud liquid water over oceans from the NOAA 15 advanced microwave sounding unit. *J. Geophys. Res.* **2001**, *106*, 2943–2953. [CrossRef]
31. Klaes, D.; Schraidt, R. The European ATOVS and AVHRR processing package (AAPP). In Proceedings of the 10th International TOVS Study Conference (ITSC X), Boulder, CO, USA, 27 January–2 February 1999.
32. Errico, R.M.; Bauer, P.; Mahfouf, J.-F. Issues regarding the assimilation of cloud and precipitation data. *J. Atmos. Sci.* **2007**, *64*, 3785–3798. [CrossRef]
33. Geer, A.J.; Bauer, P. *Enhanced Use of All-Sky Microwave Observations Sensitive to Water Vapour, Cloud and Precipitation*; ECMWF Technical Memoranda 620; ECMWF: Reading, UK, 2010.
34. Geer, A.J.; Lonitz, K.; Weston, P.; Kazumori, M.; Okamoto, K.; Zhu, Y.; Liu, E.H.; Collard, A.; Bell, W.; Migliorini, S.; et al. All-sky satellite data assimilation at operational weather forecasting centres. *Q. J. R. Meteor. Soc.* **2018**, *144*, 1191–1217. [CrossRef]
35. Lorenc, A.C. Analysis methods for numerical weather prediction. *Q. J. R. Meteor. Soc.* **1986**, *112*, 1177–1194. [CrossRef]
36. Harris, B.A.; Kelly, G. A satellite radiance-bias correction scheme for data assimilation. *Q. J. R. Meteorol. Soc.* **2001**, *127*, 1453–1468. [CrossRef]
37. Liu, Z.Q.; Zhang, F.Y.; Wu, X.B.; Xue, J. A regional atovs radiance-bias correction scheme for rediance assimilation. *Acta Meteorol. Sin.* **2007**, *65*, 113–123.
38. Isaksen, L.; Vasiljevic, D.; Dee, D.P.; Healy, S. *Bias Correction of Aircraft Data Implemented in November 2011*; ECMWF Newsletter, No. 131; ECMWF: Reading, UK, 2012; pp. 6–7.
39. Dee, D.P. Variational bias correction of radiance data in the ECMWF system. In Proceedings of the ECMWF Workshop on Assimilation of High Spectral Resolution Sounders in NWP, Reading, UK, 28 June–1 July 2004; pp. 97–112.
40. Dee, D.P. Bias and data assimilation. *Q. J. R. Meteorol. Soc.* **2005**, *131*, 3323–3343. [CrossRef]

remote sensing

MDPI

Communication

Assessing FY-3E HIRAS-II Radiance Accuracy Using AHI and MERSI-LL

Hongtao Chen and Li Guan *

Collaborative Innovation Center on Forecast and Evaluation of Meteorological Disasters, Nanjing University of Information Science &Technology, Nanjing 210044, China
* Correspondence: liguan@nuist.edu.cn

Abstract: The FY-3E/HIRAS-II (Hyperspectral Infrared Atmospheric Sounder-II), as an infrared hyperspectral instrument onboard the world's first early morning polar-orbiting satellite, plays a major role in improving the accuracy and timeliness of global numerical weather predictions. In order to assess its observation quality, the geometrically, temporally, and spatially matched scene homogeneous HIRAS-II hyperspectral observations were convolved to the channels corresponding to the Himawari-8/AHI (Advanced Himawari Imager) and FY-3E/MERSI-LL (Medium-Resolution Spectral Imager) imagers from 15 March to 21 April 2022, and their brightness temperature deviation characteristics were statistically calculated in this paper. The results show that the HIRAS-II in-orbit observed brightness temperatures are slightly warmer than the AHI observations in all the matched AHI channels (long wave infrared channel 8 to channel 16) with a mean brightness temperature bias less than 0.65 K. The bias of the atmospheric absorption channel is slightly larger than that of the window channel. A standard deviation less than 0.31 K and a correlation coefficient higher than 0.98 in all channels means that the quality of the observation is satisfactory. The thresholds chosen for the colocation approximation factors (e.g., observation geometry angle, scene uniformity, observation azimuth, and observation time) for matching the HIRAS-II with AHI contribute little and negligible uncertainty to the bias assessment, so the difference between the two observed radiations is considered to be mainly from the systematic bias of the two-instrument measurement. Compared with MERSI-LL window channel 5, the observations of both instruments are very close, with a mean bias of 0.002 K and a standard deviation of 0.31 K. The mean brightness temperature bias (HIRAS-II minus MERSI-LL) of the MERSI-LL water vapor channel 4 is 0.66 K with a standard deviation of 0.22 K. The mean brightness temperature bias of channel 6 and channel 7 is 0.63 K (the standard deviation is 0.36 K) and 0.5 K (the standard deviation is 0.3 K), respectively. The biases of channel 4 are significantly and positively correlated with the target scene temperature, and the biases of channel 6 and 7 show a U-shaped change with the increase in the scene temperature, and the biases are smallest (close to 0 K) when the scene temperature is between 250 K and 280 K. The statistical characteristics of the HIRAS-II–MERSI-LL difference vary minimally and almost constantly over a period of time, indicating that the performance of the HIRAS-II instrument is stable.

Keywords: Hyperspectral Infrared Atmospheric Sounder-II; FY-3E; Himawari-8/AHI

Citation: Chen, H.; Guan, L. Assessing FY-3E HIRAS-II Radiance Accuracy Using AHI and MERSI-LL. *Remote Sens.* **2022**, *14*, 4309. https://doi.org/10.3390/rs14174309

Academic Editors: Jie Cheng and Nicholas R. Nalli

Received: 20 July 2022
Accepted: 23 August 2022
Published: 1 September 2022

1. Introduction

Satellite-borne infrared hyperspectral atmospheric sounders can obtain global meteorological observations with high precision and high spectral resolution and have been frequently applied to retrieving atmospheric temperature and humidity profiles, data assimilation, and climate studies [1]. In addition, quality control of meteorological satellite observations is a pivotal step before using satellite data for assimilation and retrieval, and it is the fundamental basis for building long-term infrared hyperspectral datasets [2].

FY-3E, the world's first early morning polar-orbiting meteorological satellite, was successfully launched on 5 July 2021. It effectively fills a void for global satellite observations and provides 100% global satellite data coverage for numerical weather prediction (NWP) at 6-h intervals [3]. The HIRAS-II (Hyperspectral Infrared Atmospheric Sounder-II) is a continuation of the infrared hyperspectral instrument HIRAS onboard the FY-3D, and it can provide hyperspectral observations in the thermal infrared band with 3041 contiguous channels. Compared with its predecessor, the field of view (FOV) array within a field of regard (FOR) has changed from 2 × 2 to 3 × 3 with the spatial resolution increased from 16 km to 14 km at the nadir, the sensitivity is enhanced by more than 2 times with the spectral calibration accuracy increased by 30% and the radiometric calibration accuracy increased from 0.7 K to 0.5 K [4]. HIRAS-II is expected to become the reference instrument for infrared remote sensing instruments; therefore, independently assessing its data quality for radiance measurements is of great importance in improving the accuracy and timeliness of global numerical weather prediction.

The combined remote sensing and intercalibration based on the satellite-borne infrared hyperspectral atmospheric sounder and high-spatial-resolution imager has become one of the most effective means to quantify the radiometric calibration accuracy for both types of instruments. Gunshor calibrated the water vapor channels and window channels of five geostationary satellite imagers using the High-Resolution Infrared Radiation Sounder (HIRS) and the Advanced Very-High-Resolution Radiometer (AVHRR) onboard NOAA-14 [5]. Tobin used the Atmospheric Infrared Sounder (AIRS) to evaluate the radiometric accuracy of the Moderate-Resolution Imaging Spectroradiometer (MODIS) carried on the same platform [6]. Wang used the Infrared Atmospheric Sounding Interferometer (IASI) to intercalibrate the water vapor channel of the GOES-11 and GOES-12 [7]. Xu Na et al., using the hyperspectral measurements of IASI as a reference, objectively assessed the on-orbit radiometric calibration accuracy of the FY-3A Medium-Resolution Spectral Imager (MERSI) thermal infrared channel [8]. Gong used the Cross-track Infrared Sounder (CrIS) onboard the Suomi National Polar-orbiting Partnership (SNPP) satellite platform to cross-check the thermal infrared channels of the Visible Infrared Imaging Radiometer Suite (VIIRS) on the same platform [9]. Yang et al. assessed the relative bias of the HIRAS radiometric calibrations using the Metop-A/B IASI observations based on the Simultaneous Nadir Overpass (SNO) intercalibration method [1].

Accuracy assessments of satellite instrument on-orbit calibrations is necessary to ensure product consistency and interoperability, and it is also extremely important for bias correction in data assimilations [10]. However, HIRAS-II—which is on board the first early morning polar-orbiting satellite launched last year—is operational this year, and the quality of its radiance measurements has not yet been reported in the literature. The Advanced Himawari Imager (AHI) mounted on the Japanese geostationary meteorological satellite Himawari-8 is recognized as one of the most accurate imaging instruments in the world. The AHI is greatly improved over those of the MTSAT (Multi-functional Transport Satellite) series in terms of the number of bands, spatial resolution, and temporal frequency; and infrared (IR) band calibration is accurate to within 0.2 K with no significant diurnal variation [11–13]. Therefore, this paper evaluates the quality of the radiance measurements based on the spatially and temporally matched Himawari-8/AHI observations from 15 March to 21 April 2022, and also performs an intercomparison with the thermal infrared observations of MERSI-LL carried out on the same platform.

2. Data Used in the Research

The HIRAS-II is an interferometric Fourier transform spectrometer carried in a polar orbit 836 km above the ground. HIRAS-II views the ground in the conventional mode through a cross-track rotary scan mirror that provides ±50.4° ground coverage every 8 s. Each scan line observes 32 fields of regard (FORs), including 28 continuous Earth targets, 2 cold space targets, and 2 blackbody targets on the satellite. Each field of regard (FOR) includes a 3 × 3 field of view (FOV) with a spatial resolution of 14 km at the nadir. HIRAS-II covers the 3.92–15.38 μm infrared band with 3041 continuous channels at a spectral resolution of 0.625 cm^{-1}. The HIRAS-II Level 1 radiance observations from 15 March to 21

April 2022 are used in this paper. The data can be found on the Chinese Feng Yun satellite remote sensing data service network (http://data.nsmc.org.cn accessed on 11 April 2022).

The Moderate-Resolution Spectral Imager-Low Light (MERSI-LL) is an important optical instrument onboard FY-3E with microlight and infrared detection capabilities. It is equipped with one visible channel operable with low-level illumination and six infrared channels. The spatial resolution of the two infrared split-window channels is 250 m, and the remaining channels are 1000 m. The MERSI-LL Level 1 radiance observations with a spatial resolution of 1000 m from 15 to 22 March 2022 are used in this paper and can be downloaded from http://data.nsmc.org.cn (accessed on 11 April 2022).

The AHI on the geostationary satellite Himawari-8 successfully launched in October 2014 and is one of the most advanced spaceborne imagers in the world. It has 16 observation channels (3 visible, 3 near-infrared, and 10 infrared), of which the spatial resolution of the infrared channel is 2 km and the temporal resolution is 10 min. Himawari-8/AHI radiation data obtained from the Japan Earth Observation Data Center (https://www.eorc.jaxa.jp/ptree/index.html accessed on 11 April 2022) from 15 March to 21 April 2022 are analyzed.

The channel settings and performance of the thermal infrared band covered by HIRAS-II, AHI, and MESI-LL are shown in Table 1. The last row of the table (spectral coverage) specifies the central wavelength and the corresponding peak height of the weighting function (in parentheses) for each channel. The AHI has nine channels that can be completely spectrum matched with the HIRAS-II spectrum, of which channels 8, 9, and 10 are water vapor absorption channels; channels 13, 14, and 15 are window channels; and channels 11, 12, and 16 are SO_2, O_3, and CO_2 absorption channels, respectively. The weighting function peak heights of AHI channels 8, 9, 10, 12, and 16 are 300 hPa, 371 hPa, 532 hPa, 40 hPa, and 863 hPa, respectively. The weighting function heights of the remaining channels are almost near the surface. For MERSI-LL, only channels 4, 5, 6, and 7 can be completely spectrum matched. Channel 4 is a water vapor channel (the peak height of the weighting function is approximately 400 hPa), and channels 5, 6, and 7 are window channel with a central wavelength of 8.55 μm, 10.8 μm, and 12.0 μm, respectively.

Table 1. Instrument performance parameters of HIRAS-II, AHI, and MERSI-LL in the longwave infrared band.

	HIRAS-II	AHI	MERSI-LL
Satellite platform	FY-3E	Himawari-8	FY-3E
Spatial resolution/km	14 km (at nadir)	Infrared: 2 km	Infrared: 1 km
Spectral coverage	3041 channels (3.92–15.38 μm) with spectral resolution 0.625 cm^{-1}	Ch8: 6.2 μm (300 hPa) Ch9: 6.9 μm (371 hPa) Ch10: 7.3 μm (532 hPa) Ch11: 8.6 μm (window) Ch12: 9.6 μm (40 hPa) Ch13: 10.4 μm (window) Ch14: 11.2 μm (window) Ch15: 12.4 μm (window) Ch16: 13.3 μm (863 hPa)	Ch4: 7.2 μm (400 hPa) Ch5: 8.55 μm (window) Ch6: 10.8 μm (window) Ch7:12.0 μm (window)

3. Observation Data Matching and Evaluation Method

Four major steps are involved in an intercomparison of the FY-3E/HIRAS-II infrared hyperspectral observations with the Himawari-8/AHI radiance observations of the corresponding longwave infrared channel, including (1) spectral convolution, (2) observation geometry and temporal matching, (3) spatial matching, and (4) uniformity checking. The matching process is shown in Figure 1. The following will be a step-by-step detailed introduction.

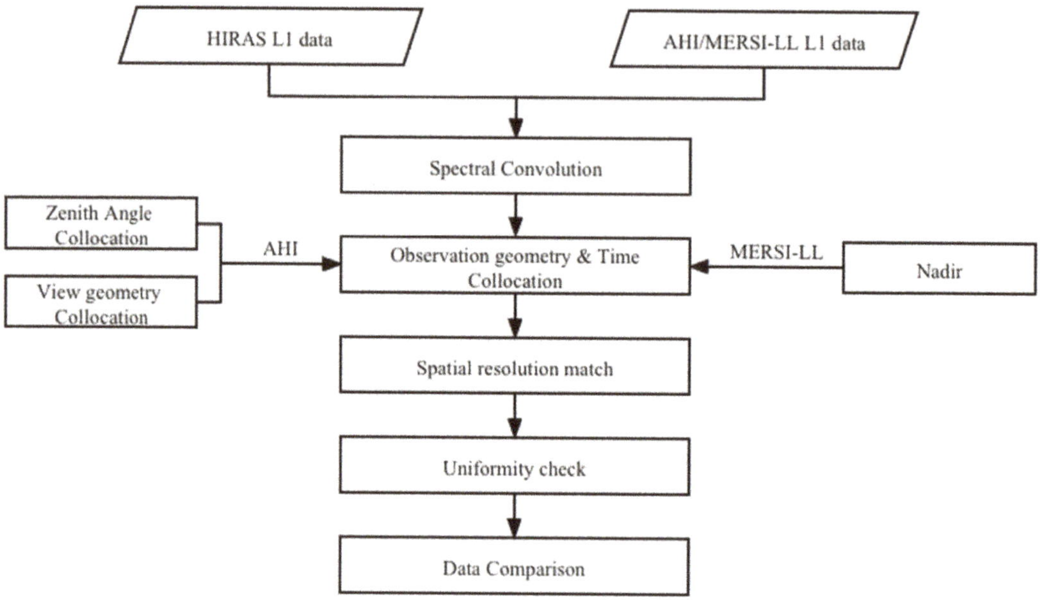

Figure 1. Intercomparison flow chart of HIRAS-II, AHI, and MERSI-LL.

3.1. Spectral Convolution

The most fundamental problem in the intercomparison between the spaceborne imager and sounder observation is radiance spectral matching. The two types of instruments with different spectral resolutions cannot be directly compared. To be compared with the imager instrument observations, the hyperspectral radiance must be convolved to match the spectral response function (SRF) of the broadband imager [14,15]. The computational formula to achieve spectral convolution is

$$L = \frac{\int_{v_1}^{v_2} R(v)S(v)dv}{\int_{v_1}^{v_2} S(v)dv} \tag{1}$$

where v is the wavenumber, L is the radiance calculated by convolution, $R(v)$ is the hyperspectral radiance at the corresponding wavenumber and $S(v)$ is the spectral response function of the imager. The simulated HIRAS-II brightness temperature spectrum using fast radiative transfer mode RTTOV under American standard atmospheric conditions, as well as the AHI and MERSI-LL spectral response functions (SRF), are given in Figure 2. The bottom horizontal coordinate is the wavelength, the upper horizontal coordinate is the wavenumber, the left vertical coordinate is the simulated brightness temperature of HIRAS-II, and the right vertical coordinate is the spectral response function. The solid line in the figure is the spectral response function of the matched AHI channels 8–16, and the dashed line denotes the spectral response function of MERSI channels 4–7.

3.2. Observation Geometry and Temporal Matching

The most important step in the intercomparison between the imager and the sounder observation is to find the consistent (same) FOV [16]. The polar-orbiting satellite FY-3E passes the geostationary satellite Himawari's nadir 140.7 E at approximately 0830 or 2030 UTC daily, the observation time difference remained within 10 min. Since the measurements of the cross-track scanning instruments are sensitive to the scan angle (satellite zenith angle), the scan angle of both instruments in the matched field of view is specified to be less than 5° to ensure that HIRAS-II and AHI observe the same scene. To minimize the difference in observation geometry and ensure that both instruments have similar observation geometry paths, the scan angle is further constrained [7,17]

$$\left|\frac{\cos\theta_1}{\cos\theta_2}-1\right|<0.002 \quad (2)$$

where θ_1 and θ_2 are the scan angles of the geostationary and polar-orbiting satellites, respectively.

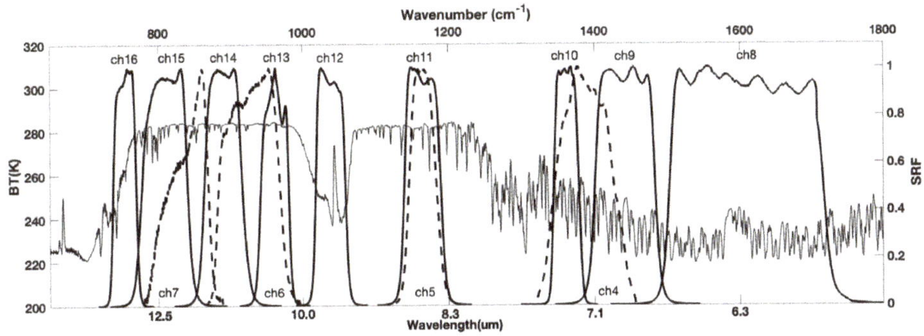

Figure 2. Simulated HIRAS-II brightness temperature spectra from RTTOV overlaid the spectral response functions of AHI (solid line) and MERSI-LL (dashed line), channel numbers for AHI are shown near the top of the graph and channel numbers for MERSI−LL are near the bottom.

3.3. Spatial Matching and Uniformity Check

The spatial field of view collocation of the FY-3E/HIRAS-II and Himawari-8/AHI in the infrared channel is shown in Figure 3, where the grid is the AHI pixel, the large circle is a HIRAS-II field of regard (FOR) centered on the field of view (FOV) to be matched (target) and the small circle is the HIRAS-II FOV within its FOR. The spatial resolution of the HIRAS-II FOV (14 km at the nadir) is coarser than that of the AHI infrared channel (2 km). Approximately 7 × 7 AHI pixels (small squares in Figure 3) fall in a HIRAS-II FOV (the small circle in Figure 3). The average of all these AHI pixel measurements is taken as the imager's measurement in the matching FOV [18]. This requires that the observation target be relatively homogeneous. Since there are many AHI pixels collocated in the HIRAS-II FOV, some of which are clear sky and some are cloudy, at the same time the underlying surface of the field of view is not homogeneous, it is necessary to check the uniformity of each HIRAS-II FOV, which is carried out by using the ratio of the standard deviation of the matched AHI radiations to its mean value

$$\mathrm{Std_{fov}}/\mathrm{Mean_{fov}} < 0.01 \quad (3)$$

where $\mathrm{Std_{fov}}$ denotes the standard deviation of the observed radiance from all the matched AHI pixels within each HIRAS-II FOV, and $\mathrm{Mean_{fov}}$ is the mean radiance of all these AHI pixels. The threshold value is set to 0.01. Only uniform scenes are selected for intercomparison to reduce the uncertainty introduced by the field of view averaging.

Furthermore, a constraint of background environmental uniformity of the HIRAS-II FOV is needed to compensate for the minor error of the spatial collocation, as well as to reduce the uncertainty due to different azimuths. The 3 × 3 HIRAS-II FOV (the large circle in Figure 3) centered on the HIRAS-II FOV to be matched is considered the background environment area.

$$\mathrm{Std_{env}}/\mathrm{Mean_{env}} < 0.05 \quad (4)$$

where $\mathrm{Std_{env}}$ denotes the standard deviation of all the matched AHI pixels' radiance within the background area, and $\mathrm{Mean_{env}}$ denotes the mean radiance of these matched AHI pixels.

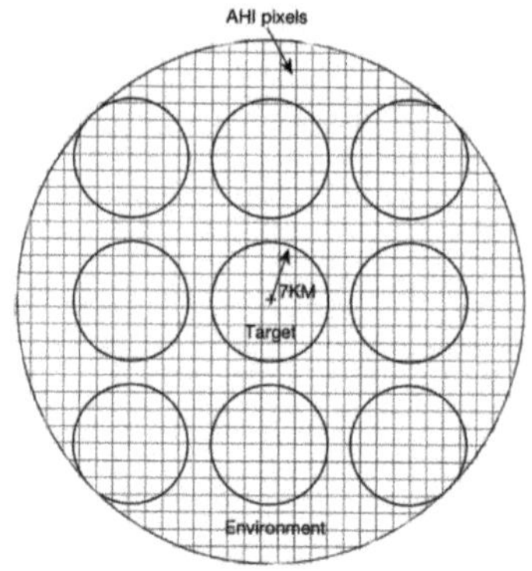

Figure 3. Spatial matching of HIRAS-II and AHI. The small square represents the AHI pixel, the small circle is the HIRAS-II FOV, and the large circle is the HIRAS-II FOR.

3.4. *Statistical Calculation*

For each collocated HIRAS-II—AHI FOV, the brightness temperature difference ($\text{BT diff} = \text{BT}_{\text{HIRAS}} - \text{BT}_{\text{AHI}}$) and the standard deviation of the brightness temperature difference ($\text{Std} = \sqrt{\sum_{i=1}^{n} (\text{BT diff}_i - \text{BT diff}_{\text{mean}})^2 / (n-1)}$) are counted (n is the number of samples).

A total of 458 pairs of samples are obtained after collocating the HIRAS-II observations with the AHIs from 15 March to 21 April 2022 based on the above matching steps.

3.5. *Matching with MERSI-LL on the Same Platform*

When the HIRAS-II observations are compared with those from the AHI nadir on the geostationary satellite, the matched fields of view are concentrated near the tropical equator, and the dynamic range of the observed brightness temperature is narrow. The HIRAS-II observations will be compared with measurements from one imager that is carried on a polar-orbiting platform to evaluate its observation accuracy on a global scale. However, FY-3E is the first early morning polar-orbiting satellite with a significant observation time difference (even more than 8 h) from the established mid-morning or afternoon orbit satellites. Therefore, the MERSI-LL imager on the same FY-3E platform is chosen to perform the HIRAS-II calibration in this paper, and the MERSI-LL channels 4–7 are spectral matched. Since HIRAS-II and MERSI-LL are on the same polar-orbiting satellite platform and are almost observed simultaneously, the matching process of the two instruments is relatively simple. Only the nadir HIRAS-II FOV is matched to ensure the same observation scene. The spectral matching, spatial matching, and uniformity checking steps are the same as those for AHI. Finally, a total of 12,395 pairs of HIRAS-II and MERSI-LL observation samples were matched over 8 days from 15 to 22 March 2022.

4. Results and Discussion

4.1. *Comparison of HIRAS-II with AHI*

There are nine Himawari-8/AHI channels (channels 8 to 16) that are spectral matched with FY-3E/HIRAS-II in the longwave infrared band. The scatter plots of the convolved HIRAS-II observed brightness temperature with AHI measurements in channel 8 to 16 are given sequentially in Figure 4. The horizontal coordinate is the HIRAS-II observed brightness temperature, the vertical axis is the AHI observation, the dashed line is the $y = x$ line, and the solid line is the linear fitting result. Because the matched samples are concentrated near the nadir of the geostationary satellite, the dynamic range of the observed brightness temperature for each channel is narrow, and the value gradually decreases as the peak height of the weighting functions of atmospheric absorption channels increases. The observations of the two instruments are very close, with correlation coefficients higher than 0.98, and the fitting lines almost coincide with the $y = x$ line in all the channels. The statistical bias and standard deviation for the HIRAS-II—AHI matching samples from 15 March to 21 April 2022 are listed in Table 2. Figure 4 and Table 2 show that all the channels have a slightly positive bias; namely, that the HIRAS-II convolved observations are slightly warmer than AHI with a maximum bias of 0.65 K (channel 9 in the water vapor wing), and the minimum is 0.22 K (window channel 14). The standard deviation of all the channel biases ranges from 0.22 to 0.31 K with small values and little difference between the channels. Water vapor absorption channels 8–10 and ozone absorption channel 12 have relatively larger biases and small standard deviations, while the biases of the window channels (such as channels 14 and 15) are relatively small, and the standard deviations are slightly large. In addition, the closer the peak height of the weighting function is to the surface, the larger the standard deviation is. This is because the value range of the observed brightness temperature of the window channels is relatively larger than that of the absorption channels, so the dispersion is larger in the window channel.

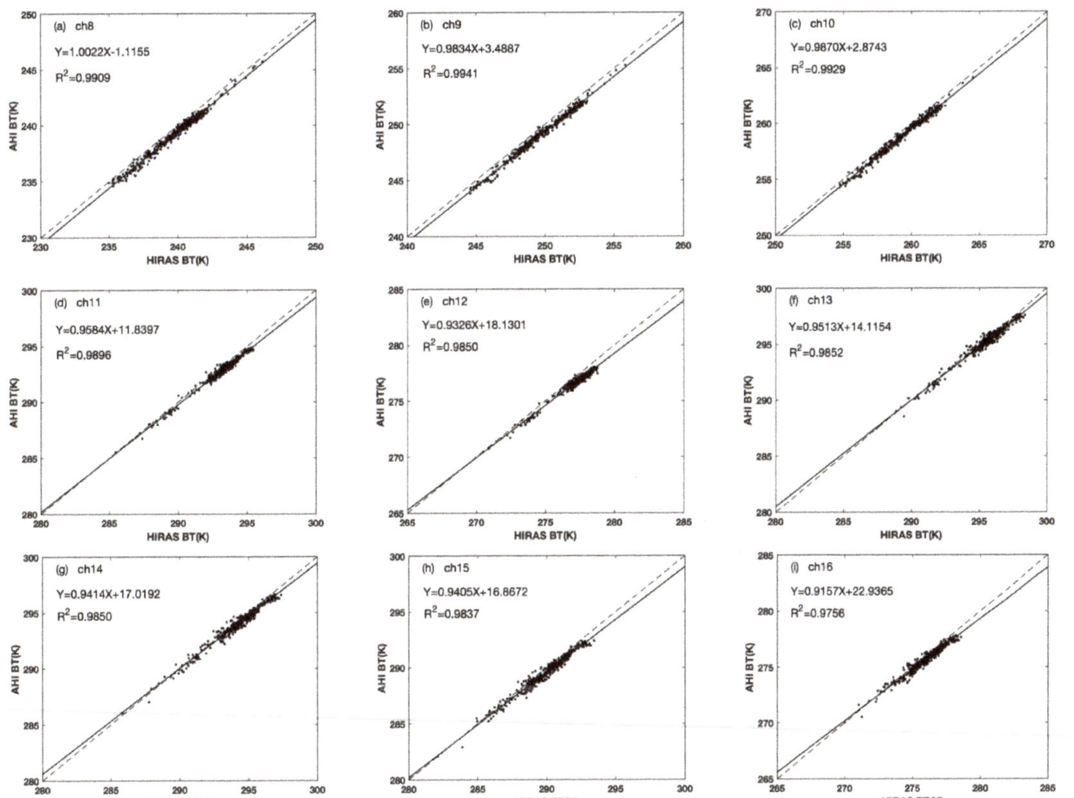

Figure 4. Scatterplot of the convolved HIRAS-II brightness temperature and the AHI brightness temperature for channels 8–16. The dashed line is $y = x$, and the solid line is the linear fitting result.

The ideal condition for the realization of the cross-calibration using the simultaneous nadir overpass (SNO) method is that the two instruments observe the same target at the same time through the same atmospheric path. However, the matching condition is appropriately approximated in the actual application to obtain enough samples, and the introduction of the matching threshold may bring uncertainty to the calibration evaluation. For the HIRAS-II and AHI cross-calibration, the observation inconsistency may be partly due to the random errors caused by the differences in observation geometry, scene uniformity, and observation time. To analyze the possible uncertainties caused by the colocation approximation factors, water vapor channel 8 with the highest peak height of the weighting function is taken as an example. The distribution of the brightness temperature differences of HIRAS-II and AHI with various matching factors: (a) observation geometry, (b) scene uniformity, (c) azimuthal difference, and (d) observation time difference are shown in Figure 5. The larger the value of the horizontal coordinate in Figure 5a denotes the larger the difference in the observation geometric viewpoint. A larger value of the x-axis in Figure 5b indicates worse uniformity within the field of view. The larger x value of Figure 5c corresponds to the larger difference in the observation azimuth of the two instruments. The solid line shows the linear fit pattern of the brightness temperature differences with the approximation factors. The brightness temperature differences are randomly and uniformly distributed with these factors and do not increase with the decrease in various matching degrees. There is no obvious linear variation characteristic with these matching factors, indicating that the influence of various matching factor differences within their threshold on observation bias can be neglected and that these reasonable matching thresholds bring little uncertainty to the bias assessment. Figure 5e shows the distribution of the brightness temperature differences with the HIRAS-II observed brightness temperature. These is also no significant scene temperature-dependent bias. The results of the remaining channels are similar and omitted. After excluding the random errors caused by the matching factors mentioned above, it can be concluded that the HIRAS-II—AHI brightness temperature differences mainly represent the systematic observation bias of the two instruments.

Table 2. Statistics of brightness temperature bias between HIRAS-II and AHI.

	AHI								
	ch8	**ch9**	**ch10**	**ch11**	**ch12**	**ch13**	**ch14**	**ch15**	**ch16**
Mean (k)	0.5780	0.6465	0.4909	0.3726	0.5465	0.2688	0.2274	0.3935	0.3259
Std (k)	0.2527	0.2398	0.2185	0.2165	0.2171	0.2881	0.2917	0.3061	0.2814
Mean HIRAS BT (k)	240.1	250.0	259.7	293.4	277.4	295.8	294.5	290.2	276.2
Correlation coefficient	0.9909	0.9941	0.9929	0.9896	0.9850	0.9852	0.9850	0.9837	0.9756

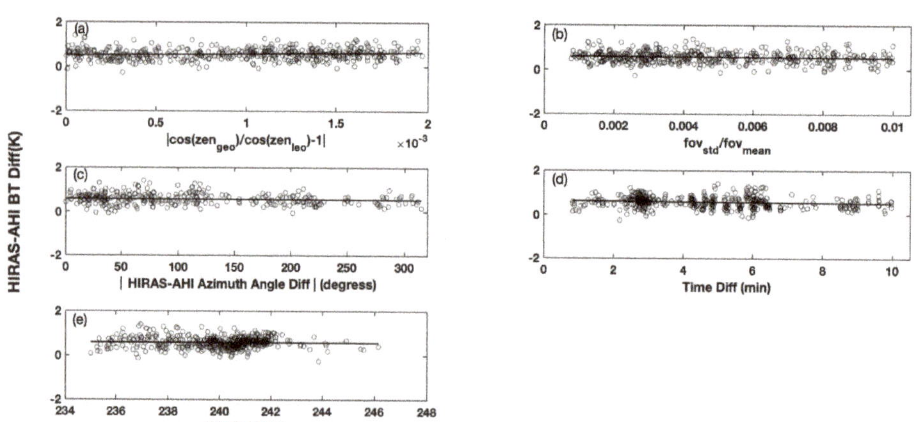

Figure 5. Brightness temperature biases between HIRAS-II and AHI of channel 8 varying with (**a**) observation geometry factor, (**b**) scene homogeneity factor, (**c**) azimuth angle factor, (**d**) observation time differences, and (**e**) HIRAS-II observations (the solid line shows the linear fitting result).

4.2. Comparison of HIRAS-II with MERSI-LL

The observed biases and standard deviations were counted based on the 12,395 pairs of samples matched by HIRAS-II and MERSI from 15–22 March 2022. The distribution of the brightness temperature differences (HIRAS-II minus MERSI-LL) with the HIRAS-II observed scene temperature for channel 4–7 are shown in the left subplot of Figure 6 with the vertical coordinates representing the brightness temperature differences and the horizontal coordinates representing the observed scene temperature. The color distinguishes the scene uniformity, and the dashed line gives the mean of the biases. The probability density distributions of the brightness temperature differences are given in the right subplot with the horizontal coordinate as the sample probability density. Channel 4 of MERSI-LL is a water vapor absorption band with a central wavelength of 7.22 μm (peak height of the weighting function 400 hPa), and channel 5–7 are window channels with a central wavelength of 8.55 μm, 10.8 μm, and 12.0 μm, respectively. The dynamic range of the channel 4 target brightness temperature is between 220 and 280 K, and the HIRAS-II measurement is slightly higher than the MERSI-LL observations (the mean bias is 0.6643 K) with a standard deviation of 0.2229 K. The dynamic range of channel 5 is slightly larger at approximately 220–300 K, and the observed brightness temperatures of HIRAS-II and MERSI are close, with a mean bias of 0.0023 K and a standard deviation of 0.3135 K. The dynamic range of channel and 7 target brightness temperature are between 210 and 310 K, and the biases of channel 6 and 7 show a U-shaped change with the increase in the scene temperature, and the biases are smallest (close to 0 K) when the scene temperature is between 250 K and 280 K. Channel 4 has smaller brightness temperature differences at lower scene temperatures (i.e., high latitudes) and relatively larger brightness temperature differences at higher scene temperatures (i.e., low latitudes), especially when the bias increases to 1.2 K near the equator. Although the channel 5 bias takes values of approximately 0 K, the bias dispersion increases as the HIRAS-II observed scene temperature increases. Both channel 6 and 7 have relatively larger brightness temperature differences at lower scene temperatures and higher scene temperatures, and the maximum value is close to 1.75 K. From the right subplot, it can be seen that the probability density distributions of the brightness temperature bias for channels 4–7 all conform to the normal distribution.

It is noteworthy that the brightness temperature differences of the water vapor channel 4 in Figure 6 are obviously positively correlated with the target scene temperature, and the window channels 6 and 7 also have an obvious scene temperature-dependence, while window channel 5 shows no scene temperature-dependent bias. At the same time, AHI water vapor channels 9 and 10—whose spectral positions are close to MERSI-LL channel 4—also do not find bias scene-dependent characteristics. Since HIRAS-II and MERSI-LL are mounted on the same platform, the scene uniformity is the only factor that introduces matching uncertainty into the intercomparison. Figure 7 shows the scatter distribution of MERSI-LL channel 4 (a), channel 5 (b), channel 6 (c), and channel 7 (d) brightness temperature differences (HIRAS-II minus MERSI-LL) with scene uniformity. A larger value of the horizontal coordinate in Figure 7 indicates worse scene uniformity, and the solid line indicates the linear fitting result. The brightness temperature differences of channel 5–7 are uniformly distributed with the scene uniformity and do not have linear variation characteristics (in Figure 7b). However, the brightness temperature differences of channel 4 show an obvious linear trend with the scene uniformity, and the biases gradually decrease as the scene uniformity worsens (in Figure 7a). Combined with the scatter color of channel 4 in Figure 6, the scene uniformity is relatively poor (yellow) in the high latitudes with a low brightness temperature, and the scene uniformity is good in the low latitudes with a high brightness temperature. This is because the underlying surface in the field of view varies greatly in the polar region when the instrument is scanning with the same spatial resolution and swath, especially the Arctic has greater underlying surface variability due to the presence of different surface types (e.g., land, snow, ocean, glacier, etc.) with higher variability in absolute temperature. Theoretically, the bias is smaller when scene uniformity is better. However, Figure 6 shows that the scene uniformity gradually improves with the increasing scene temperature, while the bias increases instead. This indicates that the scene uniformity is not the cause of the scene temperature-dependent bias.

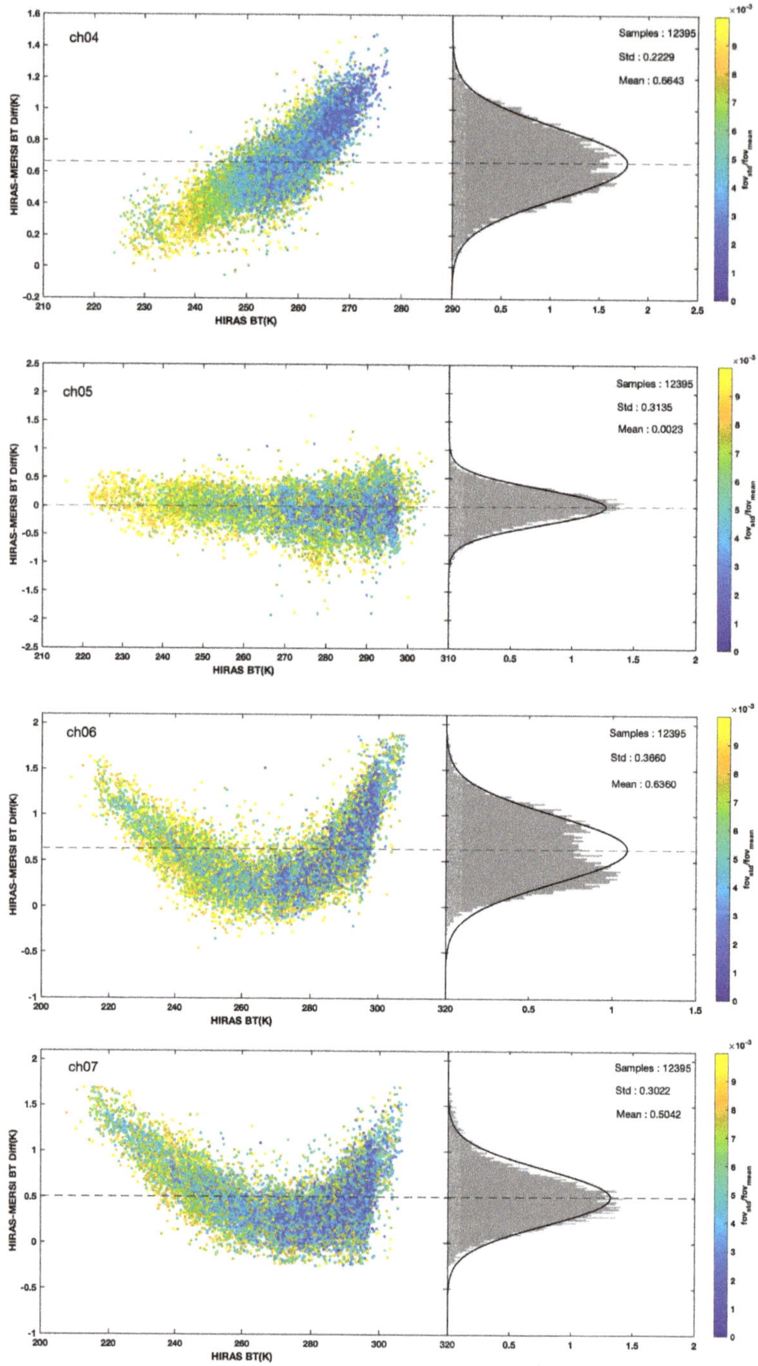

Figure 6. (left) Scatterplot of the HIRAS-II–MERSI-LL BT bias versus the scene temperature and (right) the histogram of the BT differences. The dashed line indicates the mean value. The color indicates the scene homogeneity.

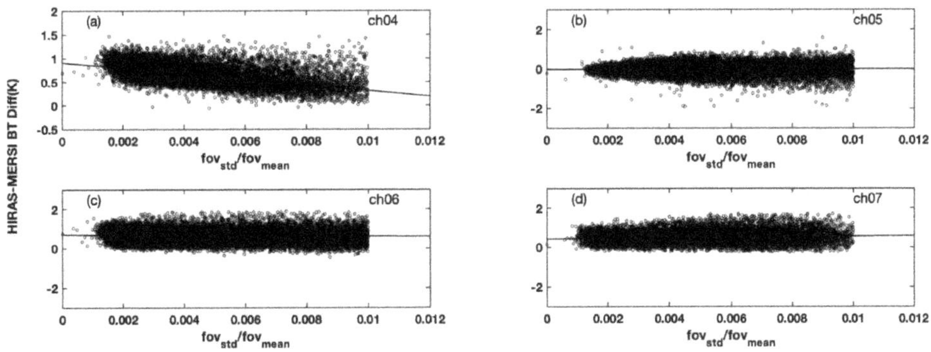

Figure 7. Scatterplot of brightness temperature biases between HIRAS-II and MERSI-LL varying with scene homogeneity of (**a**) channel 4, (**b**) channel 5, (**c**) channel 6, and (**d**) channel 7 (the solid line shows the linear fitting result).

The main factors that may cause spaceborne radiation imager calibration errors mainly include blackbody emissivity and spectral response function instrument nonlinearity. MERSI-LL channel 4 is located in the wing area of the water vapor absorption band, and a very small drift in the spectral response function can also lead to a temperature-dependent bias in the scene. However, to date, there have been no specific references about the design of the black bodies, the calibration system, and so on of these two instruments onboard FY-3E. These matters require further study in the future.

The day-to-day variations in the mean biases and standard deviations are counted based on the HIRAS-II–MERSI-LL matched samples. The daily mean biases of channel 4 range from 0.64 to 0.68 K, and the standard deviations are all approximately 0.2 K. The daily mean biases of channel 5 range from −0.02 to 0.02 K, and the standard deviations are approximately 0.3 K. The daily mean biases of channel 6 range from 0.62 K to 0.64 K, and the standard deviations are approximately 0.35 K. The daily mean biases of channel 7 range from 0.48 K to 0.51 K, and the standard deviations are approximately 0.3 K. The biases of the two instruments vary minimally and almost constantly over a period of time, indicating that the performance of the HIRAS-II instrument is stable.

5. Conclusions

To assess HIRAS-II's on-orbit observation quality, the geometrically, temporally, and spatially matched scene homogeneous HIRAS-II hyperspectral observations were convolved to the longwave infrared channels corresponding to the Himawari-8/AHI and FY-3E/MERSI-LL from 15 March to 21 April 2022, and their brightness temperature deviation characteristics were statistically calculated in this paper. The matching samples of HIRAS-II and AHI are concentrated near the equator, and the spectral matching channels are longwave infrared channel 8 to channel 16 onboard the same polar orbiting satellite platform FY-3E. The matching samples of HIRAS-II and MERSI-LL are evenly distributed all over the world with spectral matching channels 4 to channel 7. The following conclusions can be made based on this analysis:

1. The HIRAS-II on-orbit observed brightness temperatures are slightly warmer than the AHI observations, with a small positive bias in all the matched channels. The brightness temperature scatters of both observations are distributed near the $y = x$ line with a correlation coefficient higher than 0.98 in all channels. The biases of water vapor channels 8–10 and ozone absorption channel 12 are relatively large, with a maximum of 0.65 K (channel 9 in the water vapor wing), and the biases of the window channels are relatively small, with a minimum of 0.22 K (channel 14). The standard deviations for all channels are small (0.22–0.31 K) and there is little difference between the channels.

2. The thresholds chosen for the colocation approximation factors (e.g., observation geometry angle, field of view uniformity, observation azimuth, and observation time) when matching the HIRAS-II with AHI contribute little and negligible uncertainty to the bias assessment, so the difference between the two observed radiations is considered to be mainly from the systematic bias of the two instrument measurements.
3. Since HIRAS-II and MERSI-LL are mounted on the same platform, the scene uniformity is the only factor introducing matching uncertainty in the intercomparison. The mean brightness temperature bias (HIRAS-II minus MERSI-LL) of the MERSI-LL water vapor channel 4 is 0.66 K with a standard deviation of 0.22 K. To window channel 5, the observations of both instruments are very close, with a mean bias of 0.002 K and a standard deviation of 0.31 K. Both channel 6 and 7 have relatively larger brightness temperature differences at lower scene temperatures and higher scene temperatures, with a mean bias of 0.63 K (the standard deviation is 0.36 K) and 0.5 K (the standard deviation is 0.3 K), respectively.
4. The biases of MERSI-LL channel 4 are obviously positively correlated with the target scene temperature. The biases of channel 6 and 7 show a U-shaped change with the increase in the scene temperature, and the biases are smallest (close to 0 K) when the scene temperature is between 250 K and 280 K. The statistical characteristics of the HIRAS-II–MERSI-LL difference vary minimally and almost constantly over a period of time, indicating that the performance of the HIRAS-II instrument is stable.

As a final note, we just found the phenomenon of bias distribution, which is not yet fully explained due to lack of relevant references. Therefore, we will use NWP data, double-difference method to further evaluate the accuracy of HIRAS-II in future studies.

Author Contributions: Conceptualization, L.G. and H.C.; methodology, L.G.; software, H.C.; validation, H.C.; formal analysis, L.G.; investigation, H.C.; resources, L.G.; data curation, H.C.; writing—original draft preparation, H.C.; writing—review and editing, H.C.; visualization, H.C.; supervision, L.G.; project administration, L.G.; funding acquisition, L.G. All authors have read and agreed to the published version of the manuscript.

Funding: This work was supported by the National Natural Science Foundation of China under grant no. 41975028.

Data Availability Statement: The HIRAS-II and MERSI-LL Level 1 data can be obtained at the Chinese Feng Yun satellite remote sensing data service network (http://data.nsmc.org.cn accessed on 11 April 2022). Himawari-8/AHI radiation data obtained from the Japan Earth Observation Data Center (https://www.eorc.jaxa.jp/ptree/index.html accessed on 11 April 2022).

Acknowledgments: The authors would like to thank the editor and reviewers for their helpful comments on the manuscript.

Conflicts of Interest: The authors declare no conflict of interest.

References

1. Yang, T.; Hu, X.; Xu, H.; Wu, C.; Qi, C.; Gu, M. Radiation Calibration Accuracy Assessment of FY-3D Hyperspectral Infrared Atmospheric Sounder Based on Inter-Comparison. *Acta Opt. Sin.* **2019**, *39*, 337–387. [CrossRef]
2. Guan, L.; Ren, X.Y. The observation quality assessment of satellite borne microwave imager at new-added frequency over East Asia land. *Trans. Atmos. Sci.* **2018**, *41*, 554–560. [CrossRef]
3. Lu, Q.-F.; Zhou, F.; Qi, C.-L.; Hu, X.; Xu, H.-L.; Wu, C.-Q. Spectral performance evaluation of high-spectral resolution infrared atmospheric sounder onboard FY-3D. *Opt. Precis. Eng.* **2019**, *27*, 2105–2115. [CrossRef]
4. Zhang, P.; Hu, X.; Lu, Q.; Zhu, A.; Lin, M.; Sun, L.; Chen, L.; Xu, N. FY-3E: The First Operational Meteorological Satellite Mission in an Early Morning Orbit. *Adv. Atmos. Sci.* **2022**, *39*, 1–8. [CrossRef]
5. Gunshor, M.M.; Schmit, T.J.; Menzel, W.P. Intercalibration of the Infrared Window and Water Vapor Channels on Operational Geostationary Environmental Satellites Using a Single Polar-Orbiting Satellite. *J. Atmos. Ocean. Technol.* **2004**, *21*, 61–68. [CrossRef]
6. Tobin, D.C.; Revercomb, H.E.; Moeller, C.C.; Pagano, T.S. Use of Atmospheric Infrared Sounder High–Spectral Resolution Spectra to Assess the Calibration of Moderate Resolution Imaging Spectroradiometer on EOS Aqua. *J. Geophys. Res.* **2006**, *111*, D09S05. [CrossRef]

7. Wang, L.; Cao, C.; Goldberg, M. Intercalibration of GOES-11 and GOES-12 Water Vapor Channels with MetOp IASI Hyperspectral Measurements. *J. Atmos. Ocean. Technol.* **2009**, *26*, 1843–1855. [CrossRef]
8. Xu, N.; Hu, X.; Chen, L.; Zhang, Y.; Hu, J.; Sun, L. On-Orbit Radiometric Calibration Accuracy of FY-3A MERSI Thermal Infrared Channel. *Spectrosc. Spectr. Analysis* **2014**, *34*, 3429–3434. [CrossRef]
9. Gong, X.; Li, Z.; Li, Z.; Moeller, C.; Cao, C.; Wang, W.-H.; Menzel, W.P. Intercomparison Between VIIRS and CrIS by Taking Into Account the CrIS Subpixel Cloudiness and Viewing Geometry. *J. Geophys. Res. Atmos.* **2018**, *123*, 5335–5345. [CrossRef]
10. Zhou, W.; Jiang, G.-M. Cross-calibration of VIRR/FY-3A split-window channels with AIRS/AQUA and IASI/METOP-A data. *Opt. Precis. Eng.* **2015**, *23*, 1892–1902. [CrossRef]
11. Zhang, P.; Guo, Q.; Chen, B.-Y.; Feng, X. The Chinese Next-Generation Geostationary Meteorological Satellite FY-4 Compared with the Japanese Himawari-8/9 Satellite. *Adv. Meteorol. Sci. Technol.* **2016**, *6*, 72–75. [CrossRef]
12. Okuyama, A.; Takahashi, M.; Date, K.; Hosaka, K.; Murata, H.; Tabata, T.; Yoshino, R. Validation of Himawari-8/AHI Radiometric Calibration Based on Two Years of In-Orbit Data. *J. Meteorol. Soc. Jpn.* **2018**, *96*, 91–109. [CrossRef]
13. Bessho, K.; Date, K.; Hayashi, M.; Ikeda, A.; Imai, T.; Inoue, H.; Kumagai, Y.; Miyakawa, T.; Murata, H.; Ohno, T.; et al. An Introduction to Himawari-8/9—Japan's New-Generation Geostationary Meteorological Satellites. *J. Meteorol. Soc. Jpn.* **2016**, *94*, 151–183. [CrossRef]
14. Di, D. *Data Assimilation Research for Geosynchronous Interferometric Infrared Sounder Onboard FengYun-4 Satellite*; Chinese Academy of Meteorological Sciences: Beijing, China, 2019.
15. Wang, L.; Wu, X.; Goldberg, M.; Cao, C.; Li, Y.; Sohn, S.-H. Comparison of AIRS and IASI Radiances Using GOES Imagers as Transfer Radiometers toward Climate Data Records. *J. Appl. Meteorol. Climatol.* **2010**, *49*, 478–492. [CrossRef]
16. Wang, L.; Cao, C.; Ciren, P. Assessing NOAA-16 HIRS Radiance Accuracy Using Simultaneous Nadir Overpass Observations from AIRS. *J. Atmos. Ocean. Technol.* **2007**, *24*, 1546–1561. [CrossRef]
17. Xu, N.; Hu, X.-Q.; Chen, L.; Zhang, Y.-X. Cross-calibration of FY-2E/VISSR infrared window and water vapor channels with TERRA/MODIS: Cross-calibration of FY-2E/VISSR infrared window and water vapor channels with TERRA/MODIS. *J. Infrared Millim. Waves* **2012**, *31*, 319–324. [CrossRef]
18. Wang, H.-M.; Zhao, Q.-C.; Wang, Y.-M.; Zhang, Z.; Yang, X.-J.; Hu, X. Check Analysis of In-Orbit Cross-Calibration Between FY-3A/TOU and Metop-B/GOME-2. *Acta Opt. Sin.* **2017**, *37*, 320–328. [CrossRef]

remote sensing

Article

Comparison of Three Convolution Neural Network Schemes to Retrieve Temperature and Humidity Profiles from the FY4A GIIRS Observations

Shuhan Yao and Li Guan *

Collaborative Innovation Center on Forecast and Evaluation of Meteorological Disasters, Key Laboratory for Aerosol-Cloud-Precipitation of China Meteorological Administration, Nanjing University of Information Science and Technology, Nanjing 210044, China
* Correspondence: liguan@nuist.edu.cn

Abstract: FY4A/GIIRS (Geostationary Interferometric Infrared Sounder) is the first infrared hyperspectral atmospheric vertical sounder onboard a geostationary satellite. It can achieve observations of atmospheric temperature and humidity profiles with high vertical and temporal resolutions. Presently, convolutional neural network algorithms are relatively less used in the field of atmospheric profile retrieval, and different convolutional neural network approaches have different characteristics. The one-dimensional convolutional neural network scheme 1D-Net and two three-dimensional retrieval schemes U-Net 1 and U-Net 2 are used to achieve atmospheric temperature and humidity profiles under all skies based on GIIRS-observed brightness temperatures in this paper. After validation with test training data, the retrievals of different schemes derived from actual GIIRS observations and level 2 operational products were verified with ERA5 reanalysis data and radiosonde measurements in summer and winter respectively. The retrieved three-dimensional temperature and humidity fields from U-Net 1 and U-Net 2 are closer to the ERA5 reanalysis field in both distribution and value than the retrievals from the 1D-Net scheme and level 2 operational products. In particular, the inversion field of the U-Net 2 scheme is more continuous in space. Compared with radiosonde observations, the accuracy of the level 2 temperature product is the highest when the field of view is completely clear both in winter and summer month. The root mean square error (RMSE) of temperature retrieval of the two U-Net schemes is the second highest, and the RMSE and bias of the 1D-Net scheme are both large. Two U-Net schemes overestimate the temperature and humidity slightly in winter and underestimate it in summer in both clear and all sky cases. Under all sky conditions, the temperature retrieval RMSE and bias of the two U-Net schemes above 800 hPa are lower than those of the level 2 products, especially the U-Net 2 scheme with an RMSE of approximately 2.5 K. The U-Net 2 scheme bias is the smallest, with a value of approximately 0.5 K in winter. Since the level 2 product only provides the atmospheric temperature above the cloud top, it indicates that its temperature product accuracy is very low when the field of view is influenced by clouds. The humidity retrieval RMSEs of the two U-Net schemes is within 2 g/kg, better than that of the 1D-Net scheme. The retrieval accuracy of the U-Net 2 scheme is approximately 0.3 g/kg better than that of the U-Net 1 scheme below 600 hPa in winter. Level 2 does not provide humidity products. The summer humidity retrieval is worse than in winter. In general, among the three deep machine learning algorithms, 1D-Net has a large retrieval error, and the temperature and humidity from U-Net 2 have the highest accuracy. The retrieval speeds of the two U-Net schemes are nearly the same, and both are faster than that of scheme 1D-Net.

Keywords: FY4A/GIIRS; temperature and humidity profiles; convolutional neural network (CNN); U-Net

Citation: Yao, S.; Guan, L. Comparison of Three Convolution Neural Network Schemes to Retrieve Temperature and Humidity Profiles from the FY4A GIIRS Observations. *Remote Sens.* **2022**, *14*, 5112. https://doi.org/10.3390/rs14205112

Academic Editor: Ji Zhou

Received: 26 August 2022
Accepted: 10 October 2022
Published: 13 October 2022

1. Introduction

Atmospheric temperature and humidity profiles are two important parameters to study atmospheric state and play an important role in the research of atmospheric science. Accurately obtaining these parameters is of great significance to improve the accuracy of numerical weather forecast and short-term weather warning and forecast [1]. The traditional way of obtaining atmospheric temperature and humidity profiles through radiosonde observations has limited spatial representativeness due to the influence of geographical conditions and other factors and has failed to meet the needs of operational and modern meteorological development. However, the development of satellite remote sensing technology has made up for the shortage of radiosondes and provided technical support for the acquisition of global atmospheric temperature and humidity profile distributions with high spatial-temporal resolution [2].

The horizontal and vertical resolutions and accuracies of atmospheric temperature and humidity profiles must meet the requirements of numerical weather prediction systems, so it is necessary to study the remote sensing accuracy that can be achieved by using satellite technology and to study which retrieval method can obtain the optimal retrieval results. The number of setting channels for atmospheric sounding radiometers uploaded on meteorological satellite platforms determines their vertical observation resolution. Due to the small number of channels, the spectral resolution of microwave and imager instruments is low, and the weight function is too wide, so it is impossible to retrieve the fine atmospheric profile in the vertical direction. In contrast, the spaceborne infrared hyperspectral atmospheric vertical sounders have thousands of channels in the thermal infrared band with high spectral resolution and narrow weighting function and can obtain high vertical resolution three-dimensional finer observations of atmospheric temperature and humidity parameters.

The Geostationary Interferometric Infrared Sounder (GIIRS) onboard FY-4A launched on 11 December 2016 (China's new generation of quantitative remote sensing meteorological satellites with geostationary orbits) is the first infrared hyperspectral atmospheric sounder mounted on a geostationary meteorological satellite. The GIIRS can provide time-continuous, high-spectral-resolution atmospheric sounding information [3], which is used to retrieve the vertical structure of atmospheric temperature and humidity parameters with high vertical resolution. The improved vertical resolution provides higher accuracy services for numerical weather forecasting, weather monitoring, and warning [4].

To date, the traditional methods commonly used to retrieve atmospheric temperature and humidity profiles based on satellite-based infrared hyperspectral observations are statistical regression methods and physical retrieval methods. The statistical regression approaches include the eigenvector method [5–7], empirical orthogonal function expansion method [8], and least squares method. Although the statistical regression algorithm is simple to calculate and retrieval results are more stable, it does not consider the physical nature of the atmospheric radiation transmission process, and the retrieval accuracy needs to be improved. Moreover, the method relies on training samples and cannot be applied to areas where the sample information is not sufficiently representative. Although the retrieval accuracy is high, the physical retrieval method, such as the one-dimensional variational method, requires an initial field, complex physical processes to be considered, and a long computation time [9,10].

With the continuous development of artificial intelligence, machine learning algorithms have been applied to various research fields and have demonstrated a powerful ability to handle big data. Machine learning algorithms with features such as adaptive, self-organizing, and real-time learning have gradually been introduced into the field of meteorology, providing new ideas for atmospheric remote sensing. Singh et al. [11] retrieved atmospheric temperature and humidity profiles from microwave and infrared hyperspectral data, respectively, with a shallow learning neural network approach, and the results showed good agreement for all atmospheric pressure levels except for below 850 hPa. The deep learning algorithm is now also gradually applied to satellite remote sensing retrieval with the continuous development of machine learning. For example, Malmgren-Hansen et al. retrieved atmospheric temperature profiles with a convolutional neural network (CNN) algorithm based on infrared atmospheric sounding interferometer (IASI) observations and found that the CNN algorithm has higher retrieval accuracy than the linear regression method [12].

As one of the representative methods of deep machine learning (namely, hierarchical machine learning methods including multilevel nonlinear transformations), the convolutional neural network algorithm mainly integrates feature extraction into multilayer perceptions through structural reorganization and weight reduction. It unifies feature representation and regression prediction with obvious advantages compared to shallow models in feature extraction and modelling and has achieved good performance in image recognition and classification [13–15]. However, its application in the field of atmospheric parameter profile retrieval has rarely been reported in the literature. Therefore, three convolutional neural network schemes are applied to retrieve atmospheric temperature and humidity profiles based on the infrared hyperspectral GIIRS observations in this paper, namely, the traditional convolutional neural network scheme 1D-Net to retrieve the one-dimensional atmospheric temperature and humidity profiles and the U-Net 1 and U-Net 2 schemes to retrieve the three-dimensional profiles. Meanwhile, the retrieval accuracy and efficiency of each scheme are evaluated by comparison and verification.

2. Data

2.1. GIIRS Data

GIIRS is the first hyperspectral infrared instrument onboard the geostationary meteorological satellite, and two infrared spectral bands (longwavelength IR (LWIR) band of 700–1130 cm^{-1} and the medium-wavelength IR (MWIR) band of 1650–2250 cm^{-1}) with a spectral resolution of 0.625 cm^{-1} are designed to detect the atmosphere. There are 1650 channels in total, and the nadir spatial resolution is 16 km. The LWIR band is 689 channels containing the CO_2 absorption band near 15 μm, the thermal infrared window region in the 8–12 μm and the 9.6 μm O_3 absorption band. The MWIR band includes the strong water vapor absorption band (5–8 μm) centered at 6.3 μm, which can be used to retrieve atmospheric water vapor profiles, as well as the CO_2 absorption band near 4.3 μm, with a total of 961 channels. The main GIIRS instrument performance parameters are given in Table 1. GIIRS observes China and its surrounding area (3°~55°N, 66°~144°E), including 7 latitude belts from north to south, with each belt taking 15 min. Each scan belt has 59 fields of regard (FORs), and each FOR contains 128 fields of view (FOVs). Each GIIRS FOR is composed of 128 FOVs arranged in a 32 × 4 array.

Table 1. Specifications of FY4A/GIIRS.

Parameter	Performance
Spectral bandwidth	Longwave: 700–1130 cm^{-1} Mid wave: 1650–2250 cm^{-1}
Spectral resolution	0.625 cm^{-1}
Spectral channels	Longwave: 689 Mid wave: 961
Sensitivity	Longwave: 0.5–1.12 $mW/(m^2 sr\ cm^{-1})$ Mid wave: 0.1–0.14 $mW/(m^2 sr\ cm^{-1})$
Spatial resolution	16 km (the nadir)
Temporal resolution	67 min (China area)
Spectral calibration accuracy	10 ppm
Radiation calibration accuracy	1.5 K

GIIRS Level 1 (L1) radiation observed data for the whole month of February 2020 and February 2021 were used in this study, including radiance measurement, latitude, longitude, and satellite zenith angle information. At the same time, cloud mask (CLM) and atmospheric temperature profile products from GIIRS Level 2 (L2) operational products were used. The GIIRS L1 and L2 datasets are from the Chinese National Satellite Meteorological Center (http://satellite.nsmc.org.cn, accessed on 12 January 2022).

2.2. ERA5 Reanalysis Data

European Centre for Medium-Range Weather Forecasts (ECMWF) consistently generates atmospheric reanalyses of the global climate by assimilating various data from the ground, upper air, and satellite observations into the earth system model [16]. The ERA5 dataset (https://apps.ecmwf.int/datasets/, accessed on 12 January 2022) is the latest generation reanalysis dataset provided by the ECMWF, which makes a significant step forward in the assimilation system, model input, spatial resolution, output frequency, and quality level compared with its former ERA-Interim [17]. The spatial resolution is $0.25° \times 0.25°$, and the temporal resolution is 1 h. There are 37 pressure levels from 1000 hPa to 1 hPa. The atmospheric temperature and water vapor parameters of the three-dimensional grid are mainly used for validation. Since the ERA5 data and GIIRS observations are different in spatial and temporal resolution, the ERA5 temperature and water vapor profile need to be spatially interpolated to each GIIRS FOV before comparison. The ERA5 value at the nearest four grid points is selected for distance-weighted averaging. The maximum time matching difference is 1 h. For example, when the GIIRS observation time is 00 to 01 (UTC), the ERA5 reanalysis dataset at 01 (UTC) will be matched.

2.3. Radiosonde Observation Data

The radiosonde data from 89 Chinese upper-air stations in February and July 2021 are used to examine the accuracy of atmospheric temperature and humidity profile retrieval. The data are from the China Meteorological Data Service Center (CMDC, http://data.cma.cn/, accessed on 10 March 2022). With the ascent of the balloon, vertical profile data of pressure, geopotential height, temperature, dew point temperature, wind direction, and wind speed from the ground to approximately 1 hPa are provided twice daily at 00 and 12 (UTC), and measurements are transmitted back to the ground station via radio signals [16]. The dew point temperature is converted to the water vapor mixing ratio to compare with humidity retrievals. The matched difference in latitude and longitude between the GIIRS FOV and radiosonde station is less than 0.25°, and the time difference is less than 1 h.

3. Introduction of Three Convolutional Neural Network Schemes

3.1. Training Data for the One-Dimensional Scheme 1D-Net

The training dataset consists of the global training profiles from the Cooperative Institute for Meteorological Satellite Studies (CIMSS) and the corresponding GIIRS observed brightness temperature simulated by the Radiative Transfer for TOVS (RTTOV) fast radiative transfer model [18]. It includes 15,704 discrete profiles of atmospheric temperature, humidity, and ozone on a global scale with 101 pressure levels from 1100 hPa to 0.005 hPa. The sample data are representative of a large number of samples and have been applied to retrieve infrared hyperspectral atmospheric parameters many times [19]. A total of 12,528 atmospheric profiles of the training data within the range of [60°N, 60°S] are used as the training sample in our paper considering the China area covered by FY-4A geostationary satellites. The 12,528 atmospheric profiles and the simulated GIIRS brightness temperature using it as RTTOV input constitute the training sample pair. One thousand pairs were selected as independent test samples according to the principle of 1 out of 10, and the remaining pairs were used to train the algorithm model. The brightness temperatures of 225 selected GIIRS channels were used as input for training the 1D-Net scheme [20]. The input and output dimensions, the number of training samples, and the time taken to retrieve the China area of the 1D-Net network are shown in detail in Table 2 (Line 2).

Table 2. Sample parameters for network training.

Model	Input Dimensions	Training Sample Size	Training Sample Size	Retrieval Time
1D-Net	225 × 1	1 × 101	11,528 pixels	8′9″
U-Net 1	32 × 32 × 225	32 × 32 × 37	4454 images	3′27″
U-Net 2	160 × 160 × 225	160 × 160 × 37	284 images	3′9″

The observed fields of view on both sides of the nadir (satellite zenith = 0°) of a geostationary meteorological satellite have different degrees of deformation, and the deformation rate increases with increasing satellite zenith angle. The brightness temperature spectrum of GIIRS observations with different satellite zenith angles simulated from RTTOV using the U.S. Standard atmospheric profiles are shown in Figure 1. The different color lines represent the bias between the simulated brightness temperature with satellite zenith angles of 10°, 20°, 30°, 40°, 50°, 60°, and 70° minus the nadir simulation. The simulated brightness temperatures are very sensitive to the satellite zenith angle, especially the difference due to the change in satellite zenith angles in some absorption bands, which can be more than 10 K. The satellite zenith angles must be considered when retrieving the atmospheric temperature and humidity profiles. Eight sets of 1D-Net convolutional neural network retrieval models were built in this study by classifying the satellite zenith angles from 0° to 80° at 10° intervals.

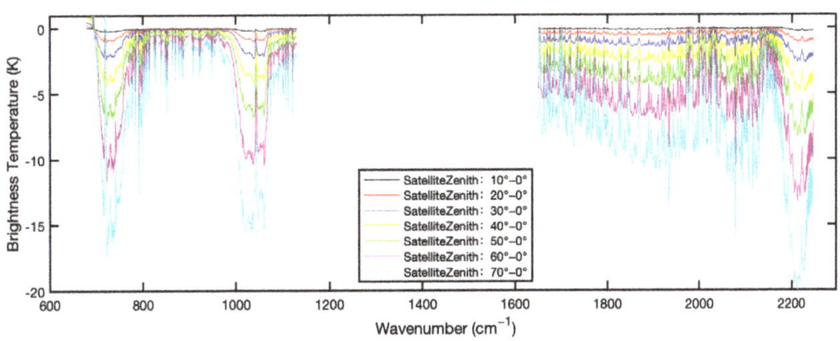

Figure 1. The simulated brightness temperature bias between different satellite zenith angles and 0° (American standard atmosphere).

When the actual observed satellite zenith angle is θ, the two nearest satellite zenith angles θ_1 and θ_2 classifications are found and the corresponding models to retrieve two sets of atmospheric profiles X_1 and X_2 are used, respectively. Then, the final retrieval parameters are obtained by linear interpolation as Equations (1)–(3):

$$X = aX_1 + bX_2 \quad (1)$$

$$a = \frac{|\theta - \theta_1|}{|\theta_2 - \theta_1|} \quad (2)$$

$$b = \frac{|\theta_2 - \theta|}{|\theta_2 - \theta_1|} \quad (3)$$

3.2. Training Data for the Three-Dimensional Scheme U-Net 1

The training dataset for the three-dimensional (3D) atmospheric temperature and humidity profile retrieval consists of the two-dimensional (horizontal) real GIIRS brightness temperature observations covering the China area and the temporal-spatial matched horizontal atmospheric temperature and humidity fields from ERA5 reanalysis data. The sample can be viewed as picture data and is correlated and continuous in horizontal space.

The input data dimension of the U-Net convolutional neural network algorithm is [H(input) × W(input) × C1(input)], and the output data dimension is [H(output) × W(output) × C2(output)], where H and W are the length and width of the sample image, respectively. The input is the GIIRS observed 225 channels (C1) brightness temperature, and the output is the atmospheric parameter of 37 vertical pressure levels (C2).

Each GIIRS observation FOR consists of 128 FOVs arranged in a 32 × 4 array. The U-Net 1 scheme spliced the observed brightness temperature of every eight consecutive FORs in the longitudinal direction to form a 32 × 32 pixel picture sample. Figure 2 shows the observed brightness temperature of the GIIRS 900 cm^{-1} channel in China from 00 to 01 (UTC) on 01 February 2020. Each white box in Figure 2 represents a sample size including 32 × 32 FOVs, where the horizontal lines represent different scan belts. The ERA5 data were spatially interpolated to the 32 × 32 pixel image. A total of 4454 pairs of samples were matched from 01 February 2020 to 10 February 2020, as detailed in Table 2.

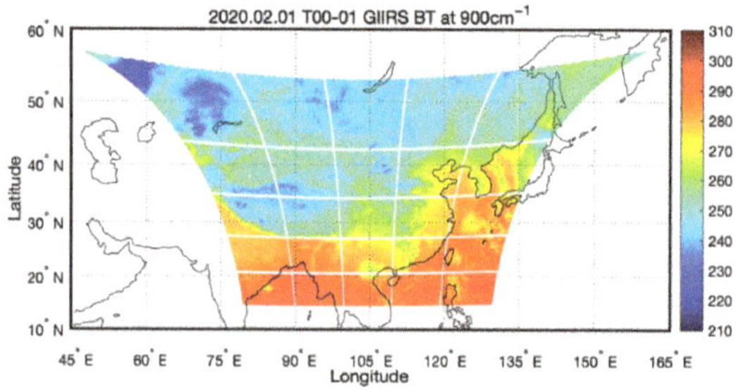

Figure 2. Example of U-Net training sample size.

3.3. *Training Data for the Three-Dimensional Scheme U-Net 2*

The U-Net 2 scheme selected the continuous GIIRS observations of the whole China area as a sample image with 160 × 160 FOVs (the full coverage area in Figure 2). There are 284 training samples for the whole month of February 2020. The input data for the U-Net 2 model are 225 channel GIIRS observations with dimensions of [160 × 160 × 225], and the output data are atmospheric parameter retrievals of 37 pressure levels with dimensions of [160 × 160 × 37].

3.4. *Model Structure and Parameter Optimization*

The structure of a traditional convolutional neural network mainly includes an input layer, convolutional layer, pooling layer, fully connected layer, and output layer. The main function of the convolutional layer is to extract features from the input image, the pooling layer is equivalent to a downsampling process, and the fully connected layer combines the local features extracted from the previous layers into the global features by nonlinear combination.

The 1D-Net model used in this study contains 1 input layer, 4 convolutional layers, 2 pooling layers, 1 fully connected layer, and 1 regression output layer. The convolutional layers and pooling layers are set alternately to form a multilayer neural network. The frame structure is shown in Figure 3. The input layer is the brightness temperature of 225 channels for each sample, which can be considered as a one-dimensional image of width 1, so the input layer size is 225 × 1. The output layer is the atmospheric temperature and humidity profiles with a size of 1 × 101. The dark part of the figure is the convolution kernel size, and the convolution operation is performed with a 1D convolution kernel. To build the optimal network, indicators such as retrieval root-mean-square error (RMSE), RMSE of network validation and network training time for test data are calculated for different parameter settings. The network optimal parameters were finally determined as follows: the convolution kernel size was 5 × 1, each pooling layer was 2 × 1 averaged pooling, the activation function was ReLU, and the training optimizer was Adam.

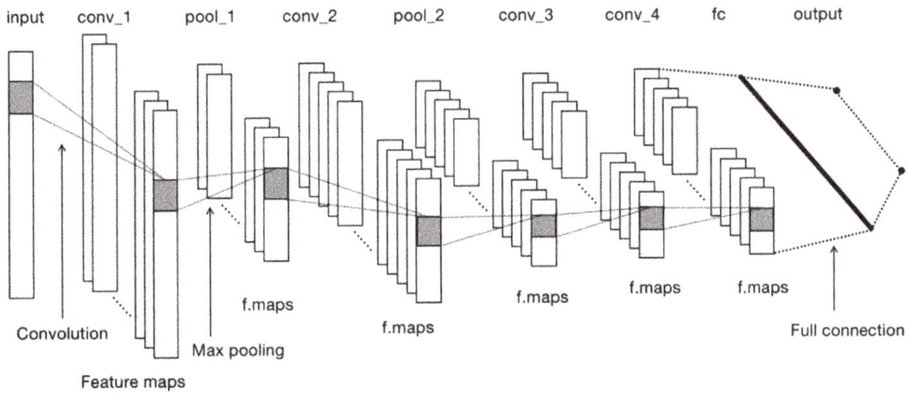

Figure 3. Frame structure for the 1D-Net model.

The U-Net convolutional neural network is a transformation of the traditional convolutional neural network and consists of two basic structure paths. The first is the contracting path, also called the encoder or analysis path, whose purpose is to capture the information in the image by convolution and pooling processes similar to the regular convolutional network. The second path is the expanding path, also known as the decoder or synthesizer path, which consists of upwards deconvolution and connecting features from the contracting path. Its purpose is to achieve precise localization of the segmented part of the image information and improve the output picture resolution. The structure of the U-Net 2 network constructed in this study is shown in Figure 4, with the contracting path on the left and the expanding path on the right. The purple arrows represent the convolution process (the convolution kernel size is 4 × 4), and each convolution process is followed by a modified linear unit called a ReLU. The grey arrows represent the crop and concatenation process, the red arrows represent the pooling process (2 × 2 maximum pooling method), and the green arrows represent the upconvolution process (2 × 2 convolution kernel). The number above the blue box indicates the number of channels in each layer, and the left box is the image size. The feature maps of the two parts are integrated using 4 crop and concatenation structures (grey arrows in Figure 4).

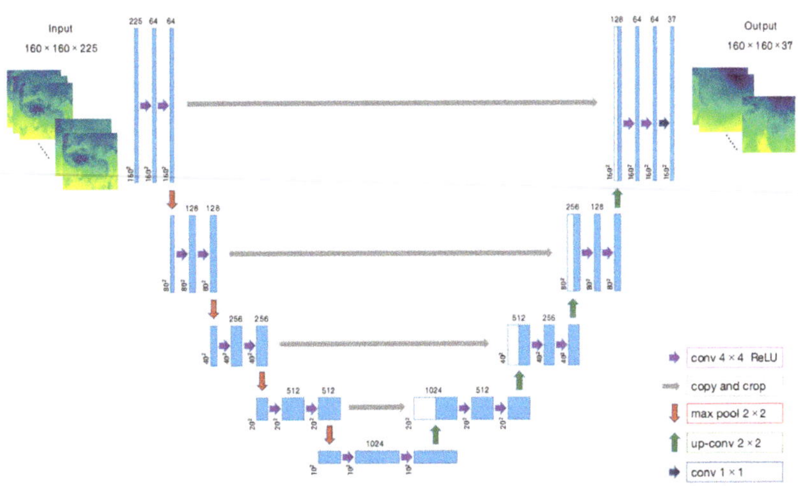

Figure 4. U-Net model structure.

4. Validation of Retrieval Results

4.1. Comparison with ERA5 Reanalysis

GIIRS Level 2 operational products downloaded from the CMDC website include CLM, temperature profiles (available only for clear sky and above the cloud top for cloudy field of view), etc., but humidity products were not released.

To test the retrieval accuracy of the above 1D-Net scheme and the two U-Net schemes, the temperature retrievals are compared with the ERA5 reanalysis fields and the GIIRS L2 operational atmospheric products. Using the GIIRS observation covering China at 00 to 01 (UTC) on 01 February 2021 as an example, the observed brightness temperature of the 900 cm^{-1} channel is shown in Figure 5. The warm areas in the figure represent the clear sky area with relatively high brightness temperature, while the cool tone areas with low values are covered with clouds. The lower the brightness temperature is, the higher the vertical cloud development height. The temperature retrievals at 1000 hPa are shown in Figure 6, where (a) is the 1D-Net scheme, (b) is the U-Net 1 scheme, (c) is the U-Net 2 scheme, and (d) and (e) are the temperature fields of GIIRS L2 and ERA5, respectively. Figure 7 illustrates the temperature retrievals at 500 hPa. The blank pixels in Figure 6e correspond to the GIIRS L2 temperature lacking FOV, and an increasing number of FOVs are missing as the height decreases because no L2 temperature product is below the cloud top under cloudy conditions. The retrieved temperature fields both at 1000 hPa and 500 hPa from the two U-Net schemes and L2 operational products are all closer to the ERA5 reanalysis field in terms of horizontal spatial distribution and values, especially for the U-Net 2 scheme, while the 1D-Net scheme is slightly worse. At 1000 hPa, the retrieval from the 1D-Net scheme is generally low (especially at high latitudes) with a large difference from ERA5. The temperatures from the two U-Net schemes are higher than those from ERA5 in the region of [60°–75°E, 50°–55°N], GIIRS L2 temperatures are underestimated relative to ERA5, the U-Net 1 retrievals are also low, and the U-Net 2 retrievals are closest to the EAR5 reanalysis in the Tibetan Plateau. The retrieved field of various schemes at 500 hPa is closer to ERA5 than that at 1000 hPa, indicating that the high-level retrieval accuracy is higher than near the surface. Since the training sample of the U-Net 1 scheme is composed of segmented small region images, traces of the segmented areas can be clearly seen in the temperature retrievals, and the retrieved temperature fields are less continuous near the region boundary. The U-Net 2 training sample is the whole China area image, so the retrieved fields are very continuous in horizontal space and are closer to the ERA5 temperature fields.

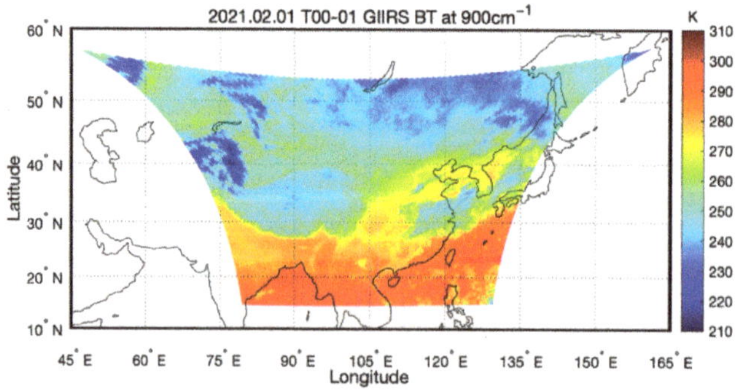

Figure 5. The GIIRS observed 900 cm^{-1} brightness temperatures (unit: K) for the China area from 00 to 01 (UTC) on 01 February 2021.

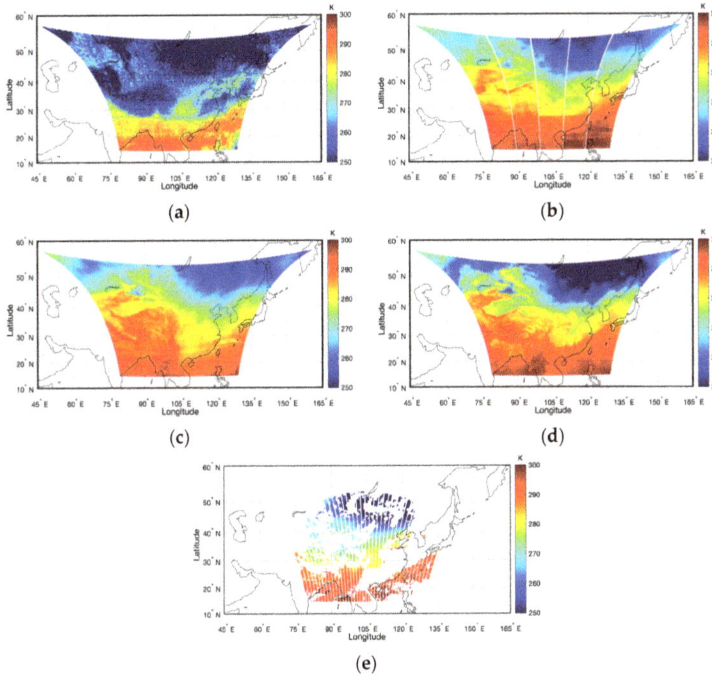

Figure 6. Temperature fields (unit: K) at 1000 hPa from 00 to 01 (UTC) on 01 February 2021: (**a**) 1D-Net; (**b**) U-Net 1; (**c**) U-Net 2; (**d**) ERA5; (**e**) Level 2 product.

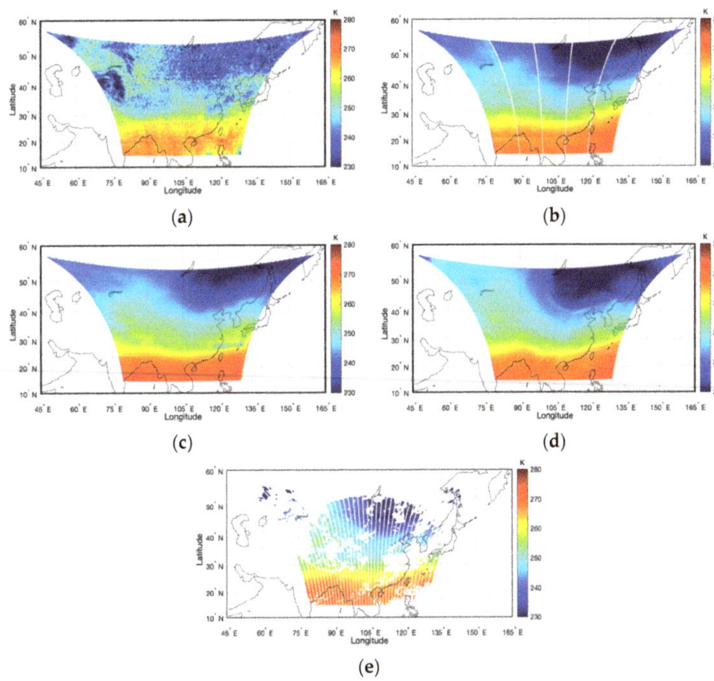

Figure 7. Same as Figure 6 except for 500 hPa: (**a**) 1D-Net; (**b**) U-Net 1; (**c**) U-Net 2; (**d**) ERA5; (**e**) Level 2 product.

The humidity retrieved results are shown in Figure 8 (at 1000 hPa) and Figure 9 (at 500 hPa). (a), (b), and (c) give the retrievals of the 1D-Net scheme, U-Net 1 scheme, and U-Net 2 scheme, respectively, and (d) is from ERA5. Humidity profiles are not provided in the GIIRS L2 products. The large value area of warm colors indicates abundant water vapor, and the cold tone corresponds to the low value dry area. The retrieved humidity fields of the two U-Net schemes are relatively close to the ERA5 reanalysis at 1000 hPa and 500 hPa, while the difference is large for the 1D-Net scheme, especially at 500 hPa. There are many clutter points of 1D-Net because this scheme only carries out retrieval of the vertical dimension for each field of view independently. The two U-Net water vapor retrievals at 500 hPa have relatively high values on the northwest side of the Tibetan Plateau, which coincides with the lower brightness temperature in Figure 5. Low brightness temperature implies cloud cover and high water vapor content. The retrieved high value areas of water vapor correspond to the low brightness temperature in Figure 5, which indicates that the retrieved results are very effective. Similarly, traces of segmented small areas can be seen in the humidity from the U-Net 1 scheme.

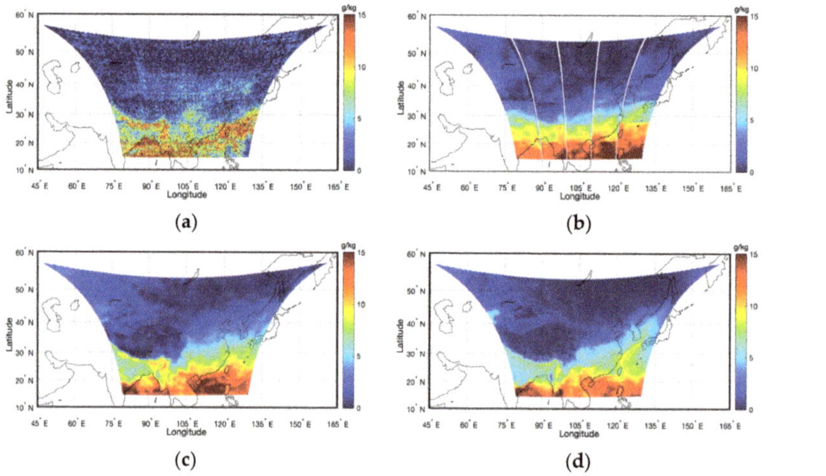

Figure 8. Water vapor mixing ratio (unit: g/kg) at 1000 hPa from 00 to 01 (UTC) on 01 February 2021: (**a**) 1D-Net; (**b**) U-Net 1; (**c**) U-Net 2; (**d**) ERA5.

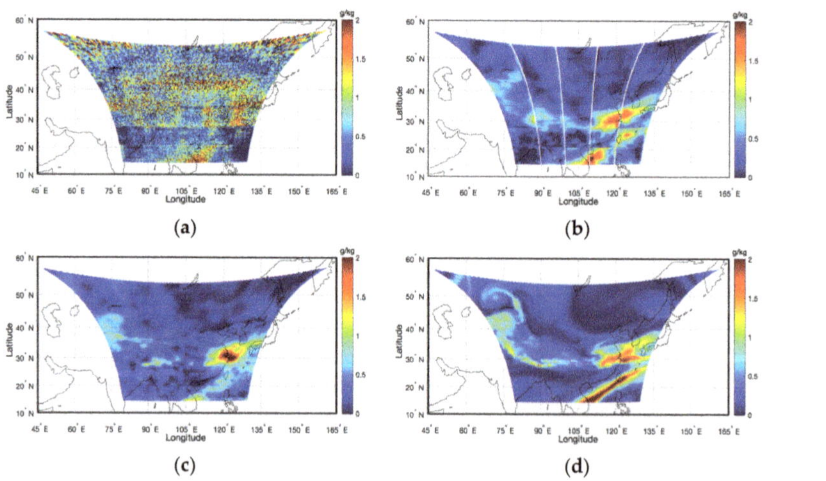

Figure 9. Same as Figure 8 except for 500 hPa: (**a**) 1D-Net; (**b**) U-Net 1; (**c**) U-Net 2; (**d**) ERA5.

To quantitative test the retrieval accuracy, the temperature and humidity ME (mean error) and RMSE (root mean square error) profiles from the three schemes and GIIRS L2 products compared with ERA5 reanalysis data for the whole February 2021 are given in Figures 10 and 11 respectively. The red lines are ME profiles, the black lines are RMSE profiles. The thicker dash-dotted line represents the 1D-Net scheme, the dashed line is the U-Net 1 scheme, the solid line is the U-Net 2 scheme, and the thinner dash-dotted line represents the L2 operational products. Figure 10 shows that most of the heights of these schemes are positive bias except for U-Net 2 from 750–150 hPa. The MEs of the two U-Net schemes are within 1 K at all pressure levels. The bias of 1D-Net scheme is significantly increased below 300 hPa. The RMSE of U-Net 1 and U-Net 2 are much smaller than 1D-Net. U-Net 2 scheme is better than that of the U-Net 1 scheme at all pressure levels. The accuracy of L2 products is close to two U-Net schemes below 700 hPa, but the RMSE increases with height. The retrieval error profiles of water vapor mixing ratio are shown in Figure 11. Humidity products are not provided by the L2 operational products. The humidity bias of two U-Net is close to 0 g/kg at all pressure levels, while 1D-Net scheme bias is larger than U-Net schemes. RMSE of all schemes decreases with height. The values of two U-Net schemes are smaller than 1D-Net. U-Net 1 humidity RMSE is lower than that of U-Net 2 above 800 hPa.

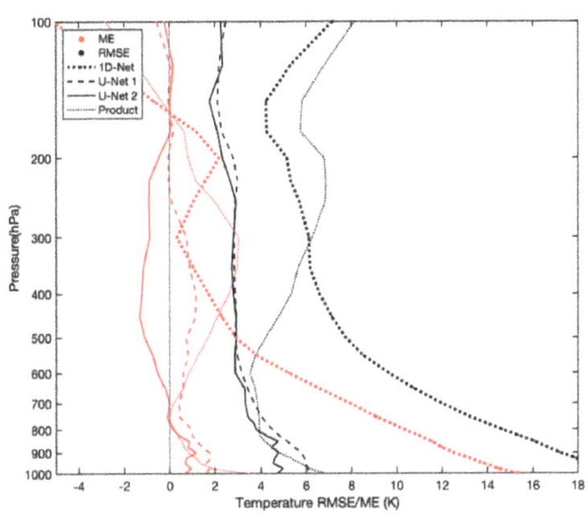

Figure 10. Temperature retrieval error profiles compared with ERA5 reanalysis field for the whole February 2021. The red lines are ME profiles, the black lines are RMSE profiles. The thicker dash-dotted line represents the 1D-Net scheme, the dashed line is the U-Net 1 scheme, the solid line is the U-Net 2 scheme, and the thinner dash-dotted line represents the L2 operational products.

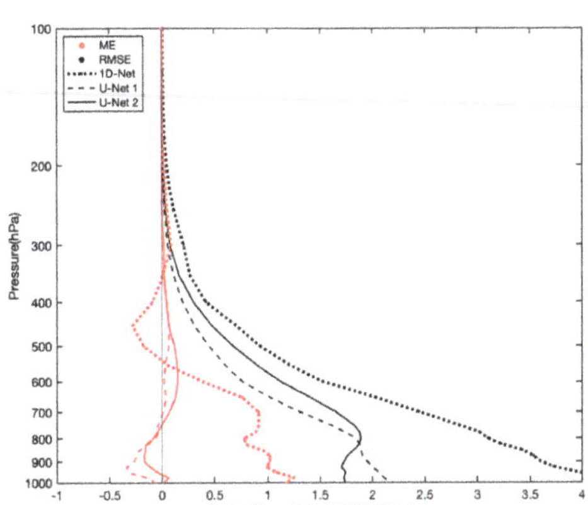

Figure 11. The same as Figure 10 except for water vapor mixing ratio.

4.2. Comparison with Radiosonde Observations

The retrievals of the three schemes are evaluated in terms of mean error and root mean square error using temporal-spatial matched radiosonde observations as true. It is divided into all sky and clear sky for the retrieval accuracy check separately based on observations from the whole month data of February 2021. The determination of clear FOV was based on the GIIRS L2 operational CLM products, which made clear sky and cloud judgements for each FOV.

The bias profile of the temperature retrieval under clear sky conditions for February 2021 is shown in Figure 12. The color and line shape denote the same as Figure 10. The sample number used to calculate the MEs and RMSEs at each pressure level is given on the right vertical coordinate. The sample number matched in each level decreases with decreasing altitude, which is because some radiosonde stations have no data at very low altitudes affected by the terrain. Figure 12 shows that the RMSEs of these scheme retrievals all slightly decrease with increasing height under clear FOVs (except near the surface), and the accuracy of the L2 operational products is higher with RMSE within 3 K. The retrieval accuracy of the two U-Net schemes is similar to that of the L2 operational products in the upper troposphere. The 1D-Net scheme RMSE is significantly larger at all pressure levels, and the U-Net 2 RMSE is slightly smaller than that of U-Net 1. The MEs of the two U-Net schemes and L2 products are close to 0 K with high accuracy above 250 hPa. L2 operational products have the smallest bias above 500 hPa, and the biases of the two U-Net schemes are smaller at heights below 500 hPa.

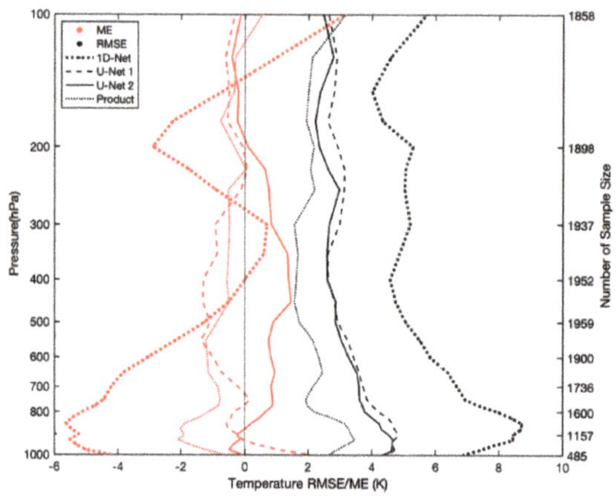

Figure 12. Temperature retrieval error profiles compared with radiosonde under clear FOVs for February 2021. The color and line shape denote the same as Figure 10.

The bias profile of the temperature retrieval under all sky conditions for February 2021 is shown in Figure 13. The red lines are the ME profiles, the black lines are the RMSE profiles. The number in the first column of the right vertical coordinate of Figure 13 represents the statistical sample size of the three convolutional neural network schemes for each pressure level matched with radiosonde observations, and the second column one represents the sample number matched with GIIRS L2 operational products, which is less than that of the convolutional neural network scheme because GIIRS L2 operational products are not retrieved below the cloud top. The RMSEs of the temperature retrieval by the two U-Net schemes are lower than those of the L2 products at almost all levels in Figure 13, and the Level 2 RMSE increases substantially above 500 hPa. The RMSEs of the two U-Net schemes are relatively large near the surface, and the accuracy of temperature gradually increases slightly with altitude above 800 hPa, with an RMSE of approximately 2.5 K. The retrieval accuracy of the U-Net 2 scheme is approximately 0.5 K better than that of the U-Net 1 scheme at all pressure levels. In terms of the temperature ME, the U-Net 2 scheme has a positive bias above 800 hPa, while the L2 operational products and U-Net 1 scheme have a negative bias, especially the bias of the L2 products above 550 hPa, which is large with the gradually increasing RMSE. The U-Net 2 scheme bias is the smallest, with a value of approximately 0.5 K. The 1D-Net scheme has a large ME and RMSE below 400 hPa.

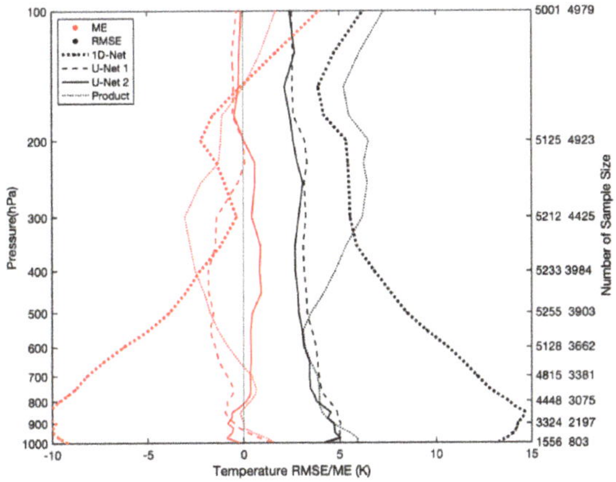

Figure 13. Same as Figure 12 except for all sky.

The ME and RMSE profiles of the retrieved humidity for all sky and clear sky FOVs for February 2021 are shown in Figures 14 and 15, respectively. Again, the sample number used to calculate the error is given for each pressure level on the right vertical coordinate. Humidity products are not provided by the L2 operational products. It can be seen from the figures that the RMSEs decrease with increasing height in both all sky conditions and clear sky conditions. The RMSE is maximum for the 1D-Net scheme at all levels, and the U-Net 1 RMSE is slightly larger than that of U-Net 2 at altitudes below 650 hPa. The bias of the 1D-Net scheme is larger than that of the U-Net schemes for almost all altitudes, the MEs of the two U-Net schemes above 650 hPa are similar (both close to 0 g/kg), and the water vapor bias of U-Net 2 is relatively small below 650 hPa.

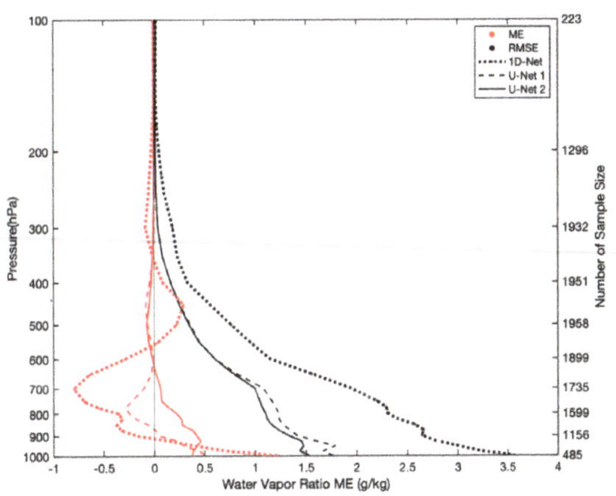

Figure 14. Water vapor mixing ratio retrieval error profiles compared with radiosonde under clear FOVs for February 2021. The color and line shape denote the same as Figure 10.

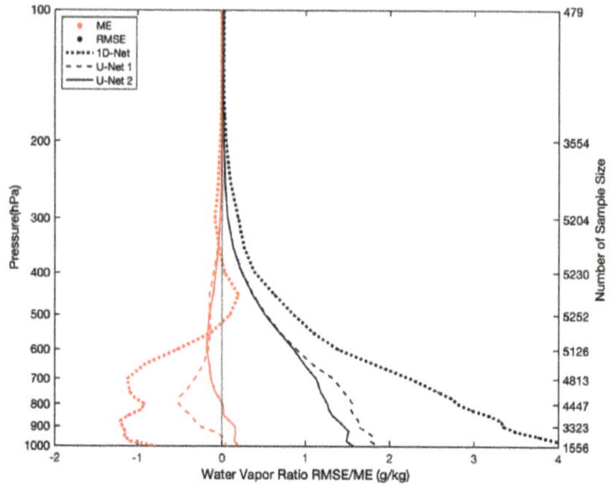

Figure 15. Same as Figure 14 except for all sky.

The GIIRS observations from July 2021 are used to further test the universality of these schemes. For summer months, the matched July 2020 data are trained to build network. The retrieval error of temperature and humidity under clear FOVs and all sky are shown in Figures 16 and 17, respectively. The U-Net 2 algorithm gives the highest retrieval accuracy in above winter month, so we just compare the U-Net 2 with L2 products in July. The solid line is U-Net 2 scheme and dashed line represents the L2 operational products. In Figure 16a, the U-Net 2 temperature is negative bias in summer while positive ones in winter (Figure 12). To bias and RMSE, the L2 products accuracy are all better than U-Net 2 for clear FOVs. The summer RMSE of L2 products is smaller than winter. To humidity (Figure 16b), the positive bias of U-Net 2 in wither (Figure 13) also changes to negative bias. This means that U-Net 2 overestimates the temperature and humidity slightly in winter and underestimates it in summer. The summer humidity RMSE is bigger than winter.

Under all sky conditions (Figure 17), the temperature retrieval accuracy of U-Net 2 is obviously higher than L2 product no matter ME or RMSE in summer and the U-Net 2 improve the temperature retrieval in the middle and lower troposphere. The summer humidity retrieval is worse than in winter.

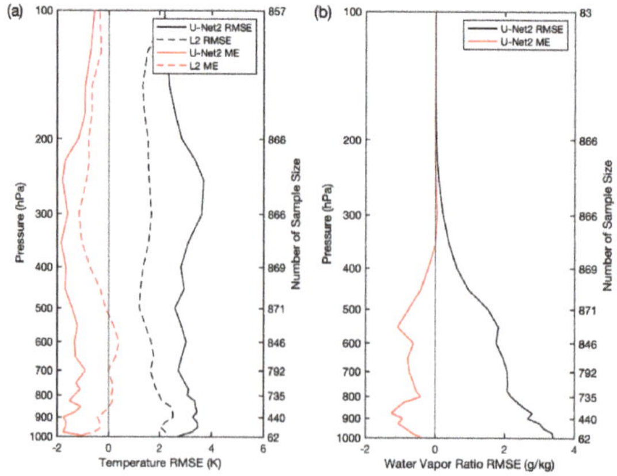

Figure 16. Temperature (**a**) and water vapor mixing ratio (**b**) retrieval error profiles compared with radiosonde under clear FOVs for July 2021. The red lines are ME profiles, the black lines are the RMSE profiles. The solid line is U-Net 2 scheme and dashed line represents the L2 operational products.

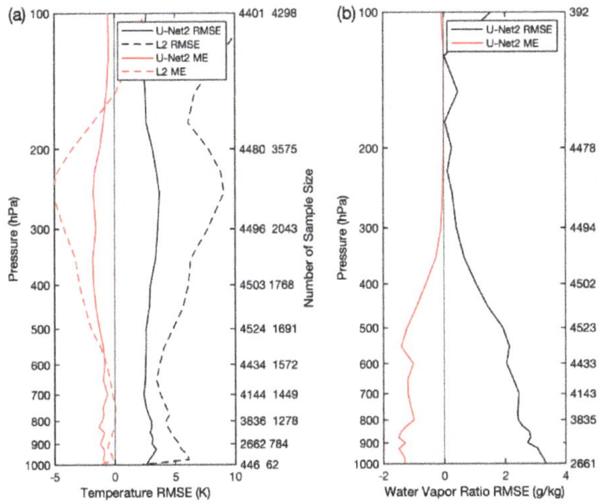

Figure 17. Same as Figure 16 except for all sky: (**a**) temperature retrieval error profiles; (**b**) water vapor mixing ratio retrieval error profiles.

4.3. Discussion of Three Convolution Neural Network Schemes

The three deep learning convolutional neural network schemes all can retrieve atmospheric temperature and humidity profiles for all the sky. The 1D-Net scheme with large retrieval bias is mainly because of just considering single FOV observations in the retrieval and fail to incorporate spatial information and feature transformations. While U-Net schemes consider the relevance from the image perspective and establish directly the relationship between the input image and output image by extracting image features and using 3D convolution. So, the U-Net temperature and humidity retrievals are more accurate and closer to the actual atmosphere, especially the U-Net 2 scheme. At the same time the retrieval fields are more continuous in horizontal distribution.

5. Conclusions

Three convolutional neural network schemes are used to retrieve one-dimensional and three-dimensional atmospheric temperature and humidity profiles, respectively, based on FY4A/GIIRS observations in this paper. The retrieval accuracy of the three schemes was examined and validated using ERA5 reanalysis fields and radiosonde observations under all sky and clear sky fields of view. The results are as follows:

(1) Compared with the ERA5 reanalysis field and the GIIRS L2 operational products, the retrieval accuracy of the 1D-Net scheme needs to be improved, and the three-dimensional atmospheric temperature and humidity field retrieved by the two U-Net schemes are closer to the ERA5 reanalysis field in both distribution and value; in particular, U-Net 2 retrieval is more continuous in horizontal space.

(2) The accuracy of L2 operational temperature products is the highest, the temperature retrieval RMSE for the two U-Net schemes is the second highest, and the RMSE and ME of the 1D-Net scheme are all larger compared with temporal-spatial matched radiosonde observations when the GIIRS field of view is completely clear. The temperature RMSE and bias of the two U-Net schemes under all sky conditions are lower than those of GIIRS L2 above 800 hPa, especially for the U-Net 2 scheme. The accuracy of the Level 2 temperature product will be reduced under the influence of clouds. The humidity RMSEs of the two U-Net schemes are within 2 g/kg, the 1D-Net scheme is worse, and humidity products are not provided by the L2 operational product.

(3) The three deep learning convolutional neural network schemes all can retrieve 3D atmospheric temperature and humidity profiles for all the sky from the perspective

of the image. The 1D-Net scheme only carries out retrieval of the vertical dimension for each field of view independently, with larger bias and discrete retrievals. The U-Net schemes use GIIRS multichannel spatial observations as input to improve the retrieval accuracy with cloud influence, and the retrieval fields are more continuous in horizontal distribution and closer to the actual atmosphere. The U-Net 2 scheme has the highest retrieval accuracy, followed by U-Net 1. The retrieval speed of the two U-Net schemes is nearly the same, faster than that of 1D-Net. The time required to retrieve the China area covered by the GIIRS is approximately 2–3 times longer than that of the U-Net schemes.

Author Contributions: Conceptualization, L.G. and S.Y.; methodology, L.G. and S.Y.; software, S.Y.; validation, L.G. and S.Y.; formal analysis, L.G. and S.Y.; investigation, L.G. and S.Y.; resources, L.G.; data curation, S.Y.; writing—original draft preparation, L.G. and S.Y.; writing—review and editing, L.G. and S.Y.; supervision, L.G.; project administration, L.G. All authors have read and agreed to the published version of the manuscript.

Funding: This work was supported by the National Natural Science Foundation of China under Grant No. 41975028.

Acknowledgments: We would like to thank the NSMC (National Satellite Meteorological Center) and the CMDC (China Meteorological Data Service Center) for sharing FY4A/GIIRS and Radiosonde data. We also thank the editor and reviewers for the comments that helped improve our manuscript.

Conflicts of Interest: The authors declare no conflict of interest.

References

1. Li, J.; Li, J.; Otkin, J.; Schmit, T.J.; Liu, C.Y. Warning Information in a Preconvection Environment from the Geostationary Advanced Infrared Sounding System-A Simulation Study Using the IHOP Case. *J. Appl. Meteorol. Climatol.* **2011**, *50*, 776–783. [CrossRef]
2. Li, J.; Fang, Z. The development of satellite meteorology-Challenges and Opportunities. *Mettir. Mon.* **2012**, *38*, 129–146.
3. Yang, J.; Zhang, Z.; Wei, C.; Lu, F.; Guo, Q. Introducing the New Generation of Chinese Geostationary Weather Satellites, Fengyun-4. *Bull. Am. Meteorol. Soc.* **2017**, *98*, 1637–1658. [CrossRef]
4. Geng, X.W.; Min, J.Z.; Yang, C.; Wang, Y.B.; Xu, D.M. Analysis of FY-4A AGRI Radiance Data Bias Characteristics and a Correction Experiment. *Chin. J. Atmos. Sci.* **2020**, *44*, 679–694. [CrossRef]
5. Jiang, D.M.; Dong, C.H.; Lu, W.S. Preliminary Study on the Capacity of High Spectral Resolution Infrared Atmospheric Sounding Instrument Using AIRS Measurements. *J. Remote Sens.* **2006**, *10*, 586–592.
6. Guan, L. Retrieving Atmospheric Profiles from MODIS/AIRS Observations. I. Eigenvector Regression Algorithms. *J. Nanjing Inst. Meteorol.* **2006**, *6*, 756–761. [CrossRef]
7. Liu, H.; Dong, C.H.; Zhang, W.J.; Zhang, P. Retrieval of clear air atmospheric temperature profiles using AIRS observation. *Acta Meteorol. Sin.* **2006**, *66*, 513–519.
8. Smith, W.L.; Weisz, E.; Kireev, S.V.; Zhou, D.K.; Li, Z.; Borbas, E.E. Dual-Regression Retrieval Algorithm for Real-Time Processing of Satellite Ultraspectral Radiances. *J. Appl. Meteorol. Climatol.* **2012**, *51*, 1455–1476. [CrossRef]
9. Zhu, L.; Bao, Y.; Petropoulos, G.P.; Zhang, P.; Lu, F.; Lu, Q.; Wu, Y.; Xu, D. Temperature and Humidity Profiles Retrieval in a Plain Area from Fengyun-3D/HIRAS Sensor Using a 1D-VAR Assimilation Scheme. *Remote Sens.* **2020**, *12*, 435. [CrossRef]
10. Xue, Q.; Guan, L.; Shi, X. One-Dimensional Variational Retrieval of Temperature and Humidity Profiles from the FY4A GIIRS. *Adv. Atmos. Sci.* **2022**, *39*, 471–486. [CrossRef]
11. Singh, D.; Bhatia, R.C. Study of Temperature and Moisture Profiles Retrieved from Microwave and Hyperspectral Infrared Sounder Data Over Indian Regions. *Indian J. Radio Space Phys.* **2006**, *35*, 286–292.
12. Malmgren-Hansen, D.; Laparra, V.; Nielsen, A.A.; Camps-Valls, G. Statistical Retrieval of Atmospheric Profiles with Deep Convolutional Neural Networks. *ISPRS J. Photogramm. Remote Sens.* **2019**, *158*, 231–240. [CrossRef]
13. Zhang, S.; Guo, Y.H.; Wang, J.J. The development of deep convolution neural network and its applications on computer vision. *Chin. J. Comput.* **2019**, *42*, 453–482.
14. Paoletti, M.E.; Haut, J.M.; Plaza, J.; Plaza, A. A New Deep Convolutional Neural Network for Fast Hyperspectral Image Classification. *ISPRS J. Photogramm. Remote Sens.* **2018**, *145*, 120–147. [CrossRef]
15. Khurana, L.; Chauhan, A.; Naved, M.; Singh, P. Speech Recognition with Deep Learning. *J. Phys. Conf. Ser.* **2021**, *1854*, 012047. [CrossRef]
16. Tan, J.; Chen, B.; Wang, W.; Yu, W.; Dai, W. Evaluating Precipitable Water Vapor Products From Fengyun-4A Meteorological Satellite Using Radiosonde, GNSS, and ERA5 Data. *IEEE Trans. Geosci. Remote Sens.* **2022**, *60*, 4106512. [CrossRef]
17. Xu, W.; Wang, W.; Wang, N.; Chen, B. A New Algorithm for Himawari-8 Aerosol Optical Depth Retrieval by Integrating Regional $PM_{2.5}$ Concentrations. *IEEE Trans. Geosci. Remote Sens.* **2022**, *60*, 4106711. [CrossRef]

18. Saunders, R.; Hocking, J.; Turner, E.; Rayer, P.; Rundle, D.; Brunel, P.; Vidot, J.; Rocquet, P.; Matricardi, M.; Geer, A.; et al. An Update on the RTTOV Fast Radiative Transfer Model (Currently at Version 12). *Geosci. Model Dev.* **2018**, *11*, 2717–2737. [CrossRef]
19. Zhu, L.; Li, J.; Zhao, Y.; Gong, H.; Li, W. Retrieval of Volcanic Ash Height from Satellite-Based Infrared Measurements: VOLCANIC ASH HEIGHT RETRIEVAL. *J. Geophys. Res. Atmos.* **2017**, *122*, 5364–5379. [CrossRef]
20. Yao, S.H.; Guan, L. Atmospheric temperature and humidity profile retrievals using machine learning algorithm based on satellite-based infrared hyperspectral observations. *Infrared Laser Eng.* **2022**, *51*, 20210707.

remote sensing

MDPI

Article

Improved Estimation of O-B Bias and Standard Deviation by an RFI Restoration Method for AMSR-2 C-Band Observations over North America

Wangbin Shen [1], Zhaohui Lin [2,3], Zhengkun Qin [1,3,*] and Xuesong Bai [1]

1 Center of Data Assimilation for Research and Application, Nanjing University of Information Science and Technology, Nanjing 210044, China
2 International Center for Climate and Environment Sciences, Institute of Atmospheric Physics, Chinese Academy of Sciences, Beijing 100029, China
3 Collaborative Innovation Center on Forecast and Evaluation of Meteorological Disasters, Nanjing University of Information Science and Technology, Nanjing 210044, China
* Correspondence: qzk_0@nuist.edu.cn; Tel.: +86-1879-580-3638

Abstract: Spaceborne microwave radiometer observations play vital roles in surface parameter retrievals and data assimilation, but widespread radio-frequency interference (RFI) signals in the C-band channel result in a lack of valuable data over large areas. Establishing repaired data based on existing observation information is crucial. In this study, Advanced Microwave Scanning Radiometer (AMSR)-2 C-band data affected by RFI were accurately repaired through the iterative principal component analysis (PCA) method in 2016 over the U.S. land area. The standard deviation (STD) and bias characteristics of the brightness temperature in the C-band vertical polarization channel were compared and analyzed before and after the restoration to verify the assimilation application prospect of the repaired data. Not only was the spatial continuity of the microwave imager observations significantly improved following restoration; the STD and bias of the observation minus background (OMB) of the restored data were basically consistent with those of the RFI-free data. The STD of OMB exhibited obvious seasonal variations, which were approximately 4.0 K from January to May and 3.0 K from June to December, whereas the biases were near zero in winter but negative (approximately −2.0 K) in summer. The surface type and terrain height also critically affected the STD and bias. The STD decreased with increasing terrain height, whereas the bias exhibited the opposite trend. The STD was largest in low-vegetation areas (4.0 K) but only approximately 2.0–3.0 K in pine forest and brush areas. These results show that the restored data have a high prospect for retrieval application and assimilation, and the STD and bias estimation results also provide a reference for land-based AMSR-2 data assimilation.

Keywords: AMSR-2; radio frequency interference; PCA iterative restoration; community radiative transfer model; bias correction

Citation: Shen, W.; Lin, Z.; Qin, Z.; Bai, X. Improved Estimation of O-B Bias and Standard Deviation by an RFI Restoration Method for AMSR-2 C-Band Observations over North America. *Remote Sens.* **2022**, *14*, 5558. https://doi.org/10.3390/rs14215558

Academic Editor: Steven C. Reising

Received: 2 September 2022
Accepted: 31 October 2022
Published: 4 November 2022

1. Introduction

A number of low-frequency microwave radiometers have been put into use (e.g., AMSR-2, Advanced Microwave Scanning Radiometer 2, etc.), which have offered opportunities for the derivation of more direct surface parameter estimations [1–5]. Modern numerical weather predictions (NWPs) rely on assimilating these satellite observations and retrievals to initialize the current state of the land surface accurately [6–11].

The continuous improvement of the assimilation effect has always been the goal of AMSR-2 data assimilation research [12,13]. During the data assimilation process, appropriate adjustment of the background field is determined by the observation error characteristics of the observation data and the background field, as well as some physical mechanisms. Due to the lack of true values, the STD of OMB is often used to characterize the observation error in data assimilation studies. Therefore, accurate STD estimations of OMB have an essential impact on the effect of data assimilation [14–16].

Bias estimation also plays a crucial role in satellite data assimilation (DA), since it is assumed that the differences between the background and observations satisfy an unbiased Gaussian distribution. In DA theory, systematic bias between satellite-observed and model-simulated radiances should be removed as a necessary condition for meeting this requirement [17,18]. Furthermore, the corrected brightness temperatures are also essential for other steps within DA, for example, cloud detection [19], which depends on the observation-minus-background (OMB) departures [18]. The proper treatment of such systematic biases is critical for the success of data assimilation systems [9,20–26].

Many studies have shown that both effective bias correction and STD estimation are significant prerequisites for successful data assimilation [9,25], but the current estimation methods mostly provide a uniform estimate over the ocean in consideration of the high spatial consistency of the ocean surface. However, the biggest difference between land and sea is the complex underlying surface characteristics of land.

In addition to large STDs caused by the artificial RFI, the variable underlying surface types over land cause considerable error in the surface emissivity. Moreover, a change in surface elevation will further complicate the simulation errors of brightness temperature caused by the surface temperatures and surface emissivity. Therefore, the assimilation of AMSR-2 data over land requires the targeted estimation of OMB standard deviations for different vegetation types and terrain heights on the basis of the current accuracy of the surface emissivity and surface temperature. Thus, the observation weight can be adjusted adaptively in the actual assimilation process and the effective assimilation of the AMSR-2 data over land can be realized.

However, the research on bias correction and STD estimation for AMSR-2 data has been restricted by RFI. AMSR2, which contains a low-frequency C-band (6.9-GHz and 7.3-GHz channels) and an X-band (10.7-GHz channel), is suitable for soil moisture monitoring [27–29]. The optimal low-frequency channel for data assimilation and retrieval using AMSR-2 is the 6.9-GHz channel, as this relatively low frequency responds to a deeper soil layer and is less attenuated by the atmosphere and vegetation than other channels [30]. However, the 6.9-GHz channel is also prone to interference by RFI signals, and the strong signal interference of RFI makes it impossible to effectively estimate the STD and bias of data from this channel, which makes the application of the channel data very difficult. Japan Aerospace Exploration Agency (JAXA) soil moisture products are mainly constructed based on the results retrieved from the 10.7-GHz channel due to the wide range of radio frequency interference (RFI) that occurs globally [28].

RFI refers to the radiation signal received by a satellite microwave radiometer that is confused by active remote sensing signals with similar bands to those used in human activities [31]. The strong signals emitted from these interfering sources conceal relatively weak thermal radiation signals from the Earth–atmosphere system, thus leading to the distortion of observations and causing significant increases in the brightness temperature of the detectors at the low-frequency band [31,32]. Numerous studies have shown that RFI is an extremely vital and nonnegligible factor in low-frequency bands (such as the C-band and the X-band), causing an anomalous bias which affects the application of microwave radiometer data [7,33,34].

An RFI filter has been used before data assimilation in a number of studies [7,9,35]. However, eliminating observational data from the low-frequency channel, which is affected by interference, inevitably causes a large amount of data to be wasted, and may also lead to a large range of observation data being lost.

To compensate for the loss of a large amount of observation data caused by RFI, Shen et al. (2019) [36] proposed an RFI data restoration method based on principal component analysis (PCA), making full use of the channel correlation and the spatial continuity of observations.

Most of the studies on AMSR-2 assimilation directly discard the data affected by RFI. Although the restored data can fill a wide range of observational data gaps, the applicability of these restored data in the assimilation process still requires further evaluation; specifically, answering the question of whether this restoration method can retain the STD and bias characteristics of the observational data is crucial for research on the follow-up of targeted bias-corrections and observational weight settings in the assimilation process. Therefore, in this paper, we used the established PCA iterative restoration method to repair RFI-affected data and then evaluated the bias and STD characteristics before and after the restoration process for different vegetation types and terrain heights. We hoped to provide more accurate bias and STD estimation results for AMSR-2 data assimilations over land.

The paper is structured as follows. In Section 2, we briefly describe the AMSR2 radiance data and the community radiative transfer model (CRTM), and give a brief introduction to the RFI detection and restoration method. In Section 3, we presents the validation of the restoration method and then compare and analyze the bias characteristics of the data before and after RFI restoration. Conclusions and discussions are summarized in Section 4.

2. Materials and Methods

2.1. AMSR-2 Brightness Temperature Observations

AMSR-2, an instrument carried on GCOM-W1, is a 14-channel, dual-polarization conically scanning passive microwave radiometer with 7 frequencies ranging from 6.9 to 89.0 GHz. This radiometer detects faint microwave emissions from the surface and atmosphere of Earth. The AMSR2 radiance observations frequencies are 6.9, 7.3, 10.65, 18.7, 23.8, 36.5, and 89.0 GHz, as listed in Table 1 [37]. The low-frequency channels below 10.65 GHz are usually used to retrieve various surface parameters, such as the soil moisture, vegetation water content, and snow thickness, as they are window channels with strong vegetation- and soil-penetrating abilities [2,3,5]. The surface incident angle of AMSR2 is maintained at 55 degrees, as this angle is less affected by sea surface winds and produces a large difference between the horizontal and vertical polarization results. The interval between the two conical scans is 1.5 s. The satellite advances approximately 10 km along the running track during this interval, and the width of one scanning line is approximately 1450 km. This scanning process can cover 99% of the world in two days.

Table 1. AMSR2 characteristics and performance.

Channel	Frequency (GHz)	Polarization	Bandwidth (MHz)	Resolution (km)	Sensitivity (K)
1/2	6.925	H/V	350	35 × 62	0.34
3/4	7.3	H/V	350	34 × 58	0.43
5/6	10.65	H/V	100	24 × 42	0.7
7/8	18.7	H/V	200	14 × 22	0.7
9/10	23.8	H/V	400	15 × 26	0.6
11/12	36.5	H/V	1000	7 × 12	0.7
13/14	89.0	H/V	3000	3 × 5	1.2

The study domain is the central and southeastern United States (30°–40°N, 260°–285°W) where C-band AMSR-2 radiance data are seriously affected by RFI. This domain also includes a variety of temperate land cover types with complex topography [38]. Performing the experiments in this domain allowed us to test the impact of the PCA iterative restoration method on changeable surface types and terrain.

To certify that this restoration method had good stability and prospects for data assimilation, it was necessary to obtain a sufficiently vast data sample to conduct RFI identification and restoration. Therefore, in this study we selected the AMSR-2 L1R-class observed brightness temperature data covering the study domain for the one-year period of 2016 (1 January to 31 December).

2.2. *Background—CRTM Simulations*

Three fast radiative transfer models have been applied worldwide: the radiative transfer for TOVS (RTTOV) [39], the community radiative transfer model (CRTM), and the advanced radiative transfer model system (ARMS) [40]. In particular, the newly developed ARMS model can be applied to the assimilation of data from the Fengyun satellites and those sensors not included in existing radiative transfer models [40,41]. The CRTM was developed by the U.S. Joint Center for Satellite Data Assimilation (JCSDA) to provide fast and accurate satellite radiance simulations and Jacobian calculations at the top of the atmosphere under all weather and surface conditions [42]. Only the CRTM model was used in this study. It can be shown that the measured radiance in this case is a weighted average of the atmospheric temperature profile.

Figure 1 showed the weighting functions calculated by the atmospheric profiles over ocean (a), and at altitudes of 1000 (b), 2000 (c) and 3000 (d) meters over land, respectively.

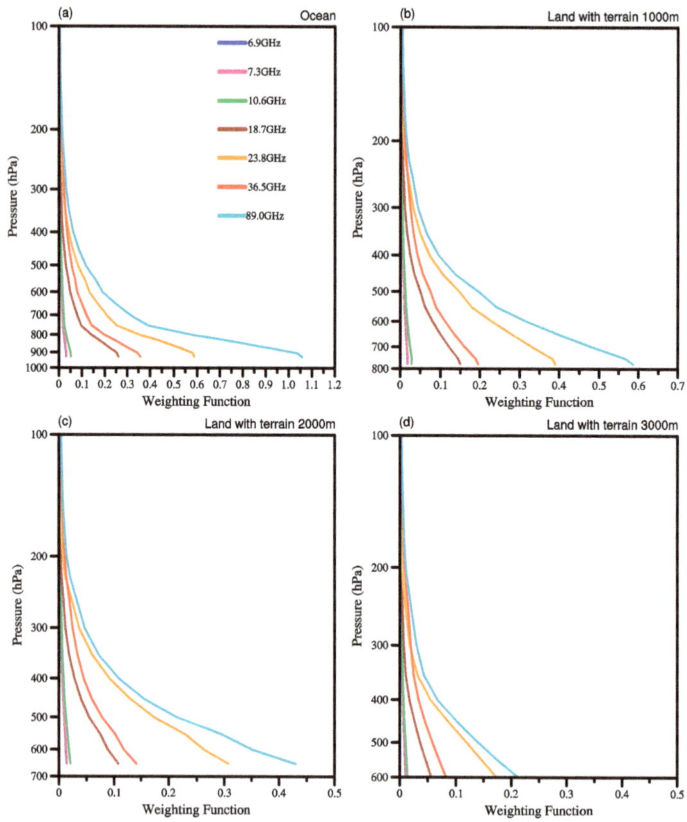

Figure 1. Weighting functions of the AMSR-2 channel 1–14 using CRTM based on the atmospheric profile over ocean (**a**) and for terrain height of 1000 m (**b**), 2000 m (**c**) and 3000 m (**d**) over land.

The weighted function $K(p)$ can be calculated as follows:

$$K(p) = {}^{d\tau}/_{dln(p)} \quad (1)$$

here τ means the atmospheric transmittance, p is for the pressure [43].

The weighting functions were calculated based on the atmospheric profiles using the CRTM. The profile information includes temperature, specific humidity and pressure profiles, as well as surface temperature and surface wind field information. It can be seen that weighting functions change little for channels with frequencies less than 10.7 GHz, but for other channels' weighting functions, the differences between the ground and the atmosphere gradually decrease with the increase of terrain height. The weighting functions of the channels with different polarization modes at the same frequency were consistent [44]. The peaks of the weighting function for each channel was located near the surface, as the microwave imager was mainly designed to improve our ability to detect surface parameters through remote sensing.

The amount of radiation detected by the microwave imager is represented by a weighted sum of surface radiation and atmospheric upward microwave radiation in different vertical layers near the ground; this value is mostly sensitive to the atmospheric temperature at the height of the maximum weighting function. The horizontal polarization channel and the vertical polarization channel with the same frequency have the same weighting function.

On the lowest-frequency channel (i.e., 6.9 GHz), the atmosphere contributes the least to the amount of observed radiation. The higher the frequency of the channel is, the wider the weighting function is. The weighting functions of the low-frequency channels are generally located inside the high-frequency channels, except for the 23.8- and 36.5-GHz channels. Thus, the brightness temperatures observed between different channels are highly correlated if the atmospheric contribution is significant [44].

2.3. Model Input—ECMWF Reanalysis Data

European Center for Medium-Range Weather Forecasting (ECMWF) hourly reanalysis data, with a horizontal resolution of 0.25×0.25 degrees and 37 vertical model levels, were used as the input for the CRTM. The input variables for CRTM include the three-dimensional atmospheric temperature, water vapor mixing ratio, and air pressure, as well as the two-dimensional surface variables of soil moisture, surface skin temperature, wind speed, and wind direction.

Hourly ECMWF liquid water path (LWP) reanalysis data with a horizontal resolution of $0.25^\circ \times 0.25^\circ$ were used to identify data collected under clear-sky conditions.

2.4. OMB Calculation Method

In this study, we used the International Geosphere-Biosphere Programme (IGBP) surface type dataset to identify the continental brightness temperature data. Among all the AMSR-2 pixels labeled as "water" in terms of their surface type, further works were carried out to eliminate the pixels within 50 km from coastlines to remove those mixed pixels with water.

Although microwave radiation is able to penetrate some non-precipitating clouds, it is basically unable to penetrate deep precipitation clouds. Even in penetrable clouds, various particles affect microwave radiation through absorption, emission and scattering effects [45,46]. To prevent effects associated with brightness temperature simulation uncertainties in cloudy areas on the bias and STD estimation, in this study we only used data obtained over continental areas under clear-sky conditions.

In order to acquire the simulated brightness temperature at AMSR-2-observed pixel locations and times, polynomial interpolation and linear interpolation were performed on the ECMWF analysis dataset in the horizontal and temporal dimensions, respectively. We processed the hourly ECMWF liquid water path (LWP) data in the same way. The brightness temperature data were considered "cloudy" data when the cloud water path value was greater than 0.01 g/kg, thus allowing us to identify data collected under clear-sky conditions. For the threshold, we referred to the study by Zou et al. (2017) [47]. The total water and ice cloud contents are close to 0.01 kg m^{-2}, which is used as the threshold to detect the cloud in Zou et al. (2017) [47].

Due to the lack of true observed values, the observation errors in the brightness temperature data are mostly estimated by obtaining the standard deviations of the OMB (observation-minus-background) values [48–51]. In satellite data assimilation, both the observations (*O*) and model simulations (*B*) are assumed to be unbiased. Therefore, STDs can be expressed as:

$$\Delta D_i = O_i - B_i$$

$$\sigma = \sqrt{\frac{\sum_{i=1}^{N}\left(\Delta D_i - \overline{\Delta D}\right)^2}{N-1}} \quad (2)$$

where O_i and B_i are the observed and simulated brightness temperature values on the same pixel, respectively, and ΔD_i means the OMB value of the pixel. $\overline{\Delta D}$ and σ represent the mean value and the standard deviations of the OMB value, respectively. N represents the counts of all the continental pixels under clear-sky conditions.

2.5. RFI Detection Method—Normalized Principal Component Analysis (NPCA)

The spatial correlations of natural-radiation-generated microwaves among different AMSR-2 instrument observation channels are often very high, as natural surfaces usually produce ultrawideband and smooth microwave radiation.

However, the brightness temperature of the low-frequency AMSR-2 channel increases significantly and abnormally in cases where RFI signals exist, resulting in weakened correlations between these RFI-affected channels and the other channels. The NPCA method, which takes advantage of the aforementioned feature, can effectively identify RFI signals through a PCA decomposition of the constructed interference coefficient matrix, using the brightness temperature difference calculated between the low-frequency channel and the high-frequency channel (low-high). On the other hand, the brightness temperature of the high-frequency channel can be strikingly reduced under the scattering effect of some natural targets (such as ice and snow), thus resulting in an inverse spectral difference gradient in continental regions covered with ice and snow. Therefore, Zou et al. (2013) [52] proposed an RFI detection method for NPCA analyses that has been shown to be effective for identifying RFI in data collected over snow- and ice-covered surfaces; this proposed method is suitable for identifying RFI over complex continental areas with mixed winter snow and RFI signals or over non-scattering surfaces in summer.

2.6. RFI Restoration Method—Iterative PCA Method

To compensate for the loss of a large amount of observation data caused by RFI, Shen et al. (2019) [36] proposed an RFI data restoration method based on principal component analysis (PCA). PCA can be used to extract observation information at different spatial scales into some independent PCA modes. The iterative PCA restoration method was established to obtain the correct brightness temperature of the RFI-affected point according to the correct observations around it.

For any observation, if the NPCA method recognizes that this observation has been affected by RFI, then on the satellite orbit where the point is located, the observation data from multiple channels for RFI-free points within the experience range of 350 km around the target point can form a repair matrix containing the target point, but the brightness temperature of the target point will be set to an initial value of 0.

PCA modes representing spatial features with different scales can be obtained through PCA decomposition of the matrix. For any data matrix B, the PCA modes correspond mathematically to the eigenvectors of the covariance matrix of B. The order of the PCA modes is determined based on the eigenvalues of the matrix corresponding to the eigenvectors. The higher-ranked modes correspond to larger eigenvalues, and larger eigenvalues correspond to spatial features with larger values of covariance. In relation to atmospheric variables, a large value of covariance often corresponds to more energy, and the energy of a large-scale weather system is generally much larger than that of a small-scale weather system. Thus, the PCA modes of meteorological variables often correspond to the weather variability features at different scales. More details can be found in Demšar et al. (2013) [53].

The brightness temperature of the target point, determined by means of a large-scale spatial structure, can be obtained by iteratively repeating the reconstruction process of the first mode. The same iterative restoration process can be performed for the rest of the PCA modes, and when all PCA modes are included, the final iterative repair results are obtained.

The proposed restoration method was used to recover observations affected by RFI with high precision [36]. The results of theoretical experiments and real data restoration experiments proved that the accuracy and effectiveness of the new method were much better than those of the Cressman method. Furthermore, the spatial continuity of observations in the recovered data were very well preserved by the new method.

3. Results

3.1. C-Band Continental RFI Characteristics

The NPCA method, described in Section 2.5, was used for RFI detection on C-band AMSR-2 data in this study. Figure 2 shows the brightness temperatures obtained by the AMSR-2 instrument in the 6.9-GHz and 10.7-GHz vertical polarization channels (Hereinafter referred to as 6.9-GHz-V and 10.7-GHz-V) over the area of the U.S. in the autumn of 2016, as well as the spatial distribution of the RFI signals identified through NPCA (Figure 2a–c). The brightness temperature of the 6.9-GHz channel was generally less than that of the 10.7-GHz channel for most of the continent, because the dielectric constant of water in soil and vegetation depends on this frequency, thus resulting in an increased surface emissivity with an increasing frequency [30]. However, the presence of an RFI signal at the 6.9-GHz frequency caused the brightness temperature of this frequency to increase abnormally, thus resulting in a spectral difference with an opposite sign to that expected. The brightness temperatures of the 6.9-GHz channel in the concentrated areas of Virginia, North Carolina, Texas, and other states were significantly higher than the brightness temperatures of the higher-frequency 10.7-GHz channel, which were far above 300 K, with notable horizontal spatial distribution discontinuities. In the identification results obtained using the NPCA method, the larger the value was, the stronger the possibility of RFI interference. As shown in Figure 2c, regions with abnormally high brightness temperatures (shown in Figure 2a) were detected as having significant RFI signals.

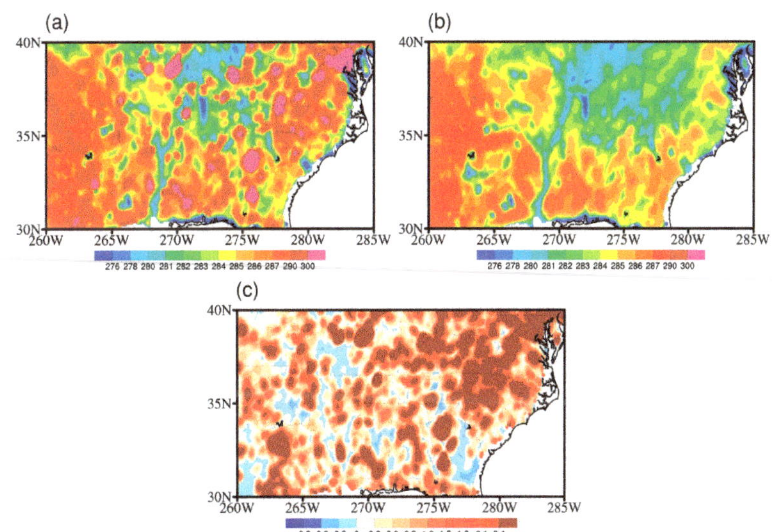

Figure 2. Spatial distributions of brightness temperatures of the 6.9-GHz-V channel (**a**) and the 10.7-GHz-V channel (**b**) over the U.S. continental area in the autumn of 2016; RFI signals identified by the NPCA for the 6.9-GHz-V channel are shown in (**c**).

The NPCA method was used for the detection of RFI signals in the horizontal and vertical AMSR-2 6.9-GHz channels over the study domain in 2016, and a daily variation curve of the proportion of the 6.9-GHz-V and 6.9-GHz-H channel scanning points affected by RFI for the land scanning points was obtained for the study domain (Figure 3). In Figure 3, the red line represents the vertical channel and the blue line represents the horizontal polarization channel. The figure shows that both the horizontal and vertical channels in the study region encountered continuous RFI signals throughout the year. In particular, the degree of interference in the vertical channel was obviously greater than that in the horizontal channel. Thirty to forty percent of the data were not available for data assimilation or retrieval applications because of RFI interference.

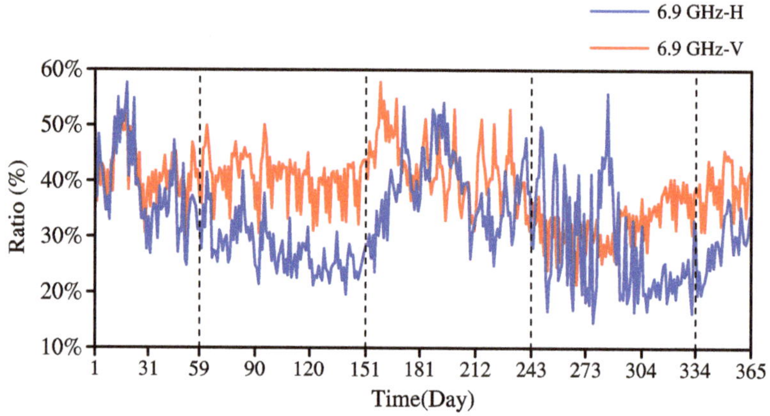

Figure 3. Daily variation curves of the proportion of pixels affected by RFI in the study domain for the 6.9-GHz-H (blue) and 6.9-GHz-V (red) channel in 2016.

3.2. RFI Restoration and Validation

Figure 4 shows the spatial distributions of the mean observed (a) and restored (b) brightness temperatures of the 6.9-GHz-V channel and the mean observed brightness temperatures of the 7.3-GHz-V (c) and 10-GHz-V (d) channels in autumn 2016. Comparing Figure 4a,b, it can be seen that those abnormally high brightness temperatures caused by RFI were well repaired. The overall geographic distribution of the brightness temperature showed good spatial continuity after this restoration, and the spatial distribution was consistent with the natural surface emission characteristics; in addition, the small-brightness temperature characteristics were restored as well.

In addition to the existing AMSR-E channel, two more channels were added to the AMSR-2 with frequencies near 6.9 GHz and 7.3 GHz. Anne et al. (2015) showed that the RFI phenomenon in the 7.3 GHz observation channel was significantly reduced in the U.S., Japan, and India, where there was severe pollution in the 6.9 GHz channel. As can be seen from Figure 4c, only a few regions showed abnormally high brightness temperatures over 300 K, such as northern West Virginia, central and eastern Alabama, and southern Kansas. However, in the corresponding region of the 6.9-GHz-V channel, there were no abnormally high brightness temperatures. The brightness temperatures of 6.9-GHz-V were generally lower than those of 10.7-GHz-V, except for the RFI-affected region. The frequencies of the 6.9-GHz channel and the 7.3-GHz channel were very close, so the brightness temperatures of the 7.3-GHz channel could be used qualitatively to verify the correctness of the repaired brightness temperatures. It can be seen that the spatial structure of the restored brightness temperature was similar to that of the 7.3-GHz channel. The low-value center in the middle of the region was well reproduced, and the spatial structures of three brightness temperature centers in the northeast of the United States, which were severely impacted by RFI, were also well restored.

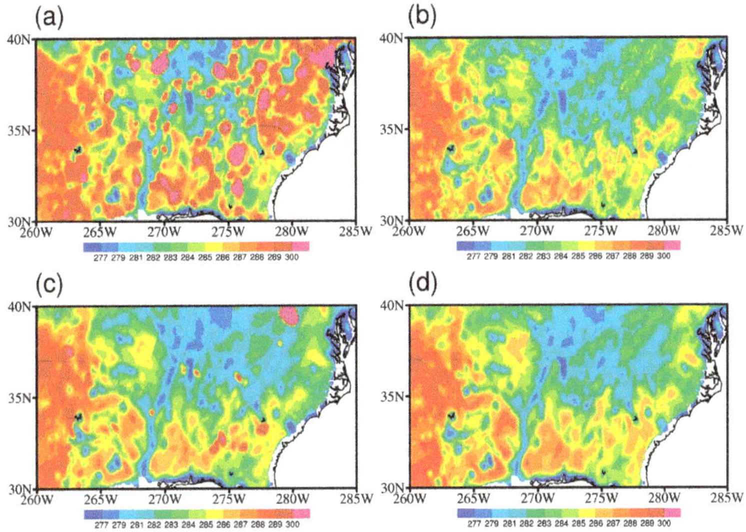

Figure 4. Spatial distributions of mean observed (**a**) and restored (**b**) brightness temperatures of the 6.9-GHz-V channel and the observed 7.3-GHz-V (**c**) and 10.7-GHz-V (**d**) channels in autumn 2016.

Figure 5 shows the distribution of the brightness temperature difference between the 6.9-GHz-V channel and the two high-frequency channels, 7.3-GHz-V (a) and 10.7-GHz-V (c), respectively. Figure 5b,d are the same as Figure 5a,c except for the restored brightness temperatures of the 6.9-GHz-V channel. RFI interference led to an abnormal increase in the brightness temperature values, resulting in the opposite spectral differences. Therefore, the larger the positive value in the spectral difference, the more affected were the values in the 6.9-GHz-V channel by RFI. As can be seen in Figure 5a,c, a large area of this region was affected by RFI, and the differences were even greater than 10 K. As can be seen in Figure 5b,d, this difference was basically within 5 K after the repair process. This indicates that the abnormal brightness temperature was well corrected, and also proves the effectiveness of the restoration method.

In consideration of the relatively high percentage of RFI signals in the 6.9-GHz-V channel (the red curve in Figure 3), in this study, we focused on the observation bias and STDs of the 6.9-GHz-V channel in the subsequent analysis.

3.3. Comparison of Simulated Brightness Temperature under Clear- and Cloudy-Sky Conditions over Ocean

The hourly cloud liquid water paths based on ERA5 reanalysis data were used to detect clear sky and cloudy data. In order to prove the accuracy of the cloud detection process and the reliability of CRTM simulation, the ocean surface area within the study area was selected for the comparison of OMB characteristics between clear-sky and cloudy areas.

Here, the monthly OMB standard deviations in clear-sky (blue line) and cloudy-sky (red line) areas were calculated separately (Figure 6). The OMB standard deviation in the cloudy area was approximately 6.0 K; this value was much larger than that obtained for the clear-sky area, with an obvious monthly difference. The largest standard deviation, reaching 7.26 K, was observed in June, whereas the smallest value was obtained for December. This may be due to the prevailing convective weather in summer, resulting in more deep clouds. However, the simulation errors for the clear-sky area were primarily reduced, as the standard deviation was maintained at around 0.9 K with a minimal standard deviation of approximately 0.6 K from June to July. The stationary standard deviation in the clear sky areas also proves the effectiveness of the cloud detection method.

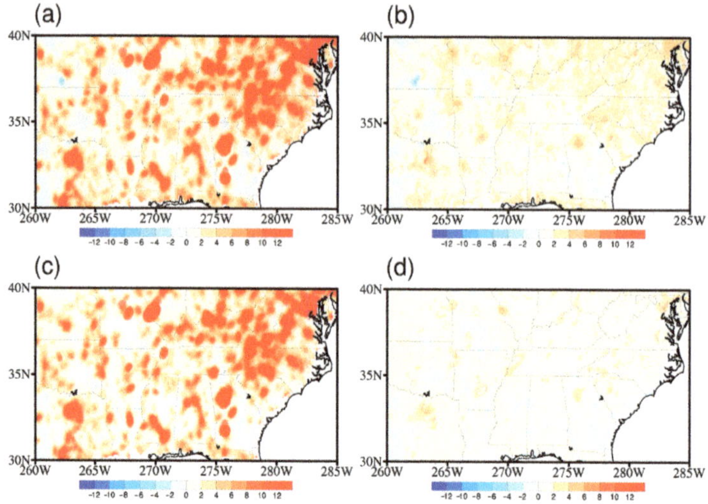

Figure 5. Spectral difference between the observed 6.9-GHz-V channel and the 7.3-GHz-V (**a**) and 10.7-GHz-V (**c**) channels, respectively, and the restored 6.9-GHz-V channel and observed 7.3-GHz-V (**b**) and 10.7-GHz-V (**d**) channels in autumn 2016.

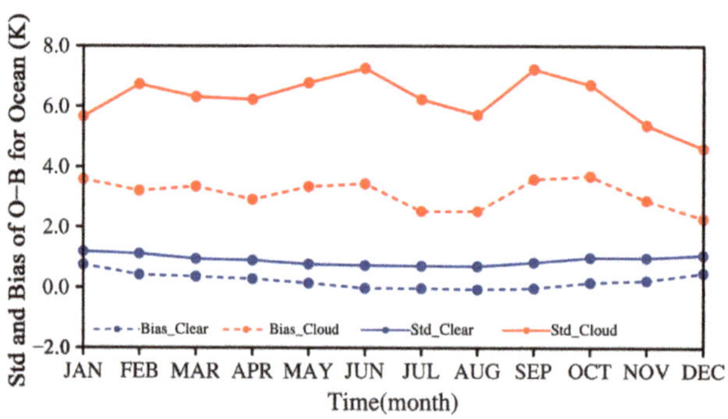

Figure 6. Monthly variations of OMB standard deviations (solid line) and bias (dotted line) for data in clear-sky (blue line) and cloudy-sky (red line) conditions over ocean from the 6.9-GHz-V channel in 2016.

It can be seen that there was also a large discrepancy between the monthly OMB bias in clear-sky (blue dotted line) and cloudy-sky (red dotted line) areas over ocean. The simulation was relatively accurate in clear-sky conditions, and the bias was basically below 1 K, with a minimum bias of zero in summer. The bias under cloudy conditions was significantly larger than that for clear-sky areas on the whole, and the bias value was basically around 3 K, with a maximum value of 3.8 K in September and October. The bias changed slightly from January to June, and was stable around 3.7 K.

3.4. Standard Deviation over Land

Figure 7 depicts the averaged OMB values before and after the restoration for the 6.9-GHz-Vchannel within the selected domain in autumn 2016. It reveals that the RFI area exhibited an obviously large bias without restoration (Figure 7a), basically exceeding 15.0 K and even exceeding 100.0 K at the maximum point. The simulation errors in the RFI-affected area were significantly reduced following the repair process (Figure 7b), with errors basically within 5.0 K, apart from some systematic deviations in high-terrain areas. Using the training data sets obtained under RFI-free conditions from AMSR-E, Wu et al. (2011) [54] developed the linear relationship between the measurements obtained at 10.7 GHz and those at 18.7 or 6.9 GHz. Then, the RFI-affected brightness temperatures were corrected based on the RFI-free measurements at 18.7 or 10.7 GHz via this linear relationship. The RFI-correction algorithm was able to produce brightness temperatures at AMSR-E frequencies with a root mean square (RMS) error of no more than 1.5 K. In this study, we focused on the 6.9-GHz-V channel of AMSR-2. The standard deviation of the OMB of this channel was 6.7 K, and it decreased to 4.0 K after restoration using the PCA iterative method.

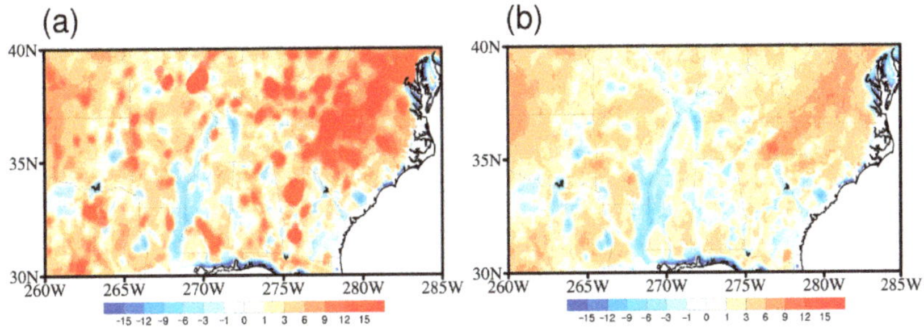

Figure 7. Averaged OMB before (**a**) and after (**b**) the restoration for the 6.9-GHz-V channel in autumn 2016.

Although it is clear that the spatial continuity of the brightness temperature data was improved through the restoration process, the impact of this restoration method on the standard deviations still needs to be further clarified in order to apply this method in data assimilation. Figure 8 shows the standard deviation and mean values of the OMB before (magenta line) and after the restoration (red line) for RFI-affected data of the 6.9-GHz-V channel. For comparison, the undisturbed data (blue line) are also shown here. The pink and gray bars in Figure 8a represent the numbers of RFI-affected and RFI-free pixels, respectively. The OMB standard deviation of the unpolluted data was approximately 4.0 K from January to May, whereas this value remained at approximately 3.0 K from June to December. The standard deviation for RFI-interfered data was significantly higher than that of the pollution-free data, with a value of approximately 8.0 K, with a minimal OMB standard deviation of 6.4 K obtained in June. From this OMB standard deviation comparison, it can be seen that the OMB STD values of RFI-affected data were significantly reduced after the restoration. The OMB STD of the restored data in each month was basically similar to that obtained from the RFI-free data; even monthly variation characteristics were also effectively reproduced in these OMB STDs.

As seen from the bias variation shown in Figure 8b, the bias of the RFI-free data was within the range of ±3.0 K. This varied obviously with the season, about 2 K in winter and −2 K in summer. From winter to summer, the bias basically showed the characteristics of a gradual decrease. The bias of RFI-affected data was significantly higher than that of the RFI-free data. The high values reached 9 K, and the low values were above 3 K. It also showed the same seasonal variation characteristics as the correct data. However, after the restoration, the bias derived for each month was very close to that obtained from the nonpolluted data, and the seasonal variation characteristics were effectively reproduced, further confirming the rationality of the restoration method. The land surface temperature had a strong impact on the simulated brightness temperatures. Some previous studies have pointed out that there are obvious seasonal biases in the surface temperature of ERA5 LST, attributed to uncertainty in land surface variables such as the leaf area index and land cover type, etc. [55]. This is a possible reason for the formation of seasonal differences in OMB biases.

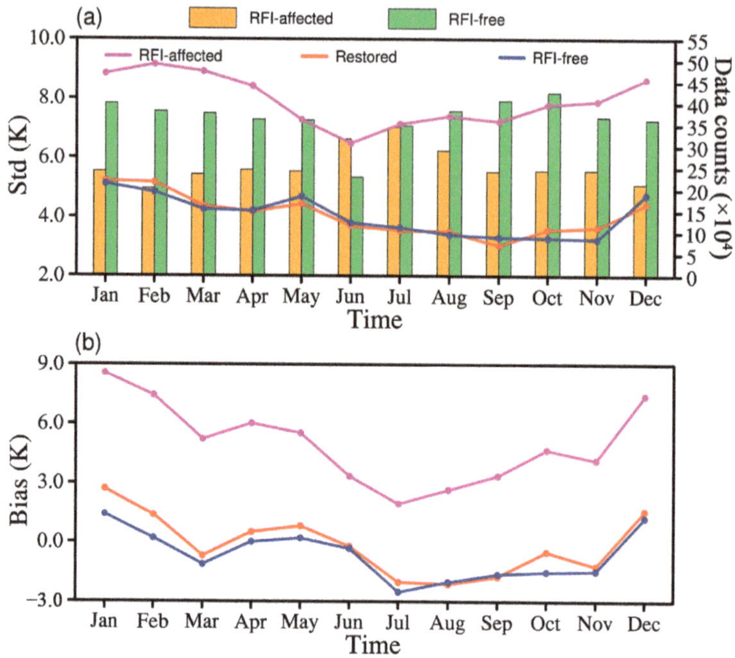

Figure 8. Monthly variations of the OMB standard deviation (**a**) and bias (**b**) obtained from the RFI-affected data before (magenta) and after (red) the restoration process and from the RFI-free data (blue) in 2016. The column bars represent the counts of considered data.

3.5. Variation Characteristics of STDs with Terrain Height and Surface Type

In contrast with the marine domain, which has uniform underlying surface properties, the underlying surfaces in land areas have two important characteristics: significant discrepancies in topographic height and changeable surface types. The STDs and bias estimation results obtained in land areas are thus inevitably affected by these two characteristics.

The biggest discrepancy between the assimilation of microwave imaging data over the land surface and the ocean is the complexity of the land surface's emissivity. In the microwave range, the land emissivity model is complicated as the land emissivity of each surface type depends on different parameters, such as soil moisture, topography, and the presence and physical properties of vegetation or snow [56]. The surface emissivity error may be significantly different for different land surface types, which will inevitably lead to inconsistency in the brightness temperature simulation bias observed over different land surface types. Therefore, it is necessary to estimate the STDs according to different surface types for the assimilation of AMSR-2 data over land. In addition, the errors of the surface temperature and wind field are much larger than those of variables in the upper atmosphere, so it is particularly important to estimate the OMB bias and STD according to the land cover type. After that, the effective bias correction and observation error specification can be achieved in the assimilation, to effectively account for the observation information of different vegetation types.

To increase the representative of the statistical results, the OMB values of the AMSR-2 6.9-GHz-V channel in the study domain were converted into grid data with a horizontal resolution of $0.25° \times 0.25°$. The spatial distributions of the standard deviations and bias before and after the restoration for 2016 within the analyzed land area are show in Figure 9. For comparison, the spatial distribution of terrain and vegetation types are also shown in the figure.

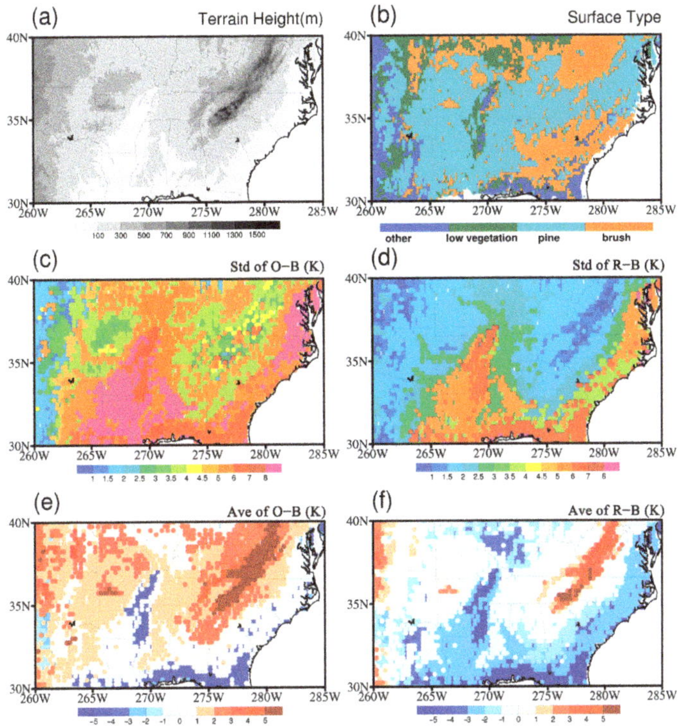

Figure 9. Spatial distributions of the terrain heights (**a**), surface types (**b**), standard deviations (**c**,**d**), and bias (**e**,**f**) before (**c**,**e**) and after the restoration (**d**,**f**) in the analyzed land area.

It can be seen from the topographical distribution shown in Figure 9a that the topography in the study domain was complex, exhibiting a large gradient that was mainly characterized by a distribution in which eastern areas were higher than western areas. The elevation of the Appalachian Mountains in the eastern study domain was relatively high, ranging from approximately 1000 to 1500 m. The elevations in the west Mississippi River Plain and the south Gulf Coast Plain were lower in comparison. As seen from the surface-type distribution (Figure 9b), the study domain mainly consisted of distributed pine trees, brush forests, and a small area of low vegetation. As seen from the STD distribution of the integral observed data in the 6.9-GHz-V channel (Figure 9c), which was abnormally large (above 4.0 K), the whole study domain was seriously affected by RFI before the restoration was applied. In the domain, the region with the largest STD—of approximately 7.0 K—was found in the Mississippi River Plain.

After the restoration of the disturbed brightness temperature data, the standard deviations characterizing this region were significantly reduced (Figure 9d). The standard deviation in the Mississippi River Plain area was approximately 3.0 K; this value was basically reduced to approximately 2.0 K in the other areas. The standard deviation in the Appalachian Mountain region basically decreased to less than 1.0 K after the restoration; this value was lower than that of the plain region because the low-frequency AMSR-2 observations are highly sensitive to soil moisture variations, which were relatively small in the mountainous region, leading to the smaller STDs obtained for this area than those obtained for the plains region. The observation bias was correspondingly large due to the strong RFI effect, as seen from its distribution (Figure 9e), with the highest mean value located in the Appalachian Mountains at approximately 8.0 K. The bias in the plain area was relatively low, with values between −3 and 3.0 K. After the restoration, the biases in most areas decreased significantly, approaching close to 0.0 K, but a positive bias was maintained in high-terrain areas, whereas the negative bias persisted in coastal and central low terrain areas.

Figure 9 shows that the spatial distribution of bias and standard deviations was very similar to that of vegetation types. In order to further clarify the impact of RFI restoration on different vegetation types, Figure 10 presents the OMB mean values and standard deviations before and after the restoration of RFI-affected data under different vegetation types.

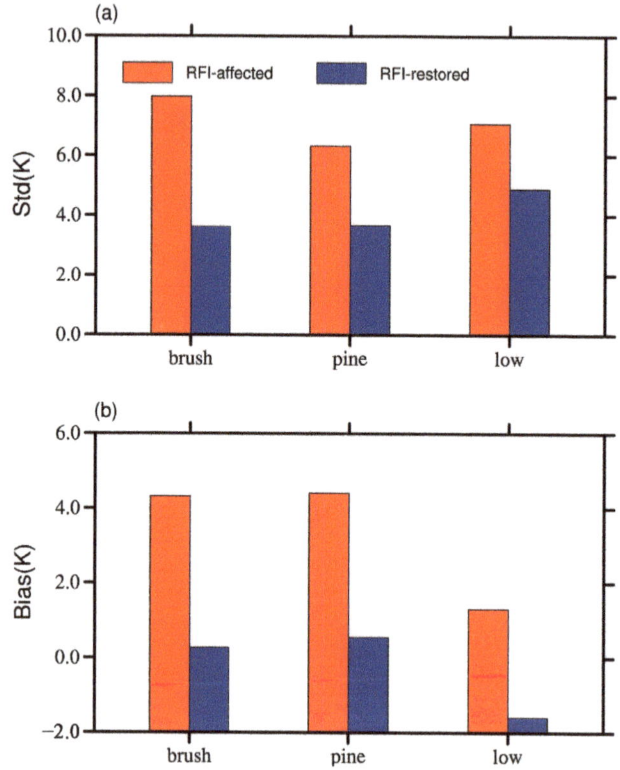

Figure 10. Comparison of standard deviation (**a**) and mean values (**b**) of OMB before (red bar) and after (blue bar) the restoration of RFI-affected data for the AMSR-2 6.9-GHz-V channel in 2016 over brush, pine forest, and low vegetation within the selected domain.

It can be seen in Figure 10a that STDs were obviously reduced after RFI restoration under all different surface types. Among these, the restoration effect of brush-covered area was the most significant, with the STD decreasing from 8.0 K to about 3.6 K. Furthermore, the STD decreased from 6.3 K to about 3.6 K within pine-forest-covered areas. The STD of low-vegetation regions was the highest after restoration, around 4.8 K. This is because increased vegetation cover and surface roughness reduce the sensitivity of microwave observations to soil moisture, leading to greater uncertainty in the background simulation [30].

The bias of the restored data was also significantly lower than before. The bias of pine and brush forest regions decreased from around 4.0 K to about 0.0 K. The bias was reduced from 1 K to −2.0 K over low-vegetation area after the accurate repair process.

In addition to vegetation types, the rapidly changing topographic height is another important feature of the land surface that is different from the ocean surface. In order to further evaluate the characteristics of data errors over land, the bias and STDs are also presented here for data under different terrain heights and different vegetation types (Figure 11). In contrast with Figure 10, the statistics here include all RFI-affected and RFI-free data, so that the statistical results can be directly applied to the actual data assimilation process.

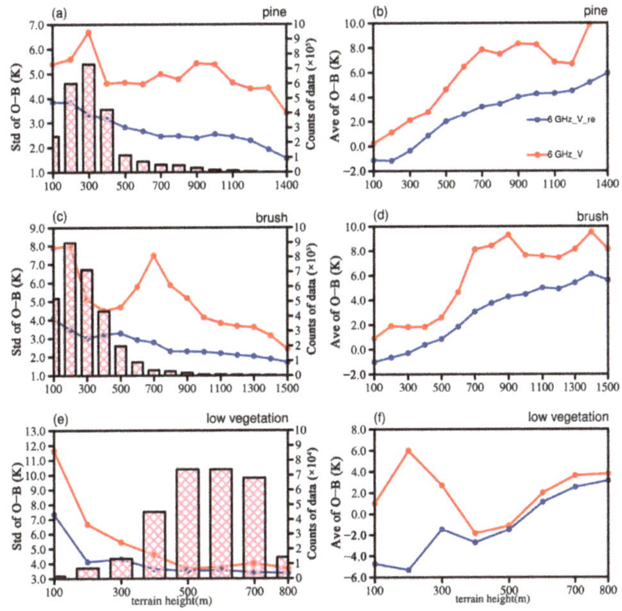

Figure 11. Variation curves of the OMB mean values (**b**,**d**,**f**) and standard deviations (**a**,**c**,**e**) characterizing the 6.9-GHz-V AMSR-2 channel with terrain height in the U.S. in 2016. The red and blue lines represent the pre- and post-repair results, respectively.

Figure 11 shows the variation characteristics of the STDs and bias obtained before and after the restoration with varying terrain heights and surface types, with pink reticulated bars indicating the amount of data processed. In this study, we analyzed three major ground types that corresponded to large amounts of data, namely, pine forests, brush regions, and low vegetation. The results revealed obvious differences in the influence of RFI on the brightness temperatures corresponding to different vegetation types under different terrain heights. Among these differences, in pine-forest- and brush-covered areas, the restoration method had an obvious improvement effect on the STD and bias values at different elevations. The STD even reached 8.0 K before the restoration, whereas this value was maintained between 2.0 and 3.0 K following the restoration, decreasing gradually with increasing terrain. The bias value obviously increased with increasing terrain height; this trend was contrary to that of the STDs. The bias value increased rapidly with increasing terrain height below 700 m. When the elevation reached heights above 700 m, the bias was reduced from 8 K to basically below 4.0 K overall following the repair process. In the area covered by low vegetation (Figure 11e,f), RFI was most serious at elevations located below 500 m, where the STD even reached 12.0 K; this value decreased to approximately 4.0 K following the restoration. In the areas with elevations over 500 m, the STDs obtained before and after the restoration were similar, both of which were approximately 4.0 K, and these barely changed with regard to terrain variations. The maximum bias obtained for elevations below 500 m before restoration was 6.0 K, and this value was gradually stabilized from −4.0 K to −2.0 K following the restoration process.

4. Discussion

The data obtained from microwave radiometer observations have important application value, especially in the case of low-frequency-channel observations, which play a crucial role in the surface parameter retrieval and data assimilation required in NWP; however, the effects of large-range RFI signals in these low-frequency channels lead to a large amount of observation data being wasted.

To obtain more effective observational data that are applicable to retrieval and assimilation tasks, an iterative PCA method was proposed to repair the RFI-affected data. Although it is clear that the spatial continuity of the brightness temperature data was improved through the restoration, the question of whether this restoration method can retain the STD and bias characteristics of the observational data is crucial for subsequent targeted bias-correction and observational weight-setting research in data assimilation.

Based on AMSR-2 observations from 1 January 2016, to 31 December 2016, in this study, we used the NPCA method to identify RFI-affected data on the C-band (6.9 GHz) in the central and southeastern United States and then applied an iterative PCA method to repair the corrupted data.

Finally, the STD and bias characteristics of the data obtained before and after the repair task and of the pollution-free data collected from the 6.9-GHz-V channel were analyzed in detail, and specifically, the variation characteristics of the STD and bias observed in land areas with varying terrain heights and surface types were further examined, thus providing a corresponding reference for subsequent data assimilation tasks involving low-frequency-channel data from AMSR-2 in land areas.

The long-term restoration results obtained herein show that the applied restoration method was not affected by the terrain height, vegetation type, or seasonal differences. Therefore, the next step will involve assimilating the restored brightness temperatures into numerical models to explore the impacts of the brightness temperature restoration process on the data assimilation.

5. Conclusions

In this study, RFI-affected AMSR-2 C-band data regarding the U.S. land area in 2016 were accurately repaired through iterative principal component analysis (PCA). The STD and bias characteristics of the brightness temperature data in the C-band vertical polarization channel were compared and analyzed before and after restoration to verify the assimilation potential of the repaired data. The main conclusions of this work are described below.

(1) The NPCA method was used to identify RFI signals in the observed brightness temperature data representing the U.S., collected from the 6.9-GHz channel for 2016. The results showed that severe RFI impacts persisted throughout the year in the U.S. The interference sources were mainly distributed in areas containing cities, such as the states of Virginia, North Carolina, and Texas. The amount of data suffering from RFI accounted for approximately 40% of the total amount of analyzed data.

(2) Based on the iterative PCA method applied herein, the disturbed brightness temperatures throughout the year were repaired. On the whole, the abnormally high brightness temperatures corresponding to RFI areas were repaired with a high level of precision. The overall brightness temperature distribution conformed to natural surface emission characteristics, maintaining good spatial continuity following the repair process, with small-scale features also being effectively recovered. At the same time, the applied restoration method was not affected by seasonal changes in brightness temperature or by variations in terrain or vegetation types and thus exhibited good stability and prospects for long-term RFI data recovery.

(3) The STD and bias in RFI-affected areas were significantly reduced following the restoration process; in addition, both of them were consistent with the corresponding values obtained from the pollution-free data, indicating that the repaired data retained the bias and STD characteristics of the observation instrument. Furthermore, in pine-forest-

and brush-covered areas, the restoration method had an obvious improvement effect. Over land, the STD decreased gradually with increasing terrain, but the trend of the bias was the opposite. These findings will be useful for subsequent data assimilation applications.

Author Contributions: Conceptualization, Z.Q. and W.S.; methodology, W.S.; software, W.S.; validation, Z.Q., W.S. and Z.L.; formal analysis, Z.L.; investigation, X.B.; resources, Z.Q.; data curation, W.S.; writing—original draft preparation, W.S.; writing—review and editing, Z.Q.; visualization, W.S.; supervision, Z.L.; project administration, Z.Q.; funding acquisition, Z.Q. All authors have read and agreed to the published version of the manuscript.

Funding: This research was jointly funded by the National Key R&D Program of China (Grant 2018YFC1507302), the National Natural Science Foundation of China (Grant 42075166), the Youth Project of the National Natural Science Foundation of China (41805076), the Natural Science Foundation of Jiangsu Province (BK20211396) and the FengYun-3 meteorological satellite engineering ground application Project (FY-3(03)-AS-11.08).

Data Availability Statement: ERA5 hourly reanalysis data: https://cds.climate.copernicus.eu/cdsapp#!/search?type=dataset&text=era5 (accessed on 2 March 2022); AMSR-2 brightness temperature data: https://gportal.jaxa.jp/gpr/search?tab=0 (accessed on 13 July 2021).

Acknowledgments: We would like to acknowledge the suggestions given by reviewers and editor. And we are grateful to the High-Performance Computing Centre of the Nanjing University of Information Science and Technology for the performed numerical calculations in this study using its blade cluster system.

Conflicts of Interest: The authors declare no conflict of interest.

References

1. Kachi, M.; Imaoka, K.; Fujii, H.; Shibata, A.; Kasahara, M.; Iida, Y.; Ito, N.; Nakagawa, K.; Shimoda, H. Status of GCOM-W1/AMSR2 development and science activities. In Proceedings of the SPIE-The International Society for Optical Engineering, Cardiff, Wales, UK, 9 October 2008.
2. Njoku, E.G.; Jackson, T.J.; Lakshmi, V.; Chan, T.K.; Nghiem, S.V. Soil moisture retrieval from AMSR-E. *IEEE Trans. Geosci. Remote Sens.* **2003**, *41*, 215–229. [CrossRef]
3. Santi, E.; Paloscia, S.; Pampaloni, P.; Pettinato, S.; Nomaki, T.; Seki, M.; Sekiya, K.; Maeda, T. Vegetation Water Content Retrieval by Means of Multifrequency Microwave Acquisitions From AMSR2. *IEEE J. Sel. Top. Appl. Earth Obs. Remote Sens.* **2017**, *10*, 3861–3873. [CrossRef]
4. Tang, Y.H.; Zhang, W.J. Land surface temperature retrieval using amsr-e data in the central tibetan plateau. In Proceedings of the 2016 IEEE International Geoscience and Remote Sensing Symposium, Beijing, China, 10–15 July 2016.
5. Dai, L.; Che, T.; Xie, H.; Wu, X. Estimation of Snow Depth over the Qinghai-Tibetan plateau based on AMSR-E and MODIS data. *Remote Sens.* **2018**, *10*, 1989. [CrossRef]
6. de Lannoy, G.J.M.; Reichle, R.H. Global Assimilation of Multiangle and Multipolarization SMOS Brightness Temperature Observations into the GEOS-5 Catchment Land Surface Model for Soil Moisture Estimation. *J. Hydrometeorol.* **2016**, *17*, 669–691. [CrossRef]
7. Muñoz-Sabater, J.; Lawrencea, H.; Albergelb, C.; Rosnaya, P.D.; Drusche, M. Assimilation of SMOS brightness temperatures in the ECMWF Integrated Forecasting System. *Q. J. R. Meteorol. Soc.* **2019**, *145*, 2524–2548. [CrossRef]
8. Wanders, N.; Karssenberg, D.; de Roo, A.; de Jong, S.M.; Bierkens, M.F.P. The suitability of remotely sensed soil moisture for improving operational flood forecasting. *Hydrol. Earth Syst. Sci.* **2014**, *18*, 2343–2357. [CrossRef]
9. Lievens, H.; Lannoy, G.; Bitar, A.A.; Drusch, M.; Dumedah, G.; Franssen, H.; Kerr, Y.H.; Tomer, S.K.; Martens, B.; Merlin, O. Assimilation of SMOS soil moisture and brightness temperature products into a land surface model. *Remote Sens. Environ.* **2016**, *180*, 292–304. [CrossRef]
10. Liang, S.; Qin, J. Data assimilation methods for land surface variable estimation. In *Advances in Land Remote Sensing: System, Modeling, Inversion and Application*; Springer: Berlin/Heidelberg, Germany, 2008; pp. 313–339.
11. Yin, J.; Zhan, X.; Zheng, Y.; Hain, C.R.; Liu, J.; Fang, L. Optimal ensemble size of ensemble Kalman filter in sequential soil moisture data assimilation. *Geophys. Res. Lett.* **2015**, *42*, 6710–6715. [CrossRef]
12. Jin, J.; Kim, M.J.; Mccarty, W.; Akella, S.; Garrett, K.; Jones, E. Assimilating GCOM-W AMSR2 Radiance Data in Future GEOS Reanalysis. In Proceedings of the AGU 2018 Fall Meeting, Washington, DC, USA, 18 December 2018.
13. Larue, F.; Royer, A.; De Sève, D.; Roy, A.; Cosme, E. Assimilation of passive microwave AMSR-2 satellite observations in a snowpack evolution model over northeastern Canada. *Hydrol. Earth Syst. Sci.* **2018**, *22*, 5711–5734. [CrossRef]
14. Geer, A.J.; Bauer, P.; Lopez, P. Direct 4D-Var assimilation of all-sky radiances. Part II: Assessment. *Q. J. R. Meteorol. Soc.* **2010**, *136*, 1886–1905. [CrossRef]

15. Yang, C.; Liu, Z.; Bresch, J.; Rizvi, S.R.H.; Huang, X.-Y.; Min, J. AMSR2 all-sky radiance assimilation and its impact on the analysis and forecast of Hurricane Sandy with a limited-area data assimilation system. *Tellus A Dyn. Meteorol. Oceanogr.* **2016**, *68*, 30917. [CrossRef]
16. Tandeo, P.; Ailliot, P.; Bocquet, M.; Carrassi, A.; Miyoshi, T.; Pulido, M.; Zhen, Y. A Review of Innovation-Based Methods to Jointly Estimate Model and Observation Error Covariance Matrices in Ensemble Data Assimilation. *Mon. Weather. Rev.* **2020**, *148*, 3973–3994. [CrossRef]
17. Dee, D.P. Bias and data assimilation. *Q. J. R. Meteorol. Soc.* **2005**, *131*, 3323–3343. [CrossRef]
18. Auligné, T.; McNally, A.P. Interaction between bias correction and quality control. *Q. J. R. Meteorol. Soc.* **2007**, *133*, 643–653. [CrossRef]
19. McNally, A.P.; Watts, P.D. A cloud detection algorithm for high-spectral-resolution infrared sounders. *Q. J. R. Meteorol. Soc.* **2003**, *129*, 3411–3423. [CrossRef]
20. Dee, D.P.; Da Silva, A.M. Data assimilation in the presence of forecast bias. *Q. J. R. Meteorol. Soc.* **1998**, *124*, 269–295. [CrossRef]
21. Eyre, J.R. Observation bias correction schemes in data assimilation systems: A theoretical study of some of their properties. *Q. J. R. Meteorol. Soc.* **2016**, *142*, 2284–2291. [CrossRef]
22. Qin, Z.; Zou, X. Impact of AMSU-A Data Assimilation over High Terrains on QPFs Downstream of the Tibetan Plateau. *J. Meteorol. Soc. Japan. Ser. II* **2019**, *97*, 1137–1154. [CrossRef]
23. Qin, Z. Adding CO_2 channel 16 to AHI data assimilation over land further improves short-range rainfall forecasts. *Tellus A Dyn. Meteorol. Oceanogr.* **2020**, *72*, 1–19. [CrossRef]
24. Xin, L.; Zou, X.; Zeng, M. An Alternative Bias Correction Scheme for CrIS Data Assimilation in a Regional Model. *Mon. Weather. Rev.* **2019**, *147*, 809–839.
25. Han, W.; Niels, B. *Constrained Adaptive Bias Correction for Satellite Radiance Assimilation in the ECMWF 4D-Var System*; ECMWF: Reading, UK, 2016.
26. Zhao, L.; Yang, Z.-L.; Hoar, T.J. Global Soil Moisture Estimation by Assimilating AMSR-E Brightness Temperatures in a Coupled CLM4–RTM–DART System. *J. Hydrometeorol.* **2016**, *17*, 2431–2454. [CrossRef]
27. Parinussa, R.M.; Holmes, T.R.H.; Wanders, N.; Dorigo, W.A.; de Jeu, R.A.M. A Preliminary Study toward Consistent Soil Moisture from AMSR2. *J. Hydrometeorol.* **2015**, *16*, 932–947. [CrossRef]
28. Kim, S.; Liu, Y.Y.; Johnson, F.M.; Parinussa, R.M.; Sharma, A. A global comparison of alternate AMSR2 soil moisture products: Why do they differ? *Remote Sens. Environ.* **2015**, *161*, 43–62. [CrossRef]
29. Zeng, J.; Li, Z.; Chen, Q.; Bi, H.; Qiu, J.; Zou, P. Evaluation of remotely sensed and reanalysis soil moisture products over the Tibetan Plateau using in-situ observations. *Remote Sens. Environ.* **2015**, *163*, 91–110. [CrossRef]
30. Parinussa, R.M.; Wang, G.; Holmes, T.R.H.; Liu, Y.Y.; Dolman, A.J.; de Jeu, R.A.M.; Jiang, T.; Zhang, P.; Shi, J. Global surface soil moisture from the Microwave Radiation Imager onboard the Fengyun-3B satellite. *Int. J. Remote Sens.* **2014**, *35*, 7007–7029. [CrossRef]
31. Njoku, E.G.; Ashcroft, P.; Chan, T.K.; Li, L. Global survey and statistics of radio-frequency interference in AMSR-E land observations. *IEEE Trans. Geosci. Remote Sens.* **2005**, *43*, 938–947. [CrossRef]
32. Zhao, J.; Zou, X.; Weng, F. WindSat Radio-Frequency Interference Signature and Its Identification Over Greenland and Antarctic. *IEEE Trans. Geosci. Remote Sens.* **2013**, *51*, 4830–4839. [CrossRef]
33. Dente, L.; Su, Z.; Wen, J. Validation of SMOS Soil Moisture Products over the Maqu and Twente Regions. *Sensors* **2012**, *12*, 9965–9986. [CrossRef]
34. Xie, X.; Wu, S.; Xu, H.; Yu, W.; He, J.; Gu, S. Ascending–Descending Bias Correction of Microwave Radiation Imager on Board FengYun-3C. *IEEE Trans. Geosci. Remote Sens.* **2019**, *57*, 3126–3134. [CrossRef]
35. Kazumori, M.; Alan, J.G.; English, S.J. *Effects of All-Sky Assimilation of GCOM-W1/AMSR2 Radiances in the ECMWF System*; ECMWF: Reading, UK, 2014.
36. Shen, W.; Qin, Z.; Lin, Z. A New Restoration Method for Radio Frequency Interference Effects on AMSR-2 over North America. *Remote Sens.* **2019**, *11*, 2917. [CrossRef]
37. Imaoka, K.; Maeda, T.; Kachi, M.; Kasahara, M.; Ito, N.; Nakagawa, K. *Status of AMSR2 Instrument on GCOM-W1*; SPIE: Bellingham, WA, USA, 2012; Volume 8528.
38. Blankenship, C.B.; Case, J.L.; Zavodsky, B.T.; Crosson, W.L. Assimilation of SMOS Retrievals in the Land Information System. *IEEE Trans. Geosci. Remote Sens.* **2016**, *54*, 6320–6332. [CrossRef] [PubMed]
39. Saunders, R.W.; Hocking, J.; Turner, E.C.; Rayer, P.J.; Rundle, D.; Brunel, P.; Vidot, J.; Roquet, P.; Matricardi, M.; Geer, A.J.; et al. An update on the RTTOV fast radiative transfer model (currently at version 12). *Geosci. Model Dev.* **2018**, *11*, 2717–2737. [CrossRef]
40. Weng, F.; Yu, X.; Duan, Y.; Yang, J.; Wang, J. Advanced Radiative Transfer Modeling System (ARMS): A New-Generation Satellite Observation Operator Developed for Numerical Weather Prediction and Remote Sensing Applications. *Adv. Atmos. Sci.* **2020**, *37*, 131–136. [CrossRef]
41. Yang, J.; Ding, S.; Dong, P.; Bi, L.; Yi, B. Advanced radiative transfer modeling system developed for satellite data assimilation and remote sensing applications. *J. Quant. Spectrosc. Radiat. Transf.* **2020**, *251*, 107043. [CrossRef]
42. Weng, F. Advances in Radiative Transfer Modeling in Support of Satellite Data Assimilation. *J. Atmos. Sci.* **2009**, *64*, 3799–3807. [CrossRef]

43. Jean-Noël, T. Satellite data assimilation in numerical weather prediction: An overview. In Proceedings of the Seminar on Recent Developments in Data Assimilation for Atmosphere and Ocean, Shinfield Park, Reading, UK, 8–12 September 2003.
44. Zou, X. Introduction to microwave imager radiance observations from polar-orbiting meteorological satellites. *Adv. Meteorol. Sci. Technol.* **2012**, *2*, 45–50. (In Chinese)
45. Greenwald, T.J.; Bennartz, R.; Lebsock, M.; Teixeira, J. An Uncertainty Data Set for Passive Microwave Satellite Observations of Warm Cloud Liquid Water Path. *J. Geophys. Res. Atmos.* **2018**, *123*, 3668–3687. [CrossRef]
46. Bennartz, R.; Watts, P.; Meirink, J.F.; Roebeling, R. Rainwater path in warm clouds derived from combined visible/near-infrared and microwave satellite observations. *J. Geophys. Res. Atmos.* **2010**, *115*, D19120. [CrossRef]
47. Zou, X.; Qin, Z.; Weng, F. Impacts from assimilation of one data stream of AMSU-A and MHS radiances on quantitative precipitation forecasts. *Q. J. R. Meteorol. Soc.* **2017**, *143*, 731–743. [CrossRef]
48. Otkin, J.A.; Potthast, R.; Lawless, A.S. Nonlinear Bias Correction for Satellite Data Assimilation Using Taylor Series Polynomials. *Mon. Weather. Rev.* **2018**, *146*, 263–285. [CrossRef]
49. Eyre, J.R. A bias correction scheme for simulated TOVS brightness temperatures. In *ECMWF Research Department Technical Memorandum No. 186*; ECMWF: Reading, UK, 1992.
50. Harris, B.A.; Kelly, G. A satellite radiance-bias correction scheme for data assimilation. *Q. J. R. Meteorol. Soc.* **2001**, *127*, 1453–1468. [CrossRef]
51. Hilton, F.; Atkinson, N.C.; English, S.J.; Eyre, J.R. Assimilation of IASI at the Met Office and assessment of its impact through observing system experiments. *Q. J. R. Meteorol. Soc.* **2009**, *135*, 495–505. [CrossRef]
52. Zou, X.; Zhao, J.; Weng, F.; Qin, Z. Detection of Radio-Frequency Interference Signal over Land from FY-3B Microwave Radiation Imager (MWRI). *Adv. Meteorol. Sci. Technol.* **2012**, *50*, 4994–5003. [CrossRef]
53. Demšar, U.; Harris, P.; Brunsdon, C.; Fotheringham, A.S.; McLoone, S. Principal Component Analysis on Spatial Data: An Overview. *Ann. Assoc. Am. Geogr.* **2013**, *103*, 106–128. [CrossRef]
54. Wu, Y.; Weng, F. Detection and correction of AMSR-E radio-frequency interference. *Acta Meteorol. Sin.* **2011**, *25*, 669–681. [CrossRef]
55. Trigo, I.F.; Boussetta, S.; Viterbo, P.; Balsamo, G.; Beljaars, A.; Sandu, I. Comparison of model land skin temperature with remotely sensed estimates and assessment of surface-atmosphere coupling. *J. Geophys. Res. Atmos.* **2015**, *120*, 12096–12111. [CrossRef]
56. Prigent, C.; Liang, P.; Tian, Y.; Aires, F.; Moncet, J.L.; Boukabara, S.A. Evaluation of modeled microwave land surface emissivities with satellite-based estimates. *J. Geophys. Res. Atmos.* **2015**, *120*, 2706–2718. [CrossRef]

remote sensing

MDPI

Article

Determining Tropical Cyclone Center and Rainband Size in Geostationary Satellite Imagery

Yanyang Hu and Xiaolei Zou *

Joint Center of Data Assimilation for Research and Application, Nanjing University of Information Science & Technology, Nanjing 210044, China; yanyang@nuist.edu.cn

* Correspondence: xzou@nuist.edu.cn

Abstract: Brightness temperature (TB) observations at an infrared channel (10.3 μm) of the Advanced Baseline Imager (ABI) on board the U. S. 16th Geostationary Operational Environmental Satellite (GOES-16) are used for determining tropical cyclone (TC) center positions and rainband sizes. Firstly, an azimuthal spectral analysis method is employed to obtain an azimuthally symmetric center of a TC. Then, inner and outer rainbands radii, denoted as R_{IR} and R_{OR}, respectively, are estimated based on radial gradients of TB observations at different azimuthal angles. The radius R_{IR} describes the size of the TC inner-core region, and the radius R_{OR} reflects the maximum radial extent of TC rainbands. Compared with the best track centers, the root mean square differences of ABI-determined centers for tropical storms and hurricanes, which totals 108 samples, are 45.35 and 29.06 km, respectively. The larger the average wavenumber-0 amplitude, the smaller the difference between the ABI-determined center and the best track center. The TB-determined R_{IR} is close but not identical to the radius of the outermost closed isobar and usually coincides with the radius where the strongest wavenumber 1 asymmetry is located. The annulus defined by the two circles with radii of R_{OR} and R_{IR} is the asymmetric area of rainbands described by azimuthal wavenumbers 1–3. In general, amplitudes of wavenumber 0 component centered on the ABI-determined center are greater than or equal to those from the best track. For a case of a 60 km distance between the ABI-determined and the best track TC center, the innermost azimuthal waves of wavenumbers 1–3 are nicely distributed along or within the radial distance R_{IR} that is determined based on the ABI-determined TC center. If R_{IR} is determined based on the best track, the azimuthal waves of wavenumbers 1–3 are found at several radial distances that are smaller than R_{IR}. The TC center positions, and rainband size radii are important for many applications, including specification of a bogus vortex for hurricane initialization and verification of propagation mechanism of vortex Rossby waves.

Keywords: geostationary satellite observations; tropical cyclone center positioning; tropical cyclone rainband size

Citation: Hu, Y.; Zou, X. Determining Tropical Cyclone Center and Rainband Size in Geostationary Satellite Imagery. *Remote Sens.* **2022**, *14*, 3499. https://doi.org/10.3390/rs14143499

Academic Editors: Yuriy Kuleshov and Hainan Gong

Received: 16 May 2022
Accepted: 20 July 2022
Published: 21 July 2022

1. Introduction

The center position and size of a tropical cyclone (TC) are essential parameters required by vortex initialization for numerical TC forecasts [1,2] and estimation of TC intensity [3,4]. Studies on the dynamics of TC structural and intensity changes also require an accurate center position and size of a TC [5–11]. For example, verifying the propagation mechanism of vortex Rossby waves by observations is usually achieved by measuring the radial scale and velocity of the TC inner spiral rainbands which rotate counterclockwise and move outward radially [7,12,13]. Guo and Tan [8] defined a TC intensity-related parameter called TC fullness, which is defined as the ratio of the difference between the radius of 17 m s^{-1} near-surface wind speed (R17) and the radius of maximum wind (RMW) divided by R17. The RMW and R17 are used to describe the sizes of the TC eye and inner-core region, respectively. Therefore, accurate TC center positions and sizes can also be applied to the calculation of TC fullness. Since TCs occur and intensify mostly over the oceans, where only a handful of in situ measurements are available, meteorological satellites have become the primary data sources for investigating TCs. Geostationary satellite imagers' brightness temperature (TB) observations in the infrared and visible channels have sufficient temporal and horizontal spatial resolution to reveal many structural features of TCs [3,4,14].

The TC center-positioning methods in numerical studies were reviewed by Ryglicki and Hart [15], who roughly divided all the methods into three categories. The first category is called the local extreme method, which defines a TC center as the geographic location where the surface pressure or the potential vorticity within the TC inner-core region reaches its minimum or maximum, respectively [16,17]. The second category is called the weighted grid points method, which takes the centroid of the properly weighted pressure or potential vorticity field as the TC center [18–20]. The third category is called the minimum azimuthal variance method, which takes the center of an annulus meeting the following conditions as the TC center: inside the annuli with varying widths centered on the position of RMW, the azimuthal variance is minimized, or the azimuthal mean of pressure or wind speed is maximized [21–23]. The last method assumes that the symmetric component of a pressure field or wind field containing a TC always dominates the asymmetric components. Observational studies of TCs mainly used reflectivity observations and derived wind field products from ground and airborne radars, as well as wind field observations from reconnaissance aircraft [13,24]. Therefore, many TC center-positioning methods are based on wind field observations around a TC, seeking a TC center position to maximize the azimuthal average tangential wind speed [24–28] or the symmetric component of the tangential wind [7] inside the annuli with different widths centered at the RMW. In addition, Corbosiero et al. [12] used the dynamical center of the vortex determined by Willoughby and Chelmow [29] as the TC center when decomposing the symmetric and asymmetric components of a TC. Guimond et al. [13] directly used the TC center position observed by the Air Force and NOAA reconnaissance aircraft when verifying the vortex Rossby waves by radar observations.

In operational analysis and forecast of TCs, abundant infrared, visible, and microwave observations provided by meteorological satellites are mainly used to locate the TC center [3,4,30–34]. The most famous TC center-positioning methods are the Advanced Dvorak Technology (ADT) for determining TC intensity and the Automated Rotational Center Hurricane Eye Retrieval (ARCHER) algorithm [33,34]. The ARCHER algorithm employs multiple data sources to locate the TC center, including TB observations at infrared, visible, 85–92-, and 37-GHz microwave channels and wind vectors from scatterometer retrievals. Hu and Zou [35] tried to locate TC centers by an azimuthal spectral analysis method using microwave TB observations at a single channel of the Advanced Technology Microwave Sounder (ATMS) and the Microwave Humidity Sounder (MHS) onboard polar-orbiting operational environmental satellites (POESs). Since a single polar-orbiting satellite can only observe the same TC twice daily at most, even during times with three possible polar orbits (morning, afternoon, and early morning) of all POESs, there would be six times global observations per 24 h. In this study, the infrared TB observations with 15 min temporal and 2 km horizontal resolutions provided by the Advanced Baseline Imager (ABI) onboard the U.S. 16th Geostationary Operational Environmental Satellite (GOES-16) are used to determine TC centers and radii of the inner and outer rainbands. Being capable of better resolving cloud top distributions than microwave radiations, GOES infrared TB observations can better capture small-scale cloud structures of TCs from space.

Aside from TC motion and intensity, TC sizes, which determine the potential impacts of TCs [36–38], as well as arial coverages and distributions of TC-induced cloud and rainfall [39,40], have received a lot of attention in research. We may name a few size parameters employed in past research: the radius of the 34 kt near-surface wind speed (R34) [8,41]; the radius of 15 m s^{-1} near-surface wind speed (R15) [42,43]; the radius of outermost closed isobar (ROCI) [44,45]; the average radius of 5 kt tangential wind (R5) at 850 hPa [46]; the average radius of 1×10^{-5} s^{-1} relative vorticity [47]; the radius of the maximum radial gradient of sea level pressure (SLP) [1]. The lack of in situ measurements makes routine operational wind radii estimation heavily dependent upon satellite observations [48]. Liu and Chan [47] measured the average radius of 1×10^{-5} s^{-1} relative vorticity as the TC size using the European Remote-Sensing Satellites-1 and wind scatter observations. Using the Quick Scatterometer (QuickSCAT) surface wind speed, Lee et al. [43] and Chan and Chan [49] measured the TC sizes as R15 and R34, respectively. Knaff et al. [46] indirectly measured the TC size (R5) by establishing a relationship between the azimuthally averaged 850 hPa tangential winds from model analyses and the radial distribution of azimuthally averaged infrared (~11 μm) TB. Lu et al. [50] measured the TC size (R34) by establishing a relationship between the R34 and the TB radial profiles derived from satellite imagery. In summary, TC sizes were generally quantified by wind or SLP data.

A TC is essentially a rotating, organized system of clouds and thunderstorms. It is natural to describe its size based on TC cloud structures. The TC size, determined from near-surface wind speeds and/or SLP data, cannot fully represent TC cloud structures. Since TC spiral rainbands are usually shown as banded low-valued TB regions, spatial variations in GOES infrared TB observations can reflect the cloud structures and cloud convection intensity of TCs [3,51–54]. A TC is usually regarded as an axisymmetric vortex [55], but asymmetric structures (non-circular TC eye, spiral rainbands, etc.) are also very important for its movement and development. The spiral rainbands are spiral band structures of convection or precipitation that occur outside the hurricane eyewall. In this study, the rainband near the TC center and of strong azimuthal symmetry in GOES infrared TB observations is called the inner core rainband, and the spiral rainband farther from the TC center often has an asymmetry distribution called the outer spiral rainband [56,57]. The TC inner-core region is typically defined as an axisymmetric region 100–200 km away from the TC center [57] or as a dynamically controlled region by cyclonic vortex circulation [58]. Since there are huge differences in TB values between the TC cloud area and its surrounding environment [59], this study, in addition to determining TC center positions, also attempts to derive the radii from the boundaries of the TC cloud area to the TC center at different azimuthal angles. Considering that the outer spiral rainbands are usually distributed in the periphery of a TC (the upwind end lies outside the TC inner-core region and the downwind end lies within the TC inner-core region), and the inner core rainbands are located within the TC inner-core region [57], it is hypothesized that the minimum and maximum radii from the TC cloud boundaries at 24 azimuthal angles of 15° to the TC center represent the radii of the inner and outer spiral rainbands from the TC center, respectively, which are called the radius of inner core rainband (R_{IR}) and the radius of outer spiral rainband (R_{OR}), respectively. Thus, R_{IR} describes the size of the TC inner-core region, and R_{OR} reflects the maximum radial extent of TC rainbands.

The article is organized as follows: Section 2 provides the instrument characteristics of ABI, the National Hurricane Center (NHC) best track data, and TC cases. Section 3 briefly describes an azimuthal spectral analysis method for TC center positioning and a step-by-step procedure for determining radii R_{IR} and R_{OR}. In Section 4, ABI-derived results on center position, R_{IR} and R_{OR} of Hurricanes Irma and Jose (2017) are given, and the relationships among the moving track, the steering flow, and TC cloud convections during the time period when Jose made a circled track are discussed. In Section 5, spatial distributions and amplitudes of symmetric and asymmetric components centered at the ABI-determined center and the best track are compared. Section 6 presents conclusions.

2. Data and TC Case Description

2.1. Instrument Characteristics of ABI and Best Track Description

The two ABIs onboard GOES-16 and GOES-17, respectively, are two same passive multi-channel imaging radiometers to provide imagery and radiometric information of Earth's surface, atmosphere, and cloud distributions in the Western Hemisphere. It is similar to the Advanced Himawari Imager (AHI) on board the Japanese satellite Himawari-8 [60]. The ABI concurrently produces a full disk for the whole Western Hemisphere every 15 min, and views the Earth with 16 different spectral bands, including two visible channels

(channels 1 and 2), four near-infrared channels (channels 3–6), and ten infrared channels (channels 7–16). The horizontal resolutions at nadir are 0.5 km for channel 2, 1 km for channels 1, 3, and 5, and 2 km for the remaining channels. Among 10 ABI infrared channels, channels 7, 11, 13, 14, and 15 are surface sensitive, channels 8, 9, and 10 are located in a water vapor absorption band, while channels 12 and 16 reside in ozone and carbon dioxide absorption bands, respectively [61]. The 10.3 μm infrared window channel 13 is less sensitive to water vapor absorption than other infrared window channels and therefore aids in cloud feature identification and estimation of cloud-top TB, whose primary application includes the identification of convective severe weather signatures and estimation of hurricane intensity [61]. The TB observations from the ABI channel 13 (10.3 μm) are used to determine TC centers and rainbands radii in this study.

The National Hurricane Center (NHC) maintains a climatology of all Atlantic tropical cyclones between 1851–2020, called the Hurricane Database2 (HURDAT2) [62]. For each TC, HURDAT2 contains six hourly estimates of its center positions, 1 min maximum sustained surface winds, minimum SLP, and the radii of 34, 50, and 64 kt wind speeds in four quadrants. The tropical cyclone extended the best track dataset (EBTRK) [63] was created by supplementing HURDAT2 with additional TC parameters, including the radius of maximum wind speed, eye diameter, pressure, and radius of the outermost closed isobar. The release time of the best track data generally lags by one year. The TC center position, the radii of 34 kt near-surface wind in four quadrants from HURDAT2, and the radius of the outermost closed isobar from the EBTRK are used in this study.

2.2. TC Case Description

Hurricanes Irma and Jose occurred in 2017 over the Atlantic Ocean and were selected as two TC examples to determine TC centers and rainband sizes. The reasons for choosing these two TC cases are as follows: Hurricane Irma had a typical northwestward track in the Northern Hemisphere, while Hurricane Jose made a clockwise loop over the southwestern Atlantic (Figure 1). Both Irma and Jose were long-lived hurricanes in September 2017. Jose reached category 4 strength on the Saffir-Simpson Hurricane Wind Scale, and Irma reached category 5 strength. Thus, the possible influences of TC intensity and motion on the accuracy of TC center positioning and TC rainband sizes can be illustrated for two TC cases with different tracks and intensity categories. The turning track of Jose is conducive to showing the hourly track by using hourly ABI observations. Criteria for TC times are the followings: all observed TCs are required to be over the ocean or before landfalling because the land surface will greatly affect the TC cloud structures in TB observations at ABI channel 13, TCs are required to be south of 40°N latitude to serve as a coarse exclusion of extratropical transition cases, and the maximum sustained wind of TCs must be greater than or equal to 50 kt (25 m s^{-1}) because organized rotation is rarely evident in the infrared imagery of ABI below the tropical storm (TS) intensity. A total of 45 times for Irma and 63 times for Jose at 6 h intervals meet the above conditions.

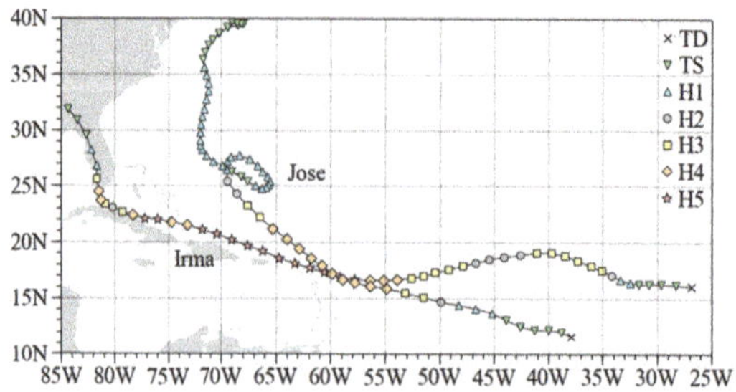

Figure 1. Best tracks (colored markers) of Hurricane Irma from 0000 UTC August 30 to 0000 UTC September 12, 2017 and Jose from 0600 UTC September 5 to 1200 UTC September 22, 2017. Intensity categories (colored markers) are shown in the legend.

3. Methods of Determining TC Center and Rainband Size

The azimuthal spectral analysis method proposed by Hu and Zou [35] was used to determine the center positions of Irma and Jose based on ABI channel-13 TB observations. It aims at obtaining a center position for which the symmetric component encompassing the center is the largest. Two steps are included in this method. The first step is to determine an initial first-guess position. Based on the fact that there is usually a region of low TB values within the eyewall and the inner core rainband close to the TC center, the first-guess position is determined as the position with the minimum value of TB observations averaged over a coarse grid of 1.5° × 1.5°. The second step is to carry out a set of azimuthal spectral analyses within circular areas of the radius R_i centered at different tryout centers. The subscript "i" of R_i is an index representing the tryout centers, which are distributed at 0.05° × 0.05° resolution and are contained in a 2° × 2° square area centered at the first-guess position. A detailed description of how to determine R_i is described in the next paragraph. Among all the tryout centers, the one which produces the largest value of the radially-mean amplitude of wavenumber 0 component (symmetric component) within the radial distance of R_i is taken as the TC center. A detailed explanation of the azimuthal spectral analysis method can be found in [35].

The azimuthal spectral analysis method for TC center positioning requires that the symmetric component of a TC within the radius R_i is always dominant. In a previous study [35] using POES microwave humidity sounder's observations, R_i was empirically set to 360 km. Taking an advantage of GOES infrared TB observations of high resolution and high cloud sensitivity, this study attempts to determine objectively the largest radial distance over the circumferences of radius R_i and smaller radii where the spectral analysis is carried out. An example is given in Figure 2 to illustrate such a procedure. Figure 2a shows the spatial distributions of ABI channel-13 TB observations at about 1200 UTC September 4, 2017, around Irma as well as the first-guess position determined from it. Variations of the observed TB values along the northwest-southeast (NW-SE) line passing through the first-guess position and the radial gradients of TB values along the NW-SE line are shown in Figure 2b. Within the radial distance of about 400 km, the TB values are relatively low and fluctuate greatly radially, reflecting the impacts of TC cloud distributions on TB. Beyond the 400 km radial distance, the TB values are relatively high and stable, which represents the variations of environmental TB distributions. Considering these features in TB observations, we propose the following method to determine objectively the largest radial distance (R_{fg}) from the first-guess position to the boundary of the TC-related cloud area. We firstly determine the inflection points where the radial gradient of TB switches from positive to negative from the center outward. If the average TB within 100 km on the radially outward side of the inflection point is greater than or equal to the average TB of the surrounding environment of a TC, the radial distance of the inflection point is regarded as the radius from the TC center to the boundary of TC related cloud area along the NW or SE direction. The average TB of the surrounding environment refers to the average TB between 800–1000 km away from the center [64]. The purpose of using average TB within the range of 100 km rather than the TB at a certain radial distance is to avoid the situation where a long and narrow high-valued TB region generated between the outer spiral rainbands is taken as the circumference separating TC from the surrounding environmental region. It can be seen from Figure 2b that the radius of the TC cloud area boundary in the northwest direction is determined as the radial distance where the inflection point of TB gradient is caused by the small-scale disturbance of TB. Obviously, the radial distance where the high-valued environment TB becomes stable is larger than the radial distance of the inflection point caused by the small-scale disturbance of TB. In order to avoid such incidents, a 50 km low pass Lanczos filter proposed by [65] was applied to obtain the filtered TB and its radial gradient, which are also shown in Figure 2b. The final radius R_{fg} is determined based on the filtered TB sequences. The two radii determined by the filtered TB observations along the NW and SE directions in Figure 2b are about 363.0 and 424.5 km, respectively. The radii of the boundaries of the TC cloud area at 24 azimuthal angles of 15° are determined by the same method mentioned above. Considering that the TC structure in the outer-core region is generally more asymmetric than that in the inner-core region [53], the smallest radius of the TC cloud area at 24 azimuthal angles is taken as the radius R_i to ensure the symmetric component is a dominant feature within R_i.

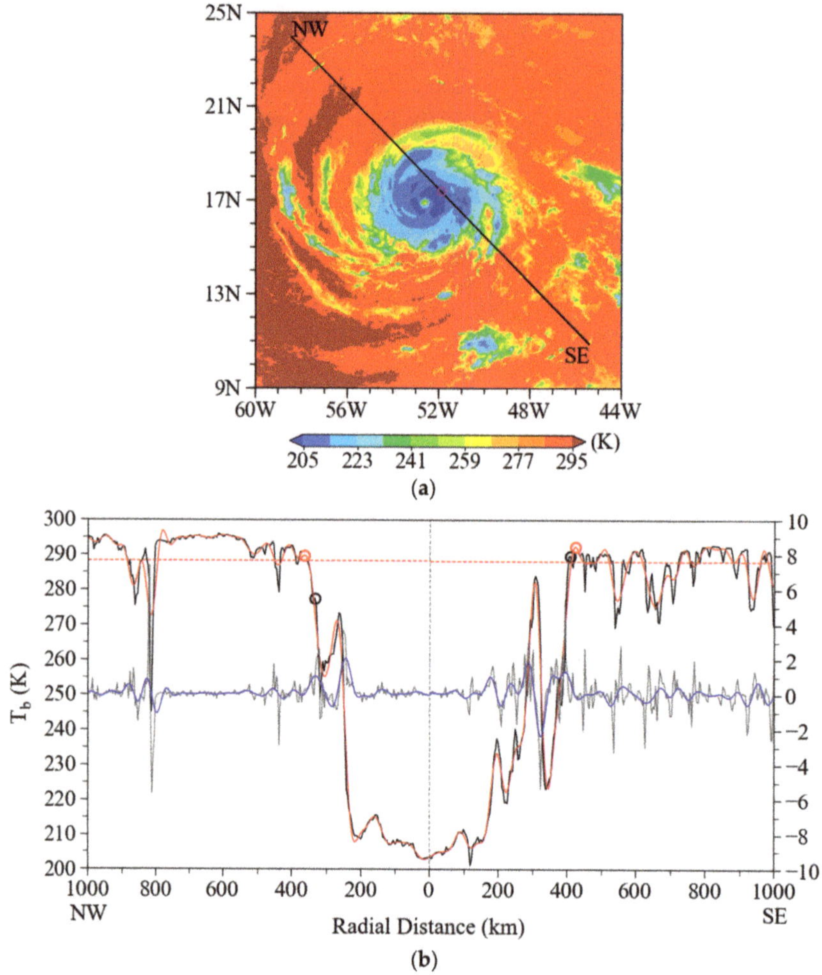

Figure 2. (**a**) Spatial distribution of ABI channel-13 brightness temperature (TB) observations (unit: K; color shading) within and around Irma at about 1200 UTC September 4, 2017 and a northwest-southeast (NW-SE) line passing through the first-guess position (magenta open circle) along which the original (black curve) and filtered (red curve) TB observations are shown in (**b**). Also shown in (**b**) are the radial gradients (unit: K) of original (gray curve) and the filtered TB observations (blue curve) along the same NW-SE line. The filtered TB value averaged over 800–1000 km distances from the first-guess position is indicated by a horizontal dashed line in red. The TB value at the boundary of TC cloud area determined by the original and filtered TB observations along the NW-SE line is indicated by black and red open circle, respectively.

Having the first-guess position and the radius R_i, an azimuthal spectral analysis is conducted at all tryout centers located in a $2° \times 2°$ square area centered on the first-guess position. The tryout center corresponding to the maximum value of the wavenumber-0 amplitude averaged over radial distances from 20 km to R_i was taken as the TC center. There are not very many TB observations along the circumferences within a radius of less than 20 km. Figure 3 shows variations of the distance between the best track center and the ABI-determined center with varying values of the largest radius for calculating the largest mean wavenumber-0 amplitude at about 1200 UTC September 4 and 1800 UTC August 30, 2017. We can see that the distance between the best track center and the ABI-determined center using the radius R_i, which is determined objectively by the ABI channel-13 TB observations, is reasonably small. At 1800 UTC August 30, 2017, the current result of the distance between the best track and the center using the objectively determined radius of R_i in this study is smaller than that of Hu and Zou [35] who used an empirical value of 360 km for R_i.

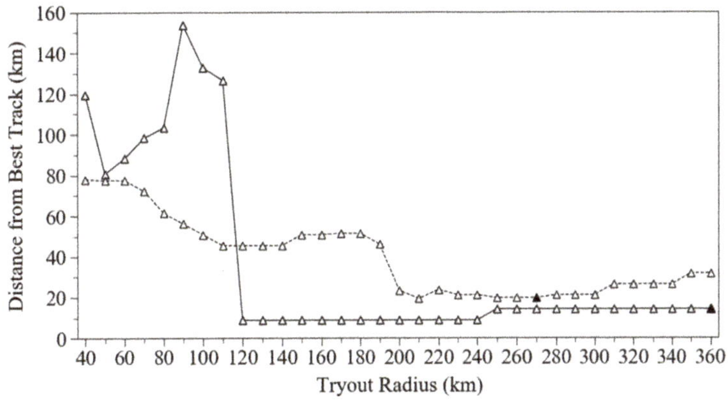

Figure 3. Variations of the distance between the best track center and the ABI-determined TC center when R_i is varied from 40 to 360 km at about 1200 UTC September 4, 2017 (solid curve) and 1800 UTC August 30, 2017 (dashed curve). The objectively determined R_i values of 270 and 360 km (solid triangles) are indicated by solid triangle symbols.

Having the TC center determined from TB observations, we proceeded to determine TC rainband sizes using the same method for determining the radius R_i. The only difference is using the TC center, instead of the first-guess position. The smallest and largest radii of the TC cloud area at 24 azimuthal angles are taken as the radius of the inner core rainband (R_{IR}) and the radius of the outer spiral rainband (R_{OR}). Two examples are provided to show the results of R_{IR} and R_{OR} for Irma at about 1200 UTC September 4, 2017, and 1800 UTC August 30, 2017. Figure 4a,b show the spatial distributions of TB observations within and around Irma, locations of the ABI-determined boundaries of the TC cloud area at 24 azimuthal angles of 15°, and two circles with radii of R_{IR} and R_{OR} at both times. It can be seen that almost all TC structures, including the outer spiral rainbands, are contained within the circumference of the radius of R_{OR}. The circle with a radius of R_{IR} describes mainly the nearly axisymmetric region with an inner core rainband. Therefore, R_{IR} describes approximately the size of the nearly axisymmetric TC inner-core region, and the R_{OR} reflects the maximum radial extent of TC rainbands. Figure 4c,d show the original and filtered TB observations and the radial gradients of filtered TB sequences along the directions in which the values of R_{IR} and R_{OR} are generated at about 1200 UTC September 4 and 1800 UTC August 30, 2017, respectively. Results from ABI-determined TC centers and the values of R_{IR} and R_{OR} at all times of Irma and Jose at 6 h intervals will be given and discussed in Section 4.

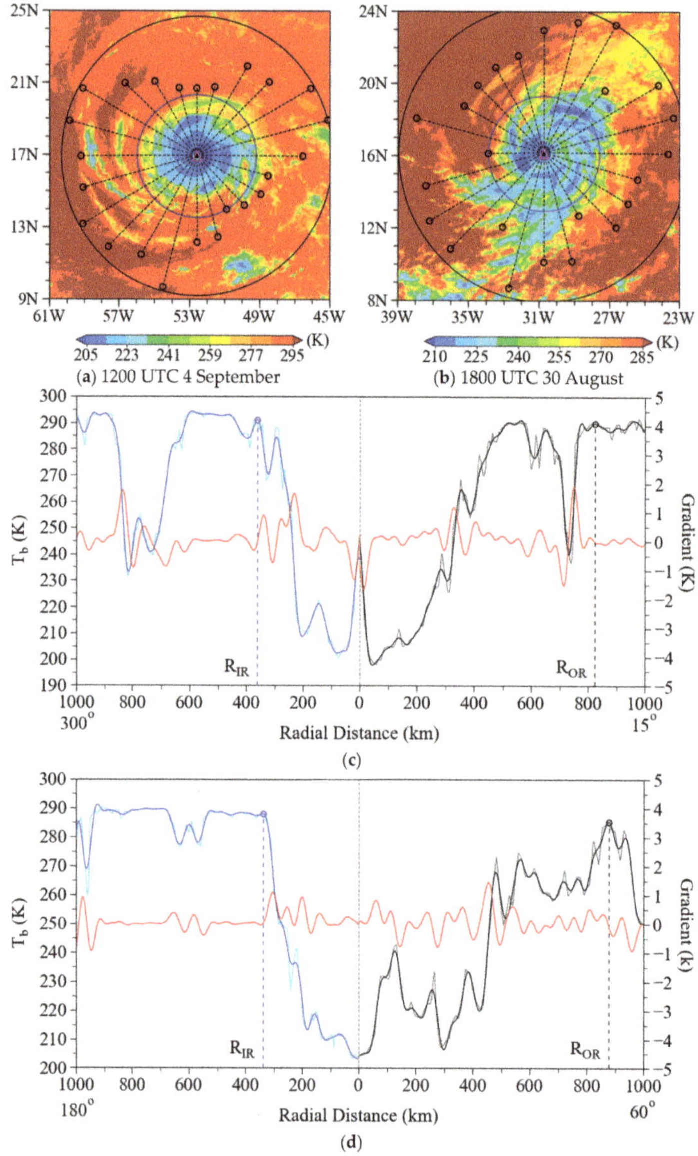

Figure 4. Spatial distributions of TB observations (color shading) around Irma, locations (small black open circles) of the ABI-determined boundaries of the TC cloud area at 24 azimuthal angles of 15° centered on the ABI-determined center (magenta open triangle), the R_{IR} (blue circle), and R_{OR} (black circle) at about (**a**) 1200 UTC September 4 and (**b**) 1800 UTC August 30, 2017. (**c**) The filtered TB sequences (blue and black curves), the original TB sequences (cyan and gray curves), and the radial gradients of filtered TB sequences (red curves) along the 300° and 15° directions counterclockwise from due east at about 1200 UTC September 4, 2017. The R_{IR} (blue dashed vertical line) and R_{OR} (black dashed vertical line) are marked. (**d**) The same as (**c**), with the exception for the time of 1800 UTC August 30, 2017.

4. Results of TC Center and Rainband Size

The TC center from the best track data refers to the center position near the surface, which usually is the location of minimum wind or minimum pressure. Although the definition of center position from the best track is quite different from the azimuthal symmetric center determined by the ABI channel-13 TB observations in this study, the best track is still the only reliable and available dataset to be compared for showing the accuracy of the ABI-determined center. The center-positioning results for Irma and Jose from ABI were compared with those from MHS/ATMS TB observations of microwave channel 18 (183.31 $\pm$ 7.0 GHz) of ATMS onboard S-NPP satellite and microwave channel 5 (190.31 GHz) of MHS onboard MetOp-A satellite using the same azimuthal center positioning method. Figure 5a shows the distances between the ABI-determined centers and ATMS channel 18- or MHS channel 5-determined TC centers from the best track centers of Irma during the time period from 1800 UTC on August 30 to 1800 UTC on September 10, 2017. Hurricane Irma was observed 31 times by the S-NPP and MetOp-A satellites. The mean distance from the best track center for the ABI-determined centers and ATMS- or MHS-determined centers are about 14.9 and 13.8 km, respectively. Therefore, compared with the best track data, the TC center-positioning differences of ABI channel-13 determined TC centers are similar to those of ATMS channel-18 or MHS channel-5 determined TC centers.

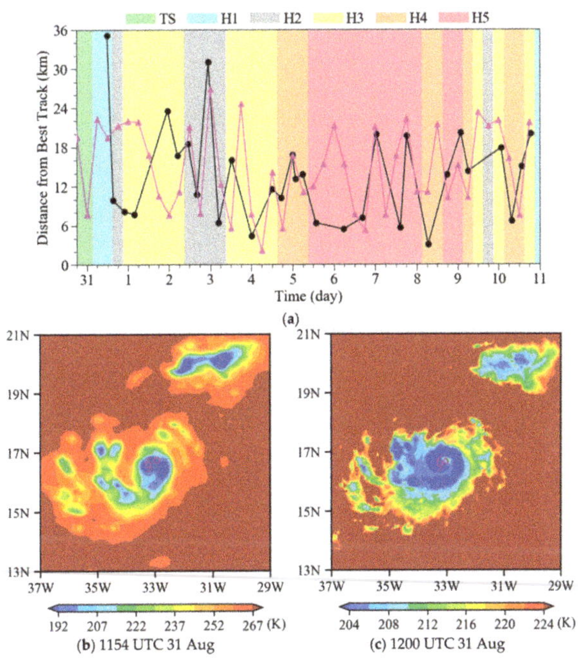

Figure 5. (**a**) Track errors of ABI channel 13-determined TC centers (solid triangles, magenta) and ATMS channel 18- or MHS channel 5-determined TC centers (solid circles, black) compared with the best track for Irma from 1800 UTC August 30 to 1800 UTC September 10, 2017. The mean distance between the best track centers for the ABI-determined centers or ATMS- or MHS-determined centers are about 14.9 and 13.8 km, respectively. Intensity categories (color shading) are shown in the legend above (**a**). (**b**,**c**) Spatial distributions of (**b**) MHS channel-5 and (**c**) ABI channel-13 TB observations (color shading; unit: K) around Hurricane Irma at about (**b**) 1154 and (**c**) 1200 UTC on August 31, 2017. The ABI-determined center (magenta open triangle), MHS-determined center (magenta open circle), and the best track center (magenta hurricane symbol) are also indicated.

However, a single polar-orbiting satellite can only observe the same TC twice daily at most, but a geostationary satellite can continuously observe it at 15 min or shorter time intervals. The horizontal resolution of ABI channel 13 is about 2 km, which is eight times higher than that of MHS/ATMS channels. This means that ABI can capture more detailed TC structures than MHS/ATMS. For example, Figure 5b,c show the spatial distributions of MHS channel-5 and ABI channel-13 TB observations around Irma, the ABI-determined center, MHS-determined center, and the best track center at about 1154 and 1200 UTC on August 31, 2017, respectively. It is seen that a clear TC eye was observed in the ABI channel-13 TB observations but not in the MHS channel-5 TB observations. The ABI-determined center is just located at the position of the TC eye and is close to the best track center. Unlike ABI infrared channel 13, microwave radiation from ATMS channel 18 and MHS channel 5 can penetrate cirrus and discern TC structures in the middle and lower troposphere [35]. To sum up, although TB observations of ABI channel 13 and ATMS channel 18/MHS channel 5 can all determine the TC centers by the azimuthal spectral analysis method, the ABI TB observations with high spatial-temporal resolutions have an advantage for the continuous center-positioning of TC.

The ABI-determined track and the best track of Irma from 1800 UTC August 30 to 1800 UTC September 10, 2017, are provided in Figure 6a. It is seen that these two tracks are quite close. Variations of R_{OR}, R_{IR} and $R_{OR} - R_{IR}$ for hurricane Irma are shown in Figure 6b. The mean values of R_{IR} and R_{OR} are about 348 and 877 km, respectively. The difference between R_{IR} and R_{OR} can be used to describe roughly the radial extent of the ring within which strong asymmetric outer spiral rainbands are located.

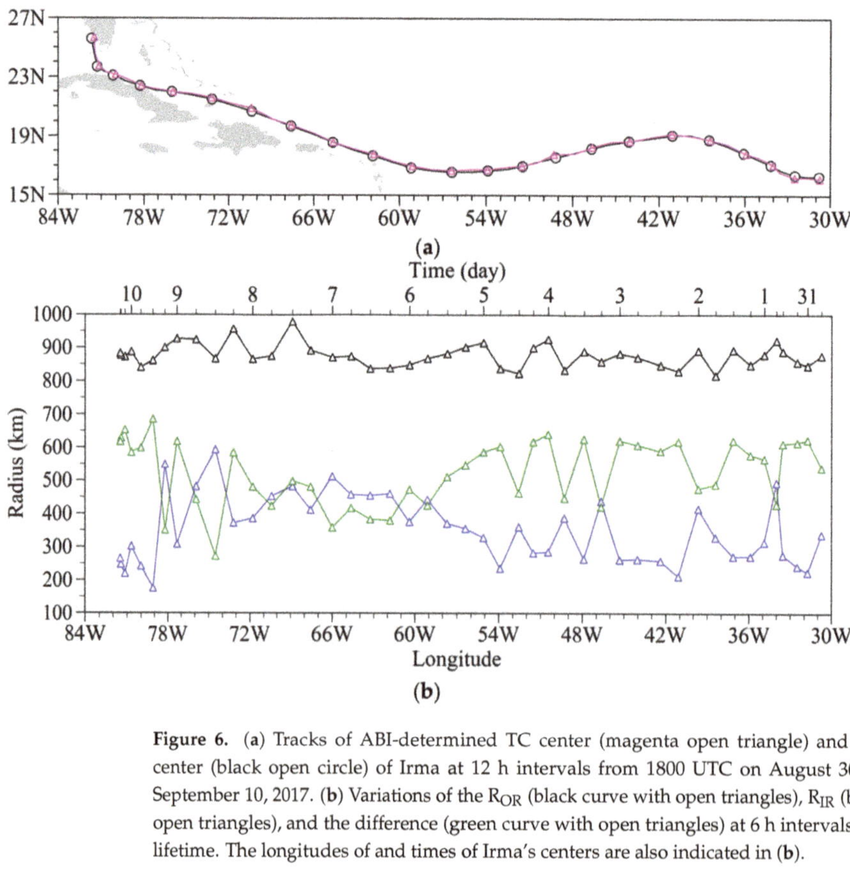

Figure 6. (**a**) Tracks of ABI-determined TC center (magenta open triangle) and the best track center (black open circle) of Irma at 12 h intervals from 1800 UTC on August 30 to 1800 UTC September 10, 2017. (**b**) Variations of the R_{OR} (black curve with open triangles), R_{IR} (blue curve with open triangles), and the difference (green curve with open triangles) at 6 h intervals during Irma's lifetime. The longitudes of and times of Irma's centers are also indicated in (**b**).

Hurricane Jose made a clockwise loop over the southwestern Atlantic during an overall northwest movement. It was a weaker hurricane than Irma on the whole. Figure 7 shows the distances between the ABI-determined centers and ATMS channel 18- or MHS channel 5-determined TC centers from the best track centers of Jose during the time period from 0600 UTC September 6 to 1800 UTC September 21, 2017. Jose was observed 42 times by the S-NPP and MetOp-A satellites. The mean distances from the best track center for the ABI-determined center and ATMS- or MHS-determined center are about 37.4 and 34.2 km, respectively. The mean distances are smaller when Jose was a category-2 or stronger hurricane than at weaker intensities. Compared with the best track, the mean distance of the ABI-determined center of Jose is larger than that of Irma, as shown in Figure 5a. Therefore, it can be inferred preliminarily that the higher the TC intensity, the smaller the difference between the ABI-determined center and the best track. Of course, a larger dataset is needed in order to confirm this. Figure 8a,b show the spatial distributions of TB observations around Jose, the locations of the ABI-determined boundaries of the hurricane cloud area at 24 azimuthal angles, R_{IR} and R_{OR} at about 1200 UTC September 12, 2017, and 1200 UTC September 15, 2017, respectively. It is seen that there is a large TB depression zone near the center of Jose without an obvious TC eye and spiral rainband structures, and the whole structure of Jose is strongly asymmetric. The circular area of radius R_{OR} contains almost all TC structures. Figure 8c shows the ABI-determined TC track and the best track of Irma from 0600 UTC September 6 to 1800 UTC September 21, 2017. Temporal variations of R_{OR} and R_{IR} and the difference between them are shown in Figure 8d. The average value of R_{IR} of Jose is about 243 km, which is smaller than the average value of R_{IR} (348 km) of a stronger hurricane like Irma. The average R_{OR} of Jose is about 873 km. Tables S1 and S2 list the best track centers, ABI-determined centers, R_{IR} and R_{OR} for Irma from 1800 UTC August 30 to 1800 UTC September 10, 2017 and Jose from 0600 UTC September 6 to 1800 UTC September 21, 2017 at 6 h intervals (a total of 108 TC times), respectively. Compared with the best track, the root mean square differences of ABI-determined centers for tropical storms and hurricanes are 45.35 and 29.06 km, respectively.

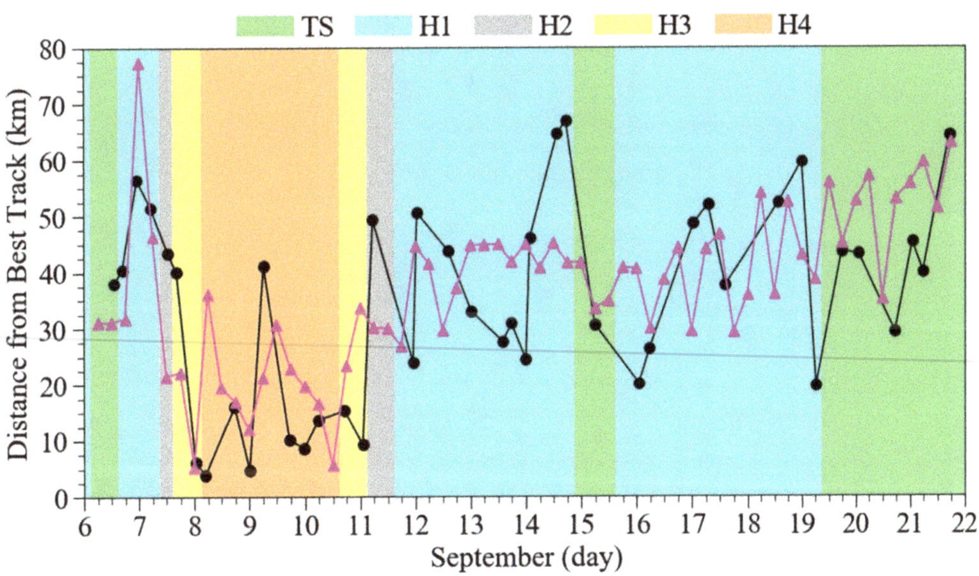

Figure 7. Same as Figure 5a except for Jose. The mean distances from the best track center for the ABI-determined centers and ATMS- or MHS-determined centers are about 37.4 and 34.2 km, respectively.

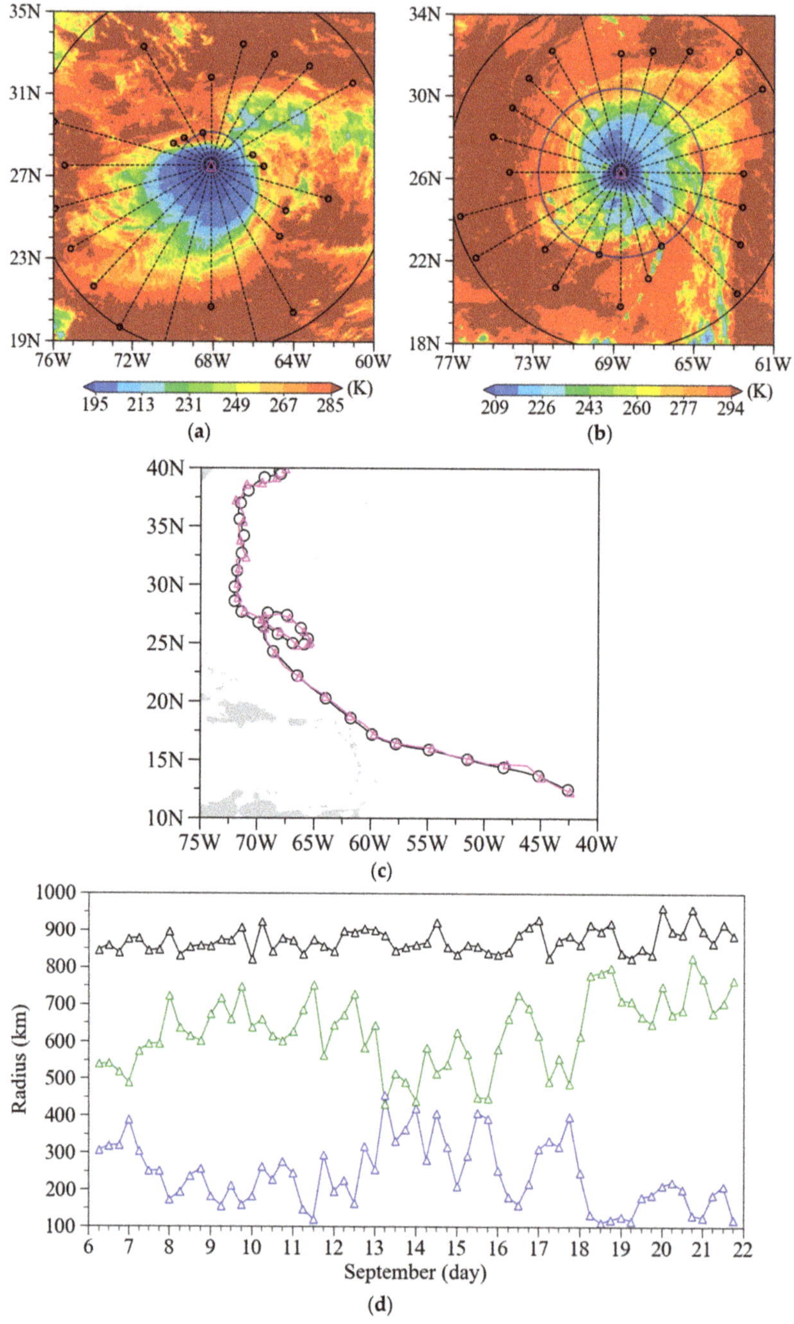

Figure 8. (**a**,**b**) Same as Figure 4a,b except for Jose at about (**a**) 1200 UTC September 12 and (**b**) 1200 UTC September 15, 2017 (**c**,**d**) Same as Figure 6a,b except for Jose from 0600 UTC September 6 to 1800 UTC September 21, 2017.

We may compare R_{OR} and R_{IR} with R34 and ROCI from the best track data. The temporal evolutions of R_{OR}, R_{IR}, R34 and ROCI are shown in Figure 9a,b for Irma and Jose, respectively. Since the best track data only provides R34 in the four quadrants of northeast, northwest, southwest, and southeast, R_{OR} and R_{IR} in Figure 9 refer to the maximum and minimum radii of the boundaries of the TC cloud area from the ABI-determined TC center in the same four quadrants. It can be seen that the R_{OR} is much larger than R34 and ROCI. For Irma (Figure 9a), R_{IR} and ROCI are nearly the same, but not identical. Both R_{IR} and ROCI are greater than R34. For Jose (Figure 9b), R_{IR} varies greatly with time and is close to ROCI and larger than R34 before 0000 UTC September 13, and then increases rapidly and is larger than R34 and ROCI between 0000 UTC September 13 and 1800 UTC September 15, and then decreases gradually and is smaller than R34 and ROCI after 1200 UTC September 18. The R34 and ROCI are measured based on near-surface wind and SLP, respectively. They are essentially different from R_{IR} and R_{OR}, which are estimated by ABI infrared TB observations of GOES-16 to describe the radial extents of TC cloud structures. Therefore, the relationship between them is not necessarily straightforward.

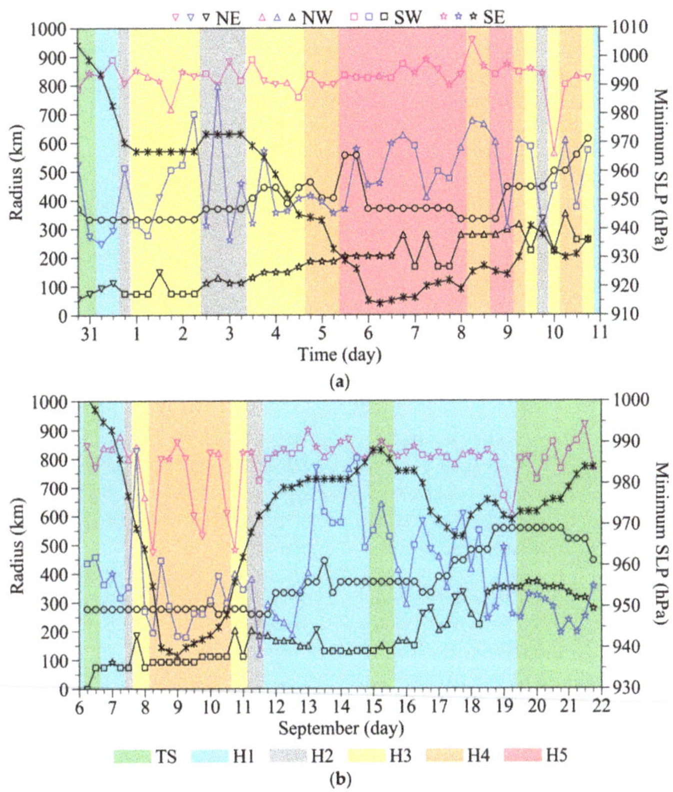

Figure 9. Temporal evolutions of R_{OR} (magenta markers indicating quadrants) and R_{IR} (blue markers indicating quadrants) along northeast (NE), northwest (NW), southwest (SW), and southeast (SE) directions, radius of the outermost closed isobar (black open circle), the radius of the 34 kt wind (black markers indicating quadrants) in the same quadrants as R_{IR}, and minimum SLP (black star markers) of (**a**) Hurricane Irma from 1800 UTC August 30 to 1800 UTC September 10, 2017 and (**b**) Jose from 0600 UTC September 6 to 1800 UTC September 21, 2017. Intensity categories are shown in color shading.

The track of ABI-determined centers at 1 h intervals and the best track at 6 h intervals from 0600 UTC 11 to 0600 UTC September 16, 2017, when Jose made a clockwise loop over the southwestern Atlantic are shown in Figure 10. The ABI-determined centers and the best track at 1800 UTC 13, 0000 UTC 14, 0600 UTC 14 and 1200 UTC 14 September are indicated for clarity purposes. It is seen that Jose oscillated around the mean path and swung occasionally in a small region, which may be related to the trochoidal oscillation of the TC track pointed out by previous studies [53,66–71]. The environmental steering flow, which is calculated as the deep-layer mean wind [72,73] based on the ERA5 winds within a circular area of 500 km radius centered at the ABI-determined center, is also shown in Figure 10. It is seen that Jose made a clockwise loop mainly under the control of the steering flow, while the small-scale oscillations are not consistent with the direction of the steering flow. Wu and Chen [74] pointed out that trochoidal oscillations associated with TC motion cannot be explained by the steering flow. Of course, the causes of the hourly oscillations of Jose require further investigation.

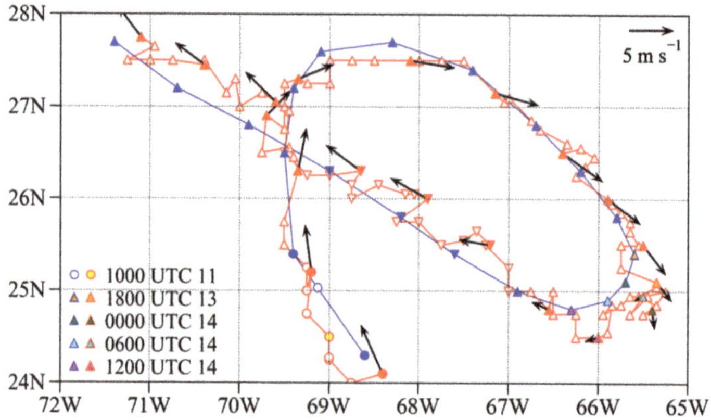

Figure 10. Track of ABI-determined centers at 1 h (red open marker) and 6 h (red solid marker) intervals and the best track (blue solid marker) of Jose from 0600 UTC September 11 to 0600 UTC September 16, 2017. Also indicated is the environmental steering flow (black arrow, the length of 5 m s^{-1}) calculated from the ERA5 wind reanalysis. The ABI-determined centers and the best track centers at 1000 UTC 11, 1800 UTC 13, 0000 UTC 14, 0600 UTC 14, and 1200 UTC 14 September 2017 are highlighted by the colors in the legend.

In order to further explore relationships among the track, the steering flow, and cloud distributions during the time period when Jose made a clockwise looping track, the spatial distributions of the ABI channel-13 TB observations, and the ERA5 geopotential height at 500 hPa within and around Jose at six selected times, along with the ABI-determined track and the steering flow of Jose at 6 h intervals, are put together in Figure 11. We see a large area of TB depression and a 500 hPa geopotential low near Jose' s center. The moving direction of Jose is consistent with the environmental steering flow. Actually, Jose became trapped between the large cyclonic circulation associated with Irma and the flow on the back side of a large mid-latitude closed low system centered off the coast of Atlantic Canada by September 12, 2017. The resulting steering pattern caused Jose to slow down considerably and make a clockwise loop over the southwestern Atlantic through September 16. Before 1200 UTC September 11, Jose moved northwest with a large area of TB depression around the TC center. After 0600 UTC September 12, there is a relatively small region of TB depression, indicating the isolated convection zone appeared northeast of the main TB depression zone. Afterward, Jose began to move northeastward. At 0000 UTC September 13, small discrete convections were separated from the southeast and south sides of the main TB depression zone, and the motion of Jose shifted from going east to southeast. At 1200 UTC September 13, an isolated convection appeared on the south side of the main TB depression zone, and Jose began to move southward. By 0600 UTC September 14, discrete small convections around the south side of the main TB depression zone dissipated, and the motion direction of Jose shifted from southward to westward. After 1800 UTC September 14, Jose restored a large TB depression around its center and kept moving northwest. Thus, it can be seen that the asymmetric convective activities within the TC circulation are associated with the TC movement to some extent, which is consistent with previous studies [75–77]. Willoughby et al. [75] and Holland and Lander [76] pointed out that the mesoscale convective systems rotating within the TC core can have an impact on the short-term motion of TC. Ritchie and Elsberry [77] acknowledged the possibility that significant changes of TC track in the order of tens to even hundreds of kilometers over several days could result from an interaction of TC with a persistent mesoscale convective system rotating within TC circulation.

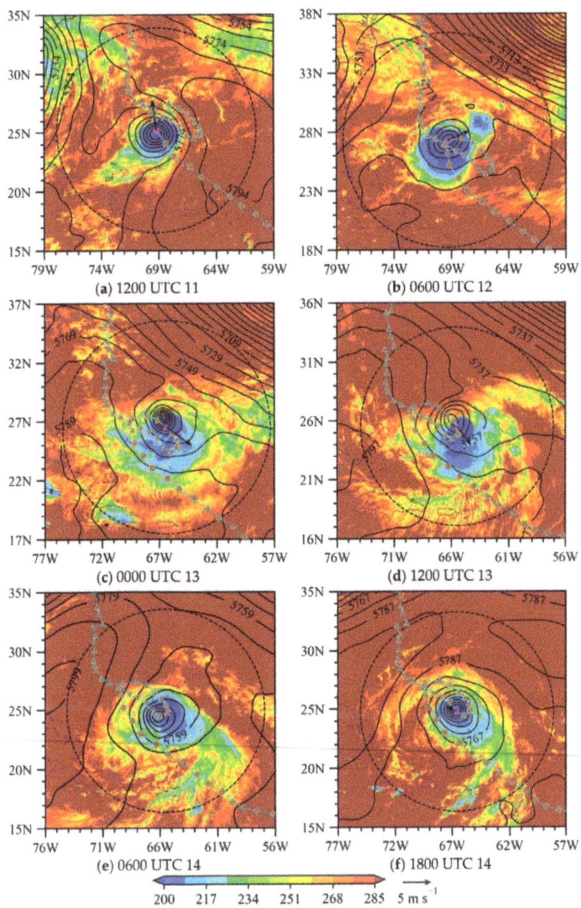

Figure 11. Spatial distributions of ABI channel 13-TB observations (color shading; unit: K) and ERA5 geopotential height (contour; unit: m) at 500 hPa at (**a**) 1200 UTC 11, (**b**) 0600 UTC 12, (**c**) 0000 and (**d**) 1200 UTC 13, (**e**) 0600 and (**f**) 1800 UTC September 14, 2017. Also shown are the ABI-determined track (cyan curve with solid markers indicating hurricane intensity) of Hurricane Jose, as well as the ABI-determined center (magenta solid marker), R_{OR} (dashed black circle), and the steering flow (black arrow, the length of 5 m s^{-1}) at the same time as TB observations.

5. Discussions of Symmetric and Asymmetric Components in ABI TB Observations

In order to compare the spatial distributions and amplitudes of symmetric and asymmetric components centered at either the ABI-determined center or the best track center, we selected a case of a 60 km distance between the ABI-determined center and the best track center. Figure 12a shows the spatial distribution of ABI channel-13 TB observations within and around Hurricane Jose, the ABI-determined cloud boundaries, the ABI-determined center, and the best track at 1000 UTC September 11, 2017. The distance between the ABI-determined center and the best track center is highlighted in Figure 10. Figure 12b compares radial variations of percentage amplitude of wavenumbers 0–3 over the sum of wavenumbers 0–10 amplitudes with the ABI-determined center and best track as assumed by Jose's centers at about 1000 UTC September 11, 2017. Taking the ABI-determined center for spectral analysis, the wavenumber-0 amplitude always accounts for the largest proportion (over 50%) of all wavenumbers' amplitudes within the 20–300 km radial distance, indicating that the symmetric component dominates the whole pattern of the TC within the 300 km radial distance. The wavenumber-1 amplitude is more than 10% beyond the radial distance of 120 km and keeps a stable value of about 20%. If the best track is used as the center, wavenumber-0 amplitudes are more than 40% larger than that of wavenumber 1 within a 20–100 km radial distance. However, the wavenumber-1 amplitude exceeds the wavenumber-0 amplitude beyond the radial distance of 100 km and keeps a stable value of about 45%. The surpassing of wavenumber-1 to wavenumber-0 amplitudes indicates that the wavenumber-1 asymmetric component dominates the entire pattern of the TC within the corresponding radial distance. By comparison, the amplitude of the wavenumber 0 component (symmetric component) centered on the ABI-determined center is always larger than that of the wavenumber 0 component centered on the best track center, but it is just the opposite for wavenumbers 1 and 3. The amplitude of the wavenumber-2 asymmetric component centered on the ABI-determined center is smaller than that of the wavenumber-2 component centered on the best track center within the radial distance of 100 km, but the amplitudes of the two are very similar outside the radial distance of 100 km.

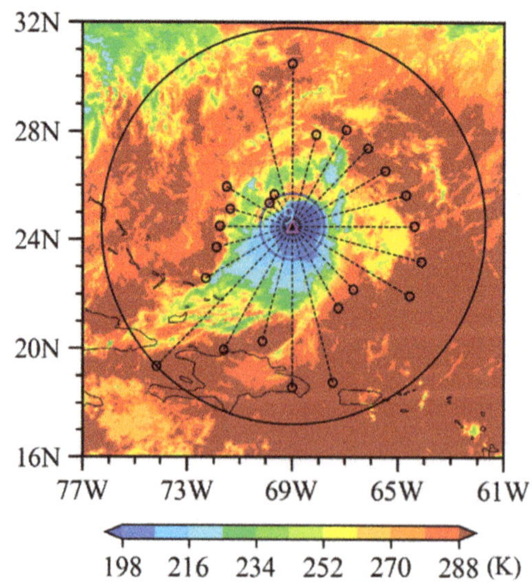

(a)

Figure 12. *Cont.*

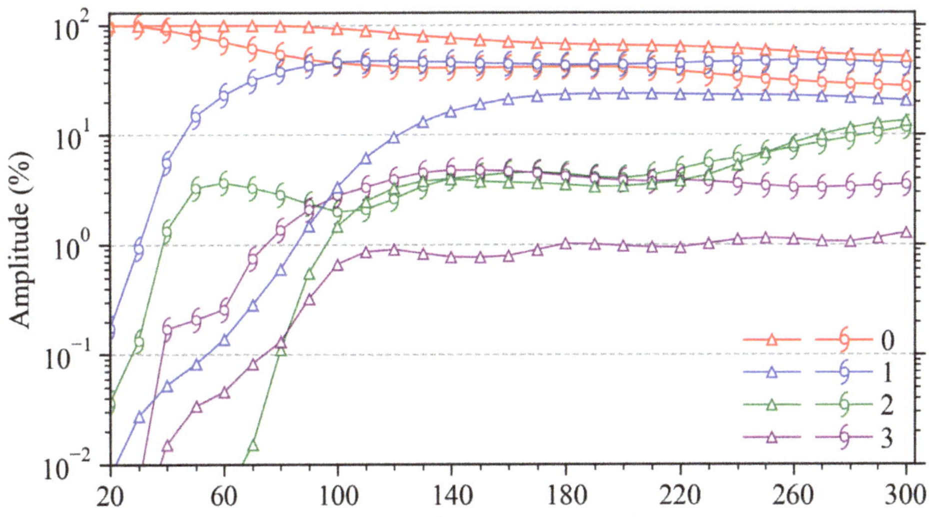

(b)

Figure 12. (**a**) Spatial distribution of ABI channel 13 TB observations (color shading; unit: K) around Hurricane Jose at 1000 UTC September 11, 2017. The ABI-determined center (magenta open triangle), the best track center (cyan hurricane symbol), the R_{IR} (blue circle), and R_{OR} (black circle) are also indicated. (**b**) Radial variations of the azimuthal wavenumbers 0–3 amplitude percentages (solid curves; %) with the ABI-determined center (open triangle) and best track (hurricane symbol) as assumed by Jose's centers at about 1000 UTC September 11, 2017.

Figure 13 shows the original TB, filtered TB, and radial gradients of the filtered TB observations along the directions where R_{IR} and R_{OR} are determined using the ABI-determined center (Figure 13a) or the best track center (Figure 13b) at about 1000 UTC September 11, 2017. Figure 14 shows the spatial distributions of wavenumbers 0–3 components of ABI channel-13 TB observations around Jose, centered on the ABI-determined center and the best-track center at 1000 UTC 11 September 2017. Also indicated in Figure 14 are two circles of radii of R_{IR} and R_{OR} determined in Figure 13. For wavenumbers 1–3 asymmetric components centered on the ABI-determined TC center, it is seen that R_{IR} coincides with the radius where the strongest wavenumber 1 asymmetry is located, and the annulus defined by the two circles of radii of R_{OR} and R_{IR} contains the main asymmetries of wavenumbers 1–3, which mainly reflect the asymmetric outer spiral rainbands. Weak wavenumbers 1–3 asymmetries also appear inside the circle at the radius of R_{IR}, which may be related to the propagation of vortex Rossby waves within the TC inner-core region [12,13]. For wavenumbers 1–3 asymmetric components centered on the best track center, the circular region with the radius of R_{IR} contains the main strong wavenumber 1 asymmetries and some strong wavenumbers 2 and 3 asymmetries. When R_{IR} is determined based on the ABI-determined TC center, the innermost azimuthal waves of wavenumbers 1–3 are nicely distributed along or within the radial distance R_{IR}. However, if R_{IR} is determined based on the best track, the azimuthal waves of wavenumbers 1–3 are found at several radial distances less than R_{IR}.

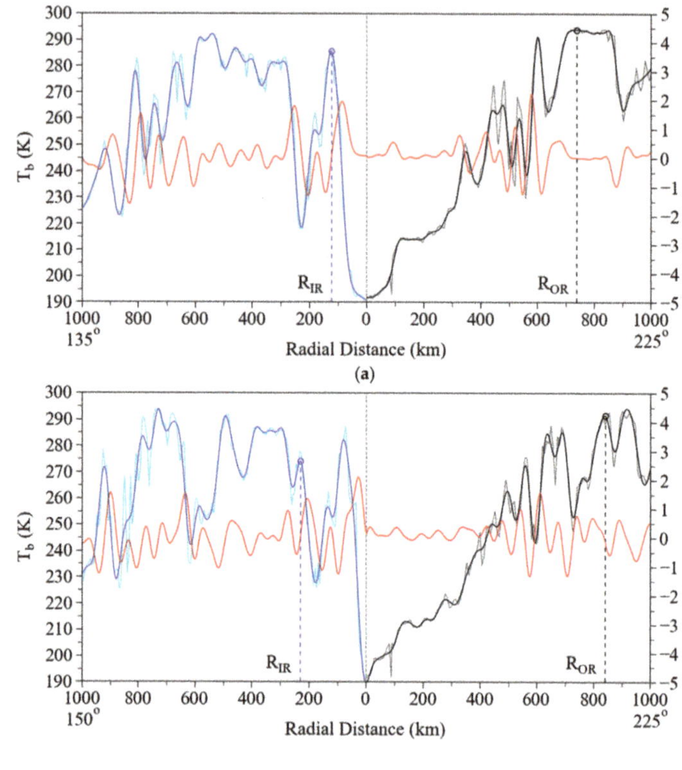

Figure 13. (**a**) Same as Figure 4c or Figure 4d except for Jose at about 1000 UTC September, 11 2017. (**b**) Same as (**a**) except for the TB sequence centered at the best track center.

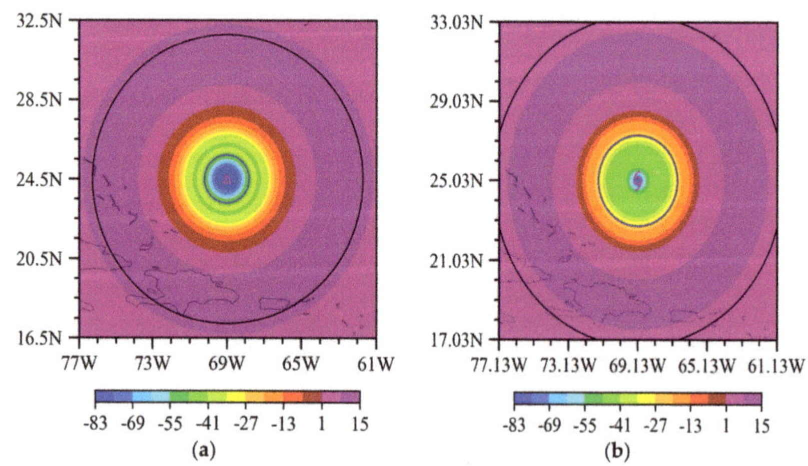

Figure 14. *Cont.*

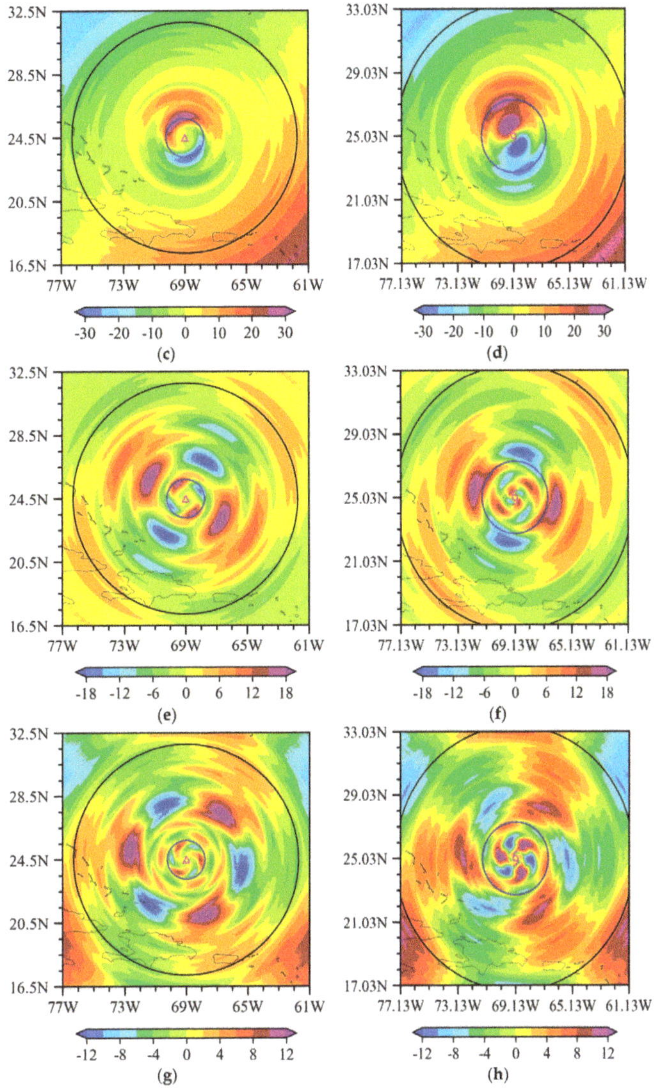

Figure 14. Spatial distributions of (**a**–**h**) wavenumbers 0–3 components (color shading; unit: K) of ABI channel-13 TB observations around Hurricane Jose centered at the ABI-determined center (magenta open triangle; left panel) and best track (magenta hurricane symbol; right panel) at 1000 UTC September 11, 2017. The R_{IR} (blue circle) and R_{OR} (black circle) are also indicated.

The average value of wavenumber-0 amplitudes and maximum wavenumbers 1, 2, and 3 amplitudes from 20 km to the outermost radius R_i of the azimuthal spectral analysis with the ABI-determined center and the best track center as the center of spectral analysis for 45 times of Hurricane Irma and 63 times of Hurricane Jose are shown in Figure 15. The distances between the ABI-determined center and the best track center are also indicated in Figure 15. It is seen that the average wavenumber-0 amplitude with the ABI-determined center are consistently larger than or equal to those with the best track center. The larger the average wavenumber-0 amplitude, the smaller center positioning difference. Specifically, the difference is less than 35 km when the average amplitude percentage of the wavenumber

0 component centered on the ABI-determined center (the best-track center) is greater than 60% (50%). The maximum amplitude of the wavenumber-1 component centered at the ABI-determined center is mostly less than or equal to that centered at the best track. Most maximum amplitudes of wavenumbers-2 or -3 components centered on the ABI-determined center and those centered on the best-track center have no obvious difference. Therefore, the amplitudes of wavenumbers 0 and 1 are sensitive to the location of the TC center, while wavenumbers 2 and 3 are fairly robust, which is consistent with the research result of [7].

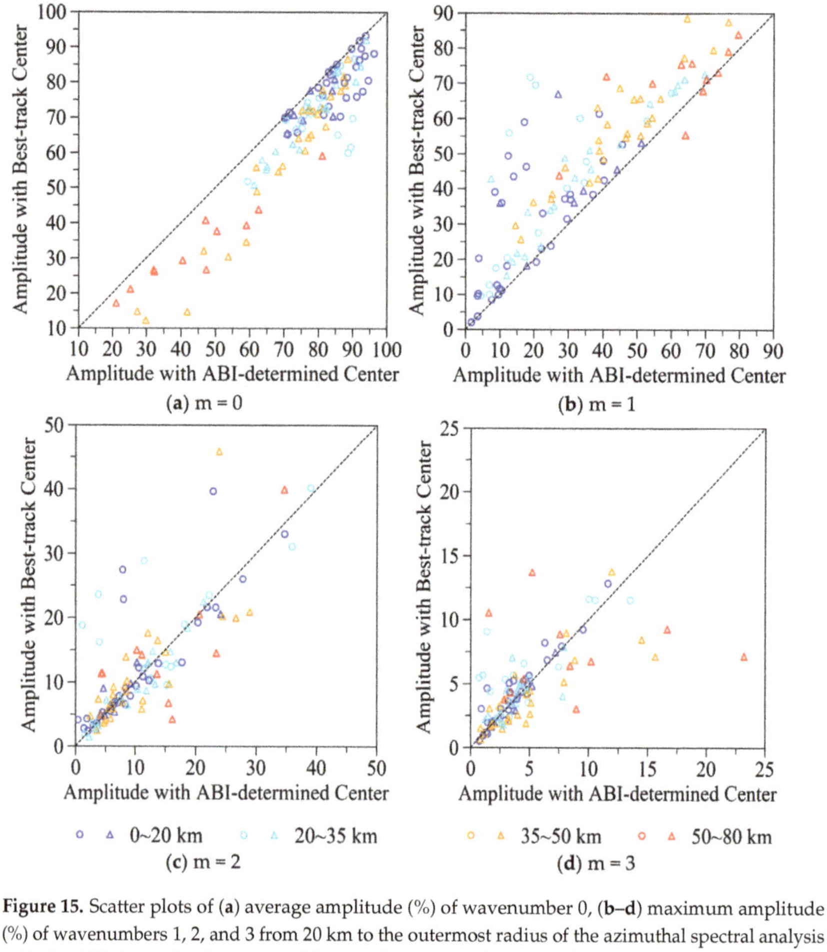

Figure 15. Scatter plots of (**a**) average amplitude (%) of wavenumber 0, (**b–d**) maximum amplitude (%) of wavenumbers 1, 2, and 3 from 20 km to the outermost radius of the azimuthal spectral analysis with the ABI-determined center and best track center as the center for the spectral decomposition for 45 times of Hurricane Irma (open circle) and 63 times of Hurricane Jose (open triangle). The differences (km) between the ABI-determined centers and best-track centers are indicated by colors.

6. Conclusions

TB observations on infrared channel 13 (10.3 μm) of ABI are used for determining the center positions and rainband sizes of Hurricanes Irma and Jose (2017). The azimuthal spectral analysis method is employed to determine the azimuthally symmetric center of TCs. With the ABI-determined azimuthally symmetric center as the TC center, the radii of inner core rainbands and outer spiral rainbands, namely R_{IR} and R_{OR}, are determined objectively based on radial gradients of TB values at different azimuthal angles in the infrared TB field observed by geostationary satellites, which describe the size of the TC inner-core region and the maximum radial extent of TC rainbands, respectively. The ABI-determined center positions of Hurricane Irma and Jose at 6 h intervals are compared with the best track in order to verify the accuracy of the azimuthal spectral analysis method for TC center positioning. The center position of Jose is determined hourly by the ABI during the period when Jose made a clockwise loop, and the hourly track is found to oscillate around the mean path and swing occasionally in a small region. The moving track, the steering flow, and TC cloud convections during the time period when Jose made a circled track are discussed, suggesting that the turning track of Jose is mainly under the control of steering flow, and mesoscale convective systems rotating within TC circulation can also influence the direction of TC movement to some extent. The R_{OR} determined by TB observations is found to be much larger than the radii of 34 kt near-surface wind and the outermost closed isobar, and R_{IR} is close but not identical to the radius of the outermost closed isobar and usually coincides with the radius where the strongest wavenumber 1 asymmetry is located. The annulus defined by the two circles with radii of R_{IR} and R_{OR} is the asymmetric area of rainbands described by azimuthal wavenumbers 1–3. Finally, the spatial distributions and amplitudes of symmetric and asymmetric components centered at the ABI-determined center and the best-track center are compared. The innermost azimuthal waves of wavenumbers 1–3 are well distributed along or within the circumference of R_{IR} that is determined based on the ABI-determined TC center. If R_{IR} is calculated using the best track, the azimuthal waves of wavenumbers 1–3 can be found at several radial distances less than R_{IR}. Besides, it is found that the larger the average wavenumber-0 amplitude, the smaller the difference between the ABI-determined center and the best-track center. The average amplitude of the wavenumber-0 (wavenumber-1) component centered on the ABI-determined center is always greater (smaller) than or equal to that centered on the best-track center.

The center position and rainband sizes of TCs estimated by the infrared TB observations from GOES-16 in this study can be used as the parameters required by specifying a bogus vortex for hurricane initialization and meeting the requirements of verifying the propagation mechanism of vortex Rossby waves by satellite observations. The azimuthal spectral analysis method for TC center positioning requires a TC intensity of 25 m s^{-1} or above in this study to ensure that the whole pattern of a TC is dominated by the symmetric component. The scenario where the asymmetric component dominates the whole pattern of a TC needs further study. The relationship between the radius of the inner core rainband measured in this study and the TC intensity has not been explored. Considering that TC intensity is usually not closely related to a single TC size parameter [8], more parameters that can describe the size of a TC, such as the TC eye diameter, should be estimated by geostationary satellites to explore the relationship between TC size and intensity in the future. The TB observations provided by geostationary satellites with high temporal and horizontal spatial resolution can fully reveal the horizontal spatial structures of TCs and the evolution of TC cloud structures over time, which lays the observation foundation for our future focus on the verification of the propagation mechanism of vortex Rossby waves in TCs by geostationary satellite observations.

Supplementary Materials: The following supporting information can be downloaded at: https://www.mdpi.com/xxx/s1, Table S1: the best track center, ABI-determined center, R_{IR} and R_{OR} of Hurricane Irma from 1800 UTC August 30 to 1800 UTC September, 10 2017; Table S2: same as Table S1 except for Hurricane Jose from 0600 UTC September 6 to 1800 UTC September 21, 2017.

Author Contributions: Conceptualization, X.Z. and Y.H.; methodology, X.Z.; software, Y.H. and X.Z.; validation, Y.H. and X.Z.; formal analysis, Y.H. and X.Z.; investigation, Y.H. and X.Z.; resources, X.Z.; data curation, Y.H.; writing—original draft preparation, Y.H.; writing—review and editing, X.Z.; visualization, Y.H.; supervision, X.Z.; project administration, X.Z.; funding acquisition, X.Z. All authors have read and agreed to the published version of the manuscript.

Funding: This research was funded by the National Key R&D Program of China (Grant 2018YFC1507004) and the Postgraduate Research and Practice Innovation Program of Jiangsu Province (Grant KYCX22_1131).

Data Availability Statement: The data that support the findings of this study are available from the author upon reasonable request.

Acknowledgments: We would like to acknowledge the suggestions given by reviewers and editor.

Conflicts of Interest: The authors declare no conflict of interest.

References

1. Zou, X.; Xiao, Q. Studies on the initialization and simulation of a mature hurricane using a variational bogus data assimilation scheme. *J. Atmos. Sci.* **2000**, *57*, 836–860. [CrossRef]
2. Park, K.; Zou, X. Toward developing an objective 4DVAR BDA scheme for hurricane initialization based on TPC observed parameters. *Mon. Weather Rev.* **2004**, *132*, 2054–2069. [CrossRef]
3. Dvorak, V.F. *Tropical Cyclone Intensity Analysis Using Satellite Data*; National Oceanic and Atmospheric Administration, National Environmental Satellite, Data, and Information Service: Washington, DC, USA, 1984; pp. 2–15.
4. Olander, T.L.; Velden, C.S. The Advanced Dvorak Technique (ADT)—Continued development of an objective scheme to estimate tropical cyclone intensity using geostationary infrared satellite imagery. *Weather Forecast.* **2007**, *22*, 287–298. [CrossRef]
5. Montgomery, M.T.; Kallenbach, R.J. A theory for vortex Rossby-waves and its application to spiral bands and intensity changes in hurricanes. *Q. J. R. Meteorol. Soc.* **1997**, *123*, 435–465. [CrossRef]
6. Schubert, W.H.; Montgomery, M.T.; Taft, R.K.; Guinn, T.A.; Fulton, S.R.; Kossin, J.P.; Edwards, J.P. Polygonal eyewalls, asymmetric eye contraction, and potential vorticity mixing in hurricanes. *J. Atmos. Sci.* **1999**, *56*, 1197–1223. [CrossRef]
7. Reasor, P.D.; Montgomery, M.T.; Marks, F.D., Jr.; Gamache, J.F. Low-wavenumber structure and evolution of the hurricane inner core observed by airborne dual-Doppler radar. *Mon. Weather Rev.* **2000**, *128*, 1653–1680. [CrossRef]
8. Guo, X.; Tan, Z.M. Tropical cyclone fullness: A new concept for interpreting storm intensity. *Geophys. Res. Lett.* **2017**, *44*, 4324–4331. [CrossRef]
9. Liang, Z.; Ding, J.; Fei, J.; Cheng, X.; Huang, X. Direct/indirect effects of aerosols and their separate contributions to Typhoon Lupit (2009): Eyewall versus peripheral rainbands. *Sci. China Earth Sci.* **2021**, *64*, 2113–2128. [CrossRef]
10. Li, X.; Cheng, X.; Fei, J.; Huang, X.; Ding, J. The modulation effect of sea surface cooling on the eyewall replacement cycle in Typhoon Trami (2018). *Mon. Weather Rev.* **2022**, *150*, 1–50. [CrossRef]
11. Ma, Z.; Fei, J. A comparison between moist and dry tropical cyclones: The low effectiveness of surface sensible heat flux in storm intensification. *J. Atmos. Sci.* **2022**, *79*, 31–49.
12. Corbosiero, K.L.; Molinari, J.; Aiyyer, A.R.; Black, M.L. The structure and evolution of Hurricane Elena (1985). Part II: Convective asymmetries and evidence for vortex Rossby waves. *Mon. Weather Rev.* **2006**, *134*, 3073–3091. [CrossRef]
13. Guimond, S.R.; Reasor, P.D.; Heymsfield, G.M.; McLinden, M.M. The dynamics of vortex Rossby waves and secondary eyewall development in Hurricane Matthew (2016): New insights from radar measurements. *J. Atmos. Sci.* **2020**, 349–374. [CrossRef]
14. Zhuo, J.Y.; Tan, Z.M. Physics-augmented deep learning to improve tropical cyclone intensity and size estimation from satellite imagery. *Mon. Weather Rev.* **2021**, *149*, 2097–2113. [CrossRef]
15. Ryglicki, D.R.; Hart, R.E. An investigation of center-finding techniques for tropical cyclones in mesoscale models. *J. Appl. Meteorol. Climatol.* **2015**, *54*, 825–846. [CrossRef]
16. Jones, S.C. The evolution of vortices in vertical shear. I: Initially barotropic vortices. *Q. J. R. Meteorol. Soc.* **1995**, *121*, 821–851. [CrossRef]
17. Frank, W.M.; Ritchie, E.A. Effects of environmental flow upon tropical cyclone structure. *Mon. Weather Rev.* **1999**, *127*, 2044–2061. [CrossRef]
18. Riemer, M.; Montgomery, M.T.; Nicholls, M.E. A new paradigm for intensity modification of tropical cyclones: Thermodynamic impact of vertical wind shear on the inflow layer. *Atmos. Chem. Phys.* **2010**, *10*, 3163–3188. [CrossRef]
19. Ryglicki, D.R.; Hart, R.E. An Investigation of Metrics Used to Determine the Center of Model Tropical Cyclones. In Proceedings of the 30th Conference on Hurricanes and Tropical Meteorology, Ponte Vedra, FL, USA, 16 April 2012.
20. Nguyen, L.T.; Molinari, J.; Thomas, D. Evaluation of tropical cyclone center identification methods in numerical models. *Mon. Weather Rev.* **2014**, *142*, 4326–4339. [CrossRef]
21. Braun, S.A. A cloud-resolving simulation of Hurricane Bob (1991): Storm structure and eyewall buoyancy. *Mon. Weather Rev.* **2002**, *130*, 1573–1592. [CrossRef]
22. Braun, S.A.; Montgomery, M.T.; Pu, Z. Highresolution simulation of Hurricane Bonnie (1998). Part I: The organization of eyewall vertical motion. *J. Atmos. Sci.* **2006**, *63*, 19–42. [CrossRef]
23. Wu, L.; Braun, S.A.; Halverson, J.; Heymsfield, G. A numerical study of Hurricane Erin (2001). Part I: Model verification and storm evolution. *J. Atmos. Sci.* **2006**, *63*, 65–86. [CrossRef]
24. Fischer, M.S.; Rogers, R.F.; Reasor, P.D. The rapid intensification and eyewall replacement cycles of hurricane irma (2017). *Mon. Weather Rev.* **2020**, *148*, 981–1004. [CrossRef]
25. Neldar, J.A.; Mead, R. A simplex method for function minimization. *Comput. J.* **1965**, *7*, 308–313. [CrossRef]
26. Marks, F.D., Jr.; Houze, R.A., Jr.; Gamache, J.F. Dual-aircraft investigation of the inner core of Hurricane Norbert. Part I: Kinematic structure. *J. Atmos. Sci.* **1992**, *49*, 919–942. [CrossRef]

27. Reasor, P.D.; Eastin, M.D. Rapidly intensifying Hurricane Guillermo (1997). Part II: Resilience in shear. *Mon. Weather Rev.* **2012**, *140*, 425–444. [CrossRef]
28. Reasor, P.D.; Rogers, R.; Lorsolo, S. Environmental flow impacts on tropical cyclone structure diagnosed from airborne Doppler radar composites. *Mon. Wea. Rev.* **2013**, *141*, 2949–2969. [CrossRef]
29. Willoughby, H.E.; Chelmow, M.B. Objective determination of hurricane tracks from aircraft observations. *Mon. Weather Rev.* **1982**, *110*, 1298–1305. [CrossRef]
30. Velden, C.S.; Olander, T.L.; Zehr, R.M. Development of an objective scheme to estimate tropical cyclone intensity from digital geostationary satellite infrared imagery. *Weather Forecast.* **1998**, *13*, 172–186. [CrossRef]
31. Olander, T.L.; Velden, C.S.; Turk, M.A. Development of the advanced objective Dvorak technique (AODT)—Current progress and future directions. In Proceedings of the 25th Conference on Hurricanes and Tropical Meteorology, San Diego, CA, USA, 28 April 2002; pp. 585–586.
32. Olander, T.; Velden, C.S.; Kossin, J. The Advanced Objective Dvorak technique (AODT): Latest upgrades and future directions. In Proceedings of the 26th Conference on Hurricanes and Tropical Meteorology, Miami, FL, USA, 3–7 May 2004; pp. 294–295.
33. Wimmers, A.J.; Velden, C.S. Objectively determining the rotational center of tropical cyclones in passive microwave satellite imagery. *J. Appl. Meteorol. Climatol.* **2010**, *49*, 2013–2034. [CrossRef]
34. Wimmers, A.J.; Velden, C.S. Advancements in objective multi-satellite tropical cyclone center fixing. *J. Appl. Meteorol. Climatol.* **2016**, *55*, 197–212. [CrossRef]
35. Hu, Y.; Zou, X. Tropical cyclone center positioning using single channel microwave satellite observations of brightness temperature. *Remote Sens.* **2021**, *13*, 2466. [CrossRef]
36. Houston, S.H.; Shaffer, W.A.; Powell, M.D.; Chen, J. Comparisons of HRD and SLOSH surface wind fields in hurricanes: Implications for storm surge and wave modeling. *Weather Forecast.* **1999**, *14*, 671–686. [CrossRef]
37. Powell, M.D.; Reinhold, T.A. Tropical cyclone destructive potential by integrated kinetic energy. *Bull. Am. Meteorol. Soc.* **2007**, *88*, 513–526. [CrossRef]
38. Irish, J.L.; Resio, D.T.; Ratcliff, J.J. The influence of storm size on hurricane surge. *J. Phys. Oceanogr.* **2008**, *38*, 2003–2013. [CrossRef]
39. Kidder, S.Q.; Knaff, J.A.; Kusselson, S.J.; Turk, M.; Ferraro, R.R.; Kuligowski, R.J. The tropical rainfall potential (TRaP) technique. Part I: Description and examples. *Weather Forecast.* **2005**, *20*, 456–464. [CrossRef]
40. Matyas, C.J. Associations between the size of hurricane rain fields at landfall and their surrounding environments. *Meteorol. Atmos. Phys.* **2010**, *106*, 135–148. [CrossRef]
41. Wu, L.; Tian, W.; Liu, Q.; Cao, J.; Knaff, J.A. Implications of the observed relationship between tropical cyclone size and intensity over the western North Pacific. *J. Clim.* **2015**, *28*, 9501–9506. [CrossRef]
42. Cocks, S.B.; Gray, W.M. Variability of the outer wind profiles of western North Pacific typhoons: Classifications and techniques for analysis and forecasting. *Mon. Weather Rev.* **2002**, *130*, 1989–2005. [CrossRef]
43. Lee, C.S.; Cheung, K.K.W.; Fang, W.T.; Elsberry, R.L. Initial maintenance of tropical cyclone size in the western North Pacific. *Mon. Weather Rev.* **2010**, *138*, 3207–3223. [CrossRef]
44. Brand, S. Very large and very small typhoon of the Western North Pacific Ocean. *J. Meteorol. Soc. Jpn.* **1972**, *50*, 332–341. [CrossRef]
45. Merrill, R.T. A Comparison of large and small tropical cyclones. *Mon. Weather Rev.* **1984**, *112*, 1408–1418. [CrossRef]
46. Knaff, J.A.; Longmore, S.P.; Molenar, D.A. An objective satellite-based tropical cyclone size climatology. *J. Clim.* **2014**, *27*, 455–476. [CrossRef]
47. Liu, K.S.; Chan, J.C.L. Size of Tropical cyclone as inferred from ERS-1 and ERS-2 Data. *Mon. Weather Rev.* **1999**, *127*, 2992–3001. [CrossRef]
48. Knaff, J.A.; Slocum, C.J.; Musgrave, K.D.; Sampson, C.R.; Strahl, B.R. Using Routinely Available Information to Estimate Tropical Cyclone Wind Structure. *Mon. Weather Rev.* **2016**, *144*, 1233–1247. [CrossRef]
49. Chan, K.T.F.; Chan, J.C.L. Size and strength of tropical cyclones as inferred from QuikSCAT data. *Mon. Weather Rev.* **2012**, *140*, 811–824. [CrossRef]
50. Lu, X.; Yu, H.; Yang, X.; Li, X. Estimating Tropical Cyclone Size in the Northwestern Pacific from Geostationary Satellite Infrared Images. *Remote Sens.* **2017**, *9*, 728. [CrossRef]
51. Dvorak, V.F. Tropical Cyclone Intensity Analysis and Forecasting from Satellite Imagery. *Mon. Weather Rev.* **1975**, *103*, 420–430. [CrossRef]
52. Kossin, J.P.; Knaff, J.A.; Berger, H.I.; Herndon, D.C.; Cram, T.A.; Velden, C.S.; Murnane, R.J.; Hawkins, J.D. Estimating hurricane wind structure in the absence of aircraft reconnaissance. *Weather Forecast.* **2007**, *22*, 89–101. [CrossRef]
53. Lajoie, F.; Walsh, K. A Technique to Determine the Radius of Maximum Wind of a Tropical Cyclone. *Weather Forecast.* **2008**, *23*, 1007–1015. [CrossRef]
54. Yang, S.; Cossuth, J. Satellite remote sensing of tropical cyclones. In *Recent Developments in Tropical Cyclone Dynamics, Prediction, and Detection*, 1st ed.; Lupo, A.R., Ed.; IntechOpen: London, UK, 2016; pp. 138–170.
55. Anthes, R.A. Tropical cyclones—Their evolution, structure and effects. *Meteorol. Monogr.* **1982**, *19*, 1–208.
56. Guinn, T.A.; Schubert, W.H. Hurricane spiral bands. *J. Atmos. Sci.* **1993**, *50*, 3380–3403. [CrossRef]
57. Willoughby, H.E. The dynamics of the tropical hurricane core. *Aust. Meteorol. Mag.* **1988**, *36*, 183–191.
58. Houze, R.A., Jr. Clouds in tropical cyclones. *Mon. Weather Rev.* **2010**, *138*, 293–344. [CrossRef]
59. Zou, X. *Atmospheric Satellite Observations*, 1st ed.; American Academic Press: Cambridge, MA, USA, 2020; pp. 235–294.

60. Bessho, K.; Coauthors. An introduction to Himawari-8/9—Japan's new-generation geostationary meteorological satellites. *J. Meteorol. Soc. Jpn.* **2016**, *94*, 151–183. [CrossRef]
61. Schmit, T.J.; Griffith, P.; Gunshor, M.M.; Daniels, J.M.; Goodman, S.J.; Lebair, W.J. A closer look at the ABI on the GOES-R series. *Bull. Am. Meteorol. Soc.* **2017**, *98*, 681–698. [CrossRef]
62. Landsea, C.W.; Franklin, J.L. Atlantic Hurricane database uncertainty and presentation of a new database format. *Mon. Weather Rev.* **2013**, *141*, 3576–3592. [CrossRef]
63. Demuth, J.L.; Demaria, M.; Knaff, J.A. Improvement of Advanced Microwave Sounding Unit Tropical Cyclone Intensity and Size Estimation Algorithms. *J. Appl. Meteorol. Climatol.* **2006**, *45*, 1573–1581. [CrossRef]
64. Pérez-Alarcón, A.; Sorí, R.; Fernández-Alvarez, J.C.; Nieto, R.; Gimeno, L. Comparative climatology of outer tropical cyclone size using radial wind profiles. *Weather Clim. Extrem.* **2021**, *33*, 100366. [CrossRef]
65. Duchon, C.E. Lanczos filtering in one and two dimensions. *J. Appl. Meteorol.* **1979**, *18*, 1016–1022. [CrossRef]
66. Jordan, H.M.; Stowell, D.J. Some small scale features of the track of Hurricane Ione. *Mon. Weather Rev.* **1955**, *83*, 210–215. [CrossRef]
67. Lawrence, M.B.; Mayfield, B.M. Satellite observations of trochoidal motion during Hurricane Belle 1976. *Mon. Weather Rev.* **1977**, *105*, 1458–1461. [CrossRef]
68. Muramatsu, T. Trochoidal motion of the eye of Typhoon 8019. *J. Meteorol. Soc. Jpn.* **1986**, *64*, 259–272. [CrossRef]
69. Hong, J.S.; Chang, P.L. The trochoid-like track in Typhoon Dujuan (2003). *Geophys. Res. Lett.* **2005**, *32*, 16801. [CrossRef]
70. Oda, M.; Nakanishi, M.; Naito, G. Interaction of an asymmetric double vortex and trochoidal motion of a tropical cyclone with the concentric eyewall structure. *J. Atmos. Sci.* **2006**, *63*, 1069–1081. [CrossRef]
71. Feng, Y.; Wu, L. Small-amplitude trochoidal oscillations in Typhoons Rammasun (2014) and Lekima (2019). *Terr. Atmos.Ocean. Sci.* **2021**, *32*, 1–10. [CrossRef]
72. Velden, C.S.; Leslie, L.M. The Basic Relationship between Tropical Cyclone Intensity and the Depth of the Environmental Steering Layer in the Australian Region. *Weather Forecast.* **1991**, *6*, 244–253. [CrossRef]
73. Wu, Y.; Zou, X. Numerical test of a simple approach for using TOMS total ozone data in hurricane environment. *Q. J. R. Meteorol. Soc.* **2008**, *134*, 1397–1408. [CrossRef]
74. Wu, L.; Chen, X. Revisiting the steering principal of tropical cyclone motion in a numerical experiment. *Atmos. Chem. Phys.* **2016**, *16*, 14925–14936. [CrossRef]
75. Willoughby, H.E.; Marks, F.D.; Feinberg, R.J. Stationary and moving convective bands in hurricanes. *J. Atmos. Sci.* **1984**, *41*, 3189–3211. [CrossRef]
76. Holland, G.J.; Lander, M. The meandering nature of TC tracks. *J. Atmos. Sci.* **1993**, *50*, 1254–1266. [CrossRef]
77. Ritchie, E.A.; Elsberry, R.L. Simulated impacts of a mesoscale convective system on the track of Typhoon Robyn in TCM-93. *Mon. Weather Rev.* **1993**, *128*, 2232–2251. [CrossRef]

remote sensing

MDPI

Article

Mitigation of Systematic Noise in F16 SSMIS LAS Channels Observations for Tropical Cyclone Applications

Huijie Dong and Xiaolei Zou *

Joint Center of Data Assimilation for Research and Application, Nanjing University of Information Science & Technology, Nanjing 210044, China; hdong@nuist.edu.cn
* Correspondence: xzou@nuist.edu.cn

Abstract: The Special Sensor Microwave Imager Sounder (SSMIS) onboard the Defense Meteorological Satellite Program (DMSP) F16, launched on 18 October 2003, was the first conical-scanning radiometer to combine the Special Sensor Microwave/Imagers (SSM/I), Special Sensor Microwave/Temperature Sounder (SSM/T), and the Special Sensor Microwave/Water Vapor Sounder (SSM/T2). Nearly 20 years of F16 SSMIS data are available to the general public, providing many opportunities to study the atmosphere at both the synoptic and decadal scales. However, data noise from complicated structures has occurred in the brightness temperature (TB) observations of lower atmospheric sounding (LAS) channels since 25 April 2013. We used a two-dimensional Fast Fourier Transform to analyze the characteristic features of data noise in cross-track and along-track directions. We found that the data noise is around 1–2 K and occurs at certain cross-track wavelengths $(\Delta\lambda)_{noise}$. A latitudinal variation was found for $(\Delta\lambda)_{noise}$. Due to noise interference, TB observations reflecting rain, clouds, tropical cyclone warm core, temperature, and water vapor distributions are not readily distinguishable, especially in channels above the middle troposphere (channels 4–7 and 24), whose dynamic TB range is smaller than low tropospheric channels 1–3. Examples are provided to show the impact of the proposed noise mitigation for conical-scanning TB observations to capture 3D structures of hurricanes directly. Once the noise in F16 SSMIS LAS channels from 25 April 2013to the present is eliminated, we may investigate the decadal change of many features of tropical cyclones derivable from these TB observations.

Keywords: F16 SSMIS LAS channels; brightness temperature observations; along-track varying cross-track data noise; tropical cyclone

Citation: Dong, H.; Zou, X. Mitigation of Systematic Noise in F16 SSMIS LAS Channels Observations for Tropical Cyclone Applications. *Remote Sens.* **2022**, *14*, 3171. https://doi.org/10.3390/rs14133171

Academic Editor: Seon Ki Park

Received: 11 May 2022
Accepted: 29 June 2022
Published: 1 July 2022

1. Introduction

The Special Sensor Microwave Imager/Sounder instrument (SSMIS) onboard the US Air Force Defense Meteorological Satellite Program (DMSP) polar-orbiting satellite F16, launched on 18 October 2003, was the successor and combination of the Special Sensor Microwave/Imagers (SSM/I), the Special Sensor Microwave/Temperature Sounder (SSM/T), and the Special Sensor Microwave/Water Vapor Sounder (SSM/T2) onboard the F10–15. As a conical scanning radiometer with a constant 45° scan angle, the SSMIS's 24 channels, whose central frequencies range from 19 to 183 GHz, are primarily designed to enhance remote-sensing capabilities of the land surface, ocean surface wind speed, cloud liquid, and rain rate (channels 12–18), measure the atmospheric radiation from the surface to about 30 hPa (channels 1–7), the upper troposphere (channel 24, ~12 hPa) and the mesosphere (channel 19, ~0.28 hPa), and obtain features sensitive to water vapor contents in the middle and lower troposphere (channels 8–11). SSMIS field-of-view (FOV) sizes are the same, and the distance between any two neighboring FOVs along a scan line does not vary. Therefore, weather-related structures are directly visible in global TB observations at channels 1–7, often called lower atmospheric sounding (LAS) channels. The F16 SSMIS TB observations have been available to the general public since 20 November 2005. As of now, we have more than 16 years of F16 SSMIS observations. Unfortunately, a significant data noise has occurred in F16 SSMIS TB observations since 25 April 2013. Our study analyzes these noise characteristics and develops an appropriate noise-detection algorithm for F16 SSMIS observations. Only then can we explore a unique opportunity to study the weather and climate of the atmosphere over 16 years using simultaneous measurements of imager channels, temperature-sounding channels, and humidity-sounding channels from F16 SSMIS.

Using satellite microwave observations to study the weather and climate of the atmosphere has been conducted more substantially using cross-track temperature sounders since the earliest Microwave Sounding Unit (MSU) onboard the National Oceanic and Atmospheric Administration (NOAA) polar-orbiting satellites TIROS-N, which was launched on 13 October 1978. The 4-channel MSU onboard TIROS-N, NOAA-6 to NOAA-14 was then replaced by the 15-channel Advanced Microwave Sounding Unit-A (AMSU-A) when NOAA-15 was launched on 13 May 1998. The AMSU-A onboard NOAA-15 to NOAA-19 and MetOp-A/B/C [1,2] was finally replaced by the Advanced Technology Microwave Sounders (ATMS) when S-NPP and NOAA-20 were launched on 28 October 2011 and 18 November 2017, respectively [3]. The global TB observations from multiple cross-track microwave temperature sounders have more than 40 years of continuous data records [4] that are routinely assimilated into NWP systems at nearly all operational centers and have contributed to significant improvements in global NWP forecast skills at the National Centers for Environmental Prediction (NCEP) [5], the European Centre of Medium-Range Weather Forecasts (ECMWF) [6,7], and China [8,9]. However, as a unique feature of cross-track radiometers, the limb effect causes the structural features of weather not to be directly visible from TB observations. This finding is not the case for TB observations from the conical scanning radiometer SSMIS.

Among different weather systems, tropical cyclones (TCs) remain of great interest to research and operational forecasts [10–13]. Due to the fast-evolving nature of their structure, track, and intensity controlled by complex, dynamic physics processes and societal impacts, satellite observations have become instrumental for investigating and predicting TCs that are mostly over oceans where conventional observations are rare. Besides direct assimilation of conical scanning microwave observations [14–17], another way to fully explore the potential values of polar-orbiting satellite microwave TB observations for TC research and forecasts is to apply TC warm core, TC center position, and inner and outer rainband sizes derivable from TB observations to vortex initialization. The microwave radiance is approximately a linear function of the atmospheric temperature at frequencies <200 GHz, larger than all SSMIS and AMSU-A channel frequencies. Based on this physical consideration, TC warm-core anomalies can be retrieved based on TB observations from these microwave instruments [18–22]. The assimilation of satellite microwaves retrieved TC warm-core temperatures improved 48-h forecasts of intensifications and vertical structures of all model state variables (e.g., temperature, water vapor mixing ratio, liquid water content mixing ratio, tangential and radial wind components, and vertical velocity) for Hurricane Florence (2018) and Typhoon Mangkhut (2018) [23]. Hu and Zou [24,25] developed an azimuthal-spectral-analysis-based center-fixing algorithm to determine the TC center position in real-time using the TC's axisymmetric structural information embedded in TB observations. The noise, if not detected and removed, prevents the application or reduces the accuracy of the TC warm-core retrieval and TC center positioning using SSMIS TB observations.

The conical-scanning node makes SSMIS LAS channels a potentially important data source complementing cross-track radiometers in revealing TC structures. The long-term F16 SSMIS data availability also allows an investigation into the decadal change of TCs. For these purposes, we aimed to remove the noise found in the F16 SSMIS LAS channels from 25 April 2013 onward. Our paper is organized as follows: Section 2 briefly describes F-16 SSMIS TB observations of LAS channels. Section 3 describes methods for analyzing and mitigating data noise. Our results are presented in Section 4, showing temporal and latitudinal dependences of data noise in F-16 SSMIS LAS channels TB observations. Methods to avoid artificial errors induced by noise mitigation due to sharp TB variations over TC heavy rainfall areas are discussed in Section 5. Our conclusions and future plans are provided in Section 6.

2. Data Description

DMSP F16 is a sun-synchronous polar-orbit satellite at an altitude of approximately 833 km, circling the Earth at 14.1 revolutions per day. The SSMIS instrument onboard F16 is a conically scanning passive microwave radiometer that collects data from the aft (forward) to the nadir for a morning ascending (descending) node orbit. It measures upwelling microwave radiation from 24 channels located in a range of frequencies from 19 to 183 GHz [26]. These 24 channels consist of the lower atmospheric sounding (LAS) channels 1–7 and 24, the environmental sensor channels 12–16, imager channels 8–11 and 17–18, and upper atmospheric sounding channels 19–23. Although only the SSMIS channels 12–18 from F16, F17, and F18 were intercalibrated to SSM/I equivalent channels to generate the so-called Fundamental Climate Data Record (FCDR) of TB data from the SSMIS sensors, the output FCDR file also contains the TB and geolocation information for all the other SSMIS channels with limited corrections and no intercalibration applied [27]. Ten years since its launch date, TB observations of the SSMIS LAS channels from DMSP F16 were contaminated by obvious noise, preventing any possible attempts to study climate change. Our study focuses on F16 SSMIS LAS channels.

Channel characteristics for the LAS channels are shown in Table 1. The LAS channels are located in the oxygen (O_2) band and provide measurements of upwelling microwave radiances responding to emissions and absorption due to O_2 rotational transitions. They provide information on atmospheric temperature. The eight LAS channels are designed to profile atmospheric temperature from the surface to the upper stratosphere near 12 hPa. Channel 1 is the window channel, and other LAS channels are sounding channels. A TB measurement from a particular channel quantifies a radiation amount from a vertical layer of the atmosphere centered around the altitude of the maximum weighting function (WF) shown in Table 1. The SSMIS antenna bore-sight is designed at a 45° angle off the nadir so that the SSMIS conically scans the Earth's surface at an incidence angle of 53°With a swath width of 1707 km. All fields-of-view (FOVs) are 27×18 km^2 in the along- and across-track directions. The sampling interval is the same as 12.5×37.5 km^2 for channels 1–7 and channel 24 in the along- and across-track directions. In 1.9 s, SSMIS can measure 60 scenes (i.e., FOVs) within a scan range of 143.2 degrees along a single scanline. Although later than MSU and much later than AMSU-A, the F16 SSMIS is the first instrument employing a conical scan geometry for temperature sounding [28].

Table 1. The LAS channel characteristics.

Channel Number	1	2	3	4	5	6	7	24
Frequency (GHz)	50.3	52.8	53.596	54.40	55.50	57.29	59.4	57.29
Peak WF (hPa)	1000	630	440	230	100	63	33	12
Bandwidth (MHz)			400			350	250	30.0
Sensitivity (K)			0.4			0.5	0.6	0.7
FOV Size ((km)^2)	27 × 18 (along track × across track)							
Sampling Interval	12.5 × 37.5 (along track × across track)							

We obtained the F16 SSMIS data used in this study from the following website: http://rain.atmos.colostate.edu/FCDR/data_access.html (accessed on 5 January 2022). Figure 1 shows the local equator crossing time (LECT) variations of F16 from its launch to 15 June 2021, the last day of available F16 SSMIS data (Figure 1). We found systematic noise in TB observations of F16 SSMIS LAS channels since 25 April 2013. For example, Figure 2 provides spatial distributions of TB observations at channels 1–7 and 24 over a small portion of an ascending swath on 5 September 2017. A systematic curvy noise pattern was seen in all TB observations of the LAS channels. The higher the channels' WF peak altitudes, the clearer the noise distributions because this part of the swath went over Hurricane Irma, whose associated cloud and precipitation affected low-level TB observations more than high-level channels. The dynamic range of TB observations for channels 4–7 and 24 (~4–5 K) are much smaller than those of channels 1–3 (~10–20 K) for the part shown in Figure 2. To search for a possible law of the curvy noise scanning pattern of the conical radiometer SSMIS, we indicated across-track distributions of 60 FOVs along several SSMIS scanlines at an interval of 25 scanlines in Figure 2e and along-track distributions of FOVs at an interval of 5 FOVs in Figure 2f. With this noise observation, we developed a method for mitigating TB noise in the F16 SSMIS LAS channels.

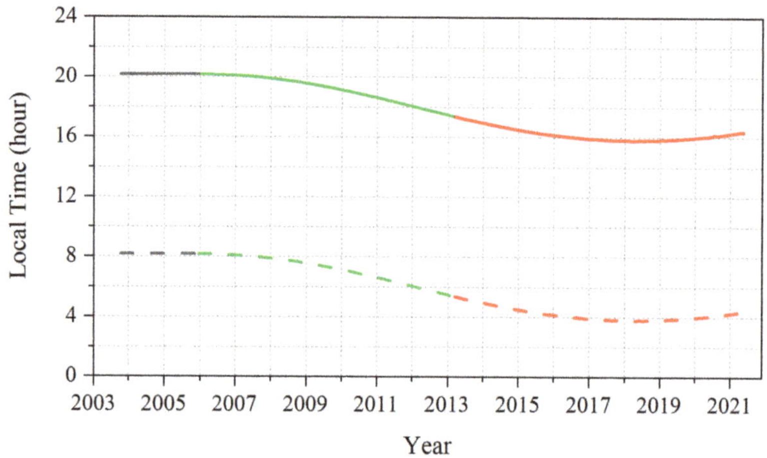

Figure 1. Variations of LECT at the ascending (solid curve) and descending (dashed curve) nodes of F16. Three time periods are also indicated: no available data (gray curve), high-quality data (green curve, from 20 November 2005 to 25 April 2013, and noisy data (red curves, from 25 April 2013 to 15 June 2021) in LAS channels TB observations.

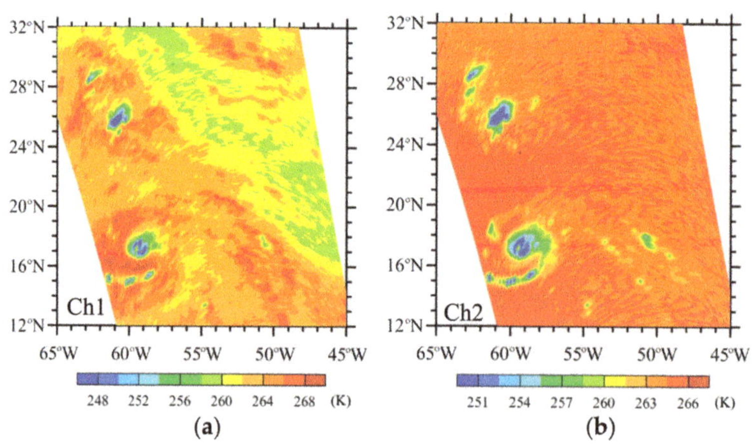

Figure 2. *Cont.*

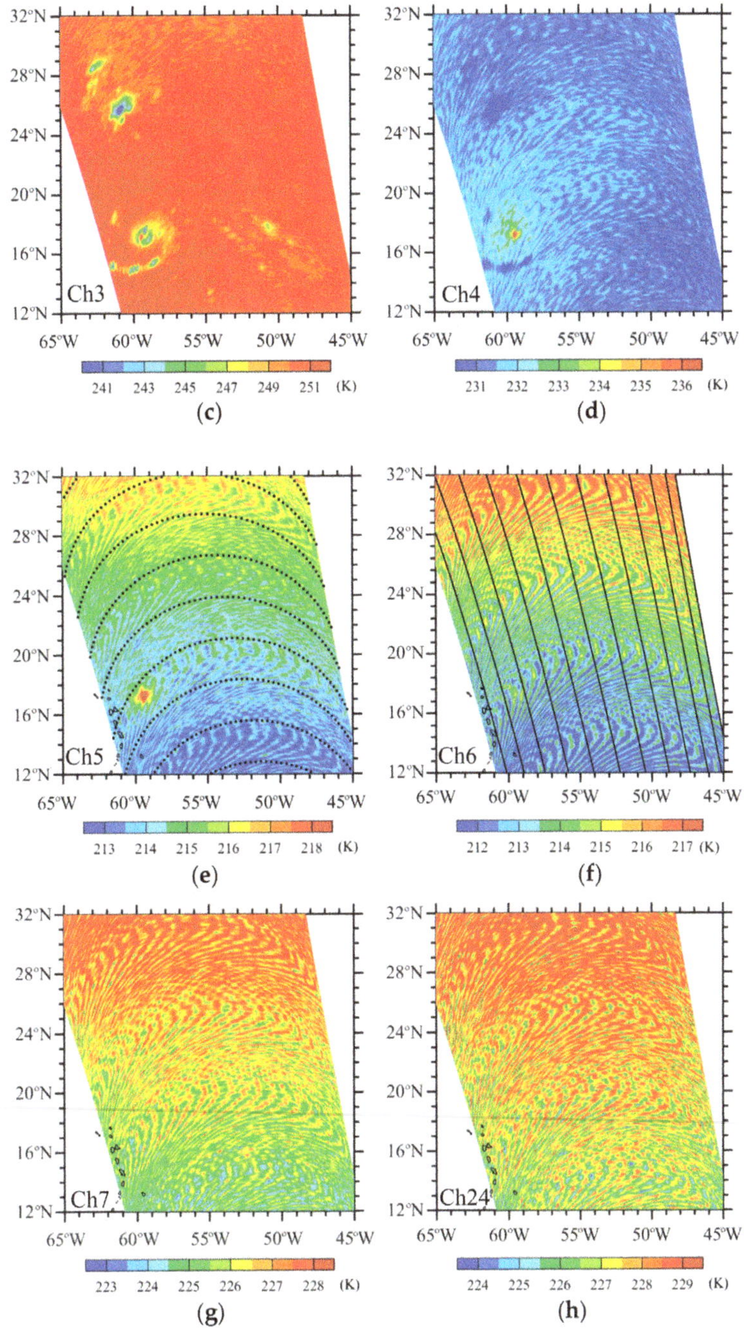

Figure 2. (**a–h**) TB observations of channels 1–7 and 24 on 5 September 2017. The across-track distributions of 60 FOVs along several SSMIS scanlines at an interval of 25 scanlines in (**e**) and the along-track distributions of FOVs at an interval of 5 FOVs in (**f**) are also shown. Hurricane Irma was located around 17.0°N, 59.7°W.

3. Method for Noise Mitigation

Using the two-dimensional (2D) discrete fast Fourier transform (FFT), we can express the TB observations over a targeted portion of swath as follows:

$$y^o_{k,l} = \sum_{m=0}^{M-1}\sum_{n=0}^{N-1} C_{m,n}e^{i(m\frac{2\pi k}{M}+n\frac{2\pi l}{N})} \tag{1}$$

where $y^o_{i,k}$ represents the TB observation at the ith FOV and the jth scanline (I = 1, 2, ..., M, j = 1, 2, ..., N); M is the total number of FOVs along a single scanline (M = 60); N is the total number of scanlines ($N \approx 300$); and $C_{m,n}$ is the amplitude of the 2D wave with wavenumbers m and n in the across- and along-track directions, respectively. The inverse Fourier transform is defined as follows:

$$C_{m,n} = \frac{1}{MN}\sum_{k=0}^{M-1}\sum_{l=0}^{N-1} y^o_{k,l}e^{-i(k\frac{2\pi m}{M}+l\frac{2\pi n}{N})} \tag{2}$$

The wavelength (λ_m) in the across-track direction is calculated from the wavenumber (m) using the formula $\lambda_m = \frac{M-1}{m} \times \Delta x_{across-track}$, where $\Delta x_{across-track}$ is equal to the across-track sampling interval of 37.5 km. Similarly, the wavelength (λ_n) in the along-track direction is calculated from the wavenumber (n) using the formula $\lambda_n = \frac{N-1}{m} \times \Delta y_{along-track}$, where $\Delta y_{along-track}$ is equal to the along-track sampling interval of 12.5 km.

In general, the TB amplitude decreases rapidly with the increasing wavenumber. If there is a sudden increase in amplitude within a range of wavenumbers $(\Delta m)_{noise}$, we can remove these wave components by setting the amplitude $C_{m,n}$ to zero when $m \in (\Delta m)_{noise}$ for all n. The reconstructed TB field is obtained by the inverse FFT:

$$\tilde{y}^o_{k,l} = \sum_{m=0}^{M-1}\sum_{n=0}^{N-1} \tilde{C}_{m,n}e^{i(m\frac{2\pi k}{M}+n\frac{2\pi l}{N})} \tag{3}$$

$$\tilde{C}_{m,n} = \begin{cases} C_{m,n} & \text{if } m \notin (\Delta m)_{noise} \\ 0, & \text{if } m \in (\Delta m)_{noise} \end{cases} \tag{4}$$

The data noise is defined as $y^o_{k,l} - \tilde{y}^o_{k,l}$. The above procedure to generate noise-mitigated TB observations ($\tilde{y}^o_{k,l}$) is applied to LAS channels 5–7 and 24.

In the presence of heavy precipitation, TB observations have outliers of an abnormally small value. Some of these low TB values sneak into extracted noise. We conducted an extra step to avoid the impacts of heavy precipitation-induced TB outliers of abnormally small value on noise mitigation. Specifically, we used an Empirical Mode Decomposition (EMD) developed by Huang et al. [29] to extract the high-frequency random noise from TB observations along a scanline. The highest frequency across-track variation, called the first intrinsic mode function (IMF), is extracted from TB observations and obtained by identifying all the local maxima and minima of $y^o_{k,l}$ (k = 1, 2, ..., M). All the local maxima are then connected with a cubic spline as the upper envelope, and all the local minima are connected with a cubic spline as the lower envelope. The upper and lower envelopes are finally averaged to obtain the local mean, denoted as $a_{1,1}(j)$. The 1st IMF of $y^o_{k,l}$ (k = 1, 2, ..., M) is defined as:

$$C^{(1)}_{k,l} = y^o_{k,l} - a_{1,l}(j) \tag{5}$$

Set

$$\hat{y}^o_{k,l} = y^o_{k,l} - C^{(1)}_{k,l} \tag{6}$$

If at some FOVs ($\{k_i\}$), values of the first IMF are greater than 1.7 times the bi-weight standard deviation, these FOVs are subtracted from TB observations:

$$\hat{y}^o_{k,l} = \begin{cases} y^o_{k,l}, & \text{if } k \notin \{k_i\} \\ y^o_{k,l} - C^{(1)}_{k,l}, & \text{if } k \in \{k_i\} \end{cases} \tag{7}$$

A 2D spectral analysis is conducted for the field of $\hat{y}^o_{k,l}$. The values subtracted on the right-hand-side of (6) are added back to the noise-mitigated TB field:

$$\tilde{\tilde{y}}^o_{k,l} = \begin{cases} \tilde{y}^o_{k,l}, & \text{if } k \notin \{k_i\} \\ \tilde{y}^o_{k,l} + C^{(1)}_{k,l}, & \text{if } k \in \{k_i\} \end{cases} \quad (8)$$

The above procedure is used to generate noise-mitigated TB observations ($\tilde{\tilde{y}}^o_{k,l}$) for LAS channels 1–4. Data noise is defined as $y^o_{k,l} - \tilde{\tilde{y}}^o_{k,l}$.

4. Results Characterizing a Systematic Noise in F16 SSMIS LAS TB Observations

Two sequential SSMIS swaths are provided in Figure 3, which shows TB observations from channel 5 on 5 September 2017. The observation times for the swath whose edges are indicated by solid black curves were from 1915 to 2057 UTC; those for the other swath with edges indicated by black dashed curves are from 2057 to 2238 UTC. There are orbital gaps in low latitudes. The global TB observations vary more than 20 K, characterized by a significant latitudinal distribution. The TBs near the south pole are below 206 K, and those near the north pole are above 226 K. The TBs near the equator are around 214 K. It is difficult to see any noise structures in Figure 3 with more than a 20 K range of TB variations and 7 K color interval.

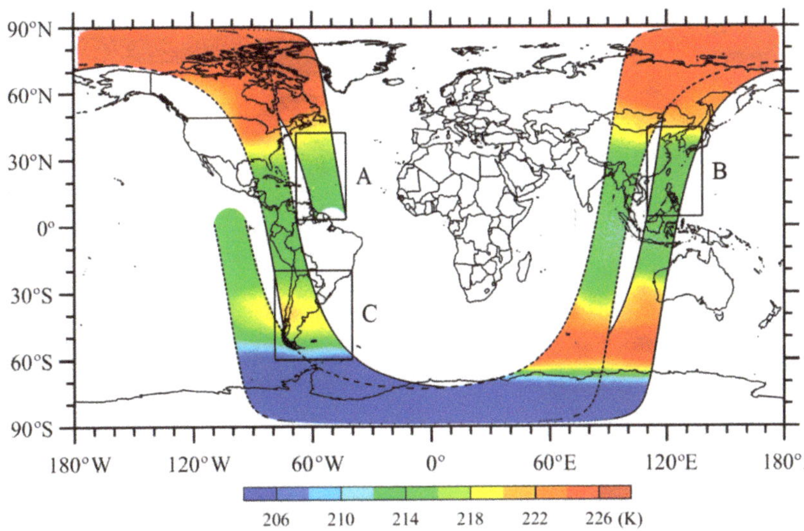

Figure 3. TB observations at channel 5 over two sequential swaths on 5 September 2017. One swath had its observation times from 1915 to 2057 UTC (edges indicated by a black solid curve) and the other swath from 2057 to 2238 UTC (edges indicated by a black dash curve). Three areas with 300 scanlines (black boxes) are also indicated: area A is at the ascending node over the western Northern Hemisphere, area B is at the descending node over the eastern Northern Hemisphere, and area C is at the ascending node over the western Southern Hemisphere.

We arbitrarily chose the following three areas of the first swath in Figure 3 for a more detailed analysis. Areas A and C are two portions at the ascending node over the western Northern and Southern Hemispheres, respectively. Area B is at the descending node over the eastern Northern Hemisphere. All three areas contain 300 scanlines. The dynamic range of channel 5 TB observations over the three local areas A, B, and C is reduced to about 5–10 K (Figure 4), and a systematic noise pattern of a curvy shape becomes visible in the distributions of TB observations. A 2D spectral analysis described in Section 3 was conducted for TB observations in these three areas. Variations of 2D amplitude with respect to wavenumbers in cross- and along-track directions are presented in two ways to provide a qualitative and quantitative look at all waves in Figures 5 and 6, respectively. The amplitudes and cross-track wavenumbers shown as the color shading and *y*-axis in Figure 5 are simply switched to the *y*-axis and curves in Figure 6. In Figure 6, amplitude variations with the along-track wavenumber are represented by a spaghetti map for all 1–150 cross-track wavenumbers in each area. A common spectral feature among TB observations over the three areas is that the amplitudes are the largest near-zero wavenumbers and decrease rapidly with wavenumbers in both across- and along-track directions. However, we also noticed amplitudes at some fixed across-track wavenumbers being higher than their neighboring wavenumbers, which may represent noise signals seen in the left panels of Figure 4. The exact across-track wavenumbers of these large-amplitude bands are different among these three areas.

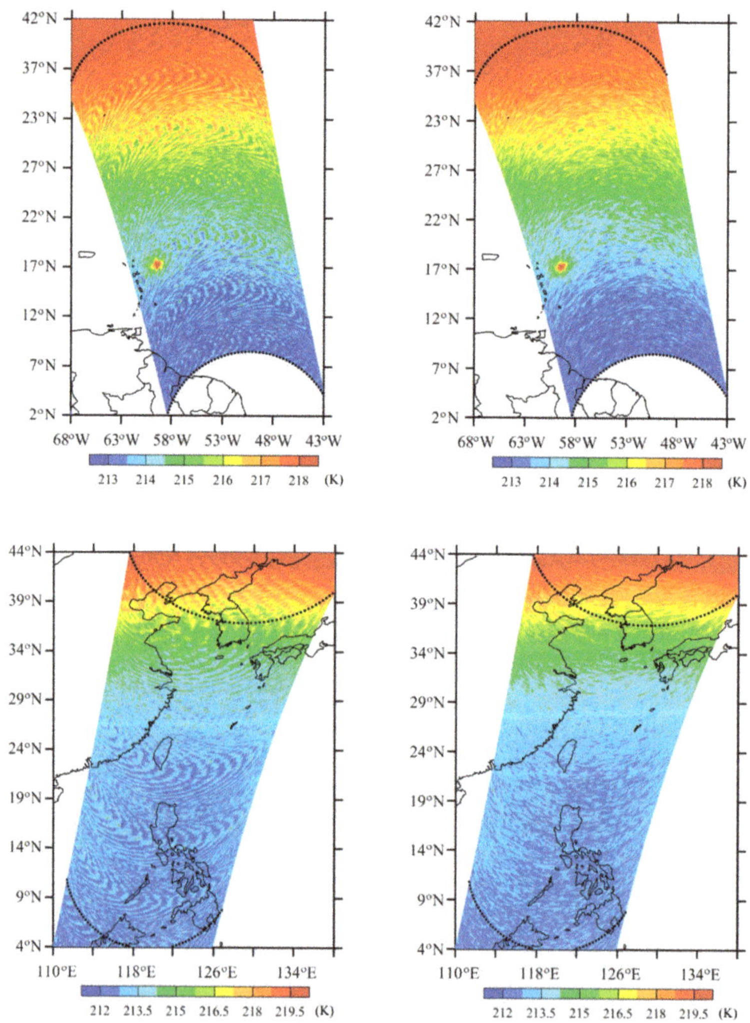

Figure 4. *Cont.*

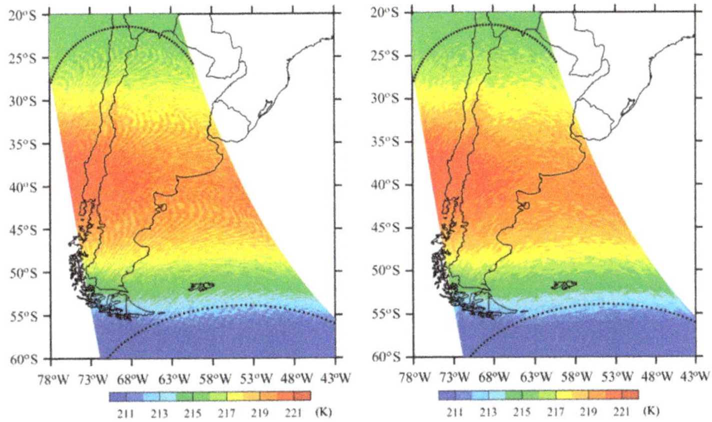

Figure 4. TB observations of channel 5 on 5 September 2017 before (**left** panels) and after noise reduction (**right** panels) in areas A (**top** panels), B (**middle** panels), and C (**bottom** panels).

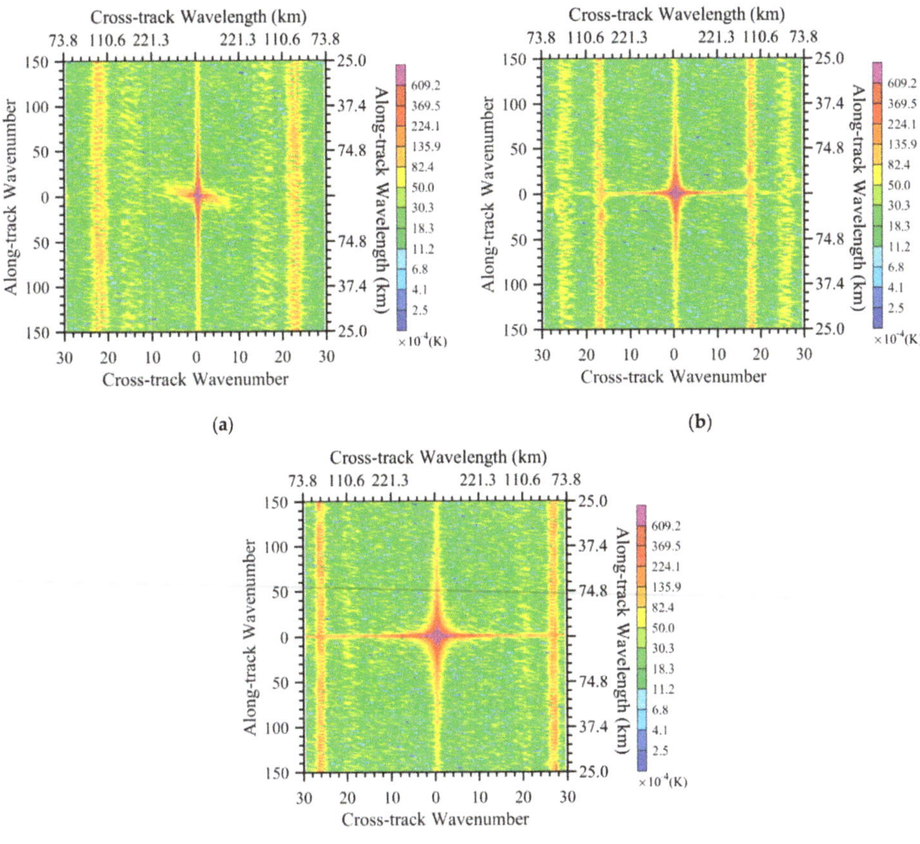

Figure 5. Variations of amplitude with respect to wavenumbers in cross- and along-track directions in areas (**a**) A, (**b**) B, and (**c**) C.

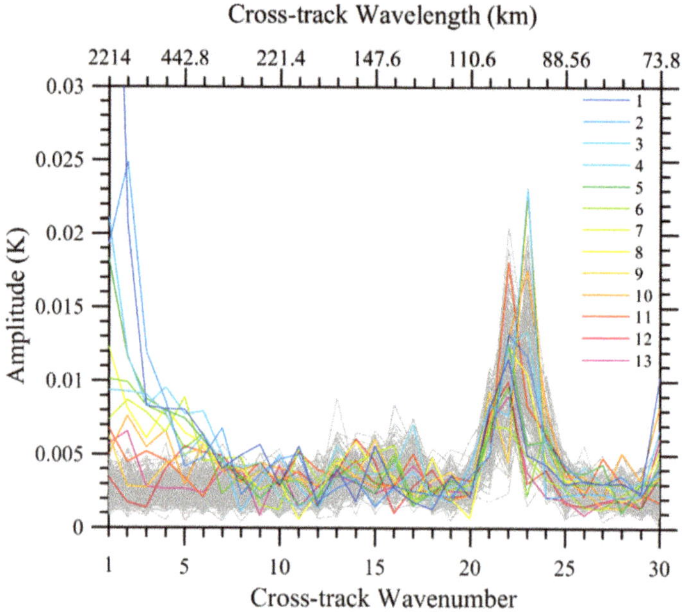

(**a**)

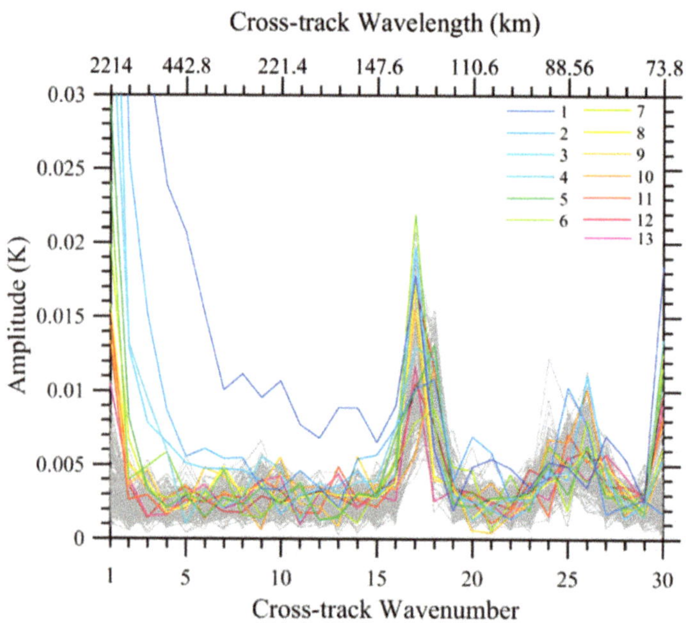

(**b**)

Figure 6. *Cont.*

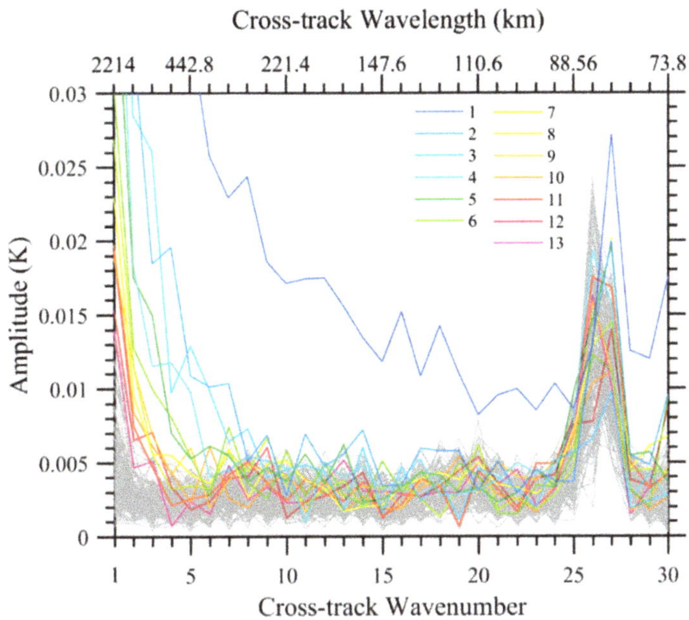

(c)

Figure 6. Variations of amplitudes with respect to the along-track wavenumbers for the cross-track wavenumbers 1–13 (color curves) and 14–150 (gray curves) over (**a**) area A, (**b**) area B, and (**c**) area C.

The detected noise, defined as the difference between the original and reconstructed TB observations, is shown in Figure 7. The curvy noise-looking pattern in the TB observations of channel 5 (left panels in Figure 4) resembles the noise detected in Figure 7 and disappears in the spatial distribution of TB observations of channel 5 over areas A, B, and C after the noise mitigation (right panels in Figure 4).

The regional dependence of data noise on across-track wavenumbers is further confirmed in Figure 8, which shows the along-track variations of 2D amplitude with respect to the cross-track wavenumber for TB observations over the two swaths in Figure 3. The across-track wavenumbers of larger amplitudes vary along both swaths compared with those of larger and smaller neighboring wavenumbers. We also found that variations of the across-track wavenumbers of data noise with respect to the observing latitude were the same for the two swaths. In other words, the F16 SSMIS LAS channels' data noise depends on the latitude of the F16 orbit. Bell et al. [28] reported that solar intrusions into the warm calibration load affect the calibration accuracy for different parts of the orbit, and the thermal cycling of an orbit may also result in a modulation of the measured TB observations by the main reflector emission. The exact root cause for this noise pattern requires further investigation and is beyond the scope of this study.

Being sensitive to the atmosphere in the lower altitudes of the middle and lower troposphere than other LAS channels in the upper troposphere and stratosphere, TB observations of LAS channels 1–4 were more strongly affected by cloud and precipitation. Figure 9 provides an example of channel-3 noise extracted by the same method as channel 5 ($y^o_{k,l} - \tilde{y}^o_{k,l}$, Figure 9a). The noise was extracted by adding a step to avoid the impacts of heavy precipitation-induced TB outliers of abnormally small value on the 2D spectrum ($y^o_{k,l} - \tilde{\tilde{y}}^o_{k,l}$, Figure 9b), and the reconstructed channel-3 TB observations (i.e., $\tilde{\tilde{y}}^o_{k,l}$) at the ascending node on 5 September 2017 (Figure 9c). The noise extracted by the same method as channel 5 (Figure 9a) was not homogeneously distributed in space, with large magnitudes in places of abnormally low TB observations. After adding a step to avoid the impacts of heavy precipitation-induced TB outliers, the noise extracted by the same method as channel 5 (Figure 9b) was homogeneously distributed in space and resembled those of the upper-level channel 5 (right panels in Figure 4).

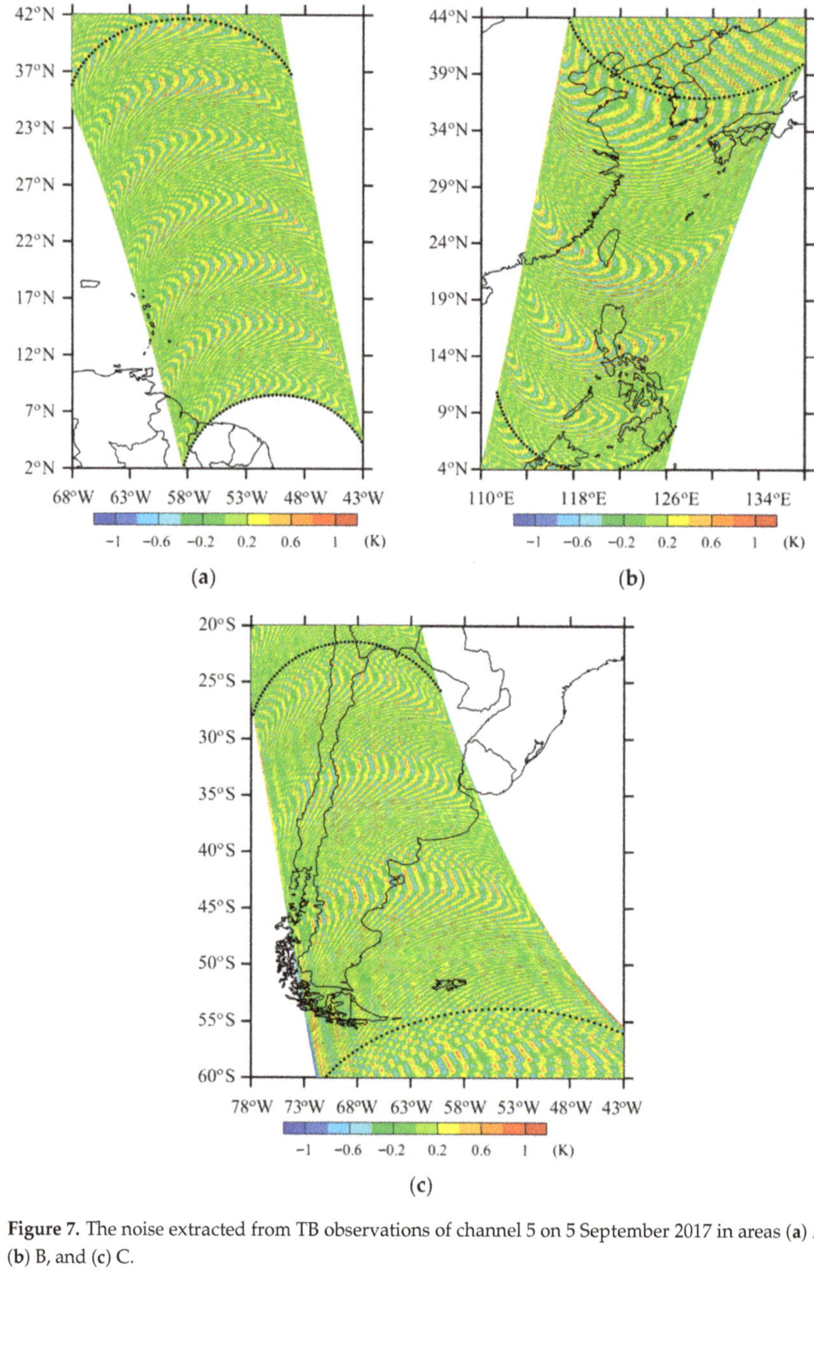

Figure 7. The noise extracted from TB observations of channel 5 on 5 September 2017 in areas (**a**) A, (**b**) B, and (**c**) C.

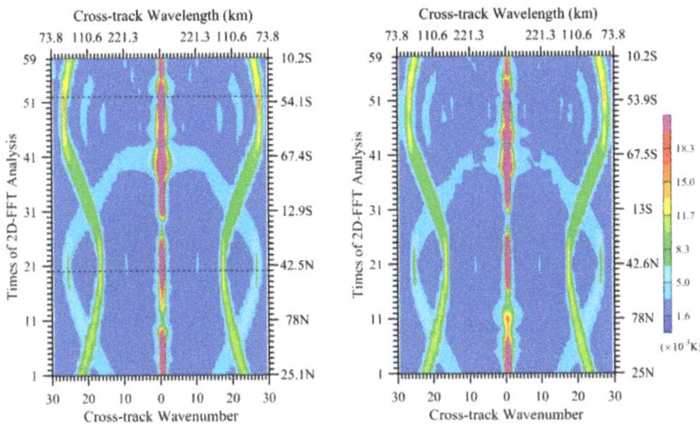

(a) (b)

Figure 8. Variations of the amplitude averaged over along-track wavenumbers 1–150 with respect to the cross-track wavenumber for TB observations over the two swaths shown in Figure 3: (**a**) the swath with its observation times from 1915 to 2057 UTC (black-curved edges) and (**b**) the other swath whose observation times were from 2057 to 2238 UTC (dash-curved edges). The 2D FFT with 300 scanlines each time is carried sequentially with an overlap of 50 scanlines between two neighboring portions of the 2D Fourier analysis domain.

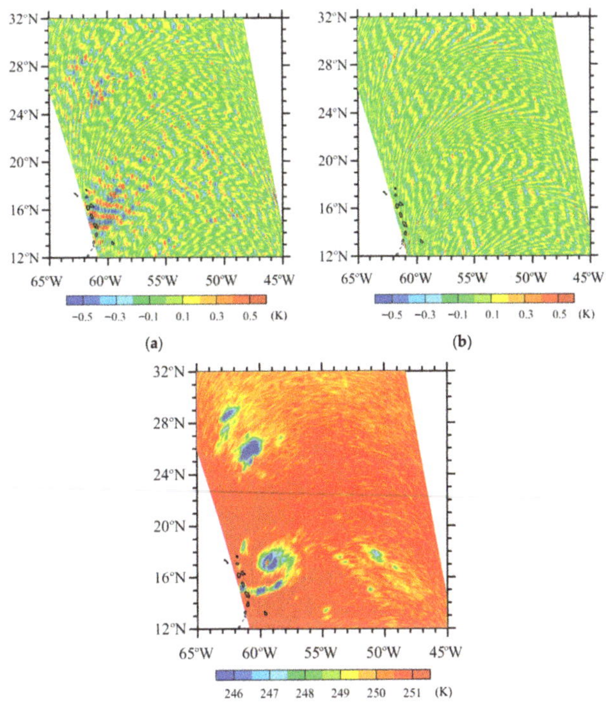

(c)

Figure 9. (**a**) Channel 3 noise at the ascending node on 5 September 2017 extracted by the same method as channel 5. (**b**) Same as (**a**) except for adding a step to avoid the impact of heavy precipitation-induced TB outliers of abnormally small value on noise mitigation. (**c**) The reconstructed channel 3 TB observations.

Figure 10 provides the spatial distribution of TB observations after noise reduction for the LAS channels 1–4, 6–7, and 24 over the same swath portion as Figure 2. The systematic curvy noise pattern seen in Figure 2 was successfully removed. What remains to be seen in TB observations besides weather signals are random errors, which are 0.4, 0.5, 0.6, and 0.7 for channels 4, 6, 7, and 24 (see Table 1). The noise detected for channels 1–4, 6–7, and 24 (Figure 11) have a curvy pattern, as seen in TB observations (Figure 2).

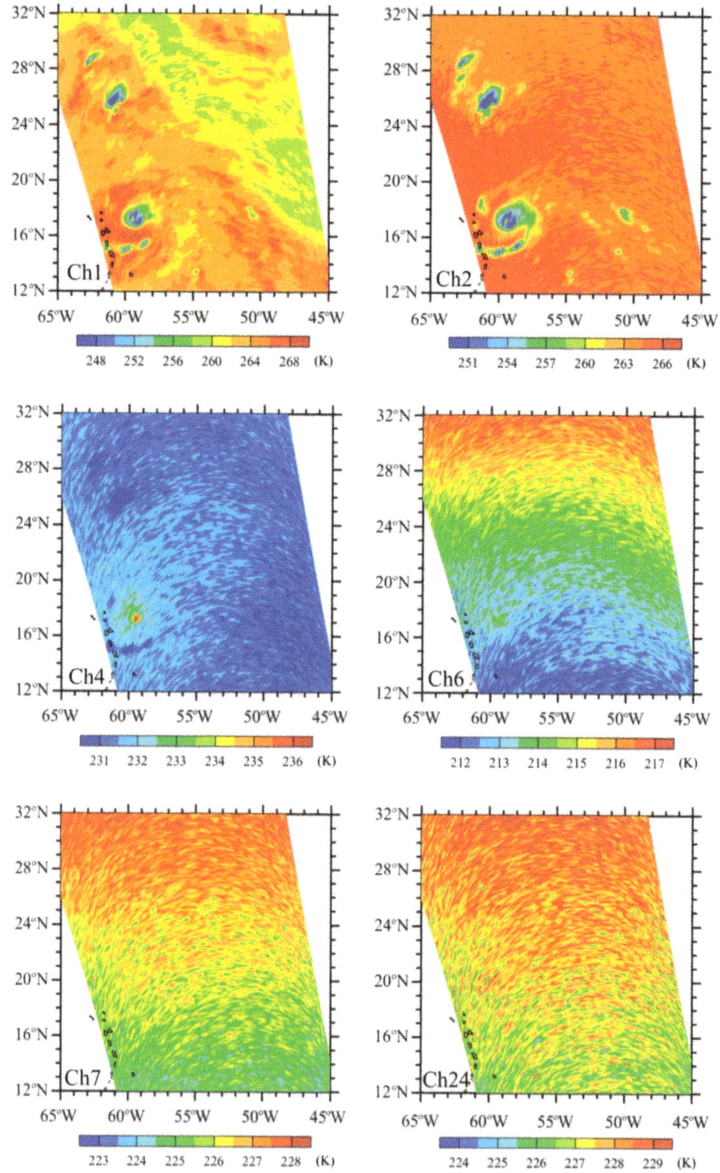

Figure 10. TB observations after noise reduction for channels 1, 2, 4, 6–7, and 24 on 5 September 2017. Hurricane Irma was located around 17.0°N, 59.7°W.

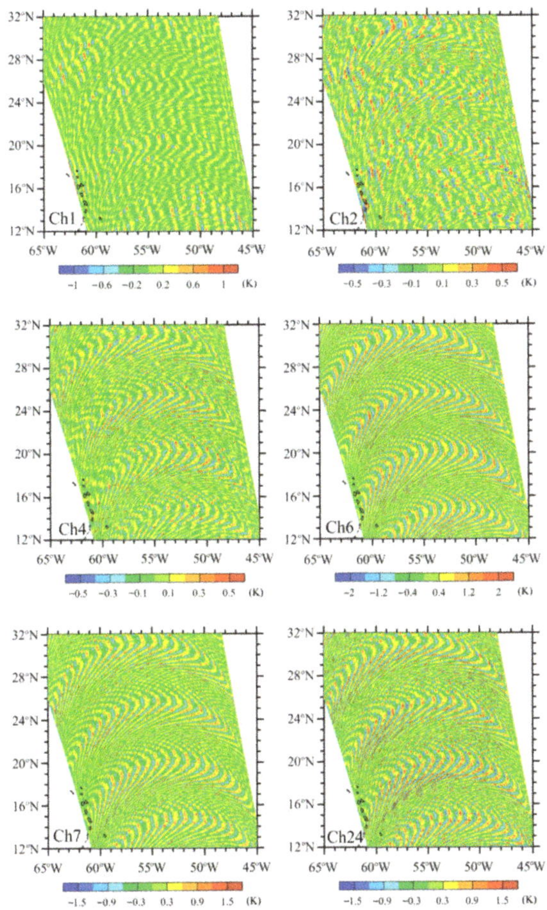

Figure 11. Detected noise at channels 1, 2, 4, 6–7, and 24.

5. Hurricane Structures Directly Revealed by SSMIS LAS Channels

Hurricanes are characterized by structured bands of cloud and precipitation. Due to the scattering effect by ice particles, TB observations in hurricane rainbands are colder than their surroundings. Therefore, a horizontal TB distribution can reveal hurricane rainband structures. We chose Hurricane Irma (2017) to illustrate TB observations from SSMIS LAS channels' capability to capture its structures.

Irma was the most intense hurricane to strike the United States since Katrina in 2005. It originated at low latitudes in the deep tropics on 30 August 2017 and rapidly intensified shortly after its formation. Irma fluctuated between hurricane categories 2 (H2) and 3 (H3) (Saffir–Simpson scale) from 1800 UTC on 31 August to 1800 UTC on 4 September 2017 while experiencing a series of eyewall replacement cycles and reaching category 5 at 1200 UTC 5 September 2017. After its first landfall in Cuba on 9 September 2017 as a category-5 hurricane, Irma made its second and third landfalls in Florida's Cudjoe Key and Marco Island at H4 and H5 intensities, respectively. It caused widespread and catastrophic property damage and many deaths.

Figure 12a shows the SSMIS TB observations at LAS channel 3 for Hurricanes Irma and Jose; the latter appeared simultaneously with Irma over the Atlantic Ocean in September 2017. Since the detected noise of channel 3 is <0.5K (Figure 13b), the TB observations after the noise deduction look the same as Figure 13a, whose color interval is 0.5 K. At 0721 UTC on 8 September, Irma reached H4 intensity and Jose, the hurricane located to the southeast of Irma, was H3. At 0000 UTC on 8 September, the maximum sustained wind speed was 161 and 121 mph for Irma and Jose, respectively. The radius of 34-kt wind speed was 296 and 222 km for Irma and Jose, respectively. In Figure 13, the west swath captured the structures of Hurricane Irma, and the east swath revealed Hurricane Jose's structure. A relatively large orbital gap exists between the two swaths. The hurricane eye is identifiable for both Irma and Jose. The observed TB values in the circularly distributed convective rainband regions are more than 6 K lower than those in their eyes, clear streaks between rainbands, and environments. The eye and rainband range of Irma are larger than those of Jose, with a fixed FOV size of 27 × 18 km^2 at a fixed interval between the neighboring FOVs of 12.5 × 37.5 km^2 (see Table 1). Such hurricane structures cannot be seen directly from TB observations of cross-track temperature sounders due to variations in their FOV size, spacing between two neighboring FOVs, and WF peak altitude along a scanline. Considering AMSU-A as an example, the altitude where the atmosphere contributes the most to an observed TB also increases with scan angle in the troposphere. Spacing between neighboring FOVs also increases with scan angle, and the across-track diameter of an FOV is 48 and 155 km at the nadir and swath edge, respectively [30]. Even the smallest AMSU-A FOV at the nadir is much coarser than the SSMIS FOV.

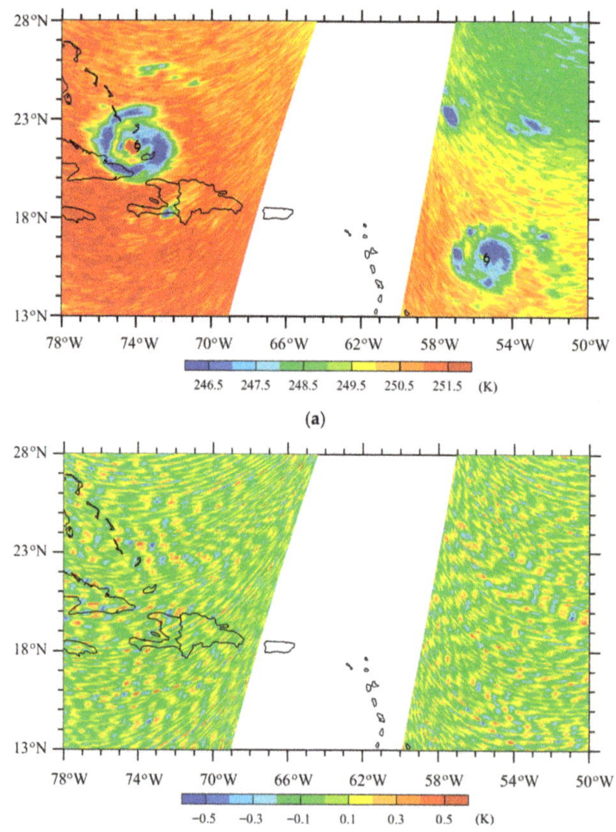

Figure 12. (**a**) TB observations and (**b**) the detected noise at channel 3 from 0721 UTC to 0910 UTC 8 on September 2017 after noise mitigation. Hurricane Irma, located at (21.6°N, 73.9°W), had a category H4 intensity, and Hurricane Jose, located at (15.9°N, 55.3°W), had a category H3 intensity.

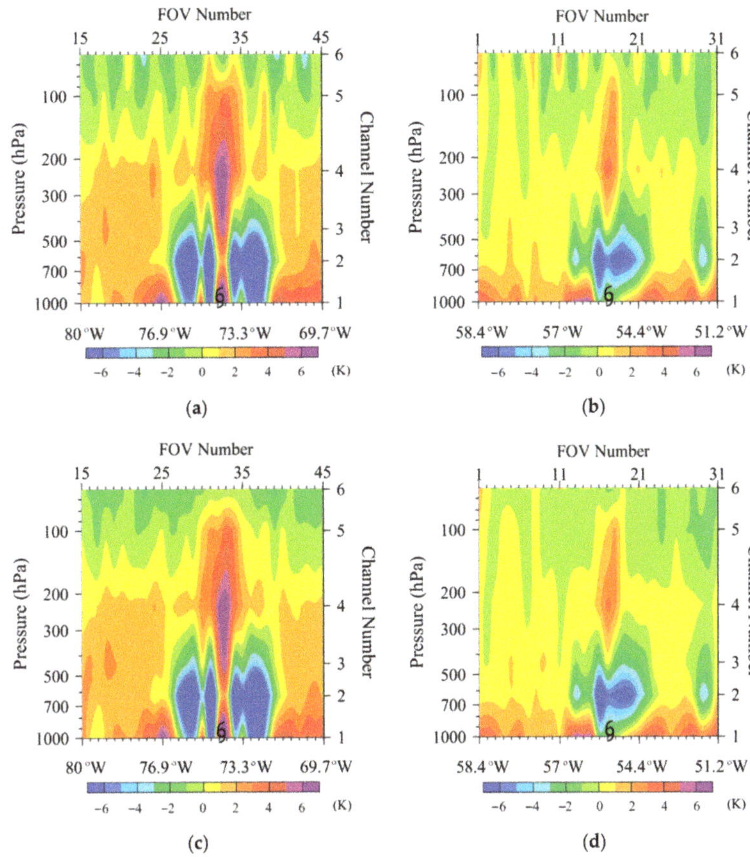

Figure 13. Cross-track cross-sections of TB anomalies at channels 1–6 along the scanline passing through the center (21.6°N, 73.9°W) of Hurricane Irma at 0906 UTC (left panels) and the center (15.9°N, 55.3°W) of Hurricane Jose at 0727 UTC (right panels) on 8 September 2017 (**a**,**b**) before and (**c**,**d**) after noise mitigation. The hurricane center position is indicated by a hurricane symbol at the bottom of each figure panel.

LAS channels were designed for vertically profiling the atmosphere and can be used to examine the vertical structures of hurricanes. Figure 13 provides vertical cross-sections of TB observations for channels 1–6 along the scanline passing through the center of Hurricane Irma located at (21.6°N, 73.9°W), and along the scanline passing through the center of Hurricane Jose at (15.9°N, 55.3°W). The mean environmental temperature within a latitude–longitude geographic box centered at the same latitude of the hurricane center but away by 1000 km was subtracted to see the warm-core structure more clearly. An unrealistic wave-like structure at a scale of about 112 km is seen in TB observations above 200 hPa without removing detected noise (Figure 13a,b). The proposed noise detection algorithm successfully removed this noise (Figure 13c,d). The warm-core maxima of both Irma and Jose are located at about 250 hPa. The Irma warm-core TB is more than 7 K, which is 4 K warmer than Jose. The warm core of Irma went to the ocean surface and extended to 70 hPa, while Jose was confined from 400 to 100 hPa. Significant scattering by precipitation size ice particles was confined below 400 and 500 hPa for Irma and Jose, respectively. Similar cross-sections of TB observations capable of revealing the vertical structures of a hurricane can be determined along any line passing through the hurricane center from the SSMIS LAS observations. However, cross-sections from AMSU-A sounding channel observations can only be determined along a line with a fixed FOV passing through the hurricane center, mainly from south to north directions, slightly tilted westward and eastward for the ascending and descending node, respectively.

As Hurricane Irma moved along its best track and the intensity changed (see Figure 14a), the vertical structure also changed, as indicated by TB observations at Irma's center from 30 August to 11 September 2017 (Figure 14b,c). The detected noise was less than 1.2 K (Figure 14d). Here, the TB anomaly is defined as the TB observations subtracted by the mean environmental temperature within a latitude–longitude geographic box centered at the same latitude of the hurricane center but away by 1000 km. Irma's warm core was maximized at about 250 hPa. As Irma's intensity increased from 31 August to 5 September 2017 (see Figure 1b in Tian and Zou [22]), the TB warm-core depth and intensity continually increased from 5 September to 8 September, during which Irma was an H5 hurricane. We noticed that the warm core extended to the surface around 0805 UTC on 5 September and 0910 UTC on 8 September due to Irma's large clear eye, as shown in Figures 13 and 14 for the 0910 UTC on 8 September. Such useful information is difficult to obtain by AMSU-A TB observations or warm-core temperature retrieval due to the AMSU-A's inability to resolve hurricane eye adequately.

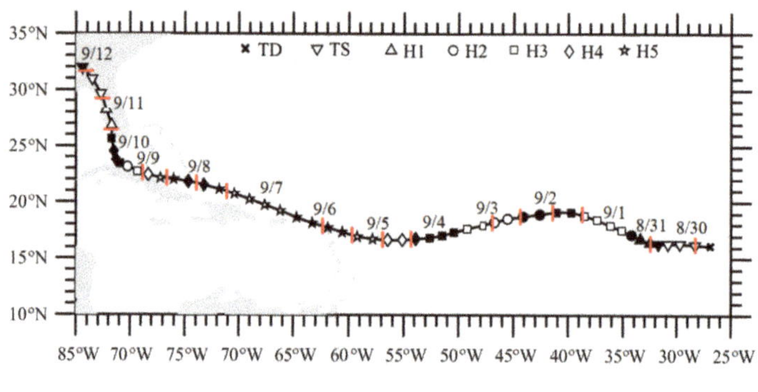

(a)

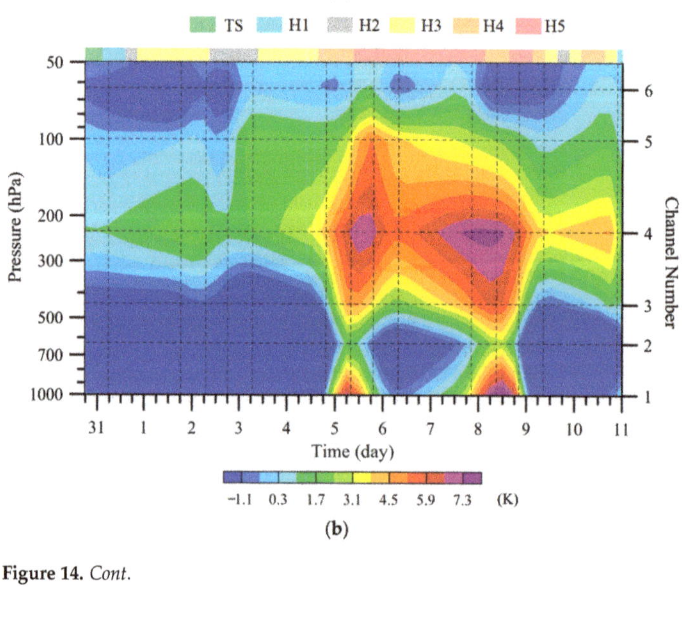

(b)

Figure 14. *Cont.*

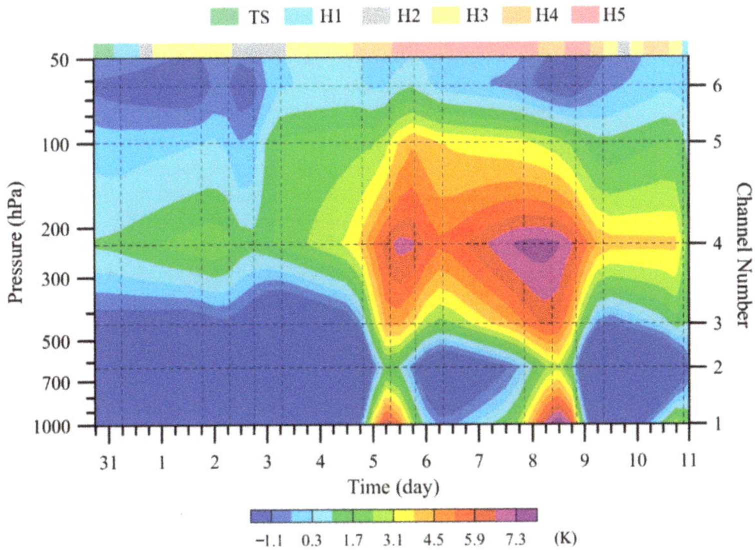

(c)

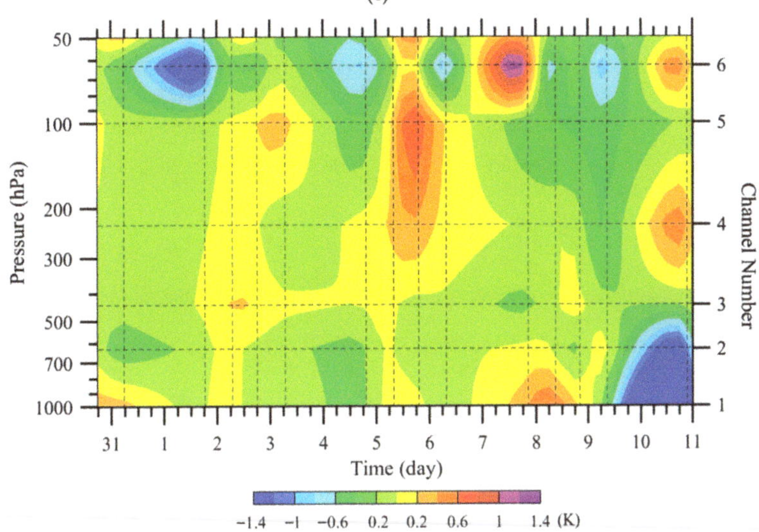

(d)

Figure 14. (**a**) The best track (black markers) of Hurricane Irma from 0000 UTC on 30 August to 0000 UTC on 12 September 2017. Different markers indicate intensity categories. The solid and hollow markers indicate different days. Red markers indicate the SSMIS observation time. (**b**,**c**) Temporal evolution of the vertical variations in SSMIS temperature anomalies at Irma's center (**b**) before and (**c**) after noise reduction from 1800 UTC on 30 August to 0000 UTC on 11 September 2017. (**d**) The detected noise. The times when Hurricane Irma's center was located within SSMIS swaths are indicated by short red lines in (**a**) and black dashed vertical lines in (**b**–**d**).

6. Conclusions

The 20-year long-term satellite observations from F16 provide opportunities and challenges for TC study. Using the conical scanning mode, the TB observations from the

LAS channels can directly reveal useful information on TC location, intensity, size, and rainband distribution. The challenge is that a systematic data noise appeared after 10 years of F16 operation. We aimed to develop an appropriate method for removing the noise so that the remaining 10 years of F16 SSMIS LAS channel observations can be used together with the first 10 years of high-quality data.

Although it appeared complicated, we found a simple 2D FFT that shows promise for the above-intended purpose. We implemented it to an SSMIS swath in a portion-by-portion manner, where a portion consisted of 300 scanlines. An extra step was added to avoid the impacts of heavy precipitation-induced TB outliers of abnormally small value in TB observations of LAS channels 1–4 during the 2D FFT analysis. For each data portion, the data noise appeared to have larger amplitudes at certain across-track wavenumbers than neighboring wavenumbers (either smaller or larger), which does not vary with the along-track wavenumbers. The across-track wavenumbers of data noise vary and depend on the latitude of the F16 orbit. The magnitude of the noise varies between 0.5 and 2 K depending on the channel number.

The TC features seen in TB observations for SSMIS LAS channels are illustrated for Hurricanes Irma and Jose, along with the impacts of the data noise. We plan to first perform an extensive validation of this technique, especially in the tropics, over a longer period and then extend to all data periods with detected noise. Finally, we will use the 20 years of F16 SSMIS observations for a more substantial study on global TCs at both the synoptic and climate scales.

Author Contributions: Conceptualization, H.D. and X.Z.; data curation, H.D.; formal analysis, H.D. and X.Z.; funding acquisition, H.D.; investigation, H.D. and X.Z.; methodology, H.D. and X.Z.; project administration, X.Z.; resources, X.Z.; writing—original draft, X.Z.; writing—review and editing, X.Z. All authors have read and agreed to the published version of the manuscript.

Funding: This research was supported by the National Key R&D Program of China (2018YFC1506702).

Conflicts of Interest: The authors declare no conflict of interest.

References

1. Mo, T. Prelaunch calibration of the advanced microwave sounding unit-A for NOAA-K. *IEEE Trans. Microw. Theory Tech.* **1996**, *44*, 1460–1469. [CrossRef]
2. Mo, T. Diurnal variation ofthe AMSU-A brightness temperatures over the Amazon rainforest. *IEEE Trans. Geosci. Remote Sens.* **2007**, *45*, 958–969. [CrossRef]
3. Kim, E.; Lyu, C.H.J.; Anderson, K.; Leslie, R.V.; Blackwell, W.J. S-NPP ATMS instrument prelaunch and on-orbit performance evaluation. *J. Geophys. Res. Atmos.* **2004**, *119*, 5653–5670. [CrossRef]
4. Zou, X. *Atmospheric Satellite Observations: Variation Assimilation and Quality Assurance*, 1st ed.; Elsevier: London, UK, 2020; pp. 210–211.
5. Derber, J.C.; Wu, W.S. The use of TOVS cloud-cleared radiances in the NCEP SSI analysis system. *Mon. Weather Rev.* **1998**, *126*, 2287–2299. [CrossRef]
6. Kelly, G.; Thépaut, J.N. Evaluation of the impact of the space component of the global observing system through observing system experiments. *ECMWF Newsl.* **2007**, *113*, 16–28.
7. Andersson, E.; Pailleux, J.; Thépaut, J.N.; Eyre, J.R.; McNally, A.P.; Kelly, G.A.; Courtier, P. Use of cloud-cleared radiances in three/four-dimensional variational data assimilation. *Q. J. R. Meteorol. Soc.* **1994**, *120*, 627–653. [CrossRef]
8. Xu, G.Q.; Chen, D.H.; Xue, J.S.; Sun, J.; Wang, S.Y. The program structure designing and optimizing tests of GRAPES physics. *Chin. Sci. Bull.* **2008**, *53*, 3470–3476. [CrossRef]
9. Li, J.; Zou, X. Impact of FY-3A MWTS radiances on prediction in GRAPES with comparison of two quality control schemes. *Front. Earth Sci.* **2014**, *1*, 251–263. [CrossRef]
10. Duan, Y.; Wan, Q.; Huang, J.; Zhao, K.; Chan, P.W. Landfalling Tropical Cyclone Research Project (LTCRP) in China. *Bull. Am. Meteorol. Soc.* **2019**, *100*, ES447–ES472. [CrossRef]
11. Li, X.; Cheng, X.; Fei, J.; Huang, X.; Ding, J. The modulation effect of sea surface cooling on the eyewall replacement cycle in Typhoon Trami (2018). *Mon. Weather Rev.* **2022**, *150*, 1417–1436. [CrossRef]
12. Liang, Z.; Ding, J.; Fei, J.; Cheng, X.; Huang, X. Direct/indirect effects of aerosols and their separate contributions to Typhoon Lupit (2009): Eyewall versus peripheral rainbands. *Sci. China Earth Sci.* **2021**, *64*, 2113–2128. [CrossRef]
13. Ma, Z.; Fei, J. A comparison between moist and dry tropical cyclones: The low effectiveness of surface sensible heat flux in storm intensification. *J. Atmos. Sci.* **2022**, *79*, 31–49. [CrossRef]

14. Amerault, C.; Zou, X. Preliminary steps in assimilating SSM/I brightness temperatures in a hurricane prediction scheme. *J. Atmos. Ocean. Technol.* **2003**, *20*, 1154–1169. [CrossRef]
15. Amerault, C.; Zou, X. Comparison of observed and model-simulated microwave radiance in hurricane environment and estimate of background error covariances for hydrometeor variables. *Mon. Weather Rev.* **2006**, *134*, 745–758. [CrossRef]
16. Amerault, C.; Zou, X.; Doyle, J. Test of an adjoint mesoscale model with explicit moist physics. Part I: Idealized tests on the cloud scales. *Mon. Weather Rev.* **2008**, *136*, 2120–2132. [CrossRef]
17. Amerault, C.; Zou, X.; Doyle, J. Assimilation of rain-affected radiances with adjoint of an explicit moist physics. *J. Appl. Remote Sens.* **2009**, *3*, 033531. [CrossRef]
18. Demuth, J.L.; DeMaria, M.; Knaff, J.A.; Vonder Haar, T.H. Evaluation of advanced microwave sounding unit tropical-cyclone intensity and size estimation algorithms. *J. Appl. Meteorol. Climatol.* **2004**, *43*, 282–296. [CrossRef]
19. Demuth, J.L.; DeMaria, M.; Knaff, J.A. Improvement of advanced microwave sounding unit tropical cyclone intensity and size estimation algorithms. *J. Appl. Meteorol. Climatol.* **2006**, *45*, 1573–1581. [CrossRef]
20. Knaff, J.A.; Seseske, S.A.; DeMaria, M.; Demuth, J.L. On the influences of vertical wind shear on symmetric tropical cyclone structure derived from AMSU. *Mon. Weather Rev.* **2004**, *132*, 2503–2510. [CrossRef]
21. Tian, X.; Zou, X. ATMS- and AMSU-A-derived hurricane warm core structures using a modified retrieval algorithm. *J. Geophys. Res. Atmos.* **2016**, *121*, 12630–12646. [CrossRef]
22. Zou, X.; Tian, X. Hurricane Warm-Core Retrievals from AMSU-A and Remapped ATMS Measurements with Rain Contamination Eliminated. *J. Geophys. Res. Atmos.* **2018**, *123*, 10815–10829. [CrossRef]
23. Tian, X.; Zou, X. A comprehensive 4D-Var vortex initialization using a nonhydrostatic axisymmetric TC model with convection accounted for. *Tellus A Dyn. Meteorol. Oceanogr.* **2019**, *71*, 1653138. [CrossRef]
24. Hu, Y.; Zou, X. Comparison of Tropical Cyclone Center Positions Determined from Satellite Observations at Infrared and Microwave Frequencies. *J. Atmos. Ocean. Technol.* **2020**, *37*, 2101–2115. [CrossRef]
25. Hu, Y.; Zou, X. Tropical Cyclone Center Positioning Using Single Channel Microwave Satellite Observations of Brightness Temperature. *Remote Sens.* **2021**, *13*, 2466. [CrossRef]
26. Northrop Grumman Corporation. Algorithm and Data User Manual for the Special Sensor Microwave Imager/Sounder (SSMIS). 2012. Available online: http://rain.atmos.colostate.edu/FCDR/Archive_Docs/SSMIS_general/ADUM_AppA_Beam_Location.pdf (accessed on 12 October 2021).
27. Berg, W. Climate Algorithm Theoretical Basis Document (C-ATBD) Fundamental Climate Data Record (FCDR) for the Special Sensor Microwave Imager/Sounder (SSMIS). 2013. Available online: https://www.ncdc.noaa.gov/sites/default/files/cdr-documentation/CDRP_ATBD_01B-17_492.pdf (accessed on 12 September 2021).
28. Bell, W.; Candy, B.; Atkinson, N.; Hilton, F.; Baker, N.; Swadley, S.D.; Campbell, W.F.; Bormann, N.; Kelly, G.; Kazumori, M. The Assimilation of SSMIS Radiances in Numerical Weather Prediction Models. *IEEE Trans. Geosci. Remote Sens.* **2008**, *46*, 884–900. [CrossRef]
29. Huang, N.E.; Shen, Z.; Long, R.S.; Wu, M.C.; Shih, H.H.; Zheng, Q.; Yen, N.; Tung, C.; Liu, H.H. The empirical mode decomposition and the Hilbert spectrum for nonlinear and non-stationary time series analysis. *Proc. R. Soc. Lond.* **1998**, *454*, 903–995. [CrossRef]
30. Qin, Z.; Zou, X.; Weng, F. Comparison between linear and nonlinear trends in NOAA-15 AMSU-A brightness temperatures during 1998–2010. *Clim. Dyn.* **2012**, *39*, 1763–1779. [CrossRef]

remote sensing

MDPI

Article

Key Factors for Improving the Resolution of Mapped Sea Surface Height from Multi-Satellite Altimeters in the South China Sea

Lei Liu [1,2], Xiaoya Zhang [2,*], Jianfang Fei [2], Zhijin Li [1,3], Wenli Shi [4], Huizan Wang [2], Xingliang Jiang [1], Ze Zhang [2] and Xianyu Lv [2]

1 Department of Atmosphere and Ocean Sciences, Institute of Atmospheric Sciences, Fudan University, Shanghai 200433, China
2 College of Meteorology and Oceanology, National University of Defense Technology, Changsha 410073, China; feijianfang17@nudt.edu.cn (J.F.)
3 Joint Institute for Regional Earth System Science and Engineering, University of California, Los Angeles, CA 90095, USA
4 Army Academy of Artillery and Air Defense, Nanjing 211131, China
* Correspondence: zxy777@nudt.edu.cn

Abstract: A satellite altimeter measures sea surface height (SSH) along the nadir track. Multiple satellite altimeters have been in orbit, and the measurements been merged for mapping mesoscale eddies of ~100 km in size in the oceans. The capability of the mapped SSH for resolving mesoscale eddies depends on mapping algorithms. A two-dimensional variational (2DVAR) algorithm was implemented to generate mapped SSH at a grid size of 1/12° in the South China Sea. A range of comparisons were performed between the mapped SSH and the commonly used AVISO (Archiving, Validation, and Interpretation of Satellite Oceanographic satellite data) mapped SSH data product at a grid size of 1/8° and 1/4°. The effective resolution, which represents the spatial scale that the data can resolve, was examined. The effective resolution of the mapped SSH using the 2DVAR algorithm is approximately 100 km, while it is 250 km with the 1/8° and 1/4° AVISO data products. The difference in the effective resolution results from the difference in the background state and thus the background error. The result suggests that the effective resolution of the mapped data could be increased by choosing a background state so that the associated errors could have a smaller decorrelation length scale.

Keywords: effective resolution; deviation from background field; merged maps; mesoscale eddies

Citation: Liu, L.; Zhang, X.; Fei, J.; Li, Z.; Shi, W.; Wang, H.; Jiang, X.; Zhang, Z.; Lv, X. Key Factors for Improving the Resolution of Mapped Sea Surface Height from Multi-Satellite Altimeters in the South China Sea. *Remote Sens.* **2023**, *15*, 4275. https://doi.org/10.3390/rs15174275

Academic Editor: Kaoru Ichikawa

Received: 25 April 2023
Revised: 25 August 2023
Accepted: 29 August 2023
Published: 31 August 2023

1. Introduction

Abundant in situ observations of oceans (e.g., Argo, drifter, and glider) have greatly promoted the progress of oceanography and led to unprecedented advancements in oceanic dynamics [1–5]. However, limited by the high cost of deploying oceanic observing platforms, the density of in situ observations is still insufficient to resolve mesoscale eddies at a scale of ca. 100 km in size over the wide-open ocean [6,7]. Mesoscale eddies have long been recognized as "synoptic systems" in the ocean [8]. With a rapid advancement of remote sensing technology, satellite altimeters, which measure sea surface height (SSH), provide most important measurements for detecting and monitoring mesoscale eddies [9].

Satellite altimeters measure sea surface height (SSH) only at nadir points, producing one-dimensional along-track SSH [10]. To resolve two-dimensional (2D) mesoscale features, it is therefore necessary to merge along-track data from multi-satellite altimeters to produce spatiotemporally continuous maps [11,12]. Currently, the most commonly used gridded altimeter maps are from the Data Unification and Altimeter Combination System (DUACS), formerly known as Archiving, Validation, and Interpretation of Satellite Oceanography (AVISO) data products. It is also called level 4 (L4) product. This L4 product was generated using an optimal interpolation (OI) algorithm. The newly reprocessed delayed-time (DT) DUACS daily global absolute dynamic topography (ADT) maps (DT2018 and DT2021) have a global spatial resolution of $1/4° \times 1/4°$ (longitude × latitude), providing an effective resolution (ER) of ca. 240 km at middle latitudes [12].

An ER represents the spatial scale that the data can resolve. We emphasize that an ER differs from the grid size. Unlike the grid size, an ER is defined as a space–time scale of the structure that the gridded data can correctly resolve. An OI method always imposes a filtering effect [13,14]. One question is then whether there is room to increase the ER beyond the AVISO data product [1,15].

The SCS is an immense marginal sea in the near-equatorial tropics. Energetic mesoscale eddies occur at a range of sizes. More than 7000 eddies were identified from the satellite altimetry data during the period 1992–2009 [16]. Based on their generation mechanisms, the eddies can be separated into four geographic regions [17]. The Kuroshio invasion is particularly considered in the surrounding area of the Luzon Strait. More than half of the eddies' radii are 100–200 km, <15% have radii exceeding 200 km, and they have a Rayleigh distribution peak of 130 km [18]. Eddies with the largest radii (100–150 km) occur in the central and western SCS. The SCS region is thus selected in this study.

A two-dimensional variational (2DVAR) method is implemented to map ADT data. ADT should be used, rather than sea level anomalies (SLA) for eddy detection [19]. The 2DVAR method was recently introduced for mapping the multi-satellite altimeter data for the East China Sea, the SCS, and the California Current system. The merged data from those areas was shown to have better ER than the AVISO data product [11,12,20].

The 2DVAR and OI methods are based on optimal estimation theory. Both methods use a background state. The associated background error covariance thus plays a key role in the methods. The background error correlation can be characterized by its correlation length scale. The main objective of this study is to illustrate that the background error decorrelation length scale is a factor that dictates the ER of the mapped data. To improve the ER, a proper background state should be carefully chosen. The background error should have a relatively smaller correlation length scale.

The ER in 2DVAR mapped data and AVISO data products are compared. The comparison is made not only with the $1/4°$ global product but also with a customized regional $1/8°$ product in the SCS, particularly provided by DUACS. Another focus is to examine whether the specially customized regional $1/8°$ product has the same performance as other regional AVISO products. Finally, this work demonstrates the robustness of the 2DVAR mapped data, in terms of its higher ER and quality. To comprehensively validate the quality of the merged 2DVAR and $1/4°$ AVISO products, we used two additional mapping datasets (the $1/4°$ AVISO and the HYCOM products), comparing their accuracy and ER. These findings on the HYCOM ER could be meaningful for future data assimilation, as the HYCOM product can distinguish 80 km-scale ocean structures.

2. Data and Methods

2.1. Datasets

Major datasets that are used in the analyses are summarized as follows.

First, the $1/4° \times 1/4°$ AVISO data product. It is also known as the DUACS-DT2021 maps. For convenience, it is called '$1/4°$ AVISO'. This is the latest version of $1/4°$ AVISO. In this version, the following improvements were implemented over the previous DUACS-DT2018 version [14]:

A. New altimetry standards and geophysical corrections were used to improve the accuracy of sea level anomaly (SLA) content. The regional mean sea level (MSL) trend and regional deviation was affected.

B. The new 'internal tide' correction was used to improve the mesoscale signal mapping.

C. The new mean sea level (non-repetitive and recent tasks) or mean profile (repetitive task) was used to improve the accuracy of SLA and regional deviation.

D. The new mean dynamic topography (MDT) was used to improve the geostrophic current and regional deviation.
E. The mesoscale signal on the L4 products were improved by using the improved mapping parameters.

Second, The 2021 customized version of the 1/8° × 1/8° grid-resolution DUACS ADT product ('1/8° AVISO').

Table 1 lists the improvements in the 1/8° regional European AVISO product compared with 1/4° global AVISO product, as well as the regional spatial ER [16,21,22].

Table 1. Variance and effective resolution (ER) of regional Europe products. The first column is the variance of the differences between 1/8° DT2021 regional Europe product and SARAL-DP/ALtika independent along-track measurements from 2016 to 2019. In parenthesis, variance reduction (in %) compared with the results obtained with the 1/4° DT2021 Global product. The second column is the spatial ER of the 1/8° DUACS-DT2018 product in the regional European seas. In parenthesis, average resolution over the basin.

European Seas	Variance [cm^2]	Effective Resolution [km]
Black Sea	14.4 (−0.94%)	100 to 150 (~130)
Mediterranean Sea	15.3 (−4.25%)	90 to 160 (~130)

The 1/8° regional European AVISO products provided substantially better spatiotemporal resolution than the global product. For the current study, 1/8° AVISO products were custom-produced for the SCS region for the first time. We hope that they will match the European Sea AVISO products in quality.

Third, the daily averaged reanalysis product of a fully 3D, multivariate, variational ocean data assimilation system (the Navy Coupled Ocean Data Assimilation (NCODA) System (HYCOM-NCODA, hereafter 'HYCOM'; U.S. Naval Oceanographic Office), obtained by averaging the data over eight periods each day. The HYCOM reanalysis product was produced by using the 3D-Var method to assimilate the satellite altimeter SSH, in two alternative ways: (1) assimilating the along-track data of the satellite altimeter directly, and (2) assimilating the merged maps of 2D horizontal analysis of SSH [23].

To evaluate the reconstruction achieved using the different models, we first used independent data, including remote sensing Multi-scale Ultra-high Resolution (MUR) Sea Surface Temperature (SST) data and in situ Global Drifter Program (GDP) drift buoy data [24,25]. In addition, the AVISO along-track L3 data—from the Sentinal-3A (S3A) and Jason-3 (J3) satellites—were used for observation-based ground truthing. Owing to their low observation error and stable operation, S3A and J3 are used primarily to analyze the merged field error and the ER and useful resolution (UR).

The selected period was from 1 April to 1 September 2018, in the SCS, covering 0°–26°N and 100°–130°E. Five altimeters were in orbit in this region and period, providing sufficiently dense observational coverage and corresponding to the customized 1/8° AVISO product provided by DUACS.

2.2. A Two-Dimensional Variational Method

2DVAR is based on the 2D variational principle. The solution is optimal in the sense of minimum error variance [11,12,20]. The solution is obtained by minimizing the cost function:

$$J(\boldsymbol{h}) = \frac{1}{2}(\boldsymbol{h}-\boldsymbol{h}_b)^T\boldsymbol{B}^{-1}(\boldsymbol{h}-\boldsymbol{h}_b) + \frac{1}{2}\sum_{s=1}^{N}(\boldsymbol{H}_s\boldsymbol{h}-\boldsymbol{h}_s^o)^T\boldsymbol{R}_s^{-1}(\boldsymbol{H}_s\boldsymbol{h}-\boldsymbol{h}_s^o) \quad (1)$$

where $\boldsymbol{h}$ is the analysis field, $\boldsymbol{h}_b$ and $\boldsymbol{h}_S^o$ are the background and observation fields of the SSH, respectively; $\boldsymbol{B}$ and $\boldsymbol{R}_s$ are the error covariance matrices of the background and observation fields, respectively; the subscript 's' denotes the time level, and 'N' is the total number of time levels of observation. $\boldsymbol{H}_s$ is the observation operator that maps the background to the observations. The superscripted 'T' indicates 'transposing matrix'.

The analytical solution that minimizes Equation (1) can be written as:

$$\boldsymbol{h}^a = \boldsymbol{h}_b + \left(\boldsymbol{B}^{-1} + \sum_{s=1}^{N} \boldsymbol{H}_s^T \boldsymbol{R}_s^{-1} \boldsymbol{h}_s^o - \boldsymbol{H}_s \right)^{-1} \sum_{Seq=1}^{N} \boldsymbol{H}_s^T \boldsymbol{R}_s^{-1} (\boldsymbol{h}_s^o - \boldsymbol{H}_s \boldsymbol{h}_b) \tag{2}$$

The AVISO products are based on the optimal interpolation principle, and the solution can be written as follows [26]:

$$h_x^a = h_x^b + \sum_{r=1}^{M} C_{xr} \left[\sum_{q=1}^{M} \boldsymbol{A}_{rq}^{-1} \left(h_q^o - h_q^b \right) \right] \tag{3}$$

where h_x^a is the SSH estimate on the regular grid x, h_x^b is the background field, h_q^o is the observation field, $\boldsymbol{A}_{rq}$ is the covariance matrix between observation points r and q, and $\boldsymbol{C}_{xr}$ is the covariance matrix between the observation and grid points. The solutions (Equations (2) and (3)) can be shown to be equivalent to each other [15].

One factor that the two methods was adapted differently is the background fields $\boldsymbol{h}_x^b$. The 2DVAR model used the merged map of the previous day, whereas DT2010, DT2014, and DT2018 (AVISO products) used mean dynamic topography (*MDT*), which is the *SSH* over a period *T* [seven years (1993–1999), 20 years (1993–2012), and 25 years (1993–2017)] above the Geoid, as follows [21]:

$$MDT_T = \langle SSH \rangle_T - Geoid \tag{4}$$

where *SSH* is the height of the surface above the ellipsoid of the reference and includes the *Geoid*. The ADT field on the initial day of the 2DVAR product was obtained by directly interpolating the observation data along a satellite track without a background field. One-day merging can eliminate the error differences between different background fields. The first day of the merging algorithm was not used in the evaluation as a product in Section 3. The 2DVAR method used 10 d of along-track data before and after the central day (21 d in total) and considered the temporal evolution error. Therefore, there was temporal correlation in the background field. The different period *T* of the background fields used in the 2DVAR and AVISO models will introduce different background errors:

$$\varepsilon_x^b = h_x^b - h_x^t \tag{5}$$

where h_x^b is the background field and h_x^t is the unknown actual value at the background state time. During mapping, the covariance ($\boldsymbol{B}$ in Equation (2) and $\boldsymbol{C}_{xr}$ in Equation (3)) is directly related to and is constructed using the background error ε_x^b. Simultaneously, the background error influences the sea level anomaly field. For example, the longer the averaged period, the closer the background field is to the actual value, and the more accurate the inter-annual and climatic-scale signals are in the SLA [26]. However, an increase in the time-averaged range of maps as the background field may not improve the covariance matrix, in turn hindering the improvement in the resolution capability of the merged maps. To demonstrate this and to analyze the influences of the background field, different $\boldsymbol{B}$ $(\boldsymbol{C}_{xr})$ values based on the above two methods were simulated. Although it cannot be obtained directly by observing its error, the estimation of $\boldsymbol{B}$ $(\boldsymbol{C}_{xr})$ can be used to solve the function equation.

In the mapping method, the background error covariance matrix $\boldsymbol{B}$ $(\boldsymbol{C}_{xr})$ and the covariance matrix of observation error ($\boldsymbol{H}_s$ in Equation (2) and $\boldsymbol{A}_{rq}$ in Equation (3)) are used as the weight parameters that determine the dissemination of information in the estimation field of the observed data. Assuming that the observation error is spatially uncorrelated, $\boldsymbol{B}$ $(\boldsymbol{C}_{xr})$ determines the propagation weight of the spatial observation data.

In data assimilation, the information propagation weight primarily represents the filtering characteristics [18]. $\boldsymbol{B}\ (\boldsymbol{C}_{xr})$ always can be decomposed as follows:

$$\boldsymbol{B} = \sum \boldsymbol{C} \sum \tag{6}$$

where $\boldsymbol{C}$ is the correlation matrix.

Estimating the matrix $\boldsymbol{B}\ (\boldsymbol{C}_{xr})$ thus requires the length scale of correlation matrix $\boldsymbol{C}$, which is the core parameter characterizing the covariance matrix $\boldsymbol{B}\ (\boldsymbol{C}_{xr})$, because the length scale extracts significant information from the matrix [27]. The auto-correlation function was used here to calculate the correlation $\boldsymbol{C}$; the distance at which the correlation dropped to 0.6 was used as the length scale of its correlation.

According to the definition of ε in the forward calculation of SSH, the simulation of ε for 2DVAR and AVISO in this study was specified as follows: 2DVAR used the merged maps of the previous day as the background fields, whereas AVISO used the 7-, 20-, and 25-year climate-averaged fields from 1993. The background error was calculated by subtracting the along-track daily data from the background field. For the linear along-track datasets, the covariance of each datasets was calculated using other data within 90 km. The covariance of all the along-track data was then averaged, and the covariance of the Gaussian distribution was obtained via ordinary least squares fitting.

The autocorrelation function of a uniform and isotropic one-dimensional spatial random field is equivalent to the power spectral density (PSD), according to the Wiener–Khinchin theorem [28]. For spatially uniform background errors, the spectral representation of the covariance matrix $\boldsymbol{B}\ (\boldsymbol{C}_{xr})$ is equivalent to the spectral representation of the correlation matrix $\boldsymbol{C}$. For homogeneous and isotropic background errors, the covariance in the spectral space is the variance of each wave number, namely PSD. The correlation based on the autocorrelation function should have physical (filtering) characteristics similar to the PSD of the background error. The degree of filtering of small-scale signals should have the same variation as the correlation: that is, these values should increase together [29].

The following section presents the relationship between the length scale of the correlation and the energy proportions of different scale errors. The Fourier transform method was used to calculate the energy spectral density which represents the PSD of the background errors. To highlight the differences in the energy spectral density between different models, the spatial distance was divided into five scales, and the portion below the minimum distinguishable scale in all models was omitted when calculating the percentages.

3. Results

3.1. Signal Proportion of Different Scales in the Background

The 2DVAR method generated the smaller correlation length scale than the AVISO method (Figure 1). For the AVISO method, the correlation length scales increased with time in years averaged for the background. This indicates that, even when a finer spatial grid is used, selecting a too-long averaged period for the background field prevents the reduction in the correlation length scale.

Around the SCS, slightly more than 25% energies of the identified signal in the background errors of the 2DVAR method were for small mesoscale signals (50 km to 150 km), and 20% energies for signals of 150 to 400 km (Figure 2). In the background errors of the AVISO method, small mesoscale (<150 km) signals had a deficient proportion (2–6%) of energies in the 7-, 20-, and 25-year averaged background. Those signals with a spatial scale >400 km accounted for 75% (7 y) and 88% (20 and 25 y) of the total energy, respectively. The 2DVAR method generated more signals that have smaller scales because it uses a day-to-day estimate, giving less weight to low-frequency longer spatial scales, whereas the AVISO methods exhibit the opposite effect at each length scale.

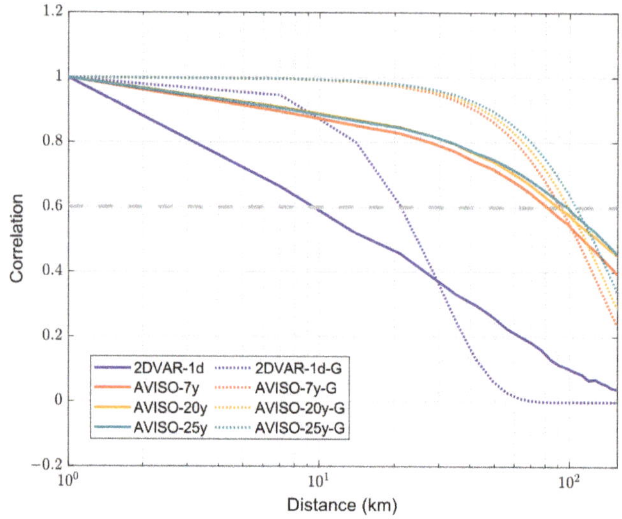

Figure 1. The correlation length scale and its Gaussian-fit distribution of the 2DVAR and AVISO maps with different background time windows.

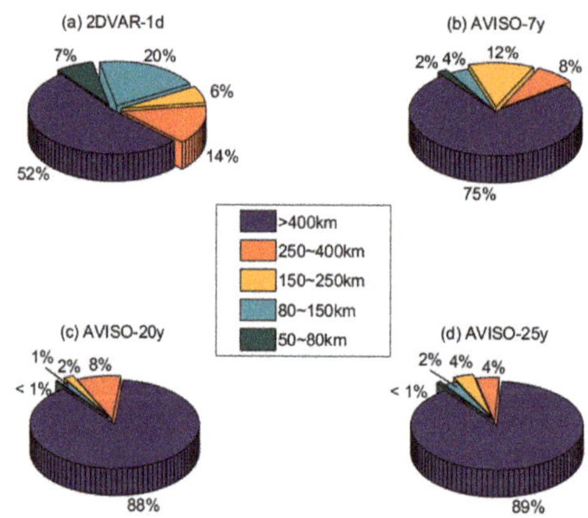

Figure 2. Proportion of error energy at different scales of background errors with the 2DVAR and AVISO methods with different background time average windows.

As the length of the background time window increases from day to year, the proportion of energies with large-scale signals remaining in the background errors increased, and the correlation length scale increases. Therefore, using a long-time-averaged background field in the mapping implementation hinders improvement of the resolution capability of the merged maps.

3.2. Evaluation of Accuracy

To validate the reconstruction of small mesoscale structure and consistency with the actual signal, different merged maps were evaluated against independent SST and drifter trajectory data and dependent along-track satellite ADT data. Tides significantly influence the coastal area of the SCS. The continental shelf, at an average depth of 200 m, was not included in the analyses in this section.

3.2.1. Remote Sensing Evaluation

Although the spatial patterns of anticyclone eddies are not entirely consistent in terms of SST, or are even orthogonal in some weaker eddies [30], SST of mesoscale eddies is often characterized by a warmer center of the anticyclone, similarly, cyclonic eddies often have colder centers. To a certain extent, SST is responsive to small mesoscale dynamic processes in the upper ocean. In this section, we compared the sea surface geostrophic current, inverted from the ADT data, with the MUR SST anomaly (SSTA) to verify the accuracy of the estimated ADT structure [31].

2DVAR single-day surface flow and temperature corresponded highly in terms of refined structural features (Figure 3a). The other three models show this correspondence only at a larger scale, with poor correspondence in terms of the same refined structure (Figure 3b–d). The geostrophic current of the 1/4° AVISO model was weaker than that of the other models. Although the 2DVAR model achieved better reconstruction of mesoscale eddies, it may induce more small-scale noise in the map the same SST.

3.2.2. In Situ Evaluation

The in situ GDP hourly drifting buoy datasets [3] is widely used to study small-scale high-frequency dynamic ocean processes. The GDP measures ocean current velocity at a depth of 15 m using a drogue that pulls a surface float underwater with each passing wave. The drogue creates strain against the lower hull of the surface float, thereby reducing the effect of wind. Here, a drifter path in the middle of the SCS, far from the coast areas, was selected to reduce the influence of tides. A drifter with a curved moving path was compared to evaluate its ability to capture the mesoscale structure in the various merged maps.

On 29 May 2018 (Figure 4), the buoy paths were parallel to the contours of the ADT gradients of the four models. On 30 and 31 May 2018 (Figure 4b,c), the buoy paths were almost parallel to the contours of the ADT gradient of the 2DVAR model and perpendicular to those of the other maps. The result of the other models are mostly inconsistent with the buoy data shown in Figure 4d, for 1 June 2018. The relationship between the buoy path and the height field gradient is similar to that in panels c–e of Figure 4. Comparing the 2DVAR height field with that of the stand-alone buoy reveals that the buoy rotated counterclockwise around a small eddy with a low center at SSH, consistent with the eddy characteristics obtained using the 2DVAR method. The two AVISO models and HYCOM have poor reconstructions of smaller-scale eddies, thus resulting in buoy trajectory crossovers with their height fields. Therefore, 2DVAR exhibits certain advantages over the other models in terms of its ability to present smaller mesoscale eddies in the ocean.

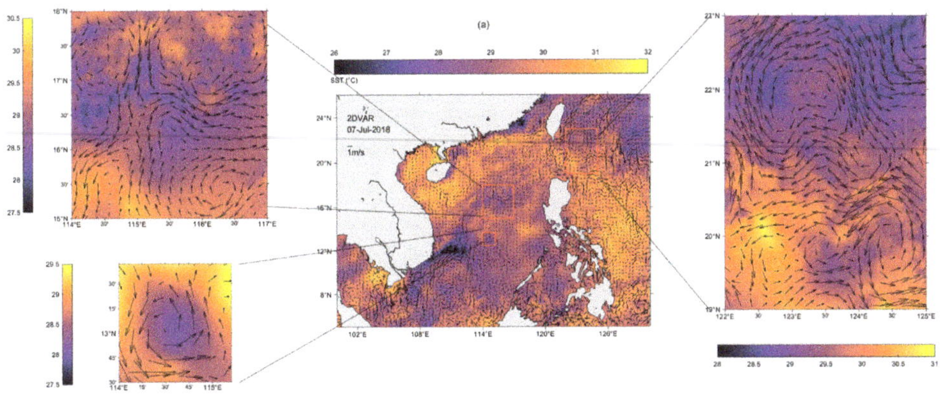

Figure 3. *Cont.*

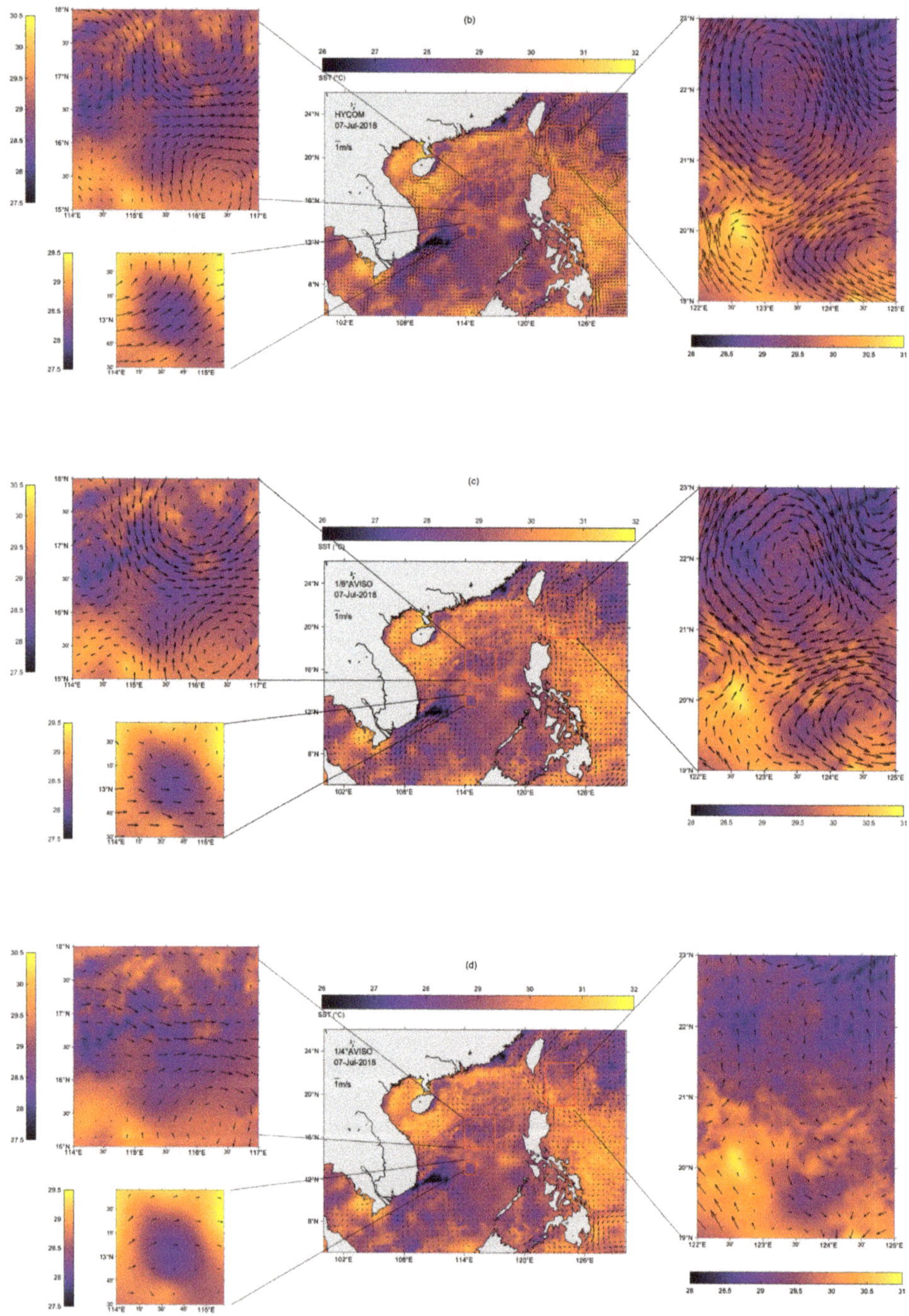

Figure 3. MUR sea surface temperature (SST) and the inversion of geostrophic current from merged absolute dynamic topography (ADT): (**a**) 2DVAR, (**b**) HYCOM, (**c**) 1/8° AVISO, and (**d**) 1/4° AVISO on 7 July 2018.

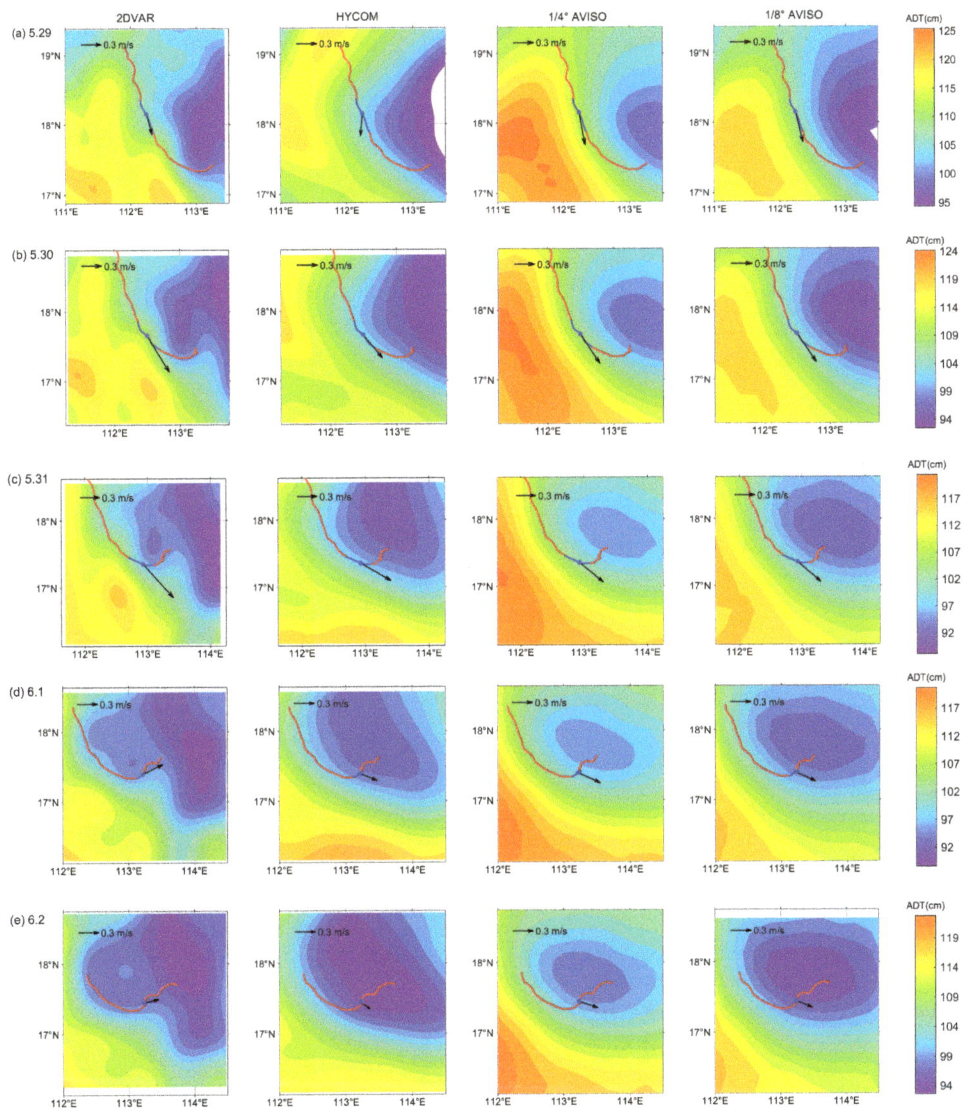

Figure 4. Absolute dynamic topography (ADT) maps of 2DVAR, HYCOM, 1/8° AVISO, and 1/4° AVISO (sorted by column) with the drifting buoy path. The days are (**a**) 29 May, (**b**) 30 May, (**c**) 31 May, (**d**) 1 June, and (**e**) 2 June 2018 (sorted by row, respectively). Blue line: drift path of the buoy on the focal days; red line, drift path during the two days before and after. The arrow indicates the geostrophic vectors from the middle moment of the focal day.

All of the models exhibited lower geostrophic velocities than the drifting buoys (Figure 5). Although the HYCOM data were the most concentrated in the regression curve, the 2DVAR regression curve illustrates a better fit, with a slope closest to 1. The 2DVAR model also had the lowest root mean square deviation (RMSD) for the entire map.

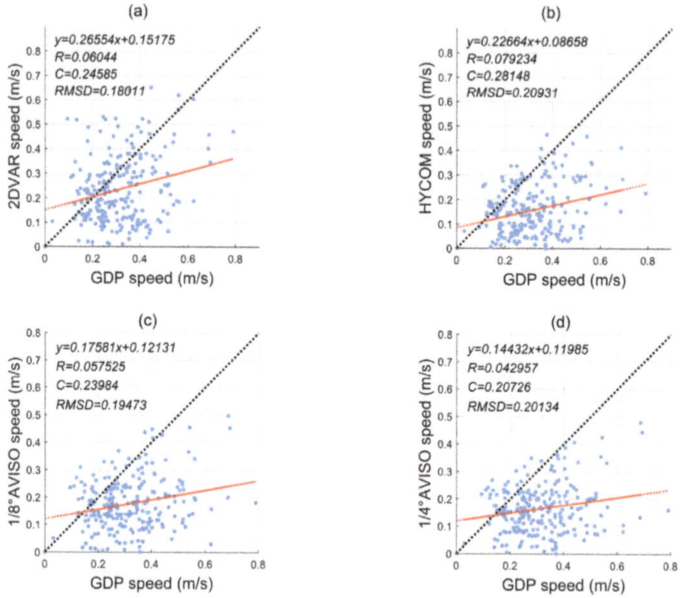

Figure 5. Geostrophic velocity scatter plot and linear regression curve of (**a**) 2DVAR, (**b**) HYCOM, (**c**) 1/8° AVISO, and (**d**) 1/4° AVISO with the drifting buoy. The slope of the dotted black dot is 1. Equation y on the upper left corner represents linear regression curve which is the dotted line colored in red. R is the correlation of linear regression coefficients. C is the correlation coefficient between the merged map and the drifting buoy data, and the root mean square deviation (RMSD) between the merged map and the drifting buoy data.

3.2.3. Along-Track Satellite Evaluation

To calculate the root mean square error (RMSE), correlations, and linear regressions of the merged maps and thus evaluate their accuracy, the L3 along-track data from the S3A and J3 satellites were used as the hypothetical true values [11]. S3A has smaller orbital spacing and higher data density satellite than J3, allowing it to cover a larger area in the full-field mapping of RMSE. To illustrate the relationship between the 2DVAR maps and those generated using the other models, an index score (S) comparing the deviations of each product was calculated [12]:

$$S = 1 - \sigma^2_{\text{2DVAR}} / \sigma^2_{\text{Others}} \quad (7)$$

where σ is the mean square error for each model; 'Others' refers to the models other than 2DVAR; and S is the degree of deviation of other models relative to 2DVAR. A positive S indicates a higher error relative to that of 2DVAR, while a negative S reflects an error less than that of 2DVAR. Table 2 lists the mean RMSE and S for the entire map, reflecting the accuracy of the merged maps.

2DVAR had the lowest RMSE for the center waters of the SCS (at only 1–2 cm), followed by the 1/4° AVISO model (at ca. 1 cm higher) (Figure 6a,c,e,g). The 1/8° AVISO model had an RMSE of ca. 5–6 cmca. double those obtained in the previous two models (Figure 6d,h). HYCOM exhibited the highest RMSE (>20 cm, relative to the S3A data) in the northern SCS (Figure 6b) and <10 cm relative to the J3 data (Figure 6f). The large RMSE obtained for the Pacific Ocean, Luzon Strait, and the Celebes Sea was due to the tidal influences in coastal areas and high eddy kinetic energy (EKE) transport from the Pacific crossing Luzon Strait to Kuroshio [32]. In summary, the mean RMSE for the entire map (Table 2) was lowest for the 2DVAR model, at ca. 0.01 cm lower than that of the AVISO global 1/4° model, and substantially lower than that of the HYCOM model. S varied in the same way as RMSE, being highest for HYCOM and lowest for 1/4° AVISO model.

Table 2. The accuracy of 4 models. The root mean square error (RMSE, in cm) for entire map compared with along-track S3A and J3 satellite L3 data, and the mean error index score S for entire map of each product compared with 2DVAR.

Experiments/ Models	RMSE [cm]		S	
	S3A	J3	S3A	J3
2DVAR	0.0299	0.0340	/	/
HYCOM	1.5946	0.5189	0.8987	0.9421
1/8° AVISO	0.0678	0.0688	0.6658	0.7070
1/4° AVISO	0.0396	0.0414	0.1125	0.2313

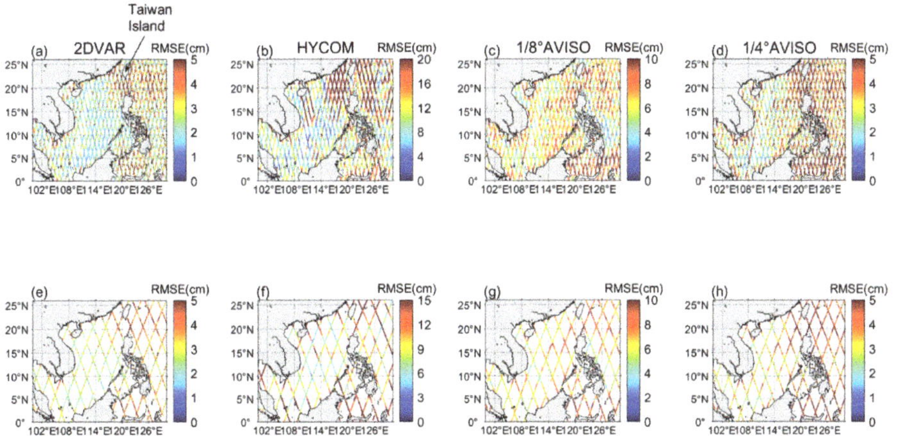

Figure 6. Root mean square error (RMSE) on the satellite track of merged absolute dynamic topography (ADT) compared with S3A (**a**–**d**) and J3 (**e**–**h**) satellite along-track data: (**a**,**e**) 2DVAR, (**b**,**f**) HYCOM, (**c**,**g**) 1/8° AVISO, and (**d**,**h**) 1/4° AVISO. Note that the color scale for each map is differ from each other for the uniform of colors.

Due to the strong influence of the Kuroshio, the zonal flow in the surface (shallower than 500 m) of the Luzon Strait is typically westward [32]. Thus, most of the high-frequency mesoscale eddies generated from the Bashi and Northern Philippines Strait were propagated from the longest channel of eddy propagation in the SCS to the southwest of the northern SCS (from the Luzon Strait to ca. 18°N, 112°E), often with a long lifecycle [33]. The high RMSE values southwest of Taiwan island in both of the AVISO models reflect the same distribution characteristic as the longest channel of eddy propagation. This result may occur because the increased presence of eddies in the channel increases the spatiotemporal variability in sea surface conditions, increasing the difficulty in reproducing smaller eddies. For the two AVISO maps, the poor scale-recognition resolution resulted in higher RMSE values in the channel. While 2DVAR model benefits from its good scale-recognition resolution, its RMSE exhibited very low consistency with the propagation channel.

The 2DVAR model exhibited a linear regression correlation coefficient (R) > 0.95 and had the highest correlation coefficient (C) (ca. 0.98) (Figure 7a,e). The correlation between the HYCOM data and the S3A hypothesized true value was only 0.64 (Figure 7b), almost 0.30 points smaller than that of the other three models. This is highly consistent with the differences in RMSE values.

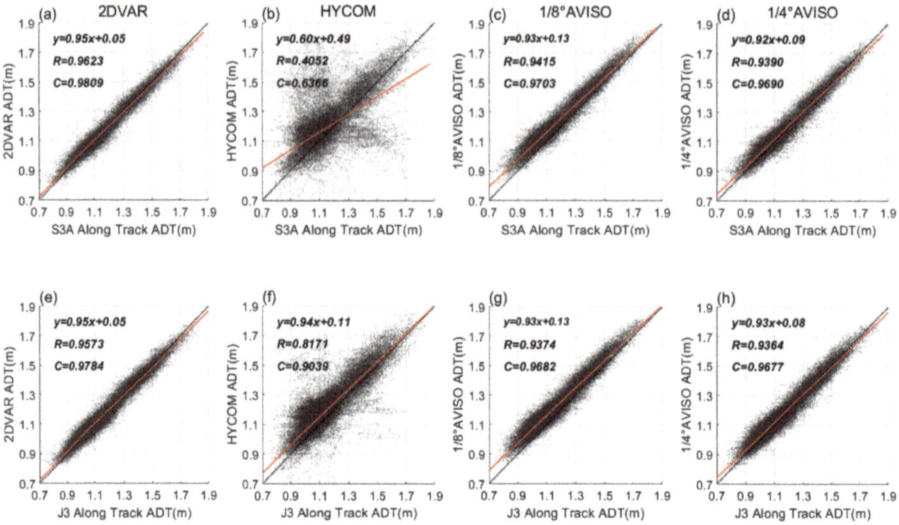

Figure 7. Correlations between the merged maps: (**a**,**e**) 2DVAR, (**b**) HYCOM, (**c**) 1/8° AVISO, (**d**) 1/4° AVISO, and S3A (**a**–**d**) and J3 (**e**–**h**) satellite along-track data. The ratio of the black dotted line is 1; *C* is the coefficient of correlation; the red dotted line is the linear regression curve; *y* represents its equation; and *R* is the coefficient of correlation for linear regression.

3.3. Evaluation of Effective Resolution

Figure 8a presents the ADT sequence of the along-track sampling points from the four merged maps and J3 satellite data for a randomly selected date (9 July 2018). The track (Figure 8b) is sorted in the sequence P1–6. Among the four merged maps, 2DVAR exhibited the best fit and best reflected the satellite signal amplitude and small-scale information, especially at P1–2. HYCOM performed the worst, with the largest deviation from satellite data, especially at P5–6. The 1/4° AVISO model deviated slightly from the satellite data. The deviation between the 1/8° AVISO and 1/4° AVISO data waveform was almost regular at every sampling point. An offset occurred because the time scale of correlation and the MDT of the customized 1/8° AVISO product had been adjusted.

Each product was a mapping of multiple satellite data, including the J3 data. Part of the merged data approximately 370 and 420 points did not match the ADT variance in J3. The reasons may not only be related to the smaller temporal scales of these processes than that used in 2DVAR or AVISO but also to their smaller spatial scales than the background error correlation coefficient scales.

ER refers to scale-recognition resolution, a parameter that relates the recognition capability of the altimeter product to the dynamic signal of ocean eddies. Scale-recognition resolution is defined as the minimum resolvable spatial scale of the signal in the merged map, and its value indicates the smallest sea-surface eddy that can be distinguished in the signal from the perspective of energy spectral density [13].

Scale-recognition resolution was calculated using the definitions of ER and UR [21] based on the wavenumber PSD of the ADT field, using the observational ADT data as a spatial sequence with distance as the independent variable. ER was obtained using the ratio of the PSD of the noise to the signal, i.e., the noise–signal ratio (NSR). UR was obtained using the ratio of the PSD between the estimated value of the merged maps and the satellite along-track signal, i.e., the signal ratio (SR). At NSR or SR of 0.5, the wavelengths (λ) corresponding to their positions are ER and UR, respectively. By comparing local phase differences, using ER reduces the large systematic error generated by the comparison between the different phases, although it may be more affected by noise. UR reflects a comparison of spatial sequence spectral amplitudes that relatively better represent the energy magnitude of the signal available within the product data. The average value of each grid point along the track during the study period was recorded to obtain the resolution along the track.

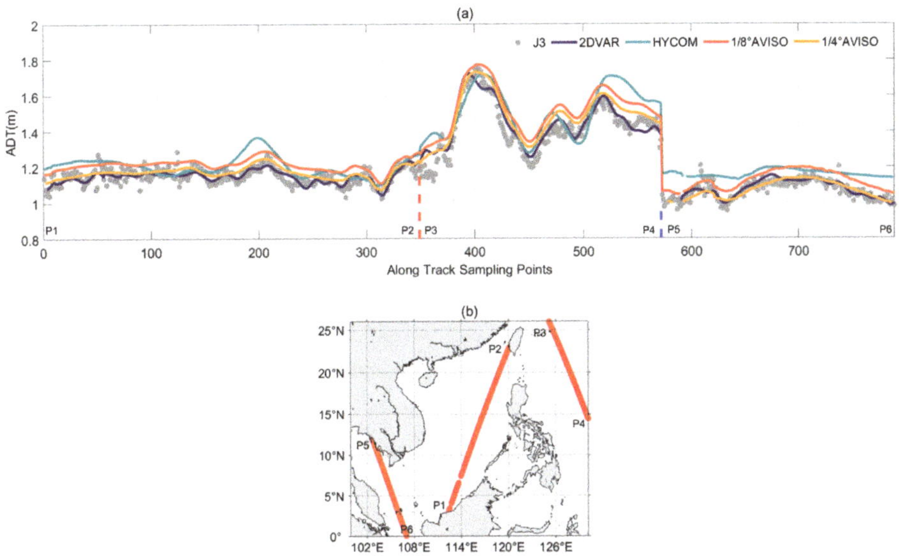

Figure 8. The absolute dynamic topography (ADT) sequence (**a**) from merged maps along the along-track points of satellite J3 on 9 July. The red and blue dashed lines are the separation position between different tracks. The J3 satellite track in the South China Sea (SCS) for 9 July (**b**); red dotted line). The sequence P1–6 indicates the direction of movement of different parts of the satellite tracks and corresponds to the sequence of the sampling fragments separated by the red and blue dashed lines.

Data for the same day as in Figure 8 (9 July 2018) was selected to display the PSD and scale-recognition resolution results. In the wavelength range of 90–400 km, the PSD value of the mapping field deviation of the 2DVAR product was much lower than that of the 1/8° and 1/4° AVISO and HYCOM models (Figure 9a). The PSD value approximately 70–200 km wavelength of the mapping field of the 2DVAR product was higher than that of the other models shown in Figure 9b. Excluding the 2DVAR model, the other three models did not differ substantially in PSD. The order of ER and UR of the same product is consistent among all products (Figure 9c,d). Relative to the two AVISO models, the 2DVAR model had ER and UR values ca. 30% lower, whereas HYCOM had slightly higher ER and UR values.

Both ER and UR of the 2DVAR model were mostly between 50 and 150 km, with the higher resolutions near the coast or islands (Figure 10a–d). The HYCOM data ER showed an extremely high value, (>200 km) and occupying a large area, with the UR mostly between 100 and 200 km (Figure 10b,f). For the two AVISO models, ER showed 150–250 km in a large area, with some being ≥250 km. For the two AVISO models, UR was concentrated approximately 200 km in most parts of the study area. For all four models, ER was much higher in the vicinity of the Philippine Islands and their coastal waters than else. Owing to the presence of anomalously fluctuating signals (noise), which could be confused with small-scale signals in the ocean, ER had limited usefulness for resolving eddies.

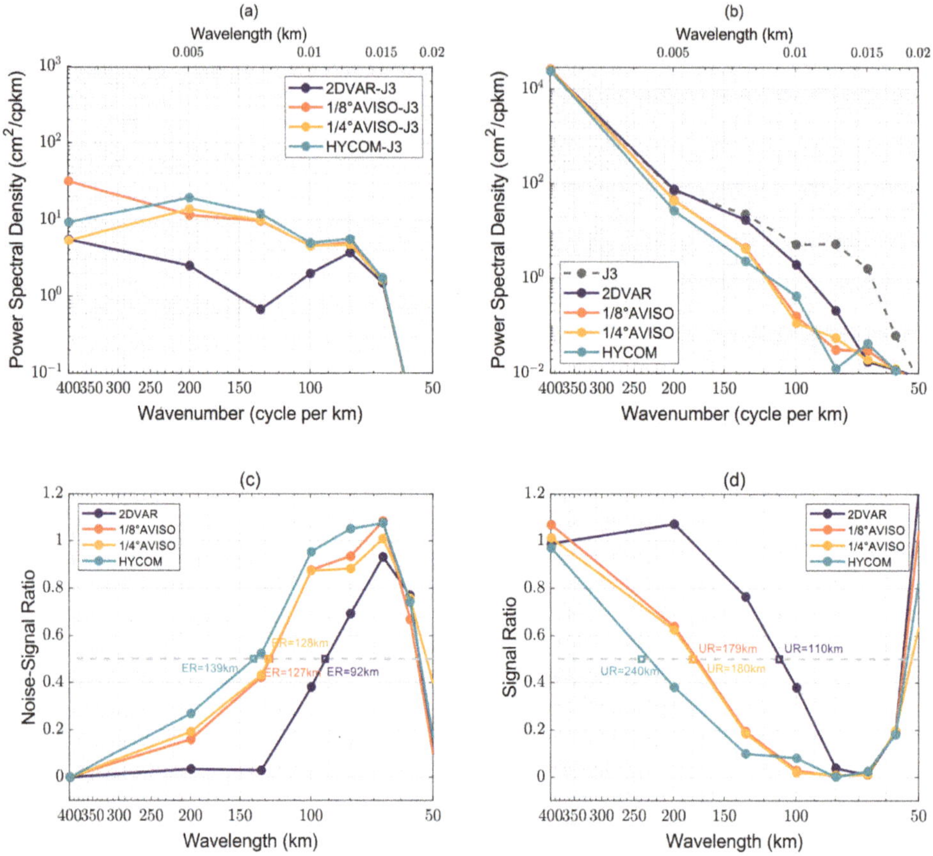

Figure 9. The power spectral density (PSD) that calculated by (**a**) subtraction and (**b**) non-subtraction of the J3 data (except for the gray line in (**a**), it was the origin data of J3 satellite) from the four merged maps (2DVAR, HYCOM, 1/8° AVISO, and 1/4° AVISO). (**c**) Effective resolution (ER) and (**d**) useful resolution (UR) of the four merged maps based on NSR and ER, respectively on 9 July.

By definition, the ER value of an area with a high error increases correspondingly, and the minimum resolved scale of the ocean eddy signal in the data increases. Comparing the error statistics described in Section 3.2.3, ER > UR for most of the regions with large RMSE. For example, for the southwestern waters regions with high EKE, around the Dongsha Islands of Taiwan Island [16], ER < UR for the 1/4° AVISO model, while for the other three models, ER > UR. Based on these findings, for zones with large RMSE values, UR is more effective than ER for identifying eddies.

Determining the ability to reconstruct eddies involves evaluating the scale of the spatial field and the time scale. Based on a previous work [20], we used the time-frequency domain and spatial wavenumber as reference elements to determine the 2D distribution of the energy spectrum of the height field. The merged maps were decomposed via a 3D Fourier transform. In terms of the period, 2DVAR captured the signals much better than the two AVISO models, within 20 d and 80–200 km (Figure 11a,c,d). The frequency wavelength PSD of HYCOM was large in the wavelength range of 200–1000 km and in the period 0–100 d (Figure 11b). The 2DVAR model exhibited a very low proportion of energy for spatial scale of <80 km (Figure 11a) relative to that in the 80–200 km range, owing to filtering by its variational merging. Although the AVISO models had an energy proportion below 80 km (Figure 11c,d), the method filtered out more energy at the 80–200 scale.

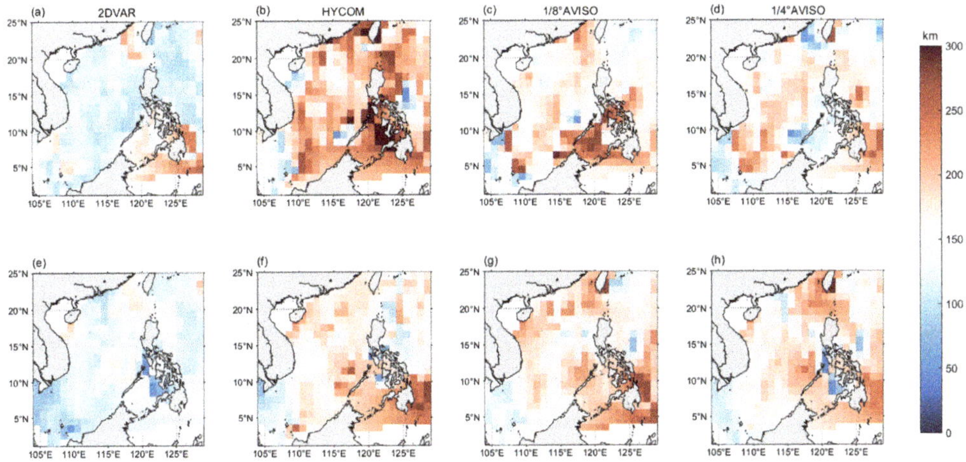

Figure 10. (**a–d**) Effective resolution (ER) and (**e–h**) useful resolution UR of the four merged models of (**a**,**e**) 2DVAR, (**b**,**f**) HYCOM, (**c**,**g**) 1/8° AVISO, and (**d**,**h**) 1/4°AVISO in the South China Sea (SCS) and nearby waters used S3A satellite data as true values; the darker the color is, the larger the ER or UR value is.

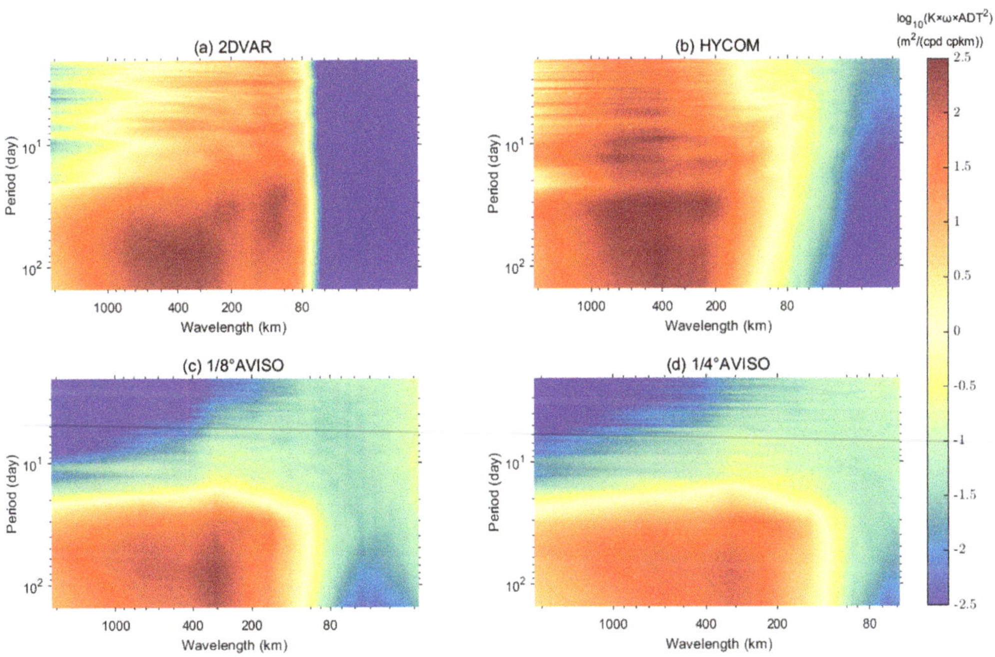

Figure 11. The frequency wavelength power spectral density (PSD) values of the four merged maps. Due to the different spatial map resolutions, the coordinate range of each spectrum was different.

The 2DVAR and AVISO models were used as the hypothetical ground truthing to examine the HYCOM data, and a parameter similar to ER was used to score the frequency wavelength PSD (*FWPSDs*), as follows [20]:

$$FWPSDs = 1 - FWPSD(ADT - ADT_{true})/FWPSD(ADT_{true}) \tag{8}$$

A PSD distribution with a score > 0.5 is considered accurate and reliable. For the 2DVAR models, the HYCOM data were reliable in time–space frequency domain > 80 km, regardless of the time span (Figure 12a). For the two AVISO models, HYCOM was considered reliable only for frequency domains > 150 km, and improved as the time scale increased (Figure 12b,c).

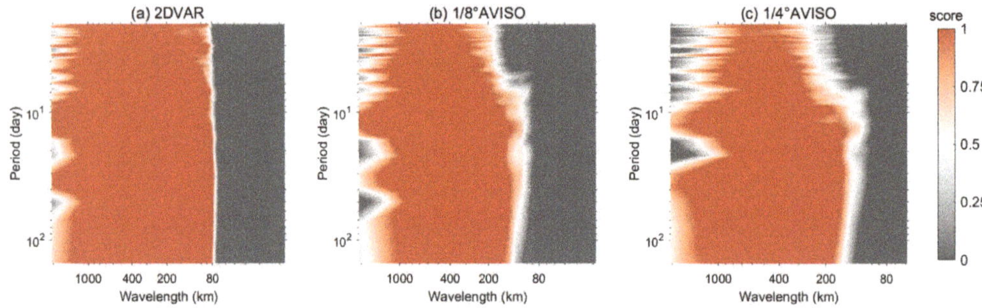

Figure 12. The frequency wavelength power spectral density (PSD) values of the three merged maps.

Based on these results, the 2DVAR merged maps have the highest quality and lowest ER among the four models. In an early study [20], 2DVAR merged maps had a higher ER than the 1/4° AVISO merged maps for the East China Sea region, especially in the open ocean and to the north of Taiwan Island, where mesoscale eddies are relatively large and long-lasting. Here, for the SCS, 2DVAR merged maps achieved better performance than 1/4° AVISO merged maps in reconstructing mesoscale oceanic structure, and their ER was as low as 130 km, or even 110 km around the middle latitudes [12]. For the California Current system, the 2DVAR merged maps could resolve smaller scales than the global 1/4° DUACS-DT2018 maps [11]. These findings together demonstrate that the 2DVAR model has advantages in terms of its merged map quality and ER, and without region-dependent.

4. Discussion

4.1. Signal Composition in Background Field and Associated Error

For the AVISO models, the background-field MDT was obtained using a multiyear-averaged SSH field. Therefore, the signal contained in the background field is strongly time-smoothed, and the error associated with the background field contained more large-scale ocean circulation signals, interannual variation, and seasonal variation. At the same time, the 2DVAR background field did not smooth the signal for any time or region, and induct an evolution error to maintain all signals in the merging processes. Therefore, the background error with the 2DVAR method comprised more small and mesoscale signals than that with the AVISO.

4.2. Filtering Effect of Correlation Coefficient Scale in Variational Method

The wavenumber energy spectral density of the background error and the characteristics of the background error covariance have similar physical (filtering) characteristics.

The filtering characteristics of the PSD associated with the background error decrease with the spatial scale, owing to the localization and intermittence of small-scale systems, and small-scale errors can only represent a certain proportion of the background error. When the spatial scale of the correlation coefficient is larger, the proportion represented by small scale error is less. For authentic ocean signals, when the proportion of the signal energy of a certain scale is lower, the error of that scale is more difficult to correct in the merged maps.

Based on these filtering characteristics, and owing to authentic ocean signals of the AVISO merged maps, it is difficult to distinguish small-scale oceanic structure, due to the larger correlation length scale with the AVISO method.

4.3. The Scale of Effective Resolution Compared with Eddy Radius

The correlation length scale was 10 km for 2DVAR and 100 km for AVISO. The 2DVAR ER was 50–150 km, while that of both AVISO models was 150–250 km.

For the eddies in the SCS, the 2DVAR ER range includes the peak wavelength of the eddy radius Rayleigh distribution, whereas AVISO fails to include the peak wavelength. The 2DVAR merged map can reconstruct mesoscale eddies more accurately than the AVISO maps.

4.4. The Restriction of HYCOM and the Advantages of 2DVAR

HYCOM reanalysis was produced using data assimilation. Altimetry data is first used to estimate temperature and salinity vertical profiles, employing the ISOP (Improved Synthetic Ocean Profiles) algorithm. The estimated temperature and salinity profiles are then assimilated into HYCOM. This assimilation method may limit the impact of altimetry data in the reanalysis:

- There is limited historical sampling data, leading to inaccurate assimilation of height field results.
- Non-steric sea surface heights in the altimeter data cannot be assimilated.
- The set of an assimilation thresholds is defined as the noise level of the satellite altimeter (currently set to 4 cm), which restricts the merging of small-scale information.

It is believed that the long background field time window of AVISO is the key factor causing its resolution to decline. The 2DVAR model has additional technical advantages [23]:

- The matrix deformation avoids inversion of the background error covariance matrix and can be minimized over the entire grid domain, and is therefore suitable for solving high-resolution problems with a large number of grid points.
- The processing methods of the background error covariance matrix and observation error covariance matrix are more flexible than those of the other models; this flexibility is convenient for simplifying and introducing dynamic constraints.
- Using the observation operator H, it is easy to merge the observation data of different properties.

Comparing with HYCOM, 2DVAR does not depend on the above assumptions such as 'noise level', directly merges along-track data, and applies the correlation length scale to supplement noise filtering, thereby retaining small and mesoscale signals. Therefore, the obtained 2DVAR merged map provides higher quality reconstructions than the map obtained using HYCOM reanalysis data.

4.5. Limitations and Future Work

Although the ER of 2DVAR product has been effectively improved, many small-scale processes still cannot be resolved due to the temporal and spatial scales. To increase the density of observation and acquire more valid information, 2DAVR introduced the evolutionary error in the observation error ($\boldsymbol{R}_s = \boldsymbol{R}_\mathrm{m} + \boldsymbol{R}_\mathrm{e}$, the observation error covariance matrix $\boldsymbol{R}_s$ consists of measurement $\boldsymbol{R}_\mathrm{m}$ and evolutionary error covariance matrices $\boldsymbol{R}_\mathrm{e}$) to address the difference between observation time and mapping time [12]. In addition, the wide-swath Surface Water and Ocean Topography (SWOT) mission was launched on 15 December 2022. As a result, the findings of this study can be extended to resolve small-scale features in maps derived using data from new multi-satellite altimeters, including SWOT data [34]. The multi-scale data merging will be tried to improve further the ER of merged maps in the future [35].

5. Conclusions

The analyses have shown that the ER of the 2DVAR mapped altimetry data is approximately 130 km, while it is approximately 250 km for both the 1/8° and 1/4° AVISO mapped data products. The factors for increasing the effective resolution was then examined. Both the 2DVAR and AVISO OI methods were formulated based on optimal estimation theory, and they are equivalent to each other [15]. The differences arise only from their implementation.

It was shown that the differences in the effective resolution result from the difference in the chosen background states and thus the associated background error. In the 2DVAR method, the mapped SSH of the previous day was used as the background state, while a 25 yeas mean used as the background state in the AVISO OI method. Thus, the background error with the 2DVAR method are relatively dominated by meso- and small-scale signals. In contrast, the background error with the AVISO OI method is the anomalies from the 25 year mean, and thus it includes large-scale signals, such as interannual and seasonal variabilities. The large-scale signals leads to a larger correlation length scale. A large correlation length scale imposes stronger filtering effect on the merged maps. Thus, the large correlation length scale is the main reason for a low effective resolution with 1/4° AVISO, even with a finer spatial grid of 1/8° AVISO and 1/12° HYCOM.

Author Contributions: Conceptualization, L.L. and J.F.; methodology, X.Z. and X.J.; validation, Z.L.; formal analysis, X.Z. and L.L.; investigation, X.Z. and L.L.; data curation, W.S., H.W. and X.L.; writing—original draft preparation, X.Z. and L.L.; writing—review and editing, X.Z., L.L. and Z.Z.; visualization, L.L. and X.Z. All authors have read and agreed to the published version of the manuscript.

Funding: This work is supported by the National Natural Science Foundation of China (42192552, 41775053 and 42022033) and the National Key Research and Development Program of China (2019YFC1510001).

Institutional Review Board Statement: Not applicable.

Informed Consent Statement: Not applicable.

Data Availability Statement: The 1/4°AVISO data, J3 and S3A Levle-3 data were downloaded from E.U. Copernicus Marine Environment Monitoring Service, available online: https://resources.marine.copernicus.eu (accessed on 22 January 2022). The MUR SST data were downloaded from Jet Propulsion Laboratory: https://coastwatch.pfeg.noaa.gov/erddap/griddap/jplMURSST41.html (accessed on 9 July 2022). The drifter data were support from GDP, available online: https://www.aoml.noaa.gov/phod/gdp/ (accessed on 30 March 2022). The 1/8°AVISO data, 2DVAR data and HYCOM data in this study is available at Zenodo (https://doi.org/10.5281/zenodo.7578466 (accessed on 7 July 2022)). Figures were made with MATLAB version 2021b, available under the MATLAB license at https://www.mathworks.com/ (accessed on 15 December 2022).

Acknowledgments: The authors are grateful to three anonymous reviewers for their constructive comments on this work and recommending several beneficial references. The authors thank Xiaolei ZOU and Xiaochun WANG for their help in improving the manuscript.

Conflicts of Interest: The authors declare no conflict of interest.

References

1. Ubelmann, C.; Dibarboure, G.; Gaultier, L.; Ponte, A.L.S.; Ardhuin, F.; Ballarotta, M.; Faugére, Y. Reconstructing Ocean Surface Current Combining Altimetry and Future Spaceborne Doppler Data. *J. Geophys. Res. Ocean.* **2021**, *126*, e2020JC016560. [CrossRef]
2. Mulet, S.; Rio, M.; Etienne, H.; Artana, C.; Cancet, M.; Dibarboure, G.; Feng, H.; Husson, R.; Picot, N.; Strub, C.P.A.P. The new CNES-CLS18 global mean dynamic topography. *Ocean Sci.* **2021**, *17*, 789. [CrossRef]
3. Elipot, S.; Lumpkin, R.; Perez, R.C.; Lilly, J.M.; Early, J.J.; Sykulski, A.M. A global surface drifter data set at hourly resolution. *J. Geophys. Res. Ocean.* **2016**, *121*, 2937–2966.

4. Abdalla, S.; Kolahchi, A.A.; Ablain, M.; Adusumilli, S.; Bhowmick, S.A.; Alou-Font, E.; Amarouche, L.; Andersen, O.B.; Antich, H.; Aouf, L.; et al. Altimetry for the future: Building on 25 years of progress. *Adv. Space Res.* **2021**, *68*, 319–363. [CrossRef]
5. Wang, G.; Wu, L.; Mei, W.; Xie, S.P. Ocean currents show global intensification of weak tropical cyclones. *Nature* **2022**, *611*, 496–500. [CrossRef] [PubMed]
6. Davis, R.; Talley, L.; Roemmich, D.; Owens, B.; Rudnick, D.; Toole, J.; Weller, R.; McPhaden, M.; Barth, J. 100 Years of Progress in Ocean Observing Systems. *Meteorol. Monogr.* **2018**, *59*, 1–46. [CrossRef]
7. Zilberman, N.; Roemmich, D.; Gille, S.; Gilson, J. Estimating the Velocity and Transport of Western Boundary Current Systems: A Case Study of the East Australian Current near Brisbane. *J. Atmos. Ocean. Technol.* **2018**, *35*, 1313–1329. [CrossRef]
8. Charney, J.; Flierl, G. Oceanic analogues of large scale atmospheric motions. In *Evolution of Physical Oceanoaraphy*; Warren, B., Wunsch, C., Eds.; MIT Press: Cambridge, MA, USA, 1981; pp. 502–546.
9. Yu, H.; Li, J.; Wu, K.; Wang, Z.; Yu, H.; Zhang, S.; Hou, Y.; Kelly, R.M. A global high-resolution ocean wave model improved by assimilating the satellite altimeter significant wave height. *Int. J. Appl. Earth Obs. Geoinf.* **2018**, *70*, 43–50. [CrossRef]
10. Detlef, S.; Anny, C. Satellite Altimetry Over Oceans and Land Surfaces. *Aeronaut. J.* **2019**, *123*, 1297–1298. [CrossRef]
11. Archer, M.R.; Li, Z.; Fu, L.L. Increasing the Space–Time Resolution of Mapped Sea Surface Height from Altimetry. *J. Geophys. Res. Ocean.* **2020**, *125*, 2019JC015878. [CrossRef]
12. Liu, L.; Jiang, X.; Fei, J.; Li, Z. Development and evaluation of a new merged sea surface height product from multi-satellite altimeters. *Chin. Sci. Bull.* **2020**, *65*, 1888–1897. [CrossRef]
13. Chelton, D.B.; Schlax, M.G.; Samelson, R.M. Global observations of nonlinear mesoscale eddies. *Prog. Oceanogr.* **2011**, *91*, 167–216. [CrossRef]
14. Chelton, D.; Dibarboure, G.; Pujol, M.I.; Taburet, G.; Schlax, M.G. *The Spatial Resolution of AVISO Gridded Sea Surface Height Fields*; OSTST: Lake Constance, Germany, 2014; pp. 28–31. Available online: https://meetings.aviso.altimetry.fr/fileadmin/user_upload/tx_ausyclsseminar/files/29Red0900-1_OSTST_Chelton.pdf (accessed on 30 March 2022).
15. Ubelmann, C.; Klein, P.; Fu, L. Dynamic Interpolation of Sea Surface Height and Potential Applications for Future High-Resolution Altimetry Mapping. *J. Atmos. Ocean. Technol.* **2015**, *32*, 177–184. [CrossRef]
16. Chen, G.; Hou, Y.; Chu, X. Mesoscale eddies in the South China Sea: Mean properties, spatiotemporal variability, and impact on thermohaline structure. *J. Geophys. Res.* **2011**, *116*, C06018. [CrossRef]
17. Wang, G.; Su, J.; Chu, P.C. Mesoscale eddies in the South China Sea observed with altimeter data. *Geophys. Res. Lett.* **2003**, *30*, 2121. [CrossRef]
18. Roberts-Jones, J.; Bovis, K.; Martin, M.J.; McLaren, A. Estimating background error covariance parameters and assessing their impact in the OSTIA system. *Remote Sens. Environ.* **2016**, *176*, 117–138. [CrossRef]
19. Pegliasco, C.; Chaigneau, A.; Morrow, R.; Dumas, F. Detection and tracking of mesoscale eddies in the Mediterranean Sea: A comparison between the Sea Level Anomaly and the Absolute Dynamic Topography fields. *Adv. Space Res.* **2021**, *68*, 401–419. [CrossRef]
20. Jiang, X.; Liu, L.; Li, Z.; Liu, L.; Sian, K.T.C.L.; Dong, C. A Two-Dimensional Variational Scheme for Merging Multiple Satellite Altimetry Data and Eddy Analysis. *Remote Sens.* **2022**, *14*, 3026. [CrossRef]
21. Taburet, G.; Pujol, M.; SL-TAC Team. QUID for Sea Level TAC DUACS Products. Available online: https://catalogue.marine.copernicus.eu/documents/QUID/CMEMS-SL-QUID-008-032-068.pdf (accessed on 5 April 2023).
22. Ballarotta, M.; Ubelmann, C.; Pujol, M.I.; Taburet, G.; Fournier, F.; Legeais, J.F.; Faugere, Y.; Delepoulle, A.; Chelton, D.; Dibarboure, G.; et al. On the resolutions of ocean altimetry maps. *Ocean Sci.* **2019**, *15*, 1091–1109. [CrossRef]
23. Lewis, J.M.; Lakshmivarahan, S.; Maryada, S.K.R. Placement of Observations for Variational Data Assimilation: Application to Burgers' Equation and Seiche Phenomenon. In *Data Assimilation for Atmospheric, Oceanic and Hydrologic Applications*; Park, S.K., Xu, L., Eds.; Springer: Cham, Switzerland, 2021; Volume IV, pp. 259–275. [CrossRef]
24. JPL MUR MEaSUREs Project. *GHRSST Level 4 MUR Global Foundation Sea Surface Temperature Analysis, Version 4.1*; PO.DAAC: Pasadena, CA, USA, 2015. [CrossRef]
25. Lumpkin, R.; Centurioni, L. Global Drifter Program Quality-Controlled 6-Hour Interpolated Data from Ocean Surface Drifting Buoys. NOAA National Centers for Environmental Information. Dataset. 2019. Available online: https://www.aoml.noaa.gov/phod/gdp/ (accessed on 30 March 2022).
26. Pujol, M.; Egrave, F.; Re, Y.; Taburet, G.; Dupuy, S.E.; Phanie; Pelloquin, C.; Ablain, M.; Picot, N. DUACS DT2014: The new multi-mission altimeter data set reprocessed over 20 years. *Ocean Sci.* **2016**, *12*, 1067–1090. [CrossRef]
27. Gaspari, G.; Cohn, S.E. Construction of correlation functions in two and three dimensions. *Q. J. R. Meteorol. Soc.* **1999**, *125*, 723–757. [CrossRef]
28. Li, Z.; Cheng, X.; Gustafson, W.I., Jr.; Vogelmann, A.M. Spectral characteristics of background error covariance and multiscale data assimilation. *Int. J. Numer. Methods Fluids* **2016**, *82*, 1035–1048. [CrossRef]
29. Daley, R. Atmospheric data assimilation. In *Cambridge Atmospheric and Space Science Series*; Cambridge University Press: Cambridge, UK, 1991.
30. Gaube, P.; Chelton, D.B.; Samelson, R.M.; Schlax, M.G.; O'Neill, L.W. Satellite Observations of mesoscale Eddy-Induced Ekman Pumping. *J. Phys. Oceanogr.* **2015**, *45*, 104–132. [CrossRef]
31. Hausmann, U.; Czaja, A. The observed signature of mesoscale eddies in sea surface temperature and the associated heat transport. *Deep. Sea Res. Part I Oceanogr. Res. Pap.* **2012**, *70*, 60–72. [CrossRef]

32. Zhang, Z.; Zhao, W.; Tian, J.; Yang, Q.; Qu, T. Spatial structure and temporal variability of the zonal flow in the Luzon Strait (Article). *J. Geophys. Res. Ocean.* **2015**, *120*, 759–776. [CrossRef]
33. Liu, Y.; Tian, F.; Chen, G. Statistical characterization of sea surface temperature over mesoscale eddies in the south China sea. *Period. Ocean. Univ. China* **2020**, *50*, 146–156. [CrossRef]
34. Morrow, R.; Fu, L.; Ardhuin, F.; Benkiran, M.; Chapron, B.; Cosme, E.; D'Ovidio, F.; Farrar, J.T.; Gille, S.; Lapeyre, G.; et al. Global Observations of Fine-Scale Ocean Surface Topography with the Surface Water and Ocean Topography (SWOT) Mission. *Front. Mar. Sci.* **2019**, *6*, 232. [CrossRef]
35. Li, Z.; Wang, J.; Fu, L. An Observing System Simulation Experiment for Ocean State Estimation to Assess the Performance of the SWOT Mission: Part 1—A Twin Experiment. *J. Geophys. Res. Ocean.* **2019**, *124*, 4838–4855. [CrossRef]

remote sensing

MDPI

Article

The Dynamics and Microphysical Characteristics of the Convection Producing the Record-Breaking Hourly Precipitation on 20 July 2021 in Zhengzhou, China

Kun Zhao [1,*,†], Xin Xu [1,†], Ming Xue [1,2], Zhe-Min Tan [1], Hao Huang [1], Ang Zhou [1], Xueqi Fan [1], Qiqing Liu [1], Chenli Wang [1], Juan Fang [1], Wen-Chau Lee [3], Qinghong Zhang [4], Fan Zhang [5], Gang Chen [6] and Ji Yang [6]

1 Key Laboratory of Mesoscale Severe Weather/Ministry of Education, and School of Atmospheric Sciences, Nanjing University, Nanjing 210023, China; zmtan@nju.edu.cn (Z.-M.T.)
2 Center for Analysis and Prediction of Storms, School of Meteorology, University of Oklahoma, Norman, OK 73072, USA
3 Earth Observing Laboratory, National Center for Atmospheric Research, Boulder, CO 80301, USA
4 Department of Atmospheric and Oceanic Sciences, School of Physics, Peking University, Beijing 100871, China
5 State Key Laboratory of Severe Weather, Chinese Academy of Meteorological Sciences, Beijing 100081, China
6 Nanjing Joint Institute for Atmospheric Sciences, Nanjing 210009, China
* Correspondence: zhaokun@nju.edu.cn
† These authors contributed equally to this work.

Abstract: An hourly rainfall of 201.9 mm fell in Zhengzhou on 20 July 2021, breaking the hourly rainfall record of mainland China and causing severe urban flooding and human casualties. This observation-based study investigates the associated convective-scale and mesoscale dynamics and microphysical processes using disdrometer and polarimetric radar observations aided by retrievals from the Variational Doppler Radar Analysis System. The synoptic flow forcing brought abundant moisture from the oceans and converged at Zhengzhou; then, the extreme rainfall was produced by a slow-moving convective storm that persisted throughout the hour over Zhengzhou. Unusually high concentrations of raindrops of all sizes (showing combined properties of maritime and continental convection) are revealed by the disdrometer data, whereas the polarimetric radar data suggest that both ice-based and warm rain processes were important contributors to the total rainfall. High precipitation efficiency was achieved with an erect updraft at the low levels, whereas enhanced easterly inflows kept the storm moving slowly. The interaction between convective-scale and mesoscale dynamics and microphysical processes within the favorable synoptic conditions led to this extremely heavy rainfall.

Keywords: extreme rainfall; dynamics and microphysics; slow-moving convection

Citation: Zhao, K.; Xu, X.; Xue, M.; Tan, Z.-M.; Huang, H.; Zhou, A.; Fan, X.; Liu, Q.; Wang, C.; Fang, J.; et al. The Dynamics and Microphysical Characteristics of the Convection Producing the Record-Breaking Hourly Precipitation on 20 July 2021 in Zhengzhou, China. *Remote Sens.* **2023**, *15*, 4511. https://doi.org/10.3390/rs15184511

Academic Editor: Angelica Tarpanelli

Received: 4 July 2023
Revised: 6 September 2023
Accepted: 7 September 2023
Published: 13 September 2023

1. Introduction

On 20 July 2021, the 24-h rainfall (0000 to 2400 LST) reached 612.9 mm (Figure 1a) in Zhengzhou (ZZ) city, Henan Province, China, approaching the city's average annual rainfall amount of ~640 mm. Between 1600 and 1700 LST, an hourly rainfall rate of 201.9 mm was observed, which broke China's previous record of 198.5 mm h^{-1} set in 1975 in the same province (known as the "75.8" rainstorm [1]). At least 302 lives were lost in this record-setting extreme rainfall that flooded ZZ, including the city's subway system and underground tunnels.

This extreme rainfall period was part of a week-long (from 17 to 23 July 2021, hereafter named as the "21.7" rainstorm) heavy rainfall episode in a broad region of Henan Province near the foothills of the Taihang and Funiu Mountains. This prolonged extreme heavy rainfall event was accompanied by abnormal synoptic conditions in East Asia (Figure 1b), including (1) the western Pacific subtropical high (WPSH) shifting northward, (2) Typhoon

In-Fa located east of Taiwan, and (3) the 200 hPa South Asian high (SAH) extending northeastward, that produced ingredients conducive to heavy rainfall in Henan [1–8]. A water vapor channel, referred to as an atmospheric river (AR) in other regions [9], was established between Typhoon In-Fa and the WPSH (indicated by the strong water vapor flux in Figure 1b) that transported abundant tropical oceanic moisture over 500 km inland. The extension of the 200 hPa SAH led to upper-level wind divergence in the vicinity of ZZ, also providing favorable upper-level conditions for the development of deep convection in ZZ.

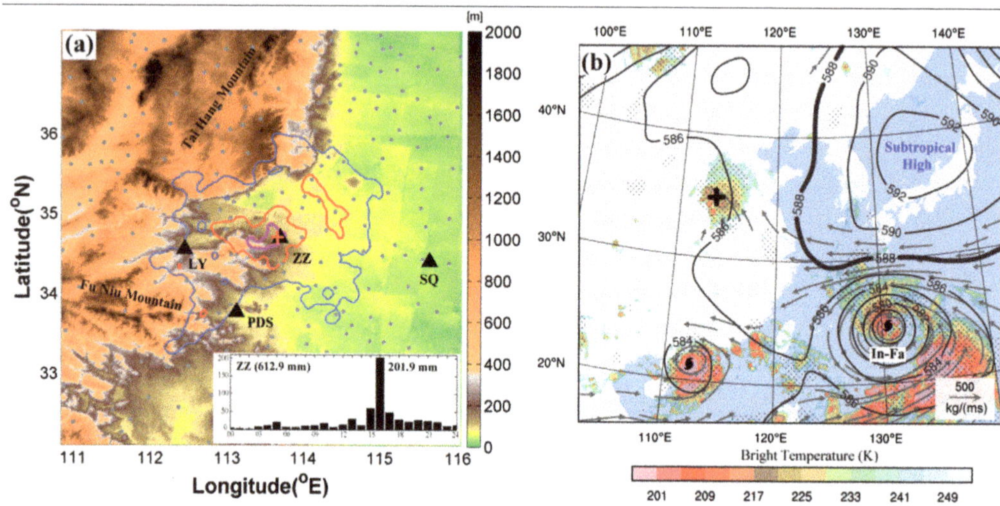

Figure 1. (**a**) Distribution map of ground instruments on the digital terrain elevation map around ZZ. Triangles denote the locations of four operational radars. Blue, red, and purple contours indicate the 24-h accumulated rainfall over 100, 250, and 400 mm on 20 July, respectively. The time series of the hourly rainfall from the ZZ national surface station (indicated by a red cross) is also shown in the bottom right corner of the figure. (**b**) Circulation fields at 0800 LST on 20 July 2021. Gray contours represent geopotential height at 500 hPa, and the bold line indicates the location of the western Pacific subtropical high (WPSH). Vectors are vertically integrated water vapor flux (in the layer of 1000–300 hPa) larger than 150 kg m^{-1} s^{-1}, and black dots indicate the wind divergence at 200 hPa that is larger than 10^{-5} s^{-1}. Shading represents the 10.8 μm infrared brightness temperature from FY-4A. The black cross indicates the location of the ZZ national surface station.

These synoptic conditions that were favorable for the heavy rainfall in Henan Province were identified by the Chinese Meteorological Administration (CMA) in their forecasting operations. Based on operational numerical models and objective forecasting techniques, the National Meteorological Center (NMC) of the CMA predicted an over 250 mm daily rainfall amount in Henan for 20 July one day in advance but failed to predict the extreme intensity and timing of the extreme rainfall in ZZ. The average hourly rainfall (Figure 1a) excluding the period of 1600–1700 LST in ZZ was about 18 mm h^{-1} throughout the day. Heavy rainfall is typically accompanied by single or multiple slow-moving convective storms over an area, whereas the basics involved in the development of flash-flood-producing storms are generally well known [10–12]. The difficulties lie in predicting the exact time, location, and amount of heavy rainfall [13–15]. To generate the extreme rainfall in ZZ, many favorable conditions needed to coexist, and dynamic, thermodynamic, and microphysical processes needed to work in synergy to efficiently convert a large amount of water vapor into precipitation. The "21.7" rainstorm event revealed the ongoing challenges in forecasting the location and amount of extreme hourly rainfall [16,17], even when the environmental conditions are correctly predicted to be very favorable.

Numerical studies were conducted to investigate the possible dynamical mechanism responsible for the production of the "21.7" extreme rainfall [6,18]. The dynamics of this extreme rainfall event were also studied by Sun et al. [19] using the four-dimensional Variational Doppler Radar Analysis System (VDRAS) [20]. Based on the combined observations of the polarimetric radar and disdrometer, the overall variability of microphysical characteristics for the heavy rainfall episode from 19 to 21 July 2021 was examined [21], whereas Yin et al. [22] explored the precipitation efficiency of the "21.7" extreme rainfall using numerical simulation. However, the detailed microphysical processes and their coupling with multiscale dynamical processes to produce the record-breaking hourly rainfall have not yet been investigated. This study focuses on the likely convective and mesoscale dynamics and microphysical processes that transformed the abundant oceanic moisture in the ZZ area to generate this record-breaking hourly rainfall between 1600 and 1700 LST on 20 July 2021. The three-dimensional wind and thermodynamic fields are obtained from the VDRAS, and the corresponding microphysical processes are deduced from the Luoyang (LY) operational polarimetric radar and a disdrometer at the ZZ national surface station.

2. Datasets and Methods

2.1. Datasets

This study used data from the operational observational network operated by the CMA, including 4 operational *S*-band Doppler radars (ZZ, LY, PDS, and SQ in Figure 1), 19 lightning detection sensors, 6900 automatic weather stations (AWSs), and a second-generation OTT particle size and velocity (PARSIVEL) disdrometer (OTT-2) located at ZZ. These radars perform volume scans at nine elevations (0.5°, 1.5°, 2.4°, 3.4°, 4.3°, 6.0°, 9.9°, 14.6°, and 19.5°) every 6 min, with an unambiguous range (velocity) of 150 km (26.5 m s^{-1}). The LY and ZZ radars can measure polarimetric variables including the differential reflectivity (Z_{DR}), differential propagation phase shift (Φ_{DP}), and specific differential phase (K_{DP}). The disdrometer can measure the number, concentration, and fall speed of raindrops within 32 bins ranging from 0 to 25 mm every minute.

2.2. Methods

The procedure for the data quality control of the OTT-2 data was introduced in Chen et al. [23]. The calculation of the raindrop size distribution (DSD), including the generalized intercept parameter (N_w) and mass-weighted mean diameter (D_m) followed Bringi et al. [24]. The radar data were firstly quality controlled to remove the ground clutter and velocity alias and correct the effect of partial beam blockage and the Z_{DR} bias [25], then bilinearly interpolated onto constant-altitude planes with a 1 km grid spacing in both the horizontal and vertical directions [26]. The ten hydrometeor types (drizzle (DZ), rain (RA), big drops (BD), ice crystals (IC), vertical aligned ice (VI), aggregates (AG), wet snow (WS), high-density graupel (HG), low-density graupel (LG), and hail (HA)) were identified from polarimetric radar data based on the fuzzy logic method presented in Dolan et al. [27], and the retrieval of ice water content (IWC) and liquid water content (LWC) mainly involved a method similar to that in Cifelli et al. [28] but with the LWCs for non-solid hydrometeors calculated from the K_{DP}. The LWC-K_{DP} relationship was fitted through the DSD data using the T-matrix method [29]. The precipitation efficiency was calculated using the rain rate divided by the total water vapor within a layer from 3 to 5 km heights [30].

The VDRAS is a four-dimensional variational data assimilation system that assimilates radar observations (i.e., radar reflectivity and radial velocity) and surface observations with the background provided by the weather research and forecasting (WRF) model [20,31]. In the VDRAS analysis, radar observations from ZZ, SQ, PDS, and LY (Figure 1) with a 6-min interval, together with surface observations of temperature and wind from the 6900 AWSs with a 5-min interval were assimilated. The VDRAS domain was centered at the ZZ radar with a horizontal resolution of 3 km and 0.4 km vertical spacing from 0900 to 2000 on 20 July 2021.

3. Results

3.1. Mesoscale and Convective Dynamical Processes

Widespread precipitation occurred in the central and northern parts of the Henan Province, with convective activities mostly in the mountainous areas west of ZZ on 20 July 2021. Figure 2a shows the horizontal winds at 0.6 km above the mean sea level (MSL) obtained from the VADRS analysis at 1200 LST, i.e., four hours before the extreme hourly rainfall in ZZ. Predominantly easterlies from the east of Henan were blocked on the southeastern slope of the Taihang Mountains and turned into northeasterlies in the form of a barrier jet [32] (see Supplementary Figure S1) that funneled into the valley towards LY, whereas east of the Funiu Mountains was southerly flows towards the Song Mountains (SM) and ZZ. Together with the easterly flows east of ZZ, a three-way low-level convergence was set up at ZZ that had the potential to produce extreme rainfall. In the lower troposphere (Figure 2b), there was a mesoscale vortex located west of ZZ [18,33]. The horizontal winds of this low-level mesovortex were primarily southerlies/southwesterlies over ZZ, which showed an increasing trend from 1000 LST, with a moderate speed of ~14 m s^{-1} at 1600 LST (see Supplement Figure S2). By 1600 LST, the low-level convergence at ZZ was strengthened, with the maximum more than doubled from 1.4×10^{-3} s^{-1} at 1200 to 3.5×10^{-3} s^{-1} (Figure 2c). Easterly winds increased toward ZZ, leading to the enhancement of convergence.

A sounding extracted from the 1200 LST VDRAS analysis 150 km southeast of ZZ (Figure 2e) exhibited a well-mixed boundary layer due to solar insolation, with a low lifting condensation level (LCL) of 685 m MSL and a moderate convective available potential energy (CAPE) of 2670 J kg^{-1}. The net low-level moisture convergence into a budget box over ZZ (red box in Figure 2a) steadily increased from 1100 LST, reaching a maximum at 1600 LST and quadrupling over the period. Most of the increase was due to the increase in inward flux through the eastern boundary and the decrease in outward flux through the western boundary (Figure 2d). The large positive flux through the southern boundary also increased somewhat, whereas the northern boundary flux remained negative but small. Because of the continuous moistening, the low-level inflow was near saturation (Figure 2e), tending to suppress the evaporative cooling of rainfall. In consequence, the storm produced a relatively weak cold pool of about 2 K, whose leading edge was near ZZ (Figure 2f).

Tracing back in time, the convective storm (labeled CS in Figure 3) that reached ZZ by 1600 LST (Figure 3c) was initiated before 1200 LST along the southern slope of the SM, ~55 km southwest of ZZ (Figure 3a). This storm was characterized by a maximum reflectivity of ~55 dBZ, with the 30 dBZ echo reaching ~6.5 km altitude and a few lightning flashes. The storm moved northeastward at a speed of ~5 m s^{-1} and reached the western edge of ZZ at around 1500 LST (Figure 3b). Because of the three-way low-level convergence over ZZ, the storm became quasi-stationary between 1600 LST and 1700 LST, producing the extreme hourly precipitation. In the west–east direction, the cold pool density current was balanced by the low-level easterly inflow, which were 6.2 m s^{-1} and -6.4 m s^{-1}, respectively, between 1600 and 1700 LST (Figure 2g). In the south–north direction, the low-level southerly and northerly winds were also nearly in balance, with mean values of 4.8 and -4.0 m s^{-1}, respectively (Figure 2h). The above balances between multiscale flows are in agreement with the numerical results of Wei et al. [18] and contributed to the quasi-stationarity of the convective storm in ZZ. The slow-moving storm reached its peak intensity (Figure 3c,d) with increased lightning. The maximum 30 dBZ echo height reached 12 km in the east–west vertical cross section through the precipitation center of the storm (Figure 3e), indicating enhanced convection intensity [34,35]. This deep, erect convection favored higher precipitation efficiency [36,37]. The VDRAS analysis showed that at 1600 LST, the low-level easterly inflows entered the storm updraft below 3 km and the air and ice aggregates exited the front (i.e., east side) of the storm above a 5 km altitude (Figure 3f). These falling ice aggregates could have helped maintain the storm by destabilizing the storm inflow [38].

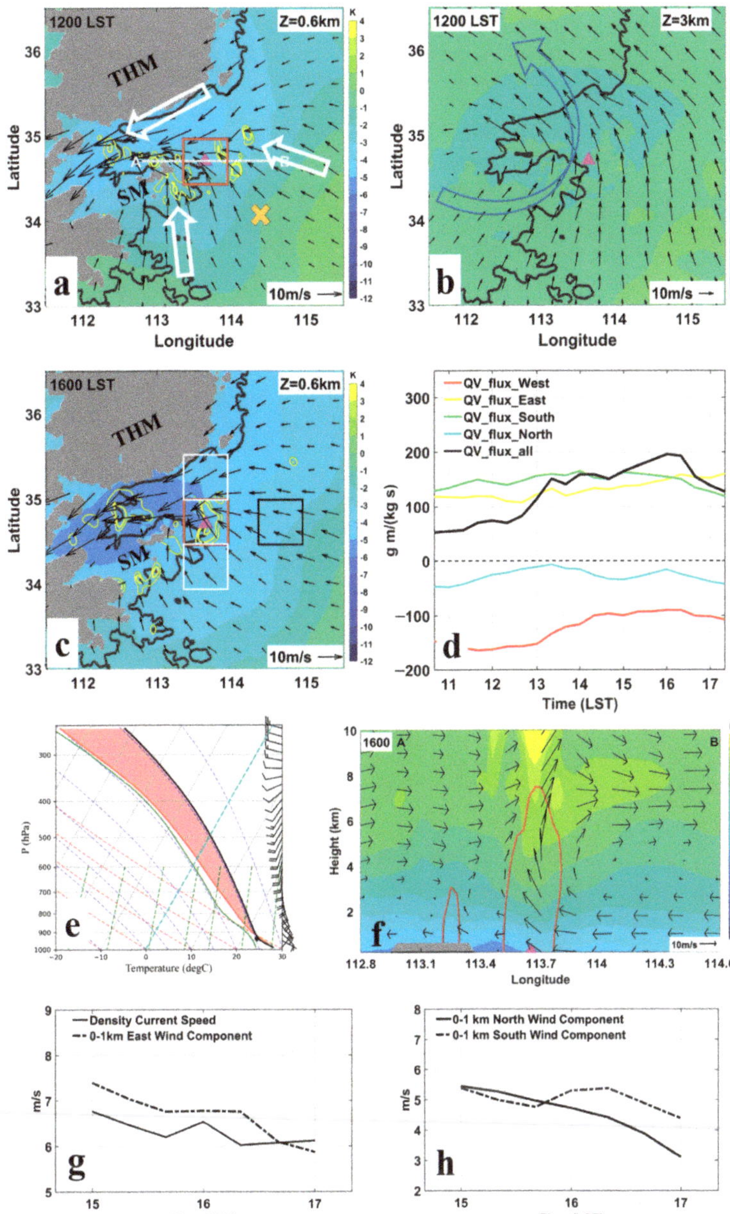

Figure 2. (**a**) Perturbation temperature (shading) and horizontal wind vectors at 0.6 km MSL in the VDRAS analysis at 1200 LST on 20 July 2021. Yellow contours are the horizontal convergence starting from -1×10^{-3} s^{-1} at an interval of -2×10^{-3} s^{-1}. Black lines are 600 m terrain elevation contours and the purple triangle indicates the location of the record-breaking hourly precipitation rain gauge station in ZZ. (**b**) As (**a**) but at 3 km. The curved arrow represents the mesoscale vortex. (**c**) As (**a**) but at 1600 LST. (**d**) 0–3 km moisture fluxes through the 4 borders of a 60 × 60 km box region covering the ZZ city (red box in (**a**,**c**)) and net flux into the box (positive into the box), based on the VADRS analysis. (**e**) Inflow environment sounding located at 150 km southeast of ZZ (yellow X in (**a**)) extracted from

the VDRAS analysis at 1200 LST with the red, green, and black solid lines representing the temperature profile, dew point profile, and the parcel ascent curve, respectively. (**f**) Perturbation temperature (shading) and wind vectors in the west–east vertical cross section through the center of ZZ city (line AB in (**c**)) in VADRAS analysis at 1600 LST. Red contours denote 40 dBZ radar reflectivity. (**g**) Evolution of cold pool density current and low-level easterlies between 0 and 1 km calculated within the red and black boxes in (**c**). (**h**) Evolution of the southerly and northerly wind components to the south and north of the storm (see the white boxes in (**c**)) over ZZ, respectively.

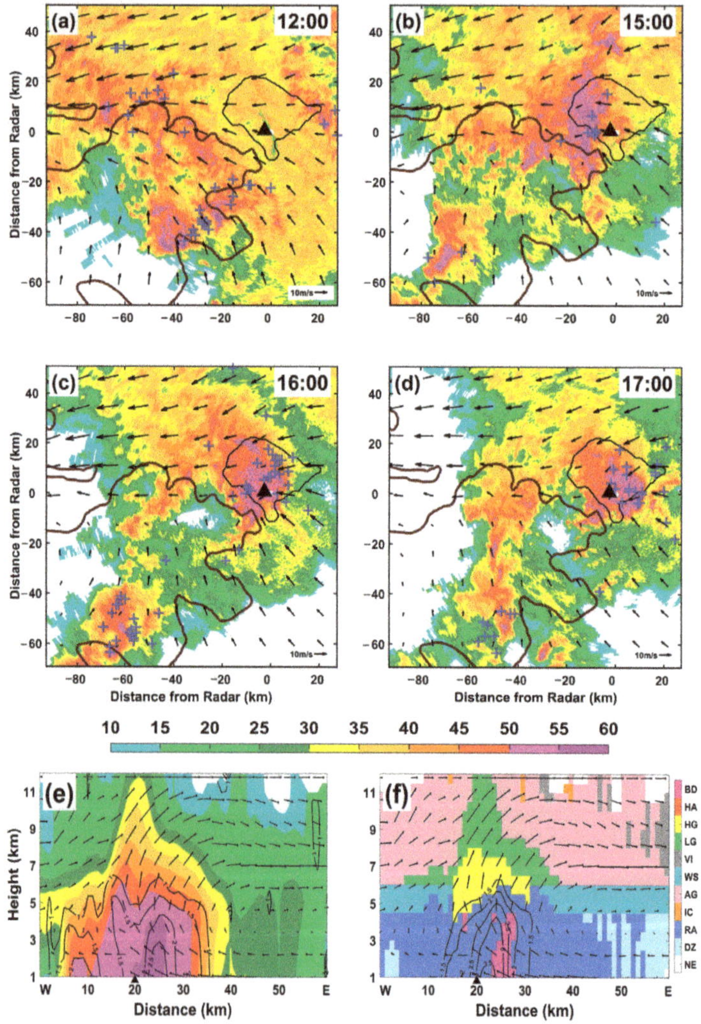

Figure 3. Observed radar reflectivity (color shading; units: dBZ) and horizontal winds at 600 m MSL at (**a**) 1200 LST, (**b**) 1500 LST, (**c**) 1600 LST, and (**d**) 1700 LST on 20 July 2021. A terrain height of 200 m is indicated by the bold brown contours. The black line represents the urban area of ZZ, and the triangle indicates the national surface station with the hourly extreme rainfall in ZZ. The blue pluses indicate that the lightning occurred 30 min before and after the radar time. SM indicates the Song Mountain. West–east cross sections of (**e**) Z_H (color shading) and Z_{DR} (contours) and (**f**) hydrometer types based on hydrometeor classification algorithm (shading) and K_{DP} (contours) across the rainfall center at 1600 LST.

3.2. Microphysical Processes in the Extreme Rainfall

3.2.1. Vertical Distributions of Hydrometeors

The hydrometeor classification algorithm (HCA [27]) based on the polarimetric radar data indicates graupels (large rain drops) above (below) the freezing level (Figure 3f). The combination of the large reflectivity Z_H (~58 dBZ), differential reflectivity Z_{DR} (~3.0 dB), and differential phase K_{DP} (~4° km^{-1}) near the surface beneath the updraft signals intense rainfall with the presence of large and oblate raindrops [39,40]. Positive Z_{DR} and K_{DP} columns were collocated with the updraft, where the region of $Z_{DR} > 1.0$ dB and $K_{DP} > 1.5° km^{-1}$ extended to a 6.5 km altitude, indicating supercooled liquid particles lifted above the freezing level (~5 km) by the strong updrafts (Figure 3e,f) [41,42]. The existence of the supercooled liquid water favored the riming process and the presence of graupel in the updraft region above the freezing level.

The time series of the Z_H, Z_{DR}, and K_{DP} fractions in specified value bins from 1200 to 2000 LST on 20 July 2021 are presented in Figure 4a–c. The polarimetric radar variables come from the 30 × 30 km box centered at the CS (labeled in Figure 3) at 2.0 km MSL. As the CS developed to the mature stage between 1600 and 1700 LST, the radar reflectivity values were frequently above 50 dBZ. About 20% of the Z_{DR} samples exceeded 2.0 dB in this hour, which was obviously higher than in any other hour, indicating abundant large raindrops formed in the CS. In addition, the fractions with a high K_{DP} value (over 3.0° km^{-1}) also peaked in the same hour, suggesting the dominance of extreme liquid water contents and instant rain rates at this vertical level. The significantly higher fractions of large polarimetric radar variables compared with other hours are consistent with the hourly extreme rainfall at the surface.

The vertical structure of the polarimetric radar variables in the region of radar-observed extreme rainfall near ZZ (K_{DP} higher than 4° km^{-1}, approximately 160 mm h^{-1} based on the radar-estimated rainfall) is illustrated in Figure 4d. More than 95% of the extreme rainfall occurred in the strong updraft region, as shown in Figure 3f. Between 12 and 8 km, Z_H increased rapidly, whereas Z_{DR} decreased slightly with decreasing height; the Z_{DR} was around 0 dB, suggesting there was aggregation in this layer [43]. Between 8 km and 5 km (the 0 °C level), the increase in both Z_H and K_{DP} indicates the existence of more supercooled liquid drops [41]. The increase in Z_H, Z_{DR}, and K_{DP} towards the 0 °C level reflects the occurrence of active riming processes, as further confirmed by the normalized frequency of hydrometeors showing that large ice particles (graupel or hail) dominate in this layer (Figure 4e). Below the 0 °C level, Z_H, Z_{DR}, and K_{DP} continuously increased toward the ground, indicating active warm rain processes of coalescence and accretion [44,45] that help increase the raindrop sizes and enhance precipitation. The enhanced Z_H and Z_{DR} within 1 km below the 0 °C level can be partly attributed to the melting of graupel and hail because of the increase in the dielectric constant and aspect ratio [46,47].

The radar-derived IWC and LWC analysis shows a rapid increase in LWC below the melting level, mainly through the accretion of cloud drops along with the auto-conversion process. The maximum LWC is about 6.2 g m^{-3} near the ground, less than twice the IWC maximum (about 3.9 g m^{-3}), suggesting that both ice and warm rain processes have important contributions to the surface extreme rainfall. This can be explained by the interactions between the kinematic and microphysical structures of the convective storm under favorable environmental conditions. A low LCL, strong low-level moisture convergence, and a deep warm cloud layer can promote the rapid and efficient conversion of inflow water vapor to cloud drops and then raindrops in the updraft through active warm rain processes [10]. Meanwhile, the strong updraft also transports water vapor and liquid water above the freezing level to promote the production of graupels and hails. Previous studies showed that upright convection has a higher precipitation efficiency than tilted convection because it allows for a more effective collection of cloud condensate by precipitation [36,48]. In this event, because the convective storm was erect, the large raindrops from melting graupels or hailstones can fall through the updraft and grow by accretion and coalescence below the melting layer and contribute to the extreme rainfall.

This suggests an optimal coupling between ice-based and warm rain processes contributed to the extreme hourly rainfall in ZZ.

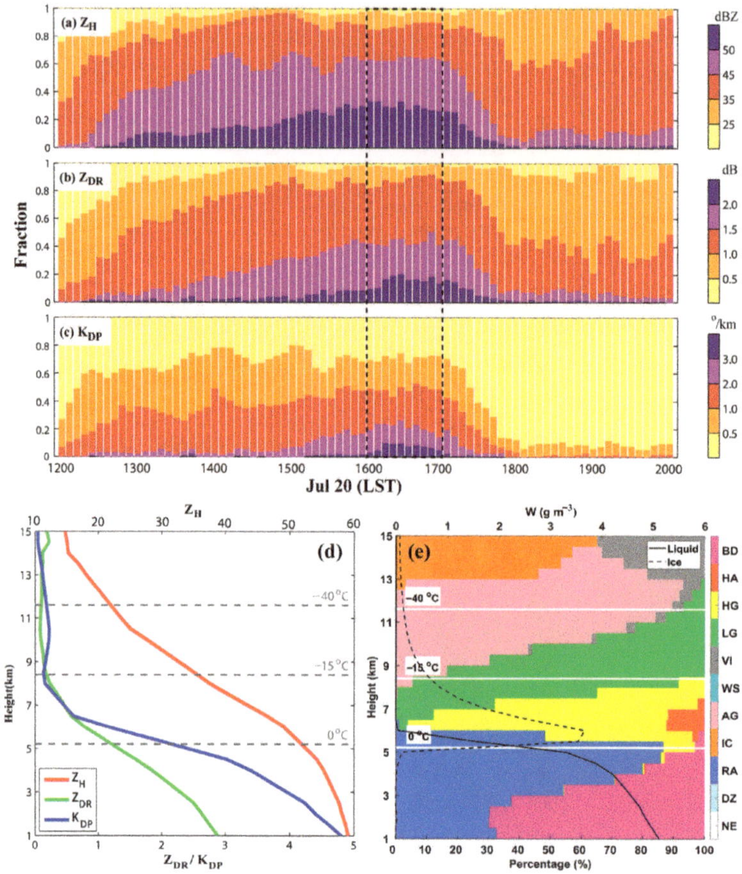

Figure 4. Time series of (**a**) Z_H, (**b**) Z_{DR}, and (**c**) K_{DP} fractions in specified bins at 2.0 km MSL in the 30 × 30 km box centered on the CS (labeled in Figure 3) from 1200 to 2000 LST. Average profiles of (**d**) polarimetric radar variables and (**e**) hydrometer types (shading), LWC (black solid), and IWC (red dashed) for K_{DP} values higher than 4° km^{-1} near the ZZ national surface station between 1600 and 1700 LST. The label for IWC/LWC is shown at the top of the panel.

3.2.2. The Raindrop Size Distribution on the Ground

The DSD is an important feature of precipitation microphysics [49]. The time series of DSD from 1 min OTT−2 observations at ZZ (Figure 5a) shows that both the drop number and maximum drop diameter increased dramatically between 1600 and 1700 LST. Note that the variations in the DSD during this period represent different parts of the convective storm moving across ZZ. The concentration of small raindrops (D < 1 mm) reached as high as 10^4 m^{-3} mm^{-1}, with the largest drop diameter being about 7 mm. With the increase in drop concentrations for all particle sizes, the hourly rain rate reached an extreme value of 201.9 mm h^{-1}. The largest contribution (66%) to total rainfall was from mid-size drops (1–3 mm), whereas large drops (D > 3 mm) were the second largest contributor (22%). On the other hand, small drops dominated the total number concentration (81.5%) but had the least contribution (12%) to the total rainfall. This is consistent with the large Z_{DR} and high K_{DP} near the surface (Figure 4a).

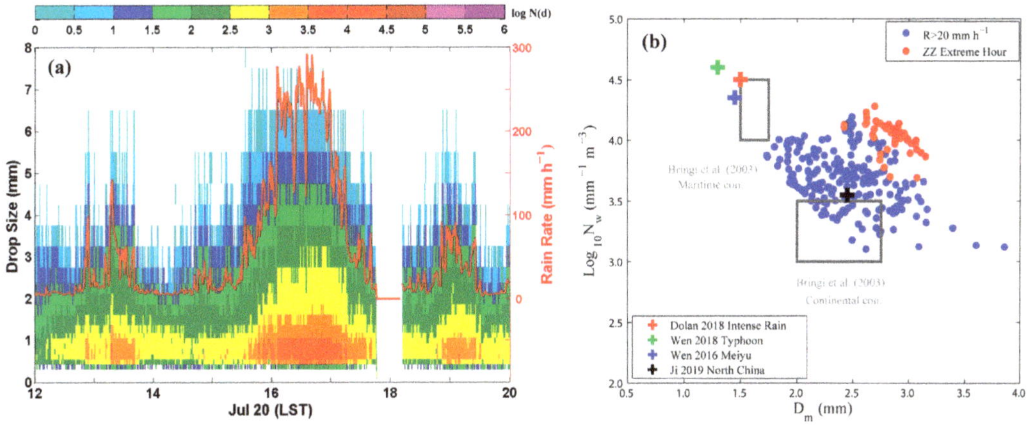

Figure 5. (**a**) Time series of 1 min DSDs from the OTT disdrometer at the ZZ national surface station. Color shading represents the DSD in logarithmic units of mm^{-1} m^{-3} and the y-axis indicates the equivalent volume diameter (mm) of raindrops; the instant rain rate is plotted as the red line. (**b**) Scatters of N_w (D_m) from the ZZ extreme hourly rainfall between 1600 and 1700 LST (red) and DSD samples with an instant rain rate over 20 mm^{-1} h^{-1} from 1200 to 2000 LST (blue), respectively. The two gray rectangles represent the maritime and continental convective clusters reported by [24]. The blue, green, black, and red crosses represent the mean values of convective rain in different regions from previous studies [50–53].

The scatter plot of $\log_{10}N_w$ (the generalized intercept parameter, indicating the raindrop concentration) and D_m (mass-weighted mean diameter) in Figure 5b illustrates the unique microphysical characteristics that distinguished the extreme hourly rainfall in ZZ (red dots) from other documented heavy rainfall events produced by typical continental or maritime convection [50–53] as well as other rainfall hours of the "21.7" rainstorm surrounding the extreme rainfall hour (blue dots). The two rectangle boxes in Figure 5b represent the characteristics of maritime and continental convection [24], dominated by warm-rain collision–coalescence processes (higher concentration of small drops) and ice-based microphysical processes (the presence of larger drops from the melting of graupel and hail), respectively [50]. Recent statistical studies have revealed that convective rainfall in South China possesses a maritime nature due to the impact of the East Asian summer monsoons or tropical cyclones [23,52,53], whereas convective rainfall in North China possesses more a continental nature being influenced by the northeast cold vortex and other mid-latitude systems [51,54]. The intense convective rainfall samples (defined as a rain rate >20 mm h^{-1} but excluding the period from 1600 to 1700 LST, i.e., blue dots in Figure 5b) show a large variation of D_m and $\log_{10}N_w$ (positioned in between these two boxes), reflecting the presence of ice processes in the ZZ rainfall. During the extreme hourly rainfall in ZZ, the concentrations of both large and small drops increased (Figure 5a), exhibiting the combined properties of maritime (high N_w) and continental (large D_m) convection. This unique characteristic of DSD indicates that both ice-based and warm-rain processes were active in producing the extreme hourly rainfall in ZZ, which is consistent with the microphysical processes inferred from the polarimetric radar observations in the previous subsection. They appear to be the results of extremely rich moisture (like the data of marine precipitation) and sufficiently large CAPE and low-level convergence forcing (that promote deep convection).

4. Conclusions

The "21.7" extreme rainfall event hit Zhengzhou, China on 20 July 2021, causing hundreds of fatalities and great economic losses. In particular, the 1 h precipitation at a national surface station in ZZ was 201.9 mm between 1600 and 1700 LST, breaking the record for hourly precipitation in mainland China. Our analyses show that the record-breaking hourly precipitation was produced by a quasi-stationary, well-organized, deep convective storm in ZZ that was fed by abundant tropical moisture via an atmospheric river between the WPSH and Typhoon In-Fa. The Taihang Mountains northwest of ZZ played a vital role in turning the environmental low-level easterlies into mountain-parallel northeasterly flows in the north, helping to block the southerly flows from the south into ZZ. The storm that moved into ZZ to produce the extreme hourly rainfall was initiated on the southern slopes of the Song Mountains southwest of ZZ.

The low-level easterly flow into ZZ was steadily enhanced in the hours preceding the extreme rainfall, whereas the airflow out of the ZZ region on its west side weakened in the same period. The flow changes led to a net moisture flux into ZZ that quadrupled in 5 h preceding the extreme hourly rainfall in ZZ. The enhanced low-level easterly flow roughly balanced the relatively weak cold pool density current and kept the storm stationary over ZZ. The rainstorm contained unusually high concentrations of small raindrops with the presence of some very large drops (about 7 mm). The rain DSDs and the polarimetric-radar-derived microphysical properties provided the first observational evidence that both oceanic (high number of raindrops, active warm rain processes) and continental (large raindrops, active ice processes) rain characteristics were active and very efficient in converting the abundant tropical moisture into the record-breaking hourly rainfall. These key dynamical and microphysical processes are summarized schematically in Figure 6.

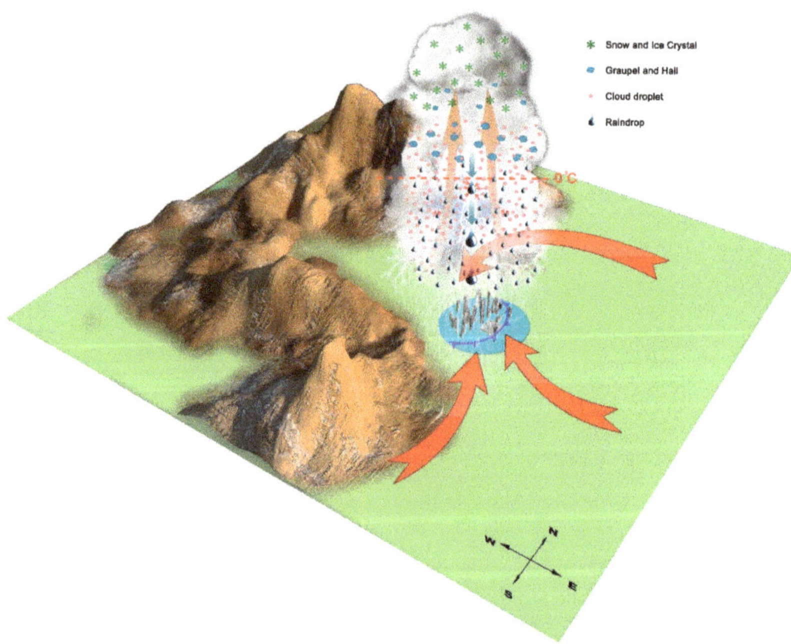

Figure 6. The conceptual model of the maintenance and precipitation microphysics of the convective storm resulting in extreme hourly rainfall in ZZ. The blue line with triangles indicates the cold pool gust front. The red arrows represent the prevalent winds.

This study provides insights into how local extreme rainfall may be better predicted by including mesoscale and convective scale processes together with the well-forecasted favorable synoptic conditions for heavy rainfall. This study also points out the unique DSD differentiating this extreme rainfall event from most other documented precipitation events in different regions of the world. We admit that an MCS should consist of both convective core and anvil cloud regions, with the latter also likely to contribute to the production of extreme rainfall. In this study, our main purpose was to investigate the key factors causing the record-breaking hourly rainfall in a local ZZ region weather station. According to radar observations, this hourly extreme rainfall was directly caused by the convective cell over ZZ. More observational and modeling studies will be conducted to investigate whether there are optimal and synergetic combinations between dynamics and microphysics in producing the unique DSD identified in the "21.7" extreme rainfall event. To be able to accurately represent within numerical weather prediction models the unique microphysical characteristics of this event and all other important ingredients that act in synergy to produce such record-breaking extreme rainfall and to provide quantitatively accurate operational forecasting with sufficient lead time remain challenging. Gaining insights and understanding of the physical processes and mechanisms involved is critical and this study represents one of the first efforts toward this goal.

Supplementary Materials: The following supporting information can be downloaded at: https://www.mdpi.com/article/10.3390/rs15184511/s1, Figure S1: (Top) Terrain elevation (shading) and VDRAS analysis horizontal winds at 0.6 km MSL (vector) at 1200 LST on 20 July 2021. (Bottom) Potential temperature (shading) and VDRAS analysis wind field in the vertical plane along line AB at 1200 LST on 20 July 2021. Red contours indicate the speed of horizontal wind normal to the vertical plane (i.e., nearly parallel to the Taihang Mountain); Figure S2: (Top) Horizontal wind vectors and speed (shading) at 3 km MSL in the VDRAS analysis from 1000 to 1600 LST on 20 July 2021. (Bottom) Time evolution of the mean horizontal wind speed over ZZ at different heights. Reference [55] is cited in Supplementary Materials.

Author Contributions: Conceptualization, K.Z.; Methodology, X.X. and G.C.; Software, H.H., X.F., Q.L. and J.Y.; Formal analysis, A.Z., Q.Z. and F.Z.; Data curation, C.W.; Writing—review & editing, M.X., Z.-M.T., J.F. and W.-C.L. All authors have read and agreed to the published version of the manuscript.

Funding: This work was primarily supported by the National Natural Science Foundation of China (grants 42025501, 4212200198, 41875053, 61827901, and 42122036), the Open Grants of the State Key Laboratory of Severe Weather (2023LASW-A01), the National Key Research and Development Program of China (Grant 2017YFC1501601, 2018YFC1506404).

Data Availability Statement: All freely available data are mentioned in the section on Data and Methods.

Conflicts of Interest: The authors declare no conflict of interest.

References

1. Ding, Y. On the study of the unprecedented heavy rainfall in Henan Province during 4–8 August 1975: Review and assessment. *Acta Meteorol. Sin.* **2015**, *73*, 411–424.
2. Shi, W.; Li, X.; Zeng, M.; Zhang, B.; Wang, H.; Zhu, K.; Zhuge, X. Multi-model comparison and high-resolution regional model forecast analysis for the "7·20" Zhengzhou severe heavy rain. *Trans. Atmos. Sci.* **2021**, *44*, 688–702. [CrossRef]
3. Sun, W.; Yu, R.; Li, J.; Yuan, W. Three-Dimensional Circulation Structure of Summer Heavy Rainfall in Central North China. *Weather Forecast.* **2015**, *30*, 238–250. [CrossRef]
4. Tao, S. *Rainstorms in China*; Science Press: Beijing, China, 1980.
5. Wu, G. *Dynamics of the Formation and Variation of Subtropical Anticyclones*; Science Press: Beijing, China, 2002.
6. Yin, J.; Gu, H.; Liang, X.; Yu, M.; Sun, J.; Xie, Y.; Li, F.; Wu, C. A Possible Dynamic Mechanism for Rapid Production of the Extreme Hourly Rainfall in Zhengzhou City on 20 July 2021. *J. Meteorol. Res.* **2021**, *36*, 6–25. [CrossRef]
7. Zhang, D.-L.; Lin, Y.; Zhao, P.; Yu, X.; Wang, S.; Kang, H.; Ding, Y. The Beijing extreme rainfall of 21 July 2012: "Right results" but for wrong reasons. *Geophys. Res. Lett.* **2013**, *40*, 1426–1431. [CrossRef]
8. Zhang, X.; Yang, H.; Wang, X.; Shen, L.; Wang, d.; Li, H. Analysis on characteristic and abnormality of atmospheric circulations of the July 2021 extreme precipitation in Henan. *Trans. Atmos. Sci.* **2021**, *44*, 672–687.
9. Ralph, F.M.; Dettinger, M.D.; Cairns, M.M.; Galarneau, T.J.; Eylander, J. Defining "Atmospheric River": How the Glossary of Meteorology Helped Resolve a Debate. *Bull. Am. Meteorol. Soc.* **2018**, *99*, 837–839. [CrossRef]
10. Doswell, C.A.; Brooks, H.E.; Maddox, R.A. Flash Flood Forecasting: An Ingredients-Based Method-ology. *Weather Forecast.* **1996**, *11*, 560–581. [CrossRef]
11. Konrad, C.E. Synoptic-Scale Features Associated with Warm Season Heavy Rainfall over the Interior Southeastern United States. *Weather Forecast.* **1997**, *12*, 557–571. [CrossRef]

12. Luo, Y.; Wu, M.; Ren, F.; Li, J.; Wong, W.-K. Synoptic Situations of Extreme Hourly Precipitation over China. *J. Clim.* **2016**, *29*, 8703–8719. [CrossRef]
13. Ding, Y. The major advances and development process of the theory of heavy rainfalls in China. *Torrential Rain Disasters* **2019**, *38*, 395–406.
14. Li, Z.; Chen, Y.; Zhang, X.; Wang, Y.; Dai, K.; Zhang, L. Development and thinking of torrential rain forecasting operation in National Meteorological Center. *Torrential Rain Disasters* **2019**, *38*, 407–415. [CrossRef]
15. Schumacher, R.S. *Heavy Rainfall and Flash Flooding*; Oxford University Press: Oxford, UK, 2017. [CrossRef]
16. Zhang, Y.; Yu, H.; Zhang, M.; Yang, Y.; Meng, Z. Uncertainties and error growth in forecasting the record-breaking rainfall in Zhengzhou, Henan on 19–20 July 2021. *Sci. China Earth Sci.* **2022**, *65*, 1903–1920. [CrossRef]
17. Zhu, K.F.; Zhang, C.; Xue, M.; Yang, N. Predictability and skill of convection-permitting ensemble forecast sys-tems in predicting the record-breaking "21·7" extreme rainfall event in Henan Province, China. *Sci. China Earth Sci.* **2022**, *65*, 1879–1902. [CrossRef]
18. Wei, P.; Xu, X.; Xue, M.; Zhang, C.; Wang, Y.; Zhao, K.; Zhou, A.; Zhang, S.; Zhu, K. On the Key Dynamical Processes Supporting the 21.7 Zhengzhou Record-breaking Hourly Rainfall in China. *Adv. Atmos. Sci.* **2023**, *40*, 337–349. [CrossRef]
19. Sun, J.; Li, R.; Zhang, Q.; Trier, S.B.; Ying, Z.; Xu, J. Mesoscale Factors Contributing to the Extreme Rainstorm on 20 July 2021 in Zhengzhou, China, as Revealed by Rapid Update 4DVar Analysis. *Mon. Weather Rev.* **2023**, *151*, 2153–2176. [CrossRef]
20. Sun, J.; Crook, N.A. Real-Time Low-Level Wind and Temperature Analysis Using Single WSR-88D Data. *Weather Forecast.* **2001**, *16*, 117–132. [CrossRef]
21. Chen, G.; Zhao, K.; Lu, Y.; Zheng, Y.; Xue, M.; Tan, Z.; Xu, X.; Huang, H.; Chen, H.; Xu, F.; et al. Variability of microphysical characteristics in the "21·7" Henan extremely heavy rainfall event. *Sci. China Earth Sci.* **2022**, *65*, 1861–1878. [CrossRef]
22. Yin, L.; Ping, F.; Mao, J.; Jin, S. Analysis on Precipitation Efficiency of the "21.7" Henan Extremely Heavy Rainfall Event. *Adv. Atmos. Sci.* **2023**, *40*, 374–392. [CrossRef]
23. Chen, G.; Zhao, K.; Wen, L.; Wang, M.; Huang, H.; Wang, M.; Yang, Z.; Zhang, G.; Zhang, P.; Lee, W.-C. Microphysical Characteristics of Three Convective Events with Intense Rainfall Observed by Polarimetric Radar and Disdrometer in Eastern China. *Remote Sens.* **2019**, *11*, 2004. [CrossRef]
24. Bringi, V.N.; Chandrasekar, V.; Hubbert, J.; Gorgucci, E.; Randeu, W.L.; Schoenhuber, M. Raindrop Size Distribution in Different Climatic Regimes from Disdrometer and Dual-Polarized Radar Analysis. *J. Atmos. Sci.* **2003**, *60*, 354–365. [CrossRef]
25. Shakti, P.C.; Maki, M.; Shimizu, S.; Maesaka, T.; Kim, D.-S.; Lee, D.-I.; Iida, H. Correction of Reflectivity in the Presence of Partial Beam Blockage over a Mountainous Region Using X-Band Dual Polarization Radar. *J. Hydrometeorol.* **2013**, *14*, 744–764. [CrossRef]
26. Mohr, C.G.; Vaughan, R.L. An Economical Procedure for Cartesian Interpolation and Display of Reflectivity Factor Data in Three-Dimensional Space. *J. Appl. Meteorol.* **1979**, *18*, 661–670. [CrossRef]
27. Dolan, B.; Rutledge, S.A.; Lim, S.; Chandrasekar, V.; Thurai, M. A Robust C-Band Hydrometeor Identification Algorithm and Application to a Long-Term Polarimetric Radar Dataset. *J. Appl. Meteorol. Clim.* **2013**, *52*, 2162–2186. [CrossRef]
28. Cifelli, R.; Petersen, W.A.; Carey, L.D.; Rutledge, S.A.; Dias, M.A.F.d.S. Radar observations of the kinematic, microphysical, and precipitation characteristics of two MCSs in TRMM LBA. *J. Geophys. Res. Atmos.* **2002**, *107*, LBA 44-1–LBA 44-16. [CrossRef]
29. Mishchenko, M.I.; Travis, L.D.; Mackowski, D.W. T-matrix computations of light scattering by nonspherical particles: A review. *J. Quant. Spectrosc. Radiat. Transf.* **1996**, *55*, 535–575. [CrossRef]
30. Wang, M.; Zhao, K.; Xue, M.; Zhang, G.; Liu, S.; Wen, L.; Chen, G. Precipitation microphysics characteristics of a Typhoon Matmo (2014) rainband after landfall over eastern China based on polarimetric radar observations. *J. Geophys. Res. Atmos.* **2016**, *121*, 12415–12433. [CrossRef]
31. Chen, X.; Zhao, K.; Sun, J.; Zhou, B.; Lee, W.-C. Assimilating surface observations in a four-dimensional variational Doppler radar data assimilation system to improve the analysis and forecast of a squall line case. *Adv. Atmos. Sci.* **2016**, *33*, 1106–1119. [CrossRef]
32. Doyle, J.D. The Influence of Mesoscale Orography on a Coastal Jet and Rainband. *Mon. Weather Rev.* **1997**, *125*, 1465–1488. [CrossRef]
33. Fu, S.; Zhang, Y.; Wang, H.; Tang, H.; Li, W.; Sun, J. On the Evolution of a Long-Lived Mesoscale Convective Vortex that Acted as a Crucial Condition for the Extremely Strong Hourly Precipitation in Zhengzhou. *J. Geophys. Res. Atmos.* **2022**, *127*, e2021JD036233. [CrossRef]
34. Xu, W. Thunderstorm Climatologies and Their Relationships to Total and Extreme Precipitation in China. *J. Geophys. Res. Atmos.* **2020**, *125*, e2020JD033152. [CrossRef]
35. Zipser, E.J.; Lutz, K.R. Vertical Profile of Radar Reflectivity of Convective Cells: A Strong Indicator of Storm Intensity and Lightning Probability? *Mon. Weather Rev.* **1994**, *122*, 1751–1759. [CrossRef]
36. Ferrier, B.S.; Simpson, J.; Tao, W.-K. Factors Responsible for Precipitation Efficiencies in Midlatitude and Tropical Squall Simulations. *Mon. Weather Rev.* **1996**, *124*, 2100–2125. [CrossRef]
37. Market, P.; Allen, S.; Scofield, R.; Kuligowski, R.; Gruber, A. Precipitation Efficiency of Warm-Season Midwestern Mesoscale Convective Systems. *Weather Forecast.* **2003**, *18*, 1273–1285. [CrossRef]
38. Parker, M.D.; Johnson, R.H. Simulated Convective Lines with Leading Precipitation. Part II: Evolution and Maintenance. *J. Atmos. Sci.* **2004**, *61*, 1656–1673. [CrossRef]
39. Doviak, R.J.; Zrnić, D.S. *Doppler Radar and Weather Observations*; Academic Press: Cambridge, MA, USA, 1993.

40. Kumjian, M. Principles and applications of dual-polarization weather radar. Part I: Description of the polarimetric radar variables. *J. Oper. Meteorol.* **2013**, *1*, 226–242. [CrossRef]
41. Kumjian, M.R.; Khain, A.P.; Benmoshe, N.; Ilotoviz, E.; Ryzhkov, A.V.; Phillips, V.T.J. The Anatomy and Physics of ZDR Columns: Investigating a Polarimetric Radar Signature with a Spectral Bin Microphysical Model. *J. Appl. Meteorol. Clim.* **2014**, *53*, 1820–1843. [CrossRef]
42. Snyder, J.C.; Ryzhkov, A.V.; Kumjian, M.R.; Khain, A.P.; Picca, J. A ZDR Column Detection Algorithm to Examine Convective Storm Updrafts. *Weather Forecast.* **2015**, *30*, 1819–1844. [CrossRef]
43. Kumjian, M.R.; Lombardo, K.A. Insights into the Evolving Microphysical and Kinematic Structure of Northeastern U.S. Winter Storms from Dual-Polarization Doppler Radar. *Mon. Weather Rev.* **2017**, *145*, 1033–1061. [CrossRef]
44. Kumjian, M.R.; Prat, O.P. The Impact of Raindrop Collisional Processes on the Polarimetric Radar Variables. *J. Atmos. Sci.* **2014**, *71*, 3052–3067. [CrossRef]
45. Wang, M.; Zhao, K.; Lee, W.-C.; Zhang, F. Microphysical and Kinematic Structure of Convective-Scale Elements in the Inner Rainband of Typhoon Matmo (2014) after Landfall. *J. Geophys. Res. Atmos.* **2018**, *123*, 6549–6564. [CrossRef]
46. Carr, N.; Kirstetter, P.E.; Gourley, J.J.; Hong, Y. Polarimetric Signatures of Midlatitude Warm-Rain Precipitation Events. *J. Appl. Meteorol. Clim.* **2017**, *56*, 697–711. [CrossRef]
47. Ryzhkov, A.V.; Kumjian, M.R.; Ganson, S.M.; Khain, A.P. Polarimetric Radar Characteristics of Melting Hail. Part I: Theoretical Simulations Using Spectral Microphysical Modeling. *J. Appl. Meteorol. Clim.* **2013**, *52*, 2849–2870. [CrossRef]
48. Zheng, H.; Zhang, Y.; Zhang, L.; Lei, H.; Wu, Z. Precipitation Microphysical Processes in the Inner Rainband of Tropical Cyclone Kajiki (2019) over the South China Sea Revealed by Polarimetric Radar. *Adv. Atmos. Sci.* **2020**, *38*, 65–80. [CrossRef]
49. Rosenfeld, D.; Ulbrich, C.W. Cloud microphysical properties, processes, and rainfall estimation opportunities. In *Radar and Atmospheric Science: A Collection of Essays in Honor of David Atlas*; Springer: Berlin/Heidelberg, Germany, 2003; pp. 237–258.
50. Dolan, B.; Fuchs, B.; Rutledge, S.A.; Barnes, E.A.; Thompson, E.J. Primary Modes of Global Drop Size Distributions. *J. Atmos. Sci.* **2018**, *75*, 1453–1476. [CrossRef]
51. Ji, L.; Chen, H.; Li, L.; Chen, B.; Xiao, X.; Chen, M.; Zhang, G. Raindrop Size Distributions and Rain Characteristics Observed by a PARSIVEL Disdrometer in Beijing, Northern China. *Remote Sens.* **2019**, *11*, 1479. [CrossRef]
52. Wen, L.; Zhao, K.; Chen, G.; Wang, M.; Zhou, B.; Huang, H.; Hu, D.; Lee, W.-C.; Hu, H. Drop Size Distribution Characteristics of Seven Typhoons in China. *J. Geophys. Res. Atmos.* **2018**, *123*, 6529–6548. [CrossRef]
53. Wen, L.; Zhao, K.; Zhang, G.; Xue, M.; Zhou, B.; Liu, S.; Chen, X. Statistical characteristics of raindrop size distributions observed in East China during the Asian summer monsoon season using 2-D video disdrometer and Micro Rain Radar data. *J. Geophys. Res. Atmos.* **2016**, *121*, 2265–2282. [CrossRef]
54. Wen, J.; Zhao, K.; Huang, H.; Zhou, B.; Yang, Z.; Chen, G.; Wang, M.; Wen, L.; Dai, H.; Xu, L.; et al. Evolution of microphysical structure of a subtropical squall line observed by a polarimetric radar and a disdrometer during OPACC in Eastern China. *J. Geophys. Res. Atmos.* **2017**, *122*, 8033–8050. [CrossRef]
55. Miranda, P.M.A.; James, I.N. Non-linear three-dimensional effects on gravity-wave drag: Splitting flow and breaking waves. *Q. J. R. Meteorol. Soc.* **1992**, *118*, 1057–1081. [CrossRef]

MDPI
St. Alban-Anlage 66
4052 Basel
Switzerland
www.mdpi.com

MDPI Books Editorial Office
E-mail: books@mdpi.com
www.mdpi.com/books

www.ingramcontent.com/pod-product-compliance
Lightning Source LLC
Chambersburg PA
CBHW061457220926
51850CB00004B/110
9783036589329